AF517021

# PESTS AND POLLINATORS OF VEGETABLE AND OILSEED CROPS

**Other related books:**

- A Manual of Practical Entomology 3rd Ed - M.M. Trigunayat
- Agricultural Insect Pests and Their Control 2nd. Ed – V.B. Awasthi
- Bastiaan Kiauta: An Odonatologist Extraordinnaire & Polymath - B. K. Tyagi
- Beekeeping: A Compressive Guide to Bees and Beekeeping - Dharam P. Abrol
- Biological and Molecular Approaches in Pest Management - Balwinder Singh et al.
- Botanical Pesticides for Pest Management - D.A. Dodia et al.
- Crop Protection Strategies: Under Climate Change Scenarios - P. Parvatha Reddy
- Emerging Crop Pest Problems: Redefining Management Strategies - P. Parvatha Reddy
- Insect Physiology (21st Century Biology And Agriculture: Textbook Series) - K.P. Sanjayan/Series Editor: Anantanarayanan Raman
- Insect, Mite and Vertebrate Pests and Their Management in Horticultural Crops - P. Parvatha Reddy
- Integrated Pest Management - A.K. Dhawan et al.
- Pest and Vector Management in the Tropics - A. Youdeowei et al.
- Pests of Forest Importance and Their Management - B. K. Tyagi et al.
- Plant Protection in Horticulture (Vol. 1-3) (Set) - P. Parvatha Reddy
- Principles of Insect Behaviour, 2nd Ed. V.B. Awasthi
- Technological Innovations in Integrated Pest Management: Biorational and Ecological Perspective - Dharam P. Abrol
- Theory and Practice of Integrated Pest Management - Ramesh Arora et al.

# PESTS AND POLLINATORS OF VEGETABLE AND OILSEED CROPS

**Dharam P. Abrol**

*Dean*
*Faculty of Agriculture*
*Sher-e-Kashmir University of*
*Agricultural Sciences & Technology of Jammu,*
*Chatha Jammu*-180009, *India*

Published by
**SCIENTIFIC PUBLISHERS (INDIA)**
Jodhpur –
5 A, New Pali Road
P.O. Box 91
Jodhpur - 342 001 INDIA





ISBN: 978-93-86652-34-8

Visit the Scientific Publishers (India) website at
http://www.scientificpub.com
Printed in India

# Dedication

I dedicate this book in the everlasting memory of my respected brother late Sh. Raj Kumar, who was a man with broad vision, open heartness, humanity, modesty and vigour. His memory encourages me to strive to the best of my ability to follow in his foot steps. I ofter this book dedicating to his pure soul and everlasting immortal memory.

# M.S. SWAMINATHAN RESEARCH FOUNDATION

**M.S. Swaminathan**
Founder Chairman
Ex-Member of Parliament (Rajya Sabha)

29 November 2017

## Foreword

Fruits and vegetables play a significant role in economic development, nutritional security, employment generation and overall economic growth of a country. They provide a dynamic tool for enhancing economic returns, creating employment avenues and ensuring ecological sustainability. The widespread use of pesticides in modern agriculture throughout the world has become necessary for the protection of the plants against insect pests and diseases to obtain higher yields to meet the food requirement of an increasing population, but the injudicious use of pesticides has resulted in contamination of agro-ecosystem and agriculture produce including nectar and pollen and caused heavy losses to the pollinators.

Most of the vegetable crops are highly cross-pollinated and depend upon insects or benefit from insect pollination for fruit set. Insect pests, on the other hand, cause major economic damage not only quantitatively but qualitatively rendering the vegetable unfit for marketing. Evidently, pest management in vegetable crops on the one hand and providing safety to the pollinators on the other is a challenging task in the context of increasing vegetable productivity without upsetting the ecological balance. The book Pests and Pollinators of Vegetable and Oilseed Crops aims to integrate and develop pest control strategies in a way to minimize their impact on beneficial insect species such as natural enemies and pollinators to enhance fruit production and quality.

The book Pests and Pollinators of Vegetable and Oilseed Crops covers interplay between pest management strategies and safety of pollinators in vegetable crops. Detailed information is provided on pests and pollinators of vegetable and oilseed crops.

The aim of this book is to fill the gap by providing the synthesis and critical analysis of different management strategies having a bearing on agriculture, sustainability and environmental protection. The proposed book provides complete information and fulfils the gap in literature, putting all the information about vegetables in one book. The present publication is an excellent reference book for students, teachers, researchers, extension functionaries and policy makers.

The book aims to promote a large, diverse, sustainable and dependable bee pollinator workforce that can meet the challenge for optimizing food production well into the twenty-first century. I congratulate the author for his efforts to provide the timely contribution.

**MS Swaminathan**

3rd Cross Road, Taramani Institutional Area, Chennai (Madras) - 600 113, India
Phone: +91-44-2254 2790, 2254 1698 Fax: +91-44-2254 1319
E-mail: founder@mssrf.res.in, swami@mssrf.res.in

# Preface

Fruits and vegetables play a significant role in economic development, nutritional security, employment generation and overall economic growth of a country. They provide a dynamic tool for enhancing economic returns, creating employment avenues and ensuring ecological sustainability. Production constraints include two diverse but mutually interlinked approaches such as attack of pests on the one hand and inadequate pollination on the other. The widespread use of pesticides in modern agriculture throughout the world has become necessary for the protection of the plants against insect pests and diseases to obtain higher yields to meet the food requirement of an increasing population, but the injudicious use of pesticides has resulted in contamination of agro-ecosystem and agriculture produce including nectar and pollen and caused heavy losses to the pollinators. Such contaminated nectar and pollen when brought to hive may cause damage to brood besides the contamination of the stored honey.

The world human population disproportionately growing at an alarming rate is a major challenge facing agriculture in the twenty-first century. The human population has already crossed the seven billion mark and is expected to rise to nine billion or more during the middle of this century requiring raising of food productivity by some 70–100 % through horizontal expansion. Production in the developing countries would have to be doubled as compared to the present. The cropping systems and pollinator demographics are changing throughout the world, and any deviation in pollinator sufficiency will prove increasingly inadequate for meeting the demands for an abundant, high-quality food supply in the twenty-first century. Evidence for decline of pollinators throughout the world is a matter of serious concern for agricultural productivity. It is manifestly clear that bee pollinators are a valuable and limited natural resource that should be conserved and encouraged at all costs. This awareness owes its origin to an apparent decline of domesticated honeybees throughout the globe due to several causes such as attack of parasitic mites such as *Varroa destructor,* which has a devastating effect. The parasite occurs now in every continent on which *A. mellifera* is kept, except Australia, and it is considered the most serious health threat to apiculture.

Most of the vegetable crops are highly cross-pollinated and depend upon insects or benefit from insect pollination for fruit set. Insect pests, on the other hand, cause major economic damage not only quantitatively but qualitatively rendering the vegetable unfit for marketing. Evidently, pest management in vegetable crops on the one hand and providing safety to the pollinators on the other is a challenging task in the context of increasing vegetable productivity without upsetting the ecological balance. The book *Pests and Pollinators of Vegetable and Oilseed Crops* aims to integrate and develop pest control strategies in a way to minimize their impact on beneficial insect species such as natural enemies and pollinators to enhance fruit production and quality.

**The book** *Pests and Pollinators of Vegetable and Oilseed Crops* covers interplay between pest management strategies and safety of pollinators in vegetable crops. The introductory chapter gives an overview of pest problems and pollinator needs of the crops followed by information on pollination scenario and mechanism of pollination in vegetable crops. Detailed information is provided on pests and pollinators of oilseed crops, Cruciferous, Solanaceous, Umbelliferous, Cucurbitaceous, Malvaceous Leguminous and Alliaceae crops.

The aim of this book is to fill the gap by providing the synthesis and critical analysis of different management strategies having a bearing on agriculture, sustainability and environmental protection. The compilation of this book is unique in the sense that it does not deal with the conventional way of discussing pest management for different crops but also takes into consideration the role of pollinators and their profitable utilization in the larger context of ecologically based pest management for management of pests on the one hand and safety of pollinators on the other.

Though number of books on pests as well as pollinators on vegetable and oilseed crops are available, they have a limited approach covering one aspect or the other. The proposed book provides complete information and fulfils the gap in literature, putting all the information about vegetables in one book.

The goal of this book is to synthesize the latest scientific literature into principles and practices that are relevant to workers in crop pollination in tropical and temperate areas. The book aims to promote a large, diverse, sustainable and dependable bee pollinator workforce that can meet the challenge for optimizing food production well into the twenty-first century.

A vast spectrum of people has helped in one way or the other in the writing of this book, which would have remained a distant dream without their active help and support. This book is the outcome of my personal experiences and the contributions of several workers which have been incorporated. I express my humble and profound thanks to all of them whose hard work has enabled me to compile the suitable information in such a manner that it would be useful to those interested in basic and applied pollination and pest management. The illustrations and figures are either original or redrawn from other sources, which have been cited individually in the figure legends. All the authors whose work has been used/referenced deserve special appreciation and heartiest acknowledgements. I am particularly thankful to Professor Dr. Raghavendra Gadagkar, Centre for Ecological Sciences, Indian Institute of Science, Bangalore, who has always been a source of inspiration, needed help, guidance and encouragement. I thank my university authorities for the excellent working atmosphere and needed encouragement for compiling such a voluminous book. I also thank Dr. Uma Shankar and Dr. Debjyoti Chatterjee for their help and support. I am also extremely thankful to Scientific Publishers Jodhpur, India who took great pains and keen interest in the publication of this book in a very impressive way. I am also highly thankful to Prof. Dr. M. S. Swaminathan, the father of green revolution in India, for writing the foreword of this book.

Last but not the least, my sincere thanks are due to my wife Professor Dr. Asha Abrol, daughter Er. Vitasta and son Er. Rajat for their endurance and help while writing this book.

**Jammu, India** **D.P. Abrol**

# Contents

# CHAPTER 1

# Introduction

Vegetables form an indispensable part of daily human diet providing carbohydrates, vitamins and minerals. They help to maintain valuable alkaline reserves in the body and provide necessary ingredients for repair and build up of human body. There are different kinds of vegetables such as edible roots, stems, leaves, fruits or seeds each of which contributes to diet in its own way (Robinson, 1990). Vegetables contribute minerals, vitamins, and fiber to the diet. Vegetables are rich source of minerals such as P, Na, K etc and poor sources of fat which makes them as excellent source of food for the people facing obese problems and help perform normal body function, maintenance and reproduction. The fibre content in vegetables helps to reduce cholesterol levels in the body.

The importance of growing vegetables lies in their fibre, trace minerals, antioxidants, vitamins, folacin, carbohydrates and protein content (Oomen and Grubben, 1978; Gormley, 1989). In most of the developing countries, there is increased emphasis to supplement the staple foods such as maize and tubers with with locally grown green vegetables such as *Amaranthus* or Chinese cabbage which increase the food's nutritional value. It is not only the green vegetable that are useful but Farinaceous vegetable consisting of starchy roots such as potatoes which contain carbohydrates and provide energy to the body are equally important. An intake of about 280 grams of vegetables per person is considered essential for maintenance of good health with intake of leafy vegetables 40 %, roots and tubers 30 % and other vegetables 30%.

Consumption of Different vegetables is good for health as they contain different nutrients for instance *Moringa* and peas are rich sources of protein, tomato spinach and peas contains calcium, cucumber rich in phosphorus , Spinach, peas, tomato, bitter gourd rich in iron, Okra, Summer squash contain iodine, Leafy vegetables, pumpkin rich in vitamin A and Vitamin K. Besides, lot of protective compounds are present in vegetables such as Cheratin in bitter gourd which is effective against diabetes. The major contribution of vegetables in human diet is in the form of fiber which is recommended from 25 to 30 g per day which is available in eggplant, broccoli, carrots, mushrooms, , Brussels sprouts, other cabbages, cauliflower, celery, spinach, fennel, green string beans, turnip, dandelion, leek, bell pepper, pumpkin, artichoke, peas, Jerusalem artichoke, white beans, black beans, kidney beans, lentils, chickpeas etc. Fibre, minerals, vitamins and antioxidant substances are the principal contributors to the nutritional value of vegetables

## Scope of Vegetable Growing

Climatic conditions of India offer a great scope for cultivation of variety of vegetables. The demand for processed vegetables is increasing thus requiring production of more vegetables. Some of the common vegetables grown in India include : Tomato (*Lycopersicon esculentum),* Brinjal *Solanum melongena* Chilli *Capsicum annuum* Bell Pepper *Capsicum frutescens* Okra *Abelmoschus esculentus* Garden Pea *Pisum sativum* French bean *Phaseolus vulgaris* Cowpea *Vigna sinensis* Dolichos bean *Dolichos lablab* Cluster bean *Cyamopsis tetragonoloba* Palak *Beta vulgaris* Spinach *Spinacia oleracea* Moringa (Drumstick) *Moringa oleifera* Amaranath *Amaranthus viridis* Lettuce *Lectuca sativa* Watermelon *Citrullus lanatus* Muskmelon *Cucumis melo* Cucumber *Cucumis sativus* Pumpkin *Cucurbita pepo* Bottle gourd *Lagenaria siceraria* Ridge gourd *Luffa acutangula* Bitter gourd *Momordica charantia* Pointed gourd *Trichosanthes dioica* Long melon *Cucumis melo var. Utilissimus.* India is rich in biodiversity of vegetables and is considered to be the centre of their origin. it is evident from the fact vegetables have been vividly described in the Indian scriptures like 'Vedas' and 'Ramayana'. Vegetables constitute an important part of agriculture and nutritional food security as they have a very short duration, high potential, economic viability, nutritional richness and on-farm and off-farm employment. India is the second largest producer of vegetables in the world. Fruits and vegetables together contribute about 92% of the total horticultural production in the country. India ranks first in the production of okra in the world (73% of world production).

India has made significant in production of vegetable which has mainly been due to development of improved vegetable varieties/hybrids coupled with their adoption by the farmers. Furthermore, the area under vegetables production has also increased (8.5 million ha), resulting in production of 146.5 million t and productivity 17.3 t/ha in the country. Major vegetable producing countries of the world during include China (48%); India (14%); USA (3%); Turkey (2.2%); and Egypt (2%). India ranks second in vegetable production in the world contributing 14 % of world vegetable production share. Per capita consumption of vegetables in India is 230 g as against 300 g recommended dietary allowance (RDA), evidently, we have shortage of about 30 million t of vegetables necessitating production of more vegetables.

## Production Constraints

Vegetable production depends upon two antagonistic approaches which influence quality and quantity of vegetable production. Attack of insect pests and diseases is a major constraint in quality and quantity of vegetable production, on an average causing 10-30 percent losses besides acting as a vector for various viral diseases which requires use of chemicals for controlling them. Insect pollination is another important input to vegetable crop production because some crops require cross-pollination by insects in order for any produce to be harvested. Lack of proper pollination is one of the most important factor responsible for low productivity in vegetables. Inspite of serious efforts at extension levels, improved management practices and introduction of new varieties desired success has not been achieved.

Evidence of emerging pollination problem has been documented by several investigators (Partap, 1998; Partap and Partap, 2000, 2001; Partap *et al.,* 2000). The use of pesticides for pest control is seriously affecting the honeybees which are most essential component for pollination. The excessive and indiscriminate use of pesticides for controlling insect-pests (Bisht *et al.,* 1980., Rana and Goyal, 1991., Zhong *et al.*, 2004) is a major problem resulting in killing of honeybees indirectly effecting pollination and crop yields.

Unfortunately, these two practices are not always compatible, as honeybees suffer heavy losses from most commonly used pesticides for killing of crop pests (Johansen, 1977., MacKenzie and Winston, 1989., Russell *et al.,*1998., Cunningham *et al*., 2002., Sundararaju, 2003, Poehling, 1989., Stark *et al*.,1995). The loss of honeybees directly affect beekeeping through loss of honey production and indirectly the crop production also incurs significant environmental, ecological and economic costs ( Bai and Reddy, 1977., Pimental *et al*., 1980., Crane and Walker, 1983., Praksah and Kumaraswami, 1984., Kasana *et al.,* 1999., Khan and Dethe, 2004).

Despite increasing efforts towards crop improvement the reliance on chemicals still continuous (Day, 1987, Rana and Goyal, 1991) and there is an urgent need of the integration of the two approaches killing crop pests and providng safety to bees for sustaining crop production.

In vegetables loss of quantity is not the only problem, but chemical method of pest management has its own ill effects on environment, human and animal healths through residual effect of chemicals in vegetables. The pesticide usage is much higher in chilli (5.13 kg a.i. /ha) followed by brinjal (4.60 kg a.i. /ha), cole crops (3.73 kg a.i. /ha) and okra (2-3 kg a.i. /ha) and if the green pods and fruits are harvested at shorter intervals and used as vegetables are prone to retain pesticide residue (Rai et al 2014). Integration of two approaches for safety and security of food is needed for sustainable production. Hence, it is imperative to develop, validate and implement multipronged approaches consisting of biological, host plant resistance/tolerance mechanical and cultural practices for managing pests of crops and providing safety to beneficial insects such as pollinating insects like honeybees. The chemicals used on the crops has more serious effect on honeybees because of their long working hours in the field for collecting nectar and pollen from flowers (Johansen, 1977, Mayer, 2003., Masoodi, 2003). The large scale killing of honeybees and other pollinators has negatibe effect on pollination process consequently effecting crop yields. Therefore for sustainable crop production safety of honeybees is a vital component in modern agriculture ((Mayer and Johansen, 1983, Smirle, 1990, Gupta and Sharma, 1998., Kremen *et al*., 2002).

This require the application of insecticides to combat the pest (Sihag, 1988) but the impact of these insecticides on the foraging activity of honeybees has not yet received much attention.

Pollination is a key ecosystem service mediated by animals that contributes to global food supply and human nutrition (Klein et al 2007, Eilers et al 2011). Recently, pollinator declines owing to habitat loss, agricultural land-use changes and other pressures [Potts et al 2010] have resulted in declines in the level and stability of seed set in insect-pollinated crops (Garibaldi LA et al. 2011., Bommarco et al 2012). In a vast majority of existing studies, crop pollination has been studied as a benefit obtained from a single context-independent interaction between a plant and its pollinators. Bos et al. (2007) however, indicated that crop pollination depends upon pollinator community interaction with other ecosystem services, such as regulation of pest pressures.

The interactions between pollination and pest control for yield indicate that plants in high resource environments have a higher ability to compensate for herbivory compared with plants in low-resource environments (Maschinski and Whitham 1989). It has been found that plants have a higher flexibility to reallocate resources to undamaged ovules following pest attack from seed predators when a higher proportion of these ovules have already been pollinated. This may be a pathway under which pollination can compensate for a lack of control of seed predating pests. There are, however, some mechanisms which speak against a compensatory interaction between pollination and control of seed predating insects.

Interactions between pollination and herbivory and its consequences for plant reproduction has so far largely been investigated from a plant evolutionary and demographic perspective with a focus on herbivory of leaves, shoots or flowers (Strauss and Irwin 2004., McCall and Irwin 2006). There is probably, no study that tested how the control of pests that attack seeds or fruits interact with pollination services for yield in a crop plant.

Herbivore attack immediately before and during flowering, and on plant reproductive parts, might be severe and irremediable for insect-pollinated plants.

The changes in floral traits such as pest infested flowers or grazed plants are not preferred by pollinators (Krupnick et al 1999., Mothershead and Marquis 2000). Moreover, herbivores that continue to consume plant parts after pollination, such as flower-feeding insects or seed predators. They might switch their foraging to selectively consume well-pollinated and seed-rich flowers, which are highly rewarding (Herrera 2000). Hence, lack of control of pests attacking plant reproductive parts may cancel positive effects of pollination, resulting in a synergistic interaction between the two services.

## Pest Problems

The major problem limiting production of vegetables is the attack of various insect pests and diseases. The losses due to pests on an average have been reported to range between 10-30 percent (Table 1.1). Insect pests not only cause direct damage to the quality and quantity of produce but also act as vector for several diseases which cause far more serious production losses and aggravate the problem still further. Very recently, there has been shift in pest problems due to changes in the cropping pattern, changes in climate, habitat alteration and use of new varieties/hybrids which have further compounded the problems of pest management. The chemical based plant protection measures have resulted in serious adverse effects on environment and human health. Most of the pests have shifted to new hosts, developed resistance to pesticides and more often resulted in secondary outbreaks. include the serpentine leaf miner (*Liriomyza trifolii*) found on tomato and cucurbitaceous crops, gall midge on brinjal, chilli and capsicum; stem fly on okra, leafhopper (*Empoasca motti*) on bitter gourd; head borer (*Hellula undalis*) on cabbage, red spider mite (*Tetranychus* spp.) on okra, brinjal, cowpea and Indian bean are some of the examples.; yellow mite (*Polyphagotarsonemus latus*) on chilli and more recently mealy bug (*Pheonacoccus solenopsis*) on okra, chilli, brinjal and tomato specially in protected conditions have intensified the severity of occurrence in different parts of the country. Similarly, many soil borne aggressive pathogens such as *Rhizoctonia, Sclerotium, Pythium* and plant parasitic nematodes are noted to threat the vegetable production system. Apart from these, the aerial pathogens such as fungal (*Alternaria, Cercospora, Colletotrichum* and mildews), bacterial (*Xanthomonas, Corynebacteria, Pseudomonaes* and *Ralstonia*) and different viruses (leaf curls, yellow mosaic) are noted to be major limiting factors for vegetable production in India. Success of management of insect pests/diseases highly depends on correct identification and selection of appropriate control measures. Among the various methods of pest control, chemical method still enjoys first choice because of its quick action and easy availability. Consequently, vegetables consume about 13-14 per cent of the total pesticide consumption in India.

Pesticide use varies in different crops, it is maximum in chilli (5.13 kg a.i. /ha) followed by brinjal (4.60 kg a.i. /ha), cole crops (3.73 kg a.i. /ha) and okra (2-3 kg a.i. /ha). Unlike cereals, the green pods and fruits harvested at shorter intervals and used as vegetables are prone to retain pesticide residue. The residue problems can only be overcome by following good management practices. The utilization of non chemical approaches such as biocontrol agents, resistant varieties, cultural and mechanical practices can help to combat the problem. However, field efficacy of bio-control agents has failed in many cases to combat the pests/disease due to evolution of new biotypes/stains/pathotypes/ races in the insect/pathogen system. It is therefore, necessary to develop, validate and implement multipronged approaches consisting of biological, host plant resistance/tolerance, cultural practices, etc. to manage the pests and diseases.

Table 1.1: *Yield losses due to major insect pests in major vegetables.*

| Crop | Pest | Yield loss (%) |
|---|---|---|
| Tomato | Fruit borer (*Helicoverpa armigera*) | 24-65 |
| | Root knot nematode (*Meloidogyne spp*) | 27.2 |
| Chilli | Thrips (*Scirtothrips dorsalis*) | 12-90 |
| | Mites (*Polyphagotarsonemus latus*) | 34 |
| | Root knot nematode (*M. incognita, M. javanica*) | 12.85 |
| Okra | Fruit borer (*H. armigera*) | 22 |
| | Leafhopper (*Amrasca biguttula biguttula*) | 54-66 |
| | Whitefly (*Bemisia tabaci*) | 54 |
| | Shoot and fruit borer (*Earias vittella*) | 23-54 |
| Brinjal | Fruit and shoot borer (*Leucinodes orbonalis*) | 11-93 |
| | Root knot nematode (*M. incognita, M. javanica*) | 16.62 |
| Cabbage | Diamond back moth (*Plutella xylostella*) | 17-99 |
| | Caterpillar (*Pieris brassicae*) | 69 |
| | Leaf webber (*Crocidolomia binotalis*) | 28-51 |
| | Cabbage borer (*Hellula undalis*) | 30-58 |
| Cucurbits | | |
| Bitter gourd | Fruit fly (*Bactrocera cucurbitae*) | 60-80 |
| Cucumber | Fruit fly (*Bactrocera cucurbitae*) | 20-39 |
| Ivy gourd | Fruit fly (*Bactrocera cucurbitae*) | 63 |
| Musk melon | Fruit fly (*Bactrocera cucurbitae*) | 76-100 |
| Snake gourd | Fruit fly (*Bactrocera cucurbitae*) | 63 |
| Sponge gourd | Fruit fly (*Bactrocera cucurbitae*) | 50 |

**Source:** *Modified from Rai et al 2014*

## Pollination Requirements in Vegetable Crops

### *Effect of pollination in improving crop productivity*

Most of the horticultural crops are self- incompatible and depend upon cross pollination for any fruit or seed production (McGregor, 1976; Free, 1993). It has been observed that it is not only the self sterile varieties that benefit from cross pollination but even the self fertile varieties produce heavier and better quality fruits if they are cross pollinated (Free, 1993). Increase in productivity of vegetables and horticultural crops shall require two inputs, the first involves the use of improved agricultural input such as quality seeds and planting material, good cultural practices, timely irrigation, efficient use of fertilizers, chemicals and new varieties, use of biotechnological techniques, such as manipulating rate of photosynthesis and biological nitrogen fixation, etc. But these practices have a success upto a limited stage after which the productivity becomes stagnant (Partap and Partap, 1997). The pollination by bees is another important input which is responsible for limiting crop yiels as more than 70 per cent of the crops are cross fertile and depend upon pollinating insects for any significant production of fruits and seeds. Different kinds of insects such as bees, flies, moths, butterflies, beetles, wasps act as pollinators. But of all these insects, honeybees are the most important pollinators which make an important improvement in the production of vegetables, fibre crops and nuts (Partap and Partap, 1997). Dependency of some horticultural crops on insects for pollination is given in the Table 1.2. These results confirm the usefulness of bee pollination and its role in increasing crop productivity and improving the quality of fruits and seeds.

### *Inadequate pollination and crop production*

The low productivity in most of the crops is due to inadequate pollination. Most of the crops donot produce any significant seeds or fruits unless pollinated by bees inspite of best management practices and introduction of new (Partap, 1998; Partap and Partap, 2000, 2001; Partap *et al.,* 2000).

### *Role of bee pollinator in crop productivity*

Honeybees are the best and most efficient pollinators of crops as they have specialized structural modifications such as hairy body and modified body parts for picking up of pollen grains. Their behaviour to work for long hours, consistency of work and floral fidelity and adaptability to work under different climatic regimes makes them most reliable pollinators. They can be placed in sufficient numbers as and when required (Free, 1993). The honey bee is undoubtedly the best known insect species that contributes most to the pollination of entomophilous crops. Boosting beekeeping shall be of benefit both to beekeepers and the horticulturist for enhancing productivity and increasing income of farmers and beekeepers. This will ensure livelihood security for both farmers and the beekeepers by generating employment opportunities and subsidiary source of income. The system of renting honeybee colonies for orchard pollination is yet another source of income for beekeepers besides selling bee hive products.

### *Insects as pollinators of vegetable crops*

Insect pollination is important to vegetable crop production because some crops require cross-pollination by insects in order for any produce to be harvested. The others benefit greatly from insect pollination. The insects involved are mostly honeybees (*Apis mellifera* L.) (Table 1.2).

## Pollination requirements of vegetables

### *Modes of Pollination*

Pollination requirements of vegetable crops vary from self pollinated to cross pollinated ones. Improvement have been observed even in self pollinated if they receive cross-pollination services from the pollinating insects including honeybees

### *Self-Pollinated Vegetable Crops*

These crops having hermaphrodite flowers reproduce by self pollination but cross pollination may occur from 5 to 50 per cent in some crops. However, the degree of cross-pollination in self-pollinated crops depends upon several factors, namely variety, environmental conditions like temperature, humidity, wind, location and abundance of pollinating insects. Self pollinated crops include :Tomato *(Lycopersicon esculentum)*, Lettuce *(Lactuca sativa)*. Parsnip *(Pastinaca sativa)*, Peas *(Pisum sativum)*, Dwarf bean *(Phaseolus vulgaris)* etc.

### *Cross-Pollinated Vegetable Crops*

The majority of the cultivated vegetable species are cross-pollinated crops. They include Cabbage (*Brassica oleracea var. capitata)*, Cauliflower *(Brassica oleracea var. botrytis)*, Broccoli, Brussel's sprouts, Knol khol Carrot Radish, Beet Spinach, Cucumber, Muskmelon Watermelon, Pumpkin, Squash and Other cucurbits such a as bitter gourd, bottle gourd, ridge gourd, sponge gourd, snake gourd, pointgourd, ash gourd etc.Amaranths) and Onion *(Allium cepa)*. In some of the vegetable crops, where cross pollination exceeds 5 percent and reaches upto 50 per cent are often referred to as cross-pollinated species such as brinjal, okra, chilli, sweet pepper.

Insect pollination is absolutely required for cucurbit crops, such as cucumber, melon, pumpkin and squash. In *Cucumis* and *Cucurbita*, the vines produce unisexual flowers, although the plants are bisexual. Pollen of these plants is typically large, spiny and oily, showing clear adaptations to insect pollination. Even though melon is self-fertile, insects are needed to effect pollination. Cross-pollination seems to be required in *Cucumis* and *Cucurbita* to obtain seeds and, in most cultivars, fruit. An exception is the long English cucumber, which forms a desirable, seedless fruit if not pollinated; if the flowers are pollinated, the fruits become misshapen (bull-necked), seeded and bitter. Honeybees are used for pollination of most cucurbit crops during the summer to avoid the production of malformed cucumbers.

Table 1.2: *Effect of bee pollination on yield increase of vegetable crops*

| Crop Plants | Yield /yield attributes | Per cent increase in yield | References |
|---|---|---|---|
| Vegetables | | | |
| Radish | Seed yield | 22-100 | Anonymous, 1997 |
| Cabbage | Seed yield | 100-300 | Anonymous, 1997 |
| Carrot | Seed yield | 9.10-135.4 | Alam *et al.*, 1995 |
| Carrot | Seed yield/umbel seed wt./umbel | 926-1397 | Alam *et al.*, 1997 |
| Turnip | Seed yield | 100-125 | Anonymous, 1996 |
| Brinjal | Seed yield | 35-67 | Anonymous, 1996 |
| Cauliflower | Seed yield | 302-1338 | Sinha and Chakrabarti, 1985 |
| Onion | Seed yield | 354-9878 | Kumar *et al.*, 1989 |
| Muskmelon | | 756-6700 | Kumar *et al.*, 1989 |
| Cucumber | Seed yield | 21.10-411 | Kumar *et al.*, 1989 |
| | Fruit setting | 127-275 | Krishansawmy, 1990 |
| | Fruit weight | 32.50-141.70 | Ahmed, 1992 |
| Squash | Fruit weight | 771.40-4800 | Alam and Quadir, 1996 |
| Bottle gourd | Fruit setting | 200-350 | Alam and Quadir, 1996 |
| Tomato | Fruit setting | 8.3-27.40 | Cribb, 1990 |

Although honeybees are the most manageable pollinators, they are not the best-suited for some vegetable crops. Pollination of squash and pumpkin is carried out effectively by the solitary squash bee *Peponapis pruinosa* (Say). Squash bees emerge from underground nests in synchrony with the blooming of squash and pumpkin. The bees are active on the newly opened flowers from just before dawn, when the flowers open for their short, morning-long life. The female squash bees gather pollen to provision their nests. They have all but finished their pollinating activity by the time honeybees begin to forage. Populations of *P. pruinosa* in the vicinity of squash and pumpkin fields are immediately beneficial to productivity.

Some crops in the pea and bean family (Leguminosae) require pollination, but most are self-fertile and self-pollinating. However, scarlet runner bean requires pollination, which is accomplished mostly by insects or sometimes by hummingbirds. Bumblebees are especially adept at pollinating scarlet runner bean and are used in Europe in greenhouse production of this crop. The importance of insect pollinators in other species and cultivars of pea and bean is difficult to judge, the available data indicating no effect or some increase in crop yield. Solanaceous vegetables, such as pepper, tomato and eggplant, usually produce fruit well when grown outdoors. Self- pollination, which is brought about by wind, agitation and gravity, seems to be effective. However, in greenhouses, pollination usually is incomplete unless assisted. Traditionally, growers ensure the transfer of pollen by touching the flower with an "electric-bee" vibrating rod. Recently, culturing of the bumblebee *Bombus terrestris* (L.) has been under taken in Europe for pollination of these solanaceous crops, especially tomato. Some seeds, apart from sprouted beans and alfalfa, also are used as vegetables. One example, sweet corn, requires cross-pollination, which occurs by wind. Newer cultivars of sunflower are self-pollinating, but the older cultivars require cross-pollination by insects. Some native bees, such as *Eumegachile pugnata* (Say), are specialist pollinators of sunflowers and may be useful on Jerusalem artichoke in some areas. Although it is assumed that many umbelliferous flowers require cross-pollination to set seed, the pollination requirements for anise, caraway, celery, coriander, dill, and fennel are incompletely understood or unknown. Generally, these plants are visited by a wide variety of insects; it is recognized that insects are necessary to bring about pollination, because the sexual parts of individual florets mature at different times.

Almost all vegetables are propagated from seed, except asparagus, garlic, Jerusalem artichoke, onion, potato, rhubarb and some herbs. Among cruciferous crops, cross-pollination by insects is the rule. It is also the generally accepted condition for onion and allied crops, for such umbelliferous crops as carrot, celery, dill, fennel, parsnip and caraway. The pollinators involved are varied and include honeybees, other bees, flies, moths and other insects. Beet, chard and spinach require cross-pollination by wind, although insects may visit the flowers of male plants for pollen, which is also the case with corn. There are special techniques for the management of insect pollinators in vegetable seed-production enclosures. The insects used include honeybees, bumblebees, orchard bees *(Osmia* spp.), leafcutting bees *(Megachile* spp.), flower flies (family Syrphidae), blow flies (family Calliphoridae), and house flies (family Muscidae). As the technology of vegetable production and improvement advances, pollination needs to be carefully considered. At present, the diversity and quality of insects as pollinators, and the botanical aspects of pollination technology for efficient cropping of vegetables are neglected areas of research.

## Conclusions

In conclusion, the declining horticultural productivity can be attributed to a number of factors, but pollination plays a crucial role. One can go for use of improved agricultural technologies, such as the use of quality seed, high yielding varieties, good agronomic practices like timely irrigation and fertilizers, but without pollination, neither fruit nor seed will be formed. Inadequate pollination is the main factor affecting crop pollination, which can be overcome by overcoming by planned pollination by supplementing crops with mamaged honeybee colonies in required numbers at desired, places. Promoting use of beekeeping for pollination of horticultural crops will be of benefit to both the beekeeper and to the farmer.

Many growers are not aware of how significant the contribution of native pollinators is to the production of their crops and farm profitability. Other challenges include active participation of researcher and extension specialist will help in expansion of use of honey bee rearing for crop pollination. It is therefore essential while attempting to control pests, safety of pollinators needs urgent attention.

## References

Abrol, D. P. 1997. *Bees and Beekeeping in India*. Kalyani publishers, Ludhiana, pp 450.

Abrol, D. P. 1993. Insect pollination and crop production in Jammu and Kashmir. *Current Science*. 65 (3): 265 – 269.

Abrol, D. P. 1996. Bee keeping in integrated rural development. *Indian Farming*. 46 (2): 17 – 22.

Abrol, D. P. and Andotra, R. S. 2003. Relative toxicity of some insecticides to *A. mellifera* L. *Journal of Asia Pacific of Entomology.* 6(2): 235-37.

Agricultural Statistics of Pakistan. 2003-2004. Govt. of Pakistan, Ministry of Food, Agri. And livestock. (Economic-wing) Islamabad, Pakistan.

Alam, M. Z. and M. A. Quadir. 1986. Role of honeybee in fruit and seed setting of bottle-gourd, *Lagenaria siceraria* (Mol.) Standl. *Punjab Veg. Grower* 21:32-34.

Alex, A. H. 1980. Honeybees aid pollination of cucumbers and cantaloupes. *Glean. Bee Cult.* 85:398-400.

Alex, A. H. 1987. Honeybees aid pollination of cucumbers and cantaloupes. *Tex. Agr. Expt. Sta. Prog. Rpt.*, 4 pp.

AOAC (Association of Official and Analytical Chemistry). 2000. Official methods of analysis (by Dr. William Horwitz) 17th Ed. Gaithersburg, MD, USA.

Asian Vegetable Research and Development Center (AVRDC) (1998) *Vegetables for Poverty Alleviation and Healthy Diets*. AVRDC,Taiwan, 29 pp.

Atsmon, D., Galun, E., and Jakob, K. B. 1965. Relative time of anthesis in pistillate and staminate cucumber flowers. *Ann. Bot.* 29: 277 - 283.

Attia, M. S., and Monger, H. M. 1950. Self-incompatibility and the production of hybrid cabbage seed. *Amer. Soc. Hort. Sci. Proc.* 56: 363 - 368.

Atwal, A. S. 1970. Biology, ecology and utilization of insects other than honeybees in the pollination of crops. Final Res. Rpt. (1965-70) OF P.L. 480 project executed at Punjab Agr. Univ., Ludhiana (India), 115 pp.

Atwal, A. S. 2000. *Essentials of Beekeeping and Pollination.* Kalyani Publishers, New Delhi, India.pp455.

Baker, S.L. and B.K. Gyawali. 1994. *Promoting Proper Pesticide Use: Obstacles and Opportunities for an Integrated Pest Management Programme in Nepal*. HMG/MOA/ Winrock Intl., Kathman-du, Nepal.pp125.

Batra, S.W.T., 1985. Bees and pollination in our changing environment.' *Apidologie,* 26: 361–370.

Battaglini, M. B.1969. The importance of honeybees for fertilizing *Cucurbita pepo. Apicolt.* 35(1): 9-12.

Bohart, G. E., and Nye, W. P. 1960. Insect pollinators of carrots in Utah. *Utah Agr. Expt. Sta. Bul.* 419, 16 pp.

Bohart, G. E., W. P. Nye and E. R. Wawrhorn. 1970. Onion pollination as affected by different levels of pollinators activity. *Utah Agric. Expt. Sta. Bull.* 482: 13-22.

Bommarco R, Lundin O, Smith HG, Rundlof M. 2012 Drastic historic shifts in bumble bee community composition in Sweden. Proc. R. Soc. B 279, 309–315. (doi:10.1098/ rspb. 2011.0647)

Bors, L. U. H. and Jasper, G. 1992. Commercial vegetable processing. Table-15.6. AVI Pub. Comp. Inc., West port, Conneticut.

Bos MM, Veddeler D, Bogdanski AK, Klein A-M, Tscharntke T, Steffan-Dewenter I, Tylianakis JM. 2007 Caveats to quantifying ecosystem services: fruit rspb. royal society publishing.org Proc R Soc B 280: 20122243 Downloaded from rspb.

royalsocietypublishing.org on February 21, 2013 abortion blurs benefits from crop pollination. Ecol. Appl. 17, 1841–1849. (doi:10.1890/06-1763.1)

Canadian Association of Professional Apiculturists (CAPA). 1995. *A guide to managing bees for crop pollination.* Scarborough: Esstee Graphics.

Caron, D. M., R. C. Lederhouse and R. A. Morse. 1975. Insect pollinators of onion in New York State. *Hortsci.* 10:273-274.

Carreck, N. L., and Williams, I. H. 1998. The economic value of bees in the UK. *Bee World, 79*, 115-123.

Chatterjea, M. N. and Shinde. R. 1998. Text book of Medical Biochemistry. Jaypee Brothers Medical Publishers: 173.

Cobley, L.C. and Steele, W.M. 1976 *An Introduction to the Botany of Tropical Plants*, 2nd Connor, I. J. 1969. Honey bee pollination requirements of hybrid cucumbers *Cucumis sativus* L. M.Sc. Thesis, Mich. State Univ. 150 pp.

Costanza et al. 1997a. The value of the world's ecosystem services and natural capital. *Nature*, *387*, 253-260.

Crane, E. and P. Walker. 1984. Pollination directory for world crops. Intl. Bee Res. Assoc., London, U.K.p55.

Cruess, W. C. 1958. Commercial fruit and vegetable products (4th Ed.). Table-45. Mcgraw-Hill Book Co. Inc. New York, Toronto, London.

Cuningham, S. A., Fitzgibben, F. and Heard, T. A. 2002. The future of pollinators for Australian Agriculture. *Agriculture.* 53(8): 893-900.

Deodikar, G.B. and Suryanarayana, M.C., 1977. Pollination in the services of increasing farm production in India. In P.K.K. Nair (ed) *Advances in Pollen Spore Research,*. New Delhi: Today and Tomorrow Printers and Publishers. pp 60-82.

Devkota, F. P. 2000. Impacts of bee pollination on the yield of broccoli (*Brassica campestris* var. *botrytis* L.) under Chitwan condition. *M. S. Thesis*, Institute of Agriculture and Animal Sciences, Rampur, Chitwan, Nepal.pp95.

Dulta, P. C., and Verma, L. R. 1987. Role of insect pollinators on yield and quality of apple fruit. *Indian Journal of Horticulture, 44*, 274-279.

Eilers EJ, Kremen C, Smith Greenleaf S, Garber AK and Klein A-M. 2011 Contribution of pollinator mediated crops to nutrients in the human food supply. PLoS ONE 6, e21363. (doi:10.1371/journal. pone.0021363)

Faegri, K. and Van der Pijl, L. 1979. *The Principles of Pollination Ecology, 3rd revised edition.* Pergamon Press, Oxford.London

Food and Agriculture Organization (FAO) (2001) *Incorporating Nutrition Considerations.*

Free, J. B. 1993. Insect pollination of crops (2 ed.). Academic Press, Harcourt Brace Jovanovich Publ., London. 684 p.

Free, J.B. 1970. *Insect Pollination of Crop Plants.* Academic Press, London; New York. 544 pp.

Garibaldi LA et al. 2011 Stability of pollination services decreases with isolation from natural areas despite honey bee visits. Ecol. Lett. 14, 1062–1072. (doi:10.1111/j.1461-0248.2011.01669.x)

Girish, P. P. 1981. Role of bees in the pollination of summer squash (*Cucurbita pepo* L.) with special preference to *Apis cerana* F. (Hymenoptera: Apidae). *M. Sc. Thesis*, Univ. Agric. Sci., Banglore, India.

Gopalan, C., Rama Sastri, B. V. and Balasubramanian, S. C. 1996. Nutritive value of Indian foods. National Institute of Nutrition. Indian Council of Medical Research, Hyderabad, India. pp.69-70

Gormley, T.R. (1989) Studies on the role of fruit and vegetables in the diet. *Professional Horticulture* 3, 66–68.

Grewal, G. S. and A. S. Sidhu. 1978.Insect pollination of some cucurbits in Punjab. *Indian J. Agric. Sci.* 48 (2): 79-83.

Gupta, J. K., Rana, B. S., and Sharma, H. K. 2000. Pollination of kiwifruit in Himachal Pradesh. In Asian Bees and Beekeeping: Progress of Research and Development. *Proceedings of the Fourth International Conference, Kathmandu* (p.274).

Hand Book on Horticulture Statistics 2014. Government of India Ministry of Agriculture Department of Agriculture and Cooperation New Delhi 1-48 pages

Hanif R, Iqbal Z, Iqbal M, Hanif S and Rasheed M 2006. Use of vegetables as nutritional food: role in human health. Journal of Agricultural and Biological Science 18(1): 18–22.

Hawthorn, L. R., Bohart, G. E., and Toole, E. H. 1956. Carrot seed yield and germination as affected by different levels of insect pollination. *Amer. Soc. Hort. Sci. Proc.* 67: 384 - 389.

Hawthorn, L. R., G. E. Bohart and E. H. Toole. 1960. Carrot seed production as affected by insect pollination.*Utah Agric. Expt. Sta. Bull.* 422. 18 p.

Herrera CM. 2000 Measuring the effects of pollinators and herbivores: evidence for nonadditivity in a perennial herb. Ecology 81: 2170–2176. (doi:10.1890/0012-9658(2000) 081[2170:MTEOPA]2.0.CO;2)

Hodson A Z and Norris, L. C.1939. A fluorometric method for determining the riboflavin content of foodstuffs. J. Biol. Chem.131: 621–630.

Howard, F. D, Macgillivray, J. H and Yamaguchi, M. 1962. Nutrient composition of fresh Californian grown vegetables. Cali. Uni. Agric. Expt. Sta. Bull 788.

Hussain, T. 1995. Food composition table for Pakistan, Department of Agric. Chemistry and Human Nut. NWFP, Agric. Univ. Peshawar: 18-30.

Hussain, T. 2001. Food composition table for Pakistan (Revised 2001). Dept. of Agric. Chemistry, NWFP Agric. Uni. Peshawar, UNICEP and Ministry of Planning and Development, Govt. of Pakistan. Islamabad. pp.14-15.

Indian Gyan. 2000. Importance of vegetables. http://www.indiangyan.com/books/healthbooks/food_that_heal/home2 (19 October, 2004).

Isaac, R. R and Johnson, W. C.1975. Collaborative study of wet and dry ashing technique for elemental analysis of plant tissue by Atomic Absorption Spectrophotometer. J. AOAC. 58: 436-440.

Jodha, N. S. and Shrestha, S. 1993. Sustainable and more productive mountain agriculture: problems and prospects. In *Mountain Environment and Development –Part 3 (Thematic Papers),* Kathmandu: ICIMOD. pp 1-65.

Kafle, G. P. 1992. Salient features of beekeeping in Nepal. In: L. R. Verma (ed.) *Honeybees in the Mountain Agriculture.* Oxford and IBH Pub. Co. Pvt. Ltd. New Delhi, India. pp 155-162.

Kenmore, P. and Krell, R., 1998. Global perspectives on pollination in agriculture and agroecosystem management. *International Workshop on the Conservation and*

*Sustainable Use of Pollinators in Agriculture, with Emphasis on Bees.* October 7-9, 1998, Sao Paulo, Brazil.

Kenmore, P., & Krell, R. 1998. Global perspectives on pollination in agriculture and agroecosystem management. *International Workshop on the Conservation and Sustainable Use of Pollinators in Agriculture, with Emphasis on Bees. Sao Paulo, Brazil.*

Kevan, P.G., E.A. Clark and V.G. Thomas. 1990. Insect pollinators and sustainable agriculture. *Am. J. Alternative Agric.* 5:13-22.

Khalil, I. A. and Rehman, S. 1977. Ber (Zizyphus Jujbe Mill) as a nutritive fruit. J. Sci. Tech. (Univ. Peshawar). 1:126-128.

Khalil, I. A. and Saleemullah. 2004. Chemistry One. Bio-analytical Chemistry. T. S.Printers, Pakistan: 41.

Khan, R. B. and Dethe, M. D. 2004 Median Lethal Time of New Pesticides to foragers of honeybee. *Pestology*. 28(1): 28-29.

Klein A-M, Vaissie`re B, Cane JH, Steffan-Dewenter I, Cunningham SA, Kremen C, and Tscharntke T. 2007 Importance of crop pollinators in changing landscapes for world crops. Proc. R. Soc. B 274, 303–313.

Krupnick GA, Weis AE, Campbell DR. 1999 The consequences of floral herbivory for pollinator service to Isomeris arborea. Ecology 80, 125–134.

Levin, M. D. (1984). Value of honeybee pollination to United States agriculture. *Am Bee J., 124*, 184-186.

Levin. M. D. 1989. Honeybee pollination of egg plant (*Solanum melongena* L.). The Thirty-first Intl. Agric. Cong., August 1987, Warsaw, Poland. Apimondia Pub. House, Bucharest. pp. 344-348.

Maschinski J and Whitham TG. 1989 The continuum of plant responses to herbivory: the influence of plant association, nutrient availability, and timing. Am. Nat. 134, 1–19. (doi:10.1086/284962)

Mayer, D. F. 2003. Bee kills from pesticides and the expert witness. *American Bee Journal.* 143 (7): 563.

McCall AC and Irwin RE. 2006 Florivory: the intersection of pollination and herbivory. Ecol. Lett. 9, 1351– 1365. (doi:10.1111/j.1461-0248.2006.00975.x)

McGregor, S.E. 1976. *Insect Pollination of Cultivated Crop Plants.* U.S. Dep. Agric., Agric. Handb. 496. 411 pp.

Mishra, R. C., J. K. Gupta and J. Kumar. 1988. Effect of modes of pollination on fruit characteristics of okra, *Abelmoschus esculentum* (L.) Moench. *Proc. Indian Natl. Sci. Acad.* 53B (2):157-160.

Mishra, R.C. 1997-98. *Perspectives in Indian Apiculture.* Agro. Botanica, HS Offset Printers, New Delhi. pp 188.

Morse, R. A., and Calderone, N. W. (2000). The value of honey bees as pollinators of U.S. crops in 2000. *Bee Culture, 128*, 1-15.

Mothershead K and Marquis RJ. 2000 Fitness impacts of herbivory through indirect effects on plantpollinator interactions in Oenothera macrocarpa. Ecology 81, 30–40.

Neupane, K. R. 2001. Foraging preference of honeybee species to selected horticultural crops. *M. Sc. Thesis.* submitted to IAAS, Rampur, Chitwan, Nepal.pp 95.

Nieuwhof, M. 1969. *Cole Crops.* 353 pp. Leonard Hill, London.

*Nudelman, N.S.* and *Nudelman.* O. 1976. Specific colorimetric determination of niacinamide in dosage forms. J. Pharm Sci. 65(1):65-7.

Oomen, H.A.P.C. and Grubben, G.J.H. 1978 *Tropical Leaf Vegetables in Human Nutrition.*

Partap, U. (1998). Successful pollination of apples in Himachal Pradesh. *Beekeeping and Development, 48*, 6-7.

Partap, U. 2000a. Foraging behaviour of *Apis cerana* on citrus (*Citrus sinensis* var. Red Junar) and its impact on fruit production. In Asian Bees and Beekeeping: Progress of Research and Development. *Proceedings of the Fourth AAA International Conference 23-28 March, 1998, Kathmandu* (p.274). New Delhi: Oxford and IBH Publishing Co. Pvt. Ltd.

Partap, U. 2000b. Pollination of strawberry by the Asian hive bee, *Apis cerana* F. In Asian Bees and Beekeeping in Asia: Progress of Research and Development. *Proceedings of the Fourth AAA International Conference 23-28 March 1998, Kathmandu* (p.274). New Delhi: Oxford and IBH Publishing Co. Pvt. Ltd.

Partap, U. 2011. *The pollination role of honeybees.* In H. Randall Hepburn, & Sarah E. Radloff (eds) *Honeybees of Asia* (pp. 227-255). Berlin: Springer Science & Business Media.

Partap, U. and Partap, T., 1997. *Managed Crop Pollination: The Missing Dimension of Mountain Agricultural Productivity.* Mountain Farming Systems' Discussion Paper Series No. MFS 97/1. Kathmandu: ICIMOD.

Partap, U., and Partap, T. 1997. *Managed Crop Pollination: The Missing Dimension of Mountain Agricultural Productivity.* Mountain Farming Systems' Discussion Paper Series No. MFS 97/1. Kathmandu, ICIMOD.

Partap, U., and Partap, T. 2000. Pollination of apples in China. *Beekeeping and Development, 54*, 6-7.

Partap, U., and Partap, T. 2002. *Warning Signals from Apple Valleys of the HKH Region: Productivity Concerns and Pollination Problems,* Kathmandu: ICIMOD, pp 106.

Partap, U., and Verma, L. R. 1992. Floral biology and foraging behaviour of Apis cerana on lettuce crop and its impact on seed production. *Progressive Horticulture, 24*, 42-47.

Partap, U., and Verma, L. R. 1994. Pollination of radish by *Apis cerana. Journal of Apicultural Research, 33*, 237-241

Partap, U., Partap, T. and Yonghua, H. 2001. Pollination failure in apple crop and farmers' management strategies. *Acta Horticulturae, 561*, 225-230.

Partap, U., Shukla, A. N., and Verma, L. R. 2000. Impact of *Apis cerana* pollination on fruit quality and yield in peach and plum in the Kathmandu valley of Nepal. In (M. Matsuka, L. R. Verma, L. R. and Partap, U. 1993. *The Asian Hive Bee, Apis cerana, as a Pollinator in Vegetable Seed Production* (p.52). Kathmandu: ICIMOD.

Pearson, O. H. 1932. Breeding plants of the cabbage group. *Calif. Agr. Expt. Sta. Bul.* 532, 22 pp.

Pesson, P., and J. Louveaux, eds. 1984. *Pollinisation et Productions Végétales.* INRA, Paris. 663 pp.

Plowright, R.C., and T.M. Laverty. 1987. Bumble bees and crop pollination in Ontario. *Proc. Entomol. Soc. Ontario* 118:155-160.

Pokhrel, S. 2006. Status and management of domesticated and wild honeybees (*Apis* spp.) in Chitwan, Nepal. Ph.D. Dissertation, TU, Institute of Agriculture and Animal Sciences, Rampur, Chitwan, Nepal. 243 pp.

Potts SG, Biesmeijer JC, Kremen C, Neumann P, Schweiger O, Kunin WE. 2010 Global pollinator declines: trends, impacts and drivers. Trends Ecol. Evol. 25, 345–353. (doi:10.1016/j.tree.2010.01.007)

Price, P. 1975. *Insect Ecology*. John Wiley and Sons, New York. USA.pp453.

Rai, N., Yadav, D. S., Rai, A. B., Rai, M., Yadav, R. K., and Sanwal, S. K. 2008. Enhancing vegetable production in north eastern hill region. *ENVIS Bulletin: Himalayan Ecology, 16*(2), 1-7.

Robinson, D. S. 1990. Food biochemistry and nutritional value. Longman scientific and technical publisher, NewYork. USA

Robinson, W.E., and R.A. Morse. 1989. The value of honeybees as pollinators of US crops. *Am. Bee J.* 129 (1): 477-487.

Royal Tropical Institute, Amsterdam and Orphan Publishing Company, Willemstad, Curacao, 140 pp.

Seppelt R, Dormann CF, Eppink FV, Lautenbach S and Schmidt S. 2011 A quantitative review of ecosystem service studies: approaches, shortcomings and the road ahead. J. Appl. Ecol. 48, 630–636. (doi:10.1111/j.1365-2664.2010.01952.x)

Sharma, D. 2007 Relative toxicity of some insecticides to adult foragers and brood of *Apis mellifera* L. *PhD Thesis*. SKUAST-Jammu. 225pp.

Sharma, K. C. 1994. Current experiences and practices in pesticide use in the Bagmati zone. *ADPI series #9*. ICIMOD, Kathmandu, Nepal.

Sharmah D., Khound Amrita, Rahman S. and Rajkumari P. 2015. Significance of Honey Bee as a Pollinator in Improving Horticultural Crop Productivity in N.E. Region, India: A Review *Asian Journal of Natural & Applied Sciences* 4(1):62-69

Shelar, D. G. and M. C. Suryanarayana. 1981. Preliminary studies on pollination of coriander (*Coriandrum sativum* L.) *Indian Bee J.* 43:110-111.

Shrestha, J. B. and K. K. Shrestha. 2000. Beekeeping in Nepal: Problems and potentials. In: *Fourth Asian Apicultural Association International Conference* Kathmandu, Nepal.In: M. Matsuka, L.R. Verma, S. Wongsiri, K. K. Shrestha, and U. Partap (eds.) *Asian Bees and Beekeeping, Progress of Research and Development*. Oxford and IBH Publishing Co. Pvt. Ltd., New Delhi, India. pp262-265.

Sihag, R. C. 1986. Insect pollination increases seed production in cruciferous and umbelliferous crops. *J. Apic. Res*. 25:121-126.

Smirnoff, N. 2000. Ascorbic acid: metabolism and functions of a multi-facetted molecule. Current Opinion in Plant Biology. 3: 229-235

Sonni Alveez. 2002. Importance of minerals and trace minerals in human nutrition. Web site: http://www.mgwater.com/impor.shtml (15 April 2004)

Sorthia, B. K. and Chari, M. S. 1985. Toxicity of some insecticides to honeybees, *Apis florea* F. and *Apis mellifera* L. *Journal of Entomological Research.* 92 (2): 195-197.

Southwick, E. E., and Southwick, L. 1992. Estimating the economic value of honey bees (Hymenoptera: Apidae) as agricultural pollinators in the United States. *Journal of Economic Entomology*, *85*, 621-633.

Srivastava, V. K. 1969. Studies on the floral biology of *Abelmoschus esculentus* (L.) Moench. *India J. Hort*. 26:165- 170.

Steekel, J. E. and Flannery, R. L. 1965. Automated determination of P, K, Ca, Na, Mg in wet digestion solution of plant tissues. In technicon symposia, Automation in analytical chemistry. pp. 116-122. 22

Strauss SY and Irwin RE. 2004 Ecological and evolutionary consequences of multispecies plant animal interactions. Ann. Rev. Ecol. Evol. Syst. 35, 435–466. (doi:10.1146/annurev.ecolsys.35. 112202.130215)

Thakur, N. S. A., Firake, D. M., Behere, G. T., Firake, P. D., and Saikia, K. 2012. Recent Scenario of Insect-pests of Guava in North East India and Their Eco-friendly Management. *Indian Journal of Hill Farming, 26*(1), 55-57.

Thapa, R. B. 2003. Pesticide pollution and integrated pest management. In: F. P. Neupane (ed.) *Integrated Pest Management in Nepal Proceedings of the National Seminar,* 25-26 September, 2002 Himalayan Resource Institute, New Baneshwor, Kathmandu, Nepal. pp. 175-194.

Thapa, R. B. 2006. Honeybees and other insect pollinators of cultivated plants- A review.*J. Inst. Agric. Anim. Sci.* 27 : 1-23.

Thapa, R.B. 2002. Insect pollinators of some crop plants in Chitwan. *NAHSON Bull*.12/13:19-21.

United Nations Food and Agriculture Organization (FAO) 1961. *Agricultural and Horticultural Seeds*. FAO Agro-studies 55, 531 pp.

Verma, L. R., and Partap, U. 1994. Foraging behaviour of *Apis cerana* on cabbage and cauliflower and its impact on seed production. *Journal of Apicultural Research, 33*, 231-236.

Verma, L.R. and Partap, U., 1994. Foraging behaviour of *Apis cerana* on cabbage and cauliflower and its impact on seed production. *J. Apic. Res.* 33: 231-236.

Verma, L.R. and Partap, U., 1993. *The Asian Hive Bee, Apis cerana, as a Pollinator in Vegetable Seed Production.* Kathmandu: ICIMOD, 52pp.

Verma, S., and Attri, P. K. 2008. Indigenous beekeeping for sustainable development in Himachal Himalaya. *Indian Journal of Traditional Knowledge*, *7*(2), 221-225.

Weinberger, K. and Msuya, J. 2004 *Indigenous Vegetables in Tanzania.* Asian Vegetable Research and Development Center (AVRDC) Technical Bulletin No. 31. AVRDC, Taiwan, 70 pp.

Williams, I. H. 1994. The dependence of crop production within the European Union on pollination by honeybees Agricultural Science Revews 6:229-257

Williams, I.H., Martin, A.P. and White, R.P. 1987. The effect of insect pollination on plant development and seed production in winter oilseed rape (*Brassica napus* L.). *Journal Agricultural Science Cambridge* 109:135-139.

Winston, M.L. and Scott, C.D. (1984). The value of bee pollination to Canadian agriculture. *Canadian Beekeeping, 11*, 134.

Zhong, H. E., Mark, L., Sheve, P., and Cate, B. 2004. Minimizing the import if the mosquito adulterine noted or honeybees, *Apis mellifera* (Hymenoptera: Apidae): Aerial Ultra volume application using a high-pressure nozzle system. *Journal of Economic Entomology*. 97 (1): 1-7.

CHAPTER

2

# Pollination in Vegetable and Oilseed Crops Productivity

Seed or fruit production in vegetables depends heavily upon proper pollination and fertilization. The pollination requirements are different for different vegetable crops depending upon their different their breeding system. In some vegetable crops, male and female flowers are borne on different plants and in others, mature at different times thus requiring insects for pollination. For instance in case of watermelon plants have separate male and female flowers. The pollen is sticky and not blown by wind, so insect pollination is necessary to transfer pollen to receptive female stigmas. Similarly, cantaloupe plants have perfect flowers and imperfect, male flowers. The perfect flowers cannot self-pollinate, and the pollen is too heavy to blow in the wind, so insect pollination is necessary. Cucumber plants have both male and female flowers, and some agent is required to transfer pollen between flowers of the same or different plant. Since the population of naturally occurring pollinators is declining due to reduction of their habitat, indiscriminate use of pesticides, and attack of pests and diseases, evidently making dependence on domesticated hive bees. The number of colonies required for pollination depends upon number of factors which include attractiveness of bees to crop, strength of colonies, natural population of pollinators available in the area and prevailing weather conditions. Pollination diversity recorded on the bloom of different important vegetable crops is discussed below:

## B. OLERACEA

Cabbage has been reported to benefit greatly from insect pollination. Radchenko (1966) recorded an increase of 300 per cent in seed crop of cabbage due to insect pollination. Rauala (1972) reported 68 per cent and 9 per cent set in cages with and without bees, respectively. Muhammad et al. (1973) reported an increase of129% seed set seed set due to presence of honeybees. Sihag (1986) reported that open pollinated plants set 978 pods per plant as compared to those isolated from insect pests through caging which produced 13 pods per plant only.

Similarly, Alam et al. (1987) recorded 9.0 and 11.5 seeds per pod in caged (without insects) and open plots. Tewari and Singh (1983) demonstrated a decrease in seed set and yield with increasing distance from three *A. cerana* colonies. In a similar study, Kumar et al (1989) recorded higher seed set and seed weight in open pollinated flowers in five cultivars of cauliflower as compared to bagged ones.

## BRASSICA OLERACEA L. (CABBAGE, CAULIFLOWER, BRUSSELS SPROUT)

*Brassica* flowers are very attractive to insects especially honeybees. Flies (Syrphidae, Calliphorinae, Muscidae) and small beetles are occasional visitors. Several investigators have reported on the insect visitors of *Brassica oleracea* (Rauala, 1972; Dhaliwal and Sharma, 1973; Muhammad et al., 1973; Adlakha and Dhaliwal, 1979; Kakkar, 1981; 1983, Tewari and Singh 1983; Varma and Joshi, 1983; Kumar et al., 1988; Mishra et al., 1988).

Free and Williams (1973) recorded *Osmia rufa*, bumblebees and many dipteran species visiting Brussels Sprout flowers. They found all the wild bees and many syrphidae carrying pollen almost equal to honeybees. However, honeybees are the most important pollinators of *B. oleracea* comprising 98 per cent of visitors in Egypt (Hussein and Abden-Aal, 1982) and 99 per cent in New Zealand (Forster et al., 1973). Honeybees, *Apis cerana, A. dorsata and A. florea* have been recored as important pollinators of cauliflower in India from different localities. In some areas solitary bees and *Eristalis* spp. have been found as important pollinators. Sharma et al(1974) recorded thirty-four species of insects, including honeybees, visiting the cauliflower bloom at Solan (Himachal Pradesh), India. Eighteen insect species belonging to fourteen families under five orders were found visiting blossom of cauliflower at Hisar, Haryana, India by Priti and Sihag (1997).

## ABELMOSCHUS ESCULENTUS L.

Honeybee and bumbles are most frequent visitors of okra (McGregor, 1976). In okra, there is a positive correlation between the number of seeds in a pod and its weight, and insect pollination increased the number of seeds present in a pod (Free, 1976). Tanda (1985) reported nineteen percent increase in yield in Punjab, India due to intensive bee pollination.

Similarly, Mishra *et al.* (1987) found open pollination was superior as compared to those isolated from insect visits. They found that that the weight and length of capsules and seed number per capsule were significantly higher in open pollinated than in bagged flowers. Mishra et al. (1987) studied pollination requirement of 9 cultivars of okra at Solan. Ants were reported as frequent visitors (44.35%) followed by *A. cerana* (21.25%). Other visitors included dipteran flies, hymenopterans and beetles. However, only *Ceratina sexmaculatus, Megachile* sp., *Xylocopa* sp. and *Bombus* spp. were important pollinators.

Abrol(1995) reported *Apis dorsata*, *Apis cerana*, *Xylocopa fenestrata*, *X. pubescens*, *Lasioglossum* sp., *Halictus* sp., *Chalicodoma flavipes* and *C. lanatum* as the important flower visitors of okra. Ants, thrips, coccinellids, *Syrphus* sp. and butterflies were also recorded on okra blossoms in very low populations. Azo'o *et al.* (2011) on the other hand reported that the mean number of matured seeds is higher in open-pollinated plants and smaller in self-pollinated plants. In India for instance, honeybees *Apis cerana* and solitary bees *Halictus* spp. appeared as the main okra pollinators (Crane, 1991; Free, 1993).

Although okra *Abelmoschus esculentus* (L.) is a self pollinated crop, but cross pollination has been reported in it (Purewal and Randhawa 1947., Frankel and Galun 1977). A number of insects visit and pollinate okra flowers (Mcgregor 1976, Frankel and Galun 1977), Choudhary et al (1973) observed *Apis cerana* and *Halictus sp* to be the chief pollinators of this crop. According to Tanda (1985) the most important insect pollinators of okra were the honeybees *A. mellifera* and *A. c. indica* which mainly foraged for nectar.

Angbanyere and Baidoo (2014) reported honeybees were the main pollinators of okra flower, however their numbers were significantly different on the different treatments. Many of the other insects found on the plant were not involved in pollination of the flowers.

There were significant differences in the numbers of pods per plant, yield per plant and the numbers of seed per pod. Halder et al (2014) reported that the effective pollinators in case of okra (*Abelmoschus esculentus* L.) are *Hermetia illucens*. wild bees, *Eurema hecabe*, *Orthellia*sp etc. Interestingly, in non-urbanized area insect pollinators' diversity was higher.

## ALLIUM SPP. (ONION)

Onion flowers are highly cross-pollinated and depend upon pollinating insects for seed production. Honeybees have been reported as the most frequent visitors of onion bloom (Treherne, 1923; Trofimec, 1940; Agati, 1952; Bohart et al., 1970; Caron et al., 1975). Other visitors to the bloom of this crop were dipteran and Solitary bees. Wild bees and dipterans (Lederhouse et al., 1968) and *Halictus farinosus* (Parker, 1982) were reported as most abundant foragers of onion bloom in New York state and Utah, respectively. However, honey bees have been reported as important pollinators of onion flowers in India (Singh and Dharamwal, 1970; Jadhav, 1981 and Rao and Suryanarayana, 1989). Kumar et al. (1985) found that three onion species, *A. cepa, A. fistulosum* and *A. cepa fistulosum* were greatly benefited by insect pollination and *A. cerana* and dipteran flies were predominant insect visitors.

In Poland, Woyke (1981), found that plots in an open field were caged without bees, caged with bees and not caged produced 2, 210 and 669 seeds per head, respectively. In India, Kumar et al. (1989) found greater set and yield and better seed germination from plots caged with bees, than from plots caged without bees and open plots (61, 90, 72 % and estimated seed yield of 73, 275, 97 kg/ha respectively on plots caged without bees, with bees and left for open pollination). Rao and Suryanarayana (1989) have reported higher seed yield in onion due to placement of bee colonies.

## DAUCUS CAROTA L. (CARROT)

Carrot flowers depend upon insects for their pollination in order to produce seed. Carrot flowers are visited by wide variety of of insects. Bohart and Nye (1960) collected 334 species belonging to 71 families on carrot flowers at Logan, Utah, USA. The pollen gathering honeybees were most valuable than nectar gatherers. In Moscow, Pankratova (1958) found that dipteran comprised 90 per cent and honeybees 9 per cent of the insect visitors to carrot bloom. But dipteran were found to collect nectar only. Galuszka and Tegrek (1987) observed that honeybees did not visit carrot crops when more attractive flower species are blooming in the vicinity. Kumar et al. (1989) recorded 20 species of bees belonging to 8 genera of 6 sub-families and 4 families on carrot flowers. Most frequent visitors were *H. splendidulus, Allodape* sp., *Nomioides* sp., *H. vachallii* and *H. himalayensis*. On the basis of pollination indices, *Halictus splendidulus* was found to be the most efficient pollinator (Kumar and Rao, 1991).

Bohart and Nye (1960) reported that pollen collector honeybees on carrot bloom literally waded across the head, swung their abdomens back and forth and scrapped the pollen from stamens with their forelegs. The nectar collectors stood higher on the flowers, moved about less and lapped droplets from the exposed nectaries. Side foraging behavior of honeybees A. mellifera and A. cerana for nectar collection was observed by Kapoor and Dhaliwal (1989) and Kumar et. al (1994) in cauliflower in Solan, Himachal Pradesh India. Similarly, Mishra et al. (1987) found nectar robbing by A. cerana in Okra flowers.

Abrol (1997) found that 95% of the flower visitors were dwarf honeybees, Apis florea with the balance of insect visitors as dipteran flies (Musca sp. and Syrphus sp.), halictine bees (*Halictus* sp.: Lasioglossum sp.) and black ants (*Camponotus compressus* Fabricius). Sinha and Chakrabarti (1992) in Karnal, India recorded dipterans (sepsid and syrphid flies) as the most abundant flower visitors constituting 90% of the total insect visitors. Ahmad and Aslam (2002) in Faisalabad, Pakistan also recorded A. florea) as the the most frequent visitors to carrot flowers.

## POLLINATION FOR HYBRID SEED PRODUCTION

Honeybees not only play an important role in the pollination of crops thereby enhancining quality and quanity of seed/ fruit production but also in specialized task like hybrid seed production. Honeybees have been used for this purpose in onion (Woyke and Dudek, 1983; Woyke, 1981), cabbage (Kubisova et al., 1987), cauliflower (Woyke, 1989; Gupta et al. 1984) carrot (Erickson and Peterson, 1979 a,b; Galuszka and Tegrek, 1987; Rodet et al., 1991) and cucumber (Ruszkowski and Bilinski, 1984)

## POLLINATION IN GREEN HOUSES/CAGES

Variety of pollinationg insects have been used for pollination under green houses, glass houses or cages which include. honeybees, bumble bees, solitary bees, syrphids and blowflies. However, Bumble bees hold promise as best pollinators under small enclosures as they can be readily obtained from flowers or by collecting their nests. They forage at very low temperatures at which honeybee activity is limited. Bumblebees are now being used to pollinate many vegetable crops in green houses. Similarly, the use of honeybees has increased in in green houses pollination. They have been used to pollinate Brussels sprout, muskmelon, onion, runner bean, tomatoes etc. in green houses or in large pollination cages (Mishra et al., 1997).

## ROLE OF HONEYBEES IN OILSEED PRODUCTION

Oilseed crops benefit greatly from insect pollination both in quantity as well as quality in terms of production of oil. The productivity of oilseed crops in india is far below the world productivity levels in spite of best agronomic practices. Of the several factors responsible for low yields lack of proper pollination at the right time the crop requires them. Failure of seed setting is attributable to the lack of proper pollination due to insufficient number of pollinators available for pollination (Rao *et al.* 1980; Free 1993; Abrol 2007, 2008, 2009). Insect pollination not improves the quality and quantity of the produce but also helps in early and uniform setting of the crop.

*Brassica rapa* L. (Cruciferae) is an important vegetable crop in Asia (Rubatzkky and Yamaguchi 2000). It has been found that more than. 95% of species of Crucifers require cross-pollination, although some cauliflower varieties are self-fruitful.

Delaplane and Mayer (2000) found that selfing in the absence of cross pollination reduces seed yield, seed size, and yield in subsequent generations. Self-incompatible plants require pollen transfer from plant to plant (Wallace *et al.* 2002). Planned pollination has been found to greatly benefit quality and quanitity of seed production in crucifers.

In male sterile oilseed rape (*B. napus*), yields of the species were increased by honey bee pollination (Westcott and Nelson 2001). Similarly, in sarson (*B. campestris*), insects pollination increase the seed yield, caused formation of well-shaped, larger grain, and more viable seed (Abrol 2009).

Many species of *Brassica,* such as rape (*Brassica napus*), sarson (*Brassica campestris* var. sarson), raya (*Brassica juncea*), white mustard (*Brassica alba*), and black mustard (*Brassica nigra* and toria (*Brassica campestris* var. toria) are cultivated widely as oilseed crops throughout the world. Most of these crops bloom during February-March for over a month. The bright yellow fragrant flowers are bright yellow and are produced in long terminal racemes. These crops are usually self-pollinated, but some degree of cross-pollination occurs in *Brassica campestris*. The brassica flowers produces a good amount of nectar and pollen and is highly attractive to honeybees and other natural insect pollinators. Cross pollination by honeybee increases yield, quality and oil content of seed. The average sugar concentration in the afternoon ranges between 40-45 % whereas in the morning it is 16 % only (Sharma 1958; Murrel and Nash 1981). Mohammed (1935) in West Punjab at Lyallpur recorded 117 visiting species of insects belonging to seven orders; *Andrena ilerda*, *Apis florea*, *Halictus* spp. comprised 82 % of insects and were the most important in this order. He also observed that stigma of the crops remained receptive for two days after opening and in 5 days their fertility was lost whereas in case of pollen viability continued for 7 days. Langridge and Goodman (1975) in Australia observed that major proportion of A. mellifera foragers were nectar collectors (72 %) followed by nectar and pollen collectors (25%) and exclusively pollen collectors whose percentage was very small (3% only). The pollen was collected between 10 am to 2 pm.

Mohammed (1935) reported that by enclosing individual plants in muslin bags; in toria, seed set was 12.3 % in brown sarson, 20.3 % and in yellow seeded sarson 91.0 %. Then he compared self pollination to cross pollination in toria. He bagged flowers for cross pollination and hand pollinated them; there was 100 % pod formation. Sihag (1986) studied seed set in toria plots caged and not caged and left open for natural cross pollination. He found that there developed 25 and 1060 pods per plant, 6.3 and 18.2 seeds per pod, and 4 and 385 seeds per plant, respectively in different treatments.Similar observations by Mishra *et al.* (1988) on sarson gave 59 and 95 % pods containing 3.5 and 12.7 seeds per pod. Observations made in field plots of both sarson and toria showed that when crop was caged to exclude insects yield was 68 g; when caged with *Apis cerena* gave 219 g and in open field near honeybee colonies where other insect pollinators were also present yield was 244 g (Latif *et al.* 1960b). In general there is quite a high degree of self incompatibility in these crops and good yields are obtained by cross pollination if with the aid of natural fauna in most places where good weather prevails and probably sufficient number of natural pollinators are available and yield can be further increased with the help of honey bee colonies to the extent of 10-30 % (Rahman 1940; Singh 1954) covering for inclement weather. Singh and Singh (1992) observed that bee pollinated plants produced 3 times heavier pods, 4 times more seed per pod, 50 times more seeds and 84 times more seeds yield per plant in field cages as compared with self pollinated plants in individual bags

Mishra *et al.* (1988) reported that % pod setting was significantly higher in plants which had access to insect pollination (95.21 %) than those which were net caged ( 70.27 %) or muslin bagged (58.81 % ). Number of seeds per pod was also significantly higher in open pollinated flowers 912.62 seeds /pod) in comparison to nylon caged (5.55 seeds/

pod) and muslin bagged (3.05 seeds/pod) in field cages as compared with self pollinated plants in individual bags than those from net caged (41.45 %) or muslin bagged (48.37 %) flowers; the latter two, however, did not differ from each other. The average weight of 100 healthy dry seeds from muslin bagged flowers was significantly higher (542.20 mg) than the weight of seeds from open pollinated (478.50 mg), and net caged flowers (459.40 mg). Per cent oil content of healthy seeds from muslin bagged (41.64 %) flowers was significantly higher than that from open pollinated (39.02 %) flowers.

Pollination by bees in rapeseed produces an increase in productivity, improving the yield and contributing to the uniformity and early establishment of the pod (Abrol, 2007). According to Sabbahi *et al.* (2005) this results from the abundant production of flowers during the flowering period, which ensures proper pollination and thus a high and relatively early production of siliques. Therefore, the absence of cross pollination, which occurs when free plants in the presence of pollinators interchange pollen, is responsible for the low production of well developed siliques.

Deodikar and Suryanarayana (1972) found enhanced yield in oilseed crops when pollinated by bees. Langridge and Goodman (1975) also found that open pollinated plots of rapeseed resulted in significant more production of seeds with better quality than the caged ones. Plots receiving unrestrained visits had significantly higher plant density, seed yield, seed/plant, weight of 1000 seeds, weight of seeds per plant, germination and % tage of oil content as compared to those isolated from insect visits.

**Table 2.1:** *Expected increase in crop yield due to bee pollination*

| Crop | Pollination requirements | Expected percent increase in yield due to cross-pollination |
|---|---|---|
| Coconut | Self pollination and possibly wind pollination appear to be the norm for this cultivar | 5 |
| Mustard rape | Mostly cross pollinated | 20 |
| Sesamum | Often cross pollinated | 20 |
| Sunflower | Cross pollinated | 20 |
| Cotton | Often cross pollinated | 10 |
| *Brassica napus* | Often cross pollinated | 50 – 72 |
| *Brassica napus* | Often cross pollinated | 35 – 51 |
| *Brassica rapa* | Often cross pollinated | 78 – 89 |
| *Brassica rapa* | Often cross pollinated | 39 – 69 |
| *Sinapis alba* | Often cross pollinated | 51 – 75 |
| Mustard | Mostly cross pollinated | 13.00 – 222.00 |
| Safflower | Often cross pollinated | 4.00 – 114.00 |
| Sunflower | Often cross pollinated | 21.00 – 3,400 |
| Sesame | Often cross pollinated | 24.00 – 40.00 |
| Niger | Often cross pollinated | 17.00 – 45.00 |
| Linseed | Mostly cross pollinated | 2.0 – 49.00 |

***Source:*** Abrol (2009)

Kevan and Eisikowitch (1990) reported that due to pollination of canola of germinability seeds increased from 83% to 96. Yield and oil content was higher in honeybee pollinated crops in comparison to self pollinated ones (Mishra and Kaushik (1992). The results showed that higher percentage of pod setting was obtained in open pollinated crops of *Brassica nigra, B.campestris, B. carinata, B. napus* and *Eruca sativa* having 80.00, 84.42, 71.90 and 70.30 per cent, respectively. Average weight of 1000 healthy seeds varied between 3.0 to 5.1 g under open pollination, whereas it was 1.8 to 3.7 g under self pollination. However, Langridge and Goodman (2003) in a trial on oilseed rape (*Brassica napus*) cv. Midas in northern Victoria, did not find any increase in yield of seed, oil content, or percentage germination from plots where bees and larger insects had access compared with plots enclosed in cages of 2.5 mm mesh to exclude these insects. The weight of 1000 seeds was slightly greater in the enclosed plots. Bees and a Syrphid, *Melangyna viridiceps*, were the predominant insects visiting the flowers. Evidently, pollination is one of the most important natural factors enhancing crop production. The studies on pollination of oil seed crops clearly reveal that cross pollination not only improves quantitatively but improves the quality of oil crops as well. Even self compatible varieties give enhanced yields and improve qualitatively when cross pollinated. Therefore, pollination in oil crops is of great significance for boosting production of oil crops in a world deficit of oil/oilseeds.

## PROBLEMS ASSOCIATED WITH BEE POLLINATION

Production of oilseed crops have two different approaches which are antagonistic to each other. The use of pesticides for the control of crop pests and pollination for quality and quantity of seed production have become essential components of modern agriculture. Unfortunately, both these approaches are not always compatible to each other, as honeybees are susceptible to many of commonly used pesticides (Johansen 1977., Russell *et al.* 1998., Cunningham *et al.* 2002; Sundararaju 2003), used for the control of insect pests (Poehling 1989; Stark *et al.*1995). The indiscriminate and excessive use of pesticides for controlling insect-pests (Bisht *et al..,* 1980, Rana and Goyal 1991., Zhong *et al.* 2004) is a major problem in plant pollinator relationship. The loss of honeybees colonies directly effect honey production and indirectly the crop production through inadequate pollination. (Bai and Reddy 1977; Pimentel *et al.* 1980; Crane and Walker 1983; Praksah and Kumaraswami 1984; Khan and Dethe 2004).

Flowering *Brassicas* attract a large number of insect pollinators especially the honey bees for nectar and pollen and also other insects such as aphids which feed on flowers, leaves, fruits thereby, causing serious economic losses (Sihag 1991; Abrol 2007, 2008, 2009). These crops are attacked by number of insect pests such as aphids, caterpillars and bugs during flowering and pod formation stage which requires the application of insecticides to combat the pest (Sihag 1986a & b; 1988, 1991, Sihag *et al.* 1999a & b). The application of pesticide poses serious problem for the foraging activity of honeybees and developing brood. Aphid control becomes a major problem during pollination when the crop is in bloom and bees are actively foraging on the flowers for nectar and pollen in the field. Insecticides presently registered for use in seed crops are considered highly toxic to bees. Thus, to avoid killing bees, the colonies are either temporarily removed from the field, or applications must be made at night. In either case, several days can pass before the bees resume normal foraging activities. Repeated applications of pesticides to control aphid populations can cause significant disruption in pollination and reduce seed yield and quality.

Bee poisoning from pesticides remains to be a serious problem for beekeepers. Mayer (2003) reported that most bee kill occurs when pesticides are applied or allowed to drift on to blooming crops or weeds. The major bee losses occur when bees pick up pesticides while foraging. The hazards of insecticidal application on flowering crops include- direct mortality, fumigative effects, repellent effect and toxicity of residues present on various floral parts and in nectar to the insect visitor. A highly toxic insecticide generally reduces the field force of a colony within a short period of time. Colonies may be reduced by one third to half in strength within 24-48 hours (Eckert and Shaw 1960), thus adversely affecting both the production and marketing segments of the honey and beekeeping industry. Poisoning of insecticides to honeybees is generally more pronounced because of their long hours of working on the crop flowers for pollen and nectar collection. Bee poisoning is more chronic in areas which lack of sufficient wild pollens and nectar plants to sustain the number of colonies required for crop pollination. Most bees colonies are placed in agricultural areas where they are subjected to killing from pesticides (Mayer and Johansen 1983). The beekeeper has little or no control of when and what pesticides are applied in the areas of his bee forage (Mayer 2003), therefore, there needs to be coordination between beekeepers and the farmers.

Furthermore, population of wild bee pollinators have considerably declined due to modern agricultural practices such as disturbed insect flower- relationship by way of disappearance destruction reduction of wastelands and uncultivated strips of land used by them for nesting. Wild bees have also been damaged by pesticides either by the food from flowers, leaves and soil used for nesting. The toxicity of a levels to both honey bees and wild bees are not always the same and even in wild bees some materials are more toxic to one species than to another.

## References

Abrol DP 1997.Impact of insect pollination on carrot seed production. Insect Environment 3: 61

Abrol, D. P. 2007. Foraging behaviour of *Apis mellifera* and *Apis cerana* as determined by the energetics of nectar production in different cultivars of *Brassica campestris* var. *toria*. Journal of Apicultural Science. 51(2):5-10.

Abrol, D. P. 2008. Plant insect interaction in crucifers *In:* Biology and Breeding of Crucifers. (Ed.) Dr. S.K.Gupta, CRC Press, Taylor and Francis, USA pp129-149.

Abrol, D. P. 2009 *Bees and Beekeeping in India*. 2nd Edition Kalyani Publishers, New Delhi. Pp720.

Abrol, D. P. 2009. *Honeybee Diseases and their Management.* 2$^{nd}$ edition. pp750. Kalyani Publishers, Ludhiana.

Adlakha, R.L. and Dhaliwal, H.S. 1979. Insect pollination of seed cauliflower (*Brassica oleracea var. botrytis*) with special reference to the role of honeybees. Indian Bee J. 41: 13-16.

Agati, G. 1952. Indagini ed osservazioni sulla biologia fiorale della cipolla. Riv. Ortoflorofruttic. ital. 36: 67-77.

Ahmad M, Aslam M 2002.Pollinators visiting carrot (*Daucus carota L.*) seed crop. Journal of Research (Science) 13: 3135.

Alam, M.Z., Quadir, M.A. and Ali, M.S. (1987b). Pollinating behaviour of honeybee. *Apis indica* F. and its influence on seed production of cauliflower. Bangladesh Horticulture 15: 25-30.

Bai, A. R. K. and Reddy, C. C. 1977. Laboratory toxicity of some insecticides to *Apis cerana indica. Journal of Apicultural Research.* 16 (3): 142, 161-162.

Bhalla, O.P., Verma A.K. and Dhaliwal, H.S. 1983 a. Insect visitors of mustard bloom (*Brassica campestris* var. *sarson*) their number and foraging behaviour under mid hill conditions. J. Ent. Res. 7: 15-17.

Bisht, D.D., Naim, M., and Mehrotra, K.N. 1980. Studies on the role of honeybees in rapeseed production. In: *Proc. 2nd Int. Conf. Apiculture in Tropical Climate.* Indian Agricultural Research Institute, New Delhi, p. 491–496.

Blackwall, F.L. 1971 a. Pod-setting and yield in the runner bean (*Phaseolus mutiflorus*). J. Royal Hort. Soc. 96: 121-130.

Bohart, G.E. and Nye, W.P. 1960. Insect pollinators of carrots in Utah. Bull. Utah agric. Exp. Stn. 419.

Bohart, G.E., Nye, W.P. and Hawthorn, L.R. 1970. Onion pollination as affected by different levels of pollinator activity. Utah Agricultural Experiment Station Bulletin. 482.

Caron, D.M, Lederhourse R.C. and Morse, R.A. 1975. Insect pollinators of onion in New York State. Hortscience 10: 273-274.

Crane, E. and Walker, P. 1983, Pollination of tropical and subtropical crops by bees. In: Crane, E., Walker, P. *The impact of pest management on bees and pollination.* Cardiff: IBRA. pp. 5-21.

Cuningham, S. A., Fitzgibben, F. and Heard, T. A. 2002. The future of pollinators for Australian Agriculture. *Agriculture.* 53(8): 893-900.

Delaplane, K. S. and Mayer, D. F. 2000. Crop pollination by bees. CABI Publishing, New. York, USA.

Dhaliwal, H.S. 1980. Ecology and relative efficiency of principal insect pollinators of *Brassica* seed crop. Ph.D. Thesis, HPKV, COA, Solan, HP, India.

Dhaliwal, H.S. and Sharma, P.L. 1973. The foraging range of the Indian honey bee on two crops. J. Apic. Res. 12: 131-143.

Erickson, E.H. and Peterson, C.E. 1979 a. Asynchrony of floral events and other differences in pollinator foraging stimuli between male fertile and male sterile carrot inbreds. J. Amer. Soc. Hort. Sci. 104: 639-643.

Erickson, E.H. and Peterson, C.E. 1979 b. Problems encountered in the pollination of cytoplasmically male sterile hybrid carrot seed parents. Proc. 4th International Symp. Poll. Maryland, 1978, 59-63.

Forster, I.W., Clinch, P.G. and Palmer-Jones, T. 1973. Observations on the pollination of chou moellier (*Brassica oleracea* L.), N.Z. J. Exp. Agric. 1:46-48.

Free J.B. 1976. Insect pollination of tropical crops. Report of the Central Association of Beekeepers.

Free J.B. and Williams, I.H. 1973. The foraging behaviour of honeybees (*Apis mellifera* L.) on Brusselss sprout (*Brassica oleracea* L.). J. Appl. Ecol. 10: 489-499.

Free, J.B. 1993. Insect pollination of crops. 3rd Edn. Academic Press, London.

Fries I. and Stark, J. 1983. Measuring the importance of honeybees in rape seed production. J. Apic. Res. 22: 272-276.

Galuszka, H. and Tegrek K. 1987. Honeybee (Apis mellifera L.) foraging in seed plantations of hybrid carrot (*Daucus carota* L.). The 31st International Apicultural Congres of Apimondia. 334-336. Bucharest: Apimondia Publishing House.

Gilbert, F.S. 1980. Flower visiting by hover flies morphology of mouth parts in relation to feeding on nectar and pollen in some urban species. Ecol. Entomol. 6: 245-265.

Gupta, J.K., Mishra, R.C. and Kumar J. 1984. Nectar sugar production and honeybee foraging activity in different cultivars of cauliflower, *Brassica oleracea* var. *botrytis*. Indian Bee Jour. 46: 21-22.

Hussein, M.H. and Abdel-Aal, S.A. 1982. Wild and honey bees as pollinators of 10 plant species in Assiut area. Egypt. Z. Angew. Ent. 93: 342-346.

Jadhav, L.D. 1981. Role of insects in the pollination of onion (*Allium cepa* L.) in the Maharashtra State, India. Indian Bee J. 43: 61-63.

Johansen, C. A. (1977). Pesticides and pollinators. *Ann. Rev. Ent.* 22, 177-192.

Johansen, C. A. and Mayer, D. F. 1990. *Pollinator Protection. A Bee and Pesticide Book.* pp150. Wicwas press, Cheshire, Connecticut.

Kakkar, K.L. 1981. Foraging behaviour of insect pollinators of cauliflower bloom. Indian J. Ecol. 8(1): 126-130.

Kakkar, K.L. 1983. Prospects of honeybees as pollinating agents of cauliflower. Proc.II Int. Conf. Apic. Trop. Climat, 29th Feb.-4th March, 1980: 545-556.

Kapil, R.P., Grewal, G.S., Kumar, S. and Atwal, A.S. 1971. Insect pollinators of rape seed and mustard. Indian J. Ent. 33: 61-66.

Kapoor, K.S. 1983. Foraging strategies and pollinating activities of *Apis cerana indica* F. and *Apis mellifera* L. on *Brassica campestris* L. var. *botrytis*. M.Sc. Thesis, HPKV, COA, Solan, India.

Kapoor, K.S. and Dhaliwal, H.S. 1989. Comparative foraging strategies of *Apis ceranaindica* and *Apis mellifera* on cauliflower Indian Bee J. 51(3):99-101.

Khan, R. B. and Dethe, M. D. 2004 Median Lethal Time of New Pesticides to foragers of honeybee. *Pestology*. 28(1): 28-29.

Koutensky, J. 1958. The results of the pollinating work of Bees. Vcelarstvi 11: 72-73.

Kubisova, S., Haslbachova, H., Strmiskova, E. and Vesely, P. 1984. Efektivnost prace vcel pri autoinkompatibilnich linii zeli (*Brassica oleracea* var. *capitata* L.) Sbornik UVTIZ Zahradnicivi 14: 300-308.

Kumar J., Gupta, J.K., Mishra, R.C. and Dogra, G.S. 1988. Pollination studies in some cultivars of cauliflower (*Brassica oleracea* var. *botrytis* L.) Indian Bee J. 50: 93-95.

Kumar, J., and Rao K.V.K. 1991. Pollinating efficiency of some bee visitors to the carrot (Daucus carota L.) crop in mid hills of Himachal Pradesh, India. Indian Bee J. 53(1-4): 34-38.

Kumar, J., Mishra, R.C. and Gupta, JK. 1985. The effect of mode of pollination on Allium species with observations on insects as pollinators. J. Apic. Res. 24(1):62-66.

Kumar, J., Rao K.V.K. and Gupta, J.K. 1993. Amount of loose pollen on the body of bees visiting spring blooming crops. J. Insect Sci. 6(2): 259-263.

Kumar, J., Rao K.V.K. and Gupta, J.K. 1994. Pollination efficiency of bees visiting blossom of *Brassica campestris* L. var. toria in mid hills of Himachal Pradesh, India. Indian Bee J. 56(3-4):202-206.

Kumar, J., Rao K.V.K., Gupta P.R. and Dogra, G.S. 1989. Temporal distribution of bees (Hymenoptera Apoidea) on spring flowering crops in mid hills of Himachal Pradesh, India.) Indian Bee J. 51(2): 55-58.

Langridge D.F., Goodman R.D., 1974. A study on pollination of sunflowers (*Helianthus annuus*). Aust J Exp Agric Anim Husb 14, 201-204.

Langridge, D.F. and Goodman, R.D. 1975. A study on pollination of oilseed rape (*Brassica campestris*). Aust. J. Exp. Agric. Anim. Hush. 15:285-288.

Langridge, D.F. and Goodman, R.D. 1975. A study on pollination of oilseed rape (*Brassica campestris*). *Austral. J. Exptl. Agric. Anim. Husband.,* 15:285–288

Latif, A., Qayyum, A., and Abbas, M. 1960. The role of *Apis indica* in the pollination of toria and sarson (*Brassica campestris* var. Toria and dichotoma). *Bee World,* 41:283–286.

Lederhouse, R.C., Caron, D.M. and Morse, R.A. 1968. Onion pollination in New York. New York's Fd Life Sci. 1: 8-9.

Levin, M. D. 1984. Value of honeybee pollination to United States agriculture. *American Bee Journal*. 124:184-186.

Mayer, D. F. 2003. Bee kills from pesticides and the expert witness. *American Bee Journal.* 143 (7): 563.

Melnichenko, A.N. 1976. Role of insect pollinators in increasing yields of agricultural plants. In: *Pollination of Entomophilous Agricultural Crops by Bees*, Kozin, R.B. (Ed.) Amerind publishing Co. Pvt. Ltd., New Delhi.pp331.

Mesquida, J. and Revard, M. 1979 a. Entomolophilous pollination of male sterile strains of winter rape seed (*Brassica napus* L.) Metzger var. *Oleifera*) and a preliminary study of alternating devices. Proc. 4th International Symp. Poll. p. 49-57. Maryland, U.S.A.

Mesquida, J. and Revard, M. 1979 b. Preliminary results on the pollination of male sterile winter rape and methods for production of hybrid seed. Information techniques No. 65 P.3-14, INRA, France.

Mishra, R.C. and Kaushik, H.D. 1992. Effect of cross- pollination on yield and oil content of *Brassica campestris* L. var. sarson with pollination efficiency of honeybees, *Apis* spp. *Ann. Entomol.,* 10:33–37.

Mishra, R.C. and Sharma, S.K. 1998. Technology for management of *Apis mellifera* in India, pp. 131-147, In Mishra, R.C. (Ed.) Perspectives in Indian Apiculture, Agro Botanica, Bikaner (India), 412 pp.

Mishra, R.C. Kumar J. and Gupta, J.K. 1988. The effect of mode of pollination on yield and oil potential of *Brassica campestris* L. var. *sarson* with observations on insect pollinators. J. Apic. Res. 27: 186-189.

Mishra, R.C., Kumar J. and Thakur S.S. 1997. Pollination for hybrid seed production and in green houses/cages. In: Perspectives in Indian Apiculture eds. R.C. Mishra p: 212-225.

Mishra, R.C., Kumar, J. and Gupta, J.K. 1987. Effect of mode of pollination on fruit characteristics of okra. *Abelmoschus esculentus* (L.) Moench. Proc. Indian. Nat. Sci. Acad. B 53: 157-160.

Mohammed, A. 1935. Pollination studies in toria *(Brassica napus* L. var. dichotsma Prain) and sarson *(Brassica campestris* L. var. sarson, Prain). *Indian J. Agric. Sci.* 5, 125-154.

Mohr, N.A. and Jay, S.C. 1988. Nectar and pollen collecting behaviour of honeybees on canola (*Brassica campestris* L. and *Brassica napus* L.). J. Apic. Res. 27: 131-136.

Muhammad, S., Gondal, A. and Manzoor-ul-Haq. 1973. Studies on the role of *Apis indica* F. in the pollination of cauliflower (*Brassica oleracea* var. *botrytis* Linn) and radish (*Raphanus sativus* Linn). S.U. Res. J. (Sci-Ser.). 7: 87-93.

Murrell, D.C. and Nash, W.T. 1981. Nectar secretion by toria (*Brassica campestris* L. v. toria) and foraging behaviour of three *Apis* species on toria in Bangladesh. J. Apic. Res. 20: 34-38.

Pankratova, E.P. 1958. Data on the biology of flowering and pollination of carrots. Dokl. TSKha 36: 118-123.

Parker, F.D. 1982. Efficiency of bees in pollinating onion flowers. J. Kansas Ent. Soc. 55: 171-176.

Pimental, D. and 11 others. 1980. Environment and pesticides: a preliminary assessment. *Oikos*. 34(2): 126-140.

Pimentel D (ed). (1980) CRC *Handbook of Energy Utilization in Agriculture.* CRC Press, Bocu Raton, Florida.

Poehling, H. M. 1989. Selective application strategies for insecticides in agricultural crops. In : *Pesticides and Non target Invertebrates.* pp151-175. Intercept, Wimborne, Dorset, U.K. Dorset, UK.

Priti and R.C. Sihag. 1997. Diversity, visitation, frequency, foraging behaviour and pollinating efficiency of insect pollinators visiting cauliflower (*Brassica oleracea* L. var. *botrytis* cv. Hazipur Local) blossoms. Indian Bee Jour. 59(4): 230-237

Radchenko, T.H. 1964. The influence of pollination on the crop and the quality of seed of winter rape. *Bdzhil'nitstvo*, 1:68–74.

Radehenko, T.G. 1966. Role of honeybees as pollinators in increasing the seed crop from cabbage and radish. Bdzhilnitstvo, 72-75.

Rahman, K.A. 1940. Insect pollinators of toria (*Brassica napus* Linn. var *dichotoma* Prain) and sarson (*Brassica campestris* Linn. var sarson Prain) at Lyallpur. Indian J. Agric. Sci. 10: 422-447.

Rahman, K.A. 1945. Progress of beekeeping in the Punjab. Bee Wld. 26: 42-44, 50-52.

Rana, B. S. and Goyal, N. P. 1991. Field toxicity of methyl demeton and dimethoate to the foragers of honeybee (*Apis cerana indica* F.) visitors to *Brassica chinensis* L. *Indian Bee Journal*. 53(1-4): 73-77.

Rao, G. M., Lazar, G. M. and Suryanarayana, M. C. (1981). Foraging behavior of honeybees in sesame (Sesamum indicum L.). *Indian bee J.* 43: 97-100.

Rao, G.M. and Suryanarayana, M.C. 1989. Effect of honeybee pollination on seed yield in onion (*Allium cepa* L.). Indian Bee J. 50: 9-11.

Rauala, T.S. 1972. Pollination studies in cauliflower (*Brassica oleracea* var. *botrytis* L.). J. Res. Punjab. Agric. Univ. 9: 580-585.

Rodet, G., Grosse, J.P.T. and Bonnet, A. 1991. Foraging behaviour of *Apis mellifera* L. on male-sterile and male fertile inbred lines of carrot (*Daucus carota* L.) in gridded enclosures. The 6th International Symposium on pollination. Tilburg. The Netherlands, August 1990. Acta Horticulturae 288: 371-375.

Rubatzkky VE, Yamaguchi M. 2000. *Sayuran Dunia: Prinsip,Produksi dan Gizi.* Bandung: Penerbit ITB.

Russell, D., Meyer, D. and Bukowski. 1998. Potential impact of micro encapsulated pesticides on New Jersey Apiaries. *American Bee Journal.* 38(3): 207-210.

Ruszkowski, A. and Bilinski, M. 1984. Proba wykorzystania trzmiek (*Bombus* Latr) I miesiarki micze Zeszyty Naukowe 28: 175-184.

Sabbahi, R., D. Oliveira & J. Marceau 2005. Influence of honey bee (Hymenoptera: Apidae) density on production of Canola (Crucifera: Brassicae). *Journal of Economic Entomology* 98:367-372.

Sharma, H.K. 1990. Floral resources and foraging characteristics of hive bees. Ph.D. Thesis, Dr Y.S. Parmar University of Horticulture and Forestry, Nauni, Solan, Himachal Pradesh. 168 p.

Sharma, P.L. 1958. Sugar concentration of nectar of some Punjab honey plants. Indian Bee J 20(7):81-91.

Sharma, P.L., Dhaliwal, H.S. and Kakar, K.L. 1974. Insect visitors and pollinators of cauliflower (*Brassica oleracea* var. *botrytis*) seed crop bloom. Him. J. Agric. Res. 2:74-78.

Sihag RC and Mishra RC. 1995. Crop pollination and *Apis cerana*. In: Kevan PG (ed). *The Asiatic Hive bee: Apiculture, Biology, and Role in Sustainable Development in Tropical and Subtropical Asia.* Ontario: Enviroquest Ltd. p 129-134.

Sihag, R. C. 1982. Effect of competition with *Parkinsonia aculeata* L. on pollination and seed production in *Medicogo saliva* L. *Indian Bee J.* 44, 89-90.

Sihag, R. C. 1986. Insect pollination increases seed production in cruciferous and umbelliferous crops. J. Apic. Res. 25: 121-126.

Sihag, R.C. 1988. Effect of pesticides and bee pollination on seed yield of some crops in India. *J. Apicult. Res*., 27:49–54.

Sihag, R.C. 1991. Ecology of European honeybee (*Apis mellifera* L.) in semi arid and sub-tropical climates. 1. Melliferous flora and over — seasoning of the colonies. *Korean J. Apicult.,* 5:31–43.

Sihag, R.C. and Rathi, A. 1994. Diversity, abundance, foraging behaviour and pollinating efficiency of different bees visiting pigeon pea (*Cajanus cajan* (L.) Mill Sp.) blossoms. Indian Bee Jour. 56(3-4):187-201.

Sihag, R.C., Khatkar, S., and Khatkar, S. 1999a. Foraging pattern of three honeybee species on eight cultivars of oilseed crops. 2. Foraging during the entire blooming period of the crops. *Int. J. Tropical Agric.,* 17:253–261.

Sihag, R.C., Khatkar, S., and Khatkar, S. 1999b. Synchronization of foraging activity of honeybees and flowering/ anthesis in oilseeds crops. *Ann. Agric. Biolog. Res.,*. 4:263–269.

Singh, J.P. and Dharmwal, S.S. 1970. The role of honey bees in seed setting on onion at Pant Nagar. Dist. Nainital, Uttar Pradesh, India. Indian Bee J. 32: 23-27.

Singh, M. P. and Singh, K. I. and Devi, C. S. 2000. Role of *Apis cerana* himalaya pollination on yield and quality of rape seed and sunflower crops. *In proceedings of 4th Asian Apicultural Association, International Conference,* Kathmandu, Nepal March 23-28, 1998

Singh, R.P. and Singh P.N. 1992. Impact of bee pollination on seed yield, carbohydrate composition and lipid composition of mustard seed. *Journal of Apicultural Research*31:128-133.

Singh, R.P. 1997 Olive In *Fruit crop pollination* (ed. L. R.Verma and K. K. Jindal), pp. 212-224. Kalyani Pub., Ludhiana, India.

Singh, S. 1954. Insect pollination and breeding of fruit varieties. Indian J Hort Sci 11:123-125.

Singh, Yogeshwar, 1997. Role of honeybees in farm production, agricultural growth and rural reconstruction in India. Indian Bee Jour. 59(1): 24-30.

Sinha SN and Ckrabarti AK 1992.Insect pollination in carrot seed crop.Seed Research 20: 37-40

Sorthia, B. K. and Chari, M. S. 1985. Toxicity of some insecticides to honeybees, *Apis florea* F. and *Apis mellifera* L. *Journal of Entomological Research.* 92 (2): 195-197.

Tanda A.S. 1984. Insect pollination in okra (*Abelmoschus esculentus*) Proc. III orient ent. Symp. Kariavattom, India. 25.

Tanda, A.S. 1985. Floral biology, pollen disposal and foraging behaviour of honeybees in okra (*Abelmoschus esculentus*).J. Apic. Res. 24: 225-227.

Tewari, G.N. and Singh, K. 1983. Studies on insect pollinators in relation to seed production in cauliflower (*Brassica oleracea* var. *botrytis* L.) Indian Bee J. 45: 54-55.

Treherne, R.C. 1923. The relation of insects to vegetable seed production. Rep. Queb. Soc. Prot. Pl. 15: 47-49.

Trofimec, N.K. 1940. The biology and fertilization in *Allium*. Vest. Sots Rasteniev. Rasten. 5:76-86.

Varma, S.K. and Joshi, N.K. 1983. Studies on the role of honey bees in the pollination of cauliflower (*Brassica oleracea var. botrytis*). Indian Bee J. 45: 57-58.

Wallace HM, Maynard GV, Trueman SJ. 2002. Insect flower visitors, foraging behaviour and their effectiveness as pollinators of *Persoonia virgata* R. Br. (Proteaceae). *Austral J Entomol* 41:55- 59.

Westcott, L., and D. Nelson. 2001. Canola pollination: an update. Bee World 82:115-129.

Woyke, H. 1989. Honeybee foraging on the flowers of cauliflower $F_1$ cultivar components. Proceedings of the XXXIst Congress of Apiculture, Warsaw, Poland, August 19-25, 1987, 360-364. Bucharest: Apimondia Publishing house.

Woyke, H.W. 1981. Some aspects of the role of the honeybee in onion seed production in Poland. Acta Horticulture, 111: 91-98.

Woyke, H.W. and Dudek, M.M. 1983. The influence of distance between male-sterile and male fertile rows on the number of honeybees visiting the flowers and on the seed yield of hybrid onion V. Symposium International Sur la Pollination, Versailles, 1983, 123-128 INRA Publications.

Zhong, H. E., Mark, L., Sheve, P., and Cate, B. 2004. Minimizing the import if the mosquito adulterine noted or honeybees, *Apis mellifera* (Hymenoptera: Apidae): Aerial Ultra volume application using a high-pressure nozzle system. *Journal of Economic Entomology*. 97 (1): 1-7.

CHAPTER

3

# Pest Problem in Vegetable and Oilseed Crops

## Insect pests of vegetables

## Major insect pests and diseases of vegetables in India

Insect pests are the major biotic constraints to vegetable production. The major pests in vegetable production include, tomato fruit borer (*Helicoverpa armigera*), brinjal shoot and fruit borer (*Leucinodes orbonalis*), chilli thrips (*Scirtothrips dorsalis*) and mite (*Polyphago tarsonemus latus*), fruit and shoot borer (*Earias* spp.) on okra, diamond back moth (*Plutella xylostella*) on cole crops, fruit fly (*Bactrocera cucurbitae*) on cucurbits. Some of the insect pests like serpentine tomato leaf miner, gall midge, okra stem fly, bitter gourd leafhopper and mites are gradually attaining the major pest status in different regions of the country. Earlier gall midge known to be a minor pest is gradually becoming a regular problem in chilli, Capsicum and brinjal, whereas *Hellula undalis* on cabbage and red spider mite on okra, brinjal, cowpea, Indian bean, etc. and more recently mealy bug, *Phenococus solenopsis* on several vegetable crops, have intensified the severity of occurrence.

The crop losses in the country due to various pests range on an average from10-30 per cent (Rai et al 2014). The extent of crop losses in vegetables due to insect pest infestation vary with the plant type, agro climatic conditions, damage potential of the pest involved and cropping season prevailing in different parts of the country. Apart from causing direct damage, many of them also act as vectors for several viral diseases which have aggravated the problem still further. In recent years due to change in cropping pattern, ecosystem and climatic changes, the pest status is also changing rapidly. Some pests have adapted to new hosts or developed resistance to pesticides. Minor pests are attaining the status of major ones coupled with secondary outbreaks. For instance, insect pests like serpentine leaf miner (*Liriomyza trifolii*) on tomato and cucurbitaceous crops, gall midge on brinjal, chilli and capsicum; leafhopper (*Empoasca motti*) on bitter gourd;stem fly on okra, head borer (*Hellula undalis*) on cabbage, red spider mite (*Tetranychus* spp.) on okra, brinjal, cowpea and Indian bean; yellow mite (*Polyphago tarsonemus latus*) on chilli and more recently mealy bug (*Pheonacoccus solenopsis*) on okra, chilli, brinjal and tomato specially in protected conditions have intensified the severity of occurrence in different parts of the country. Success of management of insect pests depends upon their correct and selection of appropriate control measures. Among the various methods of pest control, chemical method continuous to be the first choice because of its quick action and easy availability. As a consequence vegetables consume 13-14 per cent of the total pesticide consumption in India.

## Important insect pests of vegetables

Insect pests attacking different groups of vegetables are discussed under the heads:

1. Cruciferous Vegetables)
2. Solanaceous Vegetables
3. Malvaceous vegetables
4. Cucurbitaceous Vegetables
5. Leguminous vegetables
6. Pests of leafy vegetables

**B. Insect pests of oilseed crops**

**1. Cruciferous Vegetables:** Cruciferous vegetables include Cabbage (*Brassica oleracea var. capitata* L, Cauliflower (*Brassica oleracea var. botrytis L.*), Knolkhol (*Brassica oleracea var. gongylodes* L.), Radish (*Raphanas sativus* L.) and Beetroot. The most important cruciferous vegetable include

Cabbage, *Brassica oleracea var capitata* L. is cultivated on the hills and plains in winter season. The edible portion is known as 'head' which is made up of numerous thick, smooth leaves overlapping each other and covering the terminal bud. The head may be round, conical or flat. It is a rich source of vitamin A, B, C and also contains various minerals like calcium, phosphorus, potassium, sodium and iron.

Cauliflower, *Brassica oleracea var. botrytis* L. is grown in cold and moist climate. The edible portion is the white curd-like mass composed of close aggregation of flowers developed on thick branches of the inflorescences. It is rich in minerals namely, iron, magnesium, phosphorus, potassium and sodium. Vitamin A and B are also found in enough quantity.

Knol-Khol, *Brassica oleracea var. gongylodes* L. has a short stem which is swollen just above the ground level into a spherical edible portion and has a large leaf scars on its surface. It is cooked as vegetable when tender and eaten as salad. It contains calcium, magnesium, phosphorus, sodium and sulphur. It is rich in vitamin A, vitamin C and iron.

Radish (*Raphanas sativus* L.) is cultivated throughout the world and is a cool season crop. It has a white or brightly coloured tuberous tap root cultivated all over India upto 3000 m in Himalayas and other hilly regions. It is cultivated for its roots which are used as vegetables and also eaten raw as salad. Leaves are also boiled and eaten. Radish is a good source of ascorbic acid and contains a variety of mineral salts. Leaves are *also* considered to be one of the richest source of vitamin A among the Leafy vegetables.

The important pests that attack these cole crops include cabbage caterpillar, diamond-back moth, cabbage semi-looper, tobacco caterpillar, crucifer leaf-webber, cabbage borer, cabbage flea beetle. Other minor pests which appear on the crop regularly but do not cause any severe damage are cut worms, sap-sucking bugs, Bihar hairy caterpillar, flea beetles, thrips and mite etc.

**1. *Crocidolmía binotalis Zell*:** The Crucifer Leaf-Webber or Cabbage Cluster-Caterpillar (Pyralidae :Lepidoptera)

**Distribution:** India, Burma, Sri Lanka, Africa, South-East Asia, Australia.

**Host:** Cabbage, radish, mustard, turnip and other Cruciferous plants. The larva is green with red head and longitudinal red stripes on the body. Thin hairs arise on tubercles on the body. It is 2cm in length. The moths are small with light-brownish forewings.

**Life-cycle:** The pest is active with the availability of Cruciferous crops in the field. The moth lays eggs in masses of 40-100 each, on the lower surface of the leaves. The eggs hatch in 5-15 days. Young larvae feed gregariously on the leaf parenchyma. As they grow older, they start webbing the leaves and feeding on them. The larval stage is completed in 24 to 27 days in summer and in about 50 days in winter. The pupation takes place in the soil in the earthen cocoon. The adult emerges in 14-40 days. The total life-cycle is competed in 43-82 days. There may be more than one generation in a year. *Microbracon mellus* Ram. and *Apanteles crocidolmiae* Ahmed parasitize the larvae of this pest.

**Damage:** The larvae web the leaves with silken strands and feed on the lower surface of the leaves completely skeletonising them. They also feed on flower buds and bore into pods causing considerable damage to the crops.

**Control:** Remove and destroy webbed leaves with larvae within. Spray 375 ml of Malathion 50 EC or, 125 ml of Dichlorvos 100EC in 250 litres of water per acre.

**2. *Pieris brassicae* (Linnaeus):** The Cabbage caterpillar or Large White butterfly( Pieridae: Lepidoptera)

**Distribution**, Europe, India and North America.

**Host:** Cabbage, cauliflower, knol-knol, turnip, radish, toria and other cruciferous plants. Full-grown larva measures 40-50 mm in length. Newly hatched larvae arc pale yellow and turn green yellow later on. They have short hairs on the body. Medium sized butterflies are pale white with black markings. The spread wings in female measures 6.5cm in length and have two black circular dots on the dorsal side of each fore wing and while in male, the black spots are on the underside of each forewing. Males are smaller than the females.

**Life cycle:** After mating, the butterflies lay on an average 164 yellowish conical eggs in clusters of 50-90 on upper or the lower side of the leaf. The incubation period is 11-17 days in November-February and 3-7 days in March-May. They in early stages feed gregariously. There are five instars and are full grown in 2-3 weeks in March-April and in 4-6 weeks during November-February. The larvae pupate on trees away from host plant. The pupal period is 7.7-14.4 days and 20-28 days in March-April and November-February, respectively. The longevity of butterflies is 2.5-12.5 days. During the active period of breeding from October to April, the pest breeds four times.

*Apanteles glomeratus* L. parasitizes the larvae of this pest in the field.

**Damage:** The damage is caused by caterpillars only. Freshly hatched caterpillars scrape the leaf surface. The older larvae eat up the leaves from margins inwards, leaving intact the main vein. Sometimes, the leaves of entire plant are eaten up.

**Control:** Hand picking and mechanical destruction of caterpillars during early stage of attack can reduce the infestation. Spray400 ml Malathion 50 EC or, 370 ml Endosulfan 35 EC, in 250 litres of water per acre and repeat the spraying at 10-days intervals, if necessary.

**3. *Hellula undalis* Fab:** The Cabbage Borer or Cabbage Web-worm (Pyralididae: Lepidoptera): **Distribution**. The insect has a world wide distribution.

**Host:** Cabbage, cauliflower, radish, knol-khol, beet root, and the weed *Gynadropsis pentaphylla.* The full-grown larva is creamy-yellow with a pinkish tinge and has seven purplish brown longitudinal stripes. It measures 12-15mm long. The adult moth is slender, pale yellowish- brown, having grey wavy lines on the fore wings and hind wings pale dusky.

**Life-cycle:** The female lays pinkish oval eggs either singly or in groups on the undersurface of the leaves or some other parts of the plant. The incubation period is 2-3 days. The caterpillars feed in the heart of the cabbage and become full-grown in 7-12 days after passing through four moultings. The full-grown caterpillar spins a cocoon among the leaves touching the ground or even inside the larval burrows. The adult emerges in about six days. Total life-cycle is completed in 15 to 25 days average being 19 days.

**Damage:** The caterpillar in its first two instars mines the leaf along the side of veins and makes it a white papery structure with excreta in it. Later on, they feed on the leaf surface, sheltered within the silken passages. Thereafter, as they grow bigger, they bore into the heads of the cauliflower and cabbage. In case of heavy attack, almost all the plants may be found infested. The plants lose their vigour and bear deformed heads unfit for marketing.

**Control:** Collection and mechanical destruction of caterpillars in early stage of attack helps to check the infestation. Spraying 150g of Carbaryl 50 WP in 60 litres of water per acre will control this pest. Repeat spray at 10-days intervals on nursery and young plants.

**4. *Plusia orichalcea* (Fabricius)** *P. nigrisigna* (Walker): The Cabbage Semi-looper (Noctuidae Lepidoptera)

**Distribution:** India (north-westem parts), South East Asia.

**Host:** Cabbage, cauliflower, groundnut and other winter vegetables.

The caterpillars are palish-green and plump. On walking, the larvae form characteristic half-loops and hence the name 'semi-looper'. The adults of *P. orichalcea* measure about 42 mm across the spread wings and are light palish-brown with a large golden patch on each forewing. The adults of P. *nigrisigna* have dark brown dirty-white patches on the fore wings.

**Life-cycle:** During the active period in winter, the females lay eggs on leaves of host plants. The larvae feed individually biting holes of varying sizes. Pupation takes place in debris lying on the ground. The adults are very active during evening time and are seen in large numbers on flowers.

**Damage:** They are polyphagous. The larvae cause damage by biting holes into the leaves. In case of severe infestation, the plant is completely defoliated.

**Control:** Same as under cabbage for *P. brassicae*

**5. *Plutella xylostella* (Linnaeus)** (= *Plutella maculipennis* (Curtis): The Diamond-back Moth (Plutellida: Lepidoptera)

**Distribution:** Completely cosmopolitan in distribution.

**Host:** Cauliflower, cabbage and other species of Brassicae and wide range of wild and cultivated crucifcreae. The full-grown larvae are pale yellow-green with fine black hairs all over the body and measure about 8 mm in length. The moths are brown or grey, with conspicuous white spots on the fore wings. They measure about 8-12mm in length. When the wings are folded on its body at rest a diamond shaped median dorsal patch is seen and hence the name 'diamond back moth'.

**Life- cycle:** The female lays minute yellowish-white eggs singly or in batches of 2-40 on the underside of leaves. About 18-356 eggs are laid by a single females. The eggs hatch in 9.1 days at 13.9°C, 3.7 days at 17.9°C and *2.5* days at 23.4°C. The freshly hatched larvae crawl to the undersurface of the leaf, penetrate the epidermis and mine in the leaf tissue.

The three later instars feed on the underside of the leaf making windows or holes right through it. They become full-grown in 16.6, 14.3, 8.6 days at 17, 20 and25°C, respectively. Pupation takes place inside a barrel-shaped silken cocoon which is open at both the ends, constructed by the larva. The cocoon is attached to the leaf surface. The pupal period is 4-5 days at 17-25°C. The moths may live upto 20 days. During September-October, the total life is completed in 2-3 weeks. There are 8 to 12 generations in a year.

The larva is parasitized by *Voria ruratis* Fab., *Itoplectis* sp., *Apanteles sicarius and Tetrastichus sokolowskii.* The pupae are parasitized by *Brachymeria excarinata, B. plutellphaga A* larval-pupal parastoid, *Thyraeella collaris* Grov. is associated with this pest.

**Damage:** The larvae cause serious damage to crops like cabbage cauliflower and rape-seed etc. The freshly hatched larvae feed by scrapping the leaf tissues and as they grow older bite holes in them and rendered the vegetables unfit for human consumption. A heavy infestation causes retardation in the growth of cabbage and cauliflower plants. A severe infestation results in the formation of under-sized flowers in cauliflower and in cabbage head formation does not take place.

**Control:** Remove and destroy all the ramnants, stubbles, debris etc. after the harvest of the crop and plough the fields. Apply 300 ml Diazinon 20 EC or, 60ml Dichlorvos 100 EC or 400 ml Malathion 50 EC in 250 litres of water per acre and repeat spray at one week intervals.

**6. *Phyllotreta cruciferae* Goeze,** *P. choranica* Duviv., *P. birmanica* Harold, *P. onceraMaulik, P. downesi* Baly,

The Cabbagc Flea beetles (Chrysomelidae : Coleoptera)

**Distribution:** Europe, U.S.S.R. North and South America, Australia, Japan India.

**Host:** Mustard, *raya, taramira, toria, vegetables like radish,* turnip, cabbage. cauliflower, knol-knol, cotton, cereal, flowering plants, *dahlia,* sweet sultan, antirrhinum and sweet peas.

The larva is dirty white in colour with pale-while head. It measures about 5 mm in length. The adults vary in colour from shiny black to black with a green sheen to black with yellow stripes on the elytra. All species have a very stout femora, with which they jump in a flea-like manner.

**Life-cycle:** The overwintered adult beetles emerge in last week of February or in first week of March. The female lays 50 to 80 creamy white eggs, in the soil around the host plant. The oviposition period and the incubation period ranges between 25-30 and 5 to 10 days, respectively. The larva moults thrice during the total larval period of 9 to 15*days.* The larva pupates in the soil than earthen cell 0.5mm long, in the vicinity of the infested plants. The prepupal and pupal period are 2-4 days and 8-14 days. respectively. The pest completes 7-8 generations in a year.

**Damage:** The larvae live in the soil and feed upon the roots of the host plants. The adults feed on the cotyledons and leaves of young plants making innumerable round holes. The stem, the flowers and even pods may also be attacked. The old eaten away leaves dry up and the young leaves rendered unfit for consumption. The attacked plants of cabbage give a special kind of decaying odour.

**Control:** Dust DDT 5 percent @ 10kg per acre, or spray 1 kg DDT 50 WP or, Carbaryl 50 WP or, 125 ml Dichiorvos 100 EC in 200 litres of water per acre.

**7. *Spodoptera litura* (Fabricius):** The Tobacco Caterpillar or Fall Army Worm or Cluster Caterpillar or Rice Cut Worm (Noctuidae: Lepidoptera)

**Distribution:** India, Pakistan, Bangladesh, SriLanka, South EastAsia, China, Korea, Japan, Philippines, Indonesia, Australia. Pacific Islands, Hawai and Fiji.

**Host:** Pest is polyphagous. Cotton, rice, tomato, tobacco, citrus, cocoa, sweet potato, rubber, groundnut, castor, legumes, millers, sorghum, maize, banana, safflower, cabbage, chillies, and many other vegetables are attacked. The full-grown larva is stout, cylindrical, pale-greenish brown with dark markings which measures about 35-40mm in length. The moth is stout, dark with *wavy* white markings on the fore wings and white hind wings, margin having a brown colour. They are about 22 mm long and measure 40 mm across the spread wings.

**Life cycle**: The pest is active throughout the year. The fernale moth after mating at night lays about 300 eggs in clusters. These clusters are covered over by brown hairs. The incubation period is about 3-5 days. After hatching, the larvae feed gregariously together and later on they feed individually, There are six instars completing in 2-4 weeks. The last instar larvae enter the soil for pupation. The pupal period and the longevity of moths are 7-15 days and 7-10 days, respectively. The total life-cycle covers 32-60 days. There are 8 generations in a year.

**Damage:** The caterpillars feed on leaves in the nursery and in the main field and cause serious damage. In crops like tomato and chillies, they bore into fruits and spoil them. They are very active at night. Heavy infestation can seriously defoliate the crop but it is not common.

**Control:** Same as under cabbage for *P. brassicae*

**8. Cabbage aphid *Brevicoryne brassicae* (Aphididae: Homoptera):** This aphid has a very wide range of distribution.

**Nature of damage:** These insects suck sap from the tissues, the plant remain stunted in growth resulting in poor head formation. If the infestation is high the plants may completely dry up and die. When infestation occurs on seedlings, they loose their vigour, get distorted and become unfit for transplanting. The aphids produce honeydew which makes the plants sticky favouring the growth of sooty mould hindering the photosynthesis and adversely affecting the plant growth.

**Life history:** Reproduction is mostly viviparous and parthenogenetic. Eggs are pale yellow with greenish tinge. Nymphs are yellowish-green in colour while adults are darker in colour. Eggs are laid during November – December. These hatch in 20 - 22 weeks. The nymphs mature in about 2 weeks and immediately start producing young ones, without mating. A single female can produce 40 - 45 young ones during her life time. It takes 11 - 45 days to complete the life cycle and as many as 21 generations per year with favourable conditions.

**Management strategies:** (i) Cut and destroy the infested shoots mechanically. (ii) if the infestation is more than 5%, spray with 0.025% methyl demeton or 0.01% imidacloprid. Repeat the spraying after a fortnight if if infestation continuous more than 5%.

**9. Mustard Sawfly, *Athalia lugens proxima* Kulg (Tenthredinidae: Hymenoptera):** Distribution: it is one of the serious hymenopterous pest with chewing and biting habits. It is a cold weather pest found all over the Indian sub-continent. It is a major pest of almost all cruciferous plants. The peak period of activity occurs during September to December after which the activity declines. The pest is rarely seen from March to July.

**Tables 3.1:** *Minor Pests of Cabbage, Cauliflower, Knol-khol and Radish*

| Sl. No. | Scientific name | Common name | Order | Family | Distribution | Nature of damage |
|---|---|---|---|---|---|---|
| 1. | *Bagrada cruciferarum* F. | Painted bug | Hemiptera | Pentatomidae | India, Africa | Sap-sucker S.E.Asia |
| 2. | *Lipaphis erysimi* Kalt | Plant lice | Hemiptera | Aphididae | Cosmopolitan | Infest foliage, virus vector |
| 3. | *Myzus persicae* S. (Green) | Plant lice | Hemiptera | Aphididae | Cosmopolitan | Infest shoots and leaves |
| 4. | *Thrips tabaci* Lind | Thrip | Thysanoptera | Thripidae | Cosmopolitan | Scarify leaves |
| 5. | *Caliothrips indicus* (Bagnall) | Thrip | Thysanoptera | Thripidae | India | Infest foliage |
| 6. | *Diacrisia obliqua Wlk.* | Bihar hairy caterpillar | Lepidoptera | Arctidae | India | Larvae defoliate |
| 7. | *Agrotis ipsilon* (Hufn.) | Cut worm | Lepidoptera | Noctuidae | Cosmopolitan | Larvae cut worms |
| 8. | *A. segetum* (Schiff.) | Cut worm | Lepidoptera | Noctuidae | India, S.E. Asia | Larvae destroy seedlings |
| 9. | *A.c–nigrum (L.)* | Cut worm | Lepidoptera | Noctuidae | India | Larvae destroy seedlings |
| 10. | *Trichoplusiani* Hub. | Cabbage green looper | Lepidoptera | Noctuidae | India, U.S.A. Mexico | Larvae defoliate |
| 11. | *Phytomyza atricornis* (Meigen) | Pea leaf miner | Diptera | Agromyzidae | India | Larvae mine leaves |
| 12. | *Athalia lugens proxima* Klug | Mustard saw fly | Hymenoptera | Tenthredinidae | Cosmopolitan | Larvae defoliate |
| 13. | *Chaetocnema basalis* Bally | Flea beetle | Coleoptera | Alticidae | India | Adults hole leaves |
| 14. | *Tetranychus cucurbitae* Rahman and Sapra | Mite | Acarina | Tetranychidae | India | Scarify leaves |

**Nature of damage:** The eggs are found on the periphery of leaves. On hatching the grubs nibble the margins of tender leaves but later on bite holes in the leaves. Grubs are diurnal in habit feeding during early morning and evening hours.

**Life history:** A female on an average lays 35 eggs per day. The egg period lasts from 6-8 days. The newly hatched eggs are greenish-grey in colour. The grub development period takes 21-31 days. Adults are 8-12 mm long having dark coloured head with orange abdomen. Females posses straw like ovipositor, hence called as sawfly. They are active during day and mostly remain inactive during night. The pest has 3 generations in north India and up to 10 generations in southern India where winters are mild.

Management strategies include hand-picking of grubs during dawn and dusk and Spraying the crop with 0.05% monocrotophos or 0.2% carbaryl. (iii) Dusting with 5% Malathion or 10 % carbaryl is also effective.

**10. Painted Bug *Bagrada cruciferarum* Kirkaldy (Pentatomidae: Hemiptera:** Distribution: Painted bug Bagrada cruciferarum have been recorded as major pests of various cruciferous crops.

**Nature of damage:** Both nymphs and adults suck cell sap from tender plant parts causing yellowing of leaves which gradually dry up, exposing the plants to secondary invasion of bacteria and fungi.

**Life history:** Eggs are laid in batches of 2 - 12 on leaves, stems and flower buds, pale yellow when freshly laid gradually becoming pinkish-orange. Nymphs are with a mixture of black, white and orange colour, 1.5 - 4.5 mm long depending on their age. Adults are black and orange in colour bugs similar in colour pattern as nymphs – that's why they have earned the common name of painted bugs. Males are 6 - 7 mm long and females 7 - 8 mm. The mating takes place 2 - 6 days after the final nymphal moult. A single female may lay as many as 230 eggs @ 15 - 20 eggs per day. life-cycle is completed in 3 - 4 weeks and adults live for 16 -18 days with 6 - 8 generations in a year.

Management strategies include clean cultivation by removing weeds harbouring this pest and in case of severe infestation, spray the crop with 0.05% dichlorvos or 0.05%. Waiting period of at least 7 - 10 days should be there between treatment and harvest.

**2. Solanaceous Vegetables**

**Pests of brinjal (*Solanum melongena* Linnaeus):** Brinjal, *Solanwn melongena* Linnaeus or eggplant is a native of India and now cultivated throughout the tropics. The fruit is large pendant berry, ovoid or oblong, smooth in texture and usually black or purple when ripe. The fruit is eaten as a vegetable, boiled, fried or stuffed. It contains vitamins A, B and C. Brinjal crop is attacked by a number of serious pests which include shoot and fruit borer, stem borer, leaf roller, hadda beetles, lace-wing bugs. The other pests are sporadic namely, mealy bugs, thrips, leaf-miners, bud worms, leaf rollers, jassids etc.

**1. The Brinjal Shoot and Fruit Borer, *Leucinodes orbonalis* Guenee (Pyralididae: Lepidoptera)**

**Distribution:** Thailand, Laos, Malaysia, India, Burma, Sri Lanka, Congo, East and South Africa, Pakistan, Germany. Host. Brinjal, potato, bitter gourd, green pods of peas and other Solanaceous plants.

Young caterpillars are creamy white, but full-grown are light pinkish in colour and measure about 18-23 mm in length. The moth is medium sized with white wings having triangular brown and red markings on the fore wings. The moth measures about 20-22mm across the spread wings.

**Life-cycle:** The hibernating larvae start pupating in early spring season. The emergence of moths take place sometimes in March-April. A female lays about 80-120 creamy white eggs, singly or in batches of 2-4 on the underside of leaves, on green stems, flower buds or the calyces of fruits during its life span of 2-5 days. The incubation period is 3-6 days. The young caterpillars bore into the tender shoots near the growing points, into flower buds or in the fruits. There are five larval instars and are full-grown in 9-28 days. A single caterpillar may destroy as many as 4-6 fruits. It pupates on the stem or fruit in a grey tough cocoon. The pupal period is 6-17 days. During the active season, the total life-cycle is completed in 20-43 days. There are five overlapping generations in a year.

*Pristomerus testaceus* Mon., *Bracon* sp., *Shirakia schoenobii* Vier and *Iphiaulax* sp. parasitoids are associated with the larvae of this pest.

**Damage:** The larvae bore into the petiole and mid-ribs of large leaves and tender shoots and cause 'dead-heart'. In the later stage, they also bore into the flower buds and fruits. The flower buds shed due to attack. The inside of the fruit is damaged considerably and the entry hole can be plugged with excreta. Attacked fruits become unfit for consumption and marketing. As high as 70 per cent infestation in brinjal and upto 40 percent in potato has been observed.

**Control:** Remove and destroy all the affected shoots and fruits with borers inside, Avoid continuous cropping of brinjal crop. Apply Permcthrin, or 800g Carbaryl 50 WP, or 100 ml Fenvalerate 2 OEC or 200 ml Cypermcthrin IOEC or 160 ml Decamethrin 28 EC in 2501 itres of water per acre as soon as the attack starts.

**2. The Brinjal Stem Borer or Egg Plant Boring Caterpillar *Euzophera pertkella Rag. (*Pyralidae: Lepidoptera)**

**Distribution.** Widely distributed in India.

**Host:** Brinjal, potato, chillies, tomato, and *Solanum aviculare.*

Full-grown caterpillars are creamy white with a few bristly hairs and measure about 20- 22mm in length. The moth is small having pale straw yellow fore wings and whitish hind wings. The moths measure about 32mm across the spread wings.

**Life-cycle:** The insect is active from March to October and hibernates as larva in the stem of old plant from November to the beginning of March. The moths emerge sometime in second half of March and soon after mating, the females start laying eggs. The eggs are cream coloured scale-like and are laid singly or in batches on tender leaves, petiole, and branches. A single female may lay 104-363 eggs in its life span of about 7 days. The incubation period is 3-10 days. The young larvae feed for a few minutes on exposed parts of plants and then bore into the stem making longitudinal tunnels. The larvae are full-fed in 26-58 days after passing four or five stages. They pupate inside the feeding galleries after making silken cocoons or in cracks and crevices in the soil. Moth emerged in 6-8 days. The insect completes its life-cycle in *35-76* days and has 5-6 overlapping generations in a year.

*Prisromerus testaceus* Morl. and *P.euzopherae* Vier. are parasitic on the larvae of this pest.

**Damage:** The caterpillars cause the damage and feed exclusively in the main stem. It enters the main stem and makes tunnel which result either in stunting of growth or withering of plants. Its infestation is seen usually in the late stage of the crop.

**Control:** Four sprays of 200 ml Malathion 50 EC or 250 g DDT 50 WP or 125 ml of Dichlorvos 100 EC in 250 litres of water per acre should be given at 15-days intervals.

**3. The Brinjal Lace-wing Bug *Urentius sentis* Distant (Tingidae: Hemiptera**

Distribution. India (nonh-wcstcm parts).

**Host:** Brinjal.

The full-grown nymphs are about 2mm long and 1.35mm broad. They are stoutly built and pale or chraceous with grey prominent spines. The adult bugs arc straw coloured on dorsal side and black on ventral side and measure about 3 mm in length.

**Life-cycle:** The adults hibernate from November to March in cracks and crevices in the soil under the brinjal plants or other protected places. The pest is active from April to October. The bug lays 35-44 shining white nipple-shaped eggs singly in the tissue on the underside of the leaves during its life span of 30-40 days. The incubation period is 3-12 days and on hatching, the young nymphs feed gregariously on the lower surface of the leaves. They grow through five stages and transform themselves into adults in 10-23 days. There are 8 overlapping generations in a year.

**Damage:** Both nymphs and adults are destructive. The young nymphs feed gregariously on the lower surface of the leaves but the full-grown nymphs are found feeding and moving about individually on the lower and upper surface of the leaves. Adults and nymphs suck the sap from leaves and inject some toxic saliva. Owing to feeding, yellowing of leaves becomes prominent when the infestation is heavy and upto 50 per cent of the crop may be destroyed. The pest is most abundant in August-September.

Control. Apply 500 g of DDT 50 WP or, 1 kg BHC 50 WP or, 250 ml of Diazinon 2OEC or 400 ml Dimethoate 3OEC in 250 litres of water per acre.

**Table 3.2:** *Minor Pests of Brinjal*

| Sl No. | Scientific name | Common name | Order | Family | Distribution | Nature of damage |
|---|---|---|---|---|---|---|
| 1. | *Anoplocnemis. janusphasiana F.* | Bug | Herniptera | Coreidae | India | Sap-sucker, toxic saliva |
| 2. | *Nezara viriduia L.* | Greenbug | Hemiptera | Pentatornidae | Cosmopolitan | Sap-sucker, toxic saliva |
| 3. | *Aspongopus janus F.* | Pumpkin stink bug | Hemiptera | Pentatomidae | India | Sap-sucker, toxic saliva |
| 4. | *Amrasca devastans* (Distant) | Jassid | Hemiptera | Cicadelliddae | India SE Asia | Infest foliage, sap-sicker |
| 5. | *Hishimonus phycitus* (Distant) | Brown leaf hopper | Haniptera | Cicadelliddae | India | Suck sap |
| 6. | Aphisgossypii Glover | Aphid | Hemiptera | Aphididae | Cosmopolitan | Infest leaves |
| 7. | *Phenacoccus insolitus* | Brinjal mealy bug | Hemiptera | Coccidae | India | Infest foliage |
| 8. | *Phycita clientella Z.* | Leaf folder | Lepidoptera | Phycitidae | India | Larvae fold leaves |
| 9. | *Psara bipuntalis F.* | Leaf webber | Lepidoptera | Pyraustidae | India | Webs the leaves |
| 10. | *Acherontia styx W.* | Leaf roller | Lepidoptera | Sphingidae | India | Larvae defoliate |
| 11. | *Eublemma olivacea Wlk.* | Brinjal leaf roller | Lepidoptera | Noctuidae | India | Larvae roll leaves |
| 12. | *Antoba olivacea W.* | Leaf roller | Lepidoptera | Nocmidae | India | Larvae defoliate |
| 13. | *Selepta celtis M.* | Hairy caterpillar | Lepidoptera | Noctuidac | India | Feeds on leaves |

**Spotted leaf beetle (or Hadda Beetle) *Henosepilachna vigintioctopunctata* (Fabr.) (Coccinellidae: Coleoptera):**

**Distribution:** It is found in different parts of South East Asia. The insect is considered to be one of the important pests of brinjal.

**Nature of damage:** The grubs and adults scrape the leaves feeding on the epidermal which get skeletonised and gradually dry away. They affect the crop in all the stages. The brownish hemispherical beetle has 12 - 28 black spots on the elytra. The female yellowish eggs in groups of 10 - 20 on the under surface of leaves. The egg period is 2 - 4 days. The yellowish spiny grubs become full grown in 10 -35 days and pupate on the leaf or stem.. Adults emerge in a week and live for a month feeding on leaves. The total life-history takes 17 - 50 days depending on weather conditions.

Different workers have studied the host plants of hadda beetle, *Epilachna vigintioctopunctata*, an important pest of solanaceous and cucurbitaceous crops like brinjal, tomato, potato, gourd, melon and cucumber, was also found infesting *Amaranthus viridis*. The leaves of *Amaranthus viridis* were found infested with both larvae and pupae and had characteristic feeding injury. This is the new food plant record for the coccinellid (Narang and Ramzan, 1984). A check list of *E. vigintioctopunctata*, their food plants and distribution in states of India is presented for 74 species which include *Coccinella delessertii* (*E. delessertii*) on *Solanum xanthocarum* and *E. vigintioctopunctata* on numerous host (Anand *et al.,*1988 ). Among the major insect pests that attack potato, the coccinellid *E. vigintioctopunctata* is very important in Asia. *E. dodecastigma* and *E. vigintioctopuncata, E. ocellata* and *H. sparsa*( *E. sparsa*) commonly attack solanaceous plants. The peak period of infestation is generally in July-August. The pest also feeds on brinjal, tomato, tobacco, pumpkin and bitter gourd. The larvae and adults scrap the green matter from leaves and damage up to 80 per cent of plants. (Rajagopal and Trivedi, 1989). Richards and Filewood, (1990) reported the infestation of *Epilachna vigintioctopunctata* on wild cucurbit plants near *Cucurbita moschata* from Parana, Brazil.

Ramzan and Singh (1982) studied the biology of *E. vigintioctopunctata* on different host plants and reported that it successfully bred on brinjal, potato, and *Solanum nigrum*. The respective, pre-ovipositional and ovipositional periods on brinjal, potato and *S. nigrum* were 8.0, 12.7; 15.0, 15.0; and 6.0, 35.0, days, respectively. The maximum average number of eggs laid by a female was on *S. nigrum* (502) followed by brinjal (269) and minimum on potato (216). The larval period was recorded longest on *Withiana sominefera* (18 days) followed by brinjal (14.7 days), potato (12.0 days) and *S. nigrum* (11.3 days). The pre-pupal and pupal periods recorded 2.5, 5.5; 3.5, 4.7; 2.5, 4.3 and 2.5, 5.0 days, respectively, on brinjal, potato, *S. nigrum* and *W. sominefera*. It completed life cycle in 26.3 days on brinjal. Similarly, Ram and Verma (1998) recorded egg laying in *E. vigintioctopunctata* was 8-59 days with 88-560 eggs per female and the average incubation period ranged from 3-6 days and the duration of Ist, IInd, IIIrd and IVth instar lasted for 2-4, 2-4, 2-6 and 3-8 days, respectively. The pre-pupal and pupal period varied from 1-3 days and 4-10 days, respectively. As the pest hibernated in winter, pre-oviposition period varied from 4-132 days. Male beetle survived for shorter duration (7-185 days) in comparison to female (2-198 days).The life cycle was completed from 17-41 days. The pest passed 7 generations in the laboratory and the duration was found shorter during April-May.

The lady beetle, *E. vigintioctopunctata* has been reported as a serious pest of some important solanaceous medicinal plants (Mathur and Srivastava, 1964) and on vegetables in Jammu and Kashmir. Munshi *et al.* (1968) observed that both adults and grubs caused damage up to 80 per cent to the host plants by feeding on chlorophyll content on the abaxial part of the leaf and caused concentric curves. Pest remained active from July to September. Mall *et al.* (1992) recorded *E. vigintioctopunctata* from 3rd weak of August to middle of October with its maximum population 0.22 beetles per leaf in the middle of September. The corresponding average temperature and relative humidity during the period were 28°C and 80 per cent, respectively. The population was noticed to gradually decline and disappear due to fall of temperature, humidity and maturity of the crop. Veeraval and Bhaskaran (1993) concluded that grub population of *E. vigintioctopunctata* was positively correlated with relative humidity, however, the adult population was noticed to decrease with increase in temperature. Again Veeravel and Bhaskaran (1994) identified different phenological/ growth stages of brinjal viz., seedling, vegetative, flowering, bearing and senescence in relation to the occurrence of pest. The grub incidence was recorded maximum at vegetative stage, whereas, peak adult population occurred during flowering and bearing stages of the brinjal plant.

In Himachal Pradesh, India Mehta *et al.,* (1993) reported *Pedobius foveolatus* and *Tetrastichus sp.* parasitizing grubs of *E. vigintioctopunctata* on brinjal. *P. foveolatus* was also reported as an effective parasitoid of *E. vigintioctopunctata* by (Kalra, 1997). Other, potential natural enemies of *E. vigintioctopunctata* on eggs and grubs as recorded were *Tetrastichus ovulara, Pediobius foveolatus* and *Pleurotropic foveolatus,* respectively by (Hamid and Singh, 1998). Veeraval and Raghuraman, (1998) have also recorded *P. foveolatus* as potential natural enemy of *E. vigintoictopunctata.* They reported that per cent infestation of *P. foveolatus* ranged from 19.4-19.5 percent. The 3$^{rd}$ and 4$^{th}$ larval instars of *E. vigintioctopunctata* were more preferred than other instars for parasitization.

Besides parasitoids, predatory wasps also parasitize IInd and IIIrd instars grubs and cause 70 per cent parasitism from July to September (Mathur and Srivastava, 1964). A pentatomid predatory bug *Cantheonidea fuscellata* Wolf. consumed 19 larvae or pupae of *E. vigintioctopunctata* per day in the nymph stage (Sheila and Abraham, 1981). Bhagat and Munshi, (1997) reported an entomogenous fungus, *Fusarium solani* as a pathogen infecting the spotted *Epilachna* beetle (*E. vigintioctopunctata*) from the Jammu region of Jammu and Kashmir, India for the first time. The infection of beetle egg masses was observed to be about 20 per cent. Hanif and Abida (1997) found *Beavaria bassiana* effective against all stages of *E. dodecastigma* except for the eggs. The 100 percent grub, pupal and adult mortality occurred after 120, 168 and 192 hrs, respectively. Venkataraman *et al.,* (1962) evaluated the insecticidal efficacy of *Bacillus thuringensis* Barl., spore suspension on *Henosepilachna vigintioctopunctata* and recorded complete mortality of the pest within 48 hrs.

Karmakar and Bhattacharya, (2000) reported that by using mechanical methods the pest population can be maintained at levels well below the economic injury level (1.25 – 11 *E. vigintioctopunctata* Fab. / plot). In search of ecofriendly and specific action compounds, (Gupta and Dogra, 1994) reported that diflubenzuron applied at micro quantity level was effective not only as potent larvicidal compounds interfering with the molting process but also effect egg hatchability and fecundity. Sardamme *et al.,* (1986) found diflubenzuron to have both stomach and contact action against the larvae of *E. vigintioctopunctata.* Mandal and Kumar (2001) evaluated the efficacy of Monocrotophos, Quinalphos, Cypermethrin and Fenvalerate against *E., vigintioctopunctata* infesting brinjal and reported that all the insecticides reduced the *E. vigintioctopunctata* populations, with cypermethrin recording the highest reduction in the pest population at all stages of plant growth. The neem oil affects reproduction by way of reduced fertility/ fecundity of *H. vigintioctopunctata* (Tewari and Moorthy, 1985. Mishra *et al.,* 1990), on *E. sparsa* (Mishra *et.al.,* 1990) and on *E. vigintioctopunctata* (Mathur and Nigam, 1993).

Management practices include collection and destruction of affected leaves along with the eggs, grubs and adults and, spraying of carbaryl 0.1% or cypermethrin 0.025% or profenofos 0.05%.

**5. The Melon Fruit Fly *Dacus cucurbitae* (Coquillett) (= *Strumeta cucurbitae* Coq.) (Tephiitidae : Diptera):** Distribution. India, Pakistan, Burma, Malaysia, China, Formosa, Japan, East Africa, Australia, Hawaiian Islands, Mauritius, Bangladesh, Sri Lanka, Indonesia, Thailand, Sarawak, the Philippines, Taiwan, Ryukyu Isles.

**Host:** Melon, tomato, chillies, guava, citrus, pear, fig, cauliflower, cotton, sunflower, lettuce, and other cucurbits both cultivated and wild.

The maggots are legless and look as headless, dirty white wriggling creatures, thicker at one end and tapering to a point at the other. The full-grown maggot measures 9-10mm long and 2mm broad in the middle. The adult flies are reddish-brown with lemon yellow markings on the thorax and have fuscous areas on the outer margins of their wings.

**Life-cycle:** The fly hibernates as an adult during winter months and becomes active in hot weather. With the onset of monsoon, the breeding starts. The adult flies after mating, take few days for the eggs to mature inside the body of a female which start laying them within 14 days. The female makes a cavity with the help of sharp ovipositor and thrusts the white elongate eggs singly or in groups of4 to 10 into the flowers and tender fruits. After laying the eggs, the female releases a gummy secretion which cements the tissue surrounding the puncture and makes the entrance water-proof. During the life span of 14-54 days a female on an average lays 58-95 eggs. The incubation period is 1-9 days. The freshly hatched maggots bore into the pulp, forming galleries. Depending on weather, the larval period extends from 3 days in summer to 21 days in winter. The full-grown maggots come out of the fruits and drop to the ground and pupate in the soil at a depth of 1.5 to 15cm. The pupal period is 5 to 9 days in summer and may extend to 30 days in cold weather. The total life-cycle occupies 12-34 days depending on weather conditions. There are several generations in a year.

Two other allied species, *Dacus ciliatis* Loew and *Dacus dorsalis* Hendel are common in India.

**Damage:** The larvae tunnel in the fruit, contaminating them with frass and providing entry points for fungi and bacteria which cause the fruit to rot. Owing to feeding on pulp, there isa premature dropping of fruits and also make them unfit for consumption. The young fruits can be destroyed in a few days, but older fruits show less obvious symptoms but on splitting open, a mass of maggots in pulp is found. In melons the infestation reaches up to 100 percent while in other cucurbitaceous fruits, the infestation may be upto 50 per cent.

**Control:** All the fallen infested fruits should be collected and destroyed. Frequent raid ng of the soil under the vine or ploughing the infested field after the crop is harvested can help in killing the pupae. Apply the bait spray containing 20 ml Malathion 50 EC + 200 g of sugar/gur in 20 litres of water per acre. When the attack is serious, it should be repeated at weekly intervals.

**6. The Red Pumpkin Beetle *Raphidopalpa foveicollis* (Lucas) (= *Aulacophora foveicollis* Lucas) (Chiysomelidae: Coleoptera)**

**Distribution:** India, Asia, Australia, Southern Europe and Africa.

**Host:** Pumpkin, ash guard, *tinda,* Ghia, tori, melon and cucumber.

The grubs arc creamy white, with a slightly darker oval shield at the back. When full- grown, they measure about 12mm in length and 3.5 mm across the mesothorax. The beetles are oblong and 5.8mm long. Their dorsal body surface is brilliant orange red and the ventral surface is black, being clothed in short white hairs.

**Life-cycle:** The beetles resume activity as soon as the season warms up and lay about 300 oval, yellow eggs singly or in batches of 8-9 in moist soil, near the base of the plant during their life span of 60-85 days. The incubation period lasts 6-15 days. The larval and pupal periods are 13-25 and 7-17 days, respectively. The pupation takes place in the thick walled earthen chambers in the soil at a depth of about 20-25 cm. On emergence, the beetles began to feed and breed. The life-cycle is completed in 26-37 days. There are five generations from March to October.

*R. aeripennis* and *A. cincta* (yellow with dark border) species are also common in north western India.

**Damage:** The grubs Lead a subterranean life feeding on roots underground stems of creepers and on fruits laying in contact with the soil. The beetles feed on those parts of the plants which arc above the ground. They injure the cotyledons, flowers and foliage by hung holes into them. The beetles are very destructive to cucurbitaceous vegetahtes especially during March- April when the creepers arc very young.

**Control:** After harvesting the infested fields, plough deep to kifl the grubs in the soil. Apply 2.75 kg of Carbofuran 3 G per acre 3-4cm deep in the soil near the bases of the plants just after germination and irrigate or spray 75-150 g Carbaryl 50WP in 50-100 litres of water per acre.

**7. Leaf hopper *Amrasca biguttula biguttula* (= *Empoasca devastans* Dist.) (Cicadellidae: Homoptera)** It is distributed all over India. It is a polyphagous pest attacking number of crops which include okra, brinjal, beans, castor, cucurbits, hollyhock, potato, sunflower and other malvaceous plants. The nymphs and adults remain on the underside of leaves and suck sap and while feeding inject their toxic saliva. The plant brcomes stunted, the leaves crinkle, turn yellowish. Brownish or reddish colour developing along the edges of the leaves is called as 'hopper burn'. The pest is very active from September to January. The adult is a slender green insect. The eggs are laid singly inside the leaf vein on the under surface of the leaves and are yellowish in colour. A female lays 15 - 30 eggs. Egg hatch in 4 -10 days. Nymphs are green and wedge shaped. They cause extensive damage and feed mostly near the base of the leaf vein. Then they distribute themselves throughout the leaf and feed from the under surface. It takes 7-21 days to develop to adults from nymphs. They breed throughout the year. Management strategies: (i) Spraying of 0.04 % phosphamidon or 0.05% monocrotophos or 0.01% imidacloprid. (ii) Application of 5% dimethoate granules in seed furrows @ 20 kg/ha.

**Chillies (*Capsicum annum* Linnaeus):**

**1. Chilli thrips, *Scirtothrips dorsalis* Hood (Thripidae: Thysanoptera)**

**Distribution:** This a polyphagous pest, widely distributed in India, Srilanka and Thailand. Besides chillies, tomato, cotton, castor, sunflower, mango, citrus and Acacia arabica. The nymphs are like adults in shape and colour but are wingless and smaller in size. The adults are slender, yellowish brown having apically pointed wings and measure about 1 mm in length. The female having long narrow wings with the fore margin fringed with long hairs. The pest is active throughout the year except theb monsoon region. A female lays 45 to 50 eggs inside the tissues of the leaves and shoots. The incubation period, pupal period larval period and the longevity of adult thrips occupy 5 days, 2 to 4 days, 7-8 days and about 31 days, respectively. The pest completes several overlapping generations in a year. Nymphs and adults suck the cell sap from the leaves and tender parts of the plants and cause the leaves to shrivel. In case of severe infestation, there is a malformation of leaves, buds and fruits, which may damage half the crop. The attacked shoots hardly develop and leaves fall off. This thrip is also responsible for transmitting leaf curl virus disease of chillies.

**Management strategies:** The thrips can be managed by spraying with 0.03% dimethoate or monocrotophos or 0.2% carbaryl or 0.04% triazophos or 0.075% acephate or by encouraging biological control agents like, The thrips *Franklinothrips vespiformis* (Crawford) and *Erythrothrips asiaticus* R. &. M. are predaceous on this thrips in nature and their population may be encouraged by avoiding chemical sprays.

**Pests of potato *(Solanum tuberosum L.)* and Tomato *(Lycopersicon esculentum Mill.)***

Potato, *Solanum tuberosum* L. is a temperate crop and is grown in about 500,000 hectares in India. The enlarged tips of underground stem (stolons) bearing edible tubers (potatoes) are eaten in various *ways.* It is rich in carbohydrates (17%) and also contains proteins (2%), calcium, vitamin A, thiamine, riboflavin, and niacin.

Tomato, *Lycopersicon escutentum* Mill. is grown widely throughout the world. Fruits are of various shapes arid sizes, green when immature, red or yellow when ripe containing vitamin A and C. The fruit is used as a vegetable, raw or cooked, made into soup, sauce juice, ketchup, paste, puree, powder or may be canned, also used unripe in chutneys. It is a common ingredient of salad.

Insect pests of major importance include potato tuber moth, tomato fruit borer. The minor pests are cutworms, leaf hoppers, leaf-eating caterpillars and leaf-eating bugs etc.

***Gnorimoschema operculella* (Zeller)** (= *Phthorimaea operculella* Zeller)

| | | |
|---|---|---|
| *Common name* | : | The Potato Tuber Moth |
| *Order* | : | Lepidoptera |
| *Family* | : | Gelcchiidae |

**Distribution:** Almost completely cosmopolitan in distribution.

**Host:** Potato, tobacco, tomato, egg-plant (brinjal) and other Solanaceae and *Beta vulgaris.*

The caterpillars are pinkish white or greenish with dark brown heads. They measure about 20mm in length. The moth is very small, narrow winged and nocturnal and is about 13 mm across the spread wings. It is greyish-brown with mottling of dark brown.

**Life-cycle:** This pest breeds throughout the year under favourable conditions. A female lays on an average 150 to 200 eggs singly on the undersurface of the leaves or on exposed tuber of potato. The incubation period is 3-6 days. The young larvae cause silver blotches on the leaves and mines them. Subsequently, they tunnel the stems. During summer, the larval period is 14 to 21 days. The full-grown larva pupates in a silken cocoon among trash, clods of earth. The pupal period is completed in about 30 days. There are five to six generations in a year.

Another spccies of tuberrnoth attacking potato is *Gnorimoschema heliopa* (Lower).

**Damage:** The leaves have silver blotches caused by the young larvae mining in the leaves, leaf veins, petiols and stems are tunnelled followed by wilting of plants. Larger caterpillars bore the tubers which often become infested with fungi or bacteria. The insect occasionally infests potato crop in the field but is particularly serious in storage. The infestation in storage may range from 30 to 70 per cent.

**Control**

1. All the infested tubers should be removed and destroyed.
2. Spray 500g DDT 50 WP or, 1 kg Carbaryl 50 WP in 250 litres of water per acre or dust seed potato 150g of DDT 5 percent or Malathion 2 per cent per 100 kg and repeat spray 2 to 3 times at fortnightly intervals.
3. Fumigation with Carbonbisulphide@ 2-3 litres per 100 cubic meters will help in checking the infestation in storage.

**2. *Helicoverpa armigera (Hubner):*** The Tomato Fruit Borer Lepidoptera Noctuidae

For details see under tomato

**Table 3.3:** *Minor Pests of Potato and Tomato*

| S. No. | Scientific name | Common name | Order | Family | Distribution | Nature of damage |
|---|---|---|---|---|---|---|
| 1. | *Nesiodioeoris tenuis* (Reuter) | Leaf bug | Hemiptera | Miridae | India | *Infest* foliage |
| 2. | *Nezara viridula* Linn. | Green bug | Hemiptera | Pentatomidac | india | Suck the sap |
| 3. | *Amrasca bigutulla bigutulla* Ishida | Leafhopper | Hemiptera | Cicadeflidae | India | Suck the sap |
| 4. | *Trialeurodes vaporariorum* (WesL) | iiwfly | Hemiptera | Aleyrodidae | India1 Europe | Infestlcaves, viruniector |
| 5. | *Bemisiatabaci Genn.* | Whitefly | Hemiptera | Aleyrodidae | Cosmopolitan | Infcstfoliage |
| 6. | *Aulacorthum solani* Kalt. | Aphid | Hemiptera | Aphididae | Cosmopolitan | . Infestfoliage. suck sap, *virus* vector |
| 7. | *Aphis gossyii Aphid G.* | *Aphid* | Hemiptera | Aphididae | Cosmopolitan | Infestfoliage, sucksap, virus vector |
| 8. | *A. nasturtii* Kalt | Aphid | Hemiptera | Aphidiclae | India | Infest follagc, suck sap, vinis vector |
| 9. | *Macrosiphum euphorbiae* | Aphid | Hemiptera | Aphididae | Cosmopolitan | Infestfoliage,suck sap1 (Thmas) virus vector |
| 10. | *Myzus ornatus Laing* | Aphid | Hemiptera | Aphididae | India | Infestfoliage, sucksap, - virus vector |
| 11. | *Myzus persicae (Sulz.)* | Aphid | Hemiptera | Aphididae | Cosmopolitan | Infestfoli age, suck sap, virus vcctor |
| 12. | *Ferrisia virgata* (CkLL) | Mealy bug | Hemiptera | Pseudo coccidae | Pantropical | Infestioliage |

| | | | | | | |
|---|---|---|---|---|---|---|
| 13. | *Agrotis ipsilon* | Cutworim | Lepidoptcra | Noctuidae | Cosmopolitan | Feed on leavcs& Cut tender Stem |
| 14. | *Spodoptera litura* | Cutworim | Lepidoptera | Noctuidae | India | Larvae defoliaie |
| 15. | *Eumerus* | hoverfuly | Dipiera | Syrphidae | India | Infest tubers |
| 16. | *llolotrichia conferta sharp* | Beetle | Coleoptera | Melolonthidae | Jndia | Giubs attack rooLs and tubers |
| 17. | *Epilachna vigintioctopunctata* | Spotted leaf beetle | Coleoptera | Coccinellidae | Jnfia | Feedsunkaves |
| 18. | *Chalaenosoma metalicum* F. | Flea beetle | Coleoptera | Aliicidae | India | Adults feed on leaves |
| 19. | *Mylioccrus subfasciatus* | Leaf weevil | Colcoptcra | Curculionidae | India | Adults fccd on leaves |

**3. Malvaceous Vegetables:** Okra, *Abelmoschus esculantus* (L.) (Moench) is the major economically important vegetable crop highly susceptible to wide array of insect pests attack right from the seedling stage upto fruit/ seed harveststage.

More than 37 insect pests have been reported causing damage to okra fruits. The pest complexes have been categorized into sucking pests, borer insect pests, and leaf defoliators.

## Sucking insect pests

The leaf hopper, *Amrasca biguttula* Ishida is widely distributed in India and one of the most destructive pest of okra. It is found throughout the season and survives on other host crops during the off season due to its polyphagous nature. The damage is caused due to sucking of sap by adults as well, as nymphs resulting in hopper burn symptoms on the foliage. Mahal et al (1994) reported that jassids /plant hoppers cause upto 50 per cent damage in the crop wheras the seed crop damage can go upto 79 percent. Reddy and Rao (2001) identified *Amrasca biguttula* as an important pest of okra in Andhra Pradesh.

The whitefly, *Bemisia tabaci* (Gennadius), is a widely distributed polyphagous pest attacking okra among other crop host crops. The nymphs and adults not only cause damage by desaping but also transmit the yellow vein mosaic virus disease which has become a serious threat in okra cultivation. No variety has shown durable resistance against the disease. The aphid, *Aphis gossypii* Glover, attacks okra crop during the rainy season and causes damage by desaping (Arora et al.,1996).

Various species of mites have been reported to feed on okra in india.Singh (1994) reported that okra is attacked by *Eutetranychus orientalis* Klien,*Oligonychus coddeae* (Nietner), *Tetranychus cinnabarinus* (Boisd), *Tetranychus macfarlanei* Baker and Pritchard,Tetranychus *ludeni* Zacher, *Tetranychus neocaledonicus* Andre.*Tetranychus puschelii* Meyer, *Brevipalpus californicus* (Banks), *B. obovatus Dennadieu, B. Phoenicus* (Geij) and *Amblyseius sp*. The red cotton bug, *Dysdercus koenigii* (Fabricius), attacks the okra crop at the seed filling stage and causes damage by sucking the sap from the leaves as well as the developing seeds. The dusky cotton bug, *Oxycarenus laetus* Kirby, attacks okra during both spring and summer and rainy seasons during, at the maturity stage. Other sucking pests of okra include the blossom thrips, *Microcephalothrips abdominalis* (Crawford), *Frankliniella schultzei* (Tryborn) and *Haplothrips gowdei* (Franklin), which infest okra flowers (Anathakrishnan, 1971). The green plant bug, *Nezara viridula* Linn. Has also been reported on Okra (Arora et al 1996).

## Borer insect pests

Two species of spotted bollworms *Earias vittela* (Fabrius and *E. insulana* (Boisduval) have been reported on okra. The larva causes damage by boring the young shoots, flowers and the developing fruits and by feeding inside. The infested fruits become unfit for human consumption. In india *E. vittella* is more common and causes a yield loss of about 45 per cent(Srinivsan and Krishnakumar, 1983). The American bollworm, *Helicoverpa armigera* Hubner also attacks okra. Singh et al (1993) recorded a high incidence of *H. Armigera* on okra.

## Foliage feeding insects

Various species of blister beetles have been reported on okra occurring during the period from flowering to harvest. They include *Myllabris pustulata* Thunb. and *M. phalerata* (Pallas) (Sangha and Mavi 1995). Other foliage feeding pests of okra include – the green semilooper, *Anormis flava* Fabr. The grey weevil, *Myllocereus undecimpustulatus maculosus* Desb.., *M. variegatus* Boehman *and M. viridanus(Fab.).*(Srinivasan 1993), cotton shoot weevil, *Alcidodes affaber* Aurivillius(Thippeswamy et al 1980) and the chrysomelid beetle, *Plagiodera rufescens* Gyllenhal (Nath and Sharma,1996).

## Bhindi (Okra) (Hibiscus esculentus Moench)

**Okra Shoot and Fruit borer Earias vittella and *Earias insulana* (Fabricius) (Arctidae: Lepidoptera)**

They are widely distributed in North Africa, India, Pakistan, Sri Lanka, Bangladesh, Burma, Indonesia, New Guinea, and Fiji and other countries. It is a polyphagous pests having 38 hosts. It feeds on large number of malvaceous plants growing both in wild as well as cultivated. The eggs are laid usually singly on buds or flowers, and when the crop is only few weeks old, the freshly hatched larvae bore into the tender shoots and tunnel downwards, these shoots droop down, wither and ultimately die. As soon as the formation of buds, flowers and fruits begins, the larvae bore inside these and feed on inner tissues. They further continue to move from bud to bud and fruit to fruit damaging all the developing fruits. The damaged buds and flowers wither down and die wheras the affected fruits become deformed, stunted in growth and become unfit for human consumption.

The eggs are about 0.5 mm in diameter, spherical in shape aqnd light bluish in colour. They have 26-32 longitudinal ridges, the alternate ridges projecting upward toform the crown. The moths emerge at dusk. The mating takes place 2 - 3 days after emergence and oviposition starts after 1 - 5 days of mating. A female on an average lays 400 eggs (65 - 695). Incubation, larval and pupal periods last for 3 - 9, 9 - 20 (50 - 60 during winter) and 8 - 12 days, respectively. It takes 22 - 25 days to complete single life cycle which may extend up to 74 days during winter. There may be 8 - 12 generations in a year. Another fruit borer, Earias insulana (Boisduval), is also found damaging okra particularly in drier regions.

Management strategies: (i) Remove debris and all the alternate host plants from field also destroy all the infested shoots and fruits (ii) if the infestation is more, give two to three fortnightly spraying with 0.2% carbaryl or 0.005% cypermethrin.

*Trichogramma chilonis* Ishii (Ashmead), an egg parasite is very useful in reducing the incidence of the pest. Its incidence varies from 3 to 15 per cent in northern India. The exotic parasitiod *Trichogramma brasiliense* has been reported to reduce the incidence by 50 per cent. The unilative release in July-August and inoculative release afterwards are recommended. Another egg parasitiod *Trichogrammatoidea* sp. *Near guamensis* (Trichogrammatidae) is also associated with this pest. *Brachymeria nephantidis* Gahan (Chalcididae) from pupae of *Earias sp.* and parasitism ranges from 13 to 57 per cent in different months.

**Red spider mite (*Tetranychus cinnabarinus*(Boisduval), *T. urticae* Koch.) (Tetranychidae: Arachnida):** It is a polyphagous non-insect pest feeding on more than 150 host plants including vegetables like brinjal, ladies finger, cowpea, cucurbits, beans etc. These mites normally inhabit the under surface of leaves and spin the webs. The plants often remain covered with dense webs under which both the developing stages and adults damage the crops by sucking the cell sap. Besides, dust particles get adhered to webs and retard normal physiological activities of the plant leading to reduction in plant growth and yield. The affected leaves turn yellow, dry up and finally wither away giving an unhealthy appearance. Generally, spider mites are found numerous during summer months, multiply quickly and pass through a several generations in a very short period of time like aphids. These mites are soft bodied red in colour and colony forming. Larva is six legged while nymphs and adults are eight legged like spider. Size varies from 0.5-0.6 mm. Spider mite completes its life cycle within 9-14 days during active period.

## Management

Tolerant/less susceptible varieties are important component of IPM programme. They reduce the pest incidence by non-preference, antibiosis and/or tolerance. Conservation of parasitoids is very important as parasitoids like *Erythmelus empoascae, Stethymium empoascae*, are found to be active against *Amrasca bigutula bigutula.* Besides some predators such as coccinellids, syrphids, spiders, staphylinids, predatory mites and anthocorid bugs, which are predominant in the okra ecosystem needs to be conserved as they are highly prone to pesticides. Microbials are the potential source of pest control operations. Spraying of *Beauveria bassiana* 1 x $10^4$ cfu/g two times with a week interval satisfactorily managed whitefly nymphs and adult. Commercial formulations of *Bacillus thuringiensis var. kurstaki Viz.* Biolep followed by Bioasp proved to be most effective at all concentrations against *E. vitella.* The botanical insecticides did not exert any adverse effect on the natural beneficial predatory fauna such as coccinellids, syrphids, and spiders and hence found suitable in okra IPM.

**4. Cucurbitaceous Vegetables:**

**I. Pests of Gourds**

1. Fruitflies *Bactrocera cucurbitae* Coq. and *B. ciliatus* Loew (Tephritidae : Diptera)

It is a major pest of cucurbits particularly in bitter gourd and musk melon. Among the several species of fruit fly, *B. cucurbitae* is most important due its polyphagous nature and cosmopolitan distribution. The female fly makes a cavity with the help of its sharp ovipositor and inserts the white, cigar-shaped, slightly curved eggs singly or in groups into the flowers and tender fruits. The incubation period is 3-9 days. The freshly hatched maggots bore into the fruit pulp by forming serpentine galleries and contaminating them with its frass and providing entry points for saprophytic fungi and bacteria, which cause the rotting of the fruit. Due to feeding on pulp, there is premature dropping of fruits and make them unfit for consumption. The infestation varies from 50-100 per cent in different cucurbitaceous crops.

Adult flies are medium sized (4-5 mm long) with reddish brown and their wings are hyaline with a dark patch on the outer margins. Hind cross veins thickened with brown and grey spots at the apex. The full-grown maggots come out of the fruits and drop to the ground and pupate in the soil. The total life cycle occupies 14-34 days depending on weather conditions. There are several over lapping generations in a year.

**2. Snake gourd semilooper *Anadevidia* (=*Plusia*) peponis F. (Noctuidae: Lepidoptera):** The pest is reported to occur in all localities where snake gourd is grown. The caterpillar cut the leaf rolls it and live inside the roll. This pest is specific to snake gourd and in case of severe infestation the larvae defoliate the plants completely. The moths are brownish in colour having shiny brown four wings. The female moth lays white spherical eggs on the under surface of leaves of host plants. The semilooper caterpillar measures 35 – 40 mm long and has greenish body with white longitudinal lines and black tubercles with thin hairs arising on them. It pupates in a thin silken cocoon in leaf fold. The pupa is initially greenish but turns dark brown before the emergence of the adult moth. Egg, larval and pupal periods last for 4 - 5, 24 - 30 and 7 - 8 days respectively. The life history is completed six weeks. Management strategies: (i) hand picking of larvae and their destruction is recommended when they are in small numbers. (ii) Spray application of 0.02% carbaryl 0.05% monochrotophos.

**3. Red Pumpkin beetle (*Aulocophora foveicollis* Lucas):** It is a bright orange red coloured medium size (6-8 mm) beetle with black ventral surface found regularly on cucurbitaceous crops like pumpkin, cucumber, ash guard, bottle gourd, bitter gourd and melons. The grubs live in soil feeding on roots, rootlets and under ground stems and sometimes on fruits in contact with the soil. On emergence, the beetles feed on aerial parts of the plants like cotyledons, flowers and foliage and make holes on them. The damage by beetles is more detrimental during the cotyledonous stage of the plants.

The female beetle lays 250-300 oval eggs, initially yellowish pink in colour and later changes to orange. The eggs are laid singly or in batches of 8-9 in moist soil, near the base of the plant. The incubation period lasts for 5-15 days. Fully grown grubs are 10-12 mm in length. The pupation takes place in earthen chambers in the soil at a depth of 15-25 cm. It completes its life cycle in 25- 38 days.

**Spotted leaf beetle (or Hadda Beetle) *Henosepilachna vigintioctopunctata* (Fabr.) (Coccinellidae: Coleoptera)**

Bitter gourd is also subjected to ravages by a wide array of insect pests during crop season and among them *Epilachna vigintioctopunctata* is one of the major pest of bitter gourd. Nearly 80 per cent of the cucurbits are reported to be partially or completely damaged by the pest every year in India. Besides cucurbits, *E. vigintioctopunctata* also feeds on brinjal, tomato, tobacco, pumpkin and potato. Different species of hadda beetle have been reported to attack brinjal of which *E. vigintioctopunctata* is the most predominant species (Butani and Verma, 1976, Kalra, 1997). *E. vigintioctopunctata* is a phytophagus lady beetle with polyphagus habit that seems to inflict considerable damage to plant foliage. Both the adult and grub stages feed on the chlorophyll content hence skeletonize the leaves which present a lace like appearance, later dry up and fall from the plants (Butani and Verma, 1967, Kalra, 1997).

**5. Leguminous Vegetables**

**2.7.1 Spotted pod borer (*Maruca vitrata* Geyer):** It is a serious pest of leguminous crops and gaining importance due to its wide geographical distribution, extensive host range and its ability to infest the young growing plant tips, stems, flower buds, flowers, pods and seeds. Young larvae feed on the flowers and young pods by scrapping and later stage they bore the pods and feed the internal content of the fruits and make unsuitable for consumption. Infested flowers drop down prematurely. Sometime, they also feed the leaves and shoots by making galleries. In cowpea, damage due to this pest is reported up to 42 per cent.

Moths are medium sized with brownish black fore wings with conspicuous transverse white elongate marking,whereas, hind wings are translucent, pinkish white in colour. Eggs are laid singly on the flowers, flower buds, pods, tender leaves and growing points. The egg hatches within 3-4 days and larval period lasts for 11-16 days.

**2.7.2 Black bean aphid (*Aphis craccivora* Koch.):** It is soft-bodied, black coloured sucking pest usually found in colonies on the growing points of the plants, tender fruits, flower buds etc. Both the nymphs and adults suck the sap from the plants and cause stunted growth of the plants and seeds shrivelled in the developing pods. They also produce copious amount of honeydew deposited on the plants and develop the sooty mould which hinder the photosynthetic activity of the plants. Beside these, they also transmit the viral diseases like bean yellow mosaic virus, bean common mosaic virus, etc. Reproduction noticed throughout the year by means of parthenogenesis. They are often attended by the symbiotic ants which favour them for dispersal and protection against natural enemies. Total life cycle completes in 9-13 days depending on the weather conditions.

## Pests of peas *(Pisum sativum L)*

*Peas, Pisum sativum* Linnaeus cornmanly known as garden pea, is grown as a vegetable. The seeds are eaten green, either fresh or frozen or used in canning. The plant body is used as livestock feed. The seeds contains vitamin A. C and B1. and also phosphorus, iron and carbohydrates.

Major pests of garden pea include, pea leaf-miner, pea stem fly and peapod borer. The minor pests include sap sap sucking bugs, whiteflies, thrips, leaf-eating caterpillars and stem gall weevils.

**Melanagromyza phaseoli(Tryon)** (The Pea Stem Fly Agromyzidae :Diptera)

**Distribution:** India, Sri Lanka, New South Wales Philippines, East Indies, China\

**Host:** Peas, soyabean, cowpea, Lab-lab, niger, The larvae are leaf and stem miners. The adult flies are metallic black,

**Life -cycle:** The metallic flies arc active in summer and mate 2-6 days after emergence The female inserts her ovipositor into the leaf and sometimes in the stem and petiole and make an elliptical cavity under the epidermis. A female lays 14-64 elongate, oval and white eggs, the average being 33. The incubation period is 2-4 days. The maggot mines through the leaf towards the petiole and later on, reaches the stem. They pass through three instars. The larval development is completed in 6-7 days in March April and in 9-12 days in November-December. The larva pupates within its gallery. The pupal period lasts *5-9* days and 18-19 days in March-April and November-December, respectively. The longevity of female and male are 8-12 days and about 11 days respectively. The pest passes through 8-9 generations from july to April. During winter it remains inactive as a larva or as a pupa.

**Damage:** The maggots burrow into the stem causing withering and ultimate drying of affected shoots, thus, reducing the bearing capacity of the host plant. The adults also cause damage by puncturing the leaves and feed on the liquid that oozes. Owing to the feeding, the injured portion turn yellow. Seedlings are more affected than grown up plants.

**Control:** Remove and destroy all the affected branches during the initial stage of attack. Seed dressing with 2 kg of 50 per cent Phorate (Thimet) or, 8 kg of Oxydisulfoton per 100kg seed provides effective protection. Oxydisulfoton 1 kg or Phorate 2kg applied as presowing treatment is also effective.First application of 300 ml of Methyl oxydemeton 25 EC in 250 litres of water per acre should be given just after germination on the crop and repeat twice at *a*n interval of two weeks each.

**Phytomyza atricornis Meigen**

(The Pea Leaf Miner (Agromyzidae: Diptera)

**Distribution:** India, New Zealand, Italy, Australia, USSR, USA and Europe.

**Host:** Pea, mustard, cauliflower, cabbage, knol-khol, turnip, radish, safflower, lentil, berseem, hollyhock, barley, carrot, potato, linseed, poppy. The larvae are minute, slender. The adults are two winged flies having greyish black mesonotum and yellowish frons.

**Life cycle:** The pest is generally active from December to May. The rest of the period passes in the soil as a pupa. The adults emerge at the beginning of December and after mating start laying eggs singly in leaf tissue, The incubation period is 24 days. After hatching the larvae feed between the lower and upper epidemis by making zigazag tunnels. They are full-fed in about 5 days and pupate within the galleries. Pupal period lasts in 6 days. The insect completes one generation in about 13-14 days and pass through several broods during the period of its activity.

**Damage:** The larvae are the only destructive form. They make prominent whitish tunnels in the leaves which can be seen by holding the leaves against bright light. It is very rare that the mine coalesce into one another. The tunnels made by the larvae interfere with photosynthesis, and also normal development of the plants,

**Control:** Spray 400 ml of Dimethoate 30 EC in 250 litres of water per acre and repeat spray at 15-days interval. After spray 20 days is waiting period for the picking of pods.

**(C) Leafy Vegetables (*Amaranthus viridis*):**

1. Leaf caterpillar *Hymenia recurvalis* (Fb.) (Pyraustidae: Lepidoptera) Distribution:

Amaranths are one of important leafy vegetables consumed throughout Africa and Asia due to their highly nutritive content and their utilization in preparation of various foods in India.. It is an herbaceous annual plants belonging to the family Amaranthaceous. Its leaves are eaten as vegetables and seeds as cereal (Ezeh *et al.*, 2015, Kamalanathan *et al.*, 1973; Oke, 1980). One of the most important limiting factors in its profitable production is the attack of wide range of pests (Aderolu *et al.*, 2013).

More than 60 insect species associated with amaranthus have been recorded causing damage. Of these pests, *Hymenia recurvalis* Fabricius causes severe losses to amaranthus species (Navarajan, 2007, Aderolu *et al.*, 2013). The larvae web the leaves and feed on the green matter. In case of severe attack complete skeletonising and drying of leaves may occur. The moth has white wavy lines on the dark wing. The female lays approximately 156 eggs on the leaves. The greenish larva pupates in the soil in an earthen cocoon. The life cycle is completed within 3-4 weeks, with egg, larval and pupal stage of 3-4, 10-16 and 8-12 days, respectively.

Management strategies: (i) Spray application of profenofos 0.05 % or phosalone 0.07 % controls the pest. Jeyasankar and Gokilamani (2016) found that neem extract showed effective antifeedant for both the larvae of H. recurvalis and P. Basalis and may be used for controlling amaranths pests.

## Pests of onion *(Allium cepa* L.)

Onion *Atliurn cepa* Linnaeus is a widely cultivated vegetable crop found all over the world and is used as an essential supplement with various other vegetables. In India, it is grown on an area of over 1.00,000 hectares either as bulbs for drying or for pickling or as salad onion. The underground bulbs, the edible portion vary in shape and colour. Onion is rich in carbohydrates and minerals namely phosphorus, calcium and also contains proteins and vitamin C.

Onion thrips and onion fly are two major pests of economic importance. A few species of eating caterpillars, groundnut ear-wig and some beetles damaging bulbs are of minor importance.

**Thrips tabaci Lindemann** (The Onion Thrips or Cotton Thrips Thripidac Thysanoptera)

**Distribution:** Almost completely cosmopolitan. Found throughout India.

**Host:** Onion, tobacco, tomato, pyrathrum, cotton, pine apple, peas, brassicas, beet, garlic, cabbage, cauliflower, potato, cucumber and many other p1ants.

The nymphs are slender, yellowish-brown and look like adults but are wingless and slightly smaller. The adults are slender yellowish-brown and measure about 1 mm in length. The males are wingless whereas the females have long, narrow strap like wings, which are furnished with long hairs along the hind margins.

**Life cycle:** The pest is found in all kinds of vegetation, and is a major pest of onion and garlic from November to May. In October it is found on cabbage and cauliflowers. The female live for 14-30 days and lays 50-60 kidney shaped eggs singly in slits which are made in leaf tissue with its sharp ovipositor. The incubation period is 4-9 days and the nymphs start feeding on the plant juices by lacerating the leaf tissues. They are generally congregated at the base of leaf or in the flowers or onion and garlic. On cotton, cauliflowers and cabbage, they usually feed on the lower surface of the leaves. Nymphs are full-grown in 4-6 days after passing through four stages. Later on, they descend to the ground and pupate at the depth of about 25 mm. The pre-pupal period is 1-2 days and pupal period is 2-4 days. In a year, several generations are completed.

**Damage:** The damage is done by both nymphs and adults. The pest infests the crops throughout the year. As a result of its attack on the leaves of onion and garlic at first show a spotted appearance and later pale-white patches due to the drainage of sap, subsequently the leaf tips fade and gradually dry up. The plants do not form bulbs and there is no seed setting in the flowers. In cotton, the leaves become wrinkled and fall off and the plants bear very few bolls. *T. tabasi* is known to act as a vector of the streak virus disease of peas and yellow spot of pine apple in Hawaii.

**Control:** Spray 250 ml Malathion 5OEC in 250 litres of water per acre as soon as the pest appears. Waiting period is at least 7 days before harvest.

**Hylemya antique (Meigen):** The Onion Maggot (Anthomylidac Diptera)

**Distribution:** Cosmopolitan and widely distributed in U.S.A. Japan, U.S.S.R. England, France, Germany, Canada, India (northern parts)

**Host:** Onion the maggots are small, white and measure about 8 mm in length. The flies are slender,. greyish, large winged and about 6mm in length.

**Life-cycle:** White and elongate eggs are laid by female flies near the base of the plant in cracks in the soil. The incubation period is 2-7 days. After hatching, the maggots crawl upto the plant, and enter the inside of the leaf sheath and reach the bulb.It is the full-fed in 14-21 days, and then crawls out the bulb to pupate into the soil. After 14-21 days, the adults emerge and start the new generation. The third generation pest attack the onions shortly before the harvest Owing to this, the onion in storage start rotting.

**Damage:** The attacked plants become flabby and yellowish. The maggots bore into the bulbs, leaving only outer sheath. At times, it destroys 8 to 9 per cent of the plants. Several maggots at a time attack larger bulbs making cavities. The infestation is not completely distinctive to the bulbs but it leads subsequent rotting in the storage. The initial damage caused by this insect leads to the development of soft rot of onion, caused by *Bacillus carorovorus.*

**Control:** Apply 4 kg Carbaryl 4 G or, Lindane 6 G or Thimet 10G to the soil, followed by light irrigation.

**B. Insect pests of oilseed crops:** In recent years, there was a breakthrough in the production of oilseeds in India and country is heading towards self-sufficiency in oilseed production. This was possible only because of improved varieties, substantial increase in irrigation facilities, availability of fertilizers, improved cultivation technologies and above all through pest control. In India, about 24.0 million hectares (kharif 13.8 m ha and rabi 10.2 m ha) land is under cultivation of oilseed crops which includes groundnut, rapeseed, nigerseed, safflower, sunflower, mustard, sesamum, linseed, castor seed, and soya bean The annual production is *18.5* million tonnes, the average being *769* kg/ha *(kharif* 703 kg/ ha and, *rabi, 859* kg/ha).

Oilseed crops are damaged by a large number of insect pests which include aphids, bugs, sawfly, leaf-miner, stem borer, white grub, whitefly, semi-looper, capsule borer, hairy caterpillar, slug, pod caterpillar, gallfly, gallmidge and bud fly. Other pests namely jassids, cutworm, flea beetle, grasshopper, termite, ear-wig, hairy caterpillar, thrips, and mite *are* of minor importance.

Pests of Brassica *(Brassica spp.); Family - Cruciferae.*

**1 *Lipaphis erysimi* Kaltenback:** The Mustard Aphid (Aphididae: Hemiptera )

**Distribution:** Cosmopolitan in warner regions

**Host:** Oil seeds like *Brassica, toria, sarson, raya, taramira,* vegetables like cauliflower, cabbage, knol-khol,etc. The nymphs and adults are louse-like and pale-greenish in colour.

**Life-cycle:** The pest is very active from December to March when various cruciferous oil seed and vegetable crops are available in the fields. In hot weather, the pest is believed to migrate to hills but some evidence show that it survives in plains also on stray cabbage plants and on some weeds. The pest breeds parthenogenetically and the females give birth to 26-133 nymphs. The development is very fast and are full-grown in 7-10 days. As many as *45* generations are completed in a year. Cold and cloudy weather is very favourable for the multiplication of this pest. In autumn and spring, winged forms are produced and they spread from locality to locality and from field to field.

*Ischiodan scutellaris (Fab.) Syrphus serarius* (Wiedmann), *Chrysopa scelestes Banks, Coccinella septimpunctata. and Menochiles sexmaculata Fab.* are predators on mustard aphid. *Diaeretiella rapae* (Curtis) and *Lipolexis gracills* Forester parasitize the aphid. The pest is also infected by *Entomophthora coronata and Cephalosporium aphidicola.*

**Damage:** The damage is caused by nymphs and adults by sucking cell sap from leaves, stems, inflorescence or the developing pods. They are seen feeding in large numbers, often covering the entire surface. Owing to feeding on cell sap, the vitality of plants is greatly reduced. The leaves acquire a curly appearance, the flowers fail to form pods and the developing pods do not produce healthy seeds. There is one fourth or one fifth reduction in yield due to infestation in this crop.

**Control:** Spray 250 to 400 ml of Oxydemeton-methyl 25 EC or Dinethoate 30 EC or Malathion 50 EC in 80-125 litres of water per acre at 10-15 days interval. Spray 60 to 100 ml of Phosphamidon 85 WSC in 80-125 litres of water per acre at 10-15 days interval. Broadcasting of 4 kg of Phorate 100, or Aldicard 10G or 8 kg of Disulfoton *5G* or, 13kg of Carbofuran 3G followed by a light irrigation is suggested.

**2. *Athalia lugens proxima (*Klug*)*** (= *Athalia proxima* Klug) The Mustard Saw Fly Hymenoptera Tenthredinidae

**Distribution:** Indonesia, Formosa, Burma and Indian sub-continent.

**Host**. Mustard, toria, rape seed, cabbage, cauliflower, knol-khol, turnip, radish etc.

The larvae are dark green and have 8 pairs of abdominal prolegs. There are five black stripes on the back and the body has a wrinkled appearance. *A* full-fed larva measures 16-18mm in length. The adults are small orange-yellow insects with black markings on the body and have smoky wings with black veins.

**Life-cycle:** The pest is active from October to March and during summer the larvae rest in their pupal cocoons. The adults emerge from these cocoons early in October. The adult longevity is 2-8 days. The female with its saw-like ovipositor inserts 30-35 eggs singly near the leaf margin. The incubation period is 4-8 days. The larvae feed in groups of 36 on leaves during morning and evening. They are full-grown in 16-35 days after passing through seven stages. The full-fed larvae descend the plant and enter in the soil to a depth of 25-30 rnm. There, they pupate in water-proof oval cocoons made of silk. The adult emergence is in 11-31 days. The life-cycle is completed in 31-74 days. During the period of pest activity from October to March, it completes 2-3 generations. The grubs of this pest are parasitized by *Perilissus cingulator* Morby and the mortality of grubs is caused by a bacterium, *Serratia marcescens* Bizia.

**Damage:** This is one of the very few hymenopterous insects noticed to infest crops in India. The grubs, alone are destructive. They bite holes on the leaves, skeletonize them completely. Seedlings succumb to the attack while older plants do not bear seeds. Severe attack causes serious damage.

**Control:** Spray young plants with 800g of BHC 50 WP, in 50-80 litres of water per acre or dust BHC 10 percent @ 10 kg/acre. The waiting period to use the treated leaves as a vegetable is for two weeks.

**3. *Bagrada cruciferarum* Kirkaldy** (= *Bagrada picta* Fabricius) The Painted Bug or Harlequin Bug (Pentotomidae: hemiptera)

**Distribution:** East Africa, India, Sri Lanka. Pakistan, South-cast Africa, Afghanistan, Burñia, Iraq, Arabia.

**Host:** Cruciferous crops and also on rice, sugarcane, indigo and coffee. The full-grown nymphs are black with a number of brown markings and measure about 4mm long and 2.66mm broad. The adult bugs are sub-ovate, black and have a number of orange or brownish spots. They measure 3.71 mm long and 3.33 mm broad.

**Life cycle:** During the period of its activity from March to December, all the stages of painted bug can be seen in the field. It passes the winter months of January and February in the adult stage under heaps of dried oil seed plants lying in the fields. These bugs lay oval, pale yellow eggs singly or in groups of 3-8 on leaves, stalks, podes, and sometimes on the soil. A female may lay 37-102 eggs during day or night in her life span of 21-30 days. The incubation period is *3-5* days during summer and 20 days during December, The nymphs transform themselves into adults in 16-22 days during the summer and 25-34 days during the winter after passing through five stages. The total life-cycle is completed in 19-54 day. There are 9 generations in a year.

*Gryan* sp. (= *Hadrophanurus* sp.) parasitizes the eggs while *Alophora* sp. parasitizes the adults of this pest.

**Damage:** The damage is caused by nymphs as well as adults which suck cell sap from the leaves and developing pods which gradually wilt and dries up. The vigour of the plants is affected. A sort of resinous material is excreted by the nymphs and adults which spoils the pods.

**Control:** Apply BHC 10 percent dust at 10 kg/acre, 400 ml of Malathion 50 EC in 50 80 liters of water per acre once in October and again in March-April.

**Table 3.4:** *Minor Pests of Brassica*

| S. No. | Scientific name | Common name | Order | Family | Distribution | Nature of damage |
|---|---|---|---|---|---|---|
| 1. | *Empoasca binotata* Pruthi | Jassid | Hemiptera | Jassidae | India | Sap-sucker, infest leaves |
| 2. | *Bemisia tabaci* (Genn.) | White fly | Hemiptera | Aleyrodidae | India | Infest foliage |
| 3. | *Myzus persicae* (L.) | Green peach aphid | Hemiptera | Aphididac | India, S.E. Asia | Infest leaves and stem, virus vector |
| 4. | *Plutella xylostella* (L.) | Diamond – backmoth | Lepidoptera | Plutellidae | Cosmopolitan | Larvae hole leaves |
| 5. | *Pieris brassicae* (Linn.) | Cabbage butterfly | Lepidoptera | Pieridae | India, Europe, N. America | Larvae eat leaves |
| 6. | *Crocidolomia binotalis* Zell. | Leaf webber | Lepidoptera | Pyraustidae | India, Africa, S.E. Asia, Australia | Larvae defoliate |
| 7. | *Hellula undalis* (Fabr.) | Cabbage Borer | Lepidoptera | Pyraustidae | India, S.E. Asia, Australia | Larvae defoliate |
| 8. | *Diacrisia oblique* Wlk. | Bihar hairy caterpillar | Lepidoptera | Artidae | India | Larvae defoliate |
| 9. | *Agrotis ipsilon* (Rott.) | Cut worm | Lepidoptera | Noctuidae | India | Larvae are cut worms |

| | | | | | | |
|---|---|---|---|---|---|---|
| 10. | *Cirphis loreyi* Dup. | Cut worm | Lepidoptera | Noctuidae | India | Larvae are cut worms |
| 11. | *Phytomyza horticola* Gour. | Pea leaf miner | Dipetra | Agromyzidae | Cosmopolitan in old world | Larvae mine leaves |
| 12. | *Phytomyza atricornis* (Meigen) | Leaf miner | Dipetra | Agromyzidae | India | Larvae mine leaves |
| 13. | *Phyllotreta cruciferae* Goeze | Flea beetle | Coleoptera | Chrysomelidae | India, Africa | Adults hole leaves |
| 14. | *Phaedan brassicae* Baly | Flea beetle | Coleoptera | Chrysomelidae | India | Adults infest flowers |

## Future Thrust

Indian agriculture is witnessing a transformation. We have our first transgenic crop of cotton. There are some transgenic vegetable crops also in the waiting list to be given the permission to be marketed. These include tomato (Cry1Ab) to be developed against fruit borer; its glasshouse experiments are completed. Brinjal (Cry1Ab) is being transformed to combat *Leucinodes* problem; its glasshouse experiments are in progress. Cauliflower and cabbage (Cry 1H and Cry 9c) are being developed to manage Lepidopterous pests and its glasshouse experiments are underway. Apart from transgenic crop our future attention should be to reduce insecticide residues from vegetables by adopting ecofriendly IPM technology *viz.,* trap and mixed cropping and avoidance of peak activity period of pest by altering sowing and harvesting time. Farmers participation is a key factor in making IPM a success at field level, case study, indigenous knowledge, training and development of resistant varieties etc. Studies on the natural mortality factors and the role-played by each factor at different locations in different seasons and ovipositional deterrent for fruit fly *D. curcubitae* has to be intensified. In cruciferous crops detailed studies are required on the bioecology and methods of mass production of important natural enemies like *A. plutellae.* This work should be undertaken urgently so as to prevent the further loss of ecology and to continue the sustainable farming.

The perusal of literature shows that there are several non chemical methods available to reduce the pesticide usage in vegetable cultivation and produce optimization. There is a need to realize the potential of indigenous biocontrol agents and attention should be given to conserve them. Importing the exotic parasitoids and predators is also needed to be intensified for several crop pests to strengthen the natural enemy fauna. Standardization of conservation methods for parasitoids and predators and use of pheromones should be intensified.

## References

Abdul-Baki, Aref A., and John Teasdale. 1997. Sustainable Production of Fresh Market Tomatoes and Other Summer Vegetables with Organic Mulches. Farmers' Bulletin No. 2279. USDA-Agriculture Research Service, Washington, D.C. 23 p. 19 :49-52.

Abou-Elhagag, G.H. 1998, Seasonal abundance of certain cotton pests and their associated natural enemies. Assiut. J. Agric. Sci., 29 (3) : 253-267.

Abrol, D.P and Andotra, R.S. 2003. Relative toxicity of some insecticides to Apis mellifera L. Journal Asian Pacific Entomology 6(2): 235-237.

Abrol, D.P. 1989. Studies on abundance, diversity, behaviour and importance of native pollinators for crop production. Korean Journal of Apiculture. 4 (2): 25-40

Abrol, D.P. and Andotra, R.S. 2001a. Field toxicity of pesticides to honeybee Apis mellifera L. foragers. Korean Journal of Apiculture 16(1) 19-26.

Abrol, D.P. and Andotra, R.S. 2001b. Impact of pesticides on the foraging activity of honeybee *Apis mellifera* L. on *Brassica campestris var. toria*. Korean Journal of Apiculture,16(2): 89-94.

Abrol, D.P.1991. Conservation of pollinators for promotion of agricultural production in India. Journal of Animal. Morphology and Physiology 38(1x2) : 123-139.

Abrol, D.P.1997 Bees and Beekeeping in India. Kalyani Publishers Ludhiana, India, 450p. 1997.

Adams, W.M. 1990. Green Development: Environment and Sustainability in the Third World. London: Routledge.

Aderolu, I. A., Omooloye, A. A and Okelana, F. A. 2013. Occurrence, Abundance and control of the Major Insect Pests Associated with Amaranths in Ibdan, Nigeria. Journal of Entomology, Ornithology & Herpetology, 2:3-7.

Aguda, R.M. Rombach, M.C. and Shepard, B.M. 1987, Suppression of population of the brown plant hopper, Nilaparvata lugens (Homoptera : Delphacidae) in field cages by entamopathogenic fungi on rice. Korean J. Appl. Ent. 104: 167-172.

Ahlawat, D.S., Verma, A.N., Singh, H. and Singh, D. 1993. Effect of host plants on the larval and post larval development of red hairy caterpillar, Amsacta moorei Butler. Annals of Biology. 9(2): 254-256.

Alam, M.Z., Quadir, M.A. and Ali, M.S. 1987. Pollinating behaviour of honeybee Apis indica F. and its influence on seed production of cauliflower. Bangledesh Horticulutre 15: 25-30.

Albuquerque, F.A., Ros, A.B., Mendes, S.C. and Weber, L.F., 1999, Control of cotton aphid Aphis gossypii (Glover) (Homoptera : Aphididae) using different insecticides as spray. Anais-II-CongBrasileiro-de-Algodao-O-Algodao-No-SeCulo perspectivar. ParaSelulo XXI, Brasil, pp. 233-235.

Aluja, M. 1994. Bionomics and management of Anastrepha. Annu. Rev. Entomol. 39:155-178.

Ambekar, D. and Kalbhor, S. M. 1981, Mechanism of resistant to leafhopper (Amrasca biguttula biguttula) in cotton. Indian. J. Agric. Sci., 51 : 114-116.

Anais-II-Cong-Brasileiro-de-Algodao-O-Algodao-No-SeCulo-XX-perspectivar.

Anderson, L. D. 1964. The effect of sevin on honeybees. Gleanings in Bee Culture. 92(6) : 358-364, 370.

Angeles, D.E. 1991 Mangifera altissima. in: Coronel, R.E. and Verheij, E.W.M. [Eds.] Edible Fruits and Nuts. Plant Res. S.E. Asia 2 (PROSEA), Pudoc, Wageningen, the Netherlands. pp. 206-207.

Annon, 2001. Integrated pest management package for Cucurbitaceous crops (Eds.) Sipra Bhattarcharya and Mamta Madhi. Offset Press, Directorate of Extension, Ministry of Agriculture, Government of India, New Delhi, India.

Annon, 2001. Integrated pest management package for okra (Eds.) Sipra Bhattarcharya and Mamta Madhi. Offset Press, Directorate of Extension, Ministry of Agriculture, Government of India, New Delhi, India.

Annon, Yearly report of Asian Vegetable Research and Development (AVRDC).1995 Management of insects and plant diseases. Taiwan, 109-116 pp.

Anon, 2001. Integrated pest management package for cucurbitaeceous crops (Eds.) Sipra Bhattarcharya and Mamta Madhi. Offset Press, Directorate of Extension, Ministry of Agriculture, Government of India, New Delhi, India.

Anon, 2001. Pest management. In: Handbook of horticulture. (Ed.) Chadha, K.L. ICAR, New Delhi. 785-799 pp.

Anon, 2001.Integrated pest management package for onion (Eds.) Sipra Bhattarcharya and Mamta Madhi. Offset Press, Directorate of Extension, Ministry of Agriculture, Government of India, New Delhi, India.

Anon, Yearly report of Asian Vegetable Research and Development (AVRDC). 1995. Management of insects and plant diseases. Taiwan, 109-116 pp.

Anon. 1999. Green Peach Aphid And Other Early Season Aphids. Webpage, Statewide IPM Project, University of California, Division of Agriculture and Natural Resources. Adams, Sean. 1997. Seein' red: colored mulch starves nematodes. Agricultural Research. October. p. 18.

Anonymous, 1987a. Package of Practices of Vegetables. SKUAST-Kashmir. pp45.

Anonymous, 2003. National Horticulture Board, data base, Gurgoan.

Anonymous, 2005 The Hindu survey of Indian Agriculture, 2004, pp. 51-56.

Anonymous, 2005. Annual repot. Department of Agriculture, Jammu ( Jammu & Kashmir).

Anonymous, 2006, Horticultural Information Services, Monthly Bulletin, pp. 60-62.

Arneson, P.A. and Losey, J.E. 1997. Integrated Pest Mangement-Lecture Notes, Cornell University.

Arora Ramesh, Battu, G.S.,and Bath, D.S. 2000. Management of insect pests of cauliflower with biopesticides. Indian J. Ecol. 27(2): 156-162.

Ashraf, M., Khaliq, A. and Ahmad, K.F. 1993. Control of brinjal fruit borer, Leucinodes orbonalis Gn. with some insecticides. Pakistan Journal of Scientific and Industrial Research. 36(6-7): 264-266.

Asiatico, J.M. and Zobisch, J.C., 1994 Control of whitefly Bemisia tabaci on tomato with biological and chemical insecticides. Meyo. Integrads De Plaga's, 24: 1-7.

Atkin, E. L. and Kellum, D.1986. Comparative morphogenic and toxicity studies on the effects of pesticides on honeybee brood. Journal of Apicultural Research. 25(4) : 242-255.

Attri, B. S. and Sharma, P. L. 1969. Toxicity of some new pesticides of Indian honeybee (Apis cerana Indica F.). Pesticides. 3:27-29.

Atwal, A. S. and Dhaliwal,G.S. 2004. Agricultural pests of south Asia and their management. Kalyani Publishers Ludhiana, India, 498p

Atwal, A. S. and Singh, K. 1969. Comparative efficacy of Bacillus thuringiensis Berliner and some contact insecticides against Pieris brassicae L. and Plutella maculipennis (Curtis.) on cauliflower. Indian Journal of Entomology. 31(4): 361-363.

Atwal, A. S.1959. The oviposìtion behaviour of Bagrada crciferarum Kirkaldy (Pentatomidae: Heteroptera) and the influence of temperature and humidity on the speed of development of eggs. Proceedings National Institute of Sciences India 25B (2):65-67.

Aykroud, W.R., 1963, I.C.M.R. Special Report series No. 42.

Ayyar, T.V.R. 1927. The parasitic hymenoptera of economic importance noted from South India. Bull. Entomol. Res. 18: 73-78.

Badenes-Perez, F.R. and Shelton, A.M. 2006. Pest management and other agricultural practices among farmers growing cruciferous vegetables in the Central and Western highland of Kenya and the Western Himalayas of India. International Journal of Pest Management. 52(4): 303-315.

Bai, A. R. K. and Reddy, C. C. 1977. Laboratory toxicity of some insecticides to Apis cerana indica. Journal of Apicultural Research. 16(3): 142, 161-162.

Bailey, J., Hoffman, M. and Olsen, K. 1988. Black light monitoring of two avocado insect pests. California Agriculture. 42: 26-27.

Bajwa, W.I. and Kogan, M. 1998. "Compendium of IPM Definitions (CID): A Collection of IPM Definitions and their Citations in Worldwide IPM Literature." Integrated Plant Protection Center (IPPC), Oregon State University.

Bakhetia, D. R. C. 1986. Control of insect-Pests of toria, sarson and rai. Indian Farming. 36(4): 43-44.

Bakhetia, D. R. C., Brar, K. S. and Sekhon, B. S. 1986. Bioefficacy of some insecticides for the control of mustard aphid, L. erysimi (Kalt.) on rapeseed and mustard. Indian Journal of Entomology. 48(2): 137-143.

Bakhetia, D. R. C., Brar, K. S. and Sekhon, B. S. 1986. Seasonal incidence of Liphaphis erysimi (Kalt.) on the Brassica sp. in Punjab. In: Proceedings of National Symposium on Aphidology in India. Pp29-36.

Banerji, R., Misra, G. and Nigam, S.K. 1985. Role of indigenous plant material in pest management. Pesticides, 32-37 pp.

Barnes, B.N. 1990. FFTRI monitoring Manual for Orchard Research Institute, Stellenbosch, 21 pp.

Barnes, B.N. 1999. Fruit fly baiting in deciduous fruit doing the right things right. Deciduous Fruit Grower. 49: 5-7.

Barnes, B.N. and Blomefield, T.L. 1997. Goading growers towards mating disruption: the South African experience with Graphotita molesta and Cydia pomonella (Lepidoptera: Tortricidae). IOBC WPRS Bulletin. 20: 45-46.

Barnes, B.N. and Eyles, D.K. 2000. Feasibility of eradicating Ceratitis spp. fruit flies from the Western Cape of South Africa by the sterile insect technique. In: Tan, K. H. (ed.) Area wide control of fruit flies and other insect pests. Penerbit Universiti sains, Pulau Pinang, Malaysia, pp. 449-455.

Barwal, R.N. and Joshi, S. 1996. Hairy caterpillar *Euproctis scintillans* (Wlk.): a new pest of capsicum. Insect Environment. 2(1): 8-9.

Batra,H.N.1958. Bionomics of Bagrada crciferarum Kirkaldy (Pentatomidae: Heteroptera) and its occurrence as a pest of mustard seeds. Indian Journal of Entomology. 20: 130-135.

Beaument, P. 1993. Pesticides, Policy and People. London: Pesticide Trust.

Beers, E., Hull, L. and Jones, V. 1994. Sampling pest and beneficial arthropods of apple. In: Pedigo, L. and Buntin, D.(eds.) Handbook of sampling methods for arthropods in agriculture. CRC Press, Boca Raton, Florida, pp. 383-415.

Bemisia tabaci in cotton. Bot. Pesticides in Integrated Pest Management, pp. 256-260.

Benbrook, Charles M. 1996. Pest Management at the Crossroads. Consumers Union, Yonkers, NY. 272 p.

Bentley, J.W. and O'Neil, R.J. 1997. On the ethics of biological control of insect pests. Agriculture and Human Values. 14(3):283-289.

Bhagat, K.C., Masoodi, M.A. and Koul, V.K. 1988. A note on the occurrence of a coccinellid predator in Kashmir. Current Research, University of Agricultural Sciences, Bangalore. 17: 49-51.

Bhalla, S., Kapur, M. L., Verma, B. R. and Singh, R. 1997. An unusual abundance of cabbage butterfly, Pieris brassicae L. on various Brassica species on the environment of Delhi for locating sources of resistance among different cultivars. Journal of Entomological Research. 21(2): 147-151.

Bhaskaran, S., 1995, Studies on traditional pest control practices of Tamil nadu. M. Sci (Agri.) Thesis, Annamalai Univ. Annamalainagar, p. 120.

Bhat, O.K., Kaul, V. and Bhagat, K.C. 1994. Incidence of pests associated with the rhizosphere of tomato in Jammu. Annals of Plant Protection Sciences. 2(2): 23-26.

Bhatia, R. and Gupta, D. 2003. Insect and mite pest status of subtropical horticultural crops in Himachal Pradesh. Journal of Insect Science. 16(1/2): 1-8.

Bhatia, R. and Verma, A. 1993. Insect-Pest complex of cauliflower in Himachal Pradesh. Journal of Insect Science. 6(2): 297-299.

Bhatia, R. and Verma, A. K. 1995. Seasonal incidence of major insect pests of summer cabbage in Himachal Pradesh. Annals of Agricultural Research. 16(3): 278-81.

Bhatia, R. and Verma, A.K. 1994. Seasonal incidence of major insect pests associated with winter crop of cabbage in Himachal Pradesh. Annals of Agricultural Research. 15(2): 222-225.

Bhatia, R. and Verma, A.K. 1995. Seasonal incidence of major insect pests of summer cabbage in in Himachal Pradesh. Annals of Agricultural Research 1995 Vol.16 No.3 pp.278-281 ref.10.

Bhatia, R., Gupta, D. and Pathania, N. K. 1995. Host preference and population build up of key pests of cole crop. Journal of Insect Science 8(1) : 59-62.

Bhavani Sankara Rao, T., Reddy, G.P.V. Murthy, M.M.K. and Deva Prasad, V., 1991, Efficacy of neem products in the control of bhendi pest complex. Indian J. Plant Prot.,

Bindra, O. S., Singh, D. and Ramzan, M. 1975. Seasonal incidence of insect pests of cauliflower at Ludhiana. Journal of Research of Punjab Agriculture University. 14 : 66-69.

Bindra, O.S. and Mahal, M.S. 1979. Investigation on varietal resistance in okra (*Abelmoschus esculentus* L.) to the cotton jassid (Amrasca bigutttula biguttula Ishida). Indian J. Hort., 36 : 212-219.

Bioecology and control of insect pests: Proceedings of the National Symposium on growth, development and control technology of Insect pests. 170-174.

Blomefield, T.L. 1997. Managing resistance of codling moth, Cydia pomonella (L.) in South African pome fruit orchards. In: Combating Resistance, IARC-Rothamsted, Harpenden, UK. pp. 14-16.

Boavida, C., Neuenschwander, P. and Schulthess, F. 1992. Spatial distribution of Rastrococcus invadens Williams (Homoptera: Pseudococcidae) in mango trees. J. Appl. Entomol. 114:381- 391.

Brar, K. S. and Sandhu, G. S. 1981. Effectiveness of some new insecticides for the control of mustard aphid, L. erysimi as brown sarson. Pesticides 15(1): 11-13.

Bugg, Robert L., Sharad C. Phatak, and James D. Dutcher. 1990. Insects associated with cool-season cover crops in southern Georgia: Implications for pest control in truck-farm and pecan agroecosystems. Biological Agriculture and Horticulture. p. 17-45.

Bull, B.C.; Raupp, M.J.; Hardin, M.R. and Sadof, C.S. 1993. Suitability of five horticulturally important armored scale insects as hosts for an exotic predaceous lady beetle. Journal of Environmental Horticulture. 11: 28-30.

Butani, D. K., Jotwani, M. G. and Verma, S. 1977. Insect-pests of Vegetables and their control on Cole Crops. Pesticides. 11(5) : 19-24.

Butani, D.K. and Jotwani, M.G. 1984. Insects in vegetables. Periodical Expert Book Agency, Delhi,

Campbell, C.W. and Knight, A.J. 1983. Production de litchis. Communication at XIII Congreso de N.O.R.C.O.F.E.L Puerto de la Cruz. Ministerio de Agricultura Espana (En Prensa).

Capinera, J.L. 2001. Handbook of vegetable pests. Academic Press, California, USA. 729pp.

cauliflower. Pest management in Horticultural Ecosystems. 3(1): 7-12.

Chadha, K.L. and Choudhary, M.L. 2007. Report of the Working Group on Horticulture, Plantation Crops and Organic Farming for the XI Five Year Plan (2007-12), GOI, Planning Commission, January, 2007.

Chahal, B. S. and Sukhija, H. S. 1969. Spraying in mustard is a must. Progressive Farming. 5(3) :14.

Chandrashekar, N. 2002, Biological activity and field efficacy of neem seed kernal extract, nem oil and certain commercial neem products against spidermite infesting vegetable crops. M.Sc. (Agri.) Thesis. Univ. Agric. Sci., Bangalore (India).

Chandrashekarappa, B. 1995, Effect of temperature, humidity, host plant and some plant products on Amblyseius longispinosus and A. tetranychivorus (Acarina : Phytoseidae). M.Sc. (Agri.) Thesis, Univ. Agric. Sci., Bangalore (India).

Chang, K.P. 1970. A study of Anoplophora macularia, aserious insect pest of litchi. Taiwan Agriculture Quarterly. 6(3):133-143.

Chattopadhyay, N., Samui, R.P., Wadekar, S.N., Satpute, U.S. and Sarode, S.V., 1996, Sensitivity of aphid infestation to meteorological parameters of Akola, Maharashtra. Indian J. Ent., 58 (4) : 291-301.

Chaudhary, H.R. and Dadeech, 1989, Incidence of insects attacking okra and the available losses caused by them. Ann. Arid Zone, 28 (3): 305-307.

Chauhan, D.V.S., 1972 Vegetable production in India. 3rd Ed., Pub. Ram Prasad and Sons, Agra.

Chauhan, U., Bhalla, O.P. and Sharma, K.C. 1997. Biology and seasonality of the diamondback moth, Plutella xylostella (L.) (Lepidoptera: Yponomeutidae) and its parasitoids on cabbage and

Chen. Wen Rui. 2001. comparative study on efficicay of five Biopesticides for the control of Plutella xylostella. Journal of Plant Protection. 27 (6): 33-34.

Chinnaiae and Mohanasundaram, M., 1999, Evaluation of neem derivatives for their toxic effect on safety to predatory mites, Amblyseius spp. (Acarina : Phytoseidae) in cotton ecosystem. Pestology, 123 : 45-48.

Chiranjeevi, C., Reddy, I.P. Narayanamma, M., and Neeraja, G.2002. Effect of shoot clipping and insecticides on the incidence of fruit borer in brinjal. In: International Conference on Vegetables. Nov. 11-14, Bangalore, India. 271 p.

Chitra, K.C., Rao, S.J. and Rao, P.K., 1997, Efficacy of plant products for controlling aphid (Aphis gossypii). Indian J. Agric. Sci., 63 : 134-135.

Choudhary, H.R. and Bajpai, N.K. 2007. Effect of triazophos against insect pests of soybean (Glycine max Linn.) in south eastern plain zone of Rajasthan. Indian Journal of Agricultural Sciences. 77(1): 62-64.

Cochran, W.G. 1977. Sampling Techniques. New York: Wiley 3 rd ed.

Coikesen, T. and Sekeroglu, E. 1987, Effect of change in temperatures on the development of the cotton whitefly Bemisia tabaci (Genn) (Homoptera : Aleyrodidae), Turkey, Ent. Dergisi, 11 (3) : 163-168.

Crane, E. 1990. Bees and Beekeeping. Bath press Ltd. Avon. pp 68-69.

cruciferous crops. In : Verghese, T. M.(ed.). Recent Advances in Pollen Research. Allied publisher. New Delhi. pp 241-258.

Cunningham, I.C. 1984. Mango insect pests. Australian Mango Research Workshop. CSIRO, Melbourne, Australia. pp. 211-224.

Dahiya, B. and Chauhan R. 1992. Chemical control of pod borer complex in pea Pisum sativum L.

Daku, L., Norton, G.W., Pfeiffer, D.G., Luther, G.C., Pitts, C.W., Taylor, D.B., Tedeschini, J. and Uka, R. 2000. "Farmers' knowledge and attitudes towards pesticide use and olive pest management practices in Vlora, Albania: A Baseline Survey." IPM CRSP Working Paper 001-1, Virginia Polytechnic Institute and State University.

Das, S.B., Duware, S.R. and Veda, O.P. 2003, Seasonal activity of homopterans and bioagents associated with cotton in western Madhya Pradesh, ISRO National Seminar: Stress management in Oilseeds, Jan, 28-30, pp. 145-147.

David Abilash, R.B., 2000. Studies on development of IPM techniques for brinjal shoot and fruit borer L. orbonalis. M.Sc. thesis, Tamil Nadu Agricultural University, Coimbatore, India. 1-67 p.

David, B.V. 2001. Elements of economic entomology. Popular Book Depot. Chennai. 562 pp.

Day, P. K., Jana, S. K., Chakraborty, G. and Somuchoudhury, A. K., 2005. Evaluation of imidacloprid (70WS and 20SL) against sucking pest complex of okra. J. Ent. Res., 29(3) : 215-218.

Day, P. R. 1987. Crop improvement : Constraints and challenges. In. British crop Protection Conference on Weeds. B CPC Publication Survey pp14-17.

DeBach, P.1964. Biological control of insect pests and weeds. Chapman & Hall, London. 844 pp.

Dent, D. 2000. Insect pest management. 2nd edition, Ascot, CABI Publishing, 399p.

Deshmuk, S. D. and Barle, M. N., 1976, Studies on the insecticidal property of indigenous plant extracts. Indian J. Ent., 3 : 11-18.

Deshmukh, S. N. and Adarsh, H. S. 1969. Comparative toxicity of some commercial of the cabbage butterfly, Pieris brassicae L. Journal of Research of Punjab Agricultural University. Ludhiana. 8(3) : 339-42.

Devi, N. and Raj, D. 1995. Biology and parasitization of diamondback moth, Plutella xylostella L. infesting cauliflower in mid hill region of Himachal Pradesh (India). Journal of Entomological Research. 19(1): 83-89.

Devjani, P. and Singh, T. K. 2001. Efficacy of certain insecticides against insect-pests of cauliflower in Manipur. Journal of the Agricultural Science Society of North East India. 14(1): 51-56.

Dey, P.K., and Somchoudhury, A.K., 2001. Evaluation of spinosyns A+D (Spinosad 48 SC) against lepidopteran pest complex of cabbage and its effect on natural enemies of diamondback moth under field conditions of W.B. Pestology, 25(1): 54-58.

Dhaliwal, G. S. 1970. Impact of chemical control on honey bees Pesticides. 4(11): 17-19.

Dhaliwal, G. S., Gill, R. S., Dilawari, V. K., Dhaliwal, G. S., Arora, R., Randhawa, N. S. and Dhawan, A. K. 1998. Management of insect-Pest complex of cabbage with neem based insecticides. In : Proceedings of International Conference on Ecological Agriculture towards Sustainable Development. pp306-14. Chandigarh, India.

Dhanalakshmi, D.N., 2006. Studies on storability and utilization of indigenous materials on okra pests. M.Sc. (Agri.) Thesis, Univ. Agric. Sci., Dharwad (India).

Dhandapani, N., Dhivahar, P. and Murugan, M. 2002. Thiacloprid – A new molecule for the management of thrips of chillies. In: International vegetable conference, Nov.11-14, Bangalore, India. 259 p.

Dhar, T., Dey, P.K. and Sarkar, P.K., 2004. Influence of abiotic factors on population build up of red spider mite Tetranychus urticae on okra vis-à-vis evaluation of some new pesticides for their control. Pestology, 24 (9) : 34-37.

Dhawan, A.K., Simwat, G.S. and Sidhu, A.S., 1989. Effect of sowing dates on incidence of sucking pests and bollworms in Arboreum cotton. J. Res. Punjab Agric. Univ., 24 (1) : 75-85.

DPPQS, Directorate of Plant Protection, Quarantine and Storage. 2001. Integrated Pest Management Package, Government of India. (IPM Package No: 28, 31, 32, 34, 37)

Drew, R.A.I. and Hancock, D.L. 1994. The Bactrocera dorsalis complex of fruitflies (Diptera: Tephritidae: Dacinae) in Asia. Bulletin of Entomological Research. Supplement Series. 2:68.

Dugger, P. and Richter, D., 1998. Mid-season cotton aphid infestation in California; effects on cotton yield. Proceedings of Beltwide Cotton conferences San-Digeo – California, USA, 5-9, January 1998, 2 : 1056-1058.

Elwell, Henry and Anita Maas. 1995. Natural Pest and Disease Control. Natural Farming Network. Harare, Zimbabwe. 128 p.

Ezeh, A., Gedegbe, Ab.O and Gedegbe. 2015. Insect Pest occurrence on Cultivated

Amaranthus Spp in Benin City, Edo State, Nigeria. Journal of Applied Sciences and Environmental Management. 19(2):335 –339.

Falcon, L.A. and Huber, J. 1991. Biological control of codling moth. In: Tortricid Pests, their Biology, Natural Enemies and Control (van der Geest L.P.S. and Evenhuis, H.H. eds.). World Crop Pests. 5: 355-369.

Farahbakhch, G. and Moini, M. 1975. Olive pests in Iran. In: Report of Plant Pests and Diseases Research Institute. National Agricultural Research Organization, Ministry of Agriculture and Natural Resources, Tehran. pp.1-14.

Figueiredo, D. and Araujo, J. 1987. Cabbage entomofauna. Boltetin-da-sociedade Portuguessa de Entrological. 82(B-12): 1-154.

Fletcher, T.B. 1914. Some south Indian Insects. Agricultural Research Institute, New Delhi. 565pp.

Fletcher, T.B. 1921. Life history of Indian Insects, Microlepidoptera. Mem. Dep. Agric., India. 6: 1- 217.

Flint, M.L. and van den Bosch, R. 1981. Introduction to Pest Management. Plenum Press, New York, NY.

Free, J.B. 1993. Insect Pollination of Crops. 3rd edition. Academic Press, London. pp 295.

Galan-Sauco, V. 1990. Los frutales tropicales en los subtropicos. Ed. Mundi Prensa, Madrid, Spain. pp. 83-84.

Gandhale, D. N., Salunkhe, G. N. and Naik, L. M. 1982. Effectiveness of insecticides for the control of Aphids (Brevicoryne brassicae L). on cabbage. Indian Journal of Plant Protection. 10 : 85-86.

Geetha, B., Swamiappan, M. and Kempuchetty, N. 2002. Predators for the biological control of the spiraling whitefly Aleurodicus dispersus on vegetables. In: International Conference on Vegetables. Nov.11-14. Bangalore, India.261 p.

Geetha, B.,Swamiappan, M. and Kempuchetty, N. 2002. Predators for the biological control of the spiralling whitefly Aleurodicus disperses on vegetables.In: International Conference on Vegetables. Nov.11-14. Bangalore, India. 261 p.

Gill, R. S., Jaisingh and Sandhu, B. S. 1996. Imidacloprid as a seed treatment – a new approach to control cotton jassid. Nation. Sem. On Century of Cotton in India, held at Gujarat Agric. Univ., Surat, pp. E-1.

Gokulpur, R.S., 1970. Records of braconids and ichneumonids from Central India. Indian J. Ent. 32(4): 383-385.

Gold, C.S., Nankinga, C., Niere, B. and Godonou, I. 2003. IPM of banana weevil in Africa with emphasis on microbial control, p. 243-257. In: Neuenschwander, P., Borgemeister, C. and Langewald, J. (eds.), Biological control in IPM systems in Africa. Wallingford, UK, CABI Publishing, 400p.

Golez, H.G. 1991. Bionomics and control of the mango seed borer, Noorda albizonalis Hampson (Pyralidae: Lepidoptera). Acta Hortic. 291:418-424.

Gopalakrishnan,G., Krishnamoorthy, A.and Gangavisalakshi., P.N. 2002. Metarhizium anisopliae var. anisopliae: A potential entomopathogenic fungus for the management of P. xylostella L. on cabbage. In: International Conference on Vegetables. Nov. 11-14, Bangalore, India, 262 p.

Guddewar, M. B., Shukla, A., Chandra, R., pandey, S. and Saini, M. L 1994. Tabenaemontana cornoaria B. Br. Apocynaceae a potential source of botanical insecticidal insecticides. Pl. Protec. Bull., 46 : 1-5.

Gulati, R., 2004. Incidence of Tetranychus cinnabarinus (Boisd.) Infestation in different varieties of Abelmoschus esculentus L. Ann. Pl. Protec. Sci., 12 (1) : 45-47.

Gupta, G.P., Mahapatra, G.K., Kundu, S.K. and Roshan, Lal., 1998, Impact of abiotic factors on population dynamics of whitefly in cotton. Indian J. Ent., 60 (3) : 293-296.

Gupta, J. C. 1971. Studies on the efficacy of various insecticides against mustard aphid, L. erysimi (Kalt.). PANS. 17:206-208.

Gupta, J. K. and Sharma, P. L. 1998. Impact of number of bee visits on cauliflower pollination. In : Matsuka M., Verma, L.R. Wongsiri, S. Shrestha K. K. and Partap, L. (eds.) Asian Bees and Beekeeping : Progress of Research and Development. pp183-185. Apimondia Publication, Nepal.

Gupta, P. R. 1984. Bionomics of the cabbage butterfly, Pieris brassicae L. in mid hills of Himachal Pradesh. Himachal Journal of Agricultural Research. 10(1): 49-54.

Gupta, P.R. 1992. Biology of Evergestis forficalis (L.) (Pyralidae), a pest of crucifers in Himachal Pradesh. Journal of Insect Science. 5(2): 211-212.

Gupta, S.L. 1990. Key for the identity of some major lepidopterous pests of vegetables in India. Bulletin of Entomology. 31(1): 69-84.

Habibi, J., 1975, The cotton whitefly, Bemisia abaci (Genn). Bioecology and Method of Control. Ent Phytopath Appl., 38 : 13-36.

Haji, F.N.P., Barbosa, F.R., Lopes, P.R.C., Moreira, A.N., de Alencar, J.A. and Ferreira, R.C.F. 2004. Monitoring mango pests within an integrated production programme in Brazil. Acta Hort. 645: 163-165.

Hameed, S. F., Adlakha, R. L. and Giamzo, S. P. 1973. Relative toxicity of some insecticides to the workers of Apis mellifera L. Madras Agricultural Journal. 60(7) : 552-556.

Hameed, S.F., Sharma, D.D. and Agarwal, M.L. 1992. Integrated pest management in litchi. National seminar on recent developments in litchi production held at Rajindra Agril. Uni., Bihar, Pusa, May 30-31, Abstract of papers, pp 38.

Hanchinal, S. G. 1998. Effect of Metarhizium anisopliae (Metsch) sor. on Tetranychus neocaledonicus and its predator Amblyseius ovalis evans on okra. M. Sc. (Agri.) Thesis, Univ. Agric. Sci., Dharwad (India).

Hansen, J.D. and Amstrong, J.W. 1990. The failure of field sanitation to reduce infestation by mango nut weevil, Cryto mangiferae in Hawaii. Tropical pest management. 36: 359-361.

Hare, D.J. 1994. Sampling arthropod pest in citrus. In: Pedigo, L. and Buntin, D. (eds.) Handbook of sampling methods for arthropods in agriculture, CRC press, Boca Raton, Florida, pp. 417-432.

Haseeb, M. 2007. Current status of insect pest problems in guava. Acta Hort. 735:453-467.

He, D.P. 2001. An overview of integrated management of insect pests in litchi orchards of Guangdong. Acta Hort. 558: 401-405.

Hederson, C.F.and Tilton, E.W.. 1955. Tests with acaricides against brown wheat mite. Journal of Economic Entomolgy. 48: 157-161.

Hegde, M., Srinivasa, M., Biradar, D. P., Udikeri, S. S. and Khadi, B. M., 2004, Seasonal incidence of key insect pests and their natural enemies on cotton at Siruguppa. Int.

Symp. Start for Sust. Cotton prod. – A Global vision, 23-25, Karnataka, India, pp. 114-115.

Hill, D. 1975. Agricultural Insect Pests of the Tropics and Their Control. Cambridge Univ. Press, New York, NY.

Ho, T. H. and Ng, K.Y. 1970. Bacillus thuringiensis Berliner for the control of diamond back moth in west Malaysia. Malaysian Agricultural Journal. 47:313-322.

Hooda, V. S., Dhankhar, B. S., Dahiya, B. S. and Singh, R. 1999. Inheritance of resistance to leafhopper (Amrasca biguttula biguttula) on okra. Indian J. Hortic., 56 (1) : 73-76.

Horowitz, A.R. Mendelson, Z., Weintraub, P.G. and Ishaaya, 1998. Comparative toxicity of foliar and systemic application of acetamiprid and imidaloprid against the cotton whitefly, Bemisia tabaci (Hemiptera : Aleyrodidae). Bull Ent. Res., 88 (4): 437-442.

Howard, P.A., Michalenko, E.M. Sage, G.W. Jarvis, W.F. Meylan, W.M. Basu, D.K. Beauman, J.A. and Gray, D.A. 1991. Handbook of Environmental fate and Exposure Data for Organic Chemicals, Lewis Publishers: Chelsea, Michigan.

Hoyt, S.C. and Burts, E.C. 1974. Integrated control of fruit pests. Ann. Rev. of Ento. 19: 231-252.

Hua, H.1965.Life history and control of diamond back moth in Malaysia. Bulletin Division of Agriculture of Malaysia. 118 : 26.

Huang, M.D. and Li, S.X. 1989. The damage and economic threshold of citrus leafminer, Phyllocnistis citrella Stainton to citrus. In: Studies on the IPM of Citrus Insect Pests. Guangzhou, Guangdong, China. Academic Book and Periodical Press.

Huda, A.K.S. and Luck, J. 2008. Early warning of pest/diseases for selected crops using climate information and crop simulation modeling approach under climate change scenarios. Proceeding of International Symposium on Agrometeorology and Food Security, 18-21 Febuary 2008, Hyderabad, India: pp 11.

Ilyas, M., Puri, S.N. and Rote, N.B., 1991. Effect of some morphological characters of leaf on incidence of cotton whitefly. J. of Maharashtra Agric. Univ., 16 (3) : 386-388.

Jack, C. Baily and Meredith, W. R., 1983. Resistance of cotton, Gossypium hirsutum L., to natural field population of two-spotted spider mites (Acari : Tetranychidae). Environ. Entomol, 12 : 763-764.

Jain, H. K., Agnithotri, N. P. and Prasad, S. K. 1979. Insecticides residue in mustard seeds. Journal of Entomological Research. 1(2) : 222-223.

Jat, M.C. and Jeyakumar, P., 2006, Bioefficacy of botanicals and bioagens on sucking pests of cotton. Ann. Pl. Protec. Sci., 14 (1) : 8-10.

Jayakumar, R., 2002. Survey of indigenous practices for the management of pests in Raichur district and evaluation of few practices against okra. M.Sc. (Agri.) Thesis, Univ. of Agric. Sci., Dharwad (India).

Jayanthi, S.,and Alexander Jessudasan, R.W.2000. Role of Pedobius foveolatus (Hymenoptera: Eulophidae) in the biocontrol of the brinjal spotted leaf beetle: Henosepilachna viginctiotopunctata (Coleoptera: Coccinellidae). Pestology, 24(5): 33-36.

Jayaraj, S., Rangarajan, G.A.V., Murugesan, S., Santharam, G., Vijayaraghavan, S. and Thangaraj, D., 1986. Outbreak of whitefly Bemisia tabaci (Genn.) on cotton in Tamil Nadu and its management. All India group Discussion on whitefly in cotton, Guntur, April 29-30, 1986.

Jayarathnam, K. 1977. Studies on the population dynamics of Plutella xylostella (L.) (Lepidoptera: Yponomeutidae) and crop loss due to the pest in cabbage Ph. D Thesis. University of Agriculture Sciences, Banglore. pp215.

Jayaswal, A. P. and Sundaramurthy, V. T 1992. Achievements in insect management in cotton. All India Coordinated Crop Improvement Project Achievements, pp. 117-151.

Jeyarani, S, and Kennedy, J. S. 2004. Efficacy of certain biopesticides against the diamond back moth, Plutella xylostella in cauliflower. Indian Journal of Plant Protection. 32(2): 129-130.

Jeyasankar, A and Gokilamani, D 2016. Biology and eco-friendly control of Amaranths pests, Hymenia recurvalis Fabricius and Psara basalis Walker (Lepidoptera: Crambidae) Inter. J. Acad. Stud. 2(4): 218-230

Jhala, R.C., Patel, B.D. and Bharpoda, T.M 1998. Ovicidal and adulticidal efficacy of various plant extracts against red spider mite, Tetramychus cinnabarinus on okra (Abstract). Nat. Sem. On Entomol, in 21st Century, April 30th to May 2nd, 1998, Udaipur.

Johansen, C. A. 1977. Pesticides and pollinators Annual. Review of Entomology. 22: 172-199.

Joia, B.S., Gursharan Singh, Navdeep Singh and Darshan Singh. 2001. Field evaluation of cartap hydrochloride for the control of *Plutella xylostella* (L.) on cauliflower. In: Proceedings of the second national symposium on integrated pest management in horticulatural crops: New molecules, biopesticides and environment. Oct. 7-19, Bangalore, India.

Joshi, B.G., Sriprakash, K., and Murthy, P.N.S. 1993. Efficacy of neem seed kernel suspension, neem seed bitter, pongamia cake water extract, Bacillus thuringiensis Berliner, nuclear polyhedrosis virus alone and in combination against tobacco caterpillar, Spodoptera litura F. in nurseries. In: Chari, M.S. and Ramaprasad, G. (Eds), Botanical pesticides in integrated pest management. Proc. Symposium on botanical pesticides in IPM (Rajmundry, India). 154-158 pp.

Jothi D.B., Tandon, P.L. and Verghese, A. 1994. Hot water immersion as a quarantine treatment for Indian mangoes infested with the oriental fruitfly, Bactrocera dorsalis (Hendel) (Diptera:Tephritidae). FAO Plant Protection Bulletin. 42(3):158-159.

Kabir, K. H. and Mia, M. D 1987. Effectiveness of some indigenous materials as repellent against mustard aphid, Bangladesh. J. Zool., 15 : 87-88.

Kadil, M.A., El-Kabbany, S.M., Siwify, G.H. and Abdallah, M.D., 1991, Efficacy of some insecticides against the cotton whitefly, Bemisia tabaci (Genn.) with special regard to their side effect on predators. Bull. Ent. Soc. Egypt, 19 : 9-17.

Kaethner, M., 1991. No side effect of neem extract on the aphidophagous predators, Chrysoperla carnrea (Step.) and Coccinella septumpunctata L. Anzeigertus schadlingarkunde Flanzeneschutz Unweltschutz, 64 : 97-99.

Kakar, K. L. 1981 Foraging behaviour of insect pollinator of cauliflower bloom. Ind. J. Ecol. 8(1):126-130.

Kale, J. V., Wadnerkar, D. W., Zanwar, P. R. and Sangle, P. D., 2005. Bioefficacy of newer inseticides against inset pests of okra. Pestology, 29 : 9-12.

Kalra, V. K. and Gupta, D. S. 1986. Chemical Control of mustard aphid, Lipaphis erysimi (Kalt.). Indian Journal of Entomology. 48(2): 148-155.

Kamalanathan, S., Sunderajan, S., Thambura, J.S and Shanmugam, A. 1973. Co I Amaranthus. A high yielding and delicious strain. Madras Agricultural Journal. 60:355-358.

Kamat, M.N., Patel, M.K. and Dhande, G.N., 1952. Occurrence of green muscardine fungus of pyrilla in Bombay. Curr. Sci., 21 : 315.

Kapil, R. P. and Lamba, D. P. S. 1974. Toxicity of some important insecticides to Apis cerana Fab. Indian Journal of Entomology. 36(1) : 6-10.

Kapoor, K.S. and Sharma, K.C. 1995. Field and storage infestation of potato tuber moth in Kangra. Journal of the Indian Potato Association. 22(3/4): 172-173.

Katole, S. R. and Patil, P. J., 2000. Biosafety of imidacloprid and thaimethoxam as seed treatment and foliar spray to some predators. Pestology, 24(11) : 11-13.

Kaul, V. and Kesar, Y. K. 2003. Incidence and management of lepidopteran fruit borers of guava (Psidium guajava L.) in Jammu, India. J. Asia-Pacific Entomol. 6(2): 201-205.

Kaul, V., Malik, G.H., Shankar, U. and Monobrullah, Md. 2007. Incidence and management of olive psylla, Euphyllura pakistanica. Indian Journal of Entomology. 69(4): 331-340.

Kaul, V., Tiku, A.K., Shankar, U. and Monobrullah, Md. 2007. Green stink bug (Hemiptera: Pentatomidae) recorded as a new pest of olive in India. Journal of Asia-Pacific Entomology. 10(1): 81-83.

Keenan, M. 1997. In: Smith, D., Beattie, G. and Broadley, R. (eds.) Citrus pests and their natural enemies. Deptt. of Primary Industries, Q 197030, Brisbane, Australia, 272 pp.

Kevin, M., Heinz and Frank, G., Zalom, 1996. Variation in trichome. Based resistance in Bemisia tabaci (Genn.) Oviposition on tomato. J. Econ. Entomol., 88 (5) : 1494-1502.

Khalil, S.K. Bortos, J. and Taborsky, V., 1983. Effect of temperature, pH of the medium and sugar on the germination of spores, development of mycelium and sporulation of Verticillium lecanii (Zimm) Viigas. Agric. Trop. Et. Subtr., Univ., Agric. Praga, 16 : 255-274.

Khalil, S.K., Taborsky, V. and Bartos, J., 1983. Studies on Verticillium lecanii for the biological control of aphids. In : 10th Int. Cong. Pl. Prot. Brighton, England, 2 : 20-25.

Khan, M.M., Khundu, R. and Alam, M.Z., 2000. Impact of trichome density on the infestation of Aphis gossypii Glover. Int. J. Pest Mangt., 4 (3): 201-204.

Khan, R.B. and Dethe, M.D.2004 Median Lethal Time of new pesticides to foragers of honeybee. Pestology. 28(1): 28-29.

Khan, Z.R. and Agarwal, R.A., 1990. Mechanism of resistance to aphid (Aphis gossypii) in cotton. Indian J. Ent., 52 (2) : 236-240.

Khangura, J. S., Kanwar, J. S., Singh, H. and Singh, D. 1992. Pest management in litchi (Litchi chinensis Sonn.). National seminar on recent developments in litchi production held at Rajindra Agril. Uni., Bihar, Pusa, May 30-31, Abstract of papers, pp 40-41.

Khedkar, J. R. and Ukey, S. P., 2003. Efficacy of newer insecticide against jassids, Amrasca biguttula biguttula (Ishida). State level Sem. on Pest Management for Sustainble Agric., February 6-7, 2003, MAU, pp. 191-192.

Koenig, J.P., Lawson, D.S., Ngo, N., Minton, B., Ishida, C. and Moore, S., 2000. Field trial results with pymetrozine and thiamethoxam for control of cotton aphid (Aphis gossypii). Proc. Beltwide Cotton Conf. San. Antoni, USA, 4-8, January, 2000, 2 : 1335-1337.

Koral, I. T. 1986. Effectiveness of bacterial preparation. Zashcita Rastenii 3: 34-35.

Kozar, F., Brown, M.W. and Lightner, G. 1994. Spatial distribution of homopteran pests and beneficial insects in an orchard and its connection with ecological plant protection. Journal of Applied Entomology. 117: 519-529.

Kranjajic, S., and Dimic, N., Peric, P., Vuska, M., Cvetkovicd, M,. Juvtic, S. and Lazic, B. 1997. Biological control of cabbage pests. In : Proceeding of Balkan Symposium on Vegetables and Potatoes. pp119-124. Belgrade, Yugoslavia.

Kremen, C., Williams, N. M. and Thorpe, R. W. 2002. Crop pollination from native bees at risk from agricultural diversification. In: Proceedings of National Academy of science of U.S.A. 99 (26) : 16812-16816.

Krishna Kumar, N. K., Krishna Moorthy, P. N. and Eswara Reddy, S. G., 2001. Imidacloprid and thiamethoxam for the control of okra leafhopper, Amrasca biguttula biguttula (Ishida). Pest Management in Hortic. Ecosystems, 7(2) : 117-123.

Krishna Kumar, N.K., and Krishnamoorthy, P.N. 2002. Advances in the use of botanicals for the IPM of major vegetable pests. In: International Conference on Vegetables.Nov. 11-14. Bangalore, India. 233 p.

Krishnaiah, K., 1980. Methodology for assessing crop losses due to pests of vegetable. Assessment of crop losses due to pests and diseases. Proc. of Workshop held from Sept, 19-30, 1977 at U.A.S., Bangalore, pp. 259-267.

Krishnaiah, K., Jagmohan, N., and Bagle, B. G. 1975. control of major insect pests of cabbage by new insecticides. Indian Journal of Plant Protection. 1(2): 44-47.

Krishnamoorti, A., and Mani, M. 1988. Feasibility of managing H. armigera on tomato through parasitoid. In: National Workshop on Heliothis management. Feb. 18-19. TNAU, Coimbatore.

Kulkarni, K.A., and Shekharappa, 2001. Integrated management of chilli fruit borer, H.armigera. In: Proceedings of the second national symposium on integrated pest management in horticultural crops: New molecules, biopesticides and environment. October 17-19, Bangalore, India, 59-60 p.

Kumar M. and Singh, A.K., 2002. Varietal resistance of okra against cotton jassid, Amrasca biguttula biguttula under field conditions. Ann. Pl. Protec. Sci, 10 (2) : 365-410.

Kumar, K., Moorthy, T. N. and Somuchoudhury, A. K., 2005. Evaluation of imidacloprid (70 WS) and thiamethoxam (70 WS) against sucking pest complex of okra. J. Ent. Res., 29 (3) : 221-223.

Kumar, N., Suresh, J. and Anbu, S. 2007. Problems and prospects of guava cultivation under Tamil Nadu conditions. Acta Hort. 735:329-334.

Kumar, S. and Srivastava, A. S. 1986. Toxicity of field weathered deposits of certain modern insecticides to L. erysimi with special reference to the seed yield of mustard. Pesticides. 20(7) : 31-33.

Kumar, S. P. and Babu. S. 1999. Evaluation of antifeedant properties of neem azal against certain leaf feeders of brinjal. Pestology, 22(10): 62- 64.

Kumar, V. and Sharma, D. D., 1993. Bio-ecology and chemical control of spidermite, Tetranychus ludeni zacher on okra. Indian J. Plant Prot., 21(1) : 68-71.

Kumar, V. D. V. N. H., Ramanath Subramanian and Natarajan, P., 1999. Evaluation of acetamiprid – a new insecticidal compound against cotton aphids, Aphis gossypii and jassids Amrasca biguttula biguttula. Pestology, 23(3) : 29-33.

Kumashiro, B.R., Lal, P.Y., Funasaki, G.Y. and Teramoto, K.K., 1983. Efficacy of Nephasphis amicola and Encarsia haitiensis in controlling Aleurodicus dispersus in Hawaii. In : Proc. Of Hawaiian Entomological Soc., 24 : 261-269.

Lal, O. P. 1990. Evaluation of control schedule against insect-pest of cabbage seed crop in Kullu valley ( Western Himalaya). India Revista di Agricolture Subtropicale Tropicale. 84(2): 249-63.

Lal, O.P. and Sinha, S.R., 2005. Impact of imidacloprid seed treatment along with some insecticidal sprayings against insect pests of okra. Indian J. Ent., 67 (4) :328- 333.

Lefroy, H.M. 1909. Indian Insect Life- Lepidoptera - Butterflies and Moths. Vol. II, Agricultural Research Institute, Pusa, India. pp.397-786.

Lichtenberg, E., Zilberman, D. and Archibald, S.O. 1990. "Economics and Pesticides." Chapter 5 in D. Zilberman and E. Lichtenberg (Eds) "Economic Perspectives on Pesticide Use in California: a Collection of Papers." Berkley University, California, November, 1990.

Lingappa, S. K., Basavanagoud, A. S., Vastrad, A. K. and Gopali, J. K. 2000. Diamond back a real threat to vegetable production in south real threat to vegetable production in south Indian and its integrated management. IPM system in Agriculture. 7 : 235-248.

Madar, H., 2003. Bioecology and management of leafhopper, Amrasca biguttula biguttula (Ishida) (Homoptera : Cicadellidae) on sunflower. M. Sc. (Agri.) Thesis, Univ. Agric. Sci., Dharwad (India).

Mahal, M.S., Harmesh Lal, Rajwant Singh and Balraj Singh, 1993. Field resistance in okra to the cotton jassid in relation to crop age. J. Insect Sci., 6 : 60-63.

Mahamood, T., Khokhar, K. M., Banaras, M. and Ashraf, M., 1990. Effect of environmental factor on the density of leafhopper Amrasca devastans (Distant) on okra. Tropical Pest Mangt., 36 : 282-284.

Malappanavar, M. C., 2000. Bioecology and management of spiraling whitefly, Aleurodicus disperses Russell by Verticillium lecanii (Zimm.) on Guava. M. Sc. (Agri.) Thesis, Univ. Agric. Sci., Dharwad (India).

Malathi, S., Sriramulu, M.,and R. Babu, T.1999. Evaluation of certain ecofriendly insecticides against lepidopteraous pests of cabbage. Indian J. Ent. 61(2): 127-133. Tabashnik, B.E.,1986. Model for managing resistance to fenvalerate in the diamondback moth (Lepidoptea: Plutellidae), J. Econ. Entomol. 80: 1091-1099.

Mallareddy, K., Lokareddy, K., Ramesh Babu and Narasimdra Reddy, K. 2004. Efficacy of certain insecticides against pests of cabbage. Pest Management and Economic Zoology. 12(2): 153-159.

Malo, S.E. and Campbell, C.W. 1994. The guava. Horticultural Sciences Department Fact Sheet HS-4. Florida Cooperative Extension Service, Institute of Food and Agricultural Sciences, University of Florida. April 1994

Mani, C., Kumar, S. and Singh, R.N., 2003. Efficacy of acaricides and botanicals against two spotted spider mite, Tetranychus urticae Koch on okra. Ann. Pl. Protec. Sci., 11 (1) : 143-183.

Mani, M. and Krishnamoorthy, A. 1990. Evaluation of the exotic predator Cryptolaemus montrouzieri Muls. (Coccinellidae, Coleoptera) in the suppression of green shield scale Chloropulvinaria psidii (Maskell) (Coccidae, Homoptera) on guava. Entomon. 15: 45-48.

Mani, M., Dinesh, M.S.and Krishnamoorthy, A.2002. Survey and evaluation of natural enemies of the exotic spiralling whitefly, Aleurodicus dispersus on vegetables in peninsular India. In: International Conference on Vegetables. Nov. 11-14, Bangalore, India, 238 p.

Mani, M., Krishnamoorthy, A. and Singh, S.P. 1989. The impact of the predator, Cryptolaemus montrouzieri Muls. on the pesticide resistant populations of the striped mealybug Ferrisia virgata (Ckll.) on guava in India. Insect Science and its Application. 11: 167 170.

Marh, S. 2000. Mechanized delivery of beneficial insects. The IPM Practitioner. April. p. 1-5.

Masood Khan, M.A., Jagadishwar Reddy, D. and Venkateshwar Rao, S., 2001. Bio efficacy of selected insecticides against pest complex in okra. Pestology, 26 (6) : 18-23.

Masoodi, M. A. 2003. Agriculture in Jammu and Kashmir. pp275. Mohisarw book series, Kashmir, Jammu and Kashmir, India.

Masoodi, M.A. 1985. Growth response of Lymantria obfuscate Walker in relation to tannin content in different host foliage. Indian Journal of Entomology. 47(4): 422-426.

Masuda, T. and Kikuchi, O., 1992. Pathogenecity of Veticillium lecanii isolates to aphid and whitefly. Japanese J. App. Ent. And Zool., 36 : 239-245.

Mathirajan, V. G. and Regupathy, A., 2001. Persistent toxicity of Thiamethoxam to Aphis gossypii Glover and Amrasca devastans Distant on cotton. Pest Mangt. And Econ. Zool., 9 (2) : 155-160.

Mathur, A.C. and Tondon, P.L. 1974. Litchi mite can be controlled. Indian Horticulture. 19:11- 12.

Mathur, R.N. 1975: Psyllidae of the Indian subcontinent. Indian Council of Agricultural Research, New Delhi, 429 pp.

Mathur,Y. K.. and Upadhaya, K. D. 1980. Evaluation of and economics of some modern insecticides against mustard aphid, Lipaphis erysimi (Kalt.) Pesticides. 14(1): 14-15.

Matin, M. A, Saghir, S. A. and Rafi, M. A. 1992. Quantitative relationship between Pieris brassicae population and damage in cauliflower. Philippine Entomologist. 8(5) : 112-18

Mayer, D. F. 2003. Bee kills from pesticides and the expert witness. American Bee Journal. 143(7) : 563.

Mayer, D.F. and Johansen, C. A. 1983. Occurrence of honeybee (Hymenoptera: apidae) poisoning in eastern Washington. Environmental Entomology. 12(2):317-320.

McBride, J. 2000. Environmentally friendly insecticides are sugar-coated—For real. ARS News and Information. March 10.

Meade, D.L. and Bruce, D.N., 1991. The use of Verticilium lecanii against sub marginal instars of Bemisia tabaci. J. Inv. Path., 57 : 296-298.

Meenakshisundaram, K. S. and Gujar, G. T. 1988. Proteolysis of Bacillus thuringiensis sp. Kurstaki endotoxin with midgut proteases of some important Lepidopterous insects. Indian Journal of Experimental Biology. 36 : 593-598.

Miller, J.C., Hanson, P.E. and Dowell, R.V. 1987. The potential of gypsy moth as a pest of fruit and nut crops. California Agriculture. 41: 10-12.

Mishra, H. P., 2002. Field evaluation of some newer insecticides against aphids (Aphid gossypii) and jassids (Amrasca biguttula biguttula) on okra. Indian J. Ent., 64(1) : 80-84.

Mishra, R. C. and Verma, A. K. 1982. Relative toxicity of some insecticides to Apis cerana indica F. workers. Indian Bee Journal. 44(3) : 475-476

Misra, S.S. and Sharma, H.C. 1988. Assessment of crop losses due to major insect pests (white grubs and cutworms) of potato crop. Annual Scientific Report-1987, Central Potato Research Institute, India. pp.142-149.

Mohamad, R., Ibrahim, Y. and Chong, W. C. 1979. Field efficacy of several insecticides against diamond back moth, Plutella xylostella (L.). on lowlands of Malaysia. Pertanika. 2: 84-88.

Mohan, C. N. 1994. Seasonal incidence of DBM: Plutella xylostella and its parasitoids in Nilgiri. Journal of Biological Control. 8(2): 77-80.

Mohyuddin, A.I. and Mahmood, R. 1993. Integrated control of mango pests in Pakistan. Acta Hortic. 341:467-483.

Mori, P.1997 Biological control of cabbage butterfly. Vita-in-campagna 15: 5, 27.

Morin, C. 1967. Cultivo de Frutales Tropicales. Librerias ABC, Lima, Per´u. pp. 225-229.

Morse, R. A..1961. The effect of the new insecticides-sevin on honeybee. Gleaning in Bee Culture. 89 : 489-491.

Mote, U. N., Datkhile, R. V. and Pawar, S. A., 1994. A new insecticide, imidacloprid, as a seed dresser for the control of sucking pests of cotton. Pestology, 17(12) : 5-9.

Mullen, J.D. 1995. "Estimating Environmental and Human Health Benefits of Reducing Pesticide Use through Integrated Pest Management Programs." Master Thesis. Virginia Polytechnic Institute and State University, USA, 1995.

Mumtaz, M., Rahim, A. and Hashmi, A. A. 1991. Insecticide efficacy traits in cauliflower fields against Pieris brassicae L. In : Proceeding of 11th Pakistan Congresss of Zoology. pp15-18. Entomologist Research Laboratory National Agriculture Research centre, Islamabad (Pakistan).

Murray, R.C. 1991. Arthropod pests of mango in the Caribbean Islands. Rev. Crop Prot. Univ. West Indies, Kingston.

Murtaza A., Akhtar, M. and Bhatti, M.A., 1999. Some physical factors affecting resistance to cotton against whitefly, Bemisia tabaci (Genn.) in Punjab and Pakistan. Pakistan Entomol., 21 (1/2) : 67-71.

Murugan, M., and Ramachandran, K. 2000. Bioefficacy of vertimec 1.8 EC against diamondback moth, Plutella xylostella (L.) on cabbage. Pestology, 24(1): 12-14.

Mushtaque, M. 1989. Control of large cabbage butterfly in Sialkot by local parasites. Progressive Farming. 9(4): 38-40.

Mushtaque, M. and Mohyuddin, A. I. 1984. Pieris brassicae (Pieridae: Lepidoptera) a pest of crucifers and its control by parasites. Pakistan Journal of Zoology. 18(1) : 121-27.

Nair, M.R.G.K. 1970. Insects and mites of crops in India. New Jack Printing Works Private Limited, Bombay. 404pp.

Nair, R.K., 1967. Leucinodes orbonalis as a serious pest of potato at two places in Mysore state. Indian J. Ent. 29(1): 96-97.

Nandihalli, B. S., Patil, B. V. and Hugar, P. S., 1993. Population fluctuation of cotton whitefly, Bemisia tabaci (Genn.) in yellow sticky traps in relation to weather factors. Karnataka J. Agric. Sci., 6(1) : 25-29.

Nandihalli, B.S., Parameshwara Hugar and Patil, B.V., 1990. Evaluation of neem and neem products against cotton whitefly, Bemisia tabaci (Genadius). Karnataka J. Agric. Sci., 3 : 58-61.

Narkiewiez-Jodko, J. 1992. The effectiveness of Marshal 25EC carbosulfan in control of some vegetable pests. In : International Symposium on Crop Protection. pp 801-06. Research Institute of vegetable crops, Skierniewice, Poland.

Naryanan,E.S.1954. The painted bug Bagrada crciferarum Kirkaldy. Indian Farming. 4: 8-9 and 27.

Natarajan, K. and Sundaram, N., 1979. The incidence of leafhopper on cotton in relation to weather factors. Cotton Development, 8 : 41-42.

Natarajan, K. and Sundaramurthy, N.T., 1990. Effect of neem oil on cotton whitefly (Bemisia tabaci). Indian J. Agric. Sci., 60 (4) : 290-291.

Natarajan, K., Sundaramurthy, V. T. and Das, S. T., 2000. Efficacy of plant products for controlling okra aphid (Aphis gossypii) and leaf hopper (Amrasca bigutulla biguttula). Indian J. Ent., 62 (1) : 28-30.

Navarajan, A.V.P. 2007. Insect pest and their management. Indian Agricultural Research institute, Newdelhi, India, 68.

Navasero, M.V. and V.J. Calilung. 1990. Biology of the egg plant fruit and shoot borer, L. orbonalis in Philippines. Philippine Entomologist, 8(2): 749-760.

Navon, A., Cohen, M., Keren, S., Levski, S., Harel, M. and Dunkelblum, E. 1997. Trapping of the leopard moth, Zeuzera pyrina L. (Cossidae: Lepidoptera) with sex pheromone traps for monitoring the pest experiments in apple and pear orchards. Alon Hanotea. 51 (8): 358-366.

Nelson, K.C. 1994. "Participation, Empowerment, and Farmer Evaluation: A Comparative Analysis of IPM Technology Generation in Nicaragua." Agriculture and Human Values, (Spring-Summer 1994): 109-125.

Nier, T., River, F. and Bermudez, J.C., 1993. First report of Mexico on the isolation of Verticillium lecanii from whitefly and in vitro pathogenecity tests on this insect. Revista Mexican De. Micologia, 7 : 149-156.

Nijjar, G. S. 1972. Litchi cultivation. Punjab Agriculture University, Ludhiana, India.

Nimbalkar, S.A., Khodke, S.M., Taley, Y.M. and Patil, K.J., 1993. Bioefficacy of some new insecticides including neem seed extract and neem oil for control of whitefly, on cotton. Botanical pesticides in integrated pest management pp 256-260.

Nirmala, R., Ramanujam, R., Rabindra, R. J. And Rao, N. S., 2006. Effect of entomofungal pathogens on mortality of three aphid species. J. Biol. Control., 20 (1) : 89-94.

Nonita Devi, M., Singh, T.K. and Chitradevi, L., 2003. Efficacy of certain botanical insecticides against cotton aphid, Aphis gossypii (Glover) on brinjal. Pestology, 27 : 6-9.

Norton G.W. and Mullen, J. 1994. Economic Evaluation of IPM Programs: A Literature Review. Report 448-120. Blacksburg, Virginia: Cooperative Extension.

Norton, G.W., Rajotte, E.G. and Gapud, V. 1999. "Participatory Research in Integrated Pest Management: Lessons from the IPM CRSP." Agriculture and Human Values, 16 : 431-439.

Obnesorge, B., 1981. Studies on the population dynamics of the whitefly, Bemisia tabaci (Genn.) in winter months. Mitt. Dt. Ges. Allag. Angew. Ent., 3 : 324-327.

Odino, M.O., 1993. Future prospects for application of insect pathogens as a component of integrated pest management in tropical root crops. Bio. Sci and Tech., 2 (3) : 179-191.

Oke, O.L. 1980. Amaranth in Nigeria. In "Proceedings of the Second Amaranth Conference". Rodale Press Emmaus, PA, USA.

Ozgur, A.F., Sekeroglu, E., Ohnesorge, B. and Gocmen, H., 1990. Studies on host plant changes, migration and population dynamics of the cotton whitefly (Bemisia tabaci) in Cukurova (Turkey). Mitte-der. Deutsch. Gesell. Fur., Allg. Und-Ang. Ent., 7 (4-6) : 653-656.

Pakash, R. and Kumaraswami, J. 1984. Toxicity of some insecticides of the Indian bee, Apis cerana Fab. Indian Bee Journal. 46: 15-17.

Panda, P. N., Padhi, J., Nanda, U. K. and Senapati, B.1990. Pesticides and honeybees. Orissa Journal of Agricultural Research. 3(2-4): 291-294.

Panday, K., Ajay, and Bilal, S. D. 2005. Field evaluation of different insecticides against cabbage aphid, Brevicoryne brassicae L. in cold arid region of Ladakh. Annals of Plant Protection Science. 13(2) : 324-327.

Pandey, A.K., Namgyal, D., Mehdi, M., Mir, M.S. and Ahmad, S.B. 2006. A case study-major insect pest associated with different vegetable crops in cold arid region Ladakh of Jammu and Kashmir. Journal of Entomological Research. 30(2): 169-174.

Pandey, V., Patel, M.G., Chaudhari, G.B., Patel, J.R., Bhatt, B.K., Vadodaria, R.P. and Shekh, A.M. 2003. Influence of weather parameters on the population dynamics of mango hopper. Journal of Agrometeorology. 5(1): 51-59.

Pandy, N.K.,Ojha, P.K.,and Upadhyay, K.D. 1999. Studies on the bioefficacy of B.thuringiensis formulations against E. vitella Fabr. Under laboratory conditions. RAU. J. Res. 9(1): 57-61.

Paroda, R.S. 1999. For a food secure future. The Hindu survey of Indian Agriculture.

Patel, B.D., Jhala, R.C. and Patel, N.C., 1998. Effectiveness of neem based botanical pesticides against red spider mite, Tetranychus cinnabarinus on okra (Abstract). Nat. Sem. On Entomol, in 21st century, April 30th – May 2nd, Udaipur.

Patel, C.B., Rai, A.B., Patel, A.J. and Shah, A.H., 1993, Acaricidal tests of botanical pesticides in comparison to conventional aearicides/pesticides against red spider mites (Acari : Tetranychidae) on okra, brinjal and Indian bean. Indian J. Ent., 55 (2) : 184-190.

Patel, I.S. and Rote, N.B., 1995. Seasonal incidence of sucking pest complex of cotton under rainfed condition of southern Gujarat. Gujarat Agric. Univ. Res. J., 21 (1) : 127-129.

Patel, Z.P. and Patel, J.R., 1996. Effect of botanicals on behavioural response and growth of jassids, Amrasca biguttula biguttula. Indian J. Plant Prot., 24 : 28-32.

Patetta, A. and Manino, M. 1980. Research on the effect of chemicals used against Varroa jacobsoni. Apicoltore Moderno. 79(3) : 109-114.

Patil, B. V., Sreenivas, A. G., Mujibur Rahaman, S. and Bheemanna, M., 1999. Imidacloprid 70WS seed treatment against early cotton sucking pests. Pestology, 23(4) : 35-39.

Patil, I. V., Bheemanna, M., Badariprasad, P. R., Gowdar, S. B. And Kumar, V. D. V. N. H., 2001. Bioefficacy of acetamiprid 20 SP against early sucking pests in irrigated cotton. Pestology, 25(9) : 29-33.

Patil, S. B., Udikeri, S. S. and Khadi, B. M., 2004. Thiamethoxam 35FS – A new seed dresser formulation for sucking pest control in cotton crop. Pestology, 28(3) : 34-37.

Pawar, D.B., Warade, S.D., and More, T.A. 2001. Efficacy of different plant extracts against leaf hopper (Amrasca devastans Dist.) and fruit borer (E. vitella) of Okra. In:Proceedings of the second National Symposium on Integrated Pest Management (IPM) in Horticultural Crops: New Molecules, Biopesticides and Environment. October 17-19. Bangalore, India..66 p.

Pawar, D.R., Warad, S.D., Patil, S.K. and Brave, H.S., 2000. Preliminary studies on the efficacy of organic products on aphid, Aphis gossypii and leaf hopper, A. devastans is of okra. Insect. Env., 6 : 111-112.

Pedigo, L.P. 2001. Entomology and Pest Management, 4th ed. Prentice Hall, New Jersey.

Pena, J. and Johnson, F. 2001. Insect management in guava. Document ENY- 412. Florida Cooperative Extension Service, Institute of Food and Agricultural Sciences, University of Florida. November 2001

Peña, J. E. 2004. Integrated pest management and monitoring techniques for mango pests. Acta Hort. 645: 151-161.

Peng, R. K. and Christian K. 2005. Integrated pest management in mango orchards in the Northern Territory Australia, using the weaver ant, Oecophylla smaragdina, (Hymenoptera: Formicidae) as a key element. International Journal of Pest Management. 51(2): 149 – 155.

Pimental, D., H. Acquay, M. Biltonen, P. Rice, M. Silva, J. Nelson, V. Lipner, S. Giordano, A. Horowitz, and M. D'Amore. 1992. "Environmental and Economic Costs of Pesticide Use." BioScience. 42(10): 750-760.

Pokharkar, D. S. Hadapad, A. B. and Puranik, T. R 2002. Bioassay and persistence of Bacillus thuriengiensis Berliner against Plutella xylostella (Linn.) on cabbage. Annals of Plant Protection Sciences. 10(1) : 1-4.

Porrini.1991. Honeybees (Apis mellifera) and pesticides: Effect of quinalphos applied to flowering alfalfa. In: Atti XVI Congresso Nazionale Italiano di Entomologia. pp.543-553. Bari- Martna Franca.

Pradhan, S. 1969. Insect pests of crops. National Book Trust, New Delhi, India, 198pp.

Pradhan, S., Jotwani, M. G. and Rai, B. K., 1962. The neem seed deterrent to locusts. Indian J. Ent., 51 : 192-195.

Prakash, A. and Rao, J., 1986. Evaluation of plant products as antifeedants against the rice storage insects. Proc. Sym. Resid. And Envi. Pollution, pp. 201-205.

Pramanik, P. and Chatterjee, M. L. 2004. Effect of Novaluron in the population of Plutella xylostella and Spodoptera litura on cabbage. Annals of Plant Protection Sciences. 12(1) : 193-242.

Prasad, S. K. 1963. Quantitative estimation of damage caused by cabbage worm, cabbage looper, diamond backmoth and cabbage aphid. Indian Journal of Entomology. 25(3) : 242-59.

Prasad, S. K. 1978.Control of mustard aphid, Lipaphis erysimi (Kalt.) Indian Journal of Entomology. 40(4) : 401-404.

Prasada Rao, G.S.L.H.V and Beevi, S. P. 2008. Forewarning Tea MosquitoBug Helopeltis antonii Sign.(Miridae:Hemiptera) in Cashew. Proceeding of International Symposium on Agrometeorology and Food Security, 18-21 Febuary 2008, Hyderabad, India, pp 203.

Praveen, P.M., and Dhandapani, N. 2001. Ecofriendly management of major pests of tomato Lycopersicon esculantum (Mill.) in India. South Indian Horti. 49 (special):261-264.

Praveen, P.M., Dhandapani N., and Kennedy, J.S. 2001. Efficacy of Bacillus thuringiensis (Berliner) formulations for the management of tomato fruit borer, Helicoverpa armigera (Hubner).Pestology, 25:58-61.

Priti and Sihag, R. C.1997.Diversity, visitation frequency, foraging behaviour and pollination efficiency of insect pollinating cauliflower (Brassica oleracea L. var. botrytis cv. Hazipur Local) blossoms. Indian Bee Journal. 59(4):230-237.

Purohit, D., Ameta, O. P. and Sarangdevot, S. S., 2006. Seasonal incidence of major insect pests of cotton and their natural enemies. Pestology, 30 (12) : 24-29.

Puttaswamy, M. and Channabasavanna, G. P., 1980. Influence of host plants on the development, fecundity and longevity of Tetranychus ludeni Zacher (Acari : Tetramychidae). Indian J. Acar., 5 : 80-84.

Quinden, R.J., 1984. The use of Verticillium lecanii an entomopathogenic fungus to control glasshouse whitefly. In : Proc. Of a Conf. At Brighton, England, 22-25 : 9-12.

Raghuraman, M. and Gupta, G.P., 2005. Field evaluation of neonicotinoids against whitefly, Bemisia tabaci Genn. In cotton. Indian J. Ent., 67 (1) : 29-33.

Ragumoorthi, K.N. and Kumar, S., 2000. Biophysical factors of Bhendi, Abelmoschus esculentus (L.) Moench. Offering resistance to leaf hopper Amrasca biguttula biguttula (Ishida) cicadellidae : Homoptera. Pestology, 23 (12) : Rai, A. B., Sejalia, A. S., patel, C. B. And Shah, A. H., 1991, Studies on okra mite Tetranychus macfarlanei (Acari : Tetranychidae) and its chemical control. Modern Acarol. Acad., 1 : 571.

Rai A.B., Loganathan M., Halder J, Venkataravanappa V. and Naik P. S 2014. Eco-friendly Approaches for Sustainable Management of Vegetable Pests. IIVR Technical Bulletin No. 53, IIVR, Varanasi, pp. 104

Rai, K. M., Gupta, B. P. and Tripathi, G. M. 1985. Biology and control of cabbage white butterfly (Pieris brassicae Linn.) (Lepidoptera: Pieridae). Progressive Horticulture. 17(4) : 376-81.

Rai, S., Srivastava, K.M., Saxena, J.D. and Sinha, S.R. 1992. Distribution pattern of diamondback moth (Plutella xylostella L.) on cabbage and cauliflower. Indian Journal of Entomology. 54(3): 262-265.

Raj, H., Bhardwaj, M.L., Sharma, I.M. and Sharma, N.K. 1993. Performance of commercial okra (Hibiscus esculentus) varieties in relation to diseases and insect pests. Indian Journal of Agricultural Sciences. 63(11): 747-748.

Rajavel, D. D, Goplan, M. and Balasubramanian, C. 1991. Effect of insecticides in the control of major pests of cauliflower. Madras. Agricultural Journal. 78(1-4) : 100-102.

Ram, S. and Pathak, K. A. 1992. Insecticidal control of cabbage butterfly, Pieris brassicae Linn. (Pieridae: Lepidoptera) in Manipur. Indian Journal of Entomology. 54(3) : 353-55.

Ramaraju, K., Veluswamy, R. and Mohana Sundaram, M., 1995b. Evaluation of plant products for the management of Tetranychus cinnabarinus on bhendi. V. Nat. Symp. On Acarology, September 20-22, Bangalore. Ramarethinam, S., Marimuthu, S. and Murugesan, N.V., 2005, Entamopathogenic fungi, Verticillium lecanii – An overview. Pestology, 29 (12) : 9-28.

Ramesh, P.1992. Insecticidal control of cabbage butterfly, Pieris brassicae Linn. (Pieridae : Lepidoptera) in Manipur. Indian Journal of Entomology 54 : 353-55.

Rana, B. S. and Goyal, N. P. 1991. Field toxicity of methyl demeton and dimethoate to the foragers of honeybee. (Apis cerana indica Fab.) visitors to Brassica chinesis L. Indian Bee Journal. 53(1-4) : 73-77.

Rana, S. C., Singh, P. B., Pandita, V. K. and Sinha, S. N., 2006. Evaluation of insecticides as seed treatment for control of early sucking pests in seed crop of okra. Ann. Pl. Protec. Sci., 14(2) : 364-367.

Randhawa, G.S., 1974. Horticulture ; Importance of pest control. Pesticides Annual, pp. 85-87.

Randhawa, J. S., 1986. Genetics of jassid resistance in Okra (A. esculentus (L.) Moench). Indian. J. Ent., 48 : 369-371.

Rao, N.V., Reddy, A.S. and Rao, K.T., 1989. Natural enemies of cotton whitefly, Bemisia tabaci Gennadius in relation to host population and weather factors. J. Biol. Control, 3 (1) : 10-12.

Rao, S.R.K. and Lal, O.P. 1999. Distribution pattern of diamondback moth, Plutella xylostella (L.) on cabbage under Delhi conditions. Journal of Entomological Research. 23(3): 261-265.

Rao, V.P., Ghani, M.A., Sankaran, T. and Mathur, K.C. 1971. A review of the biological control of insects and other pests in South East Asia and the pacific region. CIBC Tech. communication. 6: 149.

Ratanpara, A.M., Shekh, J.R., Patel and Patel, N.M., 1994. Effect of weather parameters on brinjal leafhopper *Amrasca biguttula biguttula* (Ishida). Gujarat Agric. J., 19 (2) : 39-43.

Rataul, H. S. 1953. Study of biology of Pieris brassicae a pest of cauliflower in Punjab. MSc. Thesis. Punjab Agricultural Univerisity, Ludhiana. pp135.

Rataul, H. S. 1959. Study of biology of cabbage butterfly (Pieris brassicae L.). Indian Journal of Horticulture. 16(4) : 255-65.

Rath, A.C., Worledge, D., Koen, T.B. and Rowe, B.A., 1995. Long term field efficacy of the entomoglnous fungus, Metarhizium anisopliae against the subterranean Scarab, Adoryphorus couloni. Biocontrol Sci. and Tech., 5 : 439-451.

Raut, G. and Sanaphati, B. 1968. Biology of mustard aphid. Lipaphis erysimi in India. Annals of Entomological Society of America. 61(2) : 259-61.

Rawat, R.R. and Sadu, H.R., 1973. Estimation of losses in growth and yield of okra due to Empoasca devastans (dist). and Erias spp. Indian J. Ent., 35 : 252-254.

Regunath, P., and Gokulapalam, C.1999. IPDM in bittergourd – experiences in participatory approaches. In: Proceedings of Hort. India 277- 279 pp.

Reichert, Susan E. and Leslie Bishop. 1989. Prey control by an assemblage of generalist predators: Spiders in garden test systems. Ecology. Fall. p. 1441-1450.

Robert, H. J. U. and Wright, D. J. 1996. Multi-trophic interaction and management of the diamond back moth : a review. Bulletin of Entomological Research. 86 : 205-216.

Roger, A.. M.1961. The effect of sevin on honeybees. Journal of Economic Entomology. 54(3) : 566-567.

Roling, N. and E. van de Fliert. 1994. "Transforming Extension for Sustainable Agriculture: The Case of Integrated Pest Management in Rice in Indonesia." Agriculture and Human Values, Spring-Summer, 96- 108.

Rombach, M. C., Aguda, R. M. And Roberts, D. W., 1986. Biological control of the brown plant hopper, Nilaparvata lugens (Homoptera : Delphacidae) with dry mycelium application of Metarrhizium anisopliae. Philippines Ent., 6 : 613-619.

Rosaiah, B. 2001. Effect of different plant products against serpentine leaf miner Liriomyza trifolii on ridge gourd Luffa accutangula. Pestology, 25(4): 7-9.

Rosaiah, R., 2001a. Performance of different botanicals against the pest complex of bhendi. Pestology, 25 : 17-19.

Roy, R.S. and De, B.N. 1950. Control of litchi mite. Indian Journal of Horticulture. 7(1): 16.

Rui, C. W. 2001. Comparative study on efficacy of five Biopesticides for the control of Plutella xylostella. Journal of Plant Protection. 27(6) : 33-34.

Ruscoe, C. N. E. 1972. Growth disruption effects of an insect antifeedant. Nature New Biology. 236(66) : 159-60.

Sachan, J. N. and Gangwar, S. K. 1980. Vertical distribution of important insect-pest of cole crops in Meghalaya as influenced by the environment factors. Indian Journal of Entomology. 42 (3) : 414-421

Sachan, J. N. and Gangwar, S. K.1990. Seasonal incidence of insect-pests of cabbage. Indian Journal of Entomology. 52(1) : 111-124.

Sachan, J.N. and Srivastava,B.P.. 1972. Studies on seasonal incidence of insect pest of cabbage. Indian Journal of Entomology. 34: 123-127.

Sahu, S. and Raychaudhuri, B., 1995. Insect-plant interactions with cotton aphid, Aphis gossypii Glover and chilli Capsicum annum L. as a model. Ann. Entomol, 13 (2) : 71-86.

Saini, M. L. and Chhabra, K. S.1966. Control of mustard aphid by systemic insecticides. Plant protection Bulletin. 18(3): 4-8.

Salinas, P. J. 1977. Studies on the ecology of diamond back moth. Plutella xylostella. L. Indian Journal of Entomology. 46(4) : 385-388.

Santos, T.M., Costa, N.P., Torres, A.L. and Junior, A.L., 2004. Effect of neem extract on the cotton aphid. Pesquisa-Agro Brasiliera, 39 (11) : 1071-1076.

Sardana, H.R. Integrated pest management in vegetables. In: Training manual -2, Training on IPM for Zonal Agricultural Research Stations.May21- 26.105-118 pp..

Sarkar, P.K., Mukherjee, A.B. and Ghosh, J., 1996. Assessment of loss of bhendi against red spider mite. Env. Eco. 14 (2) : 480-481.

Sarup, P., Wadhwa, S., Dayashanker, S. and Lal, R. 1985. Testing of pesticides against cabbage aphid, Brevicoryne brassicae Linn. Indian Journal of Entomology. 28 : 369-74.

Sauco, G. V. and Menini, U.G. 1989. Litchi cultivation. FAO Plant production and protection paper. 83.

Saxena, A.P., V.K., Chandla and Raj, B.T. 1988. Ecology of Potato pests and their natural enemies. Annual Scientific Report-1987, Central Potato Research Institute, India. pp.139-140.

Senapati, B. and Khan, S.R., 1978. A note on population fluctuation of Amarasca biguthula biguthula (Ishida) on okra at Bubaneshwar. Indian J. Agric. Res., 12 (2) : 97-98.

Senapati, B. and Mohanty, G.B., 1980. A note on the population fluctuation of sucking pests on cotton. Madras Agric. J., 67 (7) : 624-630.

Shankar, U., Kaul, V. and Rai, S. 2007. New recode of anar fruit borer, Deudorix isocrates (Fabricius) on aonla. Journal of Research, SKUAST-J. 6(2): 292-295.

Sharma, D and Abrol, D.P.2004. Studies on abundance and diversity of different pollinators visiting Brassica campestris var toria and their impact on seed production. Indian bee journal 66(3-4): 109-113.

Sharma, D. and Abrol, D. P. 2005. Contact toxicity of some insecticides to honeybee, A. mellifera and A. cerana F. Journal of Asia Pacific Entomology. 8(1) : 113-115.

Sharma, D., Abrol, D. P. and Bhagat, R. M. 2005. Seasonal incidence of Brevicoryne brassicae and Pieris brassicae on cauliflower seed crop in Jammu subtropics. Journal of Research, SKUAST-Jammu. 4(2) : 214-218

Sharma, D.2005.Relative toxicity of some insecticides to brood and adult honeybee foragers (Apis mellifera L.). Ph.D thesis, SKUAST JAMMU India.

Sharma, D.D. 1985. Major pests of litchi in Bihar. Indian Farming. 35(2): 25-26.

Sharma, D.R. 2006. Management of insect pests of tropical and sub-tropical fruit crops. In:Proceedings of advance training course on Emerging trends in economic entomology, February 21st - 13th March 2006, Hisar, India, pp 123-135.

Sharma, D.R. and Singh, S. 2006. Management of insect pest of temperate fruits. In:Proceedings of advance training course on Emerging trends in economic entomology, February 21st - 13th March 2006, Hisar, India, pp 111-122.

Sharma, G.N. and Sharma, P.D., 1997b. Population dynamics of cotton leafhopper Amrasca biguttula biguttula on cotton and okra in relation to the physical factors of environment in Haryana. Annals of Biology Ludhiana, 13 (1) : 179-183.

Sharma, J. C. and Joshi, F. L. 1972. Effect of different insecticides against mustard aphid, Lipaphis erysimi (Kalt.) (Homoptera:Aphididae). N.N.KVW. Research Journal. 6(1) : 53-54.

Sharma, P. and Rishi, N., 2004. Population build up of the cotton whitefly, Bemisia tabaci Genn. In relation to weather factors at Hisar, Haryana. Pest. Mangt. And Econ. Zool, 12 (1) : 33-38.

Sharma, S. 1998. Integrated management of major insect pests of cauliflower, Brassica oleracea var. botrytis in Jammu. MSc. Thesis, SKUAST-Jammu. pp110.

Sharma, S., Verma, A. K. and Sood, P. 2001. Evaluation of bioefficacy of some new insecticides against pupae of pests infesting cole crops. Journal of Hill Research. 14(1): 54-57.

Sheela, M.S., Nisha, M.S.,and Mohandas, C., 2002. Impact of trap cropping and crop rotation for the management of root-knot nematode in vegetables. In: International Conference on Vegetables. Nov. 11-14. Bangalore, India. 245 p.

Shivalingaswamy, T.M., Satpathy, S., Singh, B. and Akhilesh Kumar. 2002. Predator-prey interaction between jassid Ammrasca devastance and staphylinid in Okra. In: International Conference on Vegetables. Nov. 11-14. Bangalore, India. 275 p.

Shivalingswami, T.M., Satpathy, S. and Banergee, M.K. 2002. Estimation of crops losses due to Insect pests in vegetables Ed. by Sarath Babu, B.,Varaprasad, K.S. Anitha, K., Prasada Rao, R.D.V.J., Chakrabarthy S.K., and Chandurkar, P.S. Resource management in plant protection.1: 24-31.

Sihag, R. C., Khatkar, S. and Khatkar, S. 1989. Foraging pattern of three honeybee species on eight cultivar of oilseed crops. 2. foraging during the entire blooming period of the crops. International Journal of Tropical Agriculture 17(1-4) : 253-261.

Sihag, R.C. 1991. Ecology of Eurpoean honeybee (Apis mellifera) is semi arid and sub tropical climate 1. Melliferous flora and over seasoning of colonies. Korean Journal of Apiculture. 5 : 31-43.

Sihag, R.C.1988. Effect of pesticides and bee pollination on seed yield of some crops in India. Journal of Apicultrual Research. 27(1) : 49-54.

Singh Dilbag.,1991. Pests of solanaceous vegetables. In: A decade of research on pests of host-crops 1980-90. Group discussion of entomologists working in coordinated projects of Hort crops held at CIHNP, Lucknow Jan. 28-29.

Singh, A., Trivedi, T.P., Sardana, H.R., Dhandapani, A. and Sabir, N. 2001. Resource Manual on validation and promotion of IPM. National Centre for IPM, New Delhi, pp.119.

Singh, G. 1991 Loss assessment, ecology and management of mango fruitfly, Dacus sp. Acta Hortic. 291:425-436.

Singh, G. and Misra, P.N. 1988. Efficacy of new insecticides for control of insect pests of spring okra (*Abelmoschus esculentus*). Indian Journal of Agricultural Sciences. 58(10): 783-785.

Singh, G.S., Sahni, M. and Shah, B.R., 1988. Biochemical and morphological basis of resistance to okra plant. Euphytica, 21 : 147-153.

Singh, H; Gupta,D.S., Yadava,T.P. and Dhawan,K.1980. Post harvest losses caused by painted bug, Bagrada cruciferarum (Kirk.) to mustard. HAU Research Journal 10(3): 407-409.

Singh, H; Rohilla, H. R. ; Ram,P and Ahlawat D. S. 1993. Outbreak of painted bug, Bagrada cruciferarum (Kirk.) on oilseeds brassicae in India. International Journal of tropical agriculture 11(2): 153-154.

Singh, J., Sohi, A. S., Biradar, D. S. and Bhaliwal, C. S., 1996. Seed dressing with imidacloprid (Gaucho 70WS) for the control of Amrasca biguttula biguttula (Ishida) on cotton. Insect Environ. 2 : 102.

Singh, J.P. 1969. Effect of thiodon on forging honeybees. Indian Bee Journal. 31: 9-10.

Singh, M., Sharma, P. L. and Dhaliwal, H. S. 1974. Toxicity of insecticides to honeybee workers *Apis cerana indica* Fab. Pesticides 8:1989.

Singh, N., Ramkishore, and Parihar, S.B.S., 2004. Preliminary efficacy of botanicals against cotton aphid, *Aphis gossypii* Glover on cotton. Insect Environ. 10 (3) : 136-137.

Singh, R. 1999. Management of insect pests of cabbage Brassica oleracea var. capitata Linn.) and persistence of chlorpyriphos. MSc Thesis. Rajindra Agricultural University, Pusa, Bihar. pp85.

Singh, R. and Jamwal, R. S. 1985. Efficacy of insecticides in controlling cabbage butterfly. Research and Development Reporter. 2(2) : 20-21.

Singh, R. P.,and Doharey, K.L. 2001. Integrated pest management in vegetables. In: Training manual -2, Training on IPM for Zonal Agricultural Research Stations. May 21-26.105-118 pp.

Singh, S.P. 1989. A note on the adult biology of Agrotis spinifera (Hubner) on potato. Bulletin of Entomology. 27(2): 191-193.

Singh, S.P. 1989. Achievements of AICRP on Biological Control. Tech.Bull. No. 2. Biological Control Centre, NCIPM, Faridabad, 20 pp.

Singh, S.P., 1997. Biological control in India. In: IPM systems in agriculture Vol 2. Biocontrol in emerging biotechnology. Ed. by Upadhyay, R.K.; Mukharjii, K.G. and Rajak, R.K. Aditya books private Ltd. New Delhi, India.

Singh, S.P., Rao, N.S., Ramani, S.and Poorani, J. (eds) (2001) Annual Report 2000-01, Project Directorate of Biological Control, Bangalore. Bangalore, India; PDBC, 218 pp.

Singh, U.C.,Singh, R., and Nagaich, K.N. 1999. Evaluation of tomato varieties against jassid (E. devastance) white fly (B. tabaci) and leaf curl. Indian J. Ent. 61(2): 173-176.

Singh, W.J.K. Singh, R,Hameed,S.F and Singh, B.1989. Field toxicity of some insecticides to Apis cerana indica Fabr. Indian Bee Journal 51(4):137.

Sinha, R. P., Yazadani, S. S. and Verma, G. D. 1990. Population dynamics of mustard aphid, Lipaphis erysimi (Kalt.) (Homoptera:Aphididae) in relation to ecological parameters. Indian Journal of Entomology. 5(3) : 387-392.

Slosser, J.E., Pinchak, W.E. and Rummel, D.R., 1998. Abiotic and biotic regulation of aphid populations in Texas rolling plains. Proceedings of Beltwide Cotton Conference San-Digeo– California, USA, 5-9, January 1998, 2 : 1065-1067.

Slosser, J.E., Pinchak, W.E. and Rummel, D.R., 1998. Biotic and abiotic regulation of Aphis gossypii Glover in West Texas dryland cotton. South-west Entomologist, 23 (1) : 31-65.

Smirle, M. J.1990. The influence of detoxifying enzymes in insecticide tolerance in honeybee. Journal of Economic Entomology. 3(3) : 715.

Smith, D, Beattie, G. and Broadley, R. 1997. Citrus pests and their natural enemies: IPM in Australia. Queensland, Department of Primary Industries, Brisbane, Australia, 272 pp.

Sokal, R.R and Rholf, F.J.1981. Biometry, 2nd Edition San Francissco

Sreelatha and Divakar, B. J., 1997. Impact of imidacloprid seed treatment on insect pest incidence of okra. Indian J. Plant Prot., 25(1) : 52-55.

Sreenivasamurthy, D., P.N. Krishna Moorthy, M. E. Raja and Shivanna. 2002. Economic assessment of pest management in brinjal and cabbage using neem products. In: International Conference on Vegetables. Nov. 11-14, Bangalore, India. 243 p.

Srinivas, P.S. and K.E. Lawande. 2002. Barrier cropping - a new method for the management of thrips in onion. In: international conference on vegetables. Nov. 11-14, Bangalore, India. 237 p.

Srinivasa Murthy, K. and Sharma, D.P., 1997. Traditional Pest management in cotton. Insect Environ. 3 : 73-74.

Srinivasa Rao and Rajendran, R., 2003. Joint action potential of neem with other plant extracts against the leaf hopper Amrasca devastance (Distant) on Okra. Pest Mangt. And Econ. Zool., 10 : 131-136.

Srinivasa, A.G., 1993. Seasonal incidence, loss estimation and management of leafhopper, Amrasca biguttula biguttula Ishida (Homoptera : cicadellidae) on cotton. M.Sc. (Agri.) Thesis, Univ. Agric. Sci., Dharwad (India).

Srinivasa, N., Mallik, B., Onkarappa, S., and Gururprasad, H., 2002. Bioefficacy of newer acaricidal molecules against chilli mite, Polyphagotarsonemus latus Banks (Acari: Tarsonemidae). In: International Vegetable Conference. Nov. 11-14, Bangalore, India, 265 p.

Srinivasan, K. 1993. Pests of vegetable crops and their control. In: Advances in Horticulture, 6: Vegetable crops Eds. K.L. Chadha and G. Kalloo. Malhotra Pub. House, New Delhi. p. 859-886.

Srinivasan, K. and Krishanakumar, N. K.1982. Pest management in cabbage, In : Annual Report of Indian Institute of Horticulture Research, Bangalore, India. pp.78-81.

Srinivasan, K. and Krishnamoorthy, P.N.1991. Indian mustard as a trap crop for management of major lepidopterous pests on cabbage. In: Trop. Pest Management, 37: 26-32. Krishnamoorthy, P.N.and Krishna Kumar, N.K 2002. Advances in the use of botanicals for the IPM of major vegetable pests. In: International Conference on Vegetables. Nov. 11-14. Bangalore, India. 233 p.

Srinivasan, K., Krishnakumar, N.K., Ramachander, P.R. and Rao, G.S.P., 1988. Seasonal pattern of leafhopper, Amrasca biguttula biguttula on okra in India. Insect. Sci. Appl., 9 (1) : 85-88.

Srinivasan, K., Krishnamoorthy, P.N. and Prasad, R. 1993. Evaluation of different trap crops for the management of fruit borer, H. armigera on tomato. In: Abs. Golden Jubillee Symp. Horticultural Research: Changing Scenario. May 24-28. Bangalore, India. 259 p.

Srinivasan, K.,and Veeresh, K.1986. The development and comparison of visual damage threshold for chemical control of Plutella xylostella and Crocidolomia binotalis on cabbage in India. Insect Sci. Appli. 7: 547- 557.

Srivastava, A.S. and Masoodi, M.A. 1985. Influence of host plants on the development and survival of *Lymantria obfuscata.* FAO Plant Protection Bulletin. 33 (2): 67-69.

Stansly, T. M., 2001. Efficacy of spinosad against cotton aphid (Aphis gossypii) by various application methods. J. Pesticide Sci., 26 (4) : 381-385.

Subhadra Acharya, H. P., Mishra, H. P. and Dash. D., 2002. Efficacy of insecticides against okra jassid, Amrasca biguttula biguttula (Ishida). Ann. Pl. Protec. Sci., 10 : 230-232.

Sugeetha, 1998. Studies on the spidermite T. macfarlanei Baker and Pritchard (Acari : Tetrnychidae) infesting okra (Abelmoschus esculentus). M.Sc. (Agri.) Thesis, Univ. Agric. Sci., Bangalore(India).

Syed, R.A., Ghani, M.A. and Murtaza, M. 1970. Studies on trypetids and their natural enemies in West Pakistan. III. Dacus (Strumeta) zonatus (Saunders). Tech. Bull. Commonw. Inst. Biol. Control. 13:1-16.

Tablot, F. V.1939. Incidence and distribution pattern of pierids. Annual Report, Research and Development Department. pp. 5-22.

Talekar, N. S. and Griggs, T. D. 1986. Diamond back moth management. In : Proceedings of First International Workshop. 11-15 March, 1985, Tainan, Taiwan AVRDC, Shanhva, Taiwan.pp471.

Tandon, P. L., Bhalla, O. P. and Verma, A. K. 1977. A note on the seasonal incidence and population fluctuations of cabbage aphid on cauliflower. Vegetable Science. 4 : 145-157.

Tandon, P.L. and Verghese, A. 1985. World list of insect, mite and other pests of mango. Tech. Doc. Indian Inst. Hortic. Res. 5. pp 22.

Terytze and terytze. 1987. The use of bacterial preparations (Bacillus thuriengiensis) for the control of harmful lepdioptera species on cabbage (Eubacteriales : Bacillacea). Archiv Fur Phytopathologie and Pflanzen schuz.23(5) : 377-387.

Tewari, G.C. and S.J. Singh. 1987. New record of baculovirus disease in L. orbonalis Guen. Curr. Sci., 56(13): 671-672.

Tewari, G.C.,and Sandana,H.R.1987. Eriborus argentiopilosus (Caweras) a new parasite of L. orbonalis. Entomon. 12(3): 227-228.

Tewari, G.N. and Sinha, K. 1983. Studies on insect pollinators in relation to seed production in cauliflower (Brassica oleracea L. var. botrytis). Indian Bee J. 43 (2-3): 57-62.

Thakur, A. K. and Kashyap, N. P. 1989. Assessment of toxicity of potential aphid controlling organophosphatic compounds against Apis mellifera on Brassica campestris. L. var. sarson Prain. Indian Bee Journal. 51(3) : 94-96.

Thakur, J.N.; Rawat, U.S. and Pawar, A.D. 1989. Investigations on the occurrence of natural enemies of San Jose scale, Quadraspidiotus perniciosus Comstock (Hemiptera: Coccidae) in Jammu & Kashmir and Himachal Pradesh. Entomon. 14: 143-146.

Thakur, J.R., Sharma, P.C. and Gupta, P.R. 1989. Occurrence of olive psylla, Euphyllura pakistanica on olive in India: A new record. Tropical Pest Management. 35 (3): 331.

Thakur, N. S. A 1996. Relationship of cabbage butterfly larval (Pieris brassicae Linn.) population on the marketable yield of cabbage. Journal of Hill Research. 9(2) : 356-58.

Thakur, S. S., and Parmar, S. 2000. Field evaluation of insecticides against Pieris brassicae Linn. infesting cabbage seed crop. Insect Environment 6(3) : 139-40.

Thakur, S.S., Chandel, K.S. and Kashyap, N.P. 1998. Field evaluation of tomato varieties against Helicoverpa armigera (Hubner) in the higher hills of Himachal Pradesh. Insect Environment. 4(2): 51-52.

Thiruvudainambi, S., S. Doraisamy, D. Veeraraghavanatham and K. Srinivasan. 2002. Chemical and non-chemical methods of managing mosaic disease of chilli. In: International conference on Vegetables, Nov. 11-14, Bangalore, India, 258 p.

Thomas, J. and Phadke, K. G. 1994. Relative toxicity of oxydemeton-methyl, chlorpyriphos and quinalphos to honeybee (A. cerana indica). Indian Journal of Agricultural Sciences. 64(3): 207-209.

Thomas, J., Phadke, K. G. and Verma, S.1994. Efficacy of some insecticides against mustard aphids, safety to predators pollinators and persistence of residues on rapeseed, mustard crop. Shaspa. 1(2):39-44.

Threhan, K.N., 1944. Distribution of whitefly in the Punjab. Indian Farming, 5 : 514-515.

Thulsiram, K., 1991. Bionomics and chemical control of the spidermite, Tetranychus macfarlanei Baker and Pritchard (Acari : Tetrany chidae) on cotton. M. Sc. (Agri.) Thesis, Univ. Agric. Sci. Dharwad (India).

Tomar, R.K.S. 1998. Efficacy and economics of biopesticides and insecticides combinations against okra shoot and fruit borer. Indian J. Ent. 60 (1): 25-28

Tripathi, R. L. and Singh, V. S. 1964. Preliminary study on residual efficacy of a few systemic insecticides against mustard aphid. Indian Oilseed Journal. 8 : 159-61.

Trivedi, T.P., Rajagopal, D. and Tandon, P.L. 1994. Assessment of losses due to tuber moth. Journal of the Indian Potato Association. 21(3/4): 207-210.

Umamaheshwari, T., Sharmila, Bharathi, C., Kanagarajan, R., Arivudainambi, S. and Selvanarayana, V., 1999. Neem formulations and castor oil – A safe way to manage okra red spider mite. J. Acarol, 14 (1&2) : 77-79.

Uthamasamy, S., Subramanian, T.R. And Jaya Raj, S., 1973. Studies on varietal resistance in Bhendi to the leaf hopper. I. Screening of varieties under field conditions. Madras Agric. J., 60 (1) : 27-31.

Vadodaria, M. P., Patel, V. G., Patel, C. J., patel, R. B. and Maisuria, I. M., 2001. Thiamethoxam (Cruiser) 70 WS : A new seed dresser against sucking pests of cotton. Pestology, 25(9) : 13-19.

Veerish, G.K. 1989. Pest problems in mango – world situation. Acta Hortic. 231:551-565.

Verma, A. N., Sharma, P. D. and Sarenima, P. U. 1969. Relative toxicity of some contact insecticides against cabbage caterpillar, Pieris brassicae. Journal of Research, Punjab Agriculture University. 6(1) : 197-99.

Vijayalakshmi, K., Subhashini, B. and Shivani, V. Koul, 1998. Plant in pest control, custard apple, Vitex, sweet flag and position nut centre for Indian Know. Syst. Chennai, pp. 1-30.

Villeirs, E.A.D.E and Matthee, W.A. 1973. The occurrence of bark borers on litchi and other fruit tree. Citrus Grower and sub-tropical Fruit Journal. 477:17-19.

Vishakantriah, M. and Gowda, B. L. N. 1975. Record of diamond back moth as a new pest of Amaranthus viridis in Karnataka. Current science 44 : 869.

Wadnekar, D.W., Zanwar, P.R., Sangle, P.D. and Bhosale, A.M., 2004. Field evaluation of thiamethoxam (Taurus 25 WG) against sucking pests of cotton. Int. Symp. On Stratg. For Sust. Cotton Pro. A Glo vision, 23-25 November 2004, Univ. Agric. Sci., Dharwad, Karnataka, pp. 190-191.

Waghulde, J.K., Patil, A.S. and Hapase, D.G., 1991. Suppression of Pyrilla purpusilla Wlk in sugarcane with Metarhizium anisopliae Bharatiya Sugar, pp. 53-58.

Watson, J. S., Hopper, B. S. and Tipton, J. D., 2003. Whitefly and the problem of sticky cotton. Span (Shell), 25 : 71-73.

Wearing, C.H. 1975. Integrated control of apple pests in New Zealand. 2. Field estimation of fifth-instar larval and pupal mortalities of codling moth by tagging with cobalt-58. New Zealand Journal of Zoology. 2: 135-149.

Whalon, M.E. and Croft, B.A. 1984. Apple IPM implementation in North America. Annu. Rev. Ent. 29: 435-70.

Whitehead, V.B. and Rust, D.J. 1972. Control of fruit piercing moth, Sorrodes partita (Fabr.) (Lepidoptera: Noctuidae). Phytophylactica. 4:9-12.

Williamson, R. C. 1999. Biorational pesticides: What are they anyway? Golf Course Management Web site.

Yadava, H.N., Mital, R.K. and Singh, G., 1967. Correlation studies between leaf midrib structure and resistance to jassids (Empoasca devastans Dist.) in cotton. Indian J. Agric. Sci., 37 (6) : 495-497.

Yadva, P. R. and Sachan, J. N.1976. Residual toxicity of some common insecticides to Lipaphis erysimi (Kalt.) (Hemiptera : Aphididae) as serious pest of cauliflower in Rajasthan. Pesticides. 2 : 39-44.

Yakomi, R.K. and Gottwald, T.R., 1998. Virulance of Verticillium lecanii isolates in aphid determined by detached leaf bioassay. Int. Inv. Path., 51 : 250-258.

Yufang, G., Hua, G,. Fang, C., Wang, H. and Fengmei, L., 1999. The incidence and control of cotton aphid in Dunhuang prefecture of Gansu province. China Cottons, 26 (12) : 25-26.

Zaz, G. M. 2001. Incidence and population build up of cabbage aphid Brevicoryne brassicae L. on cabbage and cauliflower. Applied Biological Research. 3(1/2 ) : 51-53.

Zien, S.M. 2001. B.U.G.S. Flyer. March. p. 1-3.

Zouiten, N. and Hadrami. I.E. 2001. Olive tree psyllid: State of knowledge and perspective of control. Agricultures. 4(10): 225-232.

Zuhua, S., and Shusheng, L.1998. Toxicity of insecticides commonly used in vegetable fields to the diamondback moth, Plutella xylostella and its parasite, Cotesia plutellae, in Chinese J. Biol. Control. 14(2): 53-57.

# Cruciferae

## INTRODUCTION

The most important crucifer vegetables include cabbage, cauliflower, broccoli and Brussels. All these crops are highly cross- pollinated and depend upon a variety of insects for pollination (Free, 1993, Verma and Phogat, 1994). At the same time, all these vegetables are also attacked by large number of insect pests which cause massive yield and economic losses (Waterhouse and Sands, 2001, Lamb, 1989; Sibanda et al., 2000). Evidently, crop production is simultaneously confronted by two problems which are antagonist to each other. One is protection of crop against insect pests and another safety of pollinators for pollination of crops. Evidently, there is need to explore the integrated approach protecting crop on one hand and safety of pollinators on the other. Integrated Pest Management (IPM) is a sustainable approach to managing insect pests by using biological, cultural and chemical methods. The aim is to ensure high-quality agricultural production and at the same time the reduction of environmental, economic and health risks through pesticides (Bajwa and Kogan, 2002). Strategies can include monitoring of pests and therefore an optimal timing of pesticide application, use of selective insecticides, trap crops, semiochemicals, resistant plant varieties (Picket et al., 1997) and favouring natural enemies.

## POLLINATION OF COLE CROPS

### Brassica Oleracea L., Family Cruciferae

A large number of crops belong to the plant species *B. oleracea*, known collectively as cole crops. Cabbage and broccoli are the most important of the cole crops. The cole crops are large-leaved, succulent, and low-growing, 1 to 2 feet, until the inflorescence is formed then they may reach 2 l/2 to 7 feet in height. More are biennial than annual, although most cauliflowers are annual. The plants are usually grown in cool climates or in the cooler part of the year in warm climates, and they do best under conditions of relatively high humidity.

#### *Inflorescence*

The flowers of cruciferous have four petals about ½ to 1 inch long forming a cross, hence referred to as cruciferae. The flower opens during the morning, the anthers a few hours later, so the flower is slightly protogynous. There are six stamens, two generally shorter than the style and facing toward it but leaning away, and four erect stamens generally longer than the style and also facing it. There is a single capitate stigma terminating the style. In most cultivars, nectar is secreted by two nectaries located between the bases of the short stamens and the ovary. Nieuwhof (1969) stated that there are also two inactive nectarines outside the base of the two pairs of long stamens. The nectaries secrete freely, 0.1 $cm^3$ nectar each 24 hours of the 3 days the flower is open (Pearson 1933). The flowers are highly attractive to pollinating insects for both nectar and pollen. When the seed producing acreage is large, beekeepers nearby frequently harvest a crop of excellent honey.

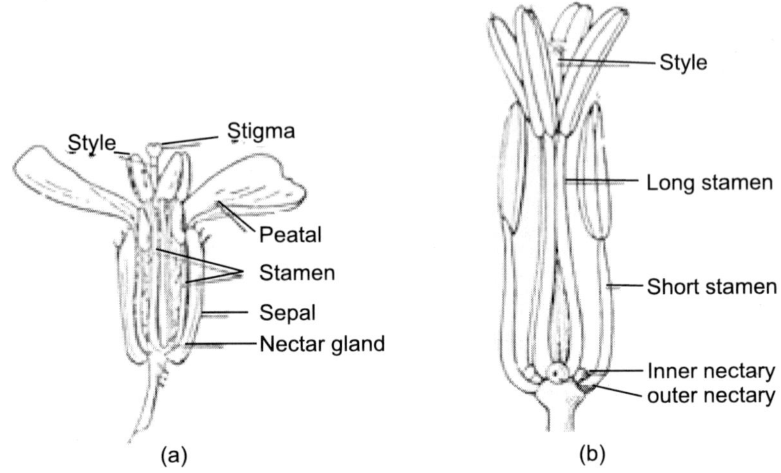


Figure 4.1. *Typical crucifer flower b Brassica napus flower*

### *Pollination Requirements*

The cole crops depend upon insects for cross pollination. Cauliflower requires cross-pollination by insects for seed setting (Nieuwhof 1963, 1969). Cauliflower is self-incompatible crop and can not utilize its own pollen. Evidently, cross pollinating insects are required to accomplish the process of pollination (Sharma et al 1974,Sihag 1985, 1986).

Anstey (1954) found that 52 percent of sprouting broccoli plants were self-incompatible, 30 percent compatible, and 18 percent somewhere in between. But even with the compatible plants, the transfer of the pollen from anthers to stigma is necessary for best seed set. Usually, plants grown in cages or otherwise isolated from pollinating agents set practically no seed even if the plants are occasionally shaken. Cross-pollinated cabbage flowers produced siliques with 10 or more seeds, but selfed flowers produced less than one seed each.

### *Pollinators*

Many kind of insects has been reported visiting cauliflower which included honey bees, wild bees, and flies. Faulkner (1962) recommended the use of Blowflies in cages where the pollination of only a few plants were involved. Pearson (1932) on the other hand regarded bees belonging to families Andrenidae, Megachilidae, and Nomadidae more important than honey bees in the pollination of cabbage. Honeybees were considered to be the most and efficient pollinators of cabbage. Skrebtsova (1964) found that in the U.S.S.R., 84 to 94 percent of the pollinating agents on cabbage were honey bees. Similarly, Radchenko (1966) reported that 85 to 100 percent of the pollinators on cabbage were honeybees.

Kakkar and Sharma (1991) reported that cauliflower *Brassica oleracea* var *botrytis* plants pollinated by honeybee, *Apis cerana indica* F. showed about 3 fold increase in yield over control with significant increase in weight and the per cent germination of the seed. Alam et al (1987) also found that open pollinated plants produced more seeds (11.5 per siliqua) compared with caged plants (9 per siliqua) and had a higher percentage of filled seed 69.70 compared with 54.2, respectively. Thus pollination is an important for higher yield of seeds, with better quality.

### *Diversity of insect pollinators visiting cauliflower bloom*

The data presented in Table 4.1 showed that cauliflower attracted insects belonging to 4 orders, 7 families 14 species visited the cauliflower bloom. Of all these insects, honeybees viz. *Apis mellifera, Apis dorsata, Apis cerana* and *Apis florea* were the dominant flower visitors and comprised more than 83.20 per cent of the total flower visiting insects. Their abundance was in the order:- *A mellifera* > *A dorsata* > *A cerana* > *A florea.* The other insect visitors included soil nesting solitary bees such as *Andrena leana* and *A. ilerda,* some beetles, butterflies, houseflies and Syrpihid flies. The latter group of insects visited the cauliflower bloom at interrupted intervals and were not considered as the reliable pollinators. Evidently, the honeybees, which frequented cauliflower in large number throughout the day constitute the most important pollinators because they are the only insects whose immature stages are reared exclusively on pollen and nectar (Kakar 1981, Sihag,.1985, Sihag, 1986, 198Abrol 1989, 1991, 1997). Priti and Sihag (1997) also reported that honeybees constituted 83.3 per cent of the pollinators on cauliflower bloom. Sharma (2005) also found honeybees as the most important pollinators of cauliflower.

**Table 4.1:** *Insect pollinator complex visiting cauliflower bloom (Source Abrol and Anil )*

| Order | Family | Species | Percentage proportion | Pollinator status |
|---|---|---|---|---|
| | Apidae | *Apis mellifera* L. | 31.20 | Major |
| | | *A. dorsata* F. | 24.20 | Major |
| | | *A. cerana* F. | 22.40 | Major |
| | | *A. florea* F. | 5.40 | Minor |
| | Andrenidae | *Andrena leaena* | 1.80 | Minor |
| | | *A. ilerda* | 1.06 | Minor |
| Coleoptera | Coccinellidae | *Coccinella septumpunctata* | 1.00 | Minor |
| Lepidoptera | Danaidae | *Danaus plexippus* | 2.0 | Minor |
| | Pieridae | *Pieris brassicae* | 1.80 | Minor |
| Diptera | Muscidae | *Musca* sp. | 3.05 | Minor |
| | Syrphidae | *Eristalis* sp. | 1.4 | Minor |
| | | *Macrosysphus* sp. | 1.8 | Minor |
| | | *Metasyrphus corollae*<br>*Episysphus balteatus* | 1.6 | Minor |

### *Impact of pollination treatments on qualitative and quantitative yield parameters of cauliflower seed*

Cauliflower crop is highly cross pollinated and depends upon pollinating insects for seed production. The data presented in Table 4.2 shows that there was considerable increase in percentage fruit set, seeds per siliqua, weight of the seeds, pod length and percentage germination open pollinated flowers as compared to those isolated from insect visits. There was an increase of 2.54 folds and 34.4 per cent in percentage of fruit set over control. Similarly, number of seeds per siliqua were 4.26 folds higher in open pollinated as compared to control. The percentage increase in seeds per siliqua was as much higher as 426 per cent. The seeds obtained from open pollinated plants were heavier than those deprived of insect visits and percentage increase over control was found to the extent of 133 per cent. The pod length in open pollinated plants was much higher as compared to the control. The percentage increase in pod length was found to be 138 per cent. Further more, seeds produced from open pollinated plants resulted in significantly higher percentage germinate on over the control. The results clearly revealed that pollinators play a significant role in the qualitative and quantitative yield attributes of cauliflower. The effectiveness of honeybees as pollinators for cauliflower seed production has been studied by different workers in open pollinated varieties. Kakkar and Sharma (1991) evaluated the role of honeybee, *Apis cerana indica* F. as pollinators of cauliflower *Brassica oleracea* var *botrytis* and reported that honeybees pollinated plants showed about 3 fold increase in yield over control. They further found that the weight and the per cent germination of the seed was also improved significantly. Alam et al (1987) in their study on pollinating behaviour of *Apis indica* F. and its influence on seed production and found that open pollinated plants produced more seeds (11.5 per siliqua) compared with caged plants (9 per siliqua) and had a higher percentage of filled seed 69.70 compared with 54.2,respectively.

Evidently, providing safety to the natural pollinators and supplementing cauliflower bloom shall result in higher yield of seeds, with better quality. The studies clearly establish that protection of the crop from the pests one hand and safety of the pollinators for pollination of the crop on other are equally important.

**Table 4.2:** *Qualitative and quantitative effect of pollination on cauliflower yield attributes.*

| Parameters | Control | Open pollination | Folds increase over control | Percentage increase over control |
|---|---|---|---|---|
| Percentage fruit set | 23.6 ± 1.4[a] | 58.0 ± 5.1 | 2.54 | 34.40 |
| Seeds/siliqua | 3.2 ± 1.6[b] | 13.66 ± 1.20 | 4.26 | 426.00 |
| Weight of 1000 seeds (gm) | 2.640[c] | 3.480 gm | 1.33 | 133.00 |
| Pod length (cm) | 3.00 ± 1.20[d] | 4.16 ± 1.30 | 1.38 | 138.00 |
| Percentage germination | 78.60 ± 1.28[e] | 83.20 ± 2.16 | 1.05 | 4.60 |

## CABBAGE AND OTHER CRUCIFER SEEDS

The flowering stem elongates after leafy growth stops. The stem has many branches, small leaves, and bright yellow or white flowers. Each flower has four petals, (1.3–2.5 cm) long, that form a cross, hence the family name Cruciferae ('cross bearing'). Each flower has six stamens; two are shorter than the style and four are longer. There is a single stigma at the tip of the style. The flower opens in the morning, but the anthers do not release pollen until a few hours later. Nectar is excreted at the bases of the short stamens and ovary. The flower is open for three days, and bees visit the blossoms for nectar and pollen.

## Crucifer Pollination Requirements

Crucifers as a group require cross-pollination (95% of them) although some cauliflower varieties are self-fruitful. Many plants are self-incompatible, and some are cross-incompatible. Selfing in the absence of cross-pollination generally reduces seed yield, seed size, and yield in subsequent generations (McGregor, 1976). Wind is not a good pollinator in *Brassica* spp. except in canola, so bees play an important role in moving pollen. This is especially true in hybrid seed production in which many parent stocks are male-sterile and seed production depends on insects moving viable pollen from male-fertile to male-sterile plants.Mcgregor (1976) recommended 5-10 colonies of honeybees per ha whereas Mayer (1986) recommended 2.5 colonies for crucifers.

## Crucifer Pollinators

Cabbage flowers are easily accessible to insects and many types of flies and bees visit them. Non-managed bees are probably important pollinators when they occur in large numbers, especially in cooler temperatures when honey bees are less likely to fly. In general, honey bees are the primary pollinator of crucifers, making up, on average, 84–100% of the pollinators in a field (McGregor, 1976). The yield is optimal when flowers receive multiple bee visits over several days; this is best achieved by providing large numbers of pollinators (Mayer, 1986).

Honey bee colonies should be brought in when there is enough bloom to attract bees. Bees tend to work down rows, and it is important in hybrid seed production for bees to cross rows and work on both type of male and female plants. It is therefore recommended to place colonies in such a way that bees leaving the hive are encouraged to forage across rows. It is important that bees do not discriminate between male and female rows in hybrid seed production. If they prefer one line over the other, they will concentrate on the preferred rows and be less apt to cross-pollinate the crop. This problem can be minimized by making sure that male and female lines have the same height, same flower colour, and bloom at the same time.

Oilseed rape is comprised of two species of *Brassica* which include *B. campestris* and *B. napus*. Rapeseed oil is naturally high in erucic acid and rapeseed meal is high in glucosinolates, both of which have limited the use of rapeseed products in human and animal food. However, plant breeders have developed varieties with low-acid, low-glucosinolate that are entirely suitable for food.

Cauliflower is a self incompatible crop (Sihag, 1985) due to which flowers cannot utilize their own pollen. This condition necessitates cross pollination which needs a pollen vector to transfer pollen from anthers of one flower to the stigma of another. Honeybees are considered to be the most important pollinator. This is because of the fact that their immature stages also depend on pollen and nectar (Kakar,1981,Crane, 1990., Free, 1993,Abrol,1997). Priti and Sihag (1997) also reported that honeybees constituted 83.3 per cent of the pollinators on cauliflower bloom. Sharma (2005) also found honeybees as the most important pollinators of cauliflower.

## IMPACT OF PESTICIDES ON BEE VISITS

Endosulfan and menazon have been reported to be least toxic to bees (Singh 1968, 1969, Hameed, et al 1973). Singh et al (1974) also found that menazon and endosulfan were least toxic against workers of honeybees (*A cerana*). Attri and Sharma (1969) and Kapil and Lamba (1974) also obtained similar results with endosulfan and menazon. Carbaryl (0.3%) was reported to be highly toxic to *A cerana indica (*Morse (1961, Roger 1961, Bai and Reddy 1977). Malathion was found to be highly toxic to bees (Carcon, 1979. Patetta and Manino, 1980). Prakash and Kumarswami (1984) found phosalone and endosulfan as least toxic to *A cerana indica,* moncrotophos and carbaryl moderately toxic, and the synthetic pyrithroids like cypermethrin, permethrin and decamethrin highly toxic to workers of honeybees.

Sihag *et al* (1989) found that as compared to endosulfan, carbaryl was four times more toxic to *A mellifera* and twenty time more toxic, to *A dorsata*. Panda *et al.* (1990) suggested that evening sprays of the phosalone on mustard had least injurious effect as compared to morning and afternoon. Dimethoate has been observed to be slightly more toxic then methyl demeton to honeybees resulting in reduction of their colony (Rana and Goyal, 1991). Thomas and Phadke (1994) reported by direct spray method, oxydemeton methyl was least toxic to bees. Thomas *et al.* (1994) found that Oxydemeton methyl was found relatively safe to bee followed by chlorpyriphos and quinalphos.

Sharma and Abrol (2005) evaluated the contact toxicity of 5 insecticides applied to foragers of *A mellfera* and *A cerana indica* and reported that the toxicity of different insecticides after 8 hours of treatment was in the order: malation >cypermethrin>demeton-s-methyl> fenvalerate >deltamethrin for *Apis mellfera* and malathion> demeton-s methyl> cypermethrin>fenvalerate> deltamethrin for *A cerana.*

## POLLINATION RECOMMENDATIONS AND PRACTICES

FAO (1961) reported that honeybees were essential for seed set in crops. Sakharov (1956) also suggested placement of honeybee colonies for quality and quantity of seed production in *Brassicas*. Different workers have recommended variable number of honeybee colonies for increased seed yields. Eckert (I959) recommended two colonies peracre whereas Odland and Noll (1950) considered one colony sufficient for the purpose. However, in general, number of colonies shall vary with strength of colonies, the size of field and competing plants near the crop.

## ECO-FRIENDLY MANAGEMENT OF INSECT PESTS OF COLE CROPS

Cole crops are extensively consumed throughout the world. One of the major problems in the production of vegetables is the attack of insect pests during different growth stages of the crop right from germination to harvest. More than 37 insect pests have been found attacking cole crops (Lal, 1975), the major ones (Mallapur, 1988).include: aphids, *Brevicoryne brassicae* (L.), *Lipaphis erysimi* (Kaltenbach) and *Myzus persicae* (Sulzer), defoliators viz., diamondback moth (DBM), *Plutella xylostella* (Linnaeus); cabbage butterfly, *Pieris brassicae* L.; cabbage semilooper, *Trichoplusia ni* Hubner; tobacco caterpillar, *Spodoptera litura* (Fabricius); cabbage leaf webber, *Crocidolomia binotalis* Zeller and head borer, *Hellula undalis* Fabricius causing damage ranging from 7 to 90 per cent resulting in yield losses from 20 to 80 per cent (Prasad, 1963). Indiscriminate use of pesticides has resulted in several problems which include resurgence of pests, insecticide resistance, killing of natural enemies and environmental pollution (Vastrad et al., 2004). Due to these reasons use of non-chemical methods needs to be encouraged (Patel et al., 2003).

## INSECT PESTS OF CAULIFLOWER

Cauliflower is attacked by a large number of insect pests at different stages of growth (Sachan and Gangwar, 1980) which cause damage from 5-100 per cent (Sachan and Gangwar, 1990). The major insect-pests include cabbage butterfly, *Pieris brassicae.*,cabbage aphid, *Brevicoryne brassicae.*, diamond back moth, *Plutella maculipennis.*, cabbage semilooper, *Trichoplusia ni.*, leaf webber, *Crocidolomia binotalis.* (Anonymous, 1987., 1993., Sharma, 1998). Important insect pests include:.

### Pieris brassicae (cabbage butterfly)

It is one of the serious pest of the crop both at nursery and main crop. Eggs are yellowish in colour laid on the leaves in clusters. The young neonates are greenish to velvety in colour. They have yellow lateral stripes with black spots. The grown up larvae are found feeding gregariously on the plant resulting on their complete defoliation.

*P. Brassicae* is found all over the world wherever cole crops are grown (Tablot 1939). Ratual (1953) found that among all cole crops, cauliflower is the most preferred host of this pest. Ratual (1959) found that under Punjab conditions the pest appeared on cruciferous crops in the beginning of October but their population was very low till January which increased thereafter resulting in extensive damage during March and April. In Himachal Pradesh, India the *P. brassicae* has been reported ro cause 86 per cent on snowball-16 variety of cauliflower (Prasad 1963).

Bhatia *et al.* (1995) recorded six regular pests viz., cabbage butterfly (*Pieris brassicae*), cabbage aphid (*Brevicoryne brassicae*), cabbage semilooper (*Plusia orichaleca*), Bihar hairy caterpillar (*Spilosoma obliqua*), cabbage worm (*Pieris rapaae*) and diamond back month (*Plutella xylostella*) infesting various cole crops in Kangra, Himachal Pradesh.

Sharma *et al.* (2005) reported that the incidence of *P. brassiace* was correlated significantly and positively with relative humidity and rainfall and negatively with temperature (maximum and minimum). The peak population of 12.03 larvae per plant was observed in the 4$^{th}$ standard week of January.

### Brevicoryne brassicae (Cabbage aphid)

*Brevicoryne brassicae* (Cabbage aphid) is another most important pest of cauliflower seed crop (Tandon *et al.* (1977, Gandhale *et al.* (1982, Sachan and Gangawar 1990). Sinha *et al.* (1990) observed that aphid population had a significant and positive correlation with temperature and relative humidity. Bhatia and Verma (1993) reported cabbage aphid as the regular pest of cauliflower besides cabbage butterfly, diamond backmoth, cutworm and semilooper.

Zaz (2001) studied the incidence and population build up of cabbage aphid *Bervicoryne brassicae* on cabbage and cauliflower and reported that the aphid was associated with the cole crops throughout their growing period from April to October with two population peaks on cauliflower (First week of June and July) and cabbage (First week of June and October). The maximum aphid population (17.4 aphids per plant) was recorded on cauliflower followed by cabbage with 52.8 aphids per plant. The per cent plant infestation was in the range of 2-25 per cent and 0-25 per cent in cabbage and cauliflower, respectively.

Sharma *et al.* (2005) carried out investigation on the fluctuations of cabbage aphid (*B. brassicae*) and reported a non-significant negative correlation with temperature (maximum and minimum), relative humidity (evening) and rainfall. They further reported a peak population of 10.30 aphids per plant during the first week of December.

## *Plutella maculipennis* Curtis (Diamond backmoth)

DBM is most serious cosmopolitan pest which causes extensive damage to the crop. The young which are white creamy in colour scrape the green tissues that turn to white patches lateron. The grown up caterpillars on the other hand make holes in the leave resulting in defoliation.

It is an important oligophogous pest of vegetable crops (Hothian, 1965., Vishakantiah and Gowda,1975., Talekar and Griggs, 1986., Robert and Wright, 1996., Lingappa *et al.* 2000). It has been reported to cause 50 to 80 per cent loss in marketable yield (Srinivasan and Krishnakumar 1982). Sachan and Srivastava (1972) reported that the incidence commences from September and gradually increases up to January reaching its peak by February. Per cent plants infested from September to October and January to March varied from 7-8 and 32 to 100, respectively. Bindra *et al.* (1975) observed the activity of diamond backmoth from October to March and reported that the pest attained peak activity during the month of October and again in March.

## *Bagrada cruciferarum* Kirk.(Painted bug)

Painted bug is a serious pest of cruciferous crops. Both nymphs and adults suck cell sap from the leaves and the developing pods which gradually wilts and dries up. The nymphs and adult bugs also excrete a sort of resinous material which spoils the pods (Narayanan, 1954., Batra, 1958., Atwal, 1959., Singh *et al.,* 1980). Atwal (1959) found that the pest was active from March to December with peak population during April and May under Punjab conditions. Batra (1958) found that the pest feeds and breeds on the mustard seed whether on the ground or on the seed causing a reduction of 30.1 and 2.7 per cent in seed weight and oil content, respectively. Singh *et al.* (1993) reported the outbreak of painted bug *Bagrada cruciferarum* on several oilseed crops like *Brassica juncea*; *Brassica carinata*; *Eruca sativa* and *Brassica campestris* from Hisar, India causing extensive damage to these crops.

## Cabbage semilooper *Trichoplusia* ni Hubner

Young caterpillars feed primarily on leaves and cause irregular holes whereas, older one completely skeletonising the leaves leaving only midrib and veins by their feeding on tender tissues. or giving them a ragged appearance. Plants thus appeared severely defoliated and stunted, producing no heads/curds.

## Tobacco caterpillar, *Spodoptera litura* (Fabricius)

The larvae are found in the form of clusters in early stages and individually in later stages. they are slender, green in colour with light wavy lines. Defoliation by *Spodoptera* results in considerable loss to the cros.

### Leaf Webber: *Crocidolomia binotalis*

It ia very serious pest of the plant and the gregarious larvae attack the tender part of the plants in its early stages and in the later stages form webs and remain inside it. The apical portion will become stunted and no formation of head/curd.

### Head borer: *Hellula undalis* Fabricius

The head borers attack the young developing heads of cabbage and curd of cauliflower. It bores deep inside the growing point/head and feed by remaining within. Faecal pellets of the pest give an indication of its presence.if the incidence is severe, huge amount of crop losses can be expected. Under severe incidence of the pest, huge crop loss can be expected.

## RADISH

*Raphanus sativus* L., family Cruciferae

Radish is grown almost entirely for the use of its succulent root as a green salad vegetable.It is a popular home garden vegetable because it is ready to harvest 3 to 6 weeks after planting.

### Plant

The radish plant first produces a relatively small rosette of leaves, compared to the cole crops, mustard and rape, the leaves being only 6 to 18 inches long, and a succulent fleshy taproot l/2 to 2 inches thick and 1 to 12 inches long, depending upon the type and cultivar. After the root growth is completed, the flowering stem elongates to a height of 2 to 3 feet. The root is harvested as soon as possible after it reaches market size. The longer it remains in the soil afterwards, the less tasty it becomes. There is no problem of shattering in the harvest of radish seed, as the pods do not dehisce. The seeds are usually harvested with standard or all-purpose combines (Hawthorn and Pollard 1954).

### Inflorescence

The white to lilac cruciferous flowers are smaller and less showy than those of mustard or rape. Each day of flowering three florets usually appear on the tip of each branch of the panicle. Each flower is capable of producing a pod 1 to 3 inches long and containing one to six seeds (Bailey 1949) or 12 seeds. Kremer (1945) stated that the flower opens during the morning with the corolla remaining fresh throughout the day or into the second day. He indicated that pollen receptivity of the flower was limited to a few hours of the day. Radish is the source of some nectar and pollen

### Pollination Requirements

The pollination of radish was studied by Crane and Mather (1943), Kremer (1945), and Radchenko (1966). The cross-pollination of radish was the main object of the study by Crane and Mather (1943) who found that the 'Icicle' and 'Scarlet Globe' cvs. were self incompatible and that crossing decreased from 30 to 40 percent at 9 inches, to 1 percent at 15 feet, and 0.1 percent at 240 feet. Radchenko (1966) stated that pollination was primarily by honey bees (77 to 94 percent of the total) and that bee pollination increased the seed crop by 22 percent and enhanced seed quality. All seemed to agree with Jones and Rosa (1928) that the radish is almost entirely insect-pollinated.

## Pollinators

Honey bees are the most important agents in the pollination of the radish (Figure 4.2). The studies by Kremer (1945) indicate that the seed yield is largely influenced by the number of honey bees visiting the radish flowers. Radchenko (1966) also reported that honey bees were the main pollinators of radish flowers, accounting for 77 to 99 percent of the total, increasing the crop by 22 percent, and enhancing the seed quality. Crane and Mather (1943) also accredited honey bees with effectively setting the seed crop, noting that the seed set was shows considerable inbreeding depression upon selfing. It is entomophilous, pollinated by wild honey bees, wild-flower flies, bumble bees, some Hymenopterans, Dipterans, Coleopterans, Lepidopterans *etc.* The seed set, numbers of seeds per siliqua and test weight were significantly higher in open pollinated radish plants compared to wind and self-pollinated plants (Verma and Phogat, 1994).

Radechenko (1964) reported that honey bees were the dominant pollinators of winter rape. They comprised of 88.00 to 97.40 per cent from the beginning of bloom to the end of blooming. The other pollinators were Andrenidae, Halictidae and Bombicidae. Kapil *et al.* (1971) reported that *Apis florea* (Fabraceous) was found to be principal visitor of mustard flowers. The other species associated with this crop were *Andrena ilerada* (Cameron), *Apis dorsata* (F.) and Halictine bees. Muhammad *et al.*, (1973) reported that honey bees comprised 77 to 94 per cent of pollinators of radish. Among the honey bees, *A. dorsata* was the highest (41%) followed by *Apis cerana* (F.) (32 %) and lowest with *A. florea* (F.) (21%). Langdridge and Goodman (1975) observed that oil seed rape (*Brassica compestris*) was visited by a large number of insects, of which honeybees were the most ( 32.9 %) followed by hover fly ( 30.7 %), blowflies ( 22.9 %), native bees ( 4.9 %) and others ( 8.8 %) at Victoria. Mesquida (1978) observed that honey bees were the most frequent visitors ( 83 %) on Kale (*Brassica oleracea* L. ) and other pollinators were bumble bees (5.4%) mainly *Bambus terrestris* (Linnaeus), *B. lapidarius* (L.) and other wild bees ( 7.2 %) Kakar (1980) reported that the highest population of *A. cerana* (42.1%) on Cauliflower bloom followed by *Eristalis* sp. (20.9%), *Ceratina* sp. (15.1%), *Halictus* sp. (5.3 %), *Lasioglossum* sp. (4.7 %) and other insects (10.9 %)

Hussein and Abdel- Aal (1982) reported the solitary bees (*Amegilla* spp., *Halictus* spp., *Nomia* spp., *Nomiodes* spp., *Megachile* spp and *Prosopis* spp.) comprised 85 per cent of flower visitor on radish and only 15 per cent were honey bees in Egypt. Bhalla *et al.* (1983) recorded 7 species of insect pollinators *viz., A. indica, Andrena reticulata* (F.), *Lasioglossum* sp., *Eristalis polymacharus* and *E. tenax* (L.) on mustard bloom. Among these, Dipteran were found most abundant visitors (47.08%) followed by honey bees (33.75 %), other insects (9.93 %) and wild bees (9.11 %) Sihag (1986) reported that nine species of Hymenopterans (*Andrena ilerada* Cameron, *A. leaena*, *A. florea, A. dorsata, Apis mellifera* L., *Halictus* sp., *Melisodes* spp., *Xylocopa fenestrata* F. and *Pithtus smargdulal*) were abundant visitors of radish for both pollen and nectar collection and six Lepidopterans (*Euploea core, catopsilia crocale* Cramer, *Papilio sp., Delias sp., Pontia sp. and Dannus sp.)* and two Dipterans (*Eristalis* sp., *Sarcophaga sp.*) were the visitors of radish flowers only for nector collection. Goyal *et al.* (1989) reported that Diptera was the most dominant group visiting carrot followed by Hymenoptera. In Diptera the family Syrphidae was predominant and in the Hymenoptera, Apidae was most abundant. *A. florea* (10.74%) was the most common, followed by *A. mellifera* and *A. cerana indica.*

Panda *et al.* (1989) reported that mustard (var.M- 27) was visited by seven species of insects. Of which, 4 species were members of Apidae (*A. dorsata, A. florea, A. cerana indica* and *Trigona* sp.,) two species of Anthoporidae (*Xylocopa sp.* and *Ceretina hieroglyphica* L.) and one species of Andrenidae (*Andrena ilerada*). Among the different flower visitors, apoidea were the most prominent flower visitors of radish comprising more than 73.82 % and among Apoidea, Indian bee, *A. cerana* was most predominant (47.94%) visitor followed by rock bee, *A. dorsata* (31.17%) and little bee, *A. florea* (20.89%) (Verma and Poghat, 1994).

Sinha *et al.* (1994) reported that mustard flowers were visited by many insect pollinators. Of these, honey bees like *A. dorsata, A. cerana indica* and *A. florea* together constituted 49 per cent and Dipterans 49.5 per cent and other pollinators such as solitary bees and Lepidopterans were less than one per cent. Priti and Sihag (1998) reported that the blossoms of carrot, *Daucus carota* L. were visited by twenty insect species during its flowering period. Hymenopterans and Dipterans constituted 57.66 and 33.31 per cent of total population respectively. The abundance of *A. florea* was maximum followed by flies and other species of honey bees. Bhatia *et al.* (1999) reported 12 species of insects on radish flowers belonging to Hymenopterans (5 species) Dipterans (4 spp.,) and Lepidopterans (2 species)

Thakur *et al.* (2000) reported that the most frequent visitor of turnip was *Eristalis tenax,* followed by *Episyrphus balteatus* Degeer and *A. cerana. E. tenax* also spent less time per flower and was rated as the most efficient pollinator of the crop. Chaudhary (2001) reported the predominance of Apoidea visiting flower of rapeseed and mustard constituting 98.5 per cent of total visitors. Among Apoidea, the social honey bees constituted 59.5 per cent of which the little bee, *A. florea* was the most abundant (42.8%) followed by rock bee, *A. dorsata* (16.6%). Solitary bees constituted 39 per cent of total visitors. Priti *et al.* (2001) reported *A. florea, A. mellifera* as the most important pollinators of radish. Kapil *et al.* (1971) observed that *A. dorsata, A. florea,* and *A. cerana indica* began foraging at 0700 to 1000 hr on rape seed and mustard. Peak foraging activity was recorded between 1230 to 1400 hr and ceased by 1700 hr. Kakar (1981) observed the commencement of foraging activity of honey bees on Cauliflower at 0800 hr and was highest between 1200 and 1400 hr of the day. *A. dorsata* foraged most rapidly visiting on an average 9.95 flowers per minute, followed by 8.05 and 7.60 by *A. mellifera* and *A. cerana indica,* respectively. Landridge and Goodman (1982) showed that bees worked readily on rape seed flower. A single bee worked on as many as 90 flowers in 47 inflorescence in 7 min 12 sec. The average time spent by single bee on one inflorescence was 8.8 sec. In 126 counts, bees visited a mean of 2.0 flowers per head with a range of 1 to 5.

Thakur *et al.* (1982) studied the foraging activity of *A. mellifera* and *A. cerana indica* on mustard at Kangra valley. *A. cerana indica* colonies, considerable activity was built up at 0900 hr when compared to *A. mellifera* which continued till 1000 hr. A decline in foraging activity was noticed soon after 1030 hr in both the honey bee species. However, there was a marginal increase between 1500 to 1530 hr, followed by gradual decline. It was observed that total foraging work was at its peak at 1200 hr. There were smaller peaks at 1400 and 1500 hr.

Sinha and Chakrabarti (1985) reported that honey bees were the major pollinators of cauliflower. Maximum activity of pollinators occurred during 1200 to 1400 hr. *A. dorsata* and *A. cerana indica* remained for longer period (1200 to 1600 hr) on the crop than *A. florea,* even with the temperature between 21.12 to 24.70 C.

Mishra *et al*, (1988) reported that the peak foraging activity of *A. mellifera*, *Eristalis* sp. and other Dipterans were observed between 1200 to 1300 hr, where as *A. cerana indica* and syrphids were observed between 1300 to 1400 hr on mustard flower at Solan, India.

Rana *et al*. (1993) conducted studies on foraging activity of *A. mellifera* and *A. cerana indica* at the hive entrance during rape seed bloom. The results revealed that the returned foragers were significantly maximum in *A. cerana indica* compared to *A. mellifera* at hive entrance. In both species, highest peak foragers wee recorded at 1200 hr and there was no much difference in the number of foragers at 0900 and 1500 hr. Chand *et al*, (1994) recorded that the foraging activity of *A. cerana indica* increased at 1000 hr and reached peak at 1100 hr on mustard bloom. Similarly, the activity of *A. dorsata* increased from 1400 hr and reached the peak at 1600 hr. The number of visits per minute by *A. cerana* was positively correlated with maximum relative humidity and minimum temperature.

Uma and Verma (1994) made observations on *A. cerana* foraging on radish flowers from 0640 to 1839 hr with peak foraging between 1100 and 1400 hr. Workers spent on an average of 4.5,5.3 and 12.8 sec. per flower, visited 8.0,9.0 and 5.0 flowers per minute and collected 11,10 and 7 mg of pollen at 0900, 1200 and 1500 hr, respectively.

Verma and Phogat (1994) reported that the foraging activity of *A. cerana, A. dorsata, A. florea* and other insects were found on radish from morning to evening with peak foraging activity of *A. cerana, A. dorsata, A. florea* and others were observed between 0800 to 1000 hr, respectively. Thereafter the activity declined as the hours of the day progressed.

Rush *et al,* (1995) reported that pollen removal by pollinators in wild radish (*Raphanus raphanistrum* L.) was extremely high with a minimum of 84 per cent of pollen produced removed in one hour. Pollen removal increased with increasing number of visits by honey bees and small native bees, but increased numbers of syrphid fly visits had no effect.

Sattigi *et al.* (1996) noticed the foraging activity of *A. cerana* throughout the day but it was at its peak between 0800 to 1100 hr in winter and 0600 to 1100 hr and 1600 to 1800 hr in summer and 0800 to 1400 hr in monsoon. The foraging activity was low during other timing of the day in different seasons.

Rana *et al*. (1997) reported that the foraging activity of *A. cerana* and *A. mellifera* were significantly higher on rape seed bloom at 1200 hr compared to 0900 hr. However, both the species were found on the crop in equal proportion from 1200 to 1500 hr. The mean numbers of *A. cerana* foragers were significantly higher than the *A. mellifera* foragers. Peak abundance and activity of all the insect pollinators were observed during mid flowering period and at 1200 to 1400 hr. Population of *A. florea* was maximum and remained active for a long duration on the biomass of carrot (Priti and Sihag, 1998).

Sharma and Singh (1999) reported that the *A. florea,* was most dominant visitor of carrot blossom compared to *A. dorsata*, however, *A. florea,* took maximum time for foraging compared to *A. dorsata*. Sihag *et al.* (1999) studied the foraging pattern of *A. dorsata*, *A. mellifera* and *A. florea* on eight cultivars of oil seed crop and reported that in oil seeds, the honey bee visitation frequency was low in the morning and reached peak between 1100 to 1300 hr and again declined in the evening.

Chandel *et al*. (2000) reported that *A. cerana* was observed to start its foraging activity at 0625 hr and *A. mellifera* at 0720 hr on toria. *A. cerana* ceased its foraging at 1730 hr and *A. mellifera* at 1740 hr. Thus, the average duration of foraging was 1118 hr and 1012 hr in *A. cerana* and *A. mellifera,* respectively. The highest number of incoming foragers of

*A. mellifera* (90.7) and *A. cerana* (274.3) was recorded at 1000 to 1030 hr at a temperature of 250 C.

Sharma and Singh (2001) noticed the foraging time of *A. florea* and *A. dorsata* in carrot ecosystem. *A. florea* spent more time (37.09 sec/flower) and visited less number of flowers (2.20/min) as compared to *A. dorsata* that visited more number of flowers (4.31/min) and spent less time (9.20 sec/flower). Shivaramu and Chithiraichelvan (2002) recorded *A. florea, Trigona* sp., *A. cerana, A. mellifera* and wasp on radish flower with peak activity between 1300 hr to 1600 hr at Chettalli. Pandey and Tripathi (2003) reported that *A. cerana indica* showed the highest frequency of visits followed by *A. dorsata, A. mellifera* and *A. florea* in mustard. Regarding the time spent of *A. dorsata* worker took the least time followed by *A. mellifera, A. cerana indica* and *A. florea.*

## Effect of bee pollination on quantitative and qualitative characters of radish seed production

Kermer (1945) reported that radish crop within 3.2 km from honey bee colonies yielded about 448 to 673 kg per hectare whereas, crop which was 3.2 km away from bee colonies yielded 224 to 336 kg per hectare.

Olsson (1952) recorded flower set of 64.7 per cent of cauliflower with 2.46 seeds per pod and 1.75 gram per pod with bees excluded plots whereas bees included plots these values were increased to 95.3, 4.08 and 2.69, respectively. Froster *et al.* (1973) observed that caging *Brassica oleracea* during flowering to exclude honey bees and bumble bees had the lower seed yield, half the weight of seeds per plant and reduced number of flowers per plant by fifth. Landridge and Goodman (1975) observed that the open pollinated plot of oil seed rape produced 50 per cent more seeds and 46 per cent more seed weight per plant than plots from which these insects had been excluded. The weight of 1000 seeds was more in enclosed plots than in open pollinated ones. Lerin (1982) reported that the number of pods and seeds per pod on the main shoot of rape was greatly increased by insect pollination. Insect pollination reduced variability in yield components on a whole plant basis and increases homogeneity of maturity.

Verma and Joshi (1983) reported that the comparison of honey bee pollination with other insects pollination on mustard bloom, the honey bee pollination increased the pod set by 74.30 per cent and number of seeds by 4.07 per pod. The weight of single seed increased by 0.19 mg.

Bisht *et al.* (1983) reported that the rape plants visited by insect pollinators yielded 472 pods per plant and 14 seeds per pod, whereas, plants caged to exclude insects yielded 37 pods per plant and 3 seeds per pod.

Sihag (1986) noticed that the pod length, number of pods per plant, number of seeds per pod, weight of the seeds per plant and seed yield per hectare were significantly higher in open pollinated plots of *Brassica compesris var.toria, Brassica var.prakash, Brassica rapa, Brassica oleracea var.botrytis* and *eruca sativa*) at Hissar.

Khan and Chaudhary (1988) reported that sarson plants pollinated by *A. cerana* had higher seed weight per plant (67.2 to 90.02 g) followed by other insect pollinators (41.72 to 56.35 g) and lowest was record in self pollination (20.46 to 34.34).

Goyal *et al.* (1989) opined that intensive bee pollination in carrot increased the yield from 59.95 to 64.42 per cent in inter-varietal trials and 70.33 to 73.62 per cent in intra-varietal trial over the control. The number of seeds per flower (2.36) per cent seed set (90.47) and per cent seed germination (77.79) in bee pollinated onion crop was significantly higher than the net caged and muslin cloth caged flowers. Seed yield was 83.81, 29.71 and 2.23 times higher under bee pollinated, open pollinated and net caged umbels; respectively than under cloth bagging (Kumar *et al.*, 1989). Prasad *et al.* (1989) reported that the plots without any pollinator had the lowest siliqua per cent (48.3%), which obtained the maximum siliqua length (92.65 mm) as compared to other treatments. The plots access to all the pollinators showed the maximum siliqua per plant (75.8%), which is closely followed by plots caged with honey bees (72.3%). The open pollinated plots had the maximum number of seeds per siliqua (10.58) with highest test weight.

Singh and Chamotre (1992) reported that average number of pods per plant, seeds per siliqua and seed yield per plant was maximum in open pollinated plants of radish (375.55, 6.15and 26.20 g respectively) compared to hand pollinated (222.60, 4.96 and 15.65 respectively) and naturally selfed (90.50, 3.99 and 5.77 g respectively) 1000 seeds weight were maximum in hand pollinated (20.64 g) compared to naturally selfed (20.44g) and open pollinated plants (15.75 g)

Verma and Poghat (1994) revealed that seed set, number of seeds per pod and 100 seed weight were maximum in open pollinated radish, intermediate in wind-pollinated plots and minimum in self-pollinated plots. Uma and Verma (1994) reported that pod set per plant (44.9%), number of seeds per pod (42.3%), 1000 seeds weight (44.5 g) and germination rate (72.7%) were significantly high in bee pollinated radish plants over open-pollination. There was no pod set in control plot.

Chand and Singh (1995) reported that the mustard plots caged without any pollinator had lowest seed yield (966 kg / ha). Whereas, the free access to all the pollinator showed the maximum yield (1620 kg / ha) followed by plots caged with honey bees (1160 kg / ha)

Priti *et al.* (2001) reported that fruit set was 81 per cent and 56.55 per cent, siliqua length 13.2 and 6.8 cm, number of seeds per pod was 8.6 and 4.8 and seed weight 1.5 and 0.9 grams in bee pollinated and self pollinated radish, respectively.

Kapila *et al.* (2002) reported that open pollinated plants of radish and greater number of grains per siliqua, seed yield per plant, 100 seed weight, seed set. And root length than the caged plants, indicating the positive effects of insect pollination on seed production and quality. Abel *et al.* (2003) observed that in *Brassica rapa* accession PI 392025 and *Brassica napus* accession PI 469944, produced significantly more seeds when pollinated by *Osmia lanigera sub sp. Lignaria* than *Osmia cornifrons* Radoszkowski.

## Pollination Recommendations and Practices

Radish is a cross-pollinated crop which depends upon a variety of insects for pollination which include wild honey bees, wild-flower flies, bumble bees, some Hymenopterans, Dipterans, Coleopterans, Lepidopterans *etc.* Kremer (1945) emphasized the need of honeybee colonies near the raddish fields. The seed set, numbers of seeds per siliqua and test weight were significantly higher in open pollinated radish plants compared to wind and self-pollinated plants (Verma and Phogat, 1994).

The crop is also attacked by number of insect pests which cause considerable damage. The important ones include Mustard sawfly, *Athalia lugens proxima* (Klug), aphid, cabbage looper and flea beetle (Srivastava *et al.*, 1972). The other insect pests repoted attacking radish are Bihar hairy caterpillar *Spilosoma obliqua* (Walker), Leaf feeder/ Plume Moth, *Pericallia ricini* (Fabricius), Leaf webber *Crocidolomia binotalis* Zeller,

Cabbage Borer *Hellula undalis* (Fabricius), Defoliator/Cabbage butterfly *Pieris brassicae* (Linnaeus), Diamond back Moth, *Plutella xylostella* (Linnaeus), Leaf feeder/Green Minor T*hysanoplusia orichalcea* (Fabricius)(Sharma 2011).

Evidently, crop production is simultaneously confronted by two problems which are antagonist to each other. One is protection of crop against insect pests and another safety of pollinators for pollination of crops. Evidently, there is need to explore the integrated approach protecting crop on one hand and safety of pollinators on the other.

## INSECT PESTS ASSOCIATED WITH RADISH

Raddish crop is also attacked by number of insect pests which cause considerable damage. Mustard sawfly, *Athalia lugens proxima* (Klug) is its most serious pest observed at seedling stage (Srivastava *et al*., 1972). Rohilla and Kumar (1991) reported that sometimes crop is damaged to such an extent that warrants resowing. The other pests attacking radddish include: aphid, cabbage looper and flea beetle. Sharma(2011) reported several insect pests attacking radish which include Bihar hairy caterpillar *Spilosoma obliqua* (Walker), Leaf feeder/Plume Moth, *Pericallia ricini* (Fabricius), Leaf webber *Crocidolomia binotalis* Zeller, Cabbage Borer *Hellula undalis* (Fabricius), Defoliator/Cabbage butterfly *Pieris brassicae* (Linnaeus), Diamond back Moth, *Plutella xylostella* (Linnaeus), Leaf feeder/ Green Minor T*hysanoplusia orichalcea* (Fabricius). The cabbage root fly drills into the plant at the top of the radish resulting in its withering followed by death.

## EVALUATION OF SOME INSECTICIDES AGAINST INSECT PESTS OF RADISH

Atwal and Singh (1969) reported that that carbaryl (0.15%) recorded the highest mortality and thuricide was the next the order of effectiveness against *P.brassicae*. Butani *et al.* (1977) found recommended that the spraying of crop with dichlorvos 0.05 per cent or malathion 0.1 per cent or dusting with carbaryl or malation 5 per cent were effective for the control of *P. Brassicae.*

Mumtaz *et al.* (1991) found that fenvalerate (0.01%) was most effective against *Pieris brassicae* on cabbage followed by malathion (0.1 %) and dichlorvos (0.01 %). Singh (1999) on the other hand reported that chlorpyriphos (0.05 %) significantly reduced the pest population.

Narkiewicz-Judko (1992) recommended carbosulfan 25 EC (0.6 kg/ha) for the control of *P. brassicae* and many other insect-pests of cabbage. Ramesh (1992) found carbaryl, endosulfan, phosphamidon dimethoate, malathion and fenetrothion for the control of cabbage butterfly and concluded that endosulfan (0.05 and 0.07 %) proved effective in minimizing the pest population, whereas, the yields were significantly higher from the plots treated with fenitrothion (0.025 and 0.05 %) followed by cabaryl (0.15 %) and endosulfan (0.07 %) as compared to other insecticides. Ram and Pathak (1992) found that carbaryl, endosulfan, phosphamidon, dimothoate, malathion and fenitrothion were effective in reducing *Pieris brassicae* infestation.

Orita (2005) reported neonicotinoids namely imidacloprid, nitenpyram, acetamiprid, and thiamethoxam were highly toxic to adults of *Aphidoletes aphidimyza*, wheras chlorfenapyr, cartap, and spinosad showed high toxicity. Similarly, pyrethroid showed high toxiocity to laevae pesticides showed high toxicity to larvae. Insect growth regulators except novaluron and acaricides except tebufenpyrad showed no toxicity.

## IMPACT OF INSECTICIDES ON BEE VISITS

Insecticides have been found to adversely affect the bee visits. Dimethoate, phosphamidon monocrotophos, oxy-demeton methyl and malathion have been reported to be more toxic than endosulfan to *A melhfera* (Sihag et al. 1999). Sharma and Abrol (2005) reported malathion to be most toxic to both *A mellfrea* and *A cerana.*

Sharma and Abrol (2005) evaluated the contact toxicity of 5 insecticides applied to foragers of *A mellfera* and *A cerana indica* and reported that the toxicity of different insecticides after 8 hours of treatment was in the order: malation >cypermethrin>demeton-s-methyl> fenvalerate >deltamethrin for *Apis mellfera* and malathion> demeton-s methyl> cypermethrin>fenvalerate> deltamethrin for *A cerana.*

Since the neonicotinoids are systemic in nature, they have an advantage of being used as seed treatment which helps in reducing direct contact to pollinator fauna unlike spray application. On other side being systemic in nature these chemical got absorbed in plant system and translocated into nectar and pollen thereby these become hazardous towards flower visiting insects.

## References

Abrol, D.P. 1989. Studies on abundance, diversity, behaviour and importance of native pollinators for crop production. Korean Journal of Apiulture. 4 (2): 25-4

Abrol, D.P.1991. Conservation of pollinators for promotion of agricultural production in India. Journal of Animal. Morphology and Physiology 38(1x2) : 123-139.

Abrol, D.P.1997 Bees and Beekeeping in India. Kalyani Publishers Ludhiana, India, 450p. 1997

Ahuja I., Rohloff J. and Bones A.M. 2010. Defense mechanisms of Brassicaceae: implications for plant-insect interactions and potential for integrated pest management. A review. Agronomy for Sustainable Development, 30: 311-348.

Alam, M.Z., Quadir, M.A. and Ali, M.S. 1987. Pollinating behaviour of honeybee Apis indica F. and its influence on seed production of cauliflower. Bangledesh Horticulutre 15: 25-30.

Anonymous, 1987. Package of Practices of Vegetables. SKUAST-Kashmir. pp45.

Anonymous, 2005. National Horticulture Board, data base, Gurgoan.

Anstey, T. H. 1954. Self-incompatibility in green sprouting broccoli (Brassica oleracea L., var. Italica plenck) 1. Its occurrence and possible use in a breeding program. Canad. Jour. Agr. Sci. 34: 59-64.

Atkinson W. T., and constable, E. E. 1937 A honey bee technique in seed production of selected cruciferous plants. Australasian beekeeper 39(6): 183-185.

Attia M. S., and monger, H. M. 1950. Self-incompatibility and the production of hybrid cabbage seed. Amer. Soc. Hort. Sci. Proc. 56: 363 - 368.

Atwal, A. S.1959. The oviposition behaviour of Bagrada crciferarum Kirkaldy (Pentatomidae: Heteroptera) and the influence of temperature and humidity on the speed of development of eggs. Proceedings National Institute of Sciences India 25B (2):65-67.

Bajwa, W. I. and M. Kogan 2002. Compendium of IPM Definitions (CID)- What is IPM and how is it defined in the Worldwide Literature? IPPC Publication, 998.

Bateman A. J. 1947-48. Contamination of seed crops: 1. Insect pollination. Jour. Genet. 48: 257-275.

Bhalla, S., Kapur, M. L, Verma, B. R and Singh, R. 1997. An unusual abundance of cabbage butterfly, Pieris brassicae L. on various Brassica species on the environment of Delhi for locating sources of resistance among different cultivars. Journal of Entomological Research. 21(2): 147-151.

Bhatia, R and Verma, A. 1993. Insect-Pest complex of cauliflower in Himachal Pradesh. Journal of Insect Science. 6(2): 297-299.

Bhatia, R., Gupta D. and Pathania, N. K. 1995. Host preference and population build up of key pests of cole crop. Journal of Insect Science 8(1) : 59-62.

Borchers E. A. 1968 yield, uniformity of heading and season of maturity of broccoli inbreds, hybrids and varieties. Amer. Soc. Hort. Sci. Proc. 93: 352 - 355.

Borchers, E. A. 1971. Hybrid broccoli seed production utilizing the m6 gene for male sterility. Amer. Soc. Hort. Sci. Proc. 96: 542-543.

Cole, K., 1959. Inheritance of male sterility in green sprouting broccoli (Brassica oleraceal)., var. Italica. Amer. Soc. Hort. Sci. Proc. 95(1): 13-14

Cole, K. 1959. Inheritance of male-sterility in green sprouting broccoli. Canad. Jour. Genet. Cytol. 1: 203-207.

Crane, E. 1990. Bees and Beekeeping. Bath press Ltd. Avon. pp 68-69.

Crane, M. B., and Mather, K. 1943. The natural cross-pollination of crop plants with particular reference to radish. Ann. Appl. Biol. 30: 301-308.

Detjen, L. R. 1927. Sterility in the common cabbage (brassica oleracea l.). Hort. Soc. N.y. Mem. 3: 277-280.

Dickson. M. H. 1970. A temperature sensitive male sterile gene in broccoli

East, E. M. 1940. The distribution of self-sterility in the flowering plants. Amer. Phil. Soc. Proc. 82: 449-518.

Faulkner, G. J. 1962. Blowflies as pollinators of brassica crops. Com. Grower [england] 3457: 807-809.

Free, J.B.1993. Insect pollination of crops. Academic Press, London U.K

Gandhale, D. N., Salunkhe, G. N and Naik, L. M.1982. Effectiveness of insecticides for the control of aphids (Brevicoryne brassicae L). on cabbage. Indian Journal of Plant Protection. 10: 85-86.

Garcia, G. M. 1954. A preliminary study of the production of cauliflower seed. Philippine jour. Agr. 19: 143-152.

Haskell, G. 1943. Spatial isolation of seed crops. Nature 152: 591-592.

Jayarathnam, K. 1977. Studies on the population dynamics of Plutella xylostella (L.) (Lepidoptera: Yponomeutidae) and crop loss due to the pest in cabbage Ph. D Thesis. University of Agriculture Sciences, Banglore. pp215.

Jenkinson, J. G., and glynne-jones, g. D. 1953. Observations on the pollination of oil rape and broccoli. Bee world 34: 173 - 177.

Johnston, T. D. 1964. Inbreeding and hybrid production in marrow-stem kale (brassica oleracea l., var. Acephala d.d.). Euphytica 13: 147-152.

Kakar, K. L., 1981, Foraging behaviour of insect pollinators of cauliflower bloom. Indian Journal of Ecology, 8 (1) : 126-130.

Kakizaki, Y. 1922. Self-sterility in chinese cabbage. Jour. Hered. 13: 374 - 376.

Kakkar and Sharma (1991) Studies on the role of honeybee Apis cerana indica F. in the pollination of cauliflower (Brassica oleracea L. var. botrytis). Indian Journal of Entomology 53:66-69.

Knott, J. E. 1949. Vegetable growing. Ed. 4, 314 pp. Lea and eebiger, philadelphia.

Kremer, J. C. 1945. Influence of honey bee habits on radish seed yield. Mich. Agr. Expt. Sta. Quart. Bul. 27: 413-420.

Kumar J., Gupta, J.K., Mishra, R.C. and Dogra, G.S. 1988. Pollination studies in some cultivars of cauliflower (Brassica oleracea var. botrytis L.) Indian Bee J. 50: 93-95.

Lal, O. P. 1975. A Compendium of insect pest of vegetables in India. Bulletin of Entomology, 16: 31-56.

Legg, P. D., and soother, E. D. 1968. Heterosis in intervarietal crosses in broccoli (brassica oleracea var. Italica). Amer. Soc. Hort. Sci. Proc. 92: 432-437.

Lerin, J. 1982. Effect of insect pollination on rape in a cage experiment, Agronomie 2(3): 249–255.

Mallapur, C. P. (1988). Studies on seasonal incidence and chemical control of cabbage pests. M. Sc. (Agri) Thesis, University of Agricultural Sciences, Dharwad. p.114.

Matin, M. A, Saghir, S. A and Rafi, M. A. 1992. Quantitative relationship between Pieris brassicae population and damage in cauliflower. Philippine Entomologist. 8(5): 112-18

Mohan, C. N. 1994. Seasonal incidence of DBM: Plutella xylostella and its parasitoids in Nilgiri. Journal of Biological Control. 8(2): 77-80.

Moore, J. F., and anstey, t. H. 1954. A study of the degree of natural selfing in green sprouting broccoli (brassica oleraceae l., var. Italica plenck) a normally cross-pollinated crop. Amer. Soc. Hort. Sci. Proc. 63: 440 - 442.

Mushtaque, M. 1989. Control of large cabbage butterfly in Sialkot by local parasites. Progressive Farming. 9(4): 38-40.

Mushtaque, M. and Mohyuddin, M. 1984. Pieris brassicae (Pieridae: Lepidoptera) a pest of crucifers and its control by parasites. Pakistan Journal of Zoology. 18(1): 121-27.

Nieuwhof, M. 1961. Male sterility in some cole crops. Euphytica 10: 351-356.

Nieuwhof, M. 1963. Pollination and contamination of brassica oleracea l. Euphytica 12: 17-26.

Nieuwhof, M. 1969. Cole crops. 353 pp. Leonard hill, london.

Odland, M. L., and noll, c. J. 1950. The utilization of cross compatibility and selfincompatibility in the production of f1 hybrid cabbage. Amer. Hort. Sci. Soc. Proc. 55, 391-402.

Oldham, C. H. 1948. Brassica crops and allied cruciferous crops. 295 pp. Lockwood, london.

Patel, P. S., Shukla, N. P. and Patel, G. M. 2003. Enhancing insecticidal properties of cow urine against sucking pests of cotton. In : Proceedings of the National Symposium on Fronterier Areas of Entomological Research, 5-7 November, 2003, pp. 385-386.

Pearson O.H. 1933. Study of the life history of brassica oleracea. Bot. Gaz. 94: 534-550.

Pearson, O. H. 1930. Observations on the type of sterility in brassica oleracea var. Capitata. Amer. Soc. Hort. Sci. Proc. 34 - 38.

Pearson, O.H. 1932. Breeding plants of the cabbage group. Calif. Agr. Expt. Sta. Bul. 532, 22 pp.

Pickett J.A., Wadhams L.J. and Woodcock C.M. 1997. Developing sustainable pest control from chemical ecology. Agr. Ecosyst. Environ. 64:149–156.

Prasad, K and Phadke, K.G.1986.Variability in the incidence of mustard aphid, lipaphis erysimi(Kaltenbach) on different varities /species of rapeseed and mustard crop. Indian J.Ent. 49(2):222-225.

Prasad, S. K. 1963. Quantitative estimation of damage to crucifers caused by cabbage worm. Cabbage looper, diamondback moth and aphids. Indian Journal of Entomology, 25 : 242-261.

Priestley G. 1954. Use of honey bees as pollinators in unheated glasshouses. New zeal. Jour. Sci. And technol. 36(3): 232 - 236.

Priti and Sihag, R. C.1997.Diversity, visitation frequency, foraging behaviour and pollination efficiency of insect pollinating cauliflower (Brassica oleracea L. var. botrytis cv. Hazipur Local) blossoms. Indian Bee Journal. 59(4):230-237.

Radchenko, T. H. 1966. [role of honey bees as pollinators in increasing the seed crop from cabbage and radish.] Bdzhil'nytstvo 2: 72-74. [in ukrainian.] Aa- 390/69.

Radchenko, T. H. 1966. [Role of honey bees as pollinators in increasing the seed crop from cabbage and radish.] Bdzhil'nitstvo 2: 72-74.

Rai, K. M., Gupta, B. P. and Tripathi, G. M. 1985. Biology and control of cabbage white butterfly (Pieris brassicae Linn.) (Lepidoptera: Pieridae). Progressive Horticulture. 17(4): 376-81.

Sachan, J. N and Gangwar, S. K. 1980. Vertical distribution of important insect-pest of cole crops in Meghalaya as influenced by the environment factors. Indian Journal of Entomology. 42 (3) : 414-421

Sachan, J. N. and Gangwar, S. K 1990. Seasonal incidence of insect-pests of cabbage. Indian Journal of Entomology. 52(1) : 111-124.

Sachan, J.N. and Srivastava, B.P. 1972. Studies on seasonal incidence of insect pest of cabbage. Indian Journal of Entomology. 34: 123-127.

Sakharov, M. K. 1956. [cabbage pollination by bees.] In krishchunas, i. V., and gubin, a. F., eds. [pollination of agricultural

Sharma, D.2005. Relative toxicity of some insecticides to brood and adult honeybee foragers (Apis mellifera L.). Ph.D thesis, SKUAST JAMMU

Sharma, P.L., Dhaliwal, H.S. and Kakar, K.L. 1974. Insect visitors and pollinators of cauliflower (Brassica oleracea var. botrytis) seed crop bloom. Him. J. Agric. Res. 2:74-78.

Sharma, S. 1998. Integrated management of major insect pests of cauliflower, Brassica oleracea var. botrytis in Jammu. MSc. Thesis, SKUAST-Jammu. pp110.

Sihag 1985 Floral biology, melittophily and pollination ecology of cultivated cruciferous crops. In : Verghese, T. M.(ed.). Recent Advances in Pollen Research. Allied publisher. New Delhi. pp 241-258.

Sihag, R.C. 1986 Insect pollination increases seed production in cruciferous and umbelliferous crops. J. Apic. Res. 25: 121-126.

Sihag, R.C. 1988. Effect of pesticides and bee pollination on seed yield of some crops in India. Journal of Apicultrual Research. 27(1) : 49-54.

Singh, M.P., Singh, K.I and Sarangthen, A.K. 1993. Studies On the effect of bee pollination on yield of rapeseed. Proc. Int. Symp. Poll. Trop. Banglore. pp. 211-212.

Sinha, R. P., Yazadani, S. S and Verma, S. S. 1990. Population dynamics of mustard aphid, Lipaphis erysimi (Kalt.) (Homoptera:Aphididae) in relation to ecological parameters. Indian Journal of Entomology. 5(3) : 387-392.

Sokal, R.R and Rholf, F.J.1981. Biometry, 2nd Edition San Francissco

Tandon, P. L., Bhalla, O. P and Verma, A. K. 1977. A note on the seasonal incidence and population fluctuations of cabbage aphid on cauliflower. Vegetable Science. 4: 145-157.

Tewari, G.N. and Singh, K. 1983. Studies on insect pollinators in relation to seed production in cauliflower (Brassica oleracea var. botrytis L.) Indian Bee J. 45: 54-55.

Thakur, N. S. A. 1996. Relationship of cabbage butterfly larval (Pieris brassicae Linn.) population on the marketable yield of cabbage. Journal of Hill Research. 9(2) : 356-58.

Varma, S.K. and Joshi, N.K. 1983. Studies on the role of honey bees in the pollination of cauliflower (Brassica oleracea var. botrytis). Indian Bee J. 45: 57-58.

Vastrad, A. S., Lingappa, S. and Basavana Goud, K. 2004. Ovicides for managing resistant populations of Diamondback Moth, Plutella xylostella L.Resistant Pest Management Newsletter, 14 : 16-17.

Verma L R and Uma Partap 1994. Foraging behaviour of Apis cerana on cauliflower and cabbage and its impact on seed production Journal of Apicultural Research 33 (4) 231-23

Zaz, G. M. 2001. Incidence and population build up of cabbage aphid *Brevicoryne brassicae* L. on cabbage and cauliflower. Applied Biological Research. 3(1/2 ) : 51-53.

*Apis dorsata* **working on raddish flowers**

**Figure 4.2: Insect pollinators of radish**

CHAPTER

5

# Solanaceous

## TOMATO

Greenhouse tomato (*Lycopersicon esculentum* Mill. (Solanaceae)) is an economically important crop. Plant yield and fruit quality is very much dependent on pollination and without pollination fruit quality is substantially reduced (Fletcher and Gregg 1907). The high value of greenhouse tomato can be attributed to the quality parameters such as roundness, size and sugar content of the tomatoes produced. Although tomato is self fertile (Fletcher and Gregg 1907) and normally pollinated by both wind and insects (Free 1970), a greenhouse environment excludes both sources of pollination. Thus, pollination is an important challenge for greenhouse tomato producers.

### The Flowers of Tomato and their Pollination

Tomato is monoecious and has hermaphroditic flowers. The tomato (*Lycopersicon esculentum*) flower is pendulous, about 3/4 in (2 cm) in diameter and perfect, containing both stamens and a pistil. There are six yellow petals and six stamens that unite at their anthers to form a cone surrounding the pistil. Inside the pistil is a five- to nine-chambered ovary. The style may be shorter than the tip of the anther cone or longer, depending on the variety. The stigma is receptive from 1–2 days before to 4–8 days after its own flower releases pollen; thus cross-pollination is possible. However, anthers release pollen inwardly towards the style, and vibration-assisted self-pollination is the norm especially in short-style varieties. In long-style varieties, the downward posture of the flower promotes self-pollination by gravity. The anther cone releases pollen around the stigma at the slightest vibration, whether from wind or insects. Tomato flowers produce pollen but little or no nectar.

Nectar secretion from tomato plants is apparently of little, if any, value in attracting bees. Schneck (1928) referred to "the absence of nectar" in the tomato flower. Neiswander (1954a, 1956) stated that the blossom "contains little or no nectar." Fink (1898) reported that bumble bees "gathered chiefly pollen" from tomato flowers. Thus, if nectar is produced, a question that should be settled, it is of little significance in the relation of insect pollination of tomatoes. The pollen is more attractive to wild bees than honey bees.

At first glance flowers seem to have only male reproductive structures because their anthers are fused forming a cone that encloses the stigma and style (Figure 5.1). The anthers of tomato are poricidal, meaning that the flower's pollen is produced within the anthers and released through small pores (Buchmann 1983, Plowright and Laverty 1987). As with poricidal anthers of other plants, tomato's anthers require agitation to release pollen (Buchmann 1983). By restricting pollen release, poricidal anthers may help to ensure that the flower's pollen is distributed to many pollinators and thus potentially many flowers (Harder and Barclay 1994). For tomato, however, pollen release alone is often adequate for pollination because the plant's flowers are self fertile and fruit quality (such as increased fruit weight) is not improved by cross-pollination (Fletcher and Gregg 1907). Regardless of whether tomato is self- or cross-pollinated, the agitation necessary for pollination is facilitated generally by insects or wind (Free 1970).

Figure 5.1. Structure of the Tomato Flower. The tomato flower looks as though it only has male reproductive structures (A,B) because the anther cone obstructs view of the female reproductive structures, the stigma, style and ovary (C: anthers removed).

## Tomato Pollination Requirements

Tomato does not produce fruit without pollination, and inadequate pollination results in low quality misshapen fruit (Fletcher and Gregg 1907). Increasing pollination activity is known to affect positively many desired characteristics of tomato fruit quality including roundness, weight and the number of seeds present (Fletcher and Gregg 1907, Sawhney and Dabbs 1978, Kevan et al. 1991, Morandin et al. 2001b). Through optimizing pollination in tomato crops, growers help to ensure that their plants are productive and that the fruit produced are of the highest quality. Unfortunately, tomato's poricidal anthers pose a pollination challenge for greenhouse growers. Greenhouse environments are artificial and exclude tomato's natural sources of agitation such as wind and pollinating insects. Thus, tomatoes within a greenhouse do not typically achieve adequate pollination without human-introduced agitation. Until the early 1990s greenhouse growers manually pollinated their tomato flowers using hand-held electric vibrators (Banda and Paxton 1991, Kevan et al. 1991). Because pollinating a tomato crop in this way was not labour-efficient, other options were examined. The use of insect pollinators within greenhouses posed an attractive and eventually cost-effective alternative to manual pollination (Velthuis and van Doorn 2006).

Tomato is self-fertile, and fruit-set is similar in self- or cross-pollinated plants (Free, 1993). The tomato flower must be shaken or vibrated in order to release pollen and self-pollinate. Natural cross-pollination in commercial varieties is low, ranging from 0.07% in California (Groenewegen *et al.*, 1994) to 12% in Mexico (Richardson and Alvarez, 1957) and is more common in long-style varieties because the stigma is more accessible to foreign pollen. Pollination can be a limiting factor in wind-free glasshouses with long-style varieties because they are not as easily self-pollinated (Wenholz, 1933).

Each seed requires one grain of pollen, and small misshapen fruit will result if the pollen transfer rate to stigmas is poor. The size and weight of fruit increase as the number of developing seeds increases. Fruit-set and the number of seeds per fruit increase the more often a flower is pollinated, up to four times (Verkerk, 1957). Male-sterility exists in tomato and is used to produce hybrid seed. This requires cross-pollination, but neither wind nor mechanical vibrators can cross-pollinate male-sterile flowers (McGregor, 1976). Thus, insects are necessary for hybrid seed production. Rate of crossing quickly decreases as the distance from the pollen source increases (Currence and Jenkins, 1942), and little pollen is transferred beyond 98 ft (30 m) from its source (Quiros and Marcias, 1978).

Various tests in greenhouses have proven that the tomato flower is not self pollinating. However, if the inflorescence is shaken, the pollen will fall from the anthers onto the stigma and fertilization will result. One pollen grain is needed for each seed, so many grains are needed on each stigma. Incomplete pollination results in misshapen fruit. Cool or cloudy weather retards pollen shedding (Stoner 1971). Growers of tomatoes in greenhouses use various types of vibrators or other devices each few days to shake the flower clusters (Beattie 1939; Bouquet 1919, 1924; Cottrell-Dormer 1945; Fletcher and Gregg 190 7; Hoffman 1958; Kerr and Kribs 1945; Lesley and Lesley 1939; Moore 1968; Neiswander 1954a, b, 1956; Ross 1963; Verkerk 1957; White 1918; Wittwer and Honma 1969). Moore (1968) obtained only 4.3 pounds of fruit from control plants in a plastic greenhouse; 6.6 pounds, from hormone treated plants; 8.8 pounds, from vibrator treated plants; and 9.8 pounds, from plants treated with both hormones and vibration. For maximum effectiveness, vibration must be repeated every 2 or 3 days.

Although Bailey and Lodeman (1895) concluded that bees in the greenhouse were of no value as pollinators of tomatoes, Neiswander (1954a, b, 1966) found that visits of honey bees increased fruit production even though the flowers had also received the shaking treatments. Marr and Hillyer ( 1968) showed that self-pollinated plants (in greenhouses) yielded less and had more misshapen fruit than crossed plants. Jones (1916) observed 1.98 percent cross-pollination of tomatoes in New Jersey and estimated that an equal undetected amount of crossing occurred. This figure has been frequently used over the years, without regard to the insect pollinator population or variety under test. He stated that he "saw no insects," although others have associated cross-pollination with insects almost exclusively. Tomato flowers in the open are usually considered to be sufficiently vibrated by wind currents to cause the pollen to fall onto the stigma and affect maximum set (Lesley and Lesley 1939).

Wind is not a factor in transferring pollen from plant to plant (Currence and Jenkins 1942); however, if the weather remains calm or if the blossom is so situated on the plant that it is not vibrated by the wind its pollination would conceivably be prevented. Under such conditions, visitation by pollinating insects would be beneficial. Cross-pollination in the field is common although the percentage is usually low (Azzam 1960; Lesley 1924; Purseglove 1968; Richardson and Alvarez 1957a; Rick 1947, 1949, 1950; Schneck 1928; Smith 1935; Soost and Rick 1957). A factor favoring self-pollination of the tomato is the relatively long time that the stigma is receptive to pollen, from 1 to 2 days before anther dehiscence to 4 to 8 days after dehiscence (Smith 1935). Another factor is the length of the stigma. If the style is short and the stigma is surrounded by dehiscing anthers, selfing after vibration is most likely. If the style is long or if it grows through the anther tube before pollen is shed, its likelihood of being cross-pollinated is increased. Jones and Rosa (1928) stated: "In some varieties, however, and probably in some flowers of all varieties, the style elongates before the anthers dehisce, thus exposing the stigma to foreign pollen." Regardless of how the pollen is applied, the more pollen (within limits) the larger the fruit (Fink 1898), and the more symmetrical it is (Hoffman 1958).

## Tomato Pollinators

Wind and insects provide the vibrating action necessary for self-pollination under field conditions. But these natural phenomena may not result in maximum potential yield. Fruit weight, fruit diameter, number of seeds, and marketable yield were increased in field-grown plants of 'Celebrity', 'Heatwave', and 'Sunny' that were regularly shaken vigorously with an air-blower device, compared to that of plants kept in ambient wind (Hanna, 1999).

For plants grown in still glasshouses it is necessary for growers to use mechanical vibrating devices or imported bees. The mechanical vibrator, also called vibrating wand or electric bee, when applied to a plant tree can shake pollen free so that it lands on stigmas. However, not all flowers mature at the same time which means that wand treatments must be repeated several times a week. Insects, especially bumble bees, can be cost-effective compared to the labour expense of mechanical pollination; plus, insect pollination is more-or-less continuous rather than intermittent. The use of insect pollinators in glasshouses, however, depends upon pest control practices that protect bees from insecticides. In an unheated glasshouse in Israel, mechanical vibrators and bumble bees had similar effects on fruit set, seed number, fruit size, and yield during autumn and early winter, but the bees were more efficient during cooler winter temperatures than were applications of the vibrator 2–3 times per week (Pressman *et al*., 1999).

Bumble bees are the most important insect pollinator of tomato, especially in glasshouses. Practical methods for rearing bumble bees year-round were developed in the 1980s, motivated by year-round demand for pollinators of glasshouse tomato. Bumble bees forage normally in glasshouses, and once released in a glasshouse they quickly discover and work tomato flowers. They work flowers when glasshouse temperatures are cool in winter and spring, and they work glasshouse flowers even when the windows are open in summer. The tomato flower is well suited for buzz-pollination, and bumble bees, unlike honey bees, are capable of buzz-pollinating.

Studies on the influence of honey bee visits to tomato flowers in polyethylene greenhouses (Spangler and Moffett, 1977) showed that *A. mellifera* erratically visited tomato (*Lycopersicon esculentum*) flowers growing in double layered polyethylene greenhouses. The percentage of flowers visited was generally low initially, but increased as the plants matured, then decreased as the plants aged. Flowers of Florida tomatoes exposed to visitation produced significantly more tomatoes by weight (22%) than flowers which were protected from visits by cheese cloth bags. Dogterom *et al.* (1998) observed that *Bombus vosnesenskii* was an effective pollinator of tomatoes in greenhouses. Bumblebee pollinated flowers produced larger fruit than non-bumble bee pollinated flowers. Higo *et al.* (2004) conducted studies on honey bee distribution and potential for supplementary pollination in commercial tomato green houses during winter. In one green house significantly larger tomatoes were produced with honey bees present compared with bumble bees alone. Quagliotti and Marletto (1987) conducted research on the pollination of runner bean (*Phaseolus coccineus* L.) for dry grain production and found that seed production in the plants to which bees had no access to be very low.

Bumble bees are very effective pollinators of glasshouse tomato and the most popular choice of growers. Bumble bee colonies in glasshouses should be fed sugar syrup in order to compensate for the lack of nectar in tomato.

A bumble bee colony has, at most, only a few hundred individuals, and colonies may need to be replaced at intervals during a season. The rearing methods are highly specialized and proprietary. For these reasons, honey bees may retain a role as secondary pollinators of glasshouse tomato, owing to their large colony populations and manageability.

In tomato glasshouses, honey bees are disorientated at first but eventually adopt a more organized foraging pattern (Cribb *et al.*, 1993). Honey bees visit relatively few tomato flowers, and because they cannot sonicate the flowers it is difficult for them to collect the pollen. Bee mortality is high, and colonies typically need supplemental sugar syrup and frames of pollen from donor colonies while they are in the glasshouses. Inspite of these drawbacks, honey bees compare favourably with mechanical vibrators, at least with varieties 'Criterium' and 'Gold Star'. The comparative performance of honey bees and vibrators observed by Cribb *et al.* (1993) differs from that found by Banda and Paxton. Thus, honey bee efficacy on glasshouse tomato may vary across growing conditions and varieties. There are numerous native bees that pollinate field tomato in the New World tropics, the native range of tomato (Free, 1993).

Pollinators can be scarce in other parts of the world (Bohart and Todd, 1961; McGregor, 1976). Although honey bees visit tomato in glasshouses when there is no other choice, they rarely visit tomato in the field. Plant breeders may do well to select for tomato varieties that are more attractive to honey bees and bumble bees. Larger, brighter flowers, functional nectaries, and anthers that allow easier access to pollen would help greatly (Cribb *et al.*, 1993). There are no recommended bee densities for field tomato. For glasshouses, growers should aim for 4–6 bumble bee colonies peracre (10–15 ha) (van Ravestijn and van der Sande, 1991).

In greenhouses, the various types of mechanical vibrators are satisfactory pollinators. In the field, the wind vibrates the plants. Neither of these methods contributes to the pollination of male-sterile plants. Only insects can serve in this capacity. They also contribute to pollination of those plants or cultivars with styles that extend beyond the stigma. Currence (1944) showed that use of hybrids could increase yields by 20 percent, and he reported finding a male-sterile plant that set good crops from artificial pollination. The value of hybrid seed has been estimated $8 per ounce (Barrows and Lucas 1942). This might be decreased if the grower could incorporate a seedling marker to aid in weeding out non hybrids (Hafen and Stevenson 1956). Others (Hojby 1958, Kerr 1955, Oba et al. 1945,

Roever 1948, and Wellington 1912) have shown the value of hybrid tomato production. Shifriss (1945) reported the production of hybrid tomato seeds, produced by the relatively inexpensive labor of college girls. Kerr (1955) associated greater numbers of seeds with larger and more rapid fruit development. Where hand pollination is impractical, insects can be used. Richardson and Alvarez (1957a, b) considered the Halictid bee (*Augochloropsis ignita* Smith) the most effective pollinator in their area. Bullard and Stevenson (1953) considered neither houseflies, blowflies, nor honey bees of value under cheese cloth cages over six plants. Azzam (1960) observed few bumble bees on tomato flowers in Puerto Rico but several hundred *Examalopsis glubosa* F. bees. It is generally known, however, that a few honey bees in such a cage do not act normally.

Fletcher and Gregg (1907) hinted that honey bees might be used to distribute tomato pollen. Lesley and Lesley (1939) indicated that "bumble bees and other insects also assist." Neiswander (1954a, b, 1956) showed that honey bees can be of value as pollinators of tomatoes. Fink (1898) considered bumble bees to be effective pollinators.

Rick (1950) suggested the use of "wild solitary bees" for cross- pollination of male-sterile tomatoes. Rick (1947) mentioned the value of insect pollinators of tomatoes and their protection from insecticides. Rick (1949) stated that at Riverside, Calif., *Anthophora urbana* Cresson was most common, but various species of solitary bees and a few species of bumble bees contributed to pollination of tomatoes. Schneck (1928) stated that bumble bees are fond of tomato flowers but that honey bees do not work them "probably because of the peculiar structure of the flower and the absence of nectar." The problem seems to be that wild pollinators in most areas are too scarce to have an impact on pollination of tomatoes from the production stand point. Occasionally, honey bees visit tomato flowers, as was demonstrated in the greenhouse.

Apparently, if they are sufficiently concentrated in a tomato-growing area, the competition could "force" them to visit tomato blossoms for pollen. Unless cultivars are found that produce nectar, there can be no insect pollination of male-sterile varieties for hybrid production. If such cultivars are found, honey bees might be practical, or useful species of wild bees might be brought from Peru, the native homeland of the tomato, to provide adequate pollination. A new look should therefore be taken at current cultivars in which there has been incorporated new germ plasm to determine if nectar is being produced or if the flower has been changed in any other way that might affect pollinating insects.

## Bumble bee pollination in tomatoes

Greenhouse tomato growers used to hire workers to manually pollinate their tomato crop by vibrating overhead wires or by using battery powered flower vibrators or air blowers; however this was costly and time consuming. Kevan et al. (1991) reported that in the early 1990s, bumble bees were identified as being capable of pollinating tomato in a greenhouse environment. Since then, bumble bee pollination in tomato greenhouses has become the standard for tomato producers world-wide. Currently, greenhouse tomato production is reliant on bumble bee pollination (Velthuis and van Doorn 2006, Morandin et al. 2001c). The greenhouse tomato industry's dependence on the bumble bee pollination system became quite clear in western Canada when shortages of their commonly used greenhouse pollinator (*Bombus occidentalis* Greene) resulted in growers needing to import an exotic species (*B. impatiens* Cresson from eastern North America) to ensure that they would still achieve pollination adequate for tomato production (Sabara and Winston 2003).

Bumble bee pollinators could be successfully used for pollinating commercial crops (Kevan et al. 1991, Dogterom et al. 1998). Studies have revealed that that bumble bee pollination is superior to manual pollination, resulting in larger yields and higher quality produce (Banda and Paxton 1991, Abak and Dasgan 2005, Palma et al. 2008). With bumble bees being less expensive and more productive than manual pollination techniques, the greenhouse tomato sector has become dependent on bumble bees to pollinate their crops (Morandin et al. 2001a,c). Today, bumble bees are the most common pollinators of greenhouse tomatoes world-wide(Velthuis and van Doorn 2006).

The species of bumble bee used for pollination depends on the geographic location of the greenhouse. Many countries discourage importing alien species for pollination (North American Pollinator Protection Campaign 2006) and have attempted to adapt their own native species (or subspecies) of bumble bees for greenhouse pollination. In eastern North America *Bombus impatiens* (Cresson) is the standard pollinator for tomato, but *B. occidentalis* (Green) originally was the pollinator used west of the Rocky Mountains, where *B. impatiens* (Cresson) is not naturally present (Velthuis and van Doorn 2006). However, commercial production of *B. occidentalis* (Greene) ceased and*B. impatiens* (Cresson) is now used throughout North America. *Bombus terrestris* L. is the most commonly used species outside of North America, being utilized throughout Europe, Asia, New Zealand and Chile (Velthuis and van Doorn 2006).

Bumble bees were selected as the standard pollinators of greenhouse tomato because of their ability to extract pollen from the poricidal anthers by buzz pollination (Plowright and Laverty 1987). Bumble bees buzz pollinate flowers by grasping the poricidal anthers with their mandibles and then vibrating their flight muscles at a highfrequency to shake pollen from the flower (Michener 1962, Buchmann 1983, Corbet et al. 1988). The pollen is released on to the venter of the bee's abdomen, where it is collected into the corbiculae and carried back to the colony. In addition, most species of bumble bees are relatively docile and do not swarm. Considering that greenhouse staff are commonly present in commercial greenhouses, it is important that they are not attacked by the pollinators.

Although the bumble bee's docility and ability to buzz pollinate make them suitable as tomato pollinators, introducing them into greenhouses creates other challenges and obstacles for tomato growers. Bumble bees require both pollen and nectar for survival: pollen for its protein and nectar for its high sugar content providing energy to forage (Heinrich 2004). Because tomato plants do not produce nectar, bumble bee colony producers provide colonies with sugar syrup in their domiciles (Ptácek 2001).

## Problems associated with use of Bumble Bee Pollination of Greenhouse Tomato

Recently, some greenhouse tomato producers have had difficulty achieving adequate pollination using bumble bees. To compensate for the reduced pollination efficiency, growers have been forced to increase the number of bumble bee colonies in their facilities; even though this tactic has had limited success. There are many possible causes for reduced pollination success, but none has been identified definitively. Suggested causes for reduced pollination success include the high prevalence of parasites in greenhouse bumble bee populations, the ability of the bumble bees to leave the greenhouse and poor attractiveness of the tomato flower.

## Parasites of Greenhouse Bumble Bees

Commercially raised bumble bees are more likely to be infected with parasites than wild bumble bees (Colla et al. 2006). Moreover, greenhouse bumble bees often drift into other colony's domiciles (Birmingham and Winston 2004), increasing their contact with other bees. Greenhouse bumble bee parasites identified in commercially raised *B. impatiens* (Cresson) and wild *Bombus* spp. include the intestinal protozoans *Crithidia bombi* (Lipa and Triggiani) and *Nosema bombi* (Fantham and Porter), as well as the tracheal mite *Locustacarus buchneri* (Stammer) (Colla et al. 2006). Those three parasites have different effects on their host and tend to affect foraging behaviour negatively (Gegear et al. 2005, Otterstatter et al. 2005, Gegear et al. 2006) which could affect the bees' ability to pollinate tomato. Recent research has shown that wild bumble bees captured near commercial greenhouses tomato production, have a higher prevalence of parasites than do bumble bee populations in places far from commercial greenhouse operations (Colla et al. 2006).

## Behavioral problem in green house foraging

Although greenhouses reduce the movement of insects between the greenhouse environment and the outside, they are not completely closed. Pollinators can, and do, leave the greenhouses through gutter ventilation to forage on plants outside the greenhouse (Morandin et al. 2001c, Sabara and Winston 2003). One suggested and partial solution to this problem was to encourage increased use of ultraviolet (UV) light transmitting coverings on the greenhouses (Morandin et al. 2001c). Bumble bees are capable of finding flowers with or without UV light (Dyer and Chittka 2004), but allowing UV transmission into greenhouses decreases the frequency of bumble bees exiting a greenhouse and increases colony activity (Morandin et al. 2001c).

Difficulties with bumble bees exiting greenhouses are more of a problem during spring and summer months when bumble bees can survive outside. In British Columbia, as much as 73% of the pollen collected by a single colony during July has been recorded as coming from plants other than tomato (Whittington et al. 2004). The amount of non-tomato pollen collected during winter, however, is dramatically reduced to 5% between September and February (Whittington et al. 2004). Considering that pollination difficulties occur mostly during the spring and summer months, tomato pollination may indeed suffer from bumble bees escaping greenhouses. Applying screens or nets to cover greenhouse vents would help restrict pollinators to forage inside the greenhouse (Sabara and Winston 2003). Although nets have been tested (Yoneda et al. 2007), tomato growers are concerned that screening vents is costly and would reduce temperature control in the greenhouse through loss of ventilation.

## Plant Attractiveness

Many floral characteristics influence the floral foraging behaviour of insects. Typical rewards for pollinators include pollen, nectar or, in some cases, oil (Kevan and Baker 1983), but pollen is the only reward presented by tomato flowers. Pollinators interested in the reward within the flower may use the advertisement to find the reward, or estimate how much reward is available. Bumble bees are adept in learning to use floral advertisements such as flower size (Blarer et al. 2002), floral display (Ishii 2006, Biernaskie and Gegear 2007), flower colour (Kunze and Gumbert 2001), floral scent (Kunze and Gumbert 2001, Laloi and Pham-Delegue 2004) or a combination of advertisements (Kunze and Gumbert

2001, Lunau 1992, Kulahci et al. 2008) to find and collect floral resources such as pollen or nectar efficiently. Prior to learning about rewards, bumble bees typically visit flowers with large displays regardless of the reward available (Makino and Sakai 2007). Because pollen is the only resource presented to tomato pollinators, any floral advertisement that indicates the presence of pollen on a tomato flower could be important to foraging decisions by bumble bees. If tomato's attractiveness to bumble bees is indeed a factor in encouraging the insects to forage, enhancing the plant's attractiveness could also increase pollination efficiency in greenhouses. Through adjusting growth conditions within greenhouses, it may be possible to grow tomatoes of high attractiveness to bumble bees.

Production of high-quality greenhouse tomatoes in partially depends on pollination success and favourable growth conditions. Modification of growth conditions could lead to the production of greenhouse tomatoes with poorer floral attractants and pollinator rewards resulting in poor pollination. When presented with poor floral attractants and rewards, bumble bees may choose to leave their greenhouses through gutter ventilation and forage on plants which are more attractive to them, resulting in poor pollination of tomato crops. This poses a serious problem for greenhouse tomato growers using bumble bees for pollination; without pollination their crops suffer in terms of both quantity and quality of produce.

## Pollination Recommendations and Practices

Because of current agricultural practices as well as the relative unattractiveness of tomato flowers to honey bees, many tomato fields are largely devoid of pollinating insects. Neiswander (1954a, 1956) concluded that honey bees should not replace vibrators in the greenhouse even though they increased production on an average of 1.1 pounds per plant on plants vibrated mechanically. There are no recommendations for supplying pollinating insects to commercial fields, although the evidence indicates that if a heavy population of insect visitors could be established the effects would be beneficial.

## Pollination of solaneceous vegetables

Proper pollination is needed for most greenhouse vegetable flowers to set fruit. Solanaceous crops such as eggplant, sweet and chilii pepper flowers are self-pollinating and require insect pollination to set fruits. Multiple pollinator visits increase size, number of seed and the consistency in the shape of the fruits. Many growers introduce honey and bumblebees that are the most common commercial pollinators in the world. This can be a tremendous labor savings over manual pollination with an electric pollinator. In open fields, the natural pollination agents like insects and wind are available, but in plastic houses, these pollination agents are limited or not available due to completely isolation by covering with plastic sheet covers and net meshes. Inadequate pollination is still a major constraint to the potential yield of eggplant, sweet and chili pepper. To maximize fruit set in different vegetable crops, honeybees and bumblebees are frequently used (Paydas *et al.*, 2000).

Natural pollinators are important for better production in commercial production. (Hayo Velthuis and Adriaan van, 2006). Adequate pollination of egg plant, sweet and chilli pepper guarantees better fruit quality and higher yields (Belletti and Quagliotti, 1995). The contribution of honeybees and bumblebees in eggplant, sweet and chilli pepper pollination is interdependent relationship. Honeybees and bumblebees visit flowers to obtain significant amount of pollen to sustain the nutritional needs of the colony (Thorp, 2000). However, most of Solanaceous plant types lack sufficient nectar, so the honey and bumblebees need to be provided with a sugar solution, which is introduced in the nuclei to increase the pollen collection and pollination by bees.

## Pepper

## Flowering

The pepper (*Capsicum annuum*) flower is 1–1.5 cm across, has five stamens, one stigma, and a bell- or wheel-shaped white corolla (Figure 5.2). The flower opens shortly after sunrise and stays open less than 1 day. The stigma is receptive before the anthers release pollen. Anthers release pollen a few hours after the flower opens, but sometimes they never release pollen at all. The stigma can be either shorter or longer than the stamens; thus the flower may not self-pollinate even if it does release pollen. In other cases, the stigma touches the anthers and self-pollination occurs. The pepper flower produces nectar at the base of the ovary. Bees visit pepper for nectar and pollen, but they are easily distracted away to more attractive blooming plants. Nectar production varies across varieties (Rabinowitch *et al.*, 1993).

The sweet pepper (*Capsicum annum* sp.) (Figure 5.3) flowers have 5-7 petals of the color white, greenish and in some varieties purple. The flowers are self-pollinating, producing plenty of pollen and nectar and lacking any scent in the sweet pepper. Occasionally there are small clusters of flowers. It has five stamens with bluish anthers and a single stigma that may vary from slightly shorter than the anthers to much longer; the corolla is somewhat bell-to wheel-shaped (Smith and Heiser, 1951). The flowers are visited by bees for both the nectar and the pollen (Markus, 1965) and high flower set by bees pollination (Nagarathnam and Rajamani, 1963).

## Pepper Pollination Requirements

Pepper is generally self-pollinating, but cross-pollination is also common. Cross-pollination is mandatory in those flowers in which the anthers never release pollen. Each seed requires one grain of pollen to be deposited on the stigma, and poorly pollinated stigmas will result in misshapen fruit. Pollination by hand, by a mechanical pollination device, or by alfalfa leafcutting bees increases fruit-set and yield even in naturally self-pollinating varieties (Rasmussen, 1985); thus, the activity of an external pollinating agent is beneficial.

The incidence of cross-pollination ranges from 5–24% (Lorenzetti and Cirica, 1974) to 58–68% (Murthy and Murthy, 1962). Such outcrossing is a liability in breeding programmes. The New Mexico (USA) Crop Improvement Association (1992) requires an isolation distance of 1 mile (1.6 km) for foundation seed and 0.25 mile (0.4 km) for certified seed. One alternative to cumbersome isolation distances is to use insect-proof plant cages that ensure self-pollination in pepper (Bosland, 1993).

Male-sterile lines give the opportunity for producing hybrid seed which is known in other crops to give superior vigour and productivity. Male-sterility occurs in certain pepper varieties (Shifriss and Frankel, 1969), but most show only partial male-sterility (Breuils and Pochard, 1975). Compared to male-fertile plants, male-sterile plants produce lower volumes of nectar with smaller amounts of sugar, and bee visitation to male-sterile flowers is 4–5 times lower than in malefertile flowers (Rabinowitch *et al.*, 1993). Rabinowitch *et al.* (1993) suggest that plant breeders could improve pollination rates by selecting for plants with attractive nectar, although the heritability of nectar characteristics in pepper is low. Hybrid seed production requires an external pollinating agent, such as honey bees, to transfer pollen from male-fertile to male-sterile lines. Anais and Torregrossa (1979) produced hybrid seed by alternating rows of male-fertile and male-sterile plants. However, honey bees cross between lines better when the varieties are interplanted in a chessboard pattern so that each row has plants of both lines (Kubi[1]ová and Háslbachová, 1991). Cold night temperatures can be a limiting factor in pepper pollination by inducing a decrease in pollen fertility. The weight and size of fruits reared from 'Latino' plants maintained at 75.2/53.6°F day/night.

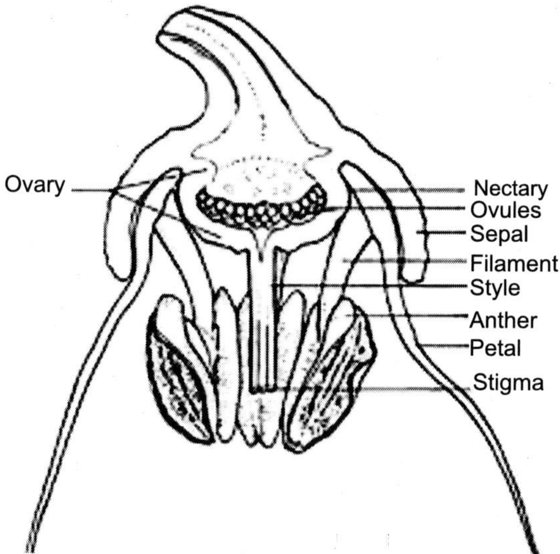


Figure 5.2: Longitudinal section of flower of bell pepper

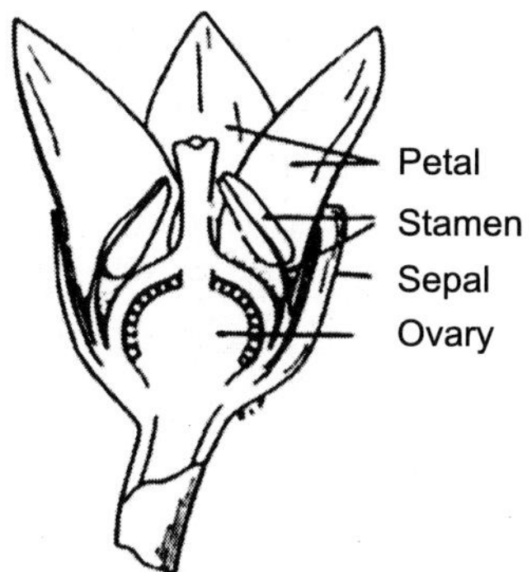


Figure 5.3: Pepper Flower

**Pepper (*Capcicum annumm*)**

## Pepper Pollinators

Several types of bees visit pepper, but honey bees are the most common pollinator. Fruit-set is satisfactory in glasshouse production in The Netherlands during summer months when windows are open and many insects are visiting the plants. Pollination can be a problem in spring and autumn, but honey bees moved into glasshouses at these times increase fruit size and weight, seed number, and number of wellshaped fruits. Honey bee pollination is standard practice in glasshouse pepper production in The Netherlands (Ruijter *et al.*, 1991). Honey bees readily visit and move between male-sterile 'Sivria 600' and male-fertile 'California Wonder' in glasshouse conditions in Czechoslovakia (Kubi'ová and Háslbachová, 1991). However, honey bees in the open concentrate on male-fertile lines to collect pollen (Breuils and Pochard, 1975).

Managed bumble bees can increase fruit weight, width, wholefruit volume, seed weight, and speed of ripening in 'Plutona' in glasshouse conditions in Ontario, Canada. Bumble bees increase speed of ripening for 'Cubico' under similar conditions (Shipp *et al.*, 1994). The increased speed of ripening for both varieties probably translates into one additional flush of fruit over the course of a growing season.

Imported horned-face bees (*Osmia cornifrons*) are known to increase pepper yield in glasshouses in Denmark early in the season when pollination is historically poor, compared to pollination by a mechanical device or by natural self-pollination. However, the bees cannot increase their populations on pepper even when glasshouses are supplemented with flowering phacelia (*Phacelia tanacetifolia*) or currant (*Ribes* spp.). By mid-June there is no difference in fruit yield between plants pollinated mechanically, by bees, or by self-pollination (Kristjansson and Rasmussen, 1991). Wind and rain do not pollinate pepper (Crane and Walker, 1984). Certain flies (*Calliphora* spp., *Lucilia* spp.) occasionally pollinate pepper (Breuils and Pochard, 1975). Unlimited visits by the drone fly *Eristalis tenax* were shown in glasshouse-grown 'Bell Boy' sweet pepper to increase the number of fruits in the 20–25 cm circumference category and in the 100–120 g weight category compared to that of plants caged to exclude pollinators (Jarlan *et al.*, 1997). There are no published recommendations for honey bee densities in pepper. For bumble bees, the data of Shipp *et al.* (1994) support a glasshouse stocking rate of one hive (containing 24–77 bees) per 1600 ft2 (150 m2) of plants.

## Egg plant

## Flowering biology

Many species of cultivated plants have particular traits, which demand different pollinators with different size and foraging behavior for a proper pollination, (e.g. eggplants) (*Solanum melongena* L.). Like most plant species of genus *Solanum*, eggplants have poricidal anthers, which require specialized behavioral and morphological adaptations by potential pollinators (Avanzi and Campos 1997). In this species, the length of stamens is about one centimeter, with very short filaments and long bright yellow anthers; each one presenting two apical pores. Effective eggplant pollinators can acquire pollen by vibrating the anthers. *S. melongena* L. is native to the tropical regions of Asia and has been cultivated for centuries by Arabs and Chinese. Its consumption is increasing mainly in Europe and the United States (Vieira *et al.* 1996), where it represents an important crop.

Eggplant (*Solanum melongena* L.) is a vegetable from Solanaceae family that originates from warm India and Chine regions [Lawande and Chavan 1998]. Wild and thorny Indian *Solanum incanum* L. or *Solanum insanum* L. Roxb. species are suspected to be the plant ancestors. The eggplant in warm climate is a perennial; its bushy and ligneous plants reach even 3m height. Under moderate climate, it is cultivated as annualplant reaching only

0.5–1 m of height. The eggplants begin flowering after forming – depending on a variety – 9 to 12 assimilation leaves, i.e. about 70–90 days after emergence.

The eggplant (*Solanum melongenae*) has a perfect flowers with a purple petals and green or purple sepals (Sambandam, 1964), flowers with five lobed corolla and yellow stamen, lack sufficient nectar. The anthers are yellow and sit on short filaments surrounding the style. The cone of stamens surrounding it, with the lowest one in the cluster being the most developed (Aguilar, 2008). The eggplant is a self-fertile and self pollinating plant however, cross-pollination through insects is possible and in many cases desirable. The eggplant doesn't self without the aid of bees or man pollination (Kakizaki, 1931).

Eggplant is a self-pollinating plant, although it can be cross-pollinated (6–10%) that results from transferring a pollen by insects such as ants, thrips, and bees (George 1985, Lawande and Chavan 1998). The eggplant flowers are hermaphrodite; they are single or arranged in inflorescences composed of 2–7 flowers each, and show positive geotropism [Konys 1993]. The flower buds develop in the corners of the same leaves as vegetative buds. At the beginning of flowering stage, eggplant flowers are very close to one another. When internodes are longer, the distances between flowers also increase (Shah and Patel 1970). Results achieved by Prasad and Prakash (1968) in study upon flowering biology of four eggplant varieties indicate that young flower buds are oval or conical in shape and green in color. In practice, the time of their opening is the same.The largest number of flowers open between 7:00 and 8:00 a.m. At 2:00 p.m. flowers begin to close and late at night they are completely closed. In the morning, flowers open again and the whole cycle repeats for about 1–3 days. Following days, flowers close only in part. The manner flowers open and close depends on the flower's age, and environmental conditions, namely air temperature and relative humidity.

The eggplant flowers are quite large 3–5 cm in diameter. The corolla consists of 5–10 accreted petals of purple, pink, or white color depending on a variety. Number of stamens is 5–7. Filaments are short, 6–20 bicameral anthers are yellow and arranged around pistil (Mc Gregor 1976). When a flower blooms, the pollen grains and germ sack are fully formed. The anthers break in upper direction 15–30 minutes after flower opening. The pollen is quite heavy. It can be spontaneously transferred for up to 1 m distance. The pollination duration depends on air temperature and humidity (Prasad and Prakash 1968).

The eggplant fruits are multi-chamber berries of different shapes. Depending on a variety, they may be: spherical, through oval, ovoid, piriform, to elongated and spiral. Black-purple fruits, in the skin of which anthocyanins are present, dominate incultivation. There are also white-colored varieties, the skin of which does not contain those pigments. At the stage of utility ripeness, fruits are light yellow to black-purple, while brown-yellow or grey-green at the physiological ripeness stage (George 1985). The skin and parenchyma color is associated with the fruit color. When the skin is colorless, fruit's parenchyma is white, yellow, or light green. When skin contains pigments, the parenchyma is usually dark – purple. Large, five-section hairy and thorny corolla strongly accreted with the ovary, which makes nursery operations, fruit harvest, and storage difficult, is characteristic for the eggplant fruits (Krug 1991). Fruit's color, sometimes its growth, depend on the flower's localization in an inflorescence. For example, side flowers which are characterized by slow growth rate amd less intensive colour set small and lighter fruits. Up to 1-25 fruits, about 30-2000 g each, may be set on a single eggplant plant.

Assuming the complexity of many factors that make difficult or impossible to self-pollination of eggplant flowers, it is recommended to apply inflorescence vibrating for cultivations under covers (Mc Gregor 1976, Gemmill-Herren and Ochieng 2008), flower harmonization (Olympios 1976, Nothmann et al. 1983, Glapś and Górecki 1989, Krug 1991, Leonardi and Romano 1997, Lawande and Chavan 1998, Kowalska 2006), and natural pollination by a bumble-bee (*Bombus terrestris*) )Mc Gregor 1976, Abak et al. 1995, Kowalska 2003a, 2003b, 2006).

## Inflorescence vibrating

Mc Gregor (1976) reported that shaking the inflorescences did not make sufficient pollen setting on pistil stigmata and was not able to pollen the eggplant flowers well. Studies of Sambandan (1964) applying the inflorescence vibrating, only 30–40% of all fruits per plant were set. Gemmill-Herren and Ochieng (2008) reported a positive influence of vibrating the eggplant inflorescences on the number of produced seeds in references to self-pollination. Object with shaken inflorescences produced fruits with significantly larger number of developed seeds as compared to those set due to self–pollination. However, discussed manner of pollination is not so efficient as pollination by insects, which is going to be the subject of further chapters.

## Natural pollinators

It is commonly assumed that bees are the most important plant pollinators all over the world (Figure 5.5). Studies upon the application of honey-bee (*Apis mellifera*) as a natural pollinator at vegetable cultivating under covers began in 1950. Although the insect was very efficient as pollinator of many plant species, it did not play its role in the case of Solanaceae plants. At present, bumble-bee *Bombus terrestris* in Europe and *Bombus impatiens* Cr. in North America are commonly used for pollinating the flowers of tomato, pepper, and eggplant grown under covers (Van Ravestijn and Van der Sande 1991, Shipp et al. 1994, Abak et al. 1995, Asada 1997, Dobromilska 1997, Dobromilskaand Fawcett 1999, Kowalska 2003b, Gemmill-Herren and Ochieng 2008). Insects applied for plant pollination (bumble-bees and bees) have biting-licking-type mouth, in which the tongue length plays an important role during pollination. *Colletes* family bees have the shortest tongues. They may visit only flowers with accessible nectar and pollen. Bumble-bees have the longest tongues (even 12 millimeters), due to which their application for pollinating in eggplant cultivation under covers gives the best results (Dylewska 1996). The activity of pollinating insects depends on the daytime and weather. At 15–25ºC, insects are the most active at noon, while in hot days (above 25ºC), they fly to gather pollen in the morning and in the afternoon, when it is cooler. When temperatures remain above 35°C for a longer time, bumble-bees remain in the hive ventilating their nest and giving up the pollen gathering. Bumble-bees, unlike other insects, visit flowers even at lower air temperatures (6–8ºC) and low light intensities, which is important at cultivating thermophilic plants, namely under more sever conditions. Moreover, they pollinate flowers regularly by moving up the plant and they never visit the same flowers (Dobromilska 1997). In sunny days, these insects start to fly about 8:00 a.m. and end about 5:00 p.m. with their maximum activity between 1:00 p.m. and 5:00 p.m. A single bumble-bee can visit up to 15 flowers per minute, and it pollinates about five of them that time. In cloudy days, bumble-bees start their flights later: about 11:00 a.m., and end before 5:00 p.m. under such conditions, a single insect visits only three flowers per minute (Abak et al. 1995). Because eggplant flowers are directed down, a bumble-bee that visits them has to hold the stamens with its mandibles very tightly. Specific bumble-bee's jingling makes a flower vibrating, and in consequence, pollen is poured out through the holes in anthers onto insect's body covered with dense hair. Bumble-bee sweeps the pollen and puts it into the baskets localized on the rear leg's shins. Then, it forms the pollen into granules called *corbiculi* (pollen baskets). When a bumble-bee releases its mandibles from the stamens, twobrown spots remains – the evidence that the flower was pollinated. Number of pollinated flowers indicates if there is an appropriate number of insects in a greenhouse or tunnel. Four to five bumble-bee families per hectare is sufficient for providing with necessary pollination (Dylewska 1996).

The yield increase and its better quality are considered to be the major benefits from bumble-bee application in vegetable pollination in a greenhouse. It was found that bumble-bees significantly reduce inputs for pollination and nursery. Biological plant protection against diseases and pests is an additional effect of the bumble-bee colony introduction for cultivations under cover (Van Ravestijn and Van der Sande 1991, Wysocka- Owczarek 1993).

In Turkey, Abak et al. (1995) carried out the experiments upon the importance of bumble-bee in eggplant and tomato cultivation in unheated greenhouse. They reported that flower pollination by insects affected the total yield increase by 23% for eggplant and by 1% for tomato. And similarly, number of fruits harvested from 1 $m^2$ increased by 22% and 6%, respectively for eggplant and tomato from plants pollinated by bumblebees. The way the flowers were pollinated had not any influence on the quality of produced fruits, whereas seed number increased by 62% and 100%, respectively after bumble-bee introduction.

Abak et al. (1995), Stępowska (1996), as well as Dobromilska (1997) found that bumble-bee makes effective fruit setting in eggplant and greenhouse pepper cultivations. They have also a positive impact on the total and commercial eggplant's and pepper's yields size. At the same time, earlier fruit ripening was observed, and fruits were heavier than those resulting from self-pollination. The first-selection fruits yieldincrease in reference to the control plants was also recorded.

Results achieved by Kowalska (2003a, 2003b, 2006) from studies upon the bumblebee efficiency in eggplant cultivated under covers revealed that flower pollination by insects caused significant increase of early and commercial yields as compared to self pollinated plants. In object where flowers were pollinated by insects, plants produced considerably more commercial fruits as compared to control ones. It should be emphasized that fruits set with a help of bumble-bee ripened earlier and were heavier than those harvested from control object. Fruits grown from flowers pollinated by a bumblebee (*Bombus terrestris*) were characterized by higher mean weight, length and diameter. In Japan, Asada and Ono (1996), as well as Asada (1997) made the comparison of the efficiency of tomato flowers pollination by four local bumble-bee species (*Bombus hypocrita hypocrita* Perez., *Bombus ignitus* Smith., *Bombus ardens ardens* Smith., *Bombus diversus diversus* Smith.) with one imported from England (*Bombus terrestris*). Experiments revealed that efficiency of tomato flower pollination by all bumble-bee species was the same and high, and fruit setting was 84–100%. Every fruit harvested within commercial yield contained large number of seeds.

Pollination ecology of *Solanum sysimbrifolium* Lamk. was studied by Babu *et al.* (1987) and found that two species of *Xylocopa* and one species of *Anthophora* are the active pollinators. All the visitors are diurnal and only larger bees are the effective pollinators. Abrol (1991 b) studied pollination of brinjal flowers (*Solanum melongena* L.)

by bumble bees. Results showed that brinjal flowers were attractive to seven species of insects belonging to four families. Abundance of various pollinating insects was in the order: bumblebees, *B. asiaticus* Morawitz, *B. albopleuralis* Friese, *B. simillimus* Smith, carpenter bee *Xylocopa valga* Gerstacker, honeybees, *A. cerana indica* F., *A. mellifera* L., and halictine bee, *Lasioglossum* spp. Bumblebees were efficient pollinators on the basis of their field behaviour, population dynamics, nectar pollen carrying capacity, and rate of flower visitation in a unit time.

Studies carried out in Africa by Amoako and Yeboah-Gyan (1991) indicate great usefulness of a honey-bee (*Apis mellifera adansonii*) under tropical climate conditions to increase the eggplant's, tomato's, and pepper's yields. Number of set fruits in relation to number of formed flowers was significantly larger from plants, that flowers of which were pollinated by insects than self-pollinated ones. The manner of flower pollination had no influence on qualitative parameters of three assessed vegetable species fruits. Gemmill-Herren and Ochieng (2008) studied the efficiency of different pollination ways on two eggplant varieties yielding. For both of them, a positive influence of bumble-bee pollination on a quality of eggplant fruits was confirmed. Fruits with significantly larger seed number were achieved in object where flowers were pollinated by insects as compared to those set due to self-pollination or inflorescence vibrating. Reference data fully confirm that pollinating insects, including bumble-bee, play very important yield-forming role at cultivating eggplants under covers. Therefore, their presence on the vegetable plantation should be considered as one of the necessary agrotechnical factors. The eggplant seed radiation to change the formation of heterostylic flowers should be also mentioned. At *Solanum khsasianum*, the seed treatment with gamma radiation caused considerable reduction of low-pistil flowers number, while increasing the number of high-pistil ones (Handique and Sarma 1995 after Chauhan and Ravindran 1979). Those results correspond with studies by Handique and Sarma (1995), who found that at *Solanum melongena* L. gamma radiation may reduce number of produced low-pistil flowers and proportionally increase high-pistil ones. It was also reported that varieties of the same species react in different ways towards various radiation intensities. The phenomenon of the high-pistil flowers number increase and reduction of low-pistil flowers resulting from gamma radiation has not been elucidated yet.

## Brinjal pests

## Pests

Among vegetables, solanaceous vegetables form an important and the biggest group being cultivated extensively through out the world as well as in our country. India is the largest producer of brinjal among the tropical and sub-tropical nations (Behra *el al.,*1997). India seems to be the original habitat of brinjal as mild varieties/ species are still available across the country. In India, this vegetable is grown under diverse agro-climatic conditions. The total area under brinjal cultivation in the country is 506.9 thousand hectares with annual production of 8372 metric tonnes (Anonymous, 2003).

Like other vegetable crops brinjal is ravaged by more than 70 insect pests Figure 5.8 whose occurrence is influenced by the prevailing weather conditions of a particular locality (Subaratnam and Butani, 1982). Among all the pests attacking brinjal, 7 major pests viz.; hadda beetle, shoot and fruit borer, stem borer, leaf roller, jassid, lace wing bug and aphids have been reported by (Butani and Jotwani, 1984). Sixteen insect pests were recorded on brinjal which cause heavy loss. The losses inflicted by these insect pests amount to 30-40 per cent and sometimes may go up as high as 100 per cent (Srivastava and Butani, 1998). Among cucurbitaceous vegetables, bitter gourd *Momordica charantia* L. is an important vegetable crop gifted with some medicinal properties and grown all over India with a cultivated area of about 0.26 lakh hectares (Gill and Tomar, 1991). In Jammu region of J and K state, the crop occupies an area of 220 hectares with an annual production of 1980 metric tonnes (Anonymous, 2002).

Integrated pest management strategies are totally safe to non target organisms, millions of untrained users, consumers and eco-system. It also helps to increase the biotic factors and thus maintains a natural pest control in a sustainable way. Since IPM evolved close observation on population dynamics, growth and related decisions about the correct control measures are to be used for controlling the pest population for sustainable crop production (Kumar, 1987).

Different workers have studied the host plants of hadda beetle, *Epilachna vigintioctopunctata*, an important pest of solanaceous and cucurbitaceous crops like brinjal, tomato, potato, gourd, melon and cucumber, was also found infesting *Amaranthus viridis*. The leaves of *Amaranthus viridis* were found infested with both larvae and pupae and had characteristic feeding injury. This is the new food plant record for the coccinellid (Narang and Ramzan, 1984). A check list of *E. vigintioctopunctata*, their food plants and distribution in states of India is presented for 74 species which include *Coccinella delessertii* (*E. delessertii*) on *Solanum xanthocarum* and *E. vigintioctopunctata* on numerous host (Anand *et al.,*1988 ). Among the major insect pests that attack potato, the coccinellid *E. vigintioctopunctata* is very important in Asia. *E. dodecastigma* and *E. vigintioctopuncata, E. ocellata* and *H. sparsa*( *E. sparsa*) commonly attack solanaceous plants. The peak period of infestation varies with region, but the peak is generally in July-August. The pest also feeds on brinjal, tomato, tobacco, pumpkin and bitter gourd. The larvae and adults scrap the green matter from leaves and damage up to 80 per cent of plants. (Rajagopal and Trivedi, 1989). Richards and Filewood, (1990), reported the infestation of *Epilachna vigintioctopunctata* on wild cucurbit plants near *Cucurbita moschata* from Parana, Brazil.

Ramzan and Singh, (1982) studied the biology of *E. vigintioctopunctata* on different host plants and reported that it successfully bred on brinjal, potato, and *Solanum nigrum*. The respective, pre-ovipositional and ovipositional periods on brinjal, potato and *S. nigrum* were 8.0, 12.7; 15.0, 15.0; and 6.0, 35.0, days, respectively. The maximum average number of eggs laid by a female was on *S. nigrum* (502) followed by brinjal (269) and minimum on potato (216). The larval period was recorded longest on *Withiana sominefera* (18 days) followed by brinjal (14.7 days), potato (12.0 days) and *S. nigrum* (11.3 days). The pre-pupal and pupal periods recorded 2.5, 5.5; 3.5, 4.7; 2.5, 4.3 and 2.5, 5.0 days, respectively, on brinjal, potato, *S. nigrum* and *W. sominefera*. It completed life cycle in 26.3 days on brinjal. Similarly, Ram and Verma (1998) recorded egg laying in *E. vigintioctopunctata* was 8-59 days with 88-560 eggs per female and the average incubation period ranged from 3-6 days and the duration of Ist, IInd, IIIrd and IVth instar lasted for 2-4, 2-4, 2-6 and 3-8 days, respectively. The pre-pupal and pupal period varied from 1-3 days and 4-10 days, respectively. As the pest hibernated in winter, pre-oviposition period varied from 4-132 days. Male beetle survived for shorter duration (7-185 days) in comparison to female (2-198 days).The life cycle was completed from 17-41 days. The pest passed 7 generations in the laboratory and the duration was found shorter during April-May.

The lady beetle, *E. vigintioctopunctata* was reported for the first time from Jammu and Kashmir as a pest of some important solanaceous medicinal plants (Mathur and Srivastava, 1964) and on vegetables. Munshi *et al.* (1968) observed that both adults and grubs caused damage up to 80 per cent to the host plants by feeding on chlorophyll content on the abaxial part of the leaf and caused concentric curves. Pest remained active from July to September. Mall *et al.* (1992) recorded *E. vigintioctopunctata* from 3rd weak of August to middle of October with its maximum population 0.22 beetles per leaf in the middle of September. The corresponding average temperature and relative humidity during the period were 28°C and 80 per cent, respectively. The population was noticed to gradually decline and disappear due to fall of temperature, humidity and maturity of the crop. Veeraval and Bhaskaran (1993) concluded that grub population of *E. vigintioctopunctata* (R2=0.3260) was positively correlated with relative humidity, however, the adult population (R2=0.1320) was noticed to decrease with increase in temperature. Again Veeravel and Bhaskaran (1994) identified different phenological/ growth stages of brinjal viz., seedling, vegetative, flowering, bearing and senescence in relation to the occurrence of pest. The grub incidence was recorded maximum at vegetative stage, whereas, peak adult population (R2=0.09) occurred during flowering and bearing stages of the brinjal plant.

In Himachal Pradesh, India Mehta *et al.,* (1993) reported *Pedobius foveolatus* and *Tetrastichus sp.* parasitizing grubs of *E. vigintioctopunctata* on brinjal. *P. foveolatus* was also reported as an effective parasitoid of *E. vigintioctopunctata* by (Kalra, 1997). Other, potential natural enemies of *E. vigintioctopunctata* on eggs and grubs as recorded were *Tetrastichus ovulara, Pediobius foveolatus* and *Pleurotropic foveolatus,* respectively by (Hamid and Singh, 1998). Veeraval and Raghuraman (1998) have also recorded *P. foveolatus* as potential natural enemy of *E. vigintoictopunctata.* They reported that per cent infestation of *P. foveolatus* ranged from 19.4-19.5 percent. The 3$^{rd}$ and 4$^{th}$ larval instars of *E. vigintioctopunctata* were more preferred than other instars for parasitization.

Besides parasitoids, predatory wasps also parasitize IInd and IIIrd instars grubs and cause 70 per cent parasitism from July to September (Mathur and Srivastava, 1964). A pentatomid predatory bug *Cantheonidea fuscellata* Wolf. consumed 19 larvae or pupae of *E. vigintioctopunctata* per day in the nymph stage (Sheila and Abraham, 1981). Bhagat and Munshi (1997) reported an entomogenous fungus, *Fusarium solani* as a pathogen infecting the spotted *Epilachna* beetle, (*E. vigintioctopunctata*) from the Jammu region of Jammu and Kashmir, India for the first time. The infection of beetle egg masses was observed to be about 20 per cent. Hanif and Abida (1997) found *Beavaria bassiana* effective against all stages of *E. dodecastigma* except for the eggs. The 100 pr cent grub, pupal and adult mortality occurred after 120, 168 and 192 hrs, respectively. Venkataraman *et al.,* (1962) evaluated the insecticidal efficacy of *Bacillus thuringensis* Barl., spore suspension on *Henosepilachna vigintioctopunctata* and recorded complete mortality of the pest within 48 hrs.

Karmakar and Bhattacharya (2000) reported that by using mechanical methods the pest population can be maintained at levels well below the economic injury level (1.25 – 11 *E. vigintioctopunctata* Fab. / plot). In search of ecofriendly and specific action compounds, (Gupta and Dogra, 1994) reported that diflubenzuron applied at micro quantity level was effective not only as potent larvicidal compounds interfering with the molting process but also effect egg hatchability and fecundity. Sardamme *et al.,* (1986) found diflubenzuron to have both stomach and contact action against the larvae of *E. vigintioctopunctata.* Mandal and Kumar (2001) evaluated the efficacy of Monocrotophos, Quinalphos, Cypermethrin and Fenvalerate against *E., vigintioctopunctata* infesting brinjal and reported that all the insecticides reduced the *E. vigintioctopunctata* populations, with cypermethrin recording the highest reduction in the pest population at all stages of plant growth. The neem oil affects reproduction by way of reduced fertility/ fecundity of *H. vigintioctopunctata* (Tewari and Moorthy, 1985 and Mishra *et al.,* 1990), on *E. sparsa* (Mishra *et.al.,* 1990) and on *E. vigintioctopunctata* (Mathur and Nigam, 1993).

## Status of biotechnological interventions

Fruits and shoot borer i.e. *Leucinodes orbonalis* being the most destructive pest in brinjal crop, efforts have been made to develop insect resistance brinjal varieties by incorporating *cry1Ac* gene in brinjal from a bacterium mainly *Bacillus thuringiensis*, commonly known as Bt brinjal. The Bt brinjal developed by M/s Maharashtra Hybrid Seeds Company Limited containing *cry1Ac* gene (Event EE1) is in the advanced stages of large scale field evaluation. MAHYCO has also transferred Bt brinjal technology to public sector institutions viz. Tamil Nadu Agricultural University (TNAU), Coimbatore, University of Agricultural Sciences (UAS), Dharwad and Indian Institute of Vegetable Research (IIVR), Varanasi, who have successfully backcrossed Event EE1 into locally adopted open pollinated brinjal varieties. The technology has been also transferred to Institute of Plant Breeding of the University of Philippines, in the Philippines and Bangladesh Agricultural Research Institute and East West Seeds Limited in Bangladesh. National Research Centre on Plant Biotechnology (NRCPB), New Delhi has also developed Bt brinjal by incorporating *cry1Fa1* gene into brinjal and transferred to the same to few seed companies for further biosafety studies and field evaluation. The Indian Institute of Horticultural Research (IIHR), Bangalore is also developing Bt brinjal using *cry1Ab* gene construct in two important brinjal varieties.

## Tomato pests

## Insect and mite pests

Insects and mites are important arthropod pests of tomatoes (Figure 5.4, 5.7). The type of damage inflicted depends upon the type of mouthparts of the damaging life stage. Many insects have chewing mouthparts, or modifications of them, and their damage appears as holes in leaves, stems or fruit, leaf mines, leaf rolls, etc. Some insects other than mites also have piercing–sucking mouthparts or modifications of them. Damage by these arthropods usually appears as speckling, spotting or distortion of leaves, stems or fruit that at times can be confused with pathogenic or nutritional causes. In addition, some insects with piercing–sucking mouthparts are important pests as vectors of many viruses that are responsible for some very serious diseases. Thus, recognition of damage can be a first step in identifying the cause and, ultimately, developing a management strategy. For this reason, pests will be presented and discussed here according to the type of damage inflicted.

Because of the large complex of arthropod pests that damage tomato plants and because of the ability of many arthropods to develop resistance to insecticides, most successful and long-term management programmes will integrate many tactics. Tomatoes fit IPM programmes well, because they can tolerate up to 30% defoliation in the lower canopy before bloom and up to 50% defoliation in the lower canopy after bloom (Keularts *et al.*, 1985). The manipulation of cultural practices probably is the oldest tactic for managing pests and often involves the use of good common sense. Some practices are adapted for specific pests but most are used for their negative impact on many pest species. Pests can be excluded from plant production facilities using finemesh screening to cover all entrances and air intakes. Row covers can be used to exclude pests from field plantings. Good sanitation within and adjacent to fields and the timely destruction of crops following the last harvest can reduce reservoirs of insects as well as the plant diseases for which they are vectors. Avoiding the use of certain plants as cover crops during fallow periods and separating new plantings in time and space from old plantings of the same or different crops can also reduce pest reservoirs. Delaying planting can create or extend crop-free periods as well as avoid early migrations of overwintering or oversummering pests. The use of ultraviolet light-reflective and coloured plastic films to cover the soil under plants can be used to interfere with the normal settling of certain insects. The film should cover as much of the soil as possible, because it is only effective as long as the insects can see it. Pesticide residue on the plastic can also reduce the effectiveness of the mulch. Growing resistant varieties can contribute to the management of insects as well as of the disease organisms that they transmit. Unfortunately, resistance to arthropod pests is not yet a reality in commercial tomato varieties. Most arthropod pests are attacked by numerous parasitoids and predators. These natural enemies can be conserved by eliminating or reducing the use of broad-spectrum pesticides and by using selective or biorational pesticides. Natural control can be augmented in some cases by the release of natural enemies reared specifically for this purpose. Certain weedy or crop areas can also serve as reservoirs or refuges of natural enemies. The application of pesticides is the method used most often to control arthropod pests. Twice-weekly scouting to ascertain the densities of pests and natural enemies is essential to the wise selection and timing of applications. Eliminating unnecessary pesticidal applications, reducing the number of applications and using selective pesticides reduces the probability of the development of pesticide resistance. Natural enemies will also be conserved and can thus contribute to reducing pest densities.

## Chewing damage: seedlings

### Crickets

Field crickets, such as *Acheta* (= *Gryllus*) *assimilis* Fabricius, may sometimes damage tomato seedlings. They are dark brown to black insects, 2–3 cm long, with long antennae and with the hind pair of legs enlarged for jumping. They migrate into the perimeters of fields from weedy, trashy or storage areas and chew seedlings off above the soil line. Aboveground plant parts may occasionally be attacked. Mole crickets, such as *Scapteriscus vicinus* Scudder, *S. borellii* Giglio-Tos, *Gryllotalpa gryllotalpa* L., *G. hexadactyla* Perty and others, are similar to field crickets but are longer (4 cm) and are lighter brown, with a claw-like front pair of legs for digging. Mole crickets also damage seedlings, chewing off plants at or just below the soil surface. They construct characteristic tunnels just below the soil surface into which excised seedlings may be pulled and wholly consumed, leaving no trace. If stems are not completely severed, plants may wilt.

If field crickets are present on immediate field perimeters, weeds should be mown and debris and trash should be cleaned and removed. Fields should be deep ploughed to destroy field cricket eggs and be kept weed free. The soil should be kept moist so that crickets remain close to the surface. Granular or bait formulations containing pesticides or products containing the entomopathogenic nematode *Steinernema scapterisci* Nguyen

& Smart may be applied. Broadspectrum soil fumigants will control mole crickets in the treated soil zone, but crickets can move back into the treated soil as the fumigant dissipates.

## Cutworms

There are many species of caterpillar pest that are generally called cutworms. Some of the most prominent species include: the variegated cutworm, *Peridroma saucia* (Hübner); the black cutworm, *Agrotis ipsilon* (Hufnagel); and the granulate cutworm, *Agrotis subterranea* (Fabricius). Larvae attack seedlings as they emerge or as transplants are set. Plants are cut at or just above the soil surface. Because larvae hide during the day, they are often overlooked. Larvae grow to about 4 cm long and vary in appearance from a dirty grey to a yellowish or dark brown with variable longitudinal or diagonal lines. Cutworm larvae appear smooth or 'greasy' to the naked eye and curl up tightly when disturbed. To control cutworms, land preparation should begin early enough to remove alternative host plants at least 10–14 days before planting. Bait formulations containing the bacterium *Bacillus thuringiensis* Berliner, or granular or bait formulations containing insecticide, should be applied if cutworm larvae are present. Broad-spectrum soil fumigants will control cutworm larvae in the treated soil zone.

## Chewing damage: foliage

## Flea beetles

Flea beetles are 3–4 mm long and shiny, with enlarged hind legs adapted for jumping. The dark brown to black beetles, primarily in the genera *Epitrix* and *Phyllotreta*, fly into tomato fields from overwintering sites. They chew small pits or holes in leaves. Damage on large plants is not usually of economic significance but damage to seedlings can result in death or stunted growth. Early plantings are usually the most affected. Larvae occur in the soil but generally are not damaging. Flea beetles in the genus *Epitrix* have also been shown to transmit eggplant mosaic virus to tomato. Seedlings can be protected from immigrating flea beetle adults with row covers of fine-mesh fabric. A single timed application of insecticide may be all that is necessary in lieu of using row covers.

## Leafminers

Four species of leafmining flies are most damaging to tomato: the vegetable leafminer, *Liriomyza sativae* Blanchard; *L. trifolii* (Burgess); *L. bryoniae* (Kaltenbach); and the pea leafminer, *L. huidobrensis* (Blanchard). Adults are 2–3 mm long and are black on the top and yellow on the heads, sides and underside. Females use their sheath-like ovipositors to make small holes in the leaf into which eggs are deposited just below the leaf surface. Females also use the ovipositor to tear larger holes, from which they feed on the exuding cell contents. The latter holes become necrotic and are often termed 'stippling'. *L. sativae*, *L. trifolii* and *L. bryoniae* deposit eggs primarily in the upper surfaces of leaves while *L. huidobrensis* deposits eggs primarily in the lower leaf surfaces. Larvae are yellowish maggots 2–5 mm long with black sickle-shaped mouth hooks that are used for scraping leaf tissue. Larvae of the former three species form serpentine mines in no particular pattern, while larvae of the latter species form mines that are generally associated with the midrib or lateral leaf veins. Larvae cut a hole at the end of the mine and drop to the soil, where they pupate under debris (Fig. 7.2).

Leafminers are generally held in control by large numbers of species of parasitic wasps that attack the larval stages; however, leafminer populations can increase to damaging levels if parasitoids are reduced or eliminated with applications of broad-spectrum insecticides targeted against other pests. Therefore, timed applications of selective insecticides with minimal impact on parasitic species are recommended.

## Loopers

Different looper species attack tomatoes in different parts of the world but some of the most common include: the cabbage looper, *Trichoplusia ni* (Hübner); the soybean looper, *Pseudoplusia includens* (Walker); the alfalfa looper, *Autographa californica* (Speyer); and the tomato looper or golden twin spot, *Chrysodeixis chalcites* (Esper). Eggs are deposited singly on the undersides of leaves. Cabbage looper eggs resemble fruitworm eggs, except that they are flatter and have a larger number of fine ridges. Young larvae leave the upper epidermis intact, forming 'windows', while older larvae chew irregular holes. Surface feeding of green fruit occurs only rarely. Larvae are up to 3.5 cm long and pale green, sometimes with white stripes on the sides or top, and they arch their bodies into a loop as they crawl.

Generally, looper populations are not large enough to result in economic loss. Applications of *B. thuringiensis* are usually effective in controlling looper larvae. Although these products contain bacterial endospores and can cause infection, much of the mortality is attributed to endotoxins that are produced by the bacteria during fermentation and which are usually enclosed in the endospores. Rotation of products based on different subspecies of *B. thurinigiensis* (i.e. *kurstaki* and *aizawei*) or products derived from transconjugation or recombinant DNA techniques will ensure that different complements of endotoxins are being applied and therefore resistance to specific endotoxins is avoided. Mass releases of the egg parasitoid, *Trichogramma pretiosum* Riley, have reduced looper populations by up to 50%.

(a)

(b)

(c)

(d)

(e)

(f)

**Figure 5.4 (a) Tomato aphid (*Lipaphis erysimi* Kaltenbach (b) American serpentine leaf miner (*Liriomyza trifoli* Burgess) (c) Tomato fruit borer (*Helicoverpa armigera Hubner*) (d) tomato mealy bug (*Phenacoccus Solenopsis* Tinsley, (e & f) Tomato white fly (*Bemisia tabaci* Gennadius)**

## Armyworms

Of the numerous species of armyworm in the genus *Spodoptera* damaging tomatoes worldwide, the most common include: the beet armyworm, *S. exigua* (Hübner); the cotton leaf worm, *S. littoralis* Boisd; the southern armyworm, *S. eridania* (Cramer) (Colour Plate 2); the yellowstriped armyworm, *S. ornithogalli* (Guenée); the western yellowstriped armyworm, *S. praefica* (Grote); the cluster caterpillar, *S. litura* (Fabricius); the fall armyworm, *S. frugiperda* (J.E. Smith); and the tomato moth or bright-line brown-eye, *Lacanobia oleracea* (L.). All species deposit eggs on the undersides of leaves in masses of 50 or more and females of most species cover the eggs with hair-like scales. Hatching larvae feed gregariously on the undersides of leaflets, leaving the upper leaf epidermis intact and forming 'windows'. Later instars disperse throughout the same and adjacent plants, eating holes through leaves and giving plants a 'shot hole' appearance. Although fruit are not preferred feeding sites, larvae will feed on the surface of fruit when they encounter them. Because of the large number of larvae present, nearly all of the fruit on a single plant may be damaged. Secondary microorganisms often enter the fruit at the feeding sites, causing the fruit to rot. Larvae of most species appear smooth to the eye and have lateral yellow, tan or white stripes. Some have black triangular markings on their backs. Mature larvae are 2.5–5 cm long, depending on species. Although most armyworm larvae are susceptible to the bacterium *B. thuringiensis*, spray applications should be applied to coincide with egg hatch for best results. *B. thuringiensis* products containing the Cry1C endotoxin are more effective (see additional comments for using *B. thuringiensis* products under the discussion for loopers, above). Products containing polyhedral occlusion bodies of the nuclear polyhedrosis virus are available for managing *S. exigua* larvae. The spinosyns, a group of compounds derived from the bacterium *Saccharopolyspora spinosa*, are effective against many species of armyworm larvae and also are less toxic to many natural enemy species.

## Fruitworms

At least three species of fruitworm attack tomatoes: the tomato fruitworm, *Helicoverpa zea* (Boddie) *Heliothis armigera* Hübner; and the tobacco budworm, *Heliothis virescens* (Fabricius). Eggs are usually deposited singly on terminal leaflets of leaves near flowers or small fruit. Hatching larvae prefer to feed on fruit but also will attack buds, flowers or stems. Ripe fruit are seldom attacked and *H. virescens* particularly seems to prefer stems. Larvae bore deeply into fruit, primarily at the stem end, and can complete their entire larval development inside a fruit that is large enough. Damaged fruit often rot due to invasion of secondary microorganisms. Larvae are similar to armyworms in size and appearance but have bristle-like spines and patches of shorter spines that are usually visible with a hand lens. Because larvae quickly bore into fruit, where they are inaccessible to insecticides, control must be targeted at the eggs or at newly hatching larvae. Eggs can be monitored on leaves immediately above or below fruit hands with open flowers or small, green fruit. Mass releases of *T. pretiosum* egg parasitoids have resulted in 50–85% parasitism. *B. thuringiensis* products are generally not as effective in controlling fruitworm larvae, because of the rapidity at which the larvae bore into fruit. Products containing the Cry1A(c) toxin are the most effective (see additional comments for using *B. thuringiensis* products under the discussion for looper control, above).

## Hornworms

The tomato hornworm, *Manduca quinquemaculata* (Haworth), and tobacco hornworm, *M. sexta* (Linnaeus), usually deposit smooth, pale green eggs on the lower surfaces of leaves. Hatching larvae are pale green and have a distinctive horn or spine on the rear end. Older larvae retain the horn but also have diagonal white stripes on their sides. Larvae grow to be up to 10 cm long and are voracious feeders, consuming entire leaves and small stems. Large pieces of fruit may be consumed, but foliage is preferred. Hornworms do not usually require control because eggs and larvae are often attacked heavily by natural enemies; however, whole plants can be defoliated if a population is left unchecked. Mass releases of the egg parasitoid *T. pretiosum* have resulted in parasitism up to 70%. Most products containing *B. thuringiensis* also are effective (see additional comments for using *B. thuringiensis* products under the discussion for looper control, above).

## Potato tuberworm

The potato tuberworm, *Phthorimaea operculella* (Zeller), is more widely distributed than the tomato pinworm but attacks tomato similarly. Eggs are deposited singly on the undersides of leaves. Most larvae form blotch mines upon hatching but eventually move to the stems. Fruit may be damaged in a way similar to that by the tomato pinworm but larvae may also bore into terminal stems, causing tip dieback. Larvae are larger than the tomato pinworm, being about 13 mm long, and are greyish white to pink. Populations of the potato tuberworm can be avoided by not planting tomatoes near infested potatoes. Insecticides should target the eggs or larvae in blotch mines.

## Tomato pinworm

The tomato pinworm, *Keiferia lycopersicella* (Walsingham), is a serious pest of tomatoes in the south-eastern and south-western USA, Mexico and some Caribbean islands. Small, reticulated white eggs about 0.5 mm in diameter are deposited singly on the undersides of leaves. Newly hatched larvae are light coloured and bore into leaves to form blotch mines. Older larvae grow to about 8 mm and are reddish purple. They roll leaves or bore into green or red fruit, usually under the calyx. Fruit damage can be difficult to detect, due to feeding under the calyx and because larvae often cover their entrance holes with silk. Infested fruit seldom rot, thus causing a potential contamination problem.

New tomato fields should be set with transplants free of tomato pinworm eggs or larvae. Old crops should be destroyed immediately upon completion of harvesting. Mass releases of the egg parasitoid *T. pretiosum* may be helpful in reducing tomato pinworm populations. Mating disruption, using applications of components of the sex-attractant pheromone, has been extremely effective in managing within-field populations. Applications of pheromone can be made when pheromone-baited sticky traps capture five moths per trap per night. Because immigrating females may already have mated, spot spraying of field perimeters with insecticide may be necessary. *B. thuringiensis* products have generally been ineffective in managing the tomato pinworm. The tomato leafminer, *Scrobipalpula absoluta* Meyrick, is similar to the tomato pinworm but most economic damage is attributed to serpentine leafmining rather than fruit feeding. It is a pest in South American countries, especially Peru, Chile, Argentina, Colombia and Brazil.

## Sucking damage: fruit

## Stink bugs

Tomatoes are attacked by a number of species of stink bug, including: the southern green stink bug, *Nezara viridula* (Linnaeus); the brown stink bug, *Euschistus servus* (Say); the consperse stink bug, *E. conspersus* Uhler; the Say stink bug, *Chlorochroa sayi* (Stål); and others. The green or brown adults are about 10–15 mm in length and shield-shaped. Small barrel-shaped eggs are deposited on leaves in clusters of ten or more. Nymphs are similar in shape to adults, only more rounded and without wings, and may be brightly coloured in black, green, orange and white. When handled, most kinds produce a strong odour. Both nymphs and adults feed on fruit, producing a lightened blotch beneath the fruit surface of green fruit that turns yellow as the fruit ripen Since stink bugs tend to increase on leguminous plants, tomatoes should not be planted near or adjacent to legume crops such as soybeans or near or adjacent to fallow fields with legume cover crops such as *Sesbania* or *Indigofera*. Some species of stink bug are predacious and so proper identification is important.

## Coreid bugs

The leaf-footed bug *Leptoglossus phyllopus* (Linnaeus), *Phthia picta* (Drury) and others are plant bugs in the family *Coreidae* that are similar to stink bugs but are longer (20 mm) and darker with parallel sides. Eggs are metallic and ovate but somewhat flattened laterally and are laid in rows, sometimes along stems, or clusters, usually on leaves. As with the stink bugs, both adults and nymphs feed on fruit, but feeding punctures usually are deeper, which results in distortion of fruit as the fruit grow and expand. Because certain leaf-footed bugs, including *L. phyllopus*, are attracted to leguminous cover crops such as beggarweed and *Crotalaria* and to solanaceous weeds such as nightshades, tomatoes should not be planted near or adjacent to fallow or weedy fields containing such plants.

## Sucking damage: foliage and flowers

## Aphids

The most common species of aphid attacking tomato plants are: the green peach aphid, *Myzus persicae* (Sulzer); the cotton or melon aphid, *Aphis gossypii* Glover; and the potato aphid, *Macrosiphum euphorbiae* (Thomas). All are pear-shaped with a posterior pair of tubercles projecting upwards and backwards from the top surface of the abdomen. The green peach aphid is about 1.5 mm long and light to dark green, while the potato aphid is larger (3 mm long) and either green or pink. *A. gossypii* is similar in size and shape to the green peach aphid but generally darker in colour. Winged and non-winged forms are all females and give birth to living nymphs. They feed on the undersides of leaves, though the potato aphid feeds on stems and petioles, causing spotting, general chlorosis and distortion of leaves. If the populations are lage stunting and wilting of plants and abscission of flowers may occur. Large populations can cause stunting and wilting of plants and abscission of flowers. Aphids feed in the phloem, extracting more plant sap than they need. The sugary excess, called honeydew, is excreted and accumulates on the upper surfaces of leaves and fruit, where it supports the growth of sooty moulds. Aphids are most damaging as vectors of plant viruses, such as potato virus Y and tobacco etch, which cause serious debilitating diseases. Aphids are not usually directly damaging, because they are attacked by a large complement of natural enemies. Insecticides should be selected for aphid control, as well as for control of other pests, that will conserve the natural enemies. Most insecticidal sprays do not afford protection against the primary transmission of virus by immigrating winged aphids, but they can be useful in reducing the amount of secondary spread within the field. Covering the soil underneath plants with aluminium plastic film, which reflects UV light, will reduce the number of alighting aphids and reduce or delay the incidence of plants with virus symptoms.

## Mites

Two groups of mites are pests of tomatoes: the spider mites in the genus *Tetranychus*, particularly *T. urticae* Koch; and the tomato russet mite, *Aculops lycopersici* (Massee). Although both groups feed on tomatoes with piercing– sucking mouthparts, they are different in appearance and in the damage they inflict.

Spider mites are 0.3–0.5 mm long and are oval. They oviposit and feed on the undersides of lower leaves. The lower surface of infested leaflets may be covered with silken threads, while the upper surface will have small chlorotic spots or stippling. Increasing populations of spider mites move to the top leaves and produce silk webbing. Plants may appear chlorotic from a distance, mimicking nutritional deficiencies. Population increase is favoured by hot, dry conditions.

Russet mites are smaller and more elongate than spider mites and detection requires at least a 14 × hand lens or dissecting microscope. Russet mites also attack the undersides of lower leaves, but the upper surfaces of damaged leaves appear silvery, chlorotic and, eventually, necrotic. As increasing populations move up the plant, stems and leaf petioles appear bronzed and lower foliage desiccates. Hot dry weather favours population increase, which may result in all foliage turning brown and desiccating. Damage may be confused with nutritional deficiencies or imbalances or with water stress.

Mites generally are kept under natural control in the field by predacious mites and numerous insect predators. Applications of pesticides for other pests can disrupt the natural control and cause outbreaks, especially in hot, dry weather. Sulphur is effective in controlling the russet mite but is less effective against spider mites. For the latter pests, other selective miticides should be used. Insecticidal soap and detergent can provide some short-term control, but thorough coverage is essential and reduced plant growth and yield can occur when detergent is applied weekly at rates above 1%.

## Thrips

Numerous species of thrips are pests of tomato. Most, especially in the genus *Frankliniella,* occur primarily in flowers and include: the western flower thrips, *F. occidentalis* (Pergande); the tobacco thrips, *F. fusca* (Hinds); the flower thrips, *F. tritici* (Fitch); and *F. schultzei* (Trybom). Other species, such as *Echinothrips americanus* Morgan, the onion thrips *Thrips tabaci* Lindeman and the greenhouse thrips *Heliothrips haemorrhoidalis* (Bouché), occur in flowers but attack foliage more commonly than the *Frankliniella* species. Thrips are minute, elongate insects 0.2–0.5 mm in length. Adults are usually yellow, brown or black with slender, feather-like wings fringed with long hairs. Damage to flower parts, especially pistils, may cause or contribute to bloom abscission and reportedly results in necrotic lines or 'zippering' and deformation or 'catfacing' of fruit. Feeding or oviposition on small fruit may cause pitting, especially on the blossom end, as the fruit develop. Oviposition by *F. occidentalis* may result in white spots surrounding the ovipositional scar. Tomato spotted wilt virus is transmitted by nine species of thrips, the most important of which include *F. occidentalis*, *F. schultzei*, *F. fusca* and *T. tabaci*. The virus can only be acquired by larvae but is transmitted by feeding of adults on leaves. Although densities of thrips greater than five per flower have resulted in an increase in bloom abscission, lower densities may actually improve pollination. Thrips migrate into tomatoes as the bloom declines in flowering plants, such as clover or fruit trees, or as weedy or grassland areas dry down. Therefore, tomatoes should be planted away from these situations in time or space. Insecticides can be helpful in reducing the secondary spread of tomato spotted wilt virus within a field but are generally ineffective in reducing the primary spread of the virus by immigrating thrips. Aluminium plastic film that reflects UV light can be used to cover the soil under plants, thereby reducing the number of alighting thrips and reducing or delaying the incidence of plants with spotted wilt symptoms.

## Whiteflies

The sweet potato whitefly, *Bemisia tabaci* (Gennadius); the silverleaf whitefly, *B. Argentifolii* Bellows & Perring (formerly strain B of *B. tabaci*); and the greenhouse whitefly, *Trialeurodes vaporariorum* (Westwood) are the most serious pests of tomatoes. The latter species is more common in the greenhouse but can reach damaging populations in the field.

Adults look like small white moths, about 1–1.5 mm in length. Adults prefer to settle and deposit eggs on the underside of foliage in the upper third of the plant. Immature whiteflies are scale-like and range in length from about 0.3 mm to 0.7 mm. The first instar nymph is the only mobile immature life stage, but it usually settles for feeding within a few millimetres of where it hatches from the egg. Subsequent nymphal instars are sessile and do not move. Adult feeding is usually of very little direct consequence, though adults of *B. argentifolii* can cause light spotting of small fruit. Adults of *Bemisia* spp. May transmit viruses, particularly geminiviruses, that cause very serious diseases. The most noteworthy is tomato yellow leaf curl virus, which reduces new growth so severely that there is little or no subsequent yield. Feeding of nymphs of *B. argentifolii* on foliage causes an irregular ripening disorder of tomato characterized by incomplete fruit ripening, especially in longitudinal streaks or sections. Also associated with nymphal feeding is an increase in internal white tissue, which seems to be especially severe in hot weather.

Because tomato plants are not only a reservoir of the whiteflies but also a major reservoir of the viruses that they transmit, it is helpful to encourage a tomato-free period by destroying old crops as soon as harvest is completed and by delaying planting of new fields. In addition, new tomato fields should be separated in time and space from old tomato fields or from fields of other infested commodities such as potatoes, cabbages, ornamentals, etc. Transplant production facilities should be located away from production fields and should have all openings covered with a fine-mesh screen to exclude migrating adults. Plastic that absorbs UV light can be used to cover plant production facilities as well as plants in the field. Yellow plastic film or aluminium plastic film that reflects UV light can be used to cover the soil under the plants. These films will result in fewer alighting whitefly adults and thus in reducing or delaying the incidence of plants with virus symptoms. Tomato cultivars resistant to tomato yellow leaf curl virus are available. Systemic insecticides such as imidacloprid can be applied to the soil for seedlings in transplant production facilities and to transplants in the field, providing not only whitefly control but also limited control of primary virus transmission by immigrating whitefly adults. This insecticide and insect growth regulators, insecticidal soap, oils and conventional insecticides can help to control whiteflies and reduce secondary spread of virus in the field.

## Pests of pepper

About 20 insect species have been recorded damaging pepper plantation (Figure 5.6). They include :

**Table 5.1:** *insect pests of chillies*

| Common name | Scientific name | Family | Order |
|---|---|---|---|
| **Major pests** | | | |
| Pollu beetle | *Longitarsus nigripennis* | Alticidae | Coleoptera |
| Top shoot borer | *Cydia hemidoxa* | Eucosmidae | Lepidoptera |
| Berry gall midge | *Cecidomyia malabarensis* | Cecidomyiidae | Diptera |
| Marginal gall thrips | *Liothrips karnyi* | Thripidae | Thysanoptera |
| **Minor pests** | | | |
| Pepper mussel scale | *Lepidosaphes piperis* | Diaspididae | Hemiptera |
| Soft scale | *Marsipococcus marsupiale* | Coccidae | Hemiptera |
| Coconut scale | *Aspidiotus destructor* | Diaspididae | Hemiptera |
| Whitefly | *Aleurocanthus piperis* | Aleyrodidae | Hemiptera |
| Wild silkworm | *Cricula trifenestrata* | Saturniidae | Lepidoptera |

## Pollu beetle

Pollu beetle (*Lanka ramakrishnai*) is the most destructive pest of black pepper and is more serious in plains and at altitudes below 300 m. The adult is a small black beetle measuring about 2.5 mm × 1.5 mm, the head and thorax being yellowish brown and the fore wings (elytra) black. Fully-grown grubs are creamy-white and measure about 5 mm in length. The adult beetles feed and damage tender leaves and spikes. The females lay eggs on tender spikes and berries. The grubs bore into and feed on the internal tissues and the infested spikes turn black and decay. The infested berries also turn black and crumble when pressed. The term pollu denotes the hollow nature of the infested berries in Malayalam. The pest infestation is more serious in shaded areas in the plantation. The pest population is higher during September-October in the field.

Regulation of shade in the plantation reduces the population of the pest in the field. Spraying quinalphos (0.05%) during June-July and September- October or quinalphos (0.05%) during July and Neemgold (0.6%) (neem-based insecticide) during August, September and October is effective for the management of the pest. Spraying needs to be made on the underside of leaves where the pest is generally seen.

## Top shoot borer

The top shoot borer (*Cydia hemidoxa*) is a serious pest in younger plantations in all black pepper areas. The adult is a tiny moth with a wing span of 10-15 mm with crimson and yellow fore wings and grey hind wings. The larvae bore into tender terminal shoots and feed on internal tissues resulting in blackening and decaying of affected shoots. Fullygrown larvae are greyish green and measure 12-15 mm in length. When successive new shoots are attacked, the growth of the vine is affected. The pest infestation is higher during July to October when numerous succulent shoots are available in the vines. Spray quinalphos (0.05%) on tender terminal shoots; repeat spraying at monthly intervals (during July-October) to protect emerging new shoots.

## Leaf gall thrips

Infestation by leaf gall thrips (*Liothrips karnyi*) is more serious at higher altitudes especially in younger vines and also in nurseries in the plains. The adults are black and measure 2.5 mm-3.0 mm in length. The larvae and pupae are creamy white. The feeding activity of thrips on leaves causes the leaf margins to curl downwards and inwards resulting in the formation of marginal leaf galls. Later the infested leaves become crinkled and malformed. In severe cases of infestation, the growth of younger vines and cuttings in the nursery is affected. Spray dimethoate (0.05%) during emergence of new flushes in young vines in the field and cuttings in the nursery.

## Scale insects

Among the various scale insects recorded on black pepper, mussel scale (*Lepidosaphes piperis*) and coconut scale (*Aspidiotus destructor*) causes serious damage to black pepper vines at higher altitudes and also to older cuttings in nurseries in the plains. Females of mussel scales are elongated (about 1 mm length) and dark brown and that of coconut scales circular (about 1 mm in diameter) and yellowish brown. Scale insects are sedentary, remaining permanently fixed to plant parts and appear as encrustations on stems, leaves and berries. They feed on plant sap and cause yellowing and wilting of infested portions; in severe cases of infestation the affected portions of vines dry up. The pest infestation is more severe during the post monsoon and summer periods. Clip off and destroy severely infested branches. Spray dimethoate (0.1%) on affected vines after harvest of produce; repeat spraying after 21 days to control the infestation completely. Initiate control measures during early stages of pest infestation. In nurseries spraying neem oil 0.3% or Neemgold 0.3% or fish oil rosin 3% is also effective in controlling the pest infestation.

## Minor pests

Leaf feeding caterpillars, especially *Synegia* sp., damage leaves and spikes of younger vines and can be controlled by spraying quinalphos (0.05%). Mealybugs, gall midges and aphids infest tender shoots especially in nurseries. Spraying dimethoate (0.05%) may be undertaken if infestations are severe. Mealybug infestation on roots can be controlled by drenching with chlorphyriphos (0.075%) and undertaking control measures against *Phytophthora* and nematode infections.

## References

Aback, K., N. Sari, M. Paksoy, O. Kaftanoglu & H. Yeninar 1995. Efficiency of bumble bees on the yield and quality of eggplant and tomato grown in unheated glasshouses. Acta Horticulturae 412:268-274.

Abak K., Guler H. Y., 1994. Polen fertility and the vegetative growth of various eggplant genotypes under low temperature greenhouse conditions. Acta Hortic., 366, 85–91.

Aguilar, A.Q., S. Kalisz and T.L. Ashman, 2008. Flower morphology and pollinator dynamics in *Solanum carolinense* (Solanaceae): Implications for the evolution of andromonoecy. Am. J. Bot., 95: 974-984.

Al-Abbadi, S.Y., 2007. Influence of different numbers of Syrian honeybee nuclei (*Apis mellifera syriaca* L.) On the production of faba bean (*Vicia faba* L.) in plastic house. Bull. Fac. Agric., Cairo Univ., 58: 266-270.

Abney T. D., Russo V. M., 1997. Factors affecting plant height and yield of eggplant. J. Sustain. Agric., 10, 37–47.

Abrol, D. P. 1987. Influence of thermal and energetic constraints on the pollination activity of carpenter bee *Xylocopa pubescens*. *Environ. Ecol.*, 5(1):90-93.

Abrol, D. P. 1991 a. Conservation of pollinators for promotion of agricultural production in India. *J. Anim. Morphol. Physiol.*, 38(1-2):123-139.

Abrol, D. P. 1991 b. Pollination of brinjal flowers [*Solanum melongena* L.] by bumble bees. *J. Anim. Morphol. Physiol.*, 38(1-2):95-98. 285

Abrol, D. P. 2005. Pollination Energetics. *Journal of Asia Pacific Entomology*, 8(1):3-14.

Abrol, D. P. and D. R. Kotwal. 1996. Insect pollinators of linseed (*Linum usitatissimum* Linn.) and their effect on yield components. *Journal of Animal Morphology and Physiology*, 43(2):157-161.

Abrol, D. P. and R. P. Kapil. 1998. Studies on Abundance and importance of insect pollinators for oilseed production. *Journal of Insect Science*, 9(2):172-174.

Al-Abbadi, S.Y., 2007. Influence of different numbers of Syrian honeybee nuclei (*Apis mellifera syriaca* L.) On the production of faba bean (*Vicia faba* L.) in plastic house. Bull. Fac. Agric., Cairo Univ., 58: 266-270.

Allen-Wardell, A.G., P. Bernhardt, R. Bitner, A. Burqez, S. Bushmann, J. Cane, P.A. Cox, V. Dalton, P. Feinsinger, M. Ingram, D. Inouye, C.E. Jones, K. Kennedy, P. Kevan and H. Andrén 1998. Effects of habitat fragmentation on birds and mammals in landscapes with different proportions of suitable habitat. Oikos 71:355-366.

Amoako J., Yeboah–Gyan K., 1991. Insect pollination of three solanaceous vegetable crops in Ghana with special reference to the role of African honey bee (*Apis mellifera adansonii*) for fruit set. Acta Hortic., 288, 255–259.

Amoako, J., K. & Yeboa-Gyan 1991. Insecto pollination of three solanaceous vegetable crops in Ghana with special reference to the hole of African honey bee (*Apis mellifera adansonii*) for fruit set. Acta Horticulturae 288:255-259.

Anand, R.K; Gupta, A.K. and Ghai, S. 1988. A check-list of Indian *Epilachna* (Coccinellidae: Coleoptera) with recorded host plants. Bulletin of Entomology (New Delhi), 29(1):121-137.

Anonymous, 2003. *Indian Horticulture Data Base.* 2003. National Horticulture Board, Ministry of Agriculture, Government of India.

Anonymous, 2003. Production Recommendation Vegetables (Jammu Division). Directorate of Extension Education, Sher-e-Kashmir University of Agricultural Sciences and Technology, Jammu.

*Apidae*): Tomato foraging behavior and pollination efficiency. Appl. Entomolog. Zool., 32, 581–586.

Armstrong, J.A., 1979. Biotic pollination mechanisms in the Australian flora: A review. N. Z. J. Bot., 17: 467-508.

Asada S., 1997. Tomato pollination with Japanese native bumblebees (*Bombus spp.)* Acta Hortic., 437, 289–291.

Asada S., Ono M., 1996. Crop pollination by Japanese Bumblebees, *Bombus spp.* (*Hymenoptera*: *Apidae*) tomato foraging behaviour and pollination efficiency. Applied Entomology Zoology 31 : 581–586.

Avanzi, M.R. & M.J.O. Campos 1997. Estrutura de guildas de polinização de *Solanum aculeatissimum* Jacq. e S. *variabile* Mart. (Solanacea). Revista Brasileira de Biologia 57(2):247-256.

Azzam, H. 1960. Natural cross-pollination of tomatoes in Puerto Rico. Amer. Soc. Hort. Sci. Caribbean Region. Proc. 4: 85-86.

Babu, G. V. S.; Bhatt, S. C. and S. C. Pandeya. 1987. Pollination Ecology of *Solanum sysimbrifolium* Lamk. *Proc. Indian Acad. Sci., Plant Sci.*, 97 (4):295-300. 287

Bailey, L. H., and Lodeman, E. G. 1895. Inforcing house miscellanies. N.Y. (Cornell) Agr. Expt. Sta. Bul. 96, pp. 327-328.

Banda, HJ and Paxton, R.J. 1991. Pollination of greenhouse tomatoes by bees. Sixth International Symposium on pollination. *Acta Horiculture* 288: 194–198.

Barrons, K. C., and Lucas, H E. 1942. The production of first generation hybrid tomato seed for commercial planting. Amer. Soc. Hort. Sci. Proc. 40: 395-404.

Bashir, C.M., 1953. Some pollination studies on chilies. Agric. Pak., 3: 125-128.

Beattie, J. H. 1939. Greenhouse tomatoes. U.S. Dept. Agr. Farmers' bul. 1431, 28 pp.

Behra, T.K.; Singh, T.S. and Kalda, T.S. 1997. Breeding for resistance to shoot and fruit borer *Leucinodes orbonalis* Guen. in egg plant (*Solanum melongena*). *Symposium on Integrated Pest Management for Sustainable Crop Production.* Dec. 2-4, 1997. Division of Entomology. Indian Agricultural Research Institute, New Delhi. pp. 39-40 (Abst.).

Belletti, P. and L. Quagliotti, 1995. Capsicum and Eggplant Newsletter. University of Turin, Plant Breeding and Seed Production, USA.

Berinjela (*Solanum melongena* L.) – mito ou realidade no combate as dislipidemias? Brazilian Journal of Pharmacognosy 16(2):252-257.

Bezerra, E.L.S. and I.C. Machado 2003. Biologia floral e sistema de polinização de *Solanum stramonifolium* jacq. (Solanaceae) em remanescente de mata Atlântica, Pernambuco. Acta Botanica Brasilica 17(2):247-257.

Bhagat, K.C. and Munshi, S.K. 1997. *Fusarium solani* (Mart.) Sacc. An entomogeneous fungus infecting the spotted *Epilachna* beetle, *Henosepilachna vigintioctopunctata* Fabr. *Pest Management and Economic Zoology*, **5**(2):139-142

Bollard, E. T., and Stevenson, E. C. 1953. Production of hybrid tomato seed. Amer. Soc. Hort. Sci. Proc. 61: 451-458.

Bortolotti, L., R. Montanari, J. Marcelino, P. Medrzycki, S. Maini & C. Porrini 2003. Effects of sub-lethal imidacloprid doses on the homing rate and foraging activity of honey bees. Bulletin of Insectology 56(1):63-67.

Bouquet, A. G. B. 1919. Pollination of tomatoes. Oreg. Agr. Expt. Sta. Bul. 158, 29 pp.

Bouquet, A. G. B. 1924. Economic results in the pollination of greenhouse tomatoes. Oreg. Agr. Expt. Sta. Cir. 55, 16 pp.

Brittain, C.A., M. Vighi, R. Bommarco, J. Setteled & S.G. Potts 2010. Impacts of a pesticide on pollinator species richness at different spatial scales. Basic and Applied Ecology 11:106-115.

Buchmann, S., E. and G. P. Nabhan. 1996. The forgotten pollinators. Island Press, Washington, D C, USA.

Buczek J., 1972. Teoretyczne podstawy zastosowania substancji wzrostowych w produkcji roślinnej. Biul. Warz., 13, 11–28.

Butani, D.K. and Jotwani, M.G. 1984. *Insects of Vegetables*. P.S. Periodical Export Book Agency, New Delhi.

Butani, D.K. and Verma, S. 1976. Pests of vegetables and their control in brinjal. *Pesticides* **11**(1): 33-35.

Campos, M.J.O., M.A. Pizano, O. Malaspina, L. Giordano, R. Leung, J. Chaud-Neto, E.G. Gomig, E.M.B. Prata, G.B. Patricio, B. Ferreira, H. Fang & E.S. Silva 2006. Manejo Agrícola e Riqueza de polinizadores. VII Encontro sobre abelhas, 7o, 2006, Ribeirão

Carvalho, C.A.L., O.M. Marques, C.A. Vidal & A.M.S. Neves 2001. Comportamento forrageiro de abelhas (Hymenoptera, Apoidea) em flores de *Solanum Paniculatum* Dunal (Solanaceae). Revista Brasileira de Zoociências 3:35-44.

CEASA-Campinas 2007. Padronização/ berinjela. Available in: www.ceasacampinas.com.br. Accessed in February 11th of 2012.

Chapagain, B., P. and Z. Weisman. 2004. Effect of potassium magnesium chloride in the fertigation solution as partial source of potassium on growth, yield and quality of green house tomato. J. Sci. Hort. 99, 279-288.

Chen, C. & H.M. LI, H. 1996. Cultivation and seed production of eggplant. Asian Vegetable

cluster and 2,4–D treatments in three eggplant cultivars. Sci. Hortic., 20, 35–44.

Corbet, S.A. 1991. Bees and the pollination of crops and wild flowers in the European community. Bee World 72:47-59.

Cottrell-dormer, w. 1945. An electric pollinator for tomatoes. Queensland jour. Agr. Sci. 2: 157-169.

Currence, T. M. 1944. A combination of semi-sterility with two simply inherited characters that can be used to reduce the cost of hybrid tomato seed. Amer. Soc. Hort. Sci. Proc. 44: 403-406.

Currence, T. M. and Jenkins, J. M., Jr. 1942. Natural crossing in tomatoes as related to distance and direction. Amer. Soc. Hort. Sci. Proc. 41 273-276.

Dobromilska R. and 1997. Plonowanie odmian papryki słodkiej w zależności od sposobu zapylania Mat. VII Ogólnopol. Zjazdu Hod. Roślin Ogrod., 11–13.

Dobromilska R., Fawceet M., 1999. Znaczenie trzmiela ziemnego w uprawie papryki i pomidora pod osłonami. Zesz. Prob. Post. Nauk Rol., 466, 493–502.

Dogterom. M. H.; Matteoni, J. A. and R. C. Plowright. 1998. Pollination of greenhouse tomatoes by the North American *Bombus vosnesenskii* (Hymenoptera: Apidae).*Journal of Economic Entomology*, 91(1):71-75.

DOS 2007 the Hashemite Kingdome of Jordan, Department of Statistic. Annual agricultural statistics.Amman, Jordan, 2007

Dylewska M., 1996. Nasze trzmiele. Karniowice. Ośrodek Doradztwa Rolniczego.

eggplant (*Solanum melongena* L.) grown in a plastic tunnel. Folia Hortic. 15/2, 77–87.

Fink, B. 1898. Pollination and reproduction of Lycopersicum esculentum. Minn. Bot. Studies 1: 636-643.

Fletcher S. W., and Gregg, O. I. 1907. Pollination of forced tomatoes. Mich. Agr. Expt. Sta. Spec. Bul. 39: 2-10.

Forni-Martins, E.R, M.C.M. Marques and M.R. Lemes 1998. Biologia floral e reprodução de *S. paniculatum* L. (Solanaceae) no estado de São Paulo, Brasil. Revista Brasileira de Botânica 21(2):117-124.

Free, J.B., 1993. Insect Pollination of Crops. 2nd Edn., Academic Press, London, UK., pp: 96-104.

Greenleaf, S.S. and C. Kremen, 2006. Wild bees enhance honey bees' pollination of hybrid sunflower. Proc. Natl. Acad. Sci. USA., 103: 13890-13895.

Hayo Velthuis, H.W. and D. Adriaan van, 2006. A century of advances in bumblebee domestication and the economic and environmental aspects of its commercialization for pollination. Apidologie, 37: 421-451.

Gemmill-Herren B., Ochieng A.O., 2008. Role of native bees and natural habitats in eggplant (*Solanum melongena*) pollination in Kenya. Agric., Ecosyst. Environ. 127, 31–36.

Gemmill-Herren, B. & A.O. Ochieng 2008. Role of native bees and natural habitats in eggplant (Solanum melongena) pollination in Kenya. Agriculture, Ecosystems and Environment 127:31-36.

George R. A. T., 1985. Vegetable seed production. London i New York, 223–229.

Gill, H.S. and Tomar, B.S. 1991. *Vegetable Statistics at a Glance.* Project Directorate on Vegetable Research (ICAR). Division of Vegetable Crops. IARI, New Delhi.

Glapś T., and Górecki R., 1989. Wpływ substancji wzrostowych na plonowanie oberżyny. Biul. Warz.m33, Supl., 125–127.

Gomig, E.G. 2007. Caracterização da fauna de abelhas silvestres com potencial de polinização de berinjela (*Solanum melongena* L.) cultivada em sistema orgânico. Trabalho de conclusão de curso (Graduação em Ecologia). Departamento de Ecologia, Instituto de Biociências, Universidade Estadual Paulista "Júlio de Mesquita Filho", Rio Claro, SP, Brasil: 31-37.

Gonçalves, M.C.R., M.F.F.M. Diniz, J.D.C. Borba & X.P. Nunes, J.M. Barbosa-Filho 2006. Berinjela (Solanum melongena L.): mito ou realidade no combate as dislipidemias? Rev. Bras. Farmacogn. 2006;16:252–257.

Górecki R. S., Espinosa-Flores A., 1996. Wpływ heterostylii kwiatów oberżyny *Solanum melongena* L. odm. Rodo na wiązanie owoców oraz liczbę nasion. Mat. Konf. Międzyn. Symp., 15–19 lipiec 1996, Skierniewice, 50–51.

Gosiewski W., Skąpski H., 1988. Pomidory szklarniowe. PWRiL, Warszawa.

Goulson, D., G.C. Lye & B. Darvill 2008. Decline and Conservation of Bumble Bees. The Annual Review of Entomology 53:191-208.


Greenleaf, S.S. and C. Kremen, 2006. Wild bees enhance honey bees' pollination of hybrid sunflower. Proc. Natl. Acad. Sci. USA., 103: 13890-13895.


Grierson, D. and M., A. Kader. 1986. Fruit ripening and quality. The tomato crop. Chapman and Hall, London. p. 240-280

Guerra, A.M.N.M., J.R.C. Neto, J.V.A.D. Marques, M.F. Pessoa & P.B. Maracajá 2007.

Gul, H. and Hamid, A. 1997. Efficacy of *Beauveria bassiana* as bio-control agent against hadda beetle *Epilachna dodecastigma* Mulsant (Coccinellidae: Coleoptera) under laboratory conditions. *Pakistan Journal of Forestry*, **47**(1/4):37-45.

Gupta, A.K. 2002. *Vegetable growing in J&K* ( Jammu Division) Division of Vegetable Science, Sher-e- Kashmir University of Agricultural Sciences and Technology, Regional Vegetable Research Station, Chatha, Jammu.

Gupta, P.R. and Dogra, G.S. 1994. Sensitivity of stages of development of *H. vigiintioctopuctata* to tropical application of diflubenzuron and penfluron. *Journal of Insect Science* 7(2):143-147.

Hafen, L., and Stevenson, E. C. 1956. Natural cross-pollination in tomato using several malesterile mutants. Amer. Soc. Hort. Sci. Proc. 68: 433-436.

Hameed, S.F. and Singh, S.P. 1998. Hand Book of Pest Management, Kalyani Publishers, New Delhi. 196p.

Handique A. K., Sarma A., 1995. Alteration of heterostyly in *Solanum melongena* L. through gamma-radiation and hormonal treatment. J. Nuc. Agric. Biol., 24, 121–126.

Hayo Velthuis, H.W. and D. Adriaan van, 2006. A century of advances in bumblebee domestication and the economic and environmental aspects of its commercialization for pollination. Apidologie, 37: 421-451.

Heard, T.A. 1999. The role of stingless bees in crop pollination. Annual Review of Ecology and Systematics 44:183-206.

Hein, L. 2009. The Economic Value of the Pollination Service, a Review Across Scales. The Open Ecology Journal 2:74-82.

Heywood, V.H. 1993. Flowering plants of the world. New York: Oxford Univ. Press, 396pp.

Higo, H. A.; Rice, N. D.; Winston, M. L. and B. Lewis. 2004. Honeybee (Hymenoptera: Apidae) distribution and potential for supplementary pollination in commercial tomato greenhouses during winter. *Journal of Economic Entomology*, 97 (2):163-170.

Hoffman, I. C. 1958. Pollination problems; here are ways to overcome poor fruit-set in tomatoes. Amer. Veg. Grower and Market Growers Jour. 6(3): 43-45.

Hojby, H. R. 1958. Utilization of hybrid vigour in tomatoes: possibilities and limitations. So. African jour. Agr. Sci. 1: 249-261.

Shipp JL, Whitfield GH and Papadopoulos AP, Effectiveness of the bumble bee, Bombus *impatiens* Cr. (*Hymenoptera:Apidae*), as a pollinator of greenhouse sweet pepper. Sci. Hort., 57, 29–39.

Kessler, A., R. Halitschke & K. Poveda 2011. Herbivory-mediated pollinator limitation: negative impacts of induced volatiles on plant–pollinator interactions. Ecology 92(9):1769-1780

Isaacs, R. and J. Tuell, 2007. Conserving Native Bees on Farmland. Michigan State Univ. Ext. Bull., UK.

Isaacs, R. and J. Tuell, 2007. Conserving Native Bees on Farmland. Michigan State Univ. Ext. Bull., UK.

Jones, D. F. 1916. Natural cross-pollination in the tomato. Science, N.S., 43: 509-510.

Kakizaki, Y., 1931. Hybrid vigor in egg-plant and its practical utilization. Genetics, 16: 1-25.

Kristjansson, K. and K. Rasmussen, 1991. Pollination of sweet pepper (*Capsicum annuum* L.) with the solitary bee (*Osmia cornifrons*) (RADOSZKOWSKI). Acta Hort., 288: 173-179.

Kwon, Y.J. and S. Saeed, 2003. Effect of temperature on the foraging activity of *Bombus terrestris* L. (Hymenoptera: Apidae) on greenhouse hot pepper (*Capsicum annuum* L.). Applied Entomol. Zool., 38: 275-280.

Kakizaki, Y., 1931. Hybrid vigor in egg-plant and its practical utilization. Genetics, 16: 1-25.

Kristjansson, K. and K. Rasmussen, 1991. Pollination of sweet pepper (*Capsicum annuum* L.) with the solitary bee (*Osmia cornifrons*) (RADOSZKOWSKI). Acta Hort., 288: 173-179.

Kwon, Y.J. and S. Saeed, 2003. Effect of temperature on the foraging activity of *Bombus terrestris* L. (Hymenoptera: Apidae) on greenhouse hot pepper (*Capsicum annuum* L.). Applied Entomol. Zool., 38: 275-280.

Kaloo, G. 1986. Tomato (Lycopersicon esculentum Miller). Allied Publishers Pvt. Ltd., New Delhi. Pp. 203-220.

Kalra, V.K. 1997. Insect pest management in brinjal and cabbage/ cauliflower In: *Advaned Training Course on Insect Pest Management.* ( Eds. J.S. Naresh; H. Singh and H.R. Rohila) Indian Council of Agricultural Research, Centre for Advanced Studies, C.C.S. H.A.U., Hissar. pp. 110-117.

Karmakar, K. and Bhattacharya, B. 2000. Performance of a local brinjal variety (*Solanum melongena* L.) and its pest management under terai agro-ecological conditions. *Environment and Ecology*, **18**(2):344-346.

Kearns, C., A. and D.W. Inouye. 1993. Techniques for Pollination Biologists, Univ. Press of Colorado, Niwot, CO

Kearns, C., D. Inouye and N. Waser 1998. Endangered mutualisms: the conservation of plant pollinator interactions. Annual Review of Ecology and Systematics 29:83-112.

Kerr E. A. and Kribs, W. 1945. Electric vibrator as an aid in greenhouse tomato production. Queensland jour. Agr. Sci. 2: 157-169.

Kerr E. A.1955. Some factors affecting earliness in the tomato. Canad. Jour. Agr. Sci. 35(3): 300-308.

Kevan, P., G. and T. P. Phillips. 2001. Economic Impacts of Pollinator Declines: An Approach to Assessing the Consequences, The Conservation Ecology v.5, i.1 Jun01 University of Guelph

Kevan, P.G. and T.P. Phillips 2001. The economic impacts of pollinator declines: an approach to assessing the consequences. Conservation Ecology 5(1): 8. [online] URL: http:// www.consecol.org/vol5/iss1/art8/

King, J. and S.L. Buchmann 2003. Floral sonication by bees: mesosomal vibration by *Bombus* and *Xylocopa*, but not *Apis* (Hymenoptera: Apidae), ejects pollen from poricidal anthers. Journal of the Kansas Entomological Society 76(2):295-305.

Klein A-M., B.E. Vaissière, J.H. Cane, I. Steffan-Dewenter, S.A. Cunningham, C. Kremen and T. Tscharntke. 2007. Importance of crop pollinators in changing landscapes for world crops. Proceedings of the Royal Society of London. Series B. Biological Sciences 274:303-313.

Konys E., 1993. Warzywa psiankowate. [w:] Gapiński M., (red.), Warzywa mało znane i zapomniane. PWRiL, Poznań, 78–82.

Kowalska G., 2003a. The influence of heterostyly, pollination method and harmonization on eggplant's (*Solanum melongena* L.) flowering and fruiting. Acta Agrobotanica. 56, 1–2, 61–76.

Kowalska G., 2003b. The effect of pollination method and flower hormonization on yielding of estrutura, organização espacial da comunidade e seletividade de espécies.

[Thesis]. [Rio Claro (SP)]: Departamento de Geografia, Universidade Estadual Paulista "Júlio de Mesquita Filho". Ciências Agrárias e Veterinárias, Universidade Estadual Paulista p.86.

Kowalska G., 2006. Eggplant (*Solanum melongena* L.) flowering and fruiting dynamics depending on pistil type as well as way of pollination and flower harmonization. Folia Hortic. 18/1, 17–29.

Kowalska, G. 2006. Eggplant (*Solanum melongena* L.) flowering and fruiting dynamics depending on pistil type as well as way of pollination and flower harmonization. Horticulturae 18(1):17-29.

Kowalska, G. 2008. Flowering biology of eggplant and procedures intensifying fruit set – review. Acta Scientiarum Polonorum, Hortorum Cultus 7(4):63-76.

Krug H., 1991. Gemüseproduktion. Verlag Paul Parey. Berlin and Hamburg, 448–452.

Kumar, J. and Parmar, B.S. 1998. Neem oil chemistry and the bio- activity. *Pesticides Researc Journal* **10**(1): 14-43.

Kumar, R. 1987. Principal and Philosophy of Integrated Pest Management for Tropical Root and Tuber Crops. pp.110-115. IIIA, Ibadan, Nigeria.

Lawande K. E., Chavan J. K., 1998. Eggplant (Brinjal). Salunkhe D. K., Kadam S. S., (red.), Handbook of vegetable science and technology. Production, composition, storage and processing. New York, 225–244.

Lenz F., 1970. Effect of fruit on sex expression in eggplant (*Solanum melongena* L.). Hortic. Res. 10, 81–82.

Lesley, J. W. 1924. Cross pollination of tomatoes: varietal differences in amount of natural cross pollination important factor in selection. Jour. Hered. 15: 233-235.

Lesley, J. W. and Lesley, M. 1939. Unfruitfulness in the tomato caused by male sterility. Jour. Agr. Res. 58: 621-630.

Losey, J., E. and M. Vaughan. 2006. The economic value of ecological services provided by insects. Biosci, 56, 311–323

Mall, N.P.; Pandey, R.S.; Singh, S.V. and Singh, S.K. 1992. Seasonal incidence of insect-pests and estimation of the losses caused by shoot and fruit borer on brinjal. *Indian Journal of Entomology*, **54**(3):241-247.

Mandal, S.K. and Kumar, B. 2001. Evaluation of certain new insecticides against *Epilachna* beetle (*Epilachna vigintioctopunctata* Fab.) on brinjal. *Environment and Ecology*, **19**(2):483-484.

Mandal, S.K.; Kumar, B. and Prasad. D 2000. Varietal resistance against *Epilachna vigintioctopunctata* Fab. on brinjal. *Journal of Applied Biology*, **10**(1):66-68

Markus, P., 1965. Investigation on cross-fertilization in red pepper for spice. Hungarian Agric. Rev., 13: 750-750.

Markus, P., 1965. Investigation on cross-fertilization in red pepper for spice. Hungarian Agric. Rev., 13: 750-750.

McGregor, S.E., 1976. Insect pollination of cultivated crop-plants. U.S.D.A. Agriculture Handbook No. 496, Version with some updated information for some crop species pp: 93–98. http://gears.tucson.ars.ag.gov/book/. Nagarathnam, A.K. and T.S. Rajamani, 1963. Studies on fruit setting in chillies (*Capsicum annum* Linn.). Madras Agric. J., 50: 138-139.

Marr, C., and Hillyer, I. G. 1968. Effect of light intensity on pollination and fertilization of field and greenhouse tomatoes. Amer. Soc. Hort. Sci. Proc. 92: 526-530.

Mathur, A.C. and Srivastava, J.B. 1964. *E. vigintioctopunctata* as a defoliator of some solanaceous medicinal plants. *Indian Journal of Entomology* 26:246.

Mathur, Y.K. and Nigam, S. 1993. Insecticidal and juvenelizing effect of neem oil ( *Azadirachta indica* A. Juss. ) against *Corcyra cephalonica* Staint. And *H. vigintioctopunctata.* In Neem and environment (eds.) R.P.Singh; M.S. Chari; A.K. Rahoja and W. Kraus. Proceedings of World neem Conference. 24-28 Feb. 1993, Bangalore, India. Oxford & IBHCO Pvt. Ltd. New Delhi, 1996 pp. 335-342.

Thakur Mehta, P.K. and Sood. A.K. 1998. Oviposition deterrent activity of some plant extracts against *Henosepilachna vigintioctopunctata* (Fabr.). *Pest Management and Economic Zoology*, **6**(2):147-150.

Mishra, B.K.; Mishra, R.R. and Patnaik, N.C. 1990. Effect of neem oil on the growth and development of *Epilachna* beetle. *Orrisa Journal of Agriculture Research* **2**(3-4): 139-172.

Montemor K.A. and D.T.M. Souza 2009. Biodiversidade de polinizadores e biologia floral em cultura de berinjela (Solanum *melongena*). Zootecnia Tropical 27(1):97-103.

Moore, E. L. 1968. Obtaining fruit set of plastic greenhouse tomatoes. Miss. Agr. Expt. Sta. Bul. 768, 8 pp.

Muller, C. H. 1940. A revision of the genus Lycopersicum. U.s. Dept. Agr. Misc. Pub. 382, 28 pp.

Munshi, G.L.; Singh, S. and Singh, R. 1968. Insect pests of agricultural importance (Brinjal) in Jammu province. Govt. Agricultural College and Research Institute, R.S. Pura, Jammu (J&K) publication pp 11.

Nagarathnam, A.K. and T.S. Rajamani, 1963. Studies on fruit setting in chillies (*Capsicum annum* Linn.). Madras Agric. J., 50: 138-139.

Neiswander, R. B. 1954a. Honey bees as pollinators of greenhouse tomatoes. Gleanings Bee Cult. 82: 610-613.

Neiswander, R. B. 1954b. Honey bees as tomato pollinators. Ohio veg. And potato growers' assoc. 39th ann. Proc., pp. 96, 98,100,102, 104, 106, 108.

Neiswander, R. B. 1956. Pollination of greenhouse tomatoes by honey bees. Jour. Econ. Ent. 49: 436-437.

Nothmann J., Koller D., 1975. Effects of low-temperature stress on fertility and fruiting of egg- -plant (*Solanum melongena* L.) in a subtropical climate. Expl. Agric., 11, 33–38.

Nothmann J., Rylski I., Spigelman M., 1983. Interactions between floral morphology, position in cluster and 2,4-D treatments in three eggplant cultivars. Scientia Hortic.,. 20: 35-44.

Oba, G. I., riner, M. E., and Scott, D. H. 1945. Experimental production of hybrid tomato seed. Amer. Soc. Hort. Sci. Proc. 46: 269-276.

Ollerton, J., R. Winfree AND S. Tarrant 2011. How many flowering plants are pollinated by animals? Oikos 120:321-326.

Olympios C. M., 1976. Effect of growth regulators on fruit–set and fruit development of the eggplant (*Solanum melongena* L.). Hort. Res., 16, 56–70.

Pal B. P., and Singh H. B., 1943. Floral characters and fruit formation in the eggplant. I. J. Genet., 3(1), 45–58.

Passam H. C., Bolmatis A., 1997. The influence of style length on the fruit set, fruit size and seed content of aubergines cultivated under high ambient temperature. Trop. Sci., 37, 221–227.

Passam, H.C., Baltas, C., Boyiatzoglou, A., Khan, E.M. 2001. Flower morphology and number of aubergine (*Solanum melongena* L.). in relation to fruit load and auxin application. Scientia. Hortic., 89, 309–316.

Patricio, G.B. Grisolia, B.B., Desuoacute I.C. and Campos, M.J.D 2012. The Importance of Bees for Eggplant Cultivations (Hymenoptera: Apidae, Andrenidae Halictidae) Sociobiology 59(13): 1038–1052.

Paxton, R., J. and H. J. Banda. 1991. Pollination of greenhouse tomatoes by bees. 6. Int. Symp. On Pollination. Acta. Hort. 288: 194-198.

Paydas, S., S. Eti, S. Sevinc, E. Yasa, K. Derin, N. Kaska and O. Kaftanoglu, 2000. Effects of different pollinators to the yield and quality of strawberries. Acta Hortic., 522: 209-215.

Ruijter, A., D. Eijnde, J. Van den and J. Steen van der, 1991. Pollination of sweet Pepper (*Capsicum annuum* L.) in greenhouses by honeybees. Acta Hort., 288: 270-274.

Paydas, S., S. Eti, S. Sevinc, E. Yasa, K. Derin, N. Kaska and O. Kaftanoglu, 2000. Effects of different pollinators to the yield and quality of strawberries. Acta Hortic., 522: 209-215.

Ruijter, A., D. Eijnde, J. Van den and J. Steen van der, 1991. Pollination of sweet Pepper (*Capsicum annuum* L.) in greenhouses by honeybees. Acta Hort., 288: 270-274.

Peet, M. M; D.H. Willits and R., Gardner.1997. Response of development and post-pollen production processes in sterile tomatoes to chronic sub-acute high temperature stress. J. of Experiment. Bot. 48: 101–111.

Plantas medicinais e hortaliças usadas para cura de doenças em residências da cidade de Mossoró – RN. Revista 2(1):70-77.

Polverente M. R., Fontes D. C. Cardoso A. I. I., 2005. Produção e qualidade de sementes de berinjela em função do horãrio de polinização manual. Bragantia, Campinas 64, 3, 467–472.

Prasad D. N., Prakash R., 1968. Floral biology of brinjal (*Solanum melongena* L.). I. J. Agric. Sci., 38, 1053–1061.

Quagliotti, L. and F. Marletto. 1987. Research on the pollination of runner bean (Phaseolus coccineus L.) for dry grain production. *Adv.Hortic. Sci.*, 1:3-49.

Rajagopal, D. and Trivedi, T.P. 1989. Status, bioecology and management of *Epilachna* beetle, *Epilachna vigintioctopunctata* (Fab.) (Coleoptera: Coccinellidae) on potato in India: a review. *Tropical Pest Management*, **35**(4):410-413.

Rajagopal, D. and Trivedi, T.P. 1989. Status, bioecology and management of *Epilachna vigintioctopunctata* (Fabr.) (Coleoptera: Coccinellidae) on potato in India: A review. *Tropical pest management* :410-413.

Rajendran, B. and Gopalan, M. 1998. Screening and grading of brinjal (*Solanum melongena*) accessions for resistance to spotted beetle (*H. vigintioctopunctata*) *Indian Journal of Agricultural Science* **68**(4): 224-225.

Ram, C and Verma, J.P. 1998. The bionomics of hadda beetle *Epilachna elatrii* Naional Seminar on Enomology in 21st century, Biodiversiy, Sustainability, Environmental safety and Human health. April 30-May 2, 1998. Rajasthan College of Agriculture, Udaipur. 198pp (Abst.).

Rangam, S.V. 1991. IPM. The role of governments and citizen groups in Integrated Pest Management in the Asia Pacific Region. pp. 13-19. CAB *International, Asian Development Bank*, Manila, Philippines, 1992.

Research and Development Center. 12pp. 1996. Disponível em: http://203.64.245.61/fulltext_pdf/eam0137.pdf>. Acesso em: 15 set. 2011.

Richards, A.J. 2001. Does low biodiversity resulting from modern agricultural practice affect crop pollination and yield? Annals of Botany 88:165-172.

Richards, A.M. and Filewood, L.W. 1990. Feeding behaviour and food preferences of the pest species comprising the *Epilachna vigintioctopunctata* (F.) complex (Col., Coccinellidae). Journal of Applied Entomology, **110**(5):501-515.

Richardson, R. W., and Alvarez, L. E. 1957a. Pollination relationships among vegetable crops in mexico. 1. Natural cross-pollination in cultivated tomatoes. Amer. Soc. Hort. Sci. Proc. 69: 366-371.

Richardson, R. W., and Alvarez, L. E. 1957b. [natural crossing in the tomato.] Agr. Tecnologia mex. 3: 18, 44-45. [in spanish, english summary.]

Rick, C. M. 1950. Male sterile tomatoes. Calif. Agr. 4(4): 7, 12.

Rick, C. M. 1947. The effect of planting design upon the amount of seed produced by male sterile tomato plants as a result of natural cross-pollination. Amer. Soc. Hort. Sci. Proc. 50: 273-284.

Rick, C. M. 1949. Rates of natural cross-pollination of tomatoes in various localities in california as measured by the fruits and seeds set on male-sterile plants. Amer. Soc. Hort. Sci. Proc. 54: 237-252.

Roever, W. E. 1948. A promising type of male-sterility for use in hybrid tomato seed production. Science 107: 506.

Romero, G.Q., P.A.A. Antiqueira and J. Koricheva 2011. A Meta-analysis of predation risk effects on pollinator behaviour. Plos One 6(6) /e20689:1-9.

Ross, R. C. 1963. Truss vibration increases tomato profits. Agr. North ireland 37 (12): 378.

Rylski I., Nothmann J., Arcan L., 1984. Differential fertility in short – styled eggplant flowers. Sci. Hortic., 22, 39–46.

Rylski, I., J. Nothmann and L. Arcan 1984. Differential fertility in short-styled eggplant flowers. Scientia Horticulturae 22:39-49.

Rylski, L., B. Aloni, L. Karni, Z. Zaidman, K. Cockshull, Y. Tuzel and A. Gul.1994. Flowering, fruit set development and fruit quality under different environmental conditions in tomato and pepper crops. Acta. hort. 366:45-55.

Sambandam C. N., 1964. Natural cross – pollinaton in egglant (*Solanum melongena* L.). Econ. Bot., 18(2), 128–131.

Saradamma, K.; Dale, D.; Amkikadevi, D. and Das, M. 1986. Bioefficacy of two insect growth regulators on spotted beetle *H. vigintioctopunctata* on bitter gourd. Proceedings III Oriental Entomology Symposium Feb. 21-24, 1984. Trivandrum 2: 133-138.

SAS, Institute, Inc.1996. SAS/STATU User's guide, ver.6, vol. 1, 4 th. ed. Gray, N. C.,U.S.A :SAS Institute, 1996, 209

Sato, S; and M. M. Peet. 2005. The effects of moderately elevated temperature stress on the timing of pollen release and germination in tomato (Lycopersicon esculentum Mill.). J.of Hort. Sci. and Biotech.80:23–28.

Schneck, H. W. 1928. Pollination of greenhouse tomatoes. N.y. (Cornell) Agr. Expt. Sta. Bull 470, 60 pp.

Schroder, R.F.W.; Athanas, M.M. and Pavan, C. 1993. *Epilachna vigintioctopunctata* (Coleoptera: Coccinellidae), new record for Western Hemisphere, with a review of host plants. *Entomological News*, **104**(2):111-112.

Shah J. J., Patel J. D., 1970. Morpho-histogenic studies in vegetative and floral buds of brinjal and chily. Phytomorphology. 20, 209–221.

Shiela, M.K. and Abraham, C.C. 1981. *Cantheconidea furcellatul* (Wolf) (Pentatomidae: Hemiptera ) as a predatory of *H. vigintioctopunctata Agricultural Research Journal of Kerala* 19:42.

Shifriss, O. 1945. Hybrid tomato. South Seedsman 8(4): 15.

Silveira, F.A., G.A.R. Melo & E.A.B. Almeida 2002. Os grupos de abelhas presentes na fauna brasileira. *In:* Silveira F.A. (ed.). Abelhas brasileiras : sistemática e identificação. Belo Horizonte: 7-182.

Smith, O. 1935. Pollination and life history studies of the tomato Lycopersicon esculentum mill. N.y. (Cornell) Agr. Expt. Sta. Mem. 184, 16 pp.

Smith, P.G. and C.B. Heiser Jr., 1951. Taxonomic and genetics studies on the cultivated peppers, (*Capsicum annuum* L.) and (*C. frutescence* L.). Am. J. Bot., 38: 362-368.

Steel, R.D.G. and J.A. Torrie, 1980. Principle and Procedure of Statistics. 1st Edn., McGraw Hill Book Co., New York, pp: 137-139.

Thorp, R.W., 2000. The collection of pollen by bees. Plant System. Evol., 222: 211-223.

Watanabe, M.E., 1994. Pollination worries rise as honeybees decline. Science, 65: 1170-1170.

Smith, P.G. and C.B. Heiser Jr., 1951. Taxonomic and genetics studies on the cultivated peppers, (*Capsicum annuum* L.) and (*C. frutescence* L.). Am. J. Bot., 38: 362-368.

Soost, R. K., and rick, C. M. 1957. Effect of varieties of pollen and ovule parents on natural cross-pollination of tomatoes. Amer. Soc. Hort. Sci. Proc. 70: 357-365.

Soud Yousef Ahmad Al-abbadi, 2009. Efficiency of Different Pollination Treatments on Solanaceae Yields Grown in Plastic House. *Journal of Biological Sciences, 9: 464-469.*

Spangler, H. G. and J. O. Moffett. 1977. Honey bee visits to tomato flowers in polyethylene greenhouses. *Am. Bee j.*, 117(9):580-582. 306

Steel, R.D.G. and J.A. Torrie, 1980. Principle and Procedure of Statistics. 1st Edn., McGraw Hill Book Co., New York, pp: 137-139.

Steffan-Dewenter, I., S.G. Potts and L. Packer 2005. Pollinator diversity and crop pollination services are at risk. Trends in Ecology and Evolution 20:651-652.

Stephenson, A.G. 1981. Flower and fruit abortion: proximate causes and ultimate functions. Annual Review of Ecology and Systematics 12:253-279.

Stępowska A., 1996. Trzmiel ziemny (*Bombus terrestris)* jako naturalny zapylacz papryki pod osłonami. Mat. Symp. Nowe rośliny i technologie w ogrodnictwie". 17–19 września 1996, Poznań, 249–252.

Stoner, A. K. 1971. Commercial production of greenhouse tomatoes. U.s. Dept. Agr., agr. Handb. 382, 32 pp.

Subaratnam, G.V. and Butani, D.K. 1982. Chemical control of insect pest complex of brinjal. *Entomon.* **7**(9): 99-100.

Sun W., Wang D., Wu Z., Zhi J., 1990. Seasonal change of fruit setting in eggplants (*Solanum melongena* L.) caused by different climatic conditions. Sci. Hortic., 44, 55–59.

Teixeira A.P. 2004. Análise de uma floresta paludosa no Município de Rio Claro, SP: florística, pod osłonami. Zesz. Prob. Post. Nauk Rol., 466, 493–502.

Tewari, G.C. and Moorthy, P.N.K. 1985. Plant extracts as antifeedants against *H. vigintioctopunctata* and their effect on their parasite. *Indian Journal of Agricultural Sciences* **55**(2): 120-124.

Thompson, H.M. and L.V. Hunt 1999. Extrapolating from Honeybees to Bumblebees in Pesticide Risk Assessment. Ecotoxicology 8(3):147-166.

Thompson, H.M. 2003. Behavioural effects of pesticides in bees - their potential for use in risk assessment. Ecotoxicology 12:317-330.

Thorp, R.W., 2000. The collection of pollen by bees. Plant System. Evol., 222: 211-223.

Watanabe, M.E., 1994. Pollination worries rise as honeybees decline. Science, 65: 1170-1170.

Tscharntke, T; A. M, Klein, A, Kruess, I, Steffan-Dewenter, and C, Thies. 2005. Landscape perspectives on agricultural intensification and biodiversity –ecosystem service management Ecol. Letter; 8:857-874.

Vaissière, B.E., B. Freitas and B. Gemmill-Herren 2011. Layout of experimental sites. *In:* Vaissière, B.E., B. Freitas & B. Gemmill-Herren (eds.) Protocol to detect and asses pollination deficits in crops: a handbook for its use. Food and Agriculture Organization of the United Nation, Rome, 29-30.

Van Ravestijn W., Van der Sande J., 1991. Use of bumblebees for the pollination of glasshouse tomatoes. Acta Hortic., 288: 204–212.

Veeraval, R. and Raghuraman, M. 1998. Studies on the development rate, host preference and seasonal occurrence of larval pupal parasitoids *Pediobius foveolatus* C. on the spotted leaf beetle *H. vigintioctopuctata* on brinjal, *National Seminar on Entomology 21st century, Biodiversity, Sustainability, Environmental safety and Human health,* April 30-May 2 1998. Rajasthan College of Agriculture, Udaipur.

Veeravel, R. and Bhaskaran, P. 1993. Influence of abiotic factors on the incidence and population build up of brinjal pests. *Pest Management and Economic Zoology* **1**: 66-69.

Veeravel, R. and Bhaskaran, P. 1994. Relationship between various growth stages and pest occurrence in brinjal. *Pest Management and Economic Zoology* **2**(2): 171-173.

Venkataraman, T.V.; Mathur, V.K. and Chander, R. 1962. Experiments on the possible use of *Bacillus thuringensis* in the control of crop pest. *Indian Journal of Entomology* **24**:274-277.

Verkerk, K. 1957. The pollination of tomatoes. Netherlands jour. Agr. Sci. 5(1): 37-54.

Verma, L R. and U. Partap.1993. The Asian hive bee, Apis cerana L. as a pollinator in vegetable seed production. ICIMOD, Kathmandu, Nepal.

Vieira, A.R., L.R. Angelocci and K. Minami 1996. Efeito do estresse hídrico no solo sobre a produção da berinjela (*Solanum Melongena* L.). Revista Brasileira de Agrometeorologia 2(4):29-33.

Wellington, R. (a.) 1912. Influence of crossing in increasing the yield of the tomato. N.Y. (geneva) agr. Expt. Sta. Bul. 346: 423 - 442.

White, t. H. 1918. Experiments with fertilizers on greenhouse crops. Md. Agr. Expt. Sta. Bul. 222, pp. 77-101.

Williams, I.H. 1994. The dependences of crop production within the European Union on pollination by honey bees. Agricultural Zoology Reviews 6:229-257.

Willians, C.S. 1995. Conserving Europe´s bees: why all the buzz? Tree 10(8):309-310.

Wittwer, S. H., and Honma, S. 1969. Greenhouse tomatoes - guidelines for successful production. Michigan state university press, east lansing, 95 pp

Wysocka-Owczarek M., 1993. Zapylanie pomidorów szklarniowych przy wykorzystaniu trzmiela ziemnego. Now. Warz., 24, 31–40.

Yang, E.C., Y.C. Chuang, Y.L. Chen, and L.H. Chang 2008. Abnormal foraging behavior induced by sublethal dosage of imidacloprid in the honey bee (Hymenoptera: Apidae). Journal of Economic Entomology 101(6):1743-1748

Zaitoun, T. and G. Vorwohl, 2003. Major pollen plant species in relation to honeybees activities in the Jordanian desert area. Int. J. Agric. Biol., 44: 411-415.

***Solanum melongena***

***Capsicum annuum***

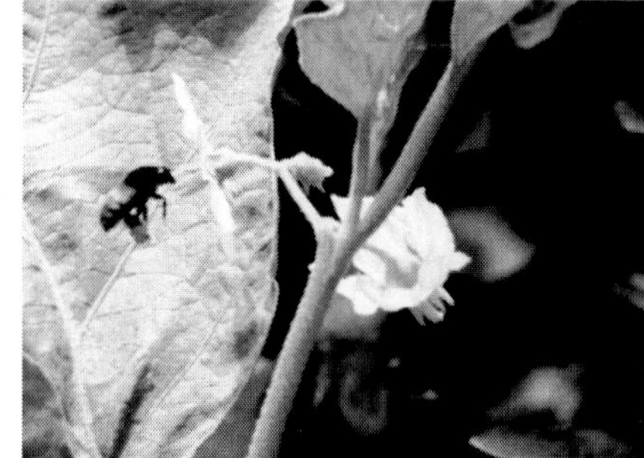

***Bombus* on brinjal**

**Figure 5.5:** Insect pollinators of brinjal and chillies

***Helicoverpa attack***

***Healthy plants*** ***Initial deformity due to insect infestation***

***Thrips and mite complex of chillies Second stage deformity caused by insects transition stage of deformation***

***Severe form of leaf curl***

**Figure 5.6: Insect pests of chillies**

***Fruit borer larvae helicoverpa***

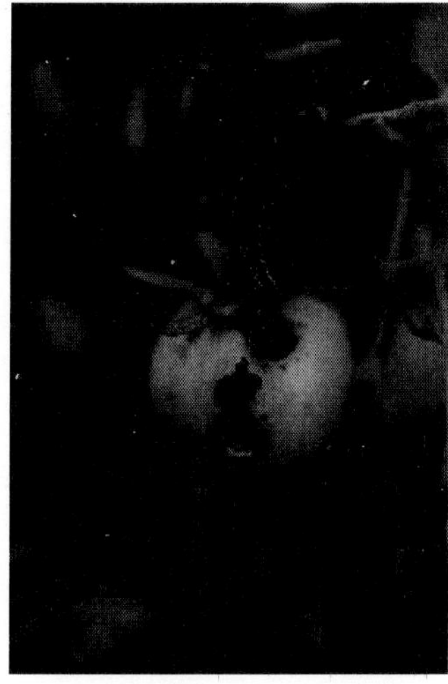

***typical damage on fruits***

*leaf miner of tomato*

*pupae on leaf surface*

*Fruit borer larvae*

*helicoverpa*

*Leaf miner of tomato*

**Figure 5.7: Insect pests of tomato**

*Drooping of infested shoots* *infested shoot with bringal fruit andshoot borer larvae*

*Management using agronet* *infested fruits with bore holes*

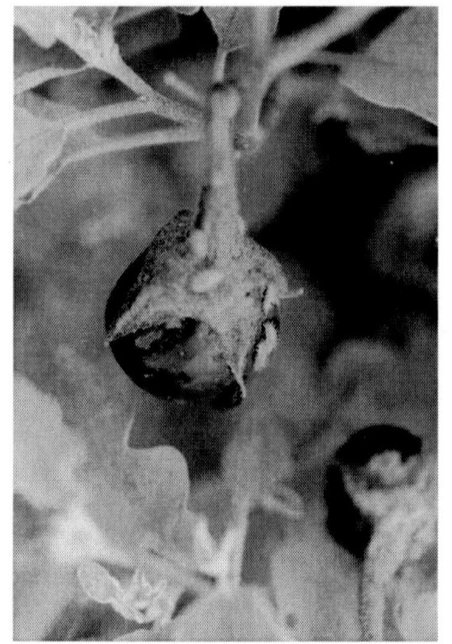

*Grubs feeding on fruits* *Seedlings infested by the grub*

***Epilachna* beetle of eggplant**

**Figure 5.8: Insect pests of brinjal**

CHAPTER

6

# Umbelliferae

## POLLINATORS OF CARROT

Carrots (*Daucus carota* L. subsp. sativus) are one of the most important vegetables grown in the world, of which production has increased recently and reached 33.7 million tons harvested from the area of 1.16 million ha in 2010 (FAO 2010). Central Asia is considered as a center of origin of cultivated carrot. Carrot *(Daucus carota* L.) is an important vegetable crop which depends heavily on cross-pollinating insects for any significant increase in seed production. The crop is protandrous with anthers shedding pollen before the stigma of the flower becomes receptive. Evidently, self pollination is not possible which necessitates the services of cross-pollinating insects for pollen transfer.

Carrot *(Daucus carota* L.) depends heavily on cross-pollinating insects for any significant increase in seed production. The crop is protandrous with anthers shedding pollen before the stigma of the flower becomes receptive. Evidently, self pollination is not possible which necessitates the services of cross-pollinating insects for pollen transfer (Goyal et al.,1989, Free, 1993). A large number of insect species have been reported visiting carrot flowers (Goyal et al. 1989). However, all insects, honeybees have been regarded as most efficient pollinators of carrot because of their specific behavioural adaptations for collection of nectar and pollen transfer (Free, 1993).

Insect pollination is a complex phenomenon which depends upon several factors. Besides, physical features of flowers such as colour, shape and odour, the diversity of environmental factors such as temperature, humidity, light, solar radiation, time of the day and nectar flow decisively shapes the behaviour of pollinating insects thereby influencing the cross-pollination and the production of the crop (Vischer and Seelay 1982, Corbet et al.,1993). Mitchell and Paton (1990) in a similar study found that differences in bill and tongue dimensions as well as in body mass influence the sugar intake rates for all three species in Australian honeyeaters.

Carrot flowers are in the form of an umbel. There is usually a primary umbel, which is produced first and then a series of increasingly smaller lower order umbels. The primary umbel is the most important for seed production. The plants will flower for between four and six weeks. The umbel is made up of groups of individual florets. The florets are presented in a number of discrete groups of flowers. The flowers around the outside of the group open first and those in the centre last. It takes about seven days for all the florets on an umbel to open. The stigma remains viable for about a week. Although carrot flowers have both male and female parts, much seed production has shifted to hybrid seed production.

The inflorescence consists of a terminal or primary compound umbel of white flowers, 5 to 6 inches across, and a system of second, third, and fourth order umbels, named in relation to their appearance on the plant below the primary umbel. The umbels decrease in size as the order number increases. The first and fourth order umbels are of little importance in seed production (Borthwick 1931). The individual flower is usually perfect (Knuth 1909), however, Braack and Kho (1958) reported that a tendency to produce only male flowers occurs with increasing frequency in the umbels of high orders. A flower normally has five functional stamens and two styles, which lead to the two locules of the ovary. Each locule contains a single ovule, thus two seeds per flower from flowers in a room free of harmful insects and supplied with flies to pollinate the flowers.

Carrot (*Daucus carota*) is a biennial or winter annual plant, which means it must go through a period of cold to produce flowers and seed. The umbrella-shaped inflorescences are borne in large groups called umbels. The primary, or king, umbel appears first near the top of the plant, followed by many second-, third-, and fourth-order umbels. Because of this developmental pattern, the plants have flowers and seed at various stages of maturity for 6–8 weeks during summer. Plants will flower for about 1 month, and flowering within a single umbel lasts about 7 days.

The individual white floret is perfect with five stamens and two styles which lead to two compartments in the ovary, each with one ovule; thus, each flower can produce two seeds. Apparently, each floret needs only two grains of pollen to fertilize the two ovules. Anthers release pollen within a floret over 1–2 days, and the stigma becomes receptive on the third or fourth day and stays receptive for a week or more. Nectar is secreted from the upper surface of the ovary. Many types of insects visit carrot for pollen and nectar.

Carrot *(D.carota)* flowers attracted a wide variety of insects belonging to order Hymenoptera, Lepidoptera and Coleoptera (Figure 6.1). In earlier studies also dipteran flies and hymenopterous insects have been recorded visiting carrot flowers (Free, 1993; Abrol, 1998, Goyal et al., (1989). Kumar et al. (1989) also reported bees as the most frequent visitor of carrot blossoms. Abrol (2010) found that insect pollinated plots produced significantly more seeds with heavier weights than those isolated from insect visits.

Nectar is secreted from a swollen disk on the upper surface of the ovary and is easily available to all types of insects. Carrot pollen is attractive to numerous insects (Bohart and Nye 1960). Gary et al. (1972) showed that carrot blossoms were much more attractive to honey bee pollen collectors than onion blossoms, as only 7 percent of the visitors to onion flowers were collecting pollen compared with 66 percent of the visitors to carrot blossoms. Flowering extends over about a month, and dehiscence within an umbel covers about 7 days. Within a floret, the anthers dehisce over a 1-to 2-day period, the stigma receptivity begins on the third or fourth day. The stigma may remain receptive a week or longer (Hawthorn and Pollard 1954, Hawthorn et al. 1960, Franklin 1953, Poole 1937).

Carrot flowers rely on insects for their pollination in order to produce quality and quantity of seed. Carrot is a highly cross-pollinated crop. The flowers are protandrous requiring cross pollination for seed production with more than 95% crossing occurring in the field (Thompson 1962). Hawthorn et al. 1960 found that for commercial seed production insect pollination is essential as s 85% increase in yield was obtained with honeybees' pollination.

Pankratova [1964] found insect visitation had a pronounced effect on carrot seed production. They found that that plots visited by insects produced 15 times more number of seeds and ten times more weight of seed as compared to that of plants covered with muslin cloth.

Whitaker et al. (1970) reported that for large scale production of seeds, where male sterile plants are used, there is a need for pollinating agents for maximum cross-pollination. Slate (1927) concluded that only 15 per cent of the carrot plants set seed from their own pollen.

Honeybees are being used extensively for pollination purposes of several crops. However in recent years, there has been a considerable decline in number of bee colonies s worldwide (Cunningham et al. 2002., Klein et al. 2007 Winfree 2008 Aizen et al. 2009.,Gallai et al. 2009) which is likely to affect the food supplies. It is therefore most essential to identify alternate pollinators of carrot crops which could be conserved and managed as a source. In earlier studies conducted worldwide a large numver of insects has been reported as pollinators of carrot (Bohart and Nye 1960 Goyal et al. 1989 Sinha and Ckrabarti (1992) Abrol 1997 Lamborn and Ollerton 2000 Ahmad and Aslam 2002). Abrol (1997) found that 95% of flower visitors to carrot were dwarf honey bees (*Apis florea* Fabricius) and other visitors very few in numbers included dipteran flies (Musca sp. and Syrphus sp.), halictine bees (*Halictus sp.: Lasioglossum* sp.) and black ants (*Camponotus compressus* Fabricius). Goyal et al. 1989 found 71 species of insects visiting carrot flowers at Ludhiana, India. They found that syrphid flies bring more numerous followed by Apidae. Ahmad and Aslam 2002 in Faisalabad, Pakistan A. florea was most predominant followed by dipterans. Sinha and Ckrabarti 1992 found yhay in Karnal India , sepsid and syrphid flies) were the most abundant flower visitors comprising of 90% of the total visitors. Lamborn and Ollerton (2000) found 20 species of insect visitors on carrot flowers, predominany included soldier beetles (*Rhagonycha fulva* (Scopoli)), syrphid flies, sawflies and various species of parasitic wasps. Bohart and Nye (1960) recorded 334 species of insect visiting carrot flowers In Utah.

Studies made by Gaffneya et al (2011) using sticky traps,water traps in Tasmania collected more than 42,000 insects which were grouped into 100 morphological categories. They found that small insects such as thrips ((<5 mm) were more than 78 % in sticky traps. The insects observed on umbels included nectar scarabs (Phyllotocus spp) which were most abundant whereas honeybee(A.*mellifera*), muscoid flies and lady bird beetles recorded in lowest numbers. However, the role of soldier, beetles, nectar scrabs as a pollinators needs to be investigated. To achieve this, lines that are male sterile or male fertile are produced. The male sterile line produces the seed and has non functional male parts. The pollen must be moved from a male fertile line, producing hybrid seed. As only the male fertile lines produce pollen, pollen-foraging insects will not normally visit both male fertile and male sterile lines unless they do so by accident. Usually only nectar foragers visit both lines and pollinate the crop. Hybrid carrot crops are usually planted with three male fertile lines and five or six male sterile lines. Some lines appear to be very attractive to insects. The difference in attractiveness was likely due to the amount of nectar being produced. Short tongued bumble bees will, however chew holes in the base of the flower to reach the nectar, without pollinating the flowers. Honey bees will then use the holes to reach the nectar. It is possible, therefore, that short tongued bumble bees may be detrimental for pollination.

Observations on insects visiting carrot flowers revealed rich faunal diversity. Abrol (2009) reported 21 species of insects belonging to 13 families, 17 genera and 4 orders visiting carrot flowers. Of all these insects, the dwarf honeybee *Apis florea* F. was the most predominant and comprised more than 94% of the total flower visiting insects. In general, *A. florea* visited less number of flowers/umbels during earlier and later part of the day. Maximum flowers/umbels/min. were visited between 1200-1400 hours. On an average, flowers visited/min ranged between 17.13-27.31 during different hours of the day with an overall average of 22.78 flowers/min. Similarly, number of umbels visited/ min ranged between 0.98-1.29 with an average of 1.14 umbels/min. He further found that the umbels left for open pollination resulted in significantly higher seed set than those excluded from insect visits. The caged umbels on an average had 8.8 per cent seed set whereas those left for open pollination resulted in 92.6 per cent. The differences were found to be highly significant statistically ($p<0.001$). Similarly, number of seeds produced per umbel also varied significantly between the two treatments ($p<0.001$). Caged umbels produced 165.42 seeds/umbel as compared to 2543.60 seeds/ umbel in case of those left for open pollination($p<0.005$). Similarly, seed weight/ umbel and weight of 1000-seed also varied significantly in both the treatments ($p<0.007$). Open pollinated plots produced significantly more seeds with heavier weights than those isolated from insect visits.

European blue blow fly *calliphora vicina* Robineau-Desvoidy, recognised as one of the potential pollinator of carrot as it occurs throughout the world and can be reared easily (Howlett). *C. vicina* has been found as effective pollinator of hybrid carrot (*Daucus carota* L.). seed yield obtained from from field-grown carrot plants caged with *C. vicina*, was similar to seed yield from uncaged plants in the presence of honey bees (*Apis mellifera* L.) Counts of *C. vicina* and honey bees on umbels revealed that honey bees were most abundant on days with maximum temperature > 25°C, while *C. vicina* was more abundant on cooler days around 20°C. *C.* This observation is more useful for crops grown in isolation cages for plant breeding purpose.

Although a number of insects visit carrot flowers and pollinate them, and some species of flies may be better pollinators than honey bees, honey bees are currently the only insect that can be managed in sufficient numbers to provide the necessary pollination services. Honey bees are introduced to pollinate carrots. They need to have large numbers of bees so they have a high demand for nectar. As the crop flowers for six weeks it is worthwhile considering introducing further hives as flowering progresses to replace any foragers that have deserted the crop. Because some hybrids appear not to be particularly attractive to bees it is important there are sufficient bees visiting the crop.

Carrot inflorescences are unspecialised in form and structure and as such are visited by a large number of insects from different morphological groupings. More than 100 grroup of small insects (< 5 mm), made up 78 % of all insects (Figure 6.2). The insects visiting umbels were nectar scarabs (*Phyllotocus spp.*) honey bees (*Apis mellifera* Linnaeus) muscoid flies and ladybirds. The composition and frequency of visitation of these insects depended upon time of the day, location and prevailing weather conditions.

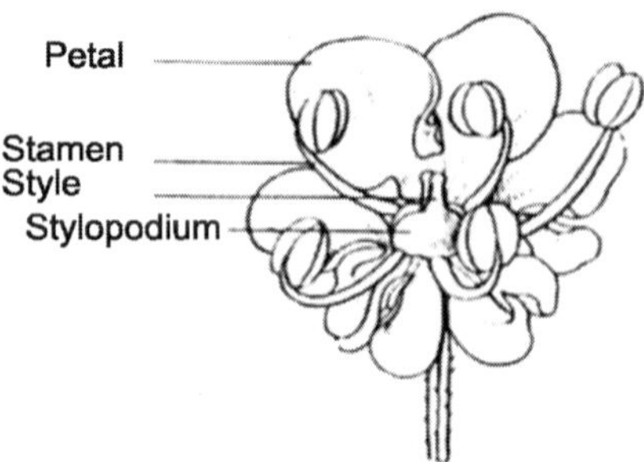


Figure 6.1. Flower of *Dacus carota* , Carrot(After Ross-Craig 1959)

In 1955 and 1956, syrphid flies (*Syritta pipiens* (Linnaeus)) and in 1957 coccinellid beetles (*Hippodamia convergens* Guérin-Méneville) were the most predominant insect visitors (Bohart and Nye 1960). Syrphid flies and *Apis spp.* were the most commonly reported visitors. Inspite of extensive use of cytoplasmically male sterile parent lines for hybrid carrot seed production worldwide, there is still incomplete documented information on the diversity of insects visiting cytoplasmically male sterile carrots in the field. (Erickson and Peterson 1979, Erickson et al. 1979, Rodet et al. 1991).

**Table 6.1:** *Effect of pollination treatments on yield components of D. carota (Abrol, 2010)*

| Parameter | No pollination (Enclosed) | Insect pollination (Open pollination) | CD at 5% |
|---|---|---|---|
| Percent fruit set | 8.8 (2.96) | 92.6 (9.62) | 10.2 |
| Number of seeds/umbel | 165.46 (50.43) | 2543.60 | 8.94 |
| Seed weight/umbel (g) | 0.42 (1.87) | 3.52 | 7.02 |
| Seed weight/1000-seed (g) | 2.21 (1.48) | 4.76 (2.18) | 0.47 |

Figures in parentheses are square root transformed values.

## Pollination Requirements

Alam et al. (1987) reported that open pollinated plots gave significantly more and heavier seeds as compared to those deprived of insect visits. Hawthorn et al. (1960) through different pollination experiments proved that insect pollinators were essential for commercial seed production. They found that in cages excluding all insects, an average of only 128 pounds of seed per 453 pounds of seed per acre when tiny insects were permitted to pollinate carrot flowers. They further found that self or mechanical pollination in the absence of pollinating insects was of little value in the commercial seed production of carrot. Slate (1927) concluded that only about 15 percent of the carrot plants set seed from their own pollen.

The relatively long flowering period of carrots is favourable for ensuring effective pollination to attract a broad spectrum of nectar and pollen to collectors, particularly honey bees. For large-scale production of seeds, however, where male-sterile plants are used, there is need for pollinating agents interested only in nectar collection that will freely cross over from the normal to the male-sterile flowers and effect maximum cross-pollination.

Most seed set occurs with pollen transferred by insects from different plants, although a small percentage of carrot plants can set seed with their own pollen. Abundant numbers of insect pollinators increase seed yield and hasten its maturity (McGregor, 1976; Free, 1993). Yields of open-pollinated carrot seed in North America are about 850 lb acre (952 kg ha), and yields for hybrid seed are about 250 lbacre (280 kg ha).

Hybrid carrots are valued for their uniform, smooth, and colourful roots. Hybrid seed production requires the transfer of pollen from pollen-donor lines to the stigmas of selected male-sterile lines. Because cross-pollination is so common, stocks are maintained by cooperating breeders who agree to isolate fields of the same stocks far from fields of different stocks.

## Pollinators

Bohart and Nye (1960) studied the insect visitors to carrot flowers. They found 334 species of insects representing 71 families visiting carrot blossoms. Most of the species of visitors were in the superfamily Apoidea, or the Ichneumonidae, Psammocharidae (Pompilidae), Sphecidae, and Vespidae families of the Hymenoptera, and the Bombyliidae, Sarcophagidae, Stratiomyidae, Syrphidae, and Tachinidae families of the Diptera. Bohart and Nye (1960) proposed an efficiency rating for the insect pollinators of carrots, based on the amount of loose pollen on the insects' body, the size of the insect, and its activity on the flower head. By multiplying this rating figure by the numbers of insects observed on the flowers, a pollination index was obtained for each species. They concluded that several genera in the Apoidea were important pollinators of carrots, but from the practical standpoint the honey bee was the only species that could be manipulated and utilized in commercial seed production.

Pankratova (1958) reported that the chief pollinators of carrots near Moscow were flies (90 percent) and bees (9 percent). The nectar collectors stand higher on the flowers, move about less, and lap up droplets from the exposed nectaries. In other species of bees, the females usually behave like pollen-collecting honey bees and the males like nectar-collecting honey bees. As shown earlier by Hawthorn et al. (1960) bees increased production of carrot seed. Both quantity and quality of carrot seeds were improved by high levels of bee pollination. Franklin (1970) reported that at one time in Parma, Idaho, the carrot fields were teeming with bees, and excellent seed crops were obtained. Then pest control methods and materials changed, competitive crops moved in, bee counts dropped, and the seed crops failed.

Commercial carrot seed production requires insect pollination. Many types of insects visit carrot flowers; however, there are not always sufficient numbers of pollinators. Wild bees and honey bees are the most important pollinators. Of these, only honey bees can be practically managed on a large scale. Hives should be placed in several areas around the field because honey bees tend to work close to their hive.

Honey bees work some varieties more readily than others which can cause large differences in seed set (Erickson and Peterson, 1979a,b). In hybrid seed production, honey bees prefer the pollenyielding male-fertile plants over the male-sterile (female) plants. A nectar-favouring pollinator that freely visits both types would be preferable in these cases. Honey bees effectively pollinate carrot, but they do not especially prefer carrot flowers. They are easily lured to more attractive flowering plants. Mayer and Lunden (1983) recommended 5 colonies of honeybees for optimum production of seeds in carrot. Levin (1986) however considered 7.4 to 10 colonies per ha.

## Pollination Recommendations and Practices

Bohart and Nye (1960) recommended to provide enough colonies of honeybees for effective pollination, avoid the presence of competing bloom. Pankratova (1957) stated that the most reliable pollinators of carrots are honey bees. He recommended transporting colonies to the field, but the number of colonies was not mentioned. Hawthorn et al. (1956) also recommended movement of colonies of honey bees to carrot fields to provide the large numbers necessary at flowering time but did not designate the number. Naturally, the number needed would be influenced by competition from other flowers, the strength and condition of the colonies, and the attractiveness of the carrot flowers. Under most conditions where carrots are grown for seed and maximum production is desired, the placement of several colonies per acre in and around the field would probably be justified. Eight bees per square yard of flowers should be striven for regardless of the number of colonies required to provide this.

## Pests of Carrot (*Daucus Carota* L., Apiaceae)

Carrot is a domesticated form of the wild carrot, *D. carota*, native to Europe and Southwest Asia. It has been selectively bred for its greatly enlarged and more palatable and edible taproot. Earlier, carrots were grown for their aromatic leaves and seeds, not their roots. Some relatives of the carrot are still grown for these uses, such as parsley, fennel, dill, and cumin. In the tropics, carrot is grown in cooler upland areas at altitudes higher than 500 m.

Carrots have few pests. The most important ones include soil arthropods that attack the root such as carrot fly, wire worms, cut worms and mole crickets. The other minor pests include leafminers, aphids and weevils. The other arthropods that occasionally attack include army worms, mites and plant bugs Although many insects feed on the foliage, they usually do not cause serious damage due to the action of natural enemies.

## Carrot fly *(Chamaepsila rosae* (F.)

Carrot fly *(Chamaepsila rosae* (F.) belongs to order siptera and family Psilidae. It is distributed in Europe and Asia having Carrot, parsnip, parsley, celery as its main hosts. Eggs are laid in the soil around the developing carrots. Upon hatching, larvae burrow into the surface of the root. The maggots feed on the outer layers of the carrot root.

**Natural enemies:** Two fungal pathogens, *Entomophthora muscae* and *Entomophthora schizophorae*, have been reported infecting adult flies. There is evidence that another fungal pathogen, *Beauveria bassiana*, has potential for controlling the larval stage. A staphylinaid ectoparasite, *Aleochara* sp., was found parasitizing the puparia.

## Green looper *Thysanoplusia orichalcea* (F.)

**Green looper *Thysanoplusia orichalcea* (F.)(** Lepidoptera::: Noctuidae) is distributed throughout Southeast Asia. The looper can be found on a wide range of cultivated crops and wild plants. The eggs are laid singly on leaves. The egg stage lasts 3–4 days. There are five larval instars lasting about 13 days. Pupation occurs in a silken cocoon att ached to and covered by the carrot leaves. Larvae feed on the leaves and occasionally they cause defoliation but usually not serious enough to reduce yields.

**Natural enemies:** Many larval parasitoids have been reared from field-collected loopers. There are also numerous predators such as carabids and pentatomids that feed on eggs and larvae. Insect pathogens play a major role in regulating looper populations, especially in carrots. Fungal pathogens such as *Nomuraea rileyi* and *Pandora gammae* are commonly found infecting larvae, and, if loopers are found in a carrot field, one can usually find the nucleopolyhedrovirus infecting that population. This is a phenomenon that the authors have observed in numerous upland locations throughout Southeast Asia.

**Other remarks:** There are two other species of loopers that may occur together with *T. orichalcea* in carrots, namely, *Chrysodeixis chalcites* and *Argyrogramma signata* (F.) (Lep. Noctinidae).

## Other pests of regional importance are:

Pea leaf miner, *Phytomyza atricornis* (Diptera: Agromyzidae); sweet potato flea beetle, *Chaetocnema basalis* (Coleoptera: Chrysomelidae); wireworms (Coleoptera: Elateridae), and grubs (Coleoptera: Scarabeidae)

## Insect Management for Carrots

Soil pests are of serious concern in cultivation of carrot. They include wireworms, cutworms, and mole crickets, all of which are sporadic. Occasional minor pests include leafminers, aphids, and weevils. Other arthropods that may occasionally cause minimal damage to carrots include armyworms, field crickets, mites, and plant bugs.

## Cutworms

Cutworms are sporadic pest of carrots and can be a problem during seedling establishment. They attack young seedlings, most actively at night, and may cut the stem off at the base. During the day, they remain hidden in debris or under the surface of the soil. Thorough soil preparation in advance of planting helps in cutworm control.

## Wireworms, *Agriotus*spp., *Melanotus* spp. and other Elateridae

Wireworms are larvae of click beetle, the adults of which make clicking sound when try to right themselves after being on their backs. The larvae are 0.5 to 1.5 inches long, slender, hard bodied and shiny. Some species can live in soil for 1 to 5 years. Eggs are usually laid singly in spring or summer at the depth of 1 to 6 inches in soil. Hatching takes place in 2 to 4 weeks. Because of the long egg-laying period, overlapping generations are present. The adults can attack the developing carrots directly, causing severe loss, or can provide entry points for pathogens that cause secondary rots.

## Tawny Mole Cricket, *Scapteriscus vicinus;* Short-winged Mole Cricket, *S. abbreviatus*

Among mole crickets only a few are pests. The tawny mole cricket damages vegetable crops by making tunnels, which are just below the surface. There is only one generation per year. Eggs are laid in chambers, from April to June about 4 to 12 inches underground. These adults overwinter and breed in the spring. For two to three months in the spring, tawny mole crickets are most commonly seen during their brief mating flights, which begin shortly after sunset. Mole crickets mainly feed on plant roots. At night, in warm, wet weather, they will also feed on stems and leaves at surface level.

The young seedlings suffer severe damage due to in and **around** tunnelling of mole crickets under the developing roots. The most serious damage occurs in carrot plants where they cut the stems of seedlings at or near the ground level. It is therefore recommended that in areas where mole crickets are expected to cause problem, incorporation of insecticides into the soil is recommended.

The various classical biocontrol agents have been evaluated for the control of mole crickets such as the tachinid fly *Ormia deplete and* steinernematid nematode *Steinernema scapterisci*. The latter has shown to be more promising where mole cricket populations are highest. It also serves as biopesticide where populations are lowest. The nematodes shows good dispersal activity even after years of application.

## Leafminers, *Liriomyza sativae* and *L. trifoli*

The adult leafminer is a small fly, about 1/8 to 1/10 of an inch long with a yellow abdomen.

The fly inserts eggs into the punctures made on the upper surface of the leaves by feeding activities. The larvae (maggots) feed between the upper and lower surface of leaves creating mines which enlarge as the larvae grow in size. Under warm weather conditions, the larva can complete its development, leaves the mine and drops to the ground to complete it pupation. The life cycle is completed in 18-21 days.

Leaf miners are a sporadic foliar pest on carrots. Infestations are more severe during growing seasons if adults migrate out of nearby crop residue.

They are a problem on seedlings during the fall carrot season. Although healthy plants can usually tolerate substantial leaf miner damage, heavy damage may cause leaf drop. The exit holes caused by feeding of aphids provides entry to pathogens. It is difficult to control leaf miners with chemicals as they are mostly protected within the leaf. Targeting small larvae results in better control. Among biocontrol agents, parasitic wasps keep the population below damaging levels.

## Melon Aphid, *Aphis gossypii;* Green Peach Aphid, *Myzus persicae*

Aphids are also a minor and sporadic pest on carrots. The green peach aphid (*Myzus persicae*) is the species most often reported on the crop whereas the melon aphid (*Aphis gossypii*), may also be found on carrots. The aphid *Hyadaphis coriandri*, is capable of colonizing carrots, but there has been no report on carrots.

Aphids are soft-bodied insects, almost egg-shaped when viewed from above. The largest melon aphids are not much more than one-sixteenth of an inch in length. The colour of the aphids varies from pale yellow to orange to dark green to almost black. Green peach aphids are larger (up to one-tenth of an inch long) and are pale yellow to medium green in colour. A small pair of tube-like structures called cornicles extends backward and upward abobe the caudal portion of aphids.

The first individuals to appear are usually with wings but then wingless aphids become more predominant and their overcrowding causes the plants to deteriorates. The wing aphids are produced again which disperse to other plants. With their needle like mouth parts, aphids suck the nutrients and water from plant tissues and at the same time inject toxins and saliva which causes the foliage to curl and deform. Aphids also deposit large amounts of honeydew which encourages the growth of black sooty mould. Lady bird beetles, lace wings and syrphids which feed on aphids help to control their populations. Occasionally, fungi will infect aphids, drastically reducing populations.

Soil inhabiting pests such as weevils and wire worms are the common pests. Army worms cause indirect damage by cutting stems or consuming foliage above ground. Besides some other insects also attack carrot plants. The insect pests attacking carrot are grouped into three categories depending upon the mode of damage:-

## Chewing insects that cut holes or entire leaves

**Caterpillars** with three pairs of legs and five pairs of prolegs

**Armyworm:** they are pale green to yellow in colour, smooth-bodied caterpillar up to 35 mm long has three dark longitudinal stripes and a tan or greenish-brown head mottled with darker brown. It feeds primarily at night on foliage and succulent stems

**Parsley worm:** This yellowish-green caterpillar up to 40 mm long has black coloured transverse bands and deep orange or yellow spots. The head is greenish yellow in colour with black stripes. The feed during the day. They have a pair of horn like scent organs which protrude from behind the head on slight disturbance.

**Yellow woolly bear:** The caterpillars are white or yellow brown covered with dense hairs on the body. They young larvae live in colonies and feed on the underside of leaves.

**Vegetable leafminer:** They are bright yellow maggots which make s-shaped leaf mines. The heavily infested leaves turn brown.

**Vegetable weevil and larva:** They are dull grayish brown weevils feed on buds and foliage during night. Both grubs' and weevils are 6.4 mm long. The larvae are legless growing up to 10 mm.

## B. Insect with needle-like mouthparts that cause foliage to be yellowed or distorted

**Aster leafhopper:** the leaf hoppers are up to 5 mm having six black spots on the head whereas nymphs are light brown instead of yellow or green.

**Tarnished plant bug:** the bugs are oval shaped, brown in colour measuring 6.4 mm long and have long legs, long antennae, and a white triangle between its "shoulders". Nymphs are yellowish-green to green with black spots on its back.

## C. Insects that feed on underground plant parts

**Vegetable weevil and larva:** The weevil is dull grayish in colour having short snout and a light V-shaped marks on wing covers, feeding primarily at night on buds and foliage. Both weevils and grubs measure approximately 6.4 mm long. The larvae are legless pale green in colour with dark mottled heads and measure about 10 mm long. The weevils mostly attack during night feeding on tap roots of plants.

**Southern potato wireworm:** The larvae have three pair of short legs and one pair of anal legs.the larvae or white cream or yellow and grow up to 17 mm long. The last abdominal segment has a closed oval notch. These wireworms form irregular holes in infested taproots

Ellis et al (1978) tested the eleven carrot cultivars to compare their susceptibility to carrot fly attack and found significant differences in the percentages of unattacked roots. Foliage height and the percentage unattacked roots were not significantly correlated.

Finch (1993) reported that improved insecticide application and pest forecasting can help to reduce the amount of pesticides applied to control the cabbage root fly (*Delia radicum*) and the carrot fly (*Psila rosae*). Growing plants partially resistant to the carrot fly can help to reduce the pesticide use.

Bolvin (1999) reported that the carrot weevil, *Listronotus oregonensis* (LeConte) (Coleoptera: Curculionidae) is a major pest of carrots. The females oviposit in the leaf petiole and the developing larvae make tunnels in the carrot roots. Egg parasitoids of the genus Anaphes (Hymenoptera: Mymaridae) are the most important natural enemies and they account for nearly 80% mortality in untreated plot. Their efficiency can be increased by managing field borders to establish secondary host species.

Uvah and Coaker(1984) reported that mixed cropping of carrots with onion reduced attack by carrot fly, *Psila rosae* Fab., on carrots and *Thrips tabaci* Lind, on onions, compared with those on carrots and onions in monoculture.

## POLLINATORS OF CORIANDER

### *Coriandrum sativum* L., family Umbelliferae

Coriander flowers are partially self-fertile and produces umbels with mostly bisexual flowers early in the season followed by male flowers. Pollen release precedes stigma receptivity so pollen from a different flower is required for seed set

The coriander flower has five irregular-shaped petals, five stamens, five sepals, and two styles. The white to pinkish flowers are in umbels. The first umbels to bloom have hermaphrodite flowers, with possibly a few staminate ones. The later umbels have only staminate flowers. The hermaphrodite flowers are completely protandrous, so that selfing is impossible. After the pollen is gone, the stigmas become receptive and are liable to crossing with other plants; however, the umbels of staminate flowers may develop in such a way that they are right over the receptive stigmas of later flowers. When these anthers dehisce, the pollen is thrown out and falls to the stigmas below in crumbling masses. In this way, some of the stigmas may be pollinated even if an insect has not brought pollen from another flower (Kerner 1897). Pollen is produced in the pinkish anthers. Nectar is freely secreted on the ovary. The blossoms are highly attractive to both pollen-collecting and nectar collecting insects

Coriander flower produces both hermaphrodite and staminate flowers. The hermaphrodite flowers are completely protandrous with anther maturity before the stigmas becomes receptive (Nemeth and Szekely, 2000) so it requires the pollen from other flowers with the help of pollinators. Ramanujam *et al.*(1964) reported that the honeybees are the major pollinators and cross-pollination is estimated to be up to 50 per cent in coriander. The studies on the pollination requirement of coriander crop was made by Glukhov (1955); Shelar and Suryanarayana (1981); Hussain and Abdel-Aal (1982); Baswana (1984); Sethi (1999); Khalid *et al.* (2008); Anonymous (2010) and Singh *et al.* (2010).

The hermaphrodite flowers are completely protandrous, so that self-fertilization is impossible; they need the intervention of pollen carriers on the receptive stigmas (McGregor, 1976; Ricciardelli D'Albore, 1986).

Coriander is a year-long herb and an excellent melliferous plant (Ricciardelli D'Albore and Intoppa, 2000) which makes it attractive to a wide variety of insect species(Free, 1993). The main attractive features include exposed nectar, abundant pollen production, zygomorphic flowers and compact umbels (Koul et al., 1989; Diederichsenn, 1996). Insects are attracted by the nectar that is secreted, especially during the period when the stigma is receptive to successful pollination (Koul et al., 1989).

### Insect visitors

Coriander flowers attract a wide variety of insects which are visitors or pollinators of coriander umbels (Figure 6.3). In India, studies have been done by Koul et al. (1989) and in Egypt El-Berry et al. (1974) who observed number of insects visting coriander flowers. Koul et al. (1989) reported that insects are attracted to the nectar secreted by stylopodium when the stigma is receptive thereby resulting in successful pollination. Chaudhary and Singh (2007) observed 34 species of insects visiting coriander flowers in Karnal Haryana, India. Insects belonging to order hymenoptera and diptera comprised of more than 96.9% of the total visitors. Among honeybee, *Apis mellifera* was most common followed by syrphid Episyrphus balteatus and some unidentified species. Honeybees foraged both for nectar and pollen wheras dipteran flies for nectar alone. *A. mellifera* started foraging early at 09.05 h, peaked from 11.00 to 16.00 h and declined drastically thereafter. The population of *A. dorsata* was maximum upto 1100 hour and observed only upto 1400 h only and therafter no bees were seen. Similarly, *Apis cerana* population was observed between 1000-1100 h in maximum numbers and seen only upto 1500 h. The syrphids and other dipteran flies had two peaks (07.00 to 11.00 h and 15.00 to 18.00 h) and were observed throughout the day.

Devi et al (2015) recorded a total of 39 insect visitors belonging to 31 genera under 18 families and 8 orders visiting the coriander bloom. Singh et al (2010) evaluated the influence of certain plant products on the insect pollinators of coriander and found that The plant origin insecticides evinced a decrease in pollinator intensity ranging from (-) 14.52 to (-) 25.06 per cent. The plots treated with neem oil (1%). gave maximum reduction of insect pollinators (25.06%).

Bendifallah (2013) found that in the El Harrach area of Algiers, the insect fauna visiting coriander flowers *Coriandrum sativum* L. flowers included Diptera (Syrphidae), Coleoptera, and Hymenoptera such as Chrysididae, Vespoidea, and Apoidea (Bees). The majority of the insect pollinators belonged to super family Apoidea comprising of four families: Apidae, Andrenidae, Halictidae, and Colletidae. The main visitors are honeybees and six species of Andrenidae: *Andrena flavipes, Andrena thoracica, Andrena lagopus, Andrena bimaculata, Andrena discors,* and *Andrena cinerea elliptica.*

## Impact of insect pollination on coriander fruit/seed production

McGregor (1976) and Glukhov (1955) reported on the role of honeybees in coriander seed production and found that that when the bees were excluded, only 49.4% of the seeds were able to set, in contrast to 68.3% when the bees were present. In Sicily, Sinacori et al. (2009) reported one study with three treatments: self-fertilization, pollination by bees, and open-pollination by the presence of all pollinators. They found that seed produced with open-pollination(1883 kg/ha) was significantly higher than with only the bees( 1 335 kg/ha) followed by absence of pollinators ;(800 kg/ha). Bogoyavlenskii and Akimenko (1966) associated seed yields with greater insect visitation. These results substantiate the significance of the bee effect noted earlier by McGregor (1976) and observed later, in other studies. In particular, in India, Abu-Hammour (2008) showed that pollination not only produced well shaped fruits but also increased the quality of the seed content

Chaudhary and Singh (2007) found that insect pollination had a significant effect on seed yield of coriander. Shelar and Suryanarayana (1981), Sihag (1986) and Lafreniere (1997) who also reported yield increase in OP and BP over WIP.

Kumar and Jaiswal (2012) found that in coriander the number of filled seeds/umbel were significantly higher in open to all insect pollinators (50.80, 46.50) followed by *A. mellifera* pollinated crops (42.40, 39.70) and lower in self-pollinated (21.60, 21.50).

## Insect pests of coriander

Coriander has only few insect pests. *Systole coriandri* whose larva damages the fruits is a specific pests of coriander. Seeds of the affected plants become unfit for consumption and sowing. Palamarja and Chotina (1953) has given a detailed account of this pest including biology of several mites , moths and beetles. Jansen (1981) also reported the serious losses caused by the chalcid fly *Systole albipennis* to fruits in Ethiopia, and reports the possibility of their biological control by the parasites *Tetrastichus spp.* and *Liondontomerus spp.* Singh and Baswana (1984) screened coriander for resistance to this pest in India. de Araujo (1986) recorded aphids *Myzus persicae* whereas (Jain and Yadav 1988) reported *Hyadaphis coriandri* on coriander crop.

**Major pest include:** Cotton Whitefly *Bemisia tabaci* Aleyrodidae Hemiptera

**Minor pests include:** Aphid *Hyadophis coriandri* Aphididae Hemiptera, Pentatomid bug *Agonoscelis nubila* Pentatomidae Hemiptera, Indigo caterpillar *Spodoptera exigua* Noctuidae Lepidoptera

The important pest of coriander is only whitefly.

**Cotton Whitefly: *Bemisia tabaci* (Hemiptera: Aleyrodidae):** The nymphs suck sap of the plants and adversely affect their growth.

Other pests which are found on coriander plants are:

**Aphid: *Hyadophis coriandri* (Hemiptera: Aphididae):** Both nymphs and adults congregate colonise on ventral surface of leaves and suck cell sap. Due to copious production of honey dew, leaves give a glistening appearance in the beginning, but later covered with sooty mould fungus. Nymphs and adults are yellowish green. A single female produces 40 to 50 young ones and they take 8 to 12 days to mature. Life cycle is completed in 14 to 21 days during summer and 6 weeks in winter.

**Pentatomid bug: *Agonoscelis nubila* (Hemiptera: Pentatomidae)**

Adult and nymphs suck the sap from leaves and stem. Heavily infested plants show stunting. Adults are yellowish. Life cycle is completed in 40 to 60 days. Spray dimethoate or quinalphos 1.5 ml/L

## References

Abrol D. P 2009. Planned honeybee pollination for improvement in horticultural crop production. Final Technical Report Horticulture Mission Mode Project-1(HTMM-1, 2.28). ICAR New Delhi pp40

Abrol D. P. 1988 Effect of climatic factors on pollination activity of alfalfa-pollinating subtropical bees Megachile nana Bingh and Megachile flavipes Spinola (Hymenoptera, Megachilidae). Acta Oecologica 9: 371-377.

Abrol DP 1997 . Impact of insect pollination on carrot seed production . Insect Environment 3 : 61 .

Abrol, D. P. 2012. Pollination Biology: Biodiversity Conservation and Agricultural Production. Springer Dordrecht HeidelbergLondon New York.

Abrol, D.P 1991 Path analysis of environmental factors influencing daily flight activity of Apis dorsata F. Acta Oecologia, Ecologia Generalis 12 : 819-824.

Abrol, D.P. 1992 Bioenergetics in bee flower interrelationship- An analysis of foraging behaviour. Korean J.Apic. 7(1): 39-66.

Abrol, D.P. 1998 Environmental factors influencing flight activity of honeybees Apis cerana F. and A. mellifera L. (Hymenoptera: Apidae). Indian Bee J. 60(2): 71-75.

Abrol, D.P. 2006. Diversity of pollinating insects visiting litchi flowers (Litchi chinensis Sonn.) and path analysis of environmental factors influencing foraging behaviour of four honeybee species. J. Apic. Res., 45(4): 180–187.

Abrol, D.P. 2006. Factors influencing flight activity of *Apis florea* F, an important pollinator of Daucus carota L Journal of Apicultural Research and Bee World 45(2): 2–6

Abrol, D.P. and Kapil, R.P. 1986. Factors affecting pollination activity of *Megachile lanata* Lepel: Proc. Indian Acad. Sci. (Anim. Sci.) 95: 757-769.

Abrol, D.P. and Kapil, R.P. 1991. Foraging strategies of honeybess and solitary bees as determined by nectar-sugar components.Proc. Indian Nat. Sci. Acad. B, 57(2): 127-132.

Abu-Hammour K. 2008 Pollination of MedicinalPlants (*Nigella sativa* and *Coriandrum sativum*) and *Cucurbita pepo* in Jordan. Doctor-Ingenieur thesis. University of Bonn. Germany. 90 pp.

Agati, G. 1952. Indagini ed osservazioni sulla biologia fiorale della cipolla. Riv. Ortoflorofruttic. ital. 36: 67-77.

Ahmad Munir and Aslam Muhammad 2002. Pollinators visiting carrot (daucus carota l.) seed crop. Journal of Research (Science), Bahauddin Zakariya University, Multan, Pakistan. 13(1): 31-35

Ahmad, M and Aslam, M. 2002. Pollinators visiting carrot (Daucus carota L.) seed crop. Journal of Research (Science), 13: 31–35.

Ahmad, S., Muhammad, T., Karim, A., Jabbar,A., and Ahmad, Gh. 2003. Effect of pollinators and insecticides on seed setting of onion (Allium cepa L.). Journal of Biological Science, 3 (3): 305-308.

Aizen, M.A., Feinsinger, P., 1994. Habitat fragmentation, native insect pollinators, and feral honey bees in Argentine "Chaco Serrano". Ecological Applications 4, 378–392.

Aizen, M.A., Garibaldi, L.A., Cunningham, S.A., Klein, A.-M., 2008. Long-term global trends in crop yield and production reveal no current pollination shortage but increasing pollinator dependency. Current Biology 18, 1572–1575.

Aizen, M.A., Garibaldi, L.A., Cunningham, S.A., Klein, A.-M., 2009. How much does agriculture depend on pollinators? Lessons from long-term trends in crop production. Annals of Botany 103, 1579–1588.

Alam, M.Z. and Quadir, M.A. 1986. Role of honey bees in fruit and seed setting of bottle gourd, Lagenaria siceraria (Mol.) Staudl.Punjab Vegetable Grower,21: 32-34.

Alam, M.Z., Quadir, M.A. and Ali, M.S. 1987. Pollination behaviour of honeybee (Apisindica) and the influence on seed production of cauliflower. Bangladesh Hortic.,15(1): 25-30.

Ali, M., Dowker, B.D., Currah, L. and Mumford, P.M. 1984. Floral biology and pollen viability of parental lines of onion hybrids.Ann. Appl. Bio., 104: 167–174.

Allen-Wardell, G., Bernhardt, P., Bitner, R., Burquez, A., Buchmann, S.L., Cox, P.A., Dalton, V., Feinsinger, P., Ingram, M., Inouye, D.W., Eugene, C., Kennedy, K., Kevan, P.G., Koopowitz, H., Medellin, R., Nabhan, G.P., Pavlik, B., Tepedino, V.J., Torchio, P.F., 1998. The potential consequences of pollinator declines on the conservation of biodiversity and stability of food crop yields. Conservation Biology 12, 8–17.

Allsopp, M.H., 2009. Bee disease threatens South African apiculture, Plant Protection News, No. 79, Agricultural Research Council: Plant Protection Research Institute.

Al-Sahaf, F. H. 2002. Effect of planting method and rose water spray on seed production in onion (Allium cepa L.). Emirate Journal of Agricultural Science, 14: 14-23. Aminpour, R. and Mortzavi Bak, A. 2004.

Altieri, M.A., 1995. Agroecology: the science of sustainable agriculture. Westview Press, Boulder, CO.

Altieri, M.A., 1999. The ecological role of biodiversity in agroecosystems. Agriculcture Ecosystems and Environment 74, 19–31.

Alves, S. B.; Vendramim, J. D.; Silveira Neto, S.; Nakano, O. 1982. Insect pollination of onion, Allium cepa L. Solo 1982 Vol. 74 No. 1/2 pp. 18-22

Amaral, E., Mitidieri, J. and Vencousky, R. 1963. Studies on the activities of Apis mellifera L. while visiting the flowers of Cucumis sativus L. Olericultura,3: 181-193.

Anderson, D.L., Sedgeley, M., Short, J.R.T. and Allwood, A.J. 1982. Insect pollination of mango in northern Australia.Australian J. agric. Res.,33: 541-548.

Aouar-Sadli M., Louadi K., Doumandji S. 2008 Pollination of the broad bean (Vicia faba L. var. major) (Fabaceae) by wild bees and honey bees (Hymenoptera: Apoidea) and its impact on the seed production in the Tizi-Ouzou area (Algeria). African Journal of Agricultural Research 3(4): 266-272.

Association of Official Seed Analysts. Rules for testing seeds. Assoc. Offic. Seed Anal. Proc. Vol. 49, no. 2. 1959.

Backman, W.W. and Waller, G.D. 1971. Honeybee responses to sugar solutions of different composition.J. apic. Res.,16: 165-169.

Baker, H.G., Cruden, R.W. and Baker, I. 1971. Minor parasitism in pollination biology and its community function: the case of Ceiba acuminate.Bio. Sci.,21:1127-1129

Banaszak J. 1980 Studies on methods of censusing the numbers of bees (Hymenoptera, Apoidea). Polish Ecological Studies 6: 355-366.

Baron, F. C. 1962. Onion seed, sample costs and production. Calif. Agr. Ext. Sen., Cost Data Sheet 22, Leaflet.

Baswana K. S. 1984 Role of insect-pollinators on seed production in coriander and fennel. South Indian Horticulture 32: 117-118.

Bell C. R. 1971 Breeding systems and floral biology of the Umbelliferae or evidence for specialization in unspecialized flowers. In: Heywood V. H. (Ed.) The biology and chemistry of the Umbelliferae. Academic Press. London. pp. 93-107.

Belluscci, S. and Marques, M.D. 2001. Circadian activity rhythm of the foragers of a eusocial bee (Scaptotrigona affdepilis, Hymenoptera, Apidae, Meliponinae) outside the nest.Biol. Rhythm Res.,32 (2): 117–124.

Benachour K., Louadi K., Terzo M. 2007 Rôle des abeilles sauvages et domestiques (Hymenoptera: Apoidea) dans la pollinisation de la fève (Vicia faba L. var. major) (Fabaceae) en région de Constantine (Algérie). Annales de la Société Entomologique de France 43(2): 213-219.

Bendifallah L Louadi K Doumandji S 2013. Bee fauna potential visitors of coriander flowers Coriandrum sativum L. (Apiaceae) in the Mitidja area (Algeria). J. Apic. Sci. 57( 2): : 59-69

Bendifallah L., Louadi K., Doumandji S., Michez D. 2011 Rôle des abeilles solitaires (Hymenoptera: Apoidea) dans la pollinisation de la fève Vicia faba L.var. Major) (Fabaceae) en région de Mitidja (Algérie). In: Proceedings of the 9 th International Conference on Pests in Agriculture. Montpellier, France. 25-27 October 2011. pp. 75-80.

Benedek P and Gall E 1972. The effect of insect pollination on seed onion with observation on the behaviour of honey bees on crop. Journal of Apiculture Research 11 : 175-180.

Benedek, P. 1976. Effect of environmental factors on the foraging rates of honey bees in red clover fields. Zeitschrift fur Angewandte Entomologie,81(1): 14-20.

Bernhardt, P., 2000. Convergent evolution and adaptive radiation of beetle-pollinated angiosperms. Plant Systematics and Evolution 222, 293–320.

Beyleveld, A.J., 1967. Ostrich farming in South Africa. Agrekon 6, 14–16.

Bezabih, G. and Gebretsadikan, K. 2014 Managed honeybees (Apis mellifera L.) increase onion (Alliun cepa) seed yield and quality.

Biesmeijer, J.C., Roberts, S.P.M., Reemer, M., Ohlemuller, R., Edwards, M., Peeters, T., Schaffers, A.P., Potts, S.G., Kleukers, R., Thomas, C.D., Settele, J., Kunin and W.E., 2006. Parallel declines in pollinators and insect-pollinated plants in Britain and the Netherlands. Science 313, 351–354.

Blanche, K.R., Cunningham, S.A., 2005. Rainforest provides pollinating beetles for Atemoya crops. Journal of Economic Entomology 98, 1193–1201.

Blanche, K.R., Ludwig, J.A., Cunningham, S.A., 2006. Proximity to rainforest enhances pollination and fruit set in orchards. Journal of Applied Ecology. 43, 1182–1187.

Bogoyavlenseii, S. H., and Akimenko, A. L. 1966. [Coriander as a nectiferous and entomophillous culture.] In Achievements of Science and Advanced Experiment in Beekeeping, pp. 119-125.

Bohart G. E and Nye W. P. 1960 Insect pollinators of carrots in Utah. Agricultural Experimental Station Bulletin. N°.49. Utah State University. Logan. UT. 16 pp.

Bohart, G. E., Nye, W. P. and Hawthorn, L. R. 1970. Onion pollination as affected by different levels of pollinator activity. Utah Agricultural Experiment Station Bulletin. p. 482.

Bohart, G.E. and Nye, W.P. 1968 "Insect pollinators of carrot in Utah", Utah Agri. Expt. Sta. Bull., 419, 16.

Bohart, G.E., W.P. Nye and Hawthorn, L.R. 1970. Onion pollination as affected by different levels of pollinator activity. Utah Agr. Exp. Sta. Bull. No. 482: 57 pp.

Boivin 1999. Integrated Management for Carrot Weevil. Integrated Pest Management Reviews 4(1):21-37.

Borneck, R. and Merle, B. 1989. Essaie d'une evaluation de l' incidence economique de l' abeillie pollinisatrice dans l' agriculture europenne. Apicata 24: 33-38.

Borthwick, H. A. 1931. Carrot seed germination. Amer. Hort. Sci. Soc. Proc. 28: 310-314.

Bose, T.K. and Som, M.G. 1990 "Vegetable crops in India", Kalyani Publishers, Naya Prakash, India, p. 409.

Boyd, J., Banzhaf, S., 2007. What are ecosystem services? The need for standardized environmental accounting units. Ecological Economics 63, 616–626.

Braak, J. P., and Kho, Y. O. 1958. Some observations on the floral biology of the carrot (Daucus carota L.). Euphytica 7(2): 131-139.

Brantjes, N.B.M. and Leemans, J.A.A.M. 1976. Silane oties (Carryophyllaceae) pollinated by nocturenal Lepidoptera and mosquitoes. Acta Botanica Neerlandica, 25: 281-285.

Brewer, J.W. 1974. Pollination requirements for watermelon seed production. J. apic. Res., 13(3): 207-212.

Brewster, J.L., 2008. Onions and other vegetable Alliums, 2nd Edition. CAB International, Oxon.

Brosi, B.J., Daily, G.C., Shih, T.M., Oviedo, F., DurÃ¡n, G., 2008. The effects of forest fragmentation on bee communities in tropical countryside. Journal of Applied Ecology 45, 773–783.

Buchmann, S. L. and Nabhan, G. P. 1996. The forgotten pollinators. Island Press, Washington D C.

Burill, R.M.; Dietz, A. 1982 The responses of honeybees to variations in solar radiation and temperature. Apidologie 12 : 318-328.

Buscher, B.E., 2008. Conservation, neoliberalism, and social science: a critical reflection on the SCB 2007 annual meeting in South Africa. Conservation Biology 22, 229–231.

Carlson, E. C. 1974 Onion varieties, honeybee visitations, and seed yield.California Agriculture 28 (9): 16-18

Caron, D. M., and Morse, R. A. 1972. Distribution and behavior of honey bees on onions. Environmental Ent. 1: 127-129.

Caron, D. M.; Lederhouse, R. C. and Morse, R. A. 1975. Insect pollinators of onion in New York State. Hortscience 10 (3) 273-274

CEPF, 2003. Ecosystem profile: The Succulent Karoo hotspot, Namibia and South Africa, Africa. Critical Ecosystem Partnership Fund.

Cervancia, C.R. and Bergonia, E.A. 1991. Insect pollination of cucumber (Cucumis sativus L.) in the Philippines.The Sixth International Symposium on Pollination, Tilburg, Netherlands, 27-31 August 1990, 278-282.

Chacoff, N.P., Aizen, M.A., 2006. Edge effects on flower-visiting insects in grapefruit plantations bordering premontane subtropical forest. Journal of Applied Ecology. 43, 18–27.

Chandel R S, Thakur R K, Bhardwaj N R and Pathania N 2004. Onion seed crop pollination a missing dimension in mountain horticulture. Acta Horticulturae 631 : 79-86.

Chandel, R.S.; Thakur, R. K.; Bhardwaj, N. 2000 R.; Pathania, N.Onion seed crop pollination: a missing dimension in mountain horticulture. Issues and advances in transplant production and stand establishment research, proceedings of the XXVI International Horticultural Congress, Toronto, Canada, 11-17 August, 2002.

Chaudhary O. P., Singh J. 2007 Diversity, temporal abundance, foraging behaviour of floral visitors and effect of different modes of pollination on coriander (Coriandrum sativum L.). Journal of Spieces and Aromatic Crops 16(1): 8-14.

Chaudhary, N. 2004. Factors influencing inter and intra specific preferences of honey bee for three concurrently flowering vegetable/ spice crops. Ph.D Thesis, CCSHAU, Hisar. 96+xxviii pp.

Chaudhary, N. and Sihag, R.C. 2003. Diversity, foraging behaviour and foraging efficiency of different pollinators of visiting onion (Alilum cepa L.) blossoms.Korean J. Apic., 18(2): 103-108.

Choi, S.Y. 1987. Foraging activity of insects on weigh onion flowers (Allium sistulosum). Korean J. Apic.,2(2): 67-74.

Chopra, R.N. 1933 "Indigenous drugs of India", The Art Press, Calcutta, p. 3, 61.

Cirudarescu, G. 1971. Pollinators of lucerne and factors influencing their effectiveness in South Eastern parts of Birse depression. Analele Universitati Bucuresti. Biologica Animale, 20: 77–81.

Clark, W.C., and Dickson, N.M., 2003. Sustainability science: the emerging research program. Proceedings of the National Academy of Sciences 100, 8059–8061.

Collison, C.H. 1973. Nectar secretion and how it affects the activity of honey bees in the pollination of hybrid pickling cucumbers, Cucumis sativus L. M.Sc. Thesis, Michigan State University, East Lansing, USA.

Collison, C.H. 1976. The interrelationship of honeybee activity, foraging behaviour, climatic conditions and flowering in the pollination of pickling cucumber, Cucumis sativus L. Ph.D. Thesis, Michigan State University, USA, 268pp.

Collison, C.H. and Martin, E.C. 1979. Behaviour of honey bees foraging on male and female flowers of Cucumis sativus. J. apic. Res.,18: 184-190.

Colville, J.F., Picker, M.D., and Cowling, R.M., 2002. Species turnover of monkey beetles Scarabaeidae: Hopliini along environmental and disturbance gradients in the Namaqualand region of the succulent Karoo, South Africa. Biodiversity and Conservation 11, 243–264.

Comin, D. 1946. Onion production. 186 pp. Orange-Judd Publishing Co., Inc., New York

Corbet, S.A. 1978. A bee's view of nectar. Bee World 57: 25-32.

Crailsheim, K. 1991. Interadult feeding of jelly in honey bees (Apis mellifera L.) colonies.J. Comp. Physiol.161B: 55-60.

Corbet, S.A., Fussell M., Ake, R., Fraser, A., Gunsen, C., Salvage, A and Smithy, K. 1993. Temperature and the pollinating activity of social bees. Ecological Entomology 18: 17-30.

Corbet, S.A.; Fussell M.; Ake, R.; Fraser, A.; Gunsen, C.; Salvage, A.; Smithy, K. 1993 Temperature and the pollinating activity of social bees. Ecological Entomology 18: 17-30.

Costanza, R., D'Arge, R., de Groot, R.S., Farber, S., Grasso, M., Hannon, B., Limburg, K., Naeem, S., O'Neill, R. V, Paruelo, J., Raskin, R.G., Sutton, P., and van den Belt, M., 1997. The value of the world's ecosystem services and natural capital. Nature 387, 253–260.

Costanza, R., D'Arge, R., de Groot, R.S., Farber, S., Grasso, M., Hannon, B., Limburg, K., Naeem, S., O'Neill, R. V, Paruelo, J., Raskin, R.G., Sutton, P., and van den Belt, M., 1998. The value of the world's ecosystem services and natural capital. Ecological Economics 25, 3–15.

Cowling, R.M. and Roux, P.W., 1987. The karoo biome: a preliminary synthesis Part 2 - vegetation and history. South African National Scientific Programmes Report, No. 142, Pretoria.

Cowling, R.M., 1986. A description of the Karoo Biome Project. South African National Scientific Programmes Report No. 122. Pretoria.

Cowling, R.M., Egoh, B., Knight, A.T., Farrell, P.J.O., Reyers, B., Rouget, M., Roux, D.J., Welz, A., Wilhelm-Rechman, A., and O'Farrell, P.J., 2008. An operational model for mainstreaming ecosystem services for implementation. Proceedings of the National Academy of Sciences 105, 9483–9488.

Cowling, R.M., Esler, K.J., Rundel, P.W., 1999. Namaqualand, South Africa – an overview of a unique winter-rainfall desert ecosystem. Plant Ecology 142, 3–21.

Cowling, R.M., Richardson, D.M. and Pierce, S.M. Eds., 1997. Vegetation of southern Africa. Cambridge University Press, Cambridge.

Cowling, R.M., Roux, P.W., and Pieterse, A.I.H., 1986. The karoo biome: a preliminary synthesis. Part 1 - physical environment. South African National Scientific Programmes Report, No. 142, Pretoria.

Crane, E. 1990. Bees and Beekeeping.Bath Press Ltd., Avon., 68-69pp.

Deodikar, G.B. and Suryanaryana, M.C. 1972. Crop yields and bee pollination.Indian Bee J., 4: 53-64.

Cremo, M.A., and Goswani, M., 1995. Divine nature: a spiritual perspective on the environmental crisis. Bhaktivedanta Book Trust International.

CSIRO 1991 . The insects of Australia: A textbook for students and research workers . 2nd edn . Carlton, Victoria , Melbourne University Press .

Cunningham, SA, Fitz Gibbon, F and Heard, TA. 2002. The future of pollinators for Australian agriculture. Australian Journal of Agricultural Research, 53: 893–900. [CrossRef]

Cupido, C.F., 2005. Assessment of veld utilisation practices and veld condition in the Little Karoo. MSc thesis, Stellenbosch University.

Currah, L., and Ockendon, D.J., 1983. Onion pollination by blowflies and honeybees in large cages. Annals of Applied Biology 103, 497–506.

Currah, L., and Ockendon, D.J., 1984. Pollination activity by blowflies and honeybees on onions in breeders' cages. Annals of Applied Biology 105, 167–176.

Daily, G.C., 1997. Nature's services: societal dependence on natural ecosystems. Island Press, Washington, D.C.

Daily, G.C., 2000. Management objectives for the protection of ecosystem services. Environmental Science & Policy 3, 333–339.

Daily, G.C., Matson, P.A., 2008. Ecosystem services: From theory to implementation. Proceedings of the National Academy of Sciences 105, 9455–9456.

De Groot, R.S., Wilson, M.A., Boumans, R.M.J., 2002. A typology for the classification, description and valuation of ecosystem functions, goods and services. Ecological Economics 41, 393– 408.

De Marco, P., and Coelho, F.M., 2004. Services performed by the ecosystem: forest remnants influence agricultural cultures' pollination and production. Biodiversity and Conservation 13, 1245–1255.

De Ruijter, A., 2002. Pollinator diversity and sustainable agriculture in the Netherlands, in: Kevan, P., Imperatriz-Fonseca, V.L. Eds., Pollinating bees: the conservation link between agriculture and nature. Ministry of Environment, Brasilia, pp. 67–70.

Dean, W.R.J., and Milton, S.J., 1995. Plant and invertebrate assemblages on old fields in the arid southern Karoo, South Africa. African Journal of Ecology 33, 1–13.

Dean, W.R.J., and Milton, S.J., 1999. The Karoo: ecological patterns and processes. Cambridge University Press, Cambridge.

Delbrassine S., Rasmont P. 1988 Contribution à l'étude de la pollinisation du Colza, Brassica napus L. var. Oleifera (Moench) Delile, en Belgique. Bulletin des Recherches Agronomiques de Gembloux 23: 123-152.

Delplane, K.S. and Mayer, D.F. 2000. Crop pollination by bees. CABI Publ., Wallingford, Oxon, UK.

Deodikar G B and Suryanarayana M C 1977 Pollination in the service of increasing farm production in India. Adv. Pollen Spore Res. 2: 67–87.

Devi M, Sharma H. K and Rana K 2015. Measuring diversity and density of insect pollinators on coriander, Coriandrum sativum L by different sampling methods International Journal of Farm Sciences 5(3) : 179-189

Devi S, Gulati R, Kanika T and Poonia A. 2015 The pollination biology of onion (Allium cepa L.)- A Review Agri. Review, 36 (1): 1-13

Devillers, J., Dore, J.C., Tisseur, M., Luzeau, S. and Maurin, G. 2004. Modelling the flight activity of Apis mellifera at the hive entrance.Computers andElectronics in Agriculture, 42(2): 87–109.

Dewey. D.R. and Lu, K.U. 1959. A correlation and path analysis of crest wheat grass seed production. Agron. J. 51 : 515 –518.

Dewey. D.R.; Lu, K.U. 1959 A correlation and path analysis of crest wheat grass seed production. Agron. J. 51 : 515 –518.

Diederichsen A. 1996 Coriander (Coriandrum sativum L.). Promoting the conservation and use of underutilized and neglected crops. 3. Inst. Plant Genetics and Crop Plant Research. Gatersleben/ International Plant Genetic Ressources Institute. Rome. 83 pp.

Donaldson, J.S., Nänni, I., Zachariades, C., and Kemper, J., 2002. Effects of habitat fragmentation on pollinator diversity and plant reproductive success in renosterveld shrublands of South Africa. Conservation Biology 16, 1267–1276.

Driver, A., Desmet, P.G., Rouget, M., Cowling, R.M., and Maze, K., 2003. Succulent Karoo ecosystem plan. Biodiversity component technical report, Cape Conservation Unit Report No CCU 1/03. Kirstenbosch, Cape Town.

Duncan, A. A. 1965. Vegetable seed productionñonion. Oreg. Agr. Ext. Serv. FS 88, Leaflet.

Duncan, D. B. Multiple range and multiple F tests. Biometrics 11(1): 1-42, 1955.

Eardley, C.D., 1996. The genus Scrapter Lepeletier & Serville Hymenoptera: Colletidae. African Entomology 4, 37–92.

Ehrlich, P.R., and Raven, P.H., 1964. Butterflies and plants: a study in coevolution. Evolution 18, 586–608.

Eickwort G. C., Ginsberg H. S. 1980 Foraging and mating behaviour in Apoidea. Annual Review of Entomology 25: 421-446.

El-Berry A R, Gawaad A A A, Moustafa M A F, El-Gayar F H, Berrry A R El, AbdelGawaad A A and Gayar F H El 1974 Pollinators other than honey bees visiting certain medicinal plants in Egypt. Zeitschrif fur Angewandte Entomologie. 76: 113–119.

El-Berry A. A., Moustafa M. A., Abdel-Gawaad A. A., El-Bialey S. 1974 Pollinators other than honey bees visiting certain vegetable plants in Egypt. Journal of Applied Entomology 77(1-4): 106-110.

Ellis P. R., Wheatley G. A. and Hardman J. A. 1978. Preliminary studies of carrot susceptibility to carrot fly attack. Annals of Applied Biology 88(1):159-170.

Emlen, J.M. 1966. The role of time and energy in food preference.American Naturalist, 100: 611-617.

Enzie J. V. 1943. Experiments in the production of carrot seed. N. Mex. Agr. Expt. Sta. Bul. 308, 11 pp.

Erickson, EH and Peterson, CE. 1979. Asynchrony of floral events and other differences in pollinator foraging stimuli between fertile and male-sterile carrot inbreds. Journal of the American Society for Horticultural Science, 104: 639–643.

Erickson, H. T., and Gabelman, W. H. 1956. The effect of distance and direction on cross-pollination in onions. Amer. Soc. Hort. Sci. Proc. 68: 351-357.

Esty, D., Levy, M.A., Srebotnjak, T., and de Sherbinin, A., 2005. 2005 Environmental sustainability index: benchmarking national environmental stewardship. Yale Center for Environmental Law and Policy, New Haven.

Ewert, E. 1942. Nectar production of cooking onions. Deut Bienenzucht 49: 234- 235. [In German.]

Ewies, M.A. and El-Sahhar, K. 1977. Observation on the behaviour of honeybee on onion and their effect on seed yield.J. Apic Res., 16: 194-196.

Fahn, A. 1949. Studies in the ecology of nectar secretation.Palest. J. Bot. Ser,4: 207-224. FAO.(2002). Production Year Book, 56.

Finch S. 1993. Integrated pest management of the cabbage root fly and the carrot fly. Crop Protection 12(6):423-430.

Fisher, B., and Turner, R.K., 2008. Ecosystem services: Classification for valuation. Biological Conservation 141, 1167–1169.

Fisher, B., Turner, R.K., and Morling, P., 2009. Defining and classifying ecosystem services for decision making. Ecological Economics 68, 643–653.

Flemion, R., and J. Olsen 1960. Lygus bugs in relation to seed production and occurrence of embryoless seeds in various Umbelliferae species. Boyce Thompson lust. Contrib. 16: 39-46.

Flemion, R., and K T. Hendrickson 1949. Further studies on the occurrence of embryoless seeds and immature embryos in the Umbelliferae. Boyce Thompson Inst. Contrib. 15: 291-297.

Franklin, D. F. 1948 some problems in carrot seed production. Seed world 63(8): 8-9,44.

Franklin, D. F. 1953. Growing carrot seed in idaho. Idaho agr. Expt. Sta. Bul. 294, 35 pp.

Franklin, D. F. 1958. Effect on hybrid onion seed production of using different ratios of male-sterile and pollen rows. Amer. Soc. Hort. Sci. Proc. 71: 435-439.

Franklin, D. F. 1970. Problems in the production of vegetable seed. In The Indispensable Pollinators, Ark. Agr. Ext. Serv. Misc. Pub. 127, pp. 112-140.

Franklin, D.F. 1970. Problems in the production of vegetable seed. In. The Indispensable Pollinators, Ark. Agr. Exp. Serv. Misc. Pub., 127: pp. 112-140.

Free J. B. 1993 Insect pollination of crops. Ed. Academic Press. Harcourt Brace Jovanich Publ. London. 684 pp.

Free, J. B. 1960. The behaviour of honey bees visiting the flowers of fruit trees. J. Anim. Ecol.,29: 385-395.

Free, J.B., 1993a. Insect pollination of crops. Academic Press, London, UK. 684 pp.

Free, J.B., 1993b. Alliaceae, in: Insect Pollination of Crops. Academic Press, London, UK, pp. 115–121.

Free, J.B., and Williams, I.H., 1972. The transport of pollen on the body hairs of honeybees Apis mellifera L. and bumblebees Bombus spp. L. Journal of Applied Ecology 9, 609–615.

Free, J.B., Williams, I.H., Longden, P.C. and Johnson, M.G. 1975. Insect pollination of sugar beet (Beta vulgaris L.) seed crops.Ann. Appl. Biol.,81: 127-134.

Free, JB. 1970. Insect pollination of crops, London: Academic Press.

Frisch, K.Von. 1950. Bees, Their Vision Chemical Sense and Language.Cornell Univ. Press, Ithaca, New York.

Gallai, N., Salles, J., Settele, J., and Vaissière, B.E., 2009. Economic valuation of the vulnerability of world agriculture confronted with pollinator decline. Ecological Economics 68, 810–821.

Gary, N. E., Witherell, P. C., and Marston, J. 1972. Foraging range and distribution of honey bees used for Carrot and onion pollination. Environmental ent. 1(1): 71 - 78.

Gary, N.E. 1975. Activities and behaviour of honey bee. In: The Hive and Honeybee. Dadant and sons, Illinois, Hamilton, 237-240 pp.

Gary, N.E., Witherell, P.C., and Marston, J.M., 1972. Foraging range and distribution of honey bees used for carrot and onion pollination. Environmental Entomology 1, 71–78.

Gary, N.E., Witherell, P.C., Lorenzen, K., and Marston, J.M., 1977. The interfield distribution of honey bees foraging on carrots, onions and safflower. Environmental Entomology 6, 637–640.

Gegear, R.J. and Thomson, J.D. 2004. Does the flower constancy of bumble bees reflect foraging economics.Ethology.,110: 1-2.

George, R.A.T., 1999. Vegetable seed production, 2nd Edition. CAB Interantional Publishing, New York.t

Gess, F.W., and Gess, S.K., 1993. Effects of increasing land utilization on species representation anddiversity of aculeate wasps and bees in the semi-arid areas of southern Africa., in: La Salle, J., Gauld, I.D. Eds., Hymenoptera and Biodiversity. CAB International, Wallingford, England, pp. 83–113.

Gess, S.K., 1992. Biogeography of the masarine wasps Hymenoptera: Vespidae: Masarinae, with particular emphasis on the southern African taxa and on correlations between masarine and forage plant distributions . Journal of Biogeography 19, 491–503.

Gess, S.K., 1996. The pollen wasps: ecology and natural history of the Masarinae. Harvard University Press, Cambridge.

Gess, S.K., 2001. The Karoo, its insect pollinators and the perils which they face. International Pollinator Initiative IPI Case Study. www.ecoport.org

Gess, S.K., and Gess, F.W., 2004a. Distributions of flower associations of pollen wasps Vespidae: Masarinae in southern Africa. Journal of Arid Environments 57, 17–44.

Gess, S.K., Gess, F.W., 2004b. A comparative overview of flower visiting by non-Apis bees in the semi-arid to arid areas of southern Africa. Journal of the Kansas Entomological Society 77, 602–618.

Gilbert, F.S. 1980. Flower visiting by hoverflies (Syrphidae). Ann. Res. Ecol. Syst., 6: 139-170.

Gill, R.J., Ramos-Rodriguez, O., and Raine, N.E., 2012. Combined pesticide exposure severely affects individual- and colony-level traits in bees. Nature 491, 105–8.

Gill, S.S., Kaushik, H.D. and Sharma, S.K. 2001. Pollination efficiency of three Apis spp. foraging on phalsa (Grewia subinaequalis D.C.).Ann. Ent., 19(1): 19-21.

Gjorgji, M. and Rukie, A. 1997. The effects of pollination and the use of bees on number and quality of onion seed 1. Balkan Symposium on vegetables and potaoes, ISHS.Acta Hortic.,462: 12.

Gladis T 1997 . Bees versus flies?—rearing methods and effectiveness of pollinators in crop germplasm regeneration . KW Richards VII International Symposium on Pollination. Lethbridge, Alberta, Canada 23–28 June, 1996 . Pp. 235 – 238 .

Glukhov M. M. 1955 Honey plants. In: Mc Gregor S. E. (Ed.) Insect pollination of cultivated crop plants. Ed. Agriculture Handbook. Serv. Rech. Agri. U.S. Gov. Printing Off. Washington. 411pp.

Glukhov M. M. 1955. [Honey Plants.] 512 pp. Izd. 6, Perer. i Dop. Moskva, Gos. Izd-vo Selkhoz Litry. [In Russian.]

Glukhov, M.M. 1955. Honey Plants. Izd-vo Seikhoz Zit-ry (In Russian): 512 PP. Izd. 6,

Goldman, R.L., Tallis, H., Kareiva, P., Daily, G.C., Goldman, L., and Dailyt, G.C., 2008. Field evidence that ecosystem service projects support biodiversity and diversify options. Proceedings of the National Academy of Sciences 105, 9445–9448.

Goyal, N.P., Singh, M. and Kandoria, J.L. 1989 "Role of insect pollination in seed production of carrot, Daucus carota Linn.", Ind. Bee J., 51, 89-93.

Goyal, N.P.; Singh, M.; Kandoria, J.L. 1989 Role of insect pollination on seed production in carrot, Daucus carota L. Indian Bee J. 51(3): 89-93.

Grace J., Nelson M. 1981 Insects and their pollen loads at a hybrid Heracleum site. New Phytologist 87: 413-423.

Greenleaf, S.S., and Kremen, C., 2006a. Wild bees enhance honey bees ' pollination of hybrid sunflower. Proceedings of the National Academy of Sciences 103, 13890–13895.

Greenleaf, S.S., and Kremen, C., 2006b. Wild bee species increase tomato production and respond differently to surrounding land use in North California. Biological Conservation 133, 81–87.

Haglar, J.R., Cohen, A.C. and Loper, G.M. 1990. Production and decomposition of onion nectar and honey bee foraging activity in Arizon.Env.Ent.,19(2): 327-331.

Hagler, J.R 1990. Honey bee (Apis mellifera L.) response to simulated onion nectars containing variable sugar and potassium concentrations. Apidologie 21:115-121.

Hagler, J.R., Cohen, A.C., and Loper, G.M., 1990. Production and composition of onion nectar and honey bee Hymenoptera: Apidae foraging activity in Arizona. Environmental Entomology 19, 327–331.

Hamilton R. P. 1946. Onions need bees. Gleanings Bee Cult. 74: 23.

Harrison, J.F. and Fewell, J.H. 2002. Environmental and genetic influences on flight metabolic rate in the honeybee, Apis mellifera. Comparative Biochem.Physio.,133(2): 323–333.

Hammert A., Harper D. A. T., Ryan P. D. 2001 P.A.S.T : Palaeontological Statistics Software Package for Education and Data Analysis. Palaeontological Electronica 4 (1): 1-9.

Harris, L.F., and Johnson, S.D., 2004. The consequences of habitat fragmentation for plant–pollinator mutualisms. International Journal of Tropical Insect Science 24, 29–43.

Haslbachova, H and Kubisova, S. 1982. Pollinating activity of honey bees (Apis mellifera) on onions (Allium cepa). I. Flight of bees and its influence on the amount of seed produced. Sbornik UVTIZ, Zahradnictvi (2): 119-126

Haslbachova, H.; Kubisova, S.; Roznovsky, M. 1981. Studies of honeybee (Apis mellifera) activity on onion (Allium cepa). Acta Universitatis Agriculturae, Brno, A, Facultas Agronomica 29 (3/4): 313-321

Hawthorn, L. R. 1951. Studies of soil moisture and spacing for seed crops of carrots AND ONIONS. U.S. Dept. Agr. Cir. 892, 26 pp.

Hawthorn, L. R., Bohart, G. E, toole, E. H., and others. 1960. Carrot seed production as affected by insect pollination. Utah agr. Expt. Sta. Bul. 422, 18 pp.

Hawthorn, L. R., Bohart, G. E., and Toole, E. H. 1956. Carrot seed yield and germination as affected by Different levels of insect pollination. Amer. Soc. Hort. Sci. Proc. 67: 384-389.

Hawthorn, L. R., G.K Bohart, and K H. Toole. Carrot seed yield and germination as affected by different levels of insect pollination. Amer. Soc. Hort. Sci. Proc. 67: 384- 389. 1956.

Hawthorn, LR, Bohart, GE and Toole, EH. 1956. Carrot seed yield and germination as affected by different levels of insect pollination. Proceedings of the American Society for Horticultural Science, 67: 384–389.

Hawthorn, R.L., Bohart, G.E., Toole, E.H., Nye, W.P. and Levin, M.D. 1960 "Carrot seed production as affected by insect pollination", Bull. Utah Agric. Exp. Sta., 422, 18.

Heal, G., 2000. Valuing ecosystem services. Ecosystems 3, 24–30.

Heard, T.A., and Exley, E., 1994. Diversity, abundance and distribution of insect visitors to macadamia flowers. Environmental Entomology 23, 91–100.

Heeger E. F. 1989 Handbuch des Arznei- und Gewürzpflanzenbaues. Unveränderte Auflage. Frankfurt. 1956 pp.

Heinrich, B. 1975 Energetics of pollination. Ann. Rev. Ecol. Syst. 6: 139-170.

Heinrich, B. and Raven, P.H. 1972. Energetics and pollination ecology.Science, 17: 597-602.

Helmholtz Association of German Research Centres. 2008. Economic Value of insect pollination. Science Daily.

Hesse, A.J., 1938. A revision of the Bombyliidae Diptera of southern Africa. Part 1. Annals of the South African Museum 34, 1–1053.

Hilton-Tailor, C., and Le Roux, A., 1989. Conservation status of the Fynbos and Karoo biomes, in: Huntley, B.J. Ed., Biotic Diversity in southern Africa. Oxford University Press, Cape Town, pp. 202–223.

Hmamouchi M. 1997 Plantes alimentaires, aromatiques, condimentaites, medicinales et toxiques au

Hoffman, M.T., Cousins, B., Meyer, T., Petersen, A., and Hendricks, H., 1999. Historical and contemporary land use and the desertification of the Karoo, in: Dean, W.R.J., Milton, S.J. Eds., The Karoo: Ecological Patterns and Processes. Cambridge University Press, Cambridge, pp. 257–273.

Howlett, B.G., Donovan, B.J., McCallum, J.A., Newstrom, L.E., and Teulon, D.A.J., 2005. Between and within field variability of New Zealand indigenous flower visitors to onions. New Zealand Plant Protection 58, 213–218.

Howlett, BG, Walker, MK, Newstrom-Lloyd, LE, Donovan, BJ and Teulon, DA. 2009. Window traps and direct observations record similar arthropod flower visitor assemblages in two mass flowering crops. Journal of Applied Entomology, 133: 553–564

Hussein M. H., Abdel-Aal S. A. 1982 Wild and honeybees as pollinators of 10 plant species in Assiut area. Egypt. Zeitschrift fur Angewandte Entomologie 93: 342-346.

Hwang, H.J., Suk, J.K. and Ha, I.J. 1998. Effect of Pollinating insect on seed yield in seed production of onion (Allium Cepa L.).J. Hortic. Sci., 40(2): 27-30.

Ilyas M. 1980 Spices in India: III. Economic Botany 34: 236-259.

Inouye, D.W. 1980. The effect of proboscis length and corolla tube lengths on patterns and rates of flower by bumble bees.Oecologia, 45: 197-201.

Ivanova K. V., Stoletova E. A. 1990 Istorija kul'tury i vnutrividovaja klassifikacija koriandra posevnogo (Coriandrum sativum L.) Sbornik Naucnych Trudov Prikladnoj Botaniki, Genetiki i Selekcii 133: 26-40.

Jablonski, B., Skowronek, J., Woyke, H. W., and Doruchowski, R. W. 1982. Biologia kwitnienia, nektarowanie, zapylanie i owocowanie meskosterylnych linii cebuli (Allim cepa L.). Pszczelnicze-Zeszyty- Naukowe, 26: 57-104.

Jadhav, L.D. 1981. Role of insects in the pollination of onion (Allium cepa L) in the Maharastra state, India.Indian bee J., 43: 61-63.

Jain, K.L. and Kapil, R.P. 1980. Foraging rhythm of megachilid bees in relation to the flowering of *Medicago sativa* L. and *Parkinsonia aculeta*. Indian Bee. J.,42: 35-38.

Jamieson, C.A. and Austin, G.H. 1956. Preference of honey bee for sugar solution. Proc.10th int. Congr.Entomol. Montreal,4: 1059-1065.

Johannsmeier, M.F., 2001. Beekeeping in South Africa, 3rd Edition. Agricultural Research Council-Plant Protection Research Institute, Handbook No. 14, Pretoria.

Johannsmeier, M.F., 2005. Beeplants of the south-Western Cape. Agicultural Research Council- Plant Protection Research Institute, Handbook No. 17., Pretoria.

Johnson, S.D., 2004. An overview of plant–pollinator relationships in southern Africa. International Journal of Tropical Insect Science 24, 45–54.

Jones O. T. and Coaker T. H. 1977. Oriented responses of carrot fly larvae, Psila rosae, to plant odours, carbon dioxide and carrot root volatiles. Physiological Entomology 2(3):189-197/

Jones, D. F., and Clarke, A. E. 1943. Inheritance of male-sterility in onion and the production of hybrid seed Amer. Soc. Hort. Sci. Proc. 43: 189-193.

Jones, H. A and Emsweller, S. L. 1934. The use of flies as onion pollinators. Amer. Soc. Hort. Sci. Proc. 31

Jones, H. A and Emsweller, S. L. 1936. A male-sterile onion. Amer. Soc. Hort. Sci. Proc. 34: 582-585.

Jones, H. A and Mann, L. K. 1964. Onions and their allies. 286 pp. Leonard Hill Book CO., London.

Jones, H. A Poole, C. F., and Emsweller, S. L. 1936. Bolting habit in the onion. Abstract in Amer. Soc. Hort. Sci. Proc. 33: 490.

Jones, H. A., and Emsweller, S. L. 1933. Methods of breeding onions. Hilgardia 7: 625-642.

Jones, H.A. 1933 Vegetable breeding at the university o California. Proc. Am. Soc. Hortic. Sci., 29:572-581. Jones, H.A. and Emsweller,S.L. 1933. Methods of breeding onions. Hilgardia, 7: 625-642.

Jones, H.A. and Clarke, A.E., 1943. Inheritance of male sterility in the onion and the production of hybrid seed. Proceedings of the American Society for Horticultural Science 43, 189–194.

Jones, H.A. and Mann, L.M., 1963. Onions and their allies, botany, cultivation and utilization. Leongard Hill, London.

Jones, H.A. and Rosa, J.T., 1928. Truck crop plants, 1st Edition. McGraw-Hill, New York.

Jula, F., Pirvu, E., and Illyes, C. 1965. The melliferous value of sainfoin (Onobrychis viciifolia) and onion (Allium cepa L.), under the soil and climatic conditions around CLUJ.] Lucr. Stiint. lnst. Agron. Cluj 21: 99-106. [In Romanian, English and Russian summaries.]

Jula, F., Privu, E. and Ilyes, G. 1965. The melliferous value of sainfoin (Onobrychis vichfolia) and onion (Allium cepa L.) under the soil and climatic condition around .Clij.Lucr.Stint.Inst.Agron.Cluj., 21: 99-106.

Jun Hwang Hae; Kyu Suh Jun; Jong Ha In and WooRyu Young 1998. Effect of pollinating insects on seed yield in seed production of onion (Allium cepa L.). RDA Journal of Horticulture Science 40 (2): 27-30

Kalmath, B. S and Sattigi, H. N 2004. Efficacy of chemicals in attracting honey bees and other insect pollinators to onion bloom. Pest Management and Economic Zoology 12(1): 101-104

Kalmath, B.S. and Sattigi, H.N. 2004. Effect of bee pollination by using attractants on the qualitative parameters of onion seed.J. Pest Manag. Econ. Zool., 12(1): 97-99

Kapil, R.P. and Brar, H.S. 1971. Foraging behaviour of Apis florea F. on Brassica campestris var toria.In: Proceedings of XXIIth International Congress of Apiculture, Moscow, 335–339.

Kapil, R.P. and Jain, K.L. 1980. Biology and utilization of insect pollinators for crop production.Final Tech. Report, Department of Zoology, Haryana Agricultural University, Hisar, India, 81p.

Kapil, R.P. and Kumar, S. 1974. Foraging activity of Apis dorsata F. on Brassica juncea Czern and Coss.J. Bombay Natural History Soc., 71(2): 327–331.

Kearns, CA, Inouye, DW and Waser, NM. 1998. Endangered mutualisms: the conservation of plant-pollinator interactions. Annual Review of Ecology and Systematics, 29: 83–112.

Kendall, D.A., Solomon, M.E., 1970. Insect pollination of fruit trees, Republication of the Agriculture Horticultural Research Station, University of Bristol.

Kevan, PG and Baker, HG. 1983. Insects as flower visitors and pollinators. Annual Review of Entomology, 28: 407–453. [CrossRef], [Web of Science ®], [CSA]

Kirtikar, K.R. and Basu, B.D. 1935 "Indian medicinal plants", Lolit Mohan Basu, Allahabad, p. 61.

Klein, A.-M., Steffan-Dewenter, I., and Tscharntke, T., 2003a. Pollination of Coffea canephora in relation to local and regional agroforestry management. Journal of Applied Ecology 40,837– 845.

Klein, A.-M., Steffan-Dewenter, I., Tscharntke, T., 2003b. Fruit set of highland coffee increases with the diversity of pollinating bees. Proceedings of the Royal Society B: Biological Sciences 270, 955–961.

Klein, A.-M., Vaissière, B.E., Cane, J.H., Steffan-Dewenter, I., Cunningham, S.A., Kremen, C. and Tscharntke, T., 2007. Importance of pollinators in changing landscapes for world crops. Proceedings of the Royal Society B: Biological Sciences 274, 303–313.

Kluser, S. an d Peduzzi, P., 2007. Global pollinator decline: a literature review. UNEP/ GRID, Geneva.

Kochetov, A.S. 2004. Use of Carpathian bees for cucumber pollination in film greenhouses. Kartofel Ovoshchi, 6: 28.

Kordakov, Z.M., 1956. Honeybees and pollination of seed plants of the common onion, in: Gubin, A.F. Ed., Pollination of Agricultural Crops. Gos. Izd-vo Selkhoz Litry, pp. 163–171.

Kordakova, Z. M. 1956. Honey Bees and pollination of seed plants of the common onion.] In Krishchunas, I. V., and Gubin, A. F., eds., Pollination of Agricultural Plants, Moskva, Gos. Izd-vo Selkhoz Lit-ry, pp. 163-171.

Kordakova, Z.M. 1956. Honey bees and pollination of seed plants of the common onion. In: Pollination of agricultural plants (eds. Krishchunas, I.V. and Gubin, A.F.). Moskva, Gos. Izd-vo Selkhoz Lit-r: 163-171.

Kosior, A., Celary, W., Olejniczak, P., Fijal, J., Król, W. and Solarz, W., Plonka, P., 2007. The decline of the bumble bees and cuckoo bees Hymenoptera: Apidae: Bombini of Western and Central Europe. Oryx 41, 79–88.

Koul A. K., Hamal I. A., Gupta S. K. 1989 Pollination mechanism in Coriandrum sativum Linn. Proceedings of the Indian Academy of Sciences Plant Science 99: 509-515.

Koul A. K., Koul P., Hamal I. A. 1986 Insects in relation to pollination of some umbellifers. Bulletin of the Botanical Survey of India 28: 39-42.

Koul, P, Koul, AK and Hamal, IA. 1989. Reproductive biology of wild and cultivated carrot (Daucus carota L.). New Phytologist, 112: 437–443.

Koul, P, Sharma, N and Koul, AK. 1993. Pollination biology of Apiaceae. Current Science, 65: 219–222.

Kremen, C., Williams, N.M., Bugg, R.L., Fray, J.P. and Thorp, R.W., 2004. The area requirements of an ecosystem service: crop pollination by native bee communities in California. Ecology Letters 7, 1109–1119.

Kremen, C., Williams, N.M., Thorp, R.W., 2002. Crop pollination from native bees at risk from agricultural intensification. Proceedings of the National Academy of Sciences 99, 16812–16816.

Kumar J, Mishra R C and Gupta J K 1989. Effect of honey bee pollination on onion (Allium cepa L.) seed production. Indian Bee Journal 51 : 3-5.

Kumar Manoj, Jaiswal Bibek Kumar 2012.Effect of honeybee (Apis mellifera L.) pollination on yield and quality in coriander Indian Journal of Entomology 74(3): 281-284.

Kumar, J., Mishra, R.C and Gupta, J K. 1985. The effect of mode of pollination on Allium species with observations on insects as pollinators. J. Apic. Res. 24(1):62-66.

Kumar, J., Mishra, R.C. and Gupta, J.K. 1985. The effect of mode of pollination on Allium species with observations on insects as pollinators J. apic. Res., 24:62-66.

Kumar, J.; Rao, K.V.K.; Gupta, P.R.; Dogra, G.S. 1989 Temporal distribution of bees (Hymenoptera: Apoidea) on spring flowering crops in mid hills of Himachal Pradesh, India. Indian Bee J. 51(2) : 55-58.

Kumar, J.; Mishra, R. C. and Gupta, J. K.1989. Effect of honey bee pollination on onion (Allium cepa L.) seed production. Indian Bee Journal 1989 51 (1): 3-5

Kumar, K. 2002. Studies on insect pollinators of sunflower (Helianthus annuus L.). M.Sc. Thesis, Haryana Agricultural University, Hisar, Haryana, India.

Kumar, R. 2004. Role of Apis melliferaL. in pollination of cucumber (Cucumis sativus L.) under poly-house conditions. M.Sc. Thesis, Haryana Agricultural University, Hisar, Haryana, India.

Kuti, J.O. and Rovelo, C. 1992. Effect of pollination techniques on fruit set and seed yield of egusi melon (Citrullus colocynthis). Report Cucurbit Genetics Cooperative,15: 80-81.

Kwon, Y.J. and Saeed, S. 2003. Effect of temperature on the foraging activity of Bombus terrestris L. (Hymenoptera: Apidae) on greenhouse hot pepper (Capsicum annuum L.) Appl. Entomol. Zool.,38(3): 275-280.

Lamborn, E and Ollerton, J. 2000. Experimental assessment of the functional morphology of inflorescences of Daucus carota (Apiaceae): testing the 'fly catcher effect'. Functional Ecology, 14: 445–454

Lamm, R.D., 2006. The culture of growth and the culture of limits. Conservation Biology 20, 269–271.

Lawrence JF , Britton EB 1991 . Coleoptera . In : The insects of Australia: A textbook for students and research workers . Vol II. 2nd edn. CSIRO Division of Entomology (ed) , Melbourne , Melbourne University Press . Pp. 1137 .

Le Baron, F.C. 1962. Onion seed sample costs and production. Calif. Agr. Ext. Sen., Cost Data Sheet 22, Leaflet.

Le Maitre, D.C., O'Farrell, P.J., Milton, S.J., Atkinson, D., De Lange, W.J., Egoh, B., Reyers, B., Colvin, C., Maherry, A. and Blignaut, J.N., 2009b. Assessment and evaluation of ecosystem services in the Succulent Karoo biome. Council for Scientific and Industrial Research, Pretoria.

Leadley, P., Pereira, H.M., Alkemade, R., Fernandez-Manjarres, J.F., Proenca, V., Scharlemann, J.P.W. and Walpole, M.J., 2010. Biodiversity scenarios: projections of 21st century - change in biodiversity and associated ecosystem services. Secretariat of the Convention on Biological Diversity, Montreal.

Lederhouse, R. C., Caron, D. M., and Morse, R. A. 1968. Onion pollination in new york. New York's Food and Life Sci. 1(3): 8-12.

Lederhouse, R.C., Caron, D.M and Morse, R.A. 1968.Onion pollination in New York. New York's Fd Life Sci. 1: 8-9.

Lederhouse, R.C., Caron, D.M. and Morse, R.A., 1972. Distribution and behavior of honey bees on onion. Environmental Entomology 1, 127–129.

Lerer, H., Bailey, W.G., Mills, P.F. and Panki, W.M. 1982. Pollination activity of Megachile rotundata (Hymenoptera; Apidae).Env.Ent.,11: 997-1000.

LI, C.C. 1956 The concept of path coefficients and its impact on population genetics. Biometrics 12: 997-1000.

Li, C.C. 1956. The concept of path coefficients and its impact on population genetics. Biometrics 12: 997-1000.

Lindsey A. H. 1984 Reproductive biology of Apiaceae. I. Floral visitors to Thaspium and Zizia and their importance in pollination. American Journal of Botany 71: 375-387.

Lorenzone M C A, Martinho M R, Freire J A H and Reis F P 1993. Effect of Pollination by honey bees on onion hybrid seed production. Presquisa-Agropecuana-Brasieira 28 : 229-235.

Losey, J. and Vaughan, M. 2006. The economic value of ecological services provided by Insects. Bioscience, 56 (4): 311-323.

Luk'janov I. A., Reznikov A. R. 1976 Koriandr. In: Smoljanova A. M., Ksendza A. T. (Eds.) Efirnomaslicnye kul'tury. Kolos. Moskva. Russia. pp. 9-57.

MacArthur, R. H. and Pianka, E. R. 1966. On the optimaluse of a patchy environment. American Naturalist, 100: 603-609.

Macgillivray, J. H. 1948. Effect of irrigation on the yield of onion seed. Amer. Soc. Hort. Sci. Proc. 51: 423-427.

Magruder, R., Wester, R. E., and others. 1941. Descriptions of types of principal american varieties of onions. U.S. Dept. Agr. Misc. Pub. 435, 87 pp.

Mann, L. K 1952. Anatomy of the garlic bulb and factors affecting bulb development. Hilgardia 21(8): 195-235.

Mann, L. K and Little, T. M. 1957. Growing garlic in California. Calif. Univ., Dept. Veg. Crops, Veg. Crops Ser. 89, 10 pp.

Mann, L. K and Woodbury, G. W. 1969. The effect of flower age, time of day and variety on pollen germination of onion, Allium cepa L. Amer. Soc. Hort. Sci. Proc. 94: 102-104.

Manning, J.C. and Goldblatt, P., 1996. The Prosoeca peringueyi Diptera: Nemestrinidae pollination guild in southern Africa: long-tongued flies and their tubular flowers. Annals of the Missouri Botanical Garden 83, 67–86.

Maroufi K., Farahani H. A., Darvishi H. H. 2010 Importance of Coriander (Coriandrum sativum L.) Between the Medicinal and Aromatic Plants. Advances in Environmental Biology 4: 433-436.

Martin, E.C. 1979. The use of bees for crop pollination. In: The Hive and Honey bee. (ed.) Dadant and Sons, 579-614 pp.

Mathias M. E. 1994 Magic, myth and medicine. Economic Botany 48: 3-7.

Mayer D. F. and Lunden, J. D. 2001. Honey bee management and wild bees for pollination of hybrid onion seed. Acta Horticulturae, 561: 275-278.

Mayer, C., 2004. Pollination services under different grazing intensities. International Journal of Tropical Insect Science 24, 95–103.

Mayer, C., Soka, G. and Picker, M.D., 2006. The importance of monkey beetle Scarabaeidae: Hopliini pollination for Aizoaceae and Asteraceae in grazed and ungrazed areas at Paulshoek, Succulent Karoo, South Africa. Journal of Insect Conservation. 10, 323–333.

Mayer, D.F. and Lunden, J.D. 2001. Honey bee management and wild bees for pollination of hybrid onion seed. Acta Hortic.,61: 275-278.

McGregor S E 1976 Insect pollination of cultivated crop plants. USDA Hand Book Washington DC.

McGregor S. E. 1976 Insect pollination of cultivated crop plants. Agriculture Handbook No. 496. Agricultural Research Service. USDA. Washington. 411 pp.

McGregor, S.E., 1976. Common vegetables for seed and fruit, in: McGregor, S.E. Ed., Insect Pollination of Cultivated Crop Plants. United States Department of Agriculture, Washington.

Meer, Q. P., Van der, and Bennekom, J. L. Van. 1968. Research in pollen distribution in onion seed fields. Euphytica 17: 216- 219.

Meer, Q. P., Van der, and Bennekom, J. L. Van. 1969. Effect of temperature on the occurrence of male sterility in onion (*Allium cepa* L.). Euphytica 18: 389394.

Meine, C., Soulé, M., Noss, R.F. and Soule, M., 2006. "A mission-driven discipline": the growth of conservation biology. Conservation Biology 20, 631–651.

Meléndez-Ramirez, V., Magaña-Rueda, S., Parra-Tabla, V., Ayala, R. and Navarro, J. 2002. Diversity of native bee visitors of cucurbit crops (Cucurbitaceae) in Yucatán, México.J Insect Conser.,6: 135-147.

Meloyan, A.A. 1975. Nectar sugar production by winter vetch (Vicia Villosa) cultivars of different origins. Zapiski Lenin Gradskogo Selskokhozyais tvennago instituta, 271: 61-63.

Mevetty, P.B.E., Pinnisch, R. and Scarth, R. 1989. The significance of floral characteristics in seed production of four summer rape cultivar.A-lines with pollination cytoplasm. Can. J. Pl. Sci., 69(3): 915-918.

Millenium Ecosystem Assessment, 2005a. Ecosystems and human well-being: biodiversity synthesis. Island Press, Washington, DC.

Millenium Ecosystem Assessment, 2005b. Ecosystems and human well-being: current state and trends. Island Press, Washington, DC.

Millenium Ecosystem Assessment, 2005c. Biodiversity regulation of ecosystem services, in: Ecosystems and human well-being: current state and trends. Island Press, Washington, DC., pp. 297–329.

Milton, S.J., Dean, W.R.J., Plessis, M.A. and Siegfried, W.R., 1994. A conceptual model of arid rangeland degradation. Bioscience 44, 70–76.

Misra, R.C., Kumar, T., Veeresh, G.K., Shaankar, R.U. and Ganeshaiah, K.N. 1993 Status of research in pollination biology in Himachel Pradesh", Proc. Intern. Symp. Poll. in Tropics, August 8-13, India, 279 - 295.

Mitchell, R.J. 1992 Testing evolutionary and ecological hypotheses using path analysis and structural equation modeling. Functional Ecology 6:123-129.

Mitchell, R.J.; Paton, D.C 1990 Effects of nectar volume and concentration on sugar intake rates of Australian honeyeaters (Meliphagidae). Oecologia 83:238246.

Moffett, J. O. 1965. Pollinating experimental onion varieties. Amer. Bee Jour. 105:378.

Mohan Rao G and Suryanarayana M C 1989. Effect of honey bee pollination on seed yield in onion (Allium cepa L.). Indian Bee Journal 51 : 9-11.

Mohr, N.A. and Jay, S.C. 1990. Nectar Production of selected cultivars of Brassicacompestris L. and Brassica napus L.J.agril. Res., 29(2): 95-100.

Mommers, J. 1977. The concentration and composition of nectar in relation to honeybee visits to fruits trees Apidologie, 8: 357-361.

Morandin, L.A. and Winston, M.L., 2005. Wild bee abundance and seed production in conventional, organic, and genetically modified Canola. Ecological Applications 15, 871–881.

Morris, H. 1968. Pollination of cucumbers. Glean. Bee Cult.,1: 14-16.

Morse R.A. and Calderone, NW. 2000. The value of honey bees as pollinators of U. S. crops. Bee Culture, 128:1–15.

Mortaza Rasekh Adel, Hussein Sadeghi Namaghi and Mojtaba hosseini 2013. Effects of insect pollinators on onion seed production quality and quantity J. Crop Prot. 2013, 2 (4): 395-402

Mucina, L. and Rutherford, M.C. Eds., 2006. The vegetation of South Africa, Lesotho and Swaziland, Strelitzia 19. South African National Biodiversity Institute, Pretoria.

Muller, H. 1983. The fertilization of flowers. Macmillian, London.

Munawar, M. S.; Shazia Raja; Shahid Niaz and Ghulam Sarwar 2011. Comparative performance of honeybees (Apis mellifera L.) and blow flies (Phormia terronovae) in onion (Allium cepa L.) seed setting. Journal of Agricultural Research (Lahore) 2011 Vol. 49 No. 1 pp. 49-56

Munawar, M.S. and Muzaffar, N. 1999. Role of honeybee (Apis mellifera L.) pollination on onion (Allium cepa) seed production.In: 36th limondia Congress September, 12-18, Vancouver-Canada, 276 pp.

Mupade, R.V. Kulkarni, S.N and Kamte, G.S.2009. Effect of bee pollination on qualitative characters of onion. Indian J Plant Prot. 37(1&2): 186-187

Myers, N., Mittermeier, R.A., Mittermeier, C.G., de Fonseca, G.A.B. and Kent, J., 2000. Biodiversity hotspots for conservation priorities. Nature 403, 853–858.

Nascimento, W. M.; Pessoa, H. B. S. V. and Araújo, M. de T.1998. Utilization of Trigona spinipes as a pollinator in onion (Allium cepa L.) breeding programmes in Brazil. Journal of Applied Seed Production 1998 Vol. 16 pp. 47-49

Nemirovich-Danchenko, E.N. 1964. Concerning the nectar yield and floral biology of cucumbers.Izv. Tomsk. Otd.Vses. Bot. Obshch.,5:127-132.

Nogueira, C.R.H. and Calmona, R.C. 1993. Insect pollination of cucumber (Cucumis sativus var. Aodai Melhorada).Naturalia Sao Paulo., 18: 77-82.

Nuenz, J.A. 1977. Nectar flow by melliferous flora and gathering flow by Apis mellifera ligustica. Journal of Insect Physiology, 23: 265–276.

Nunez, J. 1997 Circadian variation of flight activity in colonies of Apis mellifera lingustica. J.Insect Physiol. 23 : 387-392.

Nye, W. P.; Waller, G. D. and Waters, N. D. 1971 Factors affecting pollination of onions in Idaho during 1969. Journal of the American Society for Horticultural Science 96(3): 330-332

Nye, W.P. and Anderson, J.L. 1974. Insect pollinators frequenting strawberry blossoms and the effect of honey bee on yield and fruit quality.J. Am. Soc. Hortic. Sci., 99: 40–44.

O'Farrell, P.J., De Lange, W.J., Le Maitre, D.C., Reyers, B., Blignaut, J.N., Milton, S.J., Atkinson, D., Egoh, B., Maherry, A., Colvin, C., Cowling, R.M., 2011. The possibilities and pitfalls presented by a pragmatic approach to ecosystem service valuation in an arid biodiversity hotspot. Journal of Arid Environments 75, 612–623.

O'Toole, C., 1993. Diversity of native bees and agroecosystems, in: LaSalle, J., Gauld, I.D. Eds., Hymenoptera and Biodiversity. CAB International, Wallingford, UK.

Oh, H.W. and Woo, K.S. 1990. A study of foraging and pollen collecting activity of honey bee (*Apis mellifera* L.) in the spring.Korean J. Apic, 5(1): 1-22.

Oldroyd, B.P., 2007. What's killing American honey bees? PLoS Biology. 5, e168.

Osgood, C.E. 1974 Relocation of nesting populations of Megachile rotundata F. an important pollinator of alfalfa. J. Apic. Res. 13 : 67-74.

Osgood, C.E. 1974. Relocation of nesting populations of Megichile rotundata as an important pollinator of alfalfa.J. Apic. Res., 13: 63–73.

Paci, P. 1956. [research on the floral biology of the carrot.] Riv. Ortoflorofruttic. Ital. 40: 414-423.

Pankratova, E. P. 1957. The effect of bee pollination on the harvest of carrot Seed. Dokl. Tskha 30 (part 2): 332-336. [in Russian. ] aa-396/61.

Pankratova, E.P. 1964 "Data on the biology of blossoming and pollination of carrots", Dokl. TSKhA, 36, 118-123.

Parker, F.D. 1982. Efficiency of bees pollinating onion flowers.J.Kansas Ento. Soc., 55: 171-176.

Parker, F.D. 1982. Efficiency of bees in pollinating onion flowers. Journal of the Kansas Entomological Society 55, 171–176.

Parker, F.D. and Hatley, C.L. 1978. Onion pollination viability of onion pollen and pollen diversity on insects body hairs. Proc.IV Int. Symp. Poll. M D Agri. Exp. Stn. Spec. Publ. Misc., 1: 201-206.

Parker, F.D. and Hatley, C.L., 1979. Onion pollination: viability of onion pollen and pollen diversity on insect body hairs., in: Proceedings of the Fourth International Symposium on Pollination. Maryland agricultural experiment station special miscellaneous publication 1, pp. 201–206.

Pathak, C.S., 2000. Hybrid seed production in onion. Journal of New Seeds 1, 89–108.

Pedersen, M. W., F. K Todd, and F. V. Lieberman. A portable field cage. U. S. Dept. Agr. Bur. Ent. and PI. Quar. ET-289. 1950. (mimeo.)

Percival M S 1950 Pollen presentation and pollen collection. New Phytol. 46: 142–173.

Percival M S 1955 The presentation of pollen in certain angiosperms and its collection by Apis mellifera. New Phytol. 54: 353– 368.

Pereboom, J.J.M. and Biesmeijer, J.C. 2003. Thermal constraints for stingless bee foragers: the importance of body size and coloration. Oecologia, 137 (1): 42–50.

Perer-1 Dop. Maskva, Gos. Gupta, J.K., Mishra, R.C. and Kumar, J. 1984. Plectranthus as forage for Apis cerana indica F. and Apis mellifera L. Apidologie,15(1): 75-82.

Picker, M.D. and Midgley, J.J., 1996. Pollination by monkey beetles Coleoptera: Scarabaeidae: Hopliini: flower and colour preferences. African Entomology 4, 7–14.

Pierre J., Le Guen J., Esnault R., Debbagh S. et Sadiki M. 1997 Méthode d'étude de la fréquentation de diverses féveroles par les insectes pollinisateurs. In: Proceedings of Mediterranean Food Legumes. Rennes, France. 20-22 February 1988. pp. 199-206.

Pierre J., Suzo M. J., Moreno M. T., Esnault R. et Le Guen J. 1999 Diversité et efficacité de l'entomofaune pollinisatrice Hymenoptera : Apidae) de la féverole (Vicia faba L.) sur deux sites en France et en Espagne. Annales de la Société Entomologique de France 35 (n.s. suppl.): 312-318.

Pimm, S.L. and Russell, G.J., Gittleman, J.L., Brooks, T.M., 1995. The future of biodiversity. Science 269, 347–350.

Poole C. F. 1937. Improving the root vegetables. U.s. Dept. Agr. Yearbook 1937: 300-325.

Prakash V. 1990 Leafy Spices. CRC Press Inc., Boca Raton, pp. 31-32.

Prasad Rabindra; Chand Hari and Singh, R. 2000. Effect of honeybee, Apis mellifera pollination on the yield and germination of onion, Allium cepa seeds. Shashpa 2000 7 (1) 53-55

Prasad, R., Chand, H. and Singh, R. 2000. Effect of boney bee, Apis mellifera pollination on the yield and germination of onion, Allium cepa seeds.Shashpa, 7: 1, 53-55.

Priti and Sihag, R.C. 1997. Diversity, visitation frequency, foraging behaviour and pollinating efficiency of insect pollinators visiting cauliflower (Brassica aleracea) blossoms.Indian Bee J., 59(4): 230-237.

Priti, C. 1998. Abundance and Pollination efficiency of insect visitors of onion bloom. Indian Bee Journal, 60 (2): 75-78.

Purseglove J. W., Brown E. G., Green C. L. and Robbins S. R. J. 1981 Spices. Vol. 2 Longman. London. pp. 736- 788.

Pyke, G.H., Pultiam, H.R. and Charnov, E.L. 1977. Optimal foraging: Selective review of theory and tests. Qtly. Rev. Biol., 52: 137-154.

Rabindra Prasad, Hari Chand and Singh R 2000. Effect of honey bee Apis mellifera pollination on the yield and germination of onion (Allium cepa) seeds. Shashapa 7 : 53-55.

Rader, R, Howlett, BG, Cunningham, SA, Westcott, DA, Newstrom-Lloyd, LE, Walker, MK, Teulon, DAJ and Edwards, W. 2009. Alternative pollinator taxa are equally efficient but not as effective as the honeybee in a mass flowering crop. Journal of Applied Ecology, 46: 1080–1087.

Rao, G. M. and Lazar, M. 1983. Studies on bee behavior and pollination in onion (Allium cepa L.). Proceedings of the 2nd International Conference on Apiculture in Tropical Climates, New Delhi, 3: 580-589.

Rao, G. M. and Suryanarayana, M. C. 1989. Effect of honeybee pollination on seed yield in onion (Allium cepa L.). Indian Bee Journal, 50: 9-11..

Rao, G.M. and Suryanarayana, M.C. 1988. Studies on pollination of watermelon Citrullus lanatus (Thunb) Manst.Indian Bee J., 50: 5-8.

Rasekh Adel, M., Sadeghi Namaghi, H., and Hosseini, M. 2012. Biodiversity of Apoidea (Insecta: Hymenoptera) associated with onion and alfalfa fields in Mashhad and Chenaran Regions. Iranian Journal of Plant Protection Science, 43 (1): 191-199 (in Persian with abstract in English).

Rashad, S. E. 1978.Utilization of non-Apis bees as crop pollinators (EG-ARS-66); first annual report 48 pp.

Rees, W.E., 1992. Ecological footprints and appropriated carrying capacity: what urban economics leaves out. Environment and Urbanization 4, 121–130.

Ricciardelli D, Albore G and Ambrosio M 1979 Preliminary observations on pollination of Coriandrum sativum by honey bees and other insects. Apicoltore Moderno. 70: 151–157.

Ricciardelli D'Albore C. G. and Intoppa F. 2000 Flori e api. La flora visitata dalle Api e dagli altri Apoidei in Europa. Calderini Edagricole. Bologna. 110 pp.

Ricciardelli D'Albore C. G., D'Ambrosio M. 1979 Aminoacids composition variations of pollens gathered by honeybees during one year activity. In: Proceedings of the 27th International Apicultural Congress of Apimondia. Athens, Greece, 14-20 September 1979. pp.14-20.

Ricciardelli D'Albore G. C. 1986 Les insectes pollinisateurs de quelques ombellifères d'intérêt agricole et condimentaire (Angelica archangelica L., Carum carvi L., Petroselinum crispum A.W. Hill., Apium graveolens L., Pimpinella anisum L., Daucus carota L., Foeniculum vulgare Miller v. Azoricum thell.). Apidologie. 17: 107-124.

Richard W., Rusta, R.W., Vaissièreb, B.E. and Westrichc, P. 2003. Pollinator biodiversity and floral resource use in Ecballium elaterium (Cucurbitaceae), a Mediterranean endemic.Apidologie, 34: 29-42.

Richards, K.W., 1993. Non-Apis bees as crop pollinators. Revue Suisse Zoology 100, 807–822.

Ricketts, T.H., 2004. Tropical forest fragments enhance pollinator activity in nearby coffee crops. Conservation Biology 18, 1262–1271.

Ricketts, T.H., Daily, G.C., Ehrlich, P.R. and Michener, C.D., 2004. Economic value of tropical forest to coffee production. Proceedings of the National Academy of Sciences 101, 12579–12582.

Roa, G.M. and Lazer, M. 1980. Studies on beee behavoiur and pollination on Allium species with observation on insect as pollinators. J. Apic. Res., 24: 62-66.

Robinson, Richard W. 1954. Seed germination problems in the Umbelliferae. Bot. Rev. 20(9): 531-550.

Rodel, G. 1990 "Pollination of carrot: recent work", Bull. Semances, 111, 47-51.

Rodel, G., Grosse, T.J.P., and Bonnel, A. 1991 "Pollination of male sterile carrot inbred lines (Daucus carota L.) by honey bees (Apis mellifera L.)", Proc. Eucarpia on carrot breeding, June 18-20, France, 29-135.

Rodet G , Torre Grossa JP , Bonnet A 1991 . Foraging behavior of Apis mellifera L. on male-sterile and male-fertile inbred lines of carrot (Daucus carota L.) in gridded enclosures. C van Heemert and A de Ruijter VI International Symposium on Pollination, Tilburg, Netherlands . Pp. 371 – 375 .

Rodger, J.G., Balkwill, K. and Gemmill-Herren, B., 2004. African pollination studies: where are the gaps? International Journal of Tropical Insect Science 24, 5–28.

Roubik, D. W. 1995. Pollination of cultivated plants in the tropics. FAO, Rome.

Ruszkowaski, A. and Bilinski, M. 1984. Proba wykorzystania trzmiek (Bombus Later) I miesiarki lucernowki (Magachile rotundata F.) dozaphlania materialu hodowlanego ogorkow. Pszczelnicze Zeszyty Naukowe, 28: 175-184.

Saeed, A., Muhammad, T., Karim, A., Jabbar, A. and Ahmad, G. 2008. Effect of pollinators and insecticides on seed setting of onion (Allium cepa L.)Ann. Appl. Boi.,130(3): 497-506.

Saeed, S., Saggad, A., Kwon, O. and Kwon, Y. J. 2008. Fidelity of Hymenoptera and Diptera pollinators in onion (Allium cepa L.) pollination. Entomological Research, 38: 276-280.

Sajjad, A., Saeed, S. and Masood, A. 2008. Pollinator community of onion (Allium cepa L.) and its role in crop reproductive success.Pakistan J. Zoo.,40: 451-456.

Sak'ıarov, M.K. 1956. Bees and onion seed production. In: Pollination of Agriculture Plants (eds. Krishchunas, I. V. and Gubin, A. F.). Moskva, Goa Izdvo Selkhoz Lit-ry, 172-173 pp.

Sanduleac, E. 1961. The pollination of vegetable seed plants.Apiculture,14: 25-26. www.


Sarviva, J.A.P.G. 1985. Bioclimatic factors affecting flights of honey bee. *Apis mellifera* L. (Hymenoptera: Apidae) Ph.D. Thesis, Colorado State University Fort Collins. USA, xi+148pp.

Scheiner, S.M.; Mitchell, R.J and Callahan, H. 2000. Using path analysis to measure natural selection. Journal of Evolutionary Biology 13:423-433

Schiavi, M.; Masera, R.; Nervo, G.; Mazzola, P. and Soressi, G. P.1984. Different efficiency of blow-flies and honey bees in controlled pollination of onion (Allium cepa L.). 3rd Eucarpia Allium Symposium, Wageningen, 4-6 September 1984. 1984 pp. 115

Schittenhelm, S., Gladis, T.and Rao, V.R., 1997. Efficiency of various insects in germplasm regeneration of carrot, onion and turnip rape accessions. Plant Breeding 116, 369–375.

Schlessman M. A. 1982 Expression of andromonoecy and pollination of Tuberous lomatiums (Umbelliferae). Systematic Botany 7: 134-139.

Schmid-Hempel, P. 1984. The importance of handling time for flight directionality in bees. Ecol. Sociobiol. 15: 303- 309.

Schoener, T.W. 1971. Theory of feeding strategies.Ann. Rev. Ecol. Syst. 2: 369-404.

Schuphan, W. and Weiller, H. 1967 "Pollination of the carrot crop for better seed production", Qual. Plant. Mater. Veg., 15, 81-101

Seeley, T.D., Camazine, S. and Sneyd, J. 1991. Collective decision making in honey bees: How colonies choose among nectar sources. Behav. Ecol. Sociobial,28: 277-290.

Sergeeva D. S. and Sil'zenko V. M. 1984 Resistance of coriander to low temperatures. Fiziologiya i Biokhimiya Kul'turnykh Rastenii 16: 52-55.

Seyedebrahimi, S., Ebadi, R., Mobli, M. and Hatami, B. 2004. Honey bee (Apis mellifera L.) foraging on some onion cultivars and its relation with their nectar quantity and quality. J. Sci Tech. Agri. Natural Resources, 8: 1, 235-246.

Shanmugasundaram, S., 1998. Onion cultivation and seed production. Shanhua, Tainan, AVRDC.

Sharma, P.L. and Sharma, B.R. 1968 "Insect pollination in carrot crop", Indian J. Hort., 25, 216.

Sharma, S. K.; Singh, J. R. and Kaushik, H. D. 2001. Efficiency of two wild honey bees (Apis spp.) as pollinators of onion (Allium cepa L.). Haryana Journal of Horticultural Sciences 30 No. 1/2 pp. 134-135

Sharma, S.K. and Singh, J.R. 1999. Pollination efficiency of Apis dorsata F. and A. florea F. on carrot (Daucus carota L.). Indian Bee J., 61(1): 1-4.

Sharma, S.K. and Singh, J.R. 2000. Efficiency of two Apis spp. pollinating sunflower (Helianthus annus L.).Ann. Ent., 18(1): 15-17.

Sharma, S.K., Singh, J.R. and Kaushik, H.D. 2001. Efficiency of two wild honey bees (Apis spp.) as pollinators of onion (Allium cepa L.). Haryana J. hortic. Sci., 30(1&2): 134-135.

Sharma, S.K., Singh, J.R. and Ombir. 2000. Comparative efficiency of three Apis spp. on Brassica campestris var. 'sarson' B.S.H.1. Crop Res., 19(1): 125-127.

Shasha'a, N. S. 1972. Limitation studies of seed set in the onion (Allium cepa L. ) (Liliaceae). Ph. D. dissertation, Utah State Univ., Logan. 1972. [Unpublished.]

Shasha'a, N. S.; Nye, W. P. and Campbell, W. F. 1973. Path-coefficient analysis of correlation between honey bee activity and seed yield in Allium cepa L. Journal of the American Society for Horticultural Science 98 (4) : 341-347

Shelar D G and Suryanarayana M C 1981 Preliminary studies on pollination of coriander (Coriandrum sativum L.). Indian Bee J. 43: 110–111.

Shepherd, M., Buckmann, S.L., Vaughan, M. and Hoffmann-Black, S., 2003. Pollinator conservation handbook. The Xerces Society, Oregon.

Shuel, R.W. 1955. Nectar secretion.Am. Bee J., 95: 229-233. Sihag, R.C. (1984). Influence of environmental factors on the pollination activity of bees. Env. Ecol., 2:149-152.

Synge A D 1947 Pollen collection by honey bees (Apis mellifera). J. Anim. Ecol. 16: 122–138b

Sihag, R. C. 1986. Insect pollination increases seed production in cruciferous and umbelliferous crops. J. Apic. Res. 25: 121-126.

Sihag, R.C. 1985. Floral biology, mellitophily and pollination ecology of onion crops. In: Recent advances in pollen research (ed. Varghese, T.M.). Allied Publishers, New Delhi.241-268 pp.

Sihag, R.C. 1991 "Some economic estimates of use of pesticides and bee pollination for crop yield", Korean J. Apic., 6, 89-94.

Sihag, R.C. and Khatkar, S. 1999. Foraging pattern of three honeybee species on eight cultivars of oilseed crops. I. Diurnal foraging. Int. J .Trop. Agric., 17(1-4).

Silva, E. M. and Dean, B. B. 2000. Effect of nectar composition and nectar concentration on honey bee (Hymenoptera: Apidae) visitations to hybrid onion flowers. Journal of Economic Entomology 93 (4): 1216-1221

Silva, E.M., Dean, B.B. and Hiller, L.K., 2003. Honey Bee Hymenoptera: Apidae foraging in response to preconditioning with onion flower scent compounds. Journal of Economic Entomology 96, 1510–1513.

Sinacori A. and Sinacori M., Oliveri E., Carrubba A. 2009 Api e pronubi selvatici nella produzione di *Coriandrum sativum* L. Apoidea 6(3): 155-161.

Singh G. and Gangwar B. 1991 Management practices for higher productivity of coriander in India - a review. Agricultural Review Karnal 12: 15-21.

Singh H, Swaminathan R. and Hussain T.2010. Influence of certain plant products on the insect pollinators of coriander. Journal of Biopesticides 3(1): 208 -211.

Singh J P and Dharamwal S S 1970. The role of honey bees in seed setting of onion at Pant Nagar, Dist. Nainital (U P), India. Indian Bee Journal 32 : 23-27.

Singh, B. and Hameed, S.F. 1995. Physical factors and foraging activity of insects visiting rye bloosoms. In: Pollination Biology: Environment Factors and pollination (ed. Sihag, R.C.), Rajendra Scientific Publishers, Hisar. 111-116pp.

Singh, J.P and Dharmwal, S.S. 1970. The role of honey bees in seed setting on onion at Pant Nagar. Dist. Nainital, Uttar Pradesh, India. Indian Bee J. 32: 23-27.

Singh, M. 1983 Role of insect pollination in seed production of carrot, Daucus carota Linn. Thesis Abst. Punjab Agri. Univ., Ludhiana, India, 9, 327-328.

Singh, Y. 1997. Role of honeybees in farm production, agricultural growth and rural reconstruction in India. Indian Bee Journal, 59 (1): 24-30.

Sinha, SN and Ckrabarti, AK. 1992. Insect pollination in carrot seed crop. Seed Research, 20: 37–40.

Skiredj A., Elattir H. and El Fadl A. 2002 Fiches techniques des cultures aromatiques et condimentaires. L'absinthe, l'ail, le céleri, le cerfeuil, la ciboulette, la coriandre, les épinards, l'estragon, le fenouil de florence, la menthe, la niora, le persil, le sésame, la verveine. Bulletin mensuel d'information et de liaison du PNTTA. Transfert de technologie en agriculture. 24 pp.

Skrebtsov, M. F.; Skrebtsova, N. D. 1980. Pollinators of onion seed crops. Zashchita Rastenii No. 10 pp. 22

Skrebtsova, N.D. 1957. Amount of pollen on the body of bees. Pchetorodstvo, 34(4): 39-42.

Slate, W. L.1927. Report of the director. Conn. Agr. Expt. Sta. Bul. 291: 91-111.

Snedecor, G. W 1956. Statistical methods. 5th ed. Ames, Iowa. Iowa State College Press. 534 p.

Sokal, R.R and Rholf, F.J 1981. Biometry (2nd Edition) WH Freeman & Co; San Francisco.

Sonnet M. and Jacob-Remacle A. 1987 Pollinisation de la légumineuse fourragère Hedysarum coronarium L. en Tunisie. Bulletin de la Recherche Agronomique de Gembloux 22: 19-32.

Southwick E. E., and Southwick, L. 1992. Estimating the economic value of honey bees (Hymenoptera: Apidae) as agricultural pollinators in the United States. Journal of Economic Entomology, 85: 621-633.

Spurr CJ , Mendham NJ , Brown PH 2001 . The effect of Rutherglen bug (Nysius vinitor) on yield and quality of hybrid carrot seed . TRI Ltd (ed) , Proceedings of the 10th Australian Agronomy Conference 2001 . Hobart , The Regional Institute Ltd .

Stanghellini, M.S., Ambrose, J.T. and Schultheis, J.R. 1997. The effects of honeybee and bumble bee pollination on fruit set and abortion of cucumber and watermelon. Am. Bee J.,137: 386-391.

Stephen, W.A. 1970. Honey bees for cucumber pollination. Am. Bee J. 110: 132-133.

Steyn, T., 2009. Bysiekte in tot 85% van korwe. Die Burger, 14 July, p.6.

Struck, M., 1994a. Flowers and their insect visitors in the arid winter rainfall region of southern Africa: observations on permanent plots. Insect visitation behaviour. Journal of Arid Environments 28, 51–74.

Struck, M., 1994b. Flowers and their insect visitors in the arid winter rainfall region of southern Africa: observations on permanent plots. Composition of the anthophilous insect fauna. Journal of Arid Environments 28, 45–50.

Szabo, T.I. and Smith, M.V. 1972. The influence of light intensity and temperature on the activity of alfalfa leaf cutter bee. Megachile rotundata under field conditions. J. Apic. Res., 11: 157–165.

Tainton, N. Ed., 1999. Veld management in South Africa. University of Natal Press, Pietermaritzburg.

Tanaka, M. and Komochi, S.1982. Use, reproduction and storage of houseflies as pollinators for onion breeding. Research Bulletin, Hokkaido National Agricultural Experiment Station 1982 No. 135 pp. 73-81

Tepedino V. J., Bowlin W. R., Griswold T. L. 2011 Diversity and pollination value of insects visiting the flowers of a rare buckwheat (Eriogonum pelinophilum: Polygonaceae) in disturbed and " natural" areas. Journal of Pollination Ecology 4(8): 57-67.

Thapa R. B. 2006 Honeybees and other insect pollinators of cultivated plants. Journal of the Institute of Agriculture and Animal Science 27: 1-23.

Thompson, D.J. 1962 "Natural cross pollination in carrots", Proc. Amer. Soc. Hort. Sci., 20, 212-219.

Thompson, M., Vlok, J., Cowling, R.M., Cundill, S.L. and Mundau, N., 2005. A land transformation map for the Little Karoo. GeoterraImage, Pretoria.

Tilman, D., Fargione, J., Wolff, B., D'Antonio, C., Dobson, A., Howarth, R., Schindler, D., Schlesinger, W.H., Simberloff, D. and Swackhamer, D., 2001. Forecasting agriculturally driven global environmental change. Science 292, 281–284.

Tolon, B. and Duman, Y. 2003. The effect of pollination by honeybee (Apis mellifera) on onion (Allim cepa) seed production and quality. XXXVIII Apimondia Ineternational Apicultural Congres, 554.

Tornier I , Drescher W 1991 . Development of a method for mass production of the pollinator and aphid predator Episyrphus balteatus Deg. (Diptera, Syrphidae) . C van Heemert and A de Ruijter VI International Symposium on Pollination, Tilburg, Netherlands 27–31 August, 1990 . Pp. 464 – 468 .

Treherne, R.C. 1923. The relation of insects to vegetable seed production. Rep. Queb. Soc. Prot. Pl. 15: 47-49.

Trofimec, N.K. 1940. The biology and fertilization in Allium. Vest. Sots Rasteniev. Rasten. 5:76-86.

Uvah I. I. I. and Coaker T. H. 1984 Effect of mixed cropping on some insect pests of carrots and onions. Entomologia Experimentalis et Applicata 36(2): 159-167

Van Der Meer, Q. P. and van Bennekom, J. L. 1972. Influence of the environment on the percentage of self-fertilization in onion and some consequences for breeding. Euphytica, 21: 45-453.

Van der Meer, Q.P. and Van Bennekom, J.L. (1968). Research in pollen distribution in onion seed fields.Euphytica, 7: 216- 219.

Van der Walt, W.J., 1999. Changing face of South African seed trade. Journal of New Seeds 1, 81–97.

Van der Walt, W.J., 2002. A review of the South African seed industry. Journal of New Seeds 4, 57–70.

Vansell, G.H. (1934). Relation between the nectar sugar concentration in fruit blossoms and the visits of honey bee. J. econ.Ent.,27: 943-945.

Vavilov, N.I., 1951. The origin, variation, immunity and breeding of cultivated plants. Chronica Botanica 13, 1–6.

Vischer, P.K and Seeley, T.D. 1982. Foraging strategies of honeybee colonies in a temperate deciduous forest. Ecology 63: 1790-1801.

Vischer, P.K.; Seeley, T.D. 1982 Foraging strategies of honeybee colonies in a temperate deciduous forest. Ecology 63: 1790-1801.

Vitousek, P.M., Mooney, H.A., Lubchenco, J. and Melillo, J.M., 1997. Human domination of Earth's ecosystems. Science 277, 494–499.

Vlok, J., Schutte V and Ok, A.L., 2010. Plants of the Klein Karoo. Umdaus Press, Pretoria.

Voss R. E., Murray M., Bradford K., Mayberry K. S. and Miller, I. 1999. Onion Seed Production in California. University of California Division of Agriculture and Natural Resources Publication, 8008: 1-10.

Wackernagel, M., 1994. Ecological footprint and appropriated carrying capacity: a tool for planning toward sustainability. PhD thesis, The University of British Columbia.

Wäckers, FL, Romeis, J and van Rijn, P. 2007. Nectar and pollen feeding by insect herbivores and implications for multitrophic interactions. Annual Review of Entomology, 52: 301–323.

Waddington, K.D. 1982. Information used in foraging. In: The Biology of Social Insects. (eds. Breed, M.D., Michner, C.D and Evans, H.E.). Boulder Westview Press.24-27 pp.

Wallace, K.J., 2007. Classification of ecosystem services: Problems and solutions. Biological Conservation 139, 235–246.

Waller, C.D. 1970. Attracting honey bees to alfafa with citral, geraniol and anise.J. Apic. Res.,9, 9-12.

Waller, G. D., Carpenter, E. W and Ziehl, O. A. 1974. Potassium in onion nectar and its probable effects on attractiveness of onion flowers to honey bees. J. Am. Soc. Hortic. Sci 97:535–539.

Walsh, R.S., 1965. Pollination of onion plants by honeybees. New Zealand Beekeeper 27, 18–20.

Wangensteen H., Samuelsen A. B., Malterud K. E. (2004) Antioxydant activity in extracts from coriander. Food Chemistry 88: 293-297.

Waser N. M., Olerton J. 2006 Plant-pollinator Interactions. From specialization to generalization. University Chicago Press. Chicago. 441pp.

Watanabe, M.E., 1994. Pollination worries rise as honey bees decline. Science 265, 1170.

Waters, N.D., 1972. Honeybee activity in blooming onion fields in Idaho. American Bee Journal 112, 218–219.

Welch, J. E., and Grimball, E. L., Jr. 1947. Male sterility in the carrot. Science 106: 594.

Whitaker, T. W., Sherf, A. F., Lange, W. H., and others. 1970. Carrot production in the united states. U.s. Dept. Agr., agr. Handb. 375, 37 pp.

Whitehead, V.B., 1984. Distribution, biology and flower relationships of fideliid bees of southern Africa Hymenoptera, Apoidea, Fideliidae. South African Journal of Zoology 19, 87–90.

Wilkaniec, Z., Giejdasz, K. and Proszynski, G. 2004. Effect of pollination of onion seeds under isolation by the red mason bee (Osmia rufa L.) (Apoidea, Megachilidae) on the setting and quality of obtained seeds. Journal of Apicultural Science, 48 (2): 35-41.

Wilkaniec, Z., Warakomska, Z. and Giejdasz, K. 2002. Rooeliny ywicielskie pszczoy samotniczej Osmia rufa L. (Apoidea, Megachilidae) okreoelone na podstawie analizy py³ku z komór lêgowych. Zesz.Nauk. PTPN, 93: 199-205.

Williams, I.H. and Free, J.B. 1974. The pollination of onion (Allium cepa L.) to produce hybrid seed. J. Appl. Ecol., 11:409-412.

Williams, N.M., Minckley, R.L., Silveira, F.A., 2001. Variation in native bee faunas and its implications for detecting community changes. Conservation Ecology 5, 7.

Wills, A.B., North, C., 1978. Problems of hybrid seed production. Acta Horticulturae. 83, 31–36.

Winfree, R. 2008. Pollinator-dependent crops: an increasingly risky business. Current Biology, 18: R968–R969

Witter, S. and Blochtein, B. 2003. Effect of pollination by bees and other insects on the production of onion seeds. Pesquisa Agropecuaria Brasileria, 38 (12): 1399-1407.

Witter, S. and Blochtein, B., 2003. Effect of pollination by bees and other insects on the production of onion seeds. Pesquisa Agropecuária Brasileira 38, 1399–1407. World Wide Fund for Nature, 2010. Living Planet Report 2010. Gland.

Wolf, S., Lensky, Y. and Paldi, N. (1999).Genetic variability in flower attractivness of honey bee (Apis mellifera L.) within the genus Citrullus. Hort. Sci.,34(5): 860-863.

Woyke, H. W. 1981. Some aspects of the role of the honeybee in onion seed production in Poland. Acta Horticulture, 111: 91-98

Wykes, G.R. 1952. The preferences of honey bee for solutions of various sugar which occur in nectar. J. exp. Biol., 29: 511-519.

Wykes, G.R. 1953. The sugar contents of nectars. Biochem. J., 53: 294-296. Yucel, B. and Duman, I. (2005). Effect of foraging activity of honey bee (Apis mellifera L.) on onion (Allium cepa) seed production and quality.Pakistan J. Biol. Sci., 8(1): 123-126.

Zych M. 2007. On flower visitors and true pollinators: The case of protandrous Heracleum sphondylium L. (Apiaceae). Plant Systematics and Evolution 263: 159-179.

**Figure 6.1: Insect pollinators of onion, *Allium cepa***

**Figure 6.2: Insect pollinators of coriander, *Coriandrum sativum***

**Figure 6.3: Insect pollinators of carrot, *Dacus carota***

CHAPTER

7

# Cucurbitaceous Crops

## 7.1 INTRODUCTION

Cucurbits form an important source of vegetables. The family Cucurbitaceae comprises of two subfamilies (Cucurbitoideae and Zanonioideae) about 130 genera and 900 species (Jeffrey, 1964, 1980). They are most common in tropical and subtropical areas and are relatively less in temperate areas because of their host sensitivity(Tsuchiya and Gupta, 1991). Cultivation of cucurbits poses a serious challenge as they are attacked by several pests and diseases which need to be managed for quality and quantity fruit production,

Another important factor in cucurbit production is that most of the varieties are monoecious having both staminate and pistillate flowers on the same plant (Figure 7.1, 7.2). In cucurbits whether monoecious or gynoecious, pollination by insects is must as the pollen grains are large and sticky and opportunity adapted to be carried by insects rather than wind between flowers of same or different plants. The stigma remains receptive during the day but is most receptive during early morning (Seaton *et al.*, 1936). Some of the monoecious cucurbits such as cucumbers are highly dependent upon honeybees for pollinations as the flower remains open for one day only (Lord, 1985).

The value of insect pollination in production of fruits and vegetables is well known. Among vegetables, cucurbits have special pollination needs for insect pollination (Woyke and Bronikowska, 1984). The presence of large sticky pollen grains with adhesive stigma demonstrates the need for insect pollination(Anderson, 1941; Sedgley and Scholefield, 1980). Further cucubit flower remains open for one day only which leaves a very small window of opportunity for pollination to occur and if not pollinated at the right time, the flowers abort and drop. If pollination occurs but is incomplete fruits donot develop properly(Motes, 1977).

Furthermore provision for adequate poliination is needed as pollination affects the quantity and quality of the fruit. McGregor (1976) reported that honeybees are the only dependable pollinators for cucumber production. As such some growers have started renting honeybee colonies for cucumber pollination (Schultheis *et al.*, 1994; Stanghellini *et al.,* 1997).

Free (1993) reported that the Asian honeybee (*Apis cerana* F.) and dwarf honeybee (*Apis florea* F.) are most important and prevalent pollinators of cucurbits as compared to *A. mellifera* in are the prevalent ones compared to *A. mellifera* in cucurbit fields. Stanghellini *et al.* (2002 a) stated that cucumber and musk melon may be pollinated by native bees which are smaller than *A.mellifera or Bombus spp.* Connor and Martin (1969) stated that native bees can not be relied upon and majority of the investigators give credit to the honey bee in pollinating cucurbits (Thompson *et al.,* 1955, Langridge, 1952; Nevkryta, 1953; Robinsons, 1952; Sanduleac, 1959; Verdieva and Ismarlova, 1960; Wiolfenbarger, 1962; Skrebtsova,1964). However, other investigators consider ants, thrips and cucumber beetles respectively as possible pollinators of cucurbits(Tontz (1944, Annand 1926, Durham 1928). Hurd (1966) stated cucumber scarabs, meloid beetles, flies and moths as important pollinators but to lesser extent than bees

The size of the flower was another factor in phenology influencing the insect visit in all crops as in the studies of Bell (1985), Eckhart (1991); Conner and Rush (1996). The size of male (♂) and female (♀) flowers produced per plant per day increased from initial phase (IP) of the season to middle phase (MP) of the season where maximum sized flowers were observed entailing increased visit in all crops

Not only the number and size of flowers but also quality and quantiy of rewards per flower were the major factors in attracting the pollinators of flowers. The dynamics of pollinator behaviour have strong correlation with different palynological aspects such as number of pollen grains produced, removed, and deposited as well as their viability which in turn positively correlated with size of the flowers, time of the day and season.

Variability in pollinator number may influence the rate of pollen depletion from flowers. The amount of pollen remaining on flowers after anthesis suggest that a seizable proportion of the total pollen produced may not be available to bees which may be related to floral structure (Stanghellini *et al*., 2002 a). Differences in temporal pollinator density and/or pollinator diversity have been suggested in both crop and natural plant setting, as primary factors influencing pollen removal and yield response (Cane et al., 1996; Fell, 1986). Successful pollination of an insect pollinated plant depends upon not only its attractiveness to pollinators but also pollinator's ability to deposit enough compatible pollen on the flower's stigma (Utelli and Roy, 2001; Herrera, 1987). Efficiency of pollinators was judged based on the quantity of pollen grains they have deposited on stigmas (Primarck and Silander, 1975; Waser and Price, 1990). Rates of pollen deposition on stigmas by flower visitors can be measured readily, and can provide one indication of the pollination efficiency of visitors (Herrera, 1987; Snow and Roubik, 1987). Pollen deposition is considered of primary importance, as it is the very definition of pollinator effectiveness (Kearns and Inouye, 1993).

Amount of pollen deposited on stigmas, influence a flower's reproductive success (Galen and Stanton, 1989; Young and Stanton, 1990). In vine crops, the dependency of fruit set on insect pollination was well established (Free, 1993; Mcgregor, 1976). The increased insect visitation and subsequent increase in fruit set found in these studies was comparable to the results obtained by other researchers working with various vine crops (Mcgregor and Todd, 1952; Mcgregor, 1950; Cauto and Calmona, 1993).

Free (1968) also found that pollination by honeybees increased fruit set in strawberries. Similar studies on cauliflower and cabbage (Verma and Partap, 1993) have shown that bee pollination increased the yield. The effectiveness of honeybees was further demonstrated by a significant increase in fruit set by Aras *et al*. (1996) in lowbush blue berry. Flowers that had the greatest number of visits and highest cumulative durations of visits had the greatest number of fruits, which is in conformity with the studies on cucumbers by Gingras *et al*. (1999). Total abortion of female flowers in the absence of bee visitation found in these experiments confirms the results of other studies on cucumber (Rahmlow, 1970; Morris, 1968; Seaton *et al*., 1936), watermelon (Spangler and Moffett, 1979; Adlerz, 1966), cantaloupe (Iselin *et al*., 1974; Rosa, 1924) and squash (Skinner and Lovett, 1992;

Cauto *et al.,* 1990). The fact is that non pollinated cucurbit flowers, with the exception of those of parthenocarpic cultivars, will not produce fruit (Free, 1993; McGregor, 1976). The numbers of fruit set on the bagged inflorescences were almost negligible, where as there was some set on every unbagged inflorescence, indicating the effectiveness of pollinating insects (Pandey and Yadava, 1970).

The studies of Stanghellini *et al.* (1997) found that cucumber and water melon varieties depended exclusively on insect pollination as there was 100 percent abortion for all pistillate flowers that received no entomophilous visitation.

Flowers that received inadequate pollination resulted in the formation of malformed fruits as stated by Hodges and Baxendale (1995). Anderson (1941) also stated that malformed fruits in cucumbers were the result of poor pollination resulting from too few bee visits per flower. Higher frequencies of insect visit resulted in more number of maximum sized fruits in the plots at harvest which was in concordance with the studies of Free (1968) who found that pollination by honeybees increased percentage of well formed fruits in strawberry. Nye and Anderson (1974) also reported that plots of strawberry caged with honeybees produced fewer malformed fruits. So it is very clear that insect pollination is essential for fruit quality, which is in conformity with the studies on strawberry by Abrol (1989 a). Gingras *et al.* (1999) also said that the rate of pollination was correlated positively with the weight of cucumbers. Steinhauer (1970) reported that the average fresh weight of cucumbers produced per unit area were greater when cucumber fields were visited by honeybees. The same was true of pickling cucumbers (Warren, 1961). In studies on berries by Aras *et al.* (1996) plots characterized by higher number of honeybees produced berries that were more than two times the weight of the blue berries produced in plots with few or no visits. Similar results were obtained by comparing the production of cucumbers from cages with and without honeybees (Alex, 1957; Kauffeld and Williams, 1972)

This is in concordance with Hayase (1953) who stated that the seed number increased in proportion to the amount of pollen deposited on the stigma. As insect visits increased there was a highly significant increase in the number of mature seeds that developed. So efficiency of pollinators has so often determined the number of seeds the flowers set (Schemske and Horvitz, 1984; Motten, 1986; Young, 1988), According to Stanghellini *et al*., (2002 a) variable bee densities likely affect seed setting patterns also. Numerous studies have been made on the influence of honeybee visitation on seed production in cucurbit crops (Avila *et al*., 1989; Bohn and Davis, 1964; Collison, 1976; Connor, 1969; Eischen and Underwood 1991; Fischer and Pomeroy,1989; Gaye *et.al*., 1991; McGregor *et al*.,1965; Stanghellini, 1996). Increased bee visitation increased watermelon seed production (Brewer, 1974; Goff, 1937). Bee pollination also significantly increased the number of seeds per pod in the experiment of Partap and Verma (1994) on strawberries. When the quality of seeds was assessed by their germination ability, the number of viable seeds were found increasing from initial phase to middle phase and decreasing to late phase of the day and season for each crop. According to Kevan and Eisikowitch (1990) cross pollination by insects increased the germinability of the resulting seeds in Canola (*Brassica napus* L.). Similar studies on cauliflower and cabbage (Partap and Verma, 1994) have shown that bee pollination enhanced the quality of the seeds of these crops.

According to Batra (1967) bees other than *Apis* spp. are remarkably scarce at areas where constant human interference, intensive grazing and harvesting the sparse wild vegetation. An additional factor limiting the population of wild bees according to her is their competition for food with numerous active workers of *Apis* spp.

Torchio (1990) and Stephen (1955) much earlier had recommended the development of habitat management program as one method to increase numbers of native species in natural as well as agricultural ecosystems. By keeping bee hives in the fields of cucurbits we can enhance the crop yield.

Sarwar et al (2008) studied the impact of honeybee *Apis mellifera* on pollination of cucumber (*Cucumis sativus L.*) and reported that open pollinated plants and plants caged with bees significantly increased fruit set, fruit weight and circumference of the fruit as compared to control. They concluded thaty honey bees are important for quality and quantity of fruit production in cucumbers and recommended moving honeybee colonies into the field 2 to 3 days after the onset of flowering.

Pests can be managed through the use of rotation, trap cropping, mechanical barriers, such as row covers(Perring et al. 1989, Adams et al. 1990, Orozco-Santos et al. 1995, Hilje et al. 2001)., and chemical applications. Minter and Bessin (2014) advocated that Row covers can be used in place of chemical control before pollination, but are not effective when removed, pests have access to fields along with the pollinators. They can be effective only if native bees could be harnessed for use under row covers.

Cucurbits form an important and vast group of vegetable crops that comprises of many species of vines with creeping growth habit *viz.* watermelon, pumpkin, cucumber, muskmelon, chow-chow, coccinia, bittergourd, ridgegourd, ash gourd, *etc.* which are either used as salad or for cooking or pickling. Plants in the cucurbitaceae family fail to produce fruit when pollination does not occur. The male flowers tend to open first, followed by the female flowers. It is only when both the male and female flowers are open that pollination can occur.

Insects are required for pollen transfer because of the large sized pollen grain, their stickiness and the way they are released from the anthers (Lauria and Fred, 1995). The female flower borne on ovary *i.e.*, inferior ovary and the stigma is receptive throughout the day. In male flowers, an anther dehisces when the corolla expands but the pollens remains on the anther as a sticky mass. The maximum pollination occurs in the forenoon. The female flower closes in the afternoon and never reopens whether or not pollination has taken place further. The highest per cent of fruit set resulted from deposition of pollen on the stigma between 0900 to 1200 h of the day (Bailey, 1949). Of the several vegetables, cucumber and bitter gourd are important crops. Cucumber (*Cucumis sativa* L.) a native to Northern India, constitutes an important green vegetables among the cucurbitaceous crops and is fourth most important vegetable after tomato, brinjal and onion. Similarly, bitter gourd *Memordica charantia* is an other important vegetable belonging to family Cucurbitaceae, widely grown for edible fruit, in Eastern India and Southern China but high species diversity occurs in Africa (Joseph, 2005). Cucumber (*Cucumis sativas* L.) and Bittergourd (*Memordica charantia* L.) belonging to cucurbitaceae are highly cross pollinated as pollen grains being sticky and large in size, need an agent to be transfer to the pistillate flowers for fruit set. Furthermore, reproductive organs of male and female flowers occur separately on staminate and pistillate flowers. Cucumber (*Cucumis sativus*) is normally monoecious which means that there are both male and female flowers on the same plant. Male flowers open about 10 days before female flowers and outnumber female flowers at least ten to one in ordinary monoecious varieties. All of the major varieties are interfertile. Each stigma should receive several hundred grains of pollen for best fruit-set and quality (Seaton *et al.*, 1936). Among the agents, the insects especially honey bees are known to be the most efficient pollinating agents of cucumber and bittergourd (Grewal and Sidhu, 1979).

## 7.2. POLLINATION IN CUCURBITACEOUS CROPS

Majority of vegetable crops belongs to the family cucurbitaceae. But only few studies on the pollination of these crops have been carried out. Grewal and Sidhu (1978) conducted studies on insect pollinators of some cucurbits in Punjab and recorded that *Apis* spp. constituted 70.7% of the total number of bees collected from *Cucurbita pepo* in 1974 and 77.2% in 1975. Among the solitary bees *Pithitis smaragdula* (10.8%) was more abundant in 1974 and *X. fenestrata* (7.3%) in 1975. On *Cucumis melo, Apis* spp. were present in predominantly large numbers and constituted over 69% of the total bees collected. The second in importance were *Nomioides variegate* (11.6%) in 1974 and *N. minutissima* (8.6%) in 1975. *Momordica charantia* attracted fewer bees than *C. pepo* or *C. melo*. In the collection made from *M. charantia*, halictids constituted over 60% of the population of bees. *Apis* group was represented mainly by *A. florea, A. dorsata* being much less numerous on this crop. The megachilid bees visited flowers of these crops only occasionally. Ordway *et al.* (1987) conducted studies on pollen dispersal in Buffalo gourd, *Cucurbita foetidissima* by bees of the genera *Apis*, *Peponapis* and *Xenoglossa* and found that honeybees (*A. mellifera*) were the primary pollen vectors in one cultivar field, and the squash bees *Xenoglossa angustior*, *Peponapis timberlakei* and *P. pruinosa* were floral visitors in another field. Fisher and Pomeroy (1989) studied the pollination of greenhouse muskmelons by bumble bees. The bumble bee workers foraged on the melon flowers from dawn until dusk and they collected only nectar from male and female flowers. Bee visits were essential for fruit development, although the results contradicted the widely reported view that pollination must be accomplished early in the morning. Njoroge *et al.* (2004) studied the pollination ecology of *Citrullus lanatus*. It was found that this species depends heavily on the honeybee, *A. mellifera* L., as the main pollinator. Other pollen vectors identified include *Xylocopa* bees, halictid bees, hypotrigona bees, flies and beetles. The Four groups of flower visitors comprising hawkmoths (*Hippotion celerio, Agrius convolvuli*), moths (*Noctuidae* spp.), skipper butterfly (*Gorgyra johnstoni*) and honeybee (*A. mellifera*) were considered active pollinators of bottlegourd (*Lagenaria siceraria*). Hawkmoths were suspected to be the major pollinators of this plant in the locations surveyed (Morimoto *et al.*, 2004).

Willis and Kevan (1995) studied the foraging dynamics of *Peponapis pruinosa* on Pumpkin (*C. pepo*). They found that the seasonal emergence of the bees closely parallel that of the plants' anthesis. Diurnal activity is related to the availability of pollen resources on the flowers of *C. pepo*. Almost all pollen is removed from the synandria within 2 h of the start of foraging by female bees. Dag and Eisikowitch (1999) noted that ventilation of greenhouses increases honeybee foraging activity on melon, *C. melo*. Foraging activity in the open green houses was two to five fold greater than that in the closed greenhouses. The higher activity could be explained by higher flower attractivity due to a higher nectar sugar concentration both in hermaphroditic flowers and male flowers. It seems that greenhouse ventilation can improve honeybee pollination activity in melon and probably other greenhouse crops as well. Ambrose *et al.* (1995) carried out an evaluation of selected commercial bee attractants in the pollination of cucumbers (*C. sativus* L.) and watermelons [*C. lanatus* (Thunb.) Matson and Nakai]. It was found that the bee attractants didn't increase the bee activity on the subject crops when compared to an untreated control, thereby the authors emphasized the need for additional colonies of bees for pollination of cucumbers and watermelons rather than investing in commercial honeybee attractants.

Aguilar and Tabla (2000) studied the importance of conserving alternative pollinators: assessing the pollination efficiency of the squash to be the most efficient pollinator as: (1) both males and females remove and deposit almost four times as much pollen as *A. mellifera* (2) they make significantly more floral visits than *A. mellifera* and (3) their visit frequency shows a strong relationship to *C. moschata* nectar production during anthesis. Stanghellini *et al.* (2002 a) studied pollen mobilization in selected cucurbitaceous crops and putative effects of pollinator abundance on pollen depletion rates. The results suggest that total pollen production in these crops may not necessarily reflect total pollen availability to floral visitors. Most of the accessible pollen was removed shortly after anthesis, which is when these crops are most receptive to pollination. Stanghellini *et al.* (2002 b) studied diurnal activity, floral visitation and pollen deposition by honey bees and bumble bees in field grown cucumber (*Cucumis sativus*) and water melon (*C. lanatus*) and found that bumble bees (*Bombus impatiens*) were more effective than honey bees (*A. mellifera*) for all three parameters on both crops. Cauto and Calmona (1993) conducted studies on entomophilous pollination in Cucumber (*C. sativus* L. var. Aodai Melhorada), with three different treatments: areas netted without honeybees, netted with honeybees, and open pollinated. Plots visited by bee produced more fruits per square meter and heavier and bigger fruits than other plots. It was concluded that *A. mellifera* was the main pollinating agent of *C. sativus.* Vaissiere and Froissart (1996) studied the pollination of cantaloupes under spun bonded row cover by honeybees in West Africa and found that total yield, fruit size, and seed content in covered plots and control were significantly greater than the open. Covered plants produced more commercial grade fruit than control plants which received 6 insecticide applications.

Cheng (1996) studied the effects of honeybees on pollination and fruit set of cantaloupe and found that the numbers of hermaphrodite and staminate flowers present did not significantly differ between open and net caged plots, but the number of melons per plant in the caged plots was significantly less than that in the open plots. Stanghellini *et al.* (1997) studied the effects of honeybee (*A. mellifera* L.) and bumblebee (*Bombus impatiens* Cresson) pollination on fruit set and abortion of cucumber (*C. sativus* L.) and watermelon, *C. lanatus* Thunb.). They found that as the number of bee visits to flowers increased, the number of fruits aborted decreased. Abortion rates for bumble bee-visited flowers were consistently less or equal to those for honeybees at equal bee visit numbers for both crops. There was 100 percent abortion for flowers receiving no entomophilous visitation. They concluded that bumble bees could serve as a back up or alternate pollinator for honeybees for cucumber and watermelon and possibly other vine crops grown in either greenhouse or field environments.

Gingras *et al.* (1999) conducted studies on visits of honeybees and their effects on Cucumber (*C. sativus* L.) yields in the field. The rate of pollination was correlated with the cumulative duration but not with the total number of bee visits. The maximum cucumber circumference was significantly correlated with the cumulative duration of bee visits. No significant relation was found between cucumber weight and the number of cumulative duration of visits by bees. The presence of honeybees together with the number and cumulative duration of their visits to the flowers are important to pollination and influence both the quality and quantity of cucumber production. Stanghellini *et al.* (1998) conducted studies on seed production in watermelon, a comparison between two commercially available pollinators. Study showed that bee visitation level had a strong positive influence on seed set. All flowers bagged to prevent insect visitation aborted, demonstrating the need for active pollen transfer between staminate and pistillate watermelon flowers.

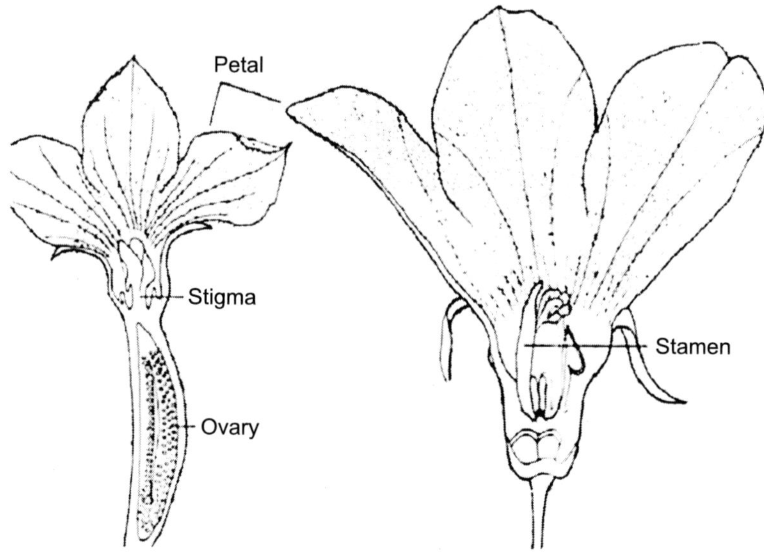


**Figure 7.1. Male and female flower of *Cucumis sativus*, cucumber (after Purseglove, 1968).**

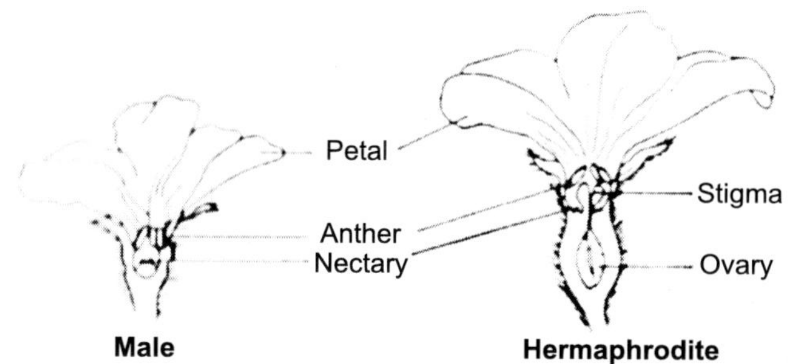


**Figure 7.2. Male and female flower of *Cucumis melo*, cucumber (after university of Arizona, 1961).**

## Pollination effect on ridge gourd

The effect of *Xylocopa fenestrata* on yield and fruit characteristics of ridge gourd was studied by Gogoi *et.al*, (2014) and reported that the flowers per plant, fruits per plant, fruit set including yield showed better result over open pollinated treatment. Rahman (2016) studied the impact of *Xylocopa* pollination ($XP_3$) treatments on ridge gourd, the number of fruits/plant was 17.81 ± 0.325 against 16.78 ± 0.58 and 5.22 ± 0.41 in open and pollinator exclusion treatments respectively (Figure 7.3). Similarly, the fruit set in Xylocopa pollination ($XP_3$) treatment was observed to be 34.91 per cent as against 32.98 and 10.31 per cent in open and pollinator exclusion treatments. The yield obtained from *Xylocopa* pollinated crop was found to have significant difference over open and pollinator exclusion treatments. The yield increase per plot with *Xylocopa* pollination treatment was 0.18±0.015 (q) whereas, 0.17 ± 0.015 (q) and 0.05 ± 0.035 (q) in open and pollinator exclusion treatments respectively. Similarly, yield increase per hectare was observed to be 71.35 ± 0.901 (q) against 53.90 ± 0.890 (q) and 20.70±1.588 (q) in open and pollinator exclusion crops respectively. Bee pollination treatments also showed 224.5 % yield increase over 160.4 % in open pollinated crop. The fruit characteristics of ridge gourd were evaluated on physical basis by measuring length and girth of the fruits. Significant differences were observed with *Xylocopa fenestrata* and open pollination over the treatment of pollination exclusion. But no significant difference was observed among open pollination and *Xylocopa* treatments (Table 7.1).

**Table 7.1:** *Effect of Xylocopa frenestrata pollination on ridge gourd (Source: Rahman 2016)*

| Treatments | Fruit set (%) | Yield/ha (q) | Yield increase over PE (%) | Fruit characteristics | |
|---|---|---|---|---|---|
| | | | | Length (cm) | Girth (cm) |
| PE | 10.31 | 20.70 ± 1.59 | | 21.5 ± 0.33 | 7.3 ± 2.3 |
| OP | 32.98 | 53.90 ± 0.90 | 160.4 | 36.0 ± 2.84 | 10.72 ± 0.58 |
| XP | 34.91 | 71.35 ± 0.90 | 224.5 | 40.31 ± 2.12 | 12.86 ± 0.44 |
| CD (0.005%) | | 2.25 | | 7.15 | 1.31 |

PE: Pollinator Exclusion, OP: Open pollination, XP: *Xylocopa* Pollination

**Figure 7.3: *Xylocopa fenestrata* foraging on ridge gourd and on cucumber (Rahman 2016)**

## Insect pollinators of cucumber and bitter gourd

Abundance of insect pollinators on cucumber and bittergourd.

Honey bees are the most important pollinator of cucumber, and numerous studies have documented their efficacy (Table 7.2). Bees collect pollen on cucumber in early morning and switch to nectar later in the morning. Pollen foraging in Maryland, USA is highest before 10 a.m. and decreases dramatically in the afternoon (Tew and Caron, 1988b). Over 80% of honey bee visits in Michigan occur between 9 a.m. and 2 p.m. (Collison, 1976). The stigma is most receptive to pollen in the morning. The first honey bee to visit a flower coats a large portion of the stigma with pollen, and over 40% of flowers receiving one bee visit will set fruit. Although there is less receptive surface on the stigma with each succeeding visit, multiple bee visits per flower (within the range of 1–20) increase fruit-set and average number of seeds per fruit (Collison, 1976). Large bee populations are especially important in gynoecious plantings where there is an enormous density of female flowers requiring bee visitation.

**Table 7.2:** *Effects of honey bees on yield in cucumber*

| Effects of honey bees | References |
|---|---|
| Increased yield in plots caged with bees > 400% compared to plots without bees | Alex (1957) |
| Increased yield in plots caged with bees > 160% compared to plots without bees | Canadian Department of Agriculture (1961) |
| Increased yield 37.5% and 47.5% in two fields with supplemental hives compared to fields without supplemental hives | Kauffeld and Williams (1972) |
| Increased weight of fruit per plant > 340% in glasshouses with bees compared to glasshouses without bees | Lemasson (1987) |
| Increased yield per plant almost 3 times higher in plots with bees compared to plots without bees | Gingras et al. (1999) |

Sommeijer et al. (1983) recorded *A. mellifera* and *Tetragona* spp. visiting flowers of *M. charantia* in Trinidad. Behera (2004) noted *A. florea, A. cerana*, and *A. dorsata* as important pollinator of *M. charantia* in India. Honey bees (Roubik, 1995) and halictids (Grewal and Sidhu, 1978) have been reported as the principal pollinators of cucurbits including *M. charantia* in Tropical America and India respectively. Lenzi et al. (2005) reported *Diabrotica speciosa* (Coleoptera, Chrysomelidae) to be the main pollinator of *M. charantia* in Brazil. Hoehn et al. (2010) found body size to influence the foraging behaviour of Pollinator in *M. charantia*. He indicated that bigger bodied bees like the *Xylocopa* spp. visited more flowers though for a shorter time when compared to smaller bees. Deyto and Cervancia (2009) recorded *A. mellifera, A. cerana, Trigona* sp. (Apidae) and *Halictus* sp. (Halictidae) foraging on *M. charantia* in Philippines. Batra (1967) reported that *Apis florea* were more in number on flowers of *cucurbita maxima* L. as compared to *A.dorsata* whereas *Lasioglossum cattulum, L. massuricum* Bluth, *Nomia oxybeloides* Smith, *Nomioides minutissima* Rossi, *N. variegate* Oliver and *N. divisa* Cameron and were in negligible numbers. Several species of solitary bees were found visiting watermelon flowers, but honey bees are reported as the principal pollinators (Goff, 1937; Brett and Sulivan, 1972). Seyman *et al.* (1969) reported that the honey bees are extremely important for the pollination of cucumber crop. Atwal (1970) recorded more than 23 species of bees visiting cucurbitaceous flowers at Ludhiana. Of the various bees, *Apis dorsata* was the most abundant species visiting the flowers followed by *A. florea, Ceratina binghami* Cockerel, *Xylocopa pubescens* Spinola, *Nomioides* sp. and halictine bees.

Prakash (2002) reported that the cucumber crop was visited by 27 insect species, of which 16 belonged to Hymenoptera and four each to Diptera, Lepidoptera and Coleoptera. The hymenopterans *viz.*, *A. dorsata, A. cerana, A. florea* and *T. iridipennis* comprised more than 82 per cent of the total insect pollinators. Nidagundi (2004) reported that among the 10 species of pollinators in bittergourd, *Apis florea* F. was the most predominant constituting 43.00 per cent of the total pollinators, followed by *A. cerana* (26.00%), *A. dorsata* (13.00%) and other pollinators (18.00%).

Mathewson (1965) reported that the native East American bee, *P. prinosa* visit exclusively squash and pumpkin flowers. However, main pollinating insects of squash were honeybees and *Trigona rufiorus* (Latreille) in Sao Paulo State, Brazil (Amaral and Mitidieri, 1966), whereas, the principal pollinators of pumpkin under field conditions in Illinois (USA) were European honeybees (*Apis mellifera* Fab), squash bees, *Bombus* sp. and *Melissodes bimaculata*, in addition to the presence of *Diabrotica* sp. and *Acalymma* sp. (Jaycox *et al.*, 1975). Cervancia and Bergonia (1990) reported that *Xylocopa chlorine*, *X. philippinesis*, *Megachile atrata* and *Apis dorsata* Fab. were the frequent visitors of pickling cucumber flowers in Philippines. Shrivastava (1990) made observations on pollination of bottlegourd (*Lagenaria siceraria* Mol.) standl.), ridgegourd (*Luffa acutangula* Roxb.), wild bittergourd (*Momordica diocia* L.) and wild pointed gourd (*Trichosanthes cucumerina* Roxb.) and found that they were pollinated by the bug, *Cyrtopeltis tenuis*, sphingid moths, pyralid moths and the ladybird beetle, *Epilachana punctata*. Shrivastava and Shrivastava (1991) studied 23 species of insects visiting cucurbitaceous crop in Rewa (India), whereas, white flower gourd/bottle gourd (*Lagenaria ciceraria* (Mol.) Standl.) was visited by *Xylocopa fenestrate* Fab. for pollen and nectar and acted as a good pollinator (Sihag, 1990; Sihag, 1993).

**Diurnal pattern of flower visitors on cucumber and bitter gourd:** Sanduleac (1959) in Rumania observed that honey bees worked on the cucurbit flowers (*C. pepo* L., *C. maxima* D. and *C.moschata* Duch ex Poir.) most intensively from 0600 to 1200 h with a maximum activity from 0800 to 0900 h. Honeybees worked on the staminate flowers more vigorously than the pistillate flowers (Figure 7.4). Bhambura (1958) recorded that *A. cerana* Fab., *A. florea* Fab. and *Melipona* sp. started collection of pollen from watermelon at 0830 h and thier activity reached the peak at 1030 h. *A. cerana* deserted the crop every day by 1200 h, but *A. florea* Fab. continued to work till sunset in Bombay. Shemetkov (1960) from Russia and Amaral and Mitidiery (1966) from Brazil reported that bees collect cucumber pollen heavily from 0800 to 1000 h and nectar from 1000 to 1200 hr of the day.

Prakash (2002) reported that the activity of *A. cerana* either in open plots or caged plots of cucumber was found to be maximum at 1000 h and lowest at 1800 h. Further, he reported that the peak pollen foraging activity of *A. cerana* was found at 1000 hr and also the time spent by different honeybees species in collection of pollen was found to be maximum between 0800 and 0900 hr in cucumber. Cervancia and Bergonia (1990) reported that common flower visitors of cucumber were *A. dorsata* F, *Xylocopa chlorinae*, *X. philippiinensis* Smith, *Megachile atrata* Smith and were most abundant from 1000 h to 1100 hours. Sattigi *et al.* (1996) reported that in general foraging activity of honey bees was noticed throughout the day, but it was at its peak between 0800 to 1100 h in winter, 0600 to 1100 h and 1600 to 1800 h in summer and 0800 to 1200 h in monsoon irrespective of the crops in transitional area. The foraging activity was low during other hours of the day in different seasons. Fakuda (1987) reported that honeybee activity on watermelon flower was highest from 0800 to 1000 h and the bees visited male flowers more frequently than female flowers. Rao and Suryanarayana (1988) stated that *A. cerana* was the principal pollinating insect and was found to be efficient pollinator than *A. florea* and *T. iridipennis*. Rapp (1981) recorded that honeybees started foraging on cucumber flowers at about 0600 h and their activity was maximum from 0900 to 1200 h and decreased in the afternoon hours.

## Role of insect pollination on quality and quantity of fruit production:

Bhambure (1958) reported that the honey bees are the most important agents for pollination of Cucurbetaceous flowers for high yield and quality fruit production. McGregor and Todd (1952) found that no melon flowers set fruit until the bees were introduced into caged melon plot, but when bees were introduced, there was a rapid fruit set resulting in the production of 184 marketable melons. But, the open melon plots produced only 145 melons as against four in plots caged without honey bees. Hence, it was opined that to get maximum quantity and better quality of melons, sufficient population of honey bees should be present throughout the flowering period (McGregor, 1950). Kauffeld and Williams (1972) and Kauffeld *et al.* (1975) reported that honey bee pollination has increased the average weight and quality of cucumbers both in open and plots caged with honey bees. The yields of muskmelon fruits were higher in plants pollinated by bees and plants in which bees were excluded set practically no fruits. Joseph (2005) and Behera et al. (2010) suggested that for a commercial fruit and seed production in *Memordica*, pollination management for this crop is essential and the use of hand pollination or the introduction of honey bee colonies in enclosures in India is recommended. Nidagundi (2004) reported that significantly highest length of fruits in bittergourd 26.10 cm as against 13.93 and 13.60 cm fruit length in open pollinated and caged plot without bees, respectively, led to pulp ratio of 0.132 as against 0.09 and 0.07 in open pollinated and caged plot without bees, respectively, highest fruit weight 129.208 as against 72.09 and 62.44 in open pollinated and caged plot without bees, respectively and yield of 118.87 q as against 68.63 and 45.23 q in open pollinated and caged plots without bees, respectively. Rao and Suryanarayana (1988) reported that there was no fruit set in watermelon plots excluded from insect pollinators. The fruit number and weight were more in honey bee pollinated crop than open pollination.

Brewer (1974b) reported the high correlation between melon weight, number of mature seeds (r=0.900) and weight of mature seeds (r=0.850), but contrary to very low correlation between melon weight and immature seeds. He also opined that more number of melons were harvested from the plots visited by bees than from those where bees were excluded. However, high population of honey bees did not improve melon weight compared with normally occurring honey bee populations. Dinesh (2003) reported that the open pollinated crop which received cacambe (10%) recorded significantly least number of dropped fruit per plant in cucumber (0.76 fruit/plant) and significantly higher yield (41.52 kg/plot as against 30.92 and 20.52 kg/plot in open pollination without spray and caged plot without bees, respectively). Thus, there was an increase in the yield of 34.28 and 101.84 per cent over open pollination without spray and caged plot without bees, respectively.

A number of workers claim to ensure cucumber pollination on yield parameters with eight visits per flower (Shemetkov, 1957; Mcgregor *et al.*, 1965 and Conner, 1969), whereas less than eight bee visits produced less number of seeds (60) per fruit with lower fruit weight (22 kg), as compared to 140 seeds per fruit which weighted 500 g, when the flowers were visited 50 times (Shemetkov, 1957). Adlerz (1960, 1961 and 1966) reported that the watermelon fruit set was significantly better after eight or more bee visits per flower and in the second year six or more bee visits per flower were significantly better than four or fewer bee visits per flower.

Verdieva and Ismailova (1960) found that 47 to 57 kg of squash yield was obtained from plots pollinated by honey bees compared with 25 to 30 kg from plots pollinated by other (Unspecified) methods. Battaglini (1969) in *C. pepo* recorded a fruit set of 61.20 per cent of the total pistillate flowers exposed to bees in comparison with fruit set of only 6.80 per cent, when flowers were caged. The cucumber yield was increased by 39 per cent with one honey bee colony per acre as compared to fields without bees (Steinhauer, 1971). Conner and Martin (1969) obtained higher yields of better quality cucumbers by caging the plants to exclude bees for first 11 days after the appearance of first pistillate flowers. They also noted that the delayed exposure of plants to bees for six days was sufficient to get optimum yield.

Kauffeld and Nelson (1982) reported that the yield of pickling cucumber was highest in the plots caged with *A. mellifera* than open plots and was lowest in control plots. Alam and Quadir (1986) reported that plots pollinated with *A. cerana* had 15 per cent fruit set and that of hand pollinated plots had 8 per cent and isolated plots had only 3 to 5 percent fruit set in bottle gourd. Cervancia and Bergonia (1990) reported that the fruit set of cucumber in bee and open pollinated plants were 75 and 58 per cent, respectively and these were significantly higher than the non-pollinated plants (33 %). Bee and open pollination also yielded heavier and uniform fruits.

Viraktamath and Anagoudar (2002) reported that maximum number of good fruits (121.60) were produced in the cucumber crop that received two sprays of attractants, followed by the crop which received sugar solution twice, one spray of Bee-Q, Bee-here and sugar solution. In contrast, lowest number of malformed fruits were recorded in the treatments that received two sprays of attractants (12.75 fruits/plot) and was at par with the caged crop with bees (13.0 fruits). However, unsprayed crop produced significantly higher number of malformed fruits (35.5 fruits) when compared to all other treatments. Brewer (1974) reported that honey bees were the adequate pollinators of watermelon in USA. Experiments in which pumpkin (*C. moschata* Duch ex Poir.) flowers were bagged to exclude bee visits and exposed to number of bee visits shown that the fruit set was 6.5 per cent after one visit and 64.5 per cent after 12 visits per flower. The pumpkin weight and number of seeds also increased with number of visits (Jaycox *et al.*, 1975).

Spangler and Moffett (1979) reported that fruit set ranges from 41 to 95 per cent of the total flowers visited by bees, whereas, no fruit was set in the plot without bees visit in melon. Grewal and Sidhu (1979) obtained 60.46 to 64.30 per cent fruit set in open pollinated *C. pepo* plots as against 41.70 and 55.00 per cent fruit set, when 20 and 30 females of *Pithitis smaragdula* (solitary bee) were enclosed respectively in caged plots. Wolfenbarger (1962) demonstrated that in the caged squash plants to exclude insects the average yield was only 19 per cent. Fisher and Pomeroy (1989) reported that pollination by Bumble bee, *Bombus terrestrus* L. yielded 90 per cent of the melons produced were export quality and weight. Ahmad (1992) reported that greater fruit weight was obtained (2.69 kg/plant) in honey bee pollinated plants compared to self-pollinated plants (2.03 kg/plant) in cucumber.

## Impact of attractants on the pollinating effectiveness of bees

"Benzene extract of pollen", also referred as "attraction fraction" has been reported as a source of stimuli which attracted bees to almond pollen. Pollen supplement containing benzene extract attracted 90 per cent bees compared to untreated check. Fructose-glucose syrup to pollen was more attractive than pollen and water. Similarly, benzene extract of pollen supplement was more attractive when mixed with fructose-glucose syrup than water alone (Doull, 1974). Dinesh (2003) reported that spraying of cacambe (10%), jaggery (10%) and Bee-Q (1.25%) had significantly influence in attracting more number of pollinators. Plots sprayed with cacambe (10%) recorded significantly more number of fruits (15.61 fruits/plant vs 7.42 and 3.34 without bees, respectively) and fruit weight (126.11 g/fruit).

Pateel and Sattagi (2007) reported that spraying of cacambe 10 per cent and Bee-Q 1.25 percent attracted maximum number of bees up to third day after first, second and third spray. Jaggery solution 10 per cent and sugar solution were next best attractants. Treatment with cacambe 10 per cent. Bee-1.25 per cent ,jaggery solution 10 percent and sugar solution 10 per cent were efficient in attracting more bees up to third day after first, second and third spray, whereas their efficacy decreased at 5th day of first, second and third spray. Nidagundi (2004) reported that spraying of cacambe @ 10 per cent, Bee-Q @ 1.25 percent and jaggery solution @ 10 per cent have significant influence in attracting more number of pollination.

According to Higo *et al.* (1995) a combination of increased recruitment of foragers nd greater time spent by foragers with increased flower visitation contributed to the enhanced pollination of blooming crops treated with Fruit boost. Viraktamath and Anagoudar (2002) reported that two applications of Bee-Q® (12.5 g/l), Bee-here® (4 ml/l) and sugar solution (10 %) on staminate flowers of *Cucumis sativa* enticed more number of bees (4.01 to 4.97 bees/flower.5 min.) up to 5 days after first and second sprays compared to unsprayed crop (3.25 to 3.59 bees). Similarly, higher visitation was recorded on pistillate flowers on the sprayed crop. Lingappa *et al.* (1999) reported that increased 21.80 and 31.80 per cent in the number of fruits formed and total yield, respectively, when Bee-Q was sprayed twice on watermelon.

Application of Bee-Q @ 12.50 and 15.00 g/l resulted in higher yield (19.56 and 19.45 t/ha respectively), maximum good fruits, minimum malformed fruits and higher size and weight in watermelon (Sattigi *et al.*, 2001a). Schultheis *et al.* (1994) evaluated two commercial bee attractants like Bee scent and Bee line on cucumber and watermelon. They found that these attractants did increase the yield and also bee visitation. Similarly, Ambrose *et al.* (1995) evaluated Bee line as honey bee feeding stimulant on watermelon and Bee scent as worker bee attractant on both cucumber and watermelon. They found that these attractants neither increased the bee activity on vine crop when compared to untreated control nor they increase the value of subsequent harvest. Allsopp and Cherry (1991) studied the attraction of *A. mellifera* to volatile compounds and they concluded that anetholes and commercial trace janapese beetle lure (10:22:11, 2 phenyl ethyl propionate:eugenol:geraniol) exposed in trace in Japanese beetle traps attracted *A.mellifera* but other floral lures and fattyacids did not attract the bees.

Williams *et al.* (1981) reported that nasonov pheromone of honey bee comprised of seven components and among these, the presence of foot print pheromone enhanced the attractiveness of synthetic nasonov pheromone. They opined that this could prove useful in attracting the honey bees to the crops for better pollination. Woodrow *et al.* (1965) screened the natural and synthetic materials as attractants and repellents for *A. mellifera* by observing responses of bees to their vapours. Out of 195 formulations tested, four were rated as weak to moderate attractants and 19 were moderate to very strong repellents *viz.,* alcohols and one fatty acid having more carbon atoms. Henning *et al.* (1992) studied behavioural responses of *A. mellifera* to primary alfalfa floral volatiles in a screened flight chamber. They found that linalool was the only compound attractive to honey bees at the optimised concentration. Two other compounds, 3-octanone and methyl salicylate were repellents. Two application of Bee scent (a liquid formulation containing 9.00 per cent pheromone and 40.00 per cent other natural attractants) was used on watermelon cultivars in Florida. Total fruit yield was increased in one farm with the treatment upto 3000 fruits/acre compared to 1500 fruits/acre without treatment and there was apparent increase in early yield in three farms. The soluble solid contents of fruit were not affected by the treatment. The number of seeds per fruit was higher with treatment on three frames (Elmstrom and Maynard, 1991).

Ortiz-Sanchoz (1993) reported the efficacy of Bee-here (Nasonov pheromone other honey bee attractant and control release formulation aids) as honey bee attractant to marrow crop (*Cucurbita pepo* L.) grown in greenhouse conditions in Almeria, Spain was tested. Honey bee counts were made on plants sprayed with recommended dose of attractant (3.00 ml/lit), plants sprayed with half the dose, plants sprayed with water and untreated plants. The bees did not exhibit preference for any experimental treatment indicating that this product being ineffective as a honey bee attractant to marrow flowers (Ortiz-Sanchoz, 1993). Looper and Rossette, (1991) conducted field trial on two adjacent fields of *Citrullus lanatus* L. in Arizona, USA, in which honey bees were introduced at a stocking density of two colonies/acre. Bee scent was sprayed over alternate 18 row strips in one field and the other field was untreated. Though, bee visitation was high on the day of application, but did not reflect in yield.

Viraktamath and Anagoudar (2002) reported that two applications of Bee-Q® (12.5 g/l), Bee-here® (4 ml/l) and sugar solution (10 %) on staminate flowers of *Cucumis sativa* enticed more number of bees (4.01 to 4.97 bees/flower.5 min.) up to 5 days after first and second sprays compared to unsprayed crop (3.25 to 3.59 bees). Similarly, higher visitation was recorded on pistillate flowers on the sprayed crop. Tsirakoglou *et al.* (1997) reported that spraying of Bee-here has no significant effect on bee visitation to kiwi fruit. Number of bees visiting kiwi fruits before spray and 4, 24, 72, 96 or 120 hr after spray was same as water treated control treatments. Feeding sugar syrup and weeding to remove wild plants did not seem to affect kiwi fruit pollen collection.

Jayaramappa et al (2011) found that in Ridge Gourd (*Luffa acutangula* L.) spraying of Fruit boost @ 0.5 ml/l and Bee-Q @ 12.5 gms/l enhanced yield parameters like number of fruits / plant to 19.00 and 17.00 fruits, when compared to 10.66 fruits / plant in open pollinated plot. Number of fruits was 21.83 and 20.83 fruits / plot, when compared to15.68 fruits / plot in open pollinated plots. Fruit length was 42.41 and 41.21cm, when compared to 32.95 cm in open pollinated plots. Fruit weight was 285.00 and 246.33 gms, when compared to 229.67 gms in open pollinated plots and yield was 154.00 and 152.00 q/ha, when compared to 130.67 q/ha in open pollinated plots were studied.

## Cucumber Pollination Requirements

Pollination is essential for fruit production in cucumbers. Fruit abortion can reach 100% in flowers bagged to exclude insect visitors (Stanghellini *et al.*, 1997), but self-pollination rates of 30–36% have been documented in the absence of insects (Jenkins, 1942; Gingras *et al.*, 1999), and a small rate of parthenocarpy is known to exist (Gustafson, 1939; Gingras *et al.*, 1999). Nevertheless, insect pollination is the norm.

All of the major varieties are interfertile. Each stigma should receive several hundred grains of pollen for best fruit-set and quality (Seaton *et al.*, 1936). Pollen grains do not necessarily need to be spread evenly on all three lobes of the stigma to form a good fruit, but it does take multiple bee visits to optimize fruit-set. The minimum number of required bee visits per flower is probably variety-dependent and has been recorded as 8–12 (Connor, 1969; Stephen, 1970; Lord, 1985), 18 (Stanghellini *et al.*, 1997), and six (Gingras *et al.*, 1999).

About 30% of the flowers in gynoecious plantings should be male in order to optimize pollination and economic return (Connor and Martin, 1971). This means that gynoecious varieties require a supplemental source of male flowers. One way this is done by pre-mixing commercial seed with around 10% monoecious seed. Cucumbers produced from early flowers on young plants often have fewer seeds and inferior shape compared to fruit produced without competition on more mature vines (Connor and Martin, 1970). In unpublished studies these authors found that older vines produce larger female flowers with more ovules. Thus, fruit quality may be improved if pollination could be delayed until vines are more mature.

Connor and Martin tested this idea with selected monoecious, gynoecious, and gynoecious hybrid varieties. Delaying pollination (with cages to exclude or include bees) for up to 11 days after the first appearance of female flowers increased the number of fruit per plant and the dollar value per acre. The increase was likely due to stronger vine and root growth in more mature plants.

It is difficult to realize the benefits of delayed pollination under field conditions. The only way to delay pollination in monoecious varieties is to delay importation of honey bee hives, but one cannot practically keep wild bees from visiting the plants. With gynoecious varieties, one can interplant male-bearing monoecious pollenizer varieties a few days after planting the gynoecious variety. That way, male flowers become available to bees only after gynoecious plants have reached a suitable level of maturity. Connor and Martin (1969a,b) recommended alternating a 40-ft (12 m)-wide strip of gynoecious variety with a narrow strip of monoecious pollenizer planted a few days after the gynoecious variety. The strip of monoecious pollenizer should probably be no wider than 3 ft (1 m) (Free, 1993). Delayed pollination in gynoecious varieties is further complicated by the fact that even gynoecious plants produce a few male flowers.

In contrast to the results of Connor and Martin (1970), delaying pollination in high-density plantings of the variety 'Napoleon' in Texas did not increase yield (Underwood and Eischen, 1992). Cucumbers are grown in Europe that set fruit parthenocarpically without pollination. Pollination is undesirable in this context because it results in misshapen, devalued fruit. Growers exclude pollinators with modified glasshouses and government-enforced restrictions on bee-keeping (Free, 1993).

## Cucumber Pollinators

Honey bees are the most important pollinator of cucumber, and numerous studies have documented their efficacy (Table 7.3). Bees collect pollen on cucumber in early morning and switch to nectar later in the morning. Pollen foraging in Maryland, USA is highest before 10 a.m. and decreases dramatically in the afternoon (Tew and Caron, 1988b). Over 80% of honey bee visits in Michigan occur between 9 a.m. and 2 p.m. (Collison, 1976). The stigma is most receptive to pollen in the morning. The first honey bee to visit a flower coats a large portion of the stigma with pollen, and over 40% of flowers receiving one bee visit will set fruit. Although there is less receptive surface on the stigma with each succeeding visit, multiple bee visits per flower (within the range of 1–20) increase fruit-set and average number of seeds per fruit (Collison, 1976). Large bee populations are especially important in gynoecious plantings where there is an enormous density of female flowers requiring bee visitation.

Cucumber growers are often interested in chemical bee attractants because honey bees are easily distracted away from cucumber to more attractive forages. Unfortunately, the research record for attractants in cucumber is not encouraging. Bee-Scent® attractant in North Carolina cucumber did not increase honey bee visitation, yield, or monetary return (Schultheis *et al.*, 1994).

Bumble bees pollinate field cucumber effectively in North Carolina (Stanghellini *et al.*, 1997). The number of aborted fruit decreases as the number of flower visits by either bumble bees or honey bees increases. The rate of abortion is lower with bumble bees when compared at an equal number of bee visits.

**Table 7.3:** *Effects of honey bees on yield in cucumber.*

| Effects of honey bees | Reference |
|---|---|
| Increased yield in plots caged with bees > 400% compared to plots without bees | Alex (1957a) |
| Increased yield in plots caged with bees > 160% compared to plots without bees | Canadian Department of Agriculture (1961) |

Increased yield 37.5% and 47.5% in two fields with supplemental hives compared to fields without supplemental hives (Kauffeld and Williams 1972). Increased weight of fruit per plant > 340% in glasshouses with bees compared to glasshouses without bees. Lemasson (1987) Increased yield per plant almost 3 times higher in plots with bees compared to plots without bees (Gingras et al. 1999). Mcgregor (1976) recommended 1-3 bee hives per acre or 2.5-7.4 hives per hectare for optimum production of cucumbers. Similarly, Atkins et al (1979) recommended 2-3 and 5-7.4 hives per acre and hectare, respectively. Levin (1986) reported 1-2 hives per acre and 2.5-5 hives per hectare in cucumber pollination. William (1994) considered 0.1 to 4 hives hives per acre and 0.3-10 hives per hectare for optimum production of cucumbers. Other estimates showed that 1 honeybee colony per 100 flowers(Mcgregor, 1976) and one colony of honeybee per 50, 000 plants(Connor, 1969). Hughes et al. (1982) reported that for gynoecious hybrids[3] 3 hives per acre or [3] 7.4 per hectare are needed.

There was significant increase in fruit set (34.25%) and yield (75.97q/ha) in bee pollination treatment over without bee (21.98%, 49.47q/ha).

**Table 7.4:** *Effect of Apis cerana pollination on yield of cucumber (Pooled data) ( Rahman 2016)*

| Treatments | Fruit set (%) | Yield/ha (q) | Yield Increase over WBP (%) | Yield Increase over OP (%) | B:C Ratio |
|---|---|---|---|---|---|
| OP | 21.98 | 49.47 ± 3.78 | 121.64 | – | |
| BP | 34.25 | 75.97 ± 4.10 | 240.36 | 53.56 | |
| WBP | 10.55 | 22.32 ± 1.26 | – | – | 3.45 |
| CD (P = 0.05) | – | 2.40 | – | – | |

OP: Open; BP= 5 *Apis cerana* colonies per ha, WBP: Without bee

**Table 7.5:** *Effect of Xylocopa fenestrata pollination on fruit set and yield of ridge gourd( Rahman 2016)*

| Treatment | Fruit set (%) | Yield/ha (q) | Yield increase in XP over WXP (%) | Fruit characteristics (cm) | |
|---|---|---|---|---|---|
| | | | | Length | Girth |
| WXP | 10.26 | 21.4 ± 1.544 | | 27.52 ± 3.25 | 8.21 ± 0.42 |
| OP | 34.42 | 71.2 ± 0.73 | 232.7 | 35.37 ± 2.31 | 10.19 ± 0.35 |
| XP3 | 36.17 | 74.8 ± 0.765 | 249.53 | 42.26 ± 3.27 | 12.58 ± 0.52 |
| S.Ed± | | 0.94 | | 2.71 | 0.63 |
| CD(0.05%) | | 2.04 | | 5.93 | 1.37 |

WXP = Without *Xylocopa* pollination, OP = Open pollination, XP3: Single, double and triple *Xylocopa fenestrata nest* on bamboo. There was significant increase in fruit set and yield of ridge gourd under *Xylocopa* treatment (36.17%, 74.8q/ha) over without bee pollination (10.26%, 21.4q/ha) Fruit characteristics viz. length and girth of the fruit were also significantly improved in *Xylocopa* pollination over without *Xylocopa* pollination.

## Squash, Pumpkin and Gourd

Squash, pumpkin, and gourd (*Cucurbita* species) plants are normally monoecious which means that there are both male and female flowers on the same plant. Male flowers outnumber female ones by 3.5:1 to 10:1. Each flower is about 3 in (7.6 1cm) wide. Each male flower occurs at the end of a slender stem and has three anthers. Each female flower occurs at the end of a short peduncle and has a thick style and a two-lobed stigma; the swollen ovary occurs at the base of the corolla and is divided into 3–5 sections. Male flowers produce nectar and pollen, and female flowers produce nectar. Female flowers produce more nectar and attract more bees compared to male flowers (Nepi and Pacini, 1993). The pollen grains are large and well suited to insect transport. Flowers open early in the morning and close around noon of the same day, never to reopen (Skinner and Lovett, 1992; Nepi and Pacini, 1993).

## Pollination Requirements

Because these plants have separate male and female flowers, some agent – usually bees – is necessary to transfer pollen from male to female flowers. Squash plants caged to exclude pollinators produced no fruit in west Tennessee, USA (Skinner and Lovett, 1992). Wind does not pollinate *Cucurbita*. Ovules are fertile only during the flowering period or on the day before in the squash variety 'Greyzini' in Italy. Pollen viability in a newly opened male flower is about 92% but by the time it closes that same morning the viability will have dropped to 75%, and by the next day it will be only 10% (Nepi and Pacini, 1993). Thus, it is important for a female flower to be pollinated as early as possible on the day it opens while pollen is still viable.

Early bee activity is even more important on hot days when flowers close early. Rate of fruit-set is similar in selfed or crossed fruit, but cross-pollination produces heavier fruit as shown in squash in India (Girish, 1981). Different varieties of the same species, and even different species, can cross each other in *Cucurbita*. Even with incompatible species the pollen of one may stimulate parthenocarpic (seedless) fruit development in another; this phenomenon does not result in contaminated seed, but it does reduce total seed yield (Free, 1993). Thus, different varieties and species of *Cucurbita* must be isolated from each other in order to produce large quantities of pure seed.

## Pollinators

Squash bees and honey bees are the most important pollinators of *Cucurbita* species. Squash bees (*Peponapis pruinosa*) are excellent pollinators of squash and pumpkin. Compared to honey bees, they make more contact with reproductive parts of a flower, work faster, and work earlier in the morning. However, squash bees are no more efficient than honey bees at setting fruit in spite of their desirable behaviours (Tepedino, 1981). Nevertheless, growers should always encourage wild squash bee population. Supplemental colonies of honey bees are unnecessary where squash bees occur in high numbers. Honey bees work *Cucurbita* flowers and effectively pollinate the crop. They are the most numerous bee visitor to summer squash in west Tennessee, and they freely visit male and female flowers. Honey bees with large pollen loads are often seen visiting female flowers. However, honey bees are generalists and they readily move to any competing bloom that offers richer reward than *Cucurbita*. This is a problem in the southeastern US where wild gallberry, a rich source of nectar, blooms in competition with early squash. This is a situation where honey bee attractants may help focus bees on the crop; however attractants have a mixed record with *Cucurbita* species (Margalith *et al.*, 1984; Loper and Roselle, 1991; Schultheis *et al.*, 1994). Another way to deal with competing bloom is to increase the density of honey bee hives so that competition for nectar and pollen forces bees to forage on the crop. In Georgia, competition from wild bloom is less of a problem in fall plantings of squash.

**Table 7.6:** *Cross-compatibility of certain Cucurbita species (Whitaker and Davis, 1962). A '+' indicates species that will cross with each other, and a '2' indicates species that will not cross.*

| | ***C. maxima* (winter squash)** | ***C. mixta* (cushaw squash)** | ***C. moschata* (pumpkin)** | ***C. pepo* (summer squash)** |
|---|---|---|---|---|
| C. maxima (winter squash) | NA | – | – | – |
| C. mixta (cushaw squash) | – | NA | – | – |
| C. moschata (pumpkin) | – | – | NA | – |
| C. pepo (summer squash) | – | – | – | NA |

NA, not applicable.

**Table 7.7:** *Effects of increasing bee visits to Cucurbita species flowers.*

| Location | Bee(s) tested | Conclusion | Reference |
|---|---|---|---|
| Illinois, USA | Honey bee | As number of bee visits to a pumpkin flower increases from 1 to 12, fruit-set increases from 6% to 64% and number of seeds per fruit increases from 273 to 366. | Jaycox et al. (1975) |
| India | Asian honey bee | As number of bee visits to a bee summer squash flower increases from 1 to 7, fruit-set increases from Girish 30% to 100%. | Tepedino (1981 |
| Utah, USA | Honey bee, | One bee visit results in 22% squash bee fruit-set in summer squash, but flowers with numerous visits have 66% fruit-set. | Tepedino (1981) |

In case of squash, pumpkin and gourd different workers have recommended different densities of honeybee for optimum production of fruits. Hughes et al (1982) recommended colony of honeybee per acre and 2.5 colonies per ha., Goebel (1984) 0.8-1.6/acre and 2-4 per ha, Levin (1986) 1-2 per acre and 2.5-5 per ha. Skinner (1995) found 1-2 and 2.5-5 colonies per acre and ha, respectively. Tepedino (1981) found that 1 squash bee per 20 flowers was sufficient for optimum quality and quantity production of fruit. Scott-Dupree et al. (1995) recommended 1–3, 2.5–7.4 colonies of honeybees per acre and ha, respectively.

Skinner and Lovett (1992) found that bumble bees were more efficient than honey bees at promoting good fruit-set in western Tennessee. Six of eight (75%) single bee flower visits by bumble bees resulted in whole fruit, whereas five of 16 (31%) single bee visits by honey bees resulted in whole fruit. However, in normal conditions and with adequate bee populations, every flower is visited many times, probably by more than one bee species, and fruit-set is very good. Fruit-set and seed number increase as the number of bee visits to flowers increases. This is due to the fact that seed formation increases as the number of pollen grains deposited on a stigma increases (Winsor *et al.*, 1987). Thus, large bee populations help ensure maximum flower visitation, pollen deposition, and crop yield. Other wild solitary bees visiting squash flowers in Tennessee, namely were *Agapostemon virescens*, *Augochlora pura*, *Dialictus* sp., *Halictus* sp., *Triepeolus remigatus*, and *Mellisodes bimaculatus*. These bees probably contribute to squash pollination, but their relative efficiency is unknown.

## *Cucurbita moschata*

Cucurbita moschata has monoecious, self-compatible unisexual flowers, which are actinomorphic with pentamerous perianths. Flowering in this species begins with opening of the male inflorescence followed by female flowers opening 2 to 3 weeks later. The brilliant yellow petals of both flowers attract pollinators. Although flowering in male and female inflorescences began at different times, the peaks of flowering in both sexes almost overlapped, favouring female receptivity (Richards, 1986; Nepi and Pacini, 1993).

Cucurbita moschata flowers that bloomed when inflorescences were 2 weeks old were usually bigger and produced more pollen and nectar, thus attracting more insect visitors. Incomplete opening, withering or shrivelling of flowers was common in the first week of flowering in both sexes. Fruit abortion was also common among inflorescences that developed at the early and later weeks of flowering. Tangmitcharoen and Owens (1997) suggested that pollination success might be greater in flowers developing early and during the peak of the flowering period than in those that develop at the end of the flowering season. This is obvious in C. moschata, where the peaks of male and female flowering almost coincide. The attractive bright yellow colour of the petals and the campanulate shape of the flowers favour insect pollination. The campanulate throat provides shade for flower visitors.

Agbagwa et al (2007) studied the pollination biology of *Cucurbita moschata* (Duch. ex Lam) Duch. ex Poir varieties and reported that the plant having both male and female flowers were pollinated by thrips and bees with nectar and pollen as major rewards for the pollinators. Several earlier investigators have reported thrips and *Apis millifera* as good pollen collectors and, as such, effective pollinators (Philippe, 1991; Nepi and Pacini, 1993; Endress, 1994; Ma et al., 1997; Tangmitcharoen and Owens, 1997; Luo & Li, 1999) on *C. moschata*. Agbagwa et al (2007) observed that thrips occurred more than *Apis mellifera* on *C. moschata* flowers although both were effective pollinators. The fact that thrips carry large pollen loads was previously demonstrated by Luo and Li (1999) and Ma et al. (1997) on Chloranthus Swartz species. The entire body of the insects collected from C. moschata flowers was covered with pollen grains. Endress (1994) summarised the features of flowers adapted to pollination by thrips and suggested that flowers that provide shelter together with a narrow entrance towards the floral centre are favoured by thrips. In addition, Luo & Li (1999) recognised scent and showy bright flowers as attractive to thrips, especially when combined. It is noteworthy that these characters outlined by Endress (1994) and Luo and Li (1999) are conspicuously present in C. moschata and most Cucurbitaceous species. As was observed, the second group of thrips that spend 6 h or more on one flower do not leave the flower until closure.

These were observed to feed on the pollen and nectar of the flower as well as to shelter in the campanulate base of the *C. moschata* flowers. Thus, the thrips are first attracted by the bright yellow and scented petals of *C. moschata*, then feed on the abundant nectar and pollen, and finally receive shelter from the day's sunlight. Although this group of thrips carries abundant pollen, the first group that fly off to other flowers before the sun becomes intense are the potential pollinators.

Bees play an essential role in the pollination of *C. moschata* as in most Cucurbitaceae (McGregor and Todd, 1952; Robinson et al., 1989; Philippe, 1991; Nepi and Pacini, 1993). The arrival of these solitary bees on *C. moschata* flowers almost coincides with the resumption of early morning sun and opening of both male and female flowers. This also corresponds to the hours of peak pollinator activity, pollen viability, and decreases in the heat of the day. In this way, a high percentage of germination is assured, a fact that has been previously observed (Gould and Gould, 1988; Nepi and Pacini, 1993). The most efficient visits are the first ones because they result in the deposition on the stigma of a large number of highly viable grains. The floral size and architecture of *C. moschata*, i.e. the large petals (7 to 12 cm long).

## *Luffa aegyptica*

*Luffa aegyptiaca* Mill (Cucurbitaceae) is a vigorous, annual climbing herb found in tropical and subtropical regions and in the same family with most of the commercially grown vine crops. L. aegyptiaca is monoecious, having separate male and female flowers on the same plant, a characteristic of many species in the family, that makes them even more dependent on pollinators, especially on bee species. Cucurbit plants do not just rely on pollinators but also require repeated visits from pollinators (Collinson 1976; Stanghellini et al. 1997; Meléndez, 2002). Furthermore, on the same plant, some bee species have been shown to be more efficient pollinators than others (Stanghellini et al. 2002). Therefore knowledge on appropriate pollinators of a plant has become important especially in the management and conservation of both the pollinators and the plants. Mensah and Kudom (2011) reported that different pollinators have different pollination efficiency on *Luffa aegyptica.* Fruits that developed from flowers that had received a single visit from *X. olivacea* had a mean weight of 428.7g and were 1.5 times heavier than fruits from flowers visited by *A. mellifera* (286.76 g). *X. olivacea* was more efficient than A. mellifera in terms of number of fruit set per single visit. This study has provided some knowledge on pollination ecology of *L. aegyptiaca*, which can be exploited to improve fruit production in commercially grown vine crops.

## Watermelon

Watermelon (*Citrullus lanatus*) plants are monoecious, bearing both male flowers and female flowers. A few varieties bear perfect flowers and male flowers. Each flower is 1 in (2.5 cm) diameter and pale green or yellow. The petals unite to form a tubular corolla enclosing three stamens that surround a short style and a three-lobed stigma. There are nectaries at the base of the corolla. Flowers open within 2 hours after sunrise. Pollen is usually released before the flower opens, but it stays on the anthers in sticky masses unless it is removed by insects. Stigmas are receptive as long as the flower is open, but receptivity is highest between 9 and 10 a.m. (Adlerz, 1966). Flowers close in the afternoon, never to reopen even if they are not pollinated. Watermelon flowers are attractive to bees for nectar and pollen, but flower density s low and bees are easily distracted away to richer resources.

## Pollination requirements

Watermelon is self-fertile, and a female flower is pollinated equally well by pollen from a male flower on the same or different plant. The pollen grains are sticky and insects are required to transfer pollen to receptive stigmas. Fruit abortion in flowers bagged to exclude insects is 100% (Stanghellini *et al.*, 1997). Each stigma needs at least 1000 pollen grains spread evenly over the three lobes in order to form a large, wellshaped fruit (Adlerz, 1966). Fruit will be misshapen and devalued if even one stigma lobe receives an insufficient amount of pollen.

Fruit-set is more likely in ovaries that are longer than 28 mm at the time of pollination compared to ovaries 20 mm or smaller (Adlerz, 1966). Plants continue to set fruit throughout a season, but fruit-set is higher in vigorously growing plants. The presence of unhar-vested cull fruit on a plant inhibits fruit-set of other female flowers; thus, regular harvesting is necessary to optimize fruit-set and yield. Triploid watermelons are grown for their highly palatable, nearly seedless fruit. Triploid seed is produced by crossing a tetraploid female parent with a selected diploid, or normal, male parent. Tetraploid female parent lines are normally self-sterile, and bees are needed to transfer pollen from the male diploid plants to the tetraploids. Tetraploid (solid colour) rows are alternated with diploid (striped) rows, and seeds are harvested from the tetraploid plants. The plant resulting from triploid seed is sterile, and its small, white seeds are easily eaten along with the fruit (Maynard and Elmstrom, 1992). Triploid plants, in turn, must be pollinated in order to form fruit.

Pollination in this case is not necessary for seed fertilization, but rather for the growth-stimulating effect provided by pollen (Kihara, 1951). Normal diploid varieties can serve as pollenizers. It is most practical to plant the pollenizer in an outside bed, next to two beds of the triploid, followed by another bed of pollenizer, and so forth in a repeating pattern. The diploid pollenizers should be well suited to the same growing regime as the triploid, visually distinguishable from the triploid, and marketable. Early yield of triploids is enhanced when early-blooming icebox watermelons are used as pollenizers, and late yield of triploids is enhanced when standard later-blooming watermelons are used as pollenizers (Maynard, 1990; Maynard and Elmstrom, 1992).

There is genetic variability among watermelon varieties for their attractiveness to bees. Bee visitation was significantly higher in the variety 'BAG' than in 'Sugar Baby' which was associated with nectar of both a higher volume and sugar concentration in 'BAG'. Moreover, this differential attraction to bees was associated with differences in seed set which was higher in 'BAG' (400 seeds per fruit) than in 'Sugar Baby' (225 seeds) (Wolf *et al.*, 1999). This study supports the idea that breeding watermelon varieties for their attraction to bees may help enhance yields.

## Pollinators

Wind and gravity contribute little or nothing to watermelon pollination. Watermelon is pollinated almost exclusively by insects. Of these, honey bees are the most abundant and widely used (Goff, 1937, 1947). Fruit-set was zero in flowers bagged to exclude bees, but 95% in nonbagged flowers in an Arizona glasshouse stocked with one honey bee colony (Spangler and Moffett, 1979).

Honey bees are reasonably efficient pollinators if they visit watermelon. Flower visitation begins early and peaks between 8 and 10 a.m. (Goff, 1937; Adlerz ( 1966). Honey bees spend 5.7–8 s per female flower (Adlerz, 1966) which is relatively lengthy and suggests that individual flowers are rich resources. Pollination efficiency may be better if the resources of individual flowers were poorer and bees were forced to visit more flowers. One way to encourage this is to increase competition for flower resources by increasing the density of bee hives in a field. Because each bee spends a long time at a flower, moving very little in the process, it takes at least eight honey bee visits per flower to adequately coat the stigma with pollen and produce a well shaped fruit (Adlerz, 1966). Seedless triploid varieties produce less pollen than regular diploid watermelon (Ambrose *et al.*, 1995). Thus, triploid varieties probably require more bee visits in order to set fruit; however, this idea is not tested.

Because honey bees are easily distracted away from watermelon, bee attractants may be helpful. However, the record for bee attractants in watermelon is not encouraging. Bee- Scent® increased honey bee visitation in Arizona for up to two days in a seedless variety and in the variety 'Picnic', but it did not increase yield in either variety (Loper and Roselle, 1991). Neither Bee-Scent® nor Beeline® improved bee visitation, yield, or monetary return in 'Royal Sweet' in North Carolina (Schultheis *et al.*, 1994). Bumble bees are efficient pollinators of watermelon in North Carolina. The number of aborted flowers decreased with increasing bee visits. It took about six honey bee visits per flower to reduce fruit abortion to a rate similar to open-pollinated flowers, but bumble bees accomplished this with only one flower visit (Stanghellini *et al.*, 1997). Concerning average seed set per fruit, it took at least 18 honey bee visits per flower, but only 12 bumble bee visits, to achieve a level comparable to that in open-pollinated flowers (Stanghellini *et al.*, 1998).

McGregor (1976) recommended 0.5–12.4 colines per ha. Atkins found 0.2-5 colonies per acre and 5–7.4 colonies per ha for optimum production water melon. Hughes et. al (1982) observed that one colony of honeybee per acre or 2.5 colonies per ha is required for production of water melon, whereas William 1984 considered 0.2-2and 0.5-5 colonies per acre and ha , respectively for this purpose. In terms of bee visits per flower Adlerz (1966) considered 8 honeybee visits per flower and 1 honeybee per 100 flowers (Cooperative Extension Service 1970) in case of watermelon.

## Pests of Cucurbits (Cucurbitaceae)

The family Cucurbitaceae, also known as the gourd family, includes many crops that are predominantly grown in the tropics such as pumpkin, bitter melon, ribbed gourd, snake gourd, watermelon, cucumber, squash, cantaloupe, and others. Cucurbits are often referred to as vine crops. Some are important crops in the drier regions of the tropics as they tolerate high temperatures. The origin of cucurbits varies depending upon the genera. For example, watermelon, cantaloupe, and honeydew melon are from Africa; cucumber, bitter gourd, and snake gourd from India; bottle gourd from Indonesia; and chayote and pumpkin from the New World (George, 2011).

Cucurbits are infested by a number of insect pests right from the germination up to harvesting stage, but a few of these *viz*., red pumpkin beetle, fruit flies and hadda beetle are of serious concern (Gupta 2004) (Figure 7.5). Red pumpkin beetle, *Aulacophora foveicollis* (Lucas) is a widely distributed polyphagous pest of cucurbit crops in India (Butani and Jotwani 1984). This beetle has also been reported from Greece, south Europe, Algeria, Cyprus, Aden, Iraq, Sri Lanka, Nepal, and Burma (Mann and Sohi 1983). The beetle is considered as one of the serious problems particularly at seedling stage in the commercial cultivation of the cucurbits. Among the cucurbits, cucumber is the most preferred by this pest while bitter gourd and sponge gourd are also equally preferred (Khan and Hajela 1987; Mehta and Sandhu, 1989). Cucurbitacins widely distributed in cucurbits, play a major role in host acceptance. Glucoside a triterpenoid of bitter gourd however acts as a feeding deterrent (Chandravadana and Pal 1983).

The beetles start feeding just after the germination and retard the growth of the seedlings due to severe foliar damage. Both the grubs and adults cause damage, the grubs which live underground are destructive to the roots. The grubs enter into the roots, and underground stem or sometimes the fruits touching the ground. The underground portion starts rotting due to attack of saprophytic fungi at the injury site and the fruits become unfit for human consumption. The adult beetles are mainly responsible for damaging the plants above the ground, attacking leaves, flowers, fruits and causing complete defoliation and sometimes the entire crop requires resowing. The beetle resumes its activity in March and remains in the field till October. The peak period of activity is from April to June while the population start declining from September onwards (Shinde and Purohit 1978; Butani and Jotwani 1984). The hadda beetle, *Henosepilachna vigintioctopunctata* (Fabricius) is also a serious pest of many plants of agricultural importance in Asia and has also been reported from Australia, Africa, East Indies and Japan (Rajgopal and Trivedi 1989; Hirano 1993). It is an important pest of solanaceous and cucurbitaceous crops in mid-hills and plains of India (Kumar and Kumar 1996). The grubs as well as adult beetles scrap and feed voraciously on the green matter of the leaf and skeletonize it (Rath *et al.* 2002; Mohasin and De 1994) in a characteristic manner leaving the upper epidermal tissue intact. The adults are fliers and damage large crop areas during their peak activity (Nagia *et al.* 1992). These usually feed on the upper surface whereas the grubs confine their attack to the lower surface of leaves. The pest eats out regular areas of the leaf tissue leaving parallel bands of uneaten portion in between. The affected leaves become lace-like in appearance, turn brown, dry up and drop prematurely resulting in retardation of plant growth and thereby reduce the bearing of the plants (Ghosh and Senapati 2001). In case of severe infestation, the plants may be defoliated. The dipteran family Tephritidae consists of over 4000 species, of which nearly 700 species belong to Dacine fruit flies (Fletcher 1987). However, 250 species are of economic importance and distributed widely in temperate, sub-tropical and tropical regions of the world (Christenson and Foote 1960). In India, fruit flies are identified as one of the ten most serious problems of the entire agriculture and because of the polyphagous nature of their larvae, many species cause high economic losses in fruits and vegetables. Of 207 species of fruit flies in India, nine are identified to be the major and economically important (Sardana *et al.* 2005).

The fruit flies constitute an important group of pests infesting cucurbit vegetables. Two species namely *Bactrocera cucurbitae* (Coquillett) and *Bactrocera tau* (Walker) commonly called as melon fruit flies are the major species found infesting cucurbits. Besides, they have been found feeding on solanaceous crops like tomato and brinjal (Kapoor and Agarwal 1983). Another fruit fly species, *Bactrocera scutellaris* (Bizzi) has also been recorded on cucurbits (Sunandita and Gupta 2007). The extent of losses caused by *B. cucurbitae* varies from 30 to 100 per cent depending on the cucurbit species and season (Dhillon *et al*. 2005). It has been reported to infest 95 per cent bitter gourd fruits in New Guinea and 90 per cent snake gourd and 60 to 87 per cent pumpkin fruits in Solomon Islands (Hollingsworth *et al*. 1997). In India Singh *et al.* (2000), however, reported fruit infestation of 31.27 per cent infestation on bitter gourd and 28.55 per cent on water melon. The melon fruit flies have more than 80 hosts and their economic importance can not be evaluated entirely from the standpoint of the direct damage to the various crops affected. Quarantine laws aimed at preventing the entry and establishment of melon flies and hence reduce the export potential of crop produce (Ronald and Jayma 2011). Generally, the female fruit flies puncture the soft and tender fruits by their sharp ovipositor and lay the eggs under fruit tissues and watery fluid oozes from the puncture. Sometimes pseudo-punctures (punctures without eggs) have also been reported on fruit skin, which reduces the market value of the produce. The eggs are laid into unopened flowers and the maggots develop into the taproot, stems and leafstalks (Weems and Heppner 2001).

After hatching, the maggots feed on the pulp of the fruits by making galleries and simultaneously the secondary infection also comes, resulting in rotting of fruits. About 50 per cent of cucurbits are partially or completely damaged by fruit flies in India (Gupta and Verma 1992). These insect pests reduce the quality as well as quantity of cucurbit fruits and adversely affect the growth of plants.

Some of the major pests of cucurbits are the melonfly, pumpkin beetles, whiteflies, and aphids.

**Fruit Fly (*Bactocera cucurbitae* Coquillett).** Order: Diptera Family: Tephritidae

This is the major pest of cucurbits. The damage by maggots results in rotting of young and ripened fruits or drying and shriveling of fruits before maturity. Sometimes even the base of the plant gets attacked and plant start wilting. The incidence is more in wet climate. It is distributed throughout in Asia and Africa (Waterhouse, 1993a). It hosts include Cucurbits, tomato, capsicum, beans, passion fruit, and eggplant. Over 125 hosts have been recorded.

It is a major pest of cucurbits particularly in bitter gourd and musk melon. Among the several species of fruit fly, *B. cucurbitae* is most important due its polyphagous nature and cosmopolitan distribution. The female fly makes a cavity with the help of its sharp ovipositor and inserts the white, cigar-shaped, slightly curved eggs singly or in groups into the flowers and tender fruits. The incubation period is 3-9 days. The freshly hatched maggots bore into the fruit pulp by forming serpentine galleries and contaminating them with its frass and providing entry points for saprophytic fungi and bacteria, which cause the rotting of the fruit. Due to feeding on pulp, there is premature dropping of fruits and make them unfit for consumption. The infestation varies from 50-100 per cent in different cucurbitaceous crops.

Adult flies are medium sized (4-5 mm long) with reddish brown and their wings are hyaline with a dark patch on the outer margins. Hind cross veins thickened with brown and grey spots at the apex. The full-grown maggots come out of the fruits and drop to the ground and pupate in the soil. The total life cycle occupies 14-34 days depending on weather conditions. There are several over lapping generations in a year.

Adults require protein for sexual maturity and in order to develop eggs. Mating is necessary for the production of fertile eggs. Each female is capable of laying over 1,000 eggs. The female inserts her ovipositor directly into the host fruit or vegetable and lays her eggs. There are three larval instars and a prepupal stage. Pupation takes place in the soil beneath the host plant (Waterhouse, 1993a). Males respond to cue lure.

**Damage:** The melon fly is the most important fruit fly pest of vegetable crops, especially of cucurbits. In addition to causing direct damage to fruit, its occurrence in a country leads to quarantine restrictions.

## Management

**Natural enemies:** The parasitoids *Fopius arisanus*, *Diachasmimorpha dausii*, *Psyttalia* (*Opius*) *fletcheri*, *Dirhinus anthracina*, *Spalangia endius*, *Spalangia hirta*, *Tetrastichus giffardianus*, and *Tetrastichus dacicida* were recorded. Predators recorded were *Chelisoches morio*, *Zelus renardii*, *Pheidole megacephala*, *Solenopsis geminata*, *Philonthus turbidus* and *Argiope* spp. Of these, only *P. fletcheri* has given moderately satisfactory control.

**IPM technology for Fruit fly management in Cucurbits**

Installation of used water bottle baited with cue lure (as MAT) saturated wood blocks (ethanol: cuelure: carbaryl in a ratio 8:1:2) @ 25 traps/ha prior to flower initiation.

Bait spray containing malathion 20 ml + 20 litres water + 500g molasses sprayed randomly on 250 plants/ha.

Use of repellent (NSKE 4%) enhanced trapping and luring in bait spots. Mean fruit fly catches were much higher in traps installed with MAT+BAT-Repellent plots (116.43/trap/week) compared to that of (75.82/trap/week) in MAT+BAT treated plot

Application of neem as a repellent increased the catch in para-pheromone traps and enhanced the luring ability of para-pheromone by 52%.

Use resistant varieties.

Soil application of neem cake @ 250 kg/ha immediately after germination and repeat at flowering followed by sprays of neem soap 1% or PNSPE 4% at 10 days interval after flowering.

Crush pumpkin 1 kg and add 100 gm jaggery and 10 ml Malathion and keep in the plot (4-6 places per acre). Adults are attracted to the fermenting pumpkin and lay eggs and get killed. Repeat the process 2-3 times in the cropping season.

Erect cuelure (para pheromone trap) 3 per acre to attract and trap male fruit flies.

Spray Carbaryl 50 WP @ 3 gm/l or Indoxacarb 0.5 ml/l.

Fine tuning of this module with addition of NSKE spray and increased diameter of bottle trap (9cm) along with use of dichlorovos 1% increased the adult trapping/killing in lure.

**Impact:** This technology helps to suppress further population build-up of insect with almost no probability of insecticide residue in and or/on the fruits.

This technology decreases fruit damage by 71.1% in cucurbits. Overall cost of pesticide application is decreased more than 10 times. Benefit is in terms of pesticide free cucurbits, safety to human health and improving quality of environment.

**Table 7.8:** *Yield losses due to fruitfly (Bactrocera cucurbitae) in cucurbitaceous vegetables in India*

| Crop | Yield loss (%) |
|---|---|
| Bitter goud | 60-80 |
| Cucumber | 20-39 |
| Ivy gourd | 63 |
| Musk melon | 76-100 |
| Snake gourd | 63 |
| Sponge gourd | 50 |

**Red Pumpkin beetle (*Aulocophora spp.*)** Order: Coleoptera Family: Chrysomelidae

**Distribution:**

*A. femoralis* – Myanmar and Vietnam

*A. fl avomarginata* – Malaysia, Brunei, and Indonesia

*A. foveicollis* – Africa, Asia, and Europe

*A. frontalis* – Thailand, Laos, Vietnam, and Singapore

*A. lewisii* – Vietnam and Malaysia

*A. similis* – South and Southeast Asia

*A. coff eae* – Malaysia

**Hosts:** Cucurbitaceous crops

**Life cycle:** It is a bright orange red coloured medium size (6-8 mm) beetle with black ventral surface found regularly on cucurbitaceous crops like pumpkin, cucumber, ash guard, bottle gourd, bitter gourd and melons. The grubs live in soil feeding on roots, rootlets and under ground stems and sometimes on fruits in contact with the soil. On emergence, the beetles feed on aerial parts of the plants like cotyledons, flowers and foliage and make holes on them. The damage by beetles is more detrimental during the cotyledonous stage of the plants.

The female beetle lays 250-300 oval eggs, initially yellowish pink in colour and later changes to orange. The eggs are laid singly or in batches of 8-9 in moist soil, near the base of the plant. The incubation period lasts for 5-15 days. Fully grown grubs are 10-12 mm in length. The pupation takes place in earthen chambers in the soil at a depth of 15-25 cm. It completes its life cycle in 25- 38 days.

Adults are known to live 100–200 days. Eggs are yellow and laid singly or in masses of two to fi ve in the soil near the base of the plant. The incubation period ranges from 8–15 days. Newly hatched grubs are pale yellow and turn orange-yellow when full-grown. Grubs bore into roots, and the larval stage lasts for 18–35 days. Pupation takes place in the soil within a cell. Pupae are elongate and pale yellow, and the pupal stage ranges from 4–14 days. The duration of life cycles varies among different species and also depends on host plants and the environment (Figs 2.4.1, 2.4.2, 2.4.3).

**Damage:** Adults cause serious damage by feeding on leaves and flowers. Seedlings may be heavily attacked to the point that they are totally destroyed. Grubs bore into the roots, causing them to swell, be discolored, misshapen, and in some cases, plant growth is retarded or plants are killed.

**Natural enemies:** Little or no information is available on the natural enemies of this pest.

**Management:** Mechanically collect and destroy the pest if incidence is low.

If the pest incidence is very severe, spray Indoxacarb 14.5 SC @ 0.5 ml/l or Cabaryl 50 WP 4g/l or 25 EC @ 2ml/l or Chlorpyriphos 20 EC2.5ml/l.

**Serpentine Leaf Miner (*Liriomyza trifolii* Burgess)**

This is an introduced pest occurring on many cucurbit vegetables. Heavy incidence is noticed in watermelon, pumpkin, cucumber, etc. However, bitter gourd seems to be resistant. A native larval parasitoid, *Hemiptarsenus vericornis* is the major parasitoid on this pest.

Soil application of neem cake @ 250 kg/ha immediately after germination.

Destroy cotyledon leaves with leaf mining at 7 days after germination.

Spray PNSPE @ 4% or neem soap 1% or neem formulation with 10000 ppm or more (2ml/l) after 15 days sowing and repeat after 15 days, if necessary.

If the incidence is high first remove all severely infected leaves and destroy. Then mix neem soap 5 gm and hostothion 1 ml/l and spray. After one week, spray neem soap 1% or PNSPE or neem formulation with 10000 ppm or more (2ml/l).

Never spray the same insecticide repeatedly.

**Cucumber moth *Palpita (Diaphania) indica* (Saunder) Order: Lepidoptera Family: Pyralidae**

**Distribution:** Asia, Africa, Australia, and Pacific

**Hosts:** Cucurbitaceous plants

**Life cycle:** Adult moths have white wings with broad black bands along the anterior margin of the forewings and on the outer margin of fore- and hind wings. Eggs are laid singly or in small clusters on the lower surface of leaves. The incubation period for eggs is about 4–5 days. The caterpillars are green with white median stripes and they feed on the underside of leaves. There are five instars, and the larval period lasts 8–19 days. Pupation takes place inside the rolled leaves, and the pupal period is about 6–7 days.

**Damage:** Caterpillars mostly feed on leaves and sometimes cause defoliation. Rarely do they attack fruits.

**Natural enemies:** The parasitoids, *Apanteles machaeralis, Apanteles taragamae, Elasmus indicus*, *Dolichogenidea stantoni*, *Goniozus sensorius*, *Phanerotoma hindecasisella*, and a nucleopolyhedrovirus (NPV), were reported attacking larvae.

**Melon thrips *Thrips palmi* Karny Order: Thysanoptera Family: Thripidae**

**Distribution:** Asia, Pacific, Caribbean, Central and South America, Florida (Capinera, 2001)

**Host plants:** Tobacco, cott on, cucurbits, crucifers, eggplant, bean, soybean, peanut, avocado, mango

**Life cycle:** Adults are pale yellow, females about 1.16 mm in length and the males 0.91 mm. Eggs are laid in young leaves or flower buds, and they hatch in about 4 days. There are two nymphal stages and in these stages they feed actively along the leaf midribs and veins. After the second instar, they undergo prepupal and pupal stages in the soil. The time taken from egg-laying to adult emergence is about 11 days in the tropics.

**Damage:** Feeding by adults and nymphal stages on plants causes flower bud drop, stunting of terminal shoots, and curled and scarred fruits. The most serious damage caused by this thrips is transmission of tospoviruses. *Peanut bud necrosis virus* on peanuts, tomato, potato, soybean, and sunflower; *Capsicum chlorosis virus* on peppers, *Iris yellow spot virus* on tomato and peanut; *Melon yellow spot virus*; *Calla lily chlorotic spot virus*; *Sunfl ower necrosis virus*, *Watermelon silver mott le virus*; and *Watermelon bud necrosis virus* are all transmitted by *T. palmi*. It is the predominant thrips vector species in tropical and subtropical Asia (Pappu *et al.*, 2009).

**Natural enemies:** The predatory bugs, *Orius mexidentex* and *Carayonocoris indicus,* were reported to feed on *T. palmi* in India. Predators recorded in Hawaii were *Franklinothrips vespiformis*, minute pirate bug, *Orius insidiosus*, *Curinus coeruleus*, *Rhinacoa forticornis*, *Paratriphleps laevisculus*, and *Ceranisus menes*. The predatory bug, *O. mexidentex,* was reported from Thailand (Kumar and Ananthakrishnan, 1984). *Orius strigicollis* and *Haplothrips brevitubus* were recorded from Japan (Kakimoto *et al.*, 2008). The fungi, *Beauveria bassiana*, *Neozygites parvispora*, *Verticillium lecanii*, and *Hirsutella* sp. are known to att ack *Thrips tabaci* (Capinera, 2001) and may also infect *T. palmi*.

**Leaf-footed bug *Leptoglossus gonagra* (F.) Hemiptera Coreidae**

**Distribution:** Southeast Asia, Pacifi c Islands

**Hosts:** Polyphagous, mostly on cucurbits and morning glory plants.

**Life cycle:** The life cycle is completed in about 40 days on fruit crops and cucurbitaceae. This pest is commonly found on bitt er gourd (*Momordica charantia*). Eggs are laid in strings of varying numbers. The egg stage lasts 4–8 days. There are five nymphal stages that last about 30 days. Adults are 17–20 mm long (Fig. 2.4.6).

**Damage:** High populations have been observed in cucumbers in Indonesia. Feeding on the young fruit causes distortion as fruit matures.

**Natural enemies:** Little is known about natural enemies of this pest.

Other pests of regional importance are the stem borer, *Melittia eurytion* (Westwood) (Lepidoptera: Sesiidae), the snake gourd semilooper, *Anadevidia* (*Plusia) peponis* (Fabricius) (Lep.: Noctuidae), and the bottle gourd plume moth, *Sphenarches caffer* Zeller (Lep.: Ptreophoridae)

## Cucumber moth or leaf eating caterpillar (*Diaphania indica* (Saunders))

It is an occasional and sporadic pest of cucumber and also becoming serious on bitter gourd and gherkins. After hatching, caterpillars feed chlorophyllous portion of the leaves by webbing them together. Sometime the flowers and young fruits are also infested. The larvae make characteristic holes on the fruits and feed inside it. The bored fruits become unfit for human consumption.

The adults are medium sized moths with whitish transparent forewings with big brown patches on the margin. Females have an anal tuft of orange coloured hairs at the caudal end. The caterpillars are elongated, bright green and bear a pair of narrow longitudinal white streaks along the mid-dorsal line. A female can lay up to 350 eggs, mostly singly or sometimes in groups on the lower surface of leaves. The egg, larval and pupal periods last for 3- 6, 9-15 and 5-12 days, respectively. The pupation takes place in the silken cocoons formed in the leaf folds. Life cycle is completed in three to five weeks.

## Red Spider Mite (*Tetranychus neocaledonicus* Andre)

This pest is serious on cucurbits during warmer climate. Colonies of mites with in silk web can be observed on ventral leaf surface when incidence is high.

## Management

Spray neem or pongamia soap at 1% on lower surface thoroughly. Alternately, spray Dimethoate 30 EC @ 2ml/l or Ethion 50 EC @ 1ml/l or Wettable Sulphur 80 WP @ 3g/l.

## Thrips (*Thrips palmi Karny*)

Both nymphs and adults feed on the young shoot tips and floral parts and is suspected vector of Tospo virus. They are serious during summer months.

## Management

Soil application of neem cake (once immediately after germination and again at flowering followed by NSPE @ 4% and neem soap 1% alternately at 10-15 days interval.

Spray any systemic insecticides like Acephate 75 SP @ lg/l or Dimethoate 30 EC @ 2ml/l.

**Leaf Eating Caterpillar (*Dipahania (=Margaronia) indica* Saund)**

Long shining caterpillars feed on leaves and fruits.

Apply neem cake to soil immediately after germination.

Spray any contact insecticides like Carbaryl 50 WP @

3g/l. Neem or pongamia soap @ 0.75% also effectively manages this pest.

Soil application of neem cake (once immediately after germination and again at flowering) followed by NSPE @ 4% and neem soap 1% alternately at 10-15 days interval.

Spray Carbaryl 50 WP @ 3g/l or Indoxacarb 0.5 ml/l.

## Root-knot Nematodes (*Meloidogyne incognita)*

The root-knot nematodes cause root galls from the initial stages of the crop growth. The larvae feed on the roots, which show typical galls, and later the entire root system shows heavy galling. The foliage becomes light yellowish; the plants become stunted and results in flower and fruit drop. In spite of irrigation the plants appear sick and drooping during daytime.

## Management

Seed treatment with bio-pesticide *Pseudomonas fluorescens* @ 10g/kg seed has proved effective.

Apply Carbofuran 3 G @ 1kg ai/ha at sowing and repeat after 45 days.

Apply 2 tons of FYM enriched with *Pochonia chlamydosporia* and *Paecilomyces lilacinus* per acre before sowing, along with 100-200 kg of neem or pongamia cake.

## Population build up studies on different insect pests of cucurbits

**Population build up of red pumpkin beetle, *Aulacophora foveicollis* (Lucas)**

According to Pavlakos (1943), Sinha and Krishna (1969) and Al Ali *et al.* (1982) *Aulacophora foveicollis* larvae and adults were most active at temperature between 27 to 32oC, but they were unable to survive at temperature exceeding 35.2oC. Shinde and Purohit (1978) studied the population build up of red pumpkin beetle in Madhya Pradesh and reported highest number (75 beetles/5 sweepings/week) in April and lowest number (30 beetles/5 sweepings/week) in October with an average sex ratio of male to female as

2:1. Pareek and Kavadia (1986) studied the seasonal incidence of this pest at two locations in Rajasthan which revealed that the beetle appeared first in the middle of March and recorded two peaks in April-May and August-September both at Udaipur (8.33 beetles/5 plants) and Jobner (6.33 and 7.00 beetles/5 plants). The population started declining in last week of October and disappeared completely during last week of November.

Chaudry and Alikhan (1990) studied the effect of temperature and relative humidity on the development and fecundity of *A. foveicollis* and reported that highest number of viable eggs per female and shortest duration of 30.40 ± 2.12 days from egg to pupal development was at 30oC and 70 per cent relative humidity and lowest number of viable eggs at 35oC and 90 per cent relative humidity. In an investigation on the fluctuation in population of red pumpkin beetle conducted by Roy and Pande (1991) in Tripura, it was revealed that the pest remained active throughout the year and its maximum (104 adults/30 leaves) and minimum (34 adults/30 leaves) populations were recorded during December and August, respectively. They further reported negative relationship between the insect population and the two abiotic factors as temperature and rainfall. Borah (1999) studied the seasonality and varietal preference of red pumpkin beetle on cucumber in Assam and recorded highest number of beetles in rainy season (June) in all the three varieties with 3.6 – 4.2 beetles/plant and 39.2 – 46.6 per cent plant damage fallowed by summer crop with 2.8 beetles/plant and 33.6 per cent plant damage and winter season with 2.1 beetles/plant and 21.1 per cent plant damage. According to Rajak (2000) the overwintering beetles become active during 7th standard metrological week (February) and recorded maximum population (28.6 beetles/5 plants) and minimum population (1.66 beetles/5 plants) during 18th (April) and 7th (February) standard metrological weeks at an average temperature of 28.8 oC and 19.0oC, respectively and population started declining with gradual increase in temperature. Among the climatic factors, temperature has significant effect and relative humidity non-significant effect on pest population. Johri and Johri (2003)a studied the seasonal biology, abundance and intensity of attack of red pumpkin beetle at Kanpur and revealed that the incubation, larval and pupal periods were lowest during pre-monsoon (7.11, 11.00 and 7.94 days) followed by monsoon (10.00, 13.00 and 9.26 days) and post-monsoon (15.00, 14.74 and 15.02 days). They observed beetle incidence was more during the period from March to September ranging from 27.70 to 47.49 per cent and the lowest as 3.92 per cent in February. The influence of weather factors *viz.*, temperature, humidity and rainfall had no significant effect on plant infestation by this pest.

## Hadda beetle, *Henosepilachna vigintioctopunctata* (Fabricius)

In Kullu Valley, Himachal Pradesh the pest was first noticed on potato in early June and increased rapidly till the third week of July. The severe infestation was coinciding with high humidity indicated its favourability for population build up of this pest (Hameed and Adlakha 1973). According to Iftekar and Khan (1980), the pest remained active from July to November on brinjal and was more prevalent at temperature of 27-29oC and relative humidity more than 80 per cent during the month of August. Singh and Kavadia (1989) also observed maximum infestation of epilachna beetle during August to October on July planted crop. Rajagopal and Trivedi (1989) reported that the peak period of pest varied with region, but it is generally in July-August. Ramzan *et al.* (1990) studied the comparative development and seasonal abundance of epilachna beetle on some solanaceous host plants and reported that egg masses were noticed from April to September, being maximum in July. The grubs commenced build up from April at temperature 25.7oC and

50 per cent relative humidity, thereafter sharply declined reaching to zero in October. The adults were found from April to November with their peak period in July on potato. In Kanpur, Uttar Pradesh, the incidence of epilachna beetle was observed from 3rd week of August to middle of October with its maximum population of 0.22 beetle per leaf in the middle of September when the average temperature and relative humidity were 28 oC and 80 per cent, respectively (Mall *et al.* 1992). Population increase of epilachna beetle was suppressed in the month of normal rainfall (≥ 300 mm) in Indonesia but unable to suppress when rainfall dropped to 50 per cent of the long term average (Inoue *et al.* 1993). Mukhopadhyay and Mandal (1994) secreened brinjal cultivars for resistance to insect pests and reported that the population of epilachna beetle was as less as 0.09 on tolerant cultivars as compared to 2.84 on highly susceptible, 2.17 to 2.03 on susceptible and 0.43 to 1.77 beetles on moderately susceptible cultivars. Raghuraman and Veeravel (1999) revealed that the epilachna beetle population was highest in February (24.2 insects/plant) and March (27.4 insects/plant). In West Bengal, the beetle was found active between April to mid-October on brinjal with highest population of 8.14 beetles per plant during mid of September and showed significant positive correlation with average temperature, relative humidity and rainfall (Ghosh and Senapati 2001). Singh and Singh (2002) reported that the epilachna beetle appeared in last week of June and remained throughout the cropping season on brinjal in Barapani, Meghalaya. According to Banerjee *et al.* (2003) the incidence of both epilachna beetle and grubs on bitter gourd in West Bengal was maximum during 1st and 2nd week of September with minimum, maximum temperature and relative humidity varying from 25 to 26oC, 31.5 to 32.7oC and 76.0 to 79.0 and 97 to 98 per cent, respectively. In Tamil Nadu, the peak incidence was recorded on brinjal during March-April (Muthukumar and Kalyanasundaram 2003). Naik *et al.* (2008) observed the initial occurrence of epilachna grubs during 3rd week of December at 22 days after transplanting of brinjal, which gradually reached the peak (17.2 grubs/plant) during 19th February and thereafter declined gradually with increase in temperature. Haseeb *et al.* (2009) noticed initial incidence of hadda beetle on brinjal during 3rd week of January and peak population during 3rd week of February. They also observed negative correlation of beetle incidence with maximum and minimum temperatures, but positive correlation with minimum and maximum humidity.

## Fruit flies, *Bactrocera* spp

A study of the effect of temperature on the activity and rate of development of *B. cucurbitae* (Coquillett) conducted by Keck (1951) and showed that the female did not deposit eggs at 55oF and no flies lived long enough to lay eggs at 100oF. The maximum number of eggs as 1813 deposited by a single female during its life period of 150 days at 80oF. According to Narayanan and Batra (1960), in severe cold the *B. cucurbitae* hide under plant debris, bushes and trees and become active in March when temperature warms up with the peak during July-August causing heavy damage to cucurbits. Nath (1966) reported that the increased relative humidity was associated with increase in level of fruit fly infestation in bottle gourd and sponge gourd, but ridge gourd was found to be less infested at high temperature and relative humidity. Bhatia and Mahto (1970) studied the effect of various temperatures under controlled conditions on the development of *B. cucurbitae* and observed that the duration of the life cycle was 36.30, 23.60, 12.61, 11.17 and 12.50 days at 15, 20, 25, 27.5 and 30oC, respectively. A temperature of 27.5 oC was found to be most favourable for the rapid development and the pupae were more sensitive to high temperature than the eggs and larvae. Lee (1976) reported that the *B. dorsalis* (Hendel) appeared early with increase in temperature and showed no activity at temperature below 14oC. The population size positively correlated with increased temperature. In a study conducted by Liu (1983) in Taiwan, temperature was found as a definite factor affecting trap catches of *B. dorsalis* and the population was observed to be lowest in late January to March which increased in April. In Hessaraghatta, Bangalore, Shukla and Prasad (1985) studied the population fluctuations of *B. dorsalis* and observed three distinct population peaks in March-April, May-June and September-October, which coincided with ripening of guava and mango fruits. Among abiotic factors, trap catches were significantly and positively correlated with maximum, minimum temperatures and maximum relative humidity. Trap catches were significantly and negatively correlated with minimum relative humidity. Shivarkar and Dumbre (1985) reported that temperature and relative humidity were important factors influencing the melon fly infestation which increased with the rise in temperature and relative humidity and *vice-versa*. Temperature was found to be the more predominant factor than relative humidity in melon fly infestation as they recorded lowest infestation from the second week of December to the first week of February due to low temperature.

An experiment carried out by Pareek and Kavadia (1986) to study the seasonal incidence of insect pests on cucurbits at Udaipur and Jonber, Rajasthan recoded that *B. cucurbitae* infestation started in the second week of April with the peak period of infestation during May-June on long melon, musk melon, bitter gourd and round gourd at both the places and high temperature and high relative humidity appeared to be favorable for the pest. In summer, the combination of high environmental temperature (above 37oC) and high humidity near plants due to frequent irrigations favored the fruit fly development, whereas during the *kharif* season, temperature (above 30oC) and relative humidity more than 70 per cent were found congenial for increase in the fly infestation. In Taiwan, Su (1986) also observed that the temperature was the most important factor affecting population size of the tephritids followed by rainfall, humidity and duration of sunshine. Vargas *et al.* (1990) carried out studies on population dynamics of the fruit fly species *viz, B. dorsalis* and *B. cucurbitae* in Hawaii for two years and observed highest *B. dorsalis* captures in December and lowest in April during first year, whereas during second year highest captures occurred in August and lowest in February. The highest number of *B. cucurbitae* was recorded in June and lowest in April during first year whereas highest captures occurred in October and lowest in February during the second year. A maximum trapping of *B. cucurbitae* was noticed during the first fortnight of October and increase in population was positively and significantly correlated with increased temperature, humidity and rainfall (Pawar *et al.* 1991). Mann (1990) reported that fruit flies passed through many overlapping generations and overwinters as larva, pupa or adult and fruit fly infestation was low during April-May, which increased rapidly from July-August on cucurbits and by the end of October causing heavy fruit infestation. In Pakistan (Faisalabad) *B. cucurbitae* activity started in March, reached to its maximum in August and progressively declined thereafter negligible during winter months. The correlation between male counts and the maximum, minimum and average temperature was positive, but negatively correlated with relative humidity (Inayatullah *et al.* 1991). According to Thakur *et al.* (1992) the mean temperature between 30 and 31oC was the most favourable for the attack of *B. cucurbitae.*

In Pakistan, Inayatullah *et al.* (1991) developed regression models to predict the fruit infestation by melon fruit fly on squash and bitter gourd based on the number of males trapped per pheromone trap and number of pupue per square feet of soil. The model indicated that at the density of 7 males per trap per day, 10 per cent fruit infestation would occur which is general economic threshold level of melon fruit fly. Similarly another model indicated that at the density of 0.74 pupa per square feet of soil, 10 per cent fruit infestation would occur. They concluded these models could be used to calculate the per cent fruit infestation at any densities of males or pupa. Population fluctuations of *B. cucurbitae* in melons and gourds were investigated by Lee *et al.* (1992) and recorded highest population densities in July and August, which decreased in September, reaching to its lowest during December to May. In south Gujarat, Kumar *et al.* (1997) studied the seasonal cyclicity of *B. correcta* (Bezzi) in mango and sapota, the major activity of the pest observed from March to June coinciding with the fruiting period of the crop. In mango, population remained considerably higher during March to June exhibiting highest peak in second fortnight of May (483.50 flies/trap), thereafter it declined gradually. In case of sapota, population started increasing considerably from first fortnight of April, attained peak in first fortnight of June (580.00 flies/trap) and afterwards showed decreasing trend in later period of the year. The population variations were positively correlated with minimum, maximum and average temperatures while other weather factors did not show any significant impact on population variations. Agarwal and Kumar (1999) studied the effect of weather factors on population dynamics of *B. zonata* (Saunders) in North Bihar and observed maximum population during the 3rd week of June (357.00 flies/trap) which declined significantly from 4th week of July reaching to its lowest number (914.3 flies/ trap) in last week of August. Fly population showed positive correlation with maximum, minimum temperature, rainfall and negative correlation with relative humidity.

According to Jalaluddin *et al.* (2001), abiotic factors and fruit ripening played an important role in regulating the *B. correcta* population. They recorded fly population (42 flies/trap) in September, gradually declined (33 flies/trap) in December reached to zero in January and then escalated to peak (803 flies/ trap) in August. The weekly trap catches were positively and significantly correlated with maximum, minimum temperatures, relative humidity and rainfall. It was concluded that adult activity was governed by combined influence of abiotic factors. Sarada *et al.* (2001) observed gradual increase in population of *Bactrocera* spp. on mango from 1st week of February (34 flies/trap) to 5th week of June (235 flies/trap) and reaching to peak during May-June, which coincided with mango fruit maturity period. The population build up was positively correlated with minimum temperature and rainfall whereas it was negative with relative humidity. In Northern Karnataka, species diversity and population dynamics of fruit flies on mango were studied by Babu and Viraktamath (2003)a who recorded four species *viz*., *B. zonata, B. correcta, B. dorsalis* and *B. cucurbitae*, among which *B. dorsalis* was found predominant. The peak catches were observed during 21st, 23rd and 46th standard week. *B. zonata* and *B. correcta* showed significant correlation whereas, *B. dorsalis* and *B. cucurbitae* had non-significant correlation with weather parameters. Another study was carried out by the same workers (Babu and Viraktamath 2003)b and they found occurrence of the same four species in cucurbit field with dominance of *B. dorsalis* and *B. cucurbitae*. Irrespective of the species, the maximum catch of flies was found during the 1st fortnight of November. In Dera Ismail Khan (Pakistan), Latif and Abdullah (2005) studied the population dynamics of *Bactrocera* spp. in mango orchard and reported that monthly mean captured fruit flies showed fluctuation over the year ranging from minimum (6 flies) during January to maximum (397 flies) during September and lowest capture rates of flies from January to March. In Bihar, *B. cucurbitae* incidence on bitter gourd commenced between 2-6 April (14th standard week) resulted in 24.6 per cent fruit infestation and the peak activity occurred during 26th standard week causing 46.3 per cent infestation, exhibited significant positive correlation with minimum temperature and relative humidity, but non-significant correlation with the maximum temperature (Mandal *et al.* 2006).

Ravikumar and Viraktamath (2006) studied the influence of weather factors on population dynamics of fruit flies for two years at Dharwad, Karnataka and reported that irrespective of the species three peaks were observed during 27th (174.33 flies/trap), 45th (128.66 flies/trap), 48th (121.17 flies/trap) standard week in 2004 and 21st (106.99 flies/trap) and 22nd standard week (114.66 flies/trap) in 2005. *B. correcta* and *B. zonata* had significant positive correlation with temperature, whereas *B. dorsalis* had significant negative correlation with relative humidity and rainfall. Three species of fruit flies namely *B. dorsalis, B. correcta* and *B. zonata* were recorded in methyl eugenol traps in mango orchard (Ranjitha and Viraktamath 2006). Irrespective of the species four peak catches were observed during 30th (120.89 flies/trap), 43th (280.28 flies/trap), 45th (361.69 flies/trap) and 52th (158.41 flies/trap) standard weeks in 2003. In 2004 two peaks were recorded during 3rd (135.38 flies/trap) and 8th (70.84 flies/trap) standard week. All weather parameters together influenced the trap catches to the tune of 51.4 per cent. Hui and Jianghong (2007) studied the population dynamics of *B. dorsalis* in Xishuangbanna, Yunnan Province, China and reported that weather parameters and host fruits were major factors comprehensively influencing the fly population fluctuation. They have also observed significant positive correlation of monthly trap captures with rainfall, mean temperature and mean maximum temperature. Hasyim *et al.* (2008) observed the effect of abiotic factors on build up of *B. tau* (Walker) and reported that the number of flies trapped with cuelure had positive and highly significant correlation with rainfall and significant correlation with temperature An investigation was carried out by Laskar and Chatterjee (2010) on the effect of meteorological factors on the population dynamics of melon fruit fly, *B. cucurbitae* in the foot hills of Himalaya and found significant positive correlation of fly incidence with minimum, maximum temperatures, minimum humidity and rainfall, but negative correlation with maximum humidity and sunshine hours.

**Use of pheromones/attractants**

Within the dipteran family Tephritidae, the genus *Bactrocera* is composed of atleast 440 species distributed primarily in tropical Asia, the South Pacific and Australia (White and Elson-Harris 1992). Males of numerous *Bactrocera* and *Dacus* species are known to be highly attracted to either methyl eugenol or cuelure (Metcalf and Metcaclf 1992). In fact, atleast 90 per cent species are strongly attracted to either of these attractants (Hardy 1979). Pheromone traps are important sampling means for early detection and monitoring of the fruit flies that have become an integrated component of integrated pest management. Cuelure and ENT 31812 lures were placed on the ground and at 2 and 5 feet above the ground to evaluate the effect on the response of *B. cucurbitae*. Both the attractants were found atleast as attractive at ground level as at higher levels and cuelure was found more attractive than ENT 31812 (Hart *et al.* 1967).

Sixty compounds related to methyl eugenol were evaluated for their attractiveness against oriental fruit fly, *B. dorsalis* and melon fruit fly, *B. cucurbitae* by Lee and Chen (1977) who reported that methyl isoeugenol, veratric acid, methyl eugenol and eugenol to be most effective attractants against *B. dorsalis* among the tested compounds. However, none of the tested chemicals was found to be significantly attractive against *B. cucurbitae*. According to Metcalf *et al.* (1983), *B. cucurbitae* was extreamly responsive to cuelure, but nonresponsive to methyl eugenol, whereas, *B. dorsalis* extremely responsive to methyl eugenol, but non-responsive to cuelure. In an experiment in melon field, commercially produced attractants Flycide C (80% cuelure content), Eugelure 20 (20%), Eugleure DB (8%), cuelure (80%) + naled cuelure (80%) + diazinon and cuelure (90%) + naled were tested against *B. cucurbitae* showed no significant difference in captured flies (Iwaizumi *et al.* 1991). A study carried out by Wong *et al.* (1991) on age related response of laboratory and wild adults of melon fly, *B. cucurbitae* to cuelure revealed that response of males increased with increase in age and corresponded with sexual maturity for each strain. They failed to eradicate the pest with male annihilation programmes against *B. cucurbitae,* which might be because of the fact that only older males, which may have already mated with gravid females, responded to cuelure. Pawar *et al.* (1991) used cuelure (sex attractant) and tephritlure (food attractant) for the monitoring of *B. cucurbitae* and found cuelure traps more efficient in trapping fruit flies as compared to tephritlure. Gazit *et al.* (1998) studied the four trap types viz., IP-McPhail trap, Frutect trap, Cylinderical trap and Ga' aton trap with three female attractant baits *viz.*, naziman, a proprietary liquid protein and a three component based synthetic attractant compound of ammonium acetate, putrescine and trimethylamine for Mediterranean fruit fly, *Ceratitis capitata* (Wiedemann). Their results ranked the trap and attractant performance as IP-McPhail trap baited with synthetic attractant > Frutect trap baited with proprietary lure > Cylinderical trap baited with synthetic attractant > IP-McPhail trap baited with naziman and Ga' aton trap baited either with synthetic attractant or naziman.

According to Vargas *et al.* (2000) methyl eugenol and cuelure were highly attractive kairomone lures to oriental fruit fly, *B. dorsalis* and melon fly, *B. cucurbitae*, respectively. They used these lures at different concentrations and found significantly highest *B. dorsalis* captures in 100 per cent methyl eugenol traps than 25, 50 and 75 per cent. However, *B. cucurbitae* captures with 25, 50 and 75 per cent cuelure were not significantly different. Babu and Viraktamath (2003)a reported that highest number of *B. dorsalis* were trapped in methyl eugenol traps followed by *B. zonata* and *B. correcta* whereas, lowest number of *B. cucurbitae* was also trapped in a mango orchard. Similarly same four species of fruit flies were recorded in methyl eugenol traps in cucurbit field by Babu and Viraktamath (2003 b). The most predominant fruit fly species was *B. dorsalis* (48%) followed by *B. cucurbitae* (21%), *B. correcta* (16%) and *B. zonata* (15%). Thomas *et al.* (2005) evaluated two parapheromones *viz.*, cuelure and methyl eugenol for their attraction to *B. cucurbitae* in a bitter gourd field and revealed that melon flies were attracted to only cuelure traps. Response of fruit flies to the traps which differed in size, shape and colour containing methyl eugenol were evaluated in mango orchard by Ranjitha and Viraktamath (2005) and observed that fruit flies showed greater response to spheres than bottles and cylinders. However, response to different colours varied among different species. Verghese *et al.* (2005) studied the comparative attractiveness of three indigenous lures/baits with three established attractants in fruit flies and reported that meyhyl eugenol attracted highest number of flies (18.25 flies/day/trap) followed by cuelure (13.5 flies/day/trap) and tulsi (5.88 flies/day/trap) whereas, flies attracted to banana, jaggery and protein hydrolysate were negligible. The number of species attracted was also higher in methyl eugenol, which attracted four species *viz.*, *B. dorsalis, B. correcta B. zonata* and *B. verbascifoliae* (Drew & Hancock) followed by ocimum with two species *viz.*, *B. dorsalis, B. correcta.* However, cuelure attracted only *B. cucurbitae*. Three species of fruit flies namely, *B. dorsalis, B. correcta* and *B. zonata* were recorded in methyl eugenol traps in guava and mango orchard by (Ranjitha and Viraktamath 2006; Ravikumar and Viraktamath 2006).

Studies on the ability of different plant extracts to attract male fruit flies carried out by Hasyim *et al.* (2007) indicated that the major compound camphor present in *Elsholtzia pubescens* (Bith) was atleast as efficient as the standard cuelure in trapping males of *B. tau* in passion fruit orchard. Singh *et al.* (2007) tested sex attractant methyl eugenol, cuelure and food attractant protein hydrolysate for attraction to fruit flies and reported that five fliy species *viz*., *B. zonata*, *B. affinis* (Hardy), *B. dorsalis, B. correcta* and *B. diversa* (Coquillett) were attracted to methyl eugenol traps and two species *viz*., *B. cucurbitae* and *B. nigrotibialis* (Perkins) to cuelure traps and two species namely, *B. cucurbitae* and *B. zonata* to protein hydrolysate traps. Vargas *et al.* (2009) evaluated various traps with methyl eugenol and cuelure for capturing fruit flies and observed that *B. dorsalis* was captured in methyl eugenol traps and *B. cucurbitae* in cuelure traps.

**Screening of cucumber varieties/genotypes for resistance against red pumpkin beetle and melon fruit flies.**

## Red pumpkin beetle, *A. foveicollis*

Benepal and Hall (1967) studied the biochemical composition of plants of *Cucurbita foetidissima* HBK. and *Cucumis melo* L. in relation to cucumber beetle feeding and reported that cucurbitacins strongly acted as attractant to field populations of spotted cucumber beetle, *Diabrotica undecimpunctata howardi* (Barber). They also observed non-consistent trend in the total free amino acids among the comparatively resistant and susceptible plants. Resistance in cucumber to three species of cucumber beetles was studied by Da Costa and Jones (1971) and found cucurbitacin B to be the most attractant to stripped cucumber beetle, *Acalymma vittata* (Fabricius), spotted cucumber beetle, *D. undecimpunctata* and banded cucumber beetle, *D. balteata* (LeConte). Vashistha and Chaudhury (1971) studied the effect of cucurbitacin on susceptibility of cucurbits and reported that higher cucurbitacin contents were associated with more susceptibility to red pumpkin beetle. Pal *et al.* (1978) evaluated 287 indigenous and exotic pumpkin germplsam for resistance to red pumpkin beetle and observed that although no entry was immune, yet rate of damage varied from 1.0 to 5.0. Low cucurbitacin content of the cotyledonary leaves was found to impart resistance to this pest and the two lines/collection numbers 596-2 and 613 contained low cucurbitacin content as 0.005 and 0.010 per cent, respectively showed less susceptibility. Grewal and Sandhu (1983) carried out laboratory studies on orientation and feeding responses of beetle to cucurbits and revealed that the beetles were not attracted to bitter gourd and wild melon. Summer squash and long melon were found more preferred and sponge gourd and pumpkin least preferred by this pest.

Sandhu and Grewal (1985) tested the cucurbits for infestation by red pumpkin beetle and reported that cucurbits exhibited higher injury ratings under multiple choice test except for the pumpkin which showed higher injury under no choice test, whereas musk melon variety Bangan Muktsar and summer squash variety Australian Green showed minor injury under no choice test. Among cucurbits, bitter gourd was found highly resistant; cucumber, musk melon and water melon moderately resistant; round melon and wild melon susceptible to this pest (Mehta and Sandhu 1989). According to Roy and Pande (1991), red pumpkin was the most preferred and sponge gourd the least out of seven cucurbits offered red pumpkin beetle. In a study on influence of cucurbitacins on the feeding activity of red pumpkin beetle under laboratory conditions, Mehta and Sandhu (1992) observed that the application of cucurbitacins on comparatively non-preferred cucurbits induced the preference for feeding by this pest and increased the leaf damage.

It was implicated that the cucurbitacins act as feeding stimulant and may play a major role in host acceptance. Dhillon (1993) also reported that purified cucurbitacin B and E acts as a feeding stimulant, but there was no correlation between the quantity of total phenols, free amino acids and reducing sugars and resistance. Seventeen musk melon varieties were evaluated for resistance to red pumpkin beetle infestation by Pareek and Kavadia (1993) at Udaipur (semi-humid) and Jobner (semi-arid) and revealed that none of the variety showed resistance, but showed significant variations. Among the varieties, Hales Best Jumbo, Jaune Canari, Faradin, Amco Sweet and Honey Dew Golden showed lower susceptibility. Sharma *et al.* (1999) carried out studies on host preference by this pest and observed highest plant damage in musk melon (15.32%) followed by bottle gourd (7.11), long melon (6.1), and ridge gourd (3.10), whereas bitter gourd was found totally free from any damage by the beetle. Borah (1999) evaluated three varieties for resistance to red pumpkin beetle and observed maximum yield in AAUC-1. In Kanpur, UP host preference of red pumpkin beetle was studied by Singh *et al.* (2000) who observed the maximum beetle intensity (0.49) was on musk melon followed by on round gourd (0.44), on cucumber, water melon and long melon 0.40, but bitter gourd was found free from infestation. Among eleven cucurbit vegetables, bitter gourd was not preferred and musk melon most preferred food by the beetle (Rajak 2001). However, Johri and Johri (2003 b) b reported highest preference on pumpkin followed by bottle gourd, cucumber and musk melon and lowest preference on ridge gourd, sponge gourd and bitter gourd.

## Melon fruit flies, *Bactrocera* spp.

Many studies have been conducted on this important aspect against melon fruit flies. Although none of the variety/genotype has been identified/developed which can be claimed to have true resistance, yet the studies have been conducted to identify varieties/cultivars which are relatively resistant to this noxious pest. Nath (1966 b) evaluated thirty cultivars of bottle gourd, sponge gourd and ridge gourd and observed that although no entry showed true resistance, yet exhibited varied degree of susceptibility. The recommended sponge gourd variety Pusa Chikni showed high susceptibility, whereas the ridge gourd variety Pusa Nasdar exhibited low susceptibility. Eighty two lines of *Cucurbita maxima* Duch. and *C. moschata* Duch. were screened for resistance to *B. cucurbitae*, the line IHR-35, 40, 79-2, 83 and 86 were found to have high resistance combined with high yield (Nath *et al.* 1976). Singh *et al.* (1977) screened various cultivars of bitter gourd for resistance to melon fly and found that all the cultivars susceptible. The per cent fruit infestation ranged from 29.4 to 48.7 with lowest damage in BG-12 and highest in BG-9 and BG-11 cultivars. Eleven cucurbit crops were screened for their relative susceptibility to the melon fruit fly which revealed highest percentage in snap melon (57% damaged fruits) and lowest in summer squash. The order of susceptibility of remaining cucurbits was bitter gourd, mush melon, water melon, citron smooth gourd, ridge gourd, bottle gourd, squash melon and pumpkin (Gupta and Verma 1978). Khandelwal and Nath (1981) screened ninety four genotypes of water melon and found J18-1 and J56-1 resistant to fruit fly. Fifty musk melon collections from different states of India and abroad screened against this pest for three consecutive years revealed that the insect damage up to 20 per cent in the first year. During second year, the collections 2, 9, 11, 41, 42, 43 and 44 retained high level of resistance, however in third year only No. 19 from this group showed high level of resistance while No. 41 and 44 remained in the resistant class (Pal *et al.* 1983). Shivarkar and Dumber (1985) screened ten cultivars of water melon against melon fly infestation and reported that none of the varieties was resistant. In Ludhiana (Punjab) twenty eight genotypes of bitter gourd were evaluated for their resistance to fruit fly and none of the genotype was free from attack of the fruit fly. However, six cultivars viz. AC-23, (10.6% infestation), ACC-23-5 (11.2%), ACC-25 (12.6%), ACC-28 (11.8%) and ACC-33 (12.9%) showed moderate resistance. Two cultivars, ACC-16-3 and Kalyanpur Sona were found highly susceptible (>20% damage) (Thakur *et al.* 1992). Later Thakur *et al.* (1994) screened ten cultivars of bitter gourd and recorded lowest fruit infestation in C-96 (9.9%) being significantly lower than other cultivars except NDBTI (10.5%) in comparison to highest in Kalyanpur Sona (17.0%). Pareek and Kavadia (1994 a) studied the relative preference of melon fruit fly on different cucurbits and reported that musk melon, round melon, bitter gourd and long melon were most preferred hosts which showed 83.12-84.71, 76.62-85.50, 73.46-78.77 and 67.41-75.95 per cent infestation, respectively. Whereas, ridge gourd, sponge gourd, cucumber, pumpkin and bottle gourd were found least preferred which showed lowest infestation 2.83-7.95, 7.82-13.41, 9.71-14.70, 11.88-23.83 and 13.08-17.52 per cent, respectively.

Pareek and Kavadia (1994 b) carried out field studies at Udaipur, Rajasthan to evaluate the varieties of musk melon for their resistance to melon fruit fly and reported that none of the varieties was found free from attack of the pest. However significant variation was observed in the degree of infestation among the varieties. The maximum damage was recorded on Durgapura Madhu (83.12-84.71%) followed by Punjab Hybrid (75.74-80.07%). The varieties Hara Gola (73.48-74.72%), Pusa Sharbati (69.20-71.97%), Arka Rajhans (71.18-75.19%), Hira Madhu (72.49-74.18%) and Punjab Sunhari (69.50-72.87%) stood close to highly susceptible ones. The rest of the varieties showed less than 60 per cent infestation and were categorized into susceptible group. They concluded that the degree of infestation of the pest might be attributed to the olfactory stimuli emitted by fruits of different varieties. The varieties having strong odour registering higher infestation than varieties with low odour. The odoriferous difference might be because of the varied concentration of the total soluble sugar in different varieties. Gupta and Verma (1995) observed slow development rate, lower fecundity and survival when flies were reared on sponge gourd in comparison to bitter gourd and cucumber. An experiment carried out by Singh *et al.* (2000) to study the host preference of melon fly on various cucurbits revealed that cucumber (21.46%), pumpkin (21.61%) and long melon (21.18%) were found least preferred hosts than bitter gourd (31.27%) which proved to be the most preferred host to this pest. Water melon (28.55%), bottle gourd (25.30%) and musk melon (23.95%) proved moderately preferred hosts. In mid hills of Himachal Pradesh out of twenty cucumber genotypes screened against melon fruit fly, none of the genotypes exhibited resistance to this pest (Ingoley *et al.* 2005). However, fly infestation varied among the genotypes and three genotypes were found moderately resistant, eight genotypes susceptible and nine as highly susceptible. The fruits of susceptible and highly susceptible genotypes were containing high amount of total free amino acids, total sugars and low amount of phenols than moderately resistant genotypes. Dhillon *et al.* (2005) screened seventeen bitter gourd genotypes under field conditions in summer and rainy seasons and observed significantly lowest infestation in IC 256185 (9.4%) and IC 248256 compared to check Pusa Do Mausmi (82.1%). They also reported that the genotypes with low infestation had low number of larvae in their fruits and found positive correlation between per cent of fruit infestation and number of larvae per fruit. Screening of thirteen bitter gourd cultivars for resistance against this pest was carried out by Gogi *et al.* (2009) in Punjab (Pakistan), who revealed that the per cent fruit infestation and larval density per fruit varied significantly in all tested genotypes. Based on per cent fruit infestation (< 20%) and larval density per fruit (3 larvae/fruit), the genotypes COL-ii and FSD-long were categorized as resistant genotypes, the genotypes COL-i –Nankana, Sahib, COL-I and GS-51 showing > 20 but < 50 per cent fruit infestation and > 3 but < 6 larvae per fruit were categorized as moderately resistant genotypes. Whereas, COL-iii, COL-Multan, COL-Vehari, Chamman, Sunder-F1, Janpuri, F1-484 and F1-485 showing >50 but < 80 per cent fruit infestation and > 6 but < 10 larvae per fruit were categorized as susceptible genotypes.

**Efficacy of insecticides/biopesticides on red pumpkin beetle, hadda beetle and fruit flies**

## Red pumpkin beetle, *A. foveicollis*

Singh and Mishra (1977) evaluated eight insecticides against *A. foveicollis* at Pantnagar, Uttrakhand and observed carbaryl @ 0.1 per cent to be the most effective (mean mortality 289.00) followed by monocrotophos @ 0.06 per cent (mean mortality 220.33). Singh *et al.* (1977) reported that the soil treatment with carbofuran @ 0.5 kg a.i. per hectare was effective in controlling the pest and protected the seedlings at its most vulnerable stage effectively. Dabi *et al.* (1980) evaluated fourteen insecticides for the control of *A. foveicollis* and reported that phosphamidon @ 0.03 per cent was the most effective throughout the observation period with 64 per cent reduction in population over control even after 15 days treatment followed by carbaryl @ 0.2 per cent and endosulfan @ 0.05 per cent. Bajwa and Mavi (1981) tested six granular insecticides for the control of *A. foveicollis* and reported that carbofuran @ 0.5 and 1 kg a.i per hectare proved quite effective in controlling the beetles up to 37 days after its application followed by carbaryl @ 1 kg a.i per hectare up to 25 days after its application. Anand and Chauhan (1983) conducted a field study at Hissar (Haryana) for the simultaneous control of *A. foveicollis*, mite, *Tetranychus cucurbitae* (Rahman & Sapra) and powdery mildew and observed that sevisulf 40:50 WP and tank mixture of carbaryl and sulphur gave good control of these pests. According to Sinha and Chakrabarti (1983), the soil application with carbofuran granules @ 0.5 kg a.i per hectare proved to be most effective and seed treatment with carbofuran WP 3 to 4 per cent equally effective against *A. foveicollis* without any adverse effect on seed germination.

Pawar *et al.* (1984) used seven insecticides for the control of *A. foveicollis* and reported that fortnightly spray of carbaryl @ 0.5 per cent was the most effective (6.75 beetles/wine) as compared to untreated check (23.00 beetles/wine). Application of phoxim and pirimiphos-methyl @ 187.5 g a.i per hectare provided effective control of *A. foveicollis* for 10 days (Mavi and Bajwa, 1984). In a field study conducted at Ludhiana, (Punjab) by Mavi and Bajwa (1985) for the control of this pest, carbaryl @ 0.05 per cent and @ 0.075 per cent was found the most potent insecticide up to 10 days after its application followed by permethrin, phoxim and pirimiphos, each @ 0.075 per cent remained effective for 4 days after their application. A field experiment was conducted by Pareek and Kavadia (1988) in two different agro-climatic regions of Rajasthan, the semi-humid Udaipur and the semi-arid Jobner which revealed that four sprays of 0.2 per cent carbaryl at 3, 5, 9 and 11 weeks after sowing of musk melon proved the most effective against *A. foveicollis*, resulting in increased yield and net profit. Mehta and Sandhu (1990) used cucurbitacin as kairomones in combination with malathion and carbaryl as poison baits for the monitoring of beetles and observed that maximum number of beetles were trapped in carbaryl poison baits than that of malathion and concluded that these baits could be used to reduce the destructive behavior of this pest. The application of carbofuran @ 1.5 kg a.i per hectare at sowing, vining and flowering stages was found to be the most effective treatment in controlling *A. foveicollis* with 84.3 per cent reduction over control after 80 days of sowing (Thomas and Jacob 1994). Chaudhary (1995) found monocrotophos @ 200 g a.i. followed by carbaryl @ 500 g a.i (spray and dust) effective during first year and cypermethrin 25 g a.i. followed by deltamethrin 10g a.i and carbaryl @ 500g a.i. (spray) per hectare during second year. Under field conditions, cypermethrin 0.1 per cent + molasses solution 1 per cent was found most effective in reducing the beetle population (8.8 beetles/5 plants) followed by cypermethrin 0.01 per cent (9.2 beetles/5 plants) and deltamethrin 0.0028 per cent (10.2 beetles/5 plants) as compared to control (18.0 beetles/5 plants) (Borah, 1997). Borah (1998) observed that application of carbofuran @ 1.5 kg a.i. at 15 days after germination to be the most effective followed by deltamethrin @ 12.5 g a.i. and decis 12.5 g a.i. per hectare at flower bud initiation stage followed by another spray at 15 days later. Khan and Jehangir (2000) studied the efficacy of different concentrations of sevin dust and found high concentration (2.0 %) to be the most effective followed by medium (1.0 %) and low (0.5 %). Khan and Wasim (2001) assessed different plant extracts and found neem extract in benzene most effective in repelling *A. foveicollis* followed by bakaion extract in benzene. Comparative efficacy of seven insecticides *viz*., neem, triazophos, chlorpyriphos, monocrotophos, abamectin, SIL-942 and Beta-cyfluthrin evaluated under field conditions against *A. foveicollis* by Babu *et al.* (2002) revealed that beta-cyfluthrin @ 18.75 g a.i. per hectare (6.86 % damaged leaves/plant) to be the most effective followed by beta-cyfluthrin @ 12.5 g a.i. (14.9 % damaged leaves/plant), monocrotophos @ 700 g a.i. (14.12 % damaged leaves/plant), neem 3ml per liter of water (15.33 % damaged leaves/plant) and SIL-942 @ 100 g a.i. (17.28 % damaged leaves/plant). Rajak and Singh (2002) evaluated various insecticides and biopesticides for the control of *A. foveicollis* and found deltamethrin followed by carbofuran and carbaryl most effective among the tested insecticides. Whereas, among biopesticides only neem powder proved to be effective against this pest. Mehmood *et al.* (2006) studied the comparartive effect of different control methods against red pumpkin beetle and observed insecticidal treatments *viz*., carbofuran and carbaryl dust more effective in killing the beetles, near the plants. In Sri Lanka, neem based formulations were also effectively used for the control of this pest in organic crop production (Rajpakse and Ratnasekera 2007). Rahaman and Prodhan (2007) studied the effect of net barrier and synthetic pesticides on *A. foveicollis* and reported zero infestation in case of net barrier and lowest infestation by the use of carbofuran. Soil treatment with carbofuran @ 500 g a.i per hectare at the time of sowing proved effective (0.93 adult/plant) followed by seed treatment with thiamethoxam @ 3 g per kg of seed + rice husk ash @ 30 kg per hectare at 15, 25, 35 and 45 days after sowing (1.26 adults/plant) (Anonymous, 2007-2008). Bioefficacy of neem based and synthetic insecticides against red pumpkin beetle was studied by Rathod *et al.* (2009) under laboratory conditions and found maximum mortality in neem based commercial formulation gronim and neem-azal-F (29.98%) and carbaryl (63.36%).

## Hadda beetle, *H. vigintioctopunctata*

David (1966) evaluated eleven insecticides against epilachna beetle on brinjal at Coimbatore, Tamil Nadu and observed carbaryl (0.1%) and trichlorophon (0.1%) most effective among tested insecticides. Hameed and Adlakha (1973) reported that under field conditions spray of dimethoate followed by carbaryl and diazinon (0.05%) gave reasonably good control of epilachna grubs and pupae within 10 days following treatments on potato. Pareek and Kavadia (1987) evaluated different insecticides against epilachna beetle on musk melon and reported that carbaryl (0.2%) was most effective followed by endosulfan (0.07%) and phosalone (0.035%). Toxaphene (camphechlor) and dicofol were least effective, whereas chlorpyriphos, dimethoate, malathion and ethion were moderately effective. According to Singh and Kavadia (1989) pests attacking brinjal including epilachna beetle were most effectively controlled with the application of disulfoton or aldicarb granules @ 1.5 kg a.i. per ha, when applied 8 days after transplanting as side dressing followed by 2-3 sprays of endosulfan @ 0.07 per cent whereas malathion and fenitrothion showed very little effectiveness against the pests. A field experiment was conducted by Reddy *et al.* (1990) to evaluate the efficacy of certain plant extracts for the control of epilachna beetle and revealed that *Azadirachta indica* Juss. and *Annona squamosa* L. were more effective in reducing the grub population at 24 hours and 3 days after spraying resulting in reduction of 88.00 and 92.99 per cent over control. Thomas and Jacob (1991) studied the bio-efficacy of carbofuran against epilachna beetle on bitter gourd and reported that the application of carbofuran @ 0.75 and 1.5 kg a.i. per ha at the time of sowing, vining and combination of sowing, vining and flowering were completely free of epilachna attack. Chitra *et al.* (1993) tested six plant products for the control of pests attacking brinjal and observed *Argemone mexicana* L. (76.18%) to be the most effective among all plant products though comparable to monocrotophos (78.82%) and endosulfan (75.46%) equally followed by *Azadirachta indica* and Neknool (formulation 0.1%) giving 69.55 and 69.02 per cent control, respectively. However, neemguard and allitin were found least effective (55.92 and 55.56%, respectively).

According to Mohasin and De (1994) application of dichlorvos (0.03%) followed by quinalphos (0.03%) and methyl-demeton (0.03%) gave better control of epilachna adults and grubs on potato. Nagia *et al.* (1992) evaluated nine insecticides against grubs and adults of epilachna beetle on brinjal and reported that carbaryl, quinalphos, endosulfan deltamethrin and fenvalerate were quite effective against grubs and adult stages of the pest, however, dimethoate, fenpropathrin and fluvalinate were effective against grubs only. Kumar and Kumar (1998) evaluated some insecticides against strains of epilachna beetle resistant to malathion and endosulfan and revealed that they did not show significant cross resistance to tested insecticides *viz.*, cypermethrin, fenvalerate monocrotophos and carbaryl and thus sparing them as effective insecticides against this pest. Abdul–Moniem *et al.* (2004) tested the commercial formulation of three natural products against the second and fourth instar grubs of *Epilachna chrysomelina* (Fabricius) under laboratory conditions and recorded highest mortality (97.5 and 89.0%) when grubs were offered squash leaves treated with spruzit flussing 1 per cent (pyrethrum) followed by neudosan (potassium soap) @ 1 per cent (94.0 and 86.0%) and Neem Azal-T/S (NSKE) @ 2 per cent (89.0 and 84.5%). Kanwar (2006) evaluated seven insecticides for the control of epilachna beetle and found cypermethrin to be the most effective resulting in maximum reduction (85.64%) one and three days after treatment whereas, the overall reduction fifteen days after treatment was maximum in carbaryl (0.1%) followed by cypermethrin (0.0075%), endosulfan (0.05%), azadirachtin (0.03%), *Beauveria bassiana* (Balsamo) (1 X 109 cfu/l), malathion and novaluron (0.007%). Anil (2008) observed highest reduction in plant infestation by epilachna beetle in lambda-cyhalothrin (0.004%) followed by agrospray oil (0.2%), deltamethrin (0.0028%) and novaluron (0.01%).

Naik *et al.* (2008) reported that four sprays of 0.1 per cent profenofos were more effective in reducing grub population (75.77%) followed by 0.015 per cent spinosad (67.51%) and 0.15 per cent azadirachtin + 0.0075 per cent spinosad (62.86%). Chandranath and Katti

(2010) evaluated six insecticides *viz.*, endosulfan, quinalphos, fenvalerate, chlropyriphos, dimethoate and NSKE and observed dimethoate to be the most effective in reducing the population (8.4 beetles/plant) as compared to control (14.90 beetles/plant). However, all the insecticides reduced the pest significantly as compared to control.

Fruit flies, *Bactrocera* spp. Kavadia *et al.* (1977) studied the effect of malathion and carbaryl on *B. cucurbitae* and found that both were better in reducing the infestation in long melon at 3, 7 and 10 days interval after treatment. However, carbaryl was found more effective than malathion. Fruit flies, *Bactrocera* spp. Kavadia *et al.* (1977) studied the effect of malathion and carbaryl on *B. cucurbitae* and found that both were better in reducing the infestation in long melon at 3, 7 and 10 days interval after treatment. However, carbaryl was found more effective than malathion. Gupta and Verma (1982) found that plots treated with fenitrothion + protein hydrolysate had lowest attack of 8.7%. However, pest incidence in fenitrothion alone (13.7%) was at par with malathion + *gur* (16.7%) as compared to control (43.3%). Doharey (1983) tested five insecticides against two fruit fly species *viz.*, *B. dorsalis* and *B. (D) cucurbitae* under cages and found that that fenitrothion even at lowest concentration of (0.03%) was most effective against both the species giving cent per cent mortality in 24 hours. The two species responded differently to decamethrin and carbaryl, as all the three concentrations (0.003, 0.004 and 0.005%) of decamethrin gave cent per cent mortality of *B. dorsalis* within 48 hours in comparison to *B. cucurbitae* which showed cent per cent mortality in 72 hours at two concentrations (0.004 and 0.005%).

An *in vitro* experiment was conducted by Wadhwani and Khan (1983) to study the effect of sub-lethal doses of toxic baits on the reproductive potential of *B. cucurbitae* which revealed that the females exposed to sodium arsenate, sodium dieldrin, malathion and sevin laid 76.54, 45.20, 52.38, 84.26 and 41.88 per cent less eggs, respectively than the normal fecundity. Singh and Srivastava (1983) reported that neem oil extract (5%) effectively prevented the oviposition in *B. cucurbitae.*

Pawar *et al.* (1984) evaluated five insecticides for the control of *B. cucurbitae* and recorded lowest infestation 15.59 per cent in treatment with monocrotophos (0.05%) followed by 23.71 per cent in carbaryl (0.5%) as compared to 70.24 per cent in control. Shivarkar and Dumbre (1985) evaluated different insecticides and found fenitrothion (0.03%) to be the most effective in reducing the fruit infestation followed by endosulfan (0.03%) and permethrin (0.01%). Ravindranath and Pillai (1986) observed that fenvalerate @ 100 g a.i per hectare was the most effective in controlling *B. cucurbitae* on bitter gourd followed by permethrin and cypermethrin @ 100 g a.i each and deltamethrin @ 15 g than malathion @ 500 g per hectare. Albrecht and Sherman (1987) tested avermectin B1 against three fruit fly species *in vitro* and found that its high dosages resulted in high mortality in all the three species. They concluded that it might be superior to malathion for the control of fruit fly species although slow acting, but was highly toxic and has definite suppressive effect on fecundity particularly in *B. cucurbitae*. Pareek and Kavadia (1988) reported that four spray applications of carbaryl (0.02%) at 3, 5, 9 and 11 weeks after sowing proved to be the most effective against fruit fly, followed by malathion (0.03%) and phosalone, which also resulted in higher benefit/cost ratios. Khan *et al.* (1992) reported that four spray applications of malathion at 10 days interval to be most effective in controlling the *B. cucurbitae* reducing infestation (4.8%) followed by cypermethrin (5.8%) and deltamethrin (7.4%) as compared to control (14%). Bhatnagar and Yadava (1992) also observed malathion the most effective resulting in minimum damage (5.7%) on bottle gourd followed by carbaryl (5.8%), quinalphos (5.9%), fenitrothion (8.2%) and dichlorovos (8.4%) over the control (14.6%). Moreover the number of maggots per infested fruit was also recorded lowest (3.8 maggots/fruit) in malathion treatment over control (11.9 maggots/per fruit). Chen *et al.* (1996) studied the deterrent effect of neem seed kernel extract on oviposition of oriental fruit fly, *B. dorsalis* and revealed that guava fruits treated with neem seed kernel extract (0.2 to 0.4%) resulted in reduction in oviposition preference (87.5 to 99.2%) over the untreated check fruits. Borah (1997) observed that cypermethrin (0.1%) + molasses solution (1.0%) at 10 days interval after 15 days of germination in cucumber resulted in the lowest fruit infestation (24.4%) followed by cypermethrin (26.6%) and deltamethrin (26.9%), but malathion was found the least effective (36.2%) over the control (40.1%).

Ranganath *et al.* (1997) tested a number of botanicals and chemical insecticides against *B. cucurbitae* on cucumber and ridge gourd in Southern Andaman, India and reported that neem oil (1.2%) was the most effective in reducing the damage in cucumber and recorded 6.2 per cent infestation over 39.0 per cent in the control. Ragumoorthi *et al.* (1998) tested five neem formulations against Moringa fruit fly, *Gotina distigma* (Meigon) and observed that nimbecidine (1500 ppm) was the most effective with high anti-ovipostion effect as 15.33 eggs per female were observed compared to 54.67 in untreated check. Hassan (1998) tested neem seed kernel extract on persimmon for its efficacy against developing stages of the Queensland fruit fly and found it significantly effective @ 120-140 mg per liter of water against 1st and 2nd instar larvae. Application of carbofuran @ 1.5 kg a.i. per hectare at 15 days interval after germination proved most effective followed by deltamethrin 12.5 g a.i. per hectare. (Borah 1998). Mehta *et al.* (2000) observed deltamethrin (0.0028%) and acephate (0.05%) to be most effective in reducing fruit infestation by *B. cucurbitae* and recorded lowest number of maggots per cucumber fruit. However, highest fruit infestation and maggot population was observed in malathion treatment. According to Khan and Khattak (2000), three sprays of dipterex and malathion @ 1.0 per cent reduced the fruit fly infestation to the extent of 21.67 and 26.00 per cent, respectively. Babu *et al.* (2002) observed abamectin (15 and 20 g a.i./ha) and neem (5 ml/liter of water) most effective against *B. cucurbitae* reducing fruit infestation by 82.64, 76.52 and 70.55 per cent, respectively followed by SIL-942 (60 g a.i./ha), beta-cyfluthrin (18.75 g a.i./ha) and monocrotophos (700 g a.i./ ha). Rajapakse and Ratnasekera (2007) also reported that neem based products provided excellent control of *B. cucurbitae*. Oke (2008) evaluated two insecticides *viz.*, deltamethrin and lambda-cyhalothrin for the control of *B. cucurbitae* on cucumber and observed both the insecticides effective against this pest. However, lambda-cyhalothrin was found to be better in respect of reducing the oviposition marks and number of pupae than that of deltamethrin. Sharma and Sinha (2009) evaluated seven insecticides for the control of *B. cucurbitae* and revealed that three foliar sprays of alphamethrin (20 g a.i/ha) at fortnight interval proved most effective followed by emamectin benzoate (15 g a.i/ha) and neem ban (1 ml/liter of water). In Nepal, application of locally made botanical pesticide Jholmal, a mixture of leaf extract of many plants including neem was found superior in terms of quality, size, yield and reduced fruit fly infestation in squash (Sapkota *et al.* 2010).

## References

Abdul-Moniem ASH, Gomma AA, Dimetry NZ, Wetzel T and Volkmar C 2004. Laboratory evaluation of certain natural compounds against the melon ladybird beetle, Epilachna chrysomelina F. attacking cucurbit plants. Archives of Phytopathology and Plant Protection 37: 71-81

Abe M and Matsuda K 2005. Chemical factors influencing the feeding preference of three Aulacophora leaf beetle species (Coleoptera : Chrysomelidae). Applied Entomology and Zoology 40 (1): 161-168

Adams, R. G., R. A. Ashley, and M. J. Brennan. 1990. Row covers for excluding insect pests from broccoli and summer squash plantings. J. Econ. Entomol. 83: 948Ð954.

Adlerz, W. C. 1966. Honey bee visit numbers and watermelon pollination. J. Econ. Entomol. 59: 28Ð30.

Agarwal ML and Kumar P 1999. Effect of weather parameters on population dynamics of peach fruit fly, Bactrocera zonata (Saunders). Entomon 24(1): 81-84 Al Ali AS,

Agbagwa I O., Ndukwu B C., Mensah S I.. 2007. Floral Biology, Breeding System, and Pollination Ecology of Cucurbita moschata (Duch. ex Lam) Duch. ex Poir. Varieties (Cucurbitaceae) from Parts of the Niger Delta, Nigeria. Turk J Bot. 31 : 451-458

Agbagwa IO and Ndukwu BC 2004. The value of morpho-anatomical features in the systematics of Cucurbita L. (Cucurbitaceae) species in Nigeria. Afr J Biotechnol 3: 541-546

Aguilar, M. A. C and V. P. Tabla. 2000. Importance of conserving alternative pollinators: assessing the pollination efficiency of the squash bee, Peponapis limitaris in Cucurbita moschata (Cucurbitaceae). Journal of Insect Conservation, 4(3):203-210.

Al Neamy IK and Alwan MS 1982. Studies on the biology and host preference of Aulacophora foveicollis Lucas. Zeitschrift für Angewandte Entomologie 84(1): 82-86.

Albrecht CP and Sherman M 1987. Lethal and sublethal effects of avermectin B1 on three fruit fly species (Diptera : Tephritidae). Journal of Economic Entomology 80(2): 344-347

Aluja M, Hurtado HC, Liedo P, Cabrera M, Castillo F, Guillen J and Rios E 1996. Seasonal populations and ecological implications for management of Anastrepha fruit flies (Diptera : Tephritidae) in commercial mango orchard in Southern Maxico. Journal of Economic Entomology 89(3): 654-667

Anand RK and Chauhan R 1983. Simultaneous control of red pumpkin beetle (Rapidopalpa foveicollis Lucas), mite (Tetranychus cucurbitae Rahman and Sapra) and powdery mildew (Sphaerotheca fuliginae Schlecht) of bottle gourd with mixture of pesticides. Indian Journal of Agricultural Sciences 53(9): 851-854

Anil 2008. Bioefficacy and presistance of insecticides against insect pests of brinjal (Solanum melongena L.). M. Sc. Thesis, Department of Entomology, CSK Himachal Pradesh Krishi Vishvavidyalaya, Palampur, India. p. 33 192

Anjum S, Rafaq M and Yazdani SS 2000. Studies on seasonal activity and control of fruit flies (Dacus spp.) on mango (Mangifera indica L.) at Faisalabad, Pakistan. Arab Journal of Plant Protection 18 (1): 121-123

Anonymous 2007-2008. Management of red pumpkin beetle and leaf minor in cucumber/ bottle gourd. XXVI AICRP (VC), Annual Report 308-309.

Anonymous 2008. Package and practices for vegetable production. Directorate of Extension Education, CSK Himachal Pradesh Krishi Vishvavidyalaya, Palampur, India. p 44-50

Babu KS and Viraktamath S 2003a. Species diversity and population dynamics of fruit flies on mango in Northern Karnataka. Pest Management and Economic Zoology 11(2): 103-110.

Babu KS and Viraktamath S 2003b. Population dynamics of fruit flies on cucurbits in North Karnataka. Pest Management and Economic Zoology 11(1): 53-57

Babu PG, Reddy DG, Jadhav DR, Chiranjeevi C and Khan MAM 2002. Comparative efficacy of selected insecticides against pests of water melon. Pesticide Research Journal 14(1): 57-62

Bajwa DS and Mavi GS 1981. Evaluation of granular insecticides with respect to the control of red pumpkin beetle, Aulacophora foveicollis (Lucas) on musk melon. Journal of Research (PAU) 18(3): 292-299

Banerjee M, Konar A, Jha S and Sahoo S 2003. Incidence of epilachna spp. on bitter gourd. In: Proceedings National Symposium on Frontier Areas of Entomological Research (B Subrahmanyam and VV Ramamurthy eds), New Delhi, India November 5-7, 200. p 137

Benepal PS and Hall CV 1967. Biochemical composition of plants of Cucurbita foetidissima HBK and Cucumis melo L. as related to cucumber beetle feeding. Proceedings of American Society for Horticultural Science 91: 353-360

Bhalla OP and Verma AK 1991. Integrated pest management studies on cauliflower and cabbage in temperate regions of Himachal Pradesh. Final Report, ICAR Research Sheme., Department of Entomology Dr YS Parmar University of Horticulture and Forestry, Nauni (HP), India. p 32 193

Bhatia SK and Mahto Y 1970. Influence of temperature on the speed of development of melon fly, Dacus cucurbitae Coquillett (Diptera : Tephritidae). Indian Journal of Agricultural Sciences, 40(9): 821-828

Bhatnagar KN and Yadava SRS 1992. An insecticidal trial for reducing the damage of some cucurbitaceous fruits due to Dacus cucurbitae Coq. Indian Journal of Entomology 54(1): 66-69 Borah RK 1997. Effect of insecticides on pest incidence of cucumber (Cucumis sativus) in hill zone of Assam. Indian Journal of Agricultural Sciences 67(8): 332-333

Borah RK 1998. Evaluation of insecticidal schedule for the control of red pumpkin beetle and fruit fly of red pumpkin in the hill zone of Assam. Indian Journal of Entomology 60(4): 417-419

Borah RK 1999. Seasonality and varietal preference red pumpkin beetle (Raphidopalpa foveicollis) on cucumber (Cucumis sativus) in Assam. Indian Journal of Agricultural Sciences 69(3): 180-181

Butani DK and Jatwani MG 1984. Insects in vegetables. Periodical Expert Book Agency, Delhi, p 84

Cane, J. H., B. J. Sampson, and S. A. Miller. 2011. Pollination value of male bees: The specialist bee Peponapis pruinosa (Apidae) at summer squash (Cucurbita pepo). Environ. Entomol. 40: 614Ð620.

Caudle, B. 2010. Development of organic melon production methods to control bacterial wilt, pp. 44Ð45. 2010 Fruit and Vegetable Research Report, University of Kentucky College of Agriculture Cooperative Extension Service, Publication PR-608.

Cauto, R. H. N. and R. C. Calmona. 1993. Entomophilous pollination in Cucumber (Cucumis sativus L. var. Aodai Melhorada). Naturalia (Sao Paulo), 18:77-82.

Celli G and Giordani G 1981. Importanza economica e biologica dell'Ape (Apis mellifera L.) per la fruttificazione del Melone (Cucumis melo L.) in coltura protetta. Bol Ist Entomol Univ Studi Bologna 36: 91-114.

Chandranath HT and Katti P 2010. Management of epilachna beetle on ashwagandha. Karnataka Journal of Agricultural Sciences 23(1): 171

Chandravadana MV and Pal AB 1983. Triterpenoid feeding deterrent of Raphidopalpa foveicollis L. Current Science 52(2): 87-88

Chaudhary RN 1995. Management of red pumpkin beetle, Aulacophora foveicollis (Lucas) using polythene cages on cucurmber. Pest Management in Horticultural Ecosystems 1(1): 55-57

Chaudry ZA and Alikhan MA 1990. Effects of temperature and relative humidity on the development and the fecundity of the red pumpkin beetle, Aulacophora foveicollis Lucas (Chrysomelidae: Coleoptera). Indian Journal of Entomology 52 (2): 274-278

Chen CC, Dong WJ, Cheng LL and Hou RF 1996. Deterrent effect of neem seed kernel extract on oviposition of the oriental fruit fly (Diptera : Tephritidae) in guava. Journal of Economic Entomology 89(2): 462-466

Cheng, Chi-Tung. 1996. Effects honeybees on pollination and fruit set of cantaloupe. Chin. J. Entomol., 16(1):25-33.

Chitra KC, Rao SJ, Rao PKA and Nagaliah K 1993. Field evaluation of certain plant products in the control of brinjal pest complex. Indian Journal of Entomology 55(3): 237-240

Christenson LD and Foote RH 1960. Biology of fruit flies. Annual Review Entomology 5: 171-192

Coolong, T., R. Bessin, S. Wright, J. Strang, and K. Seebold. 2011. Vegetable Production Guide for Commercial Growers, ID-36, p. 132. University of Kentucky College of Agriculture, Lexington KY.

Couto RHN, Pereira JMS and Couto LA 1990. Effects of pollination in Cucurbita pepo (summer squash). Cientifica 18: 21-23.

Da Costa CP and Jones CM 1971. Resistance in cucumber Cucumis sativus L. to three species of cucumber beetles. Hort Science 6 (4): 340-342

Dabi RK, Qureshi QG and Sharma SK 1980. Field evaluation of insecticidal sprays against Aulacophora foveicollis Lucas. Pesticides 14(3): 7-8

Dag, A. and D. Eisikowitch. 1999. Ventilation of greenhouses increases honey bee foraging activity on melon, Cucumis melo. Journal Of Apicultural Research, 38(3-4):169-175.

David AL 1966. Trial with insecticides for the control of brinjal leaf beetle (Epilachna vigintioctopunctata Fabr.) and fruit borer (Leucinodes orbonalis Guen.). Indian Journal of Agricultural Sciences 36(4): 153-160

Delaplane, K. S., and D. F. Mayer. 2000. Crop pollination by bees, p. 344. CAB International, Oxon, United Kingdom.

Dhillon MK, Naresh JS, Singh R and Sharma NK 2005. Reaction of different bitter gourd (Momordica charantia L.) genotypes to melon fruit fly, Bactrocera cucurbitae (Coquillett). Indian Journal of Plant Protection 33(1): 55-59

Dhillon MK, Singh R, Naresh JS and Sharma HC 2005. The melon fruit fly, Bactrocera cucurbitae: A review of its biology and management. Journal of Insect Science 40(1): 1-16

Dhillon NPS 1993. The lack of relationship between bitterness and resistance of cucurbits to red pumpkin beetle (Aulacophora foveicollis). Plant Breeding 110(1): 73-76 Doharey KL 1983. Efficacy of some insecticides against fruit flies. Indian Journal of Entomology 45(4): 465-469

Drew RAI, Zalucki MP and Hopper GHS 1984. Ecological studies of eastern Australian fruit flies (Diptera : Tephritidae) in their endemic habitat I. Temperal variation in abundance. Oecologia 64 (4): 267-272

Droege, S. 2010. The very handy manual: how to catch and identify bees and manage a collection. (http://pollinators. nbii.gov/documents/Handy%20Bee%20Manual.pdf).

Dubois M, Gills KA, Hamilton JK and Smith F 1956. Colorimetric methods of determinations of sugars and related substances. Analytical Chemistry 28(3): 350-356

Endress PK 1994. Diversity and Evolutionary Biology of Tropical Flowers. New York: Cambridge University Press.

Estes JR, Amos BB and Sullivan JR 1983. Pollination from two perspectives: the agricultural and biological sciences. In: Jones CE & Little RJ, eds. Handbook of Experimental Pollination Biology. New York: Van Norstrand Reinhold, 173-183.

Faegri K and Van der Pijl L 1979. The Principles of Pollination Ecology. Pergamon Press, Oxford.

Fischer, R. M. and N. Pomeroy. 1989. Pollination of greenhouse musk melons by Bumble bees (Hymenoptera: Apidae). J. Econ. Entomol., 82(4):1061-1066.

Fleming R and Retnakaran A 1985. Evaluating single treatment data using Abbott,s formula with reference to insecticides. Journal of Economic Entomology 78(6): 1179-1183

Fletcher BS 1974. The ecology of natural population of the Queensland fruit fly, Dacus tryoni iv. The immigration and emigration of adults. Australian Journal of Zoology 21 (4): 541-565

Fletcher BS 1987. The biology of Dacine fruit flies. Annual Review of Entomology 32: 115-144

Free, J. B. 1993. Cucurbitacae, pp. 190–208. In Insect pollination of crops, 2nd ed, vol. 21. Academic, London, United Kingdom.

Gazit Y, Rossler Y, Epsky ND and Heath RR 1998. Trapping females of the Mediterranean fruit fly (Diptera : Tephritidae) in Israel: Comparison of lures and trap type. Journal of Economic Entomology 91(6): 1355-1359

Ghosh SK and Senapati SK 2001. Biology and seasonal incidence of Henosepilachna vigintioctopunctata Fabr. on brinjal under terai region of West Bengal. Indian Journal of Agricultural Research 35(3): 149-154



Ghosh SK and Senapati SK 2001. Biology and seasonal incidence of Henosepilachna vigintioctopunctata Fabr. on brinjal under terai region of West Bengal. Indian Journal of Agricultural Research 35(3): 149-154



Gill LS 1988. Taxonomy of Flowering Plants. Onitsha, Nigeria: Africana-FEP. 338 pp.

Gingras, D.; Gingras, J and D.De Oliveira. 1999. Visits of Honeybees (Hymenoptera :Apidae) and their effects on Cucumber yields in the field. Journal of Economic Entomology, 92 (2):435-438.

Gogi MD, Ashfaq M, Arif MJ and Khan MA 2009. Screening of bitter gourd ((Momordica charantia) germplasm for source of resistance against melon fruit fly, (Bactrocera cucurbitae) in Pakistan. International Journal of Agriculture and Biology 11(6): 746-750

Gogi MD, Ashfaq M, Arif MJ, Sarfraz RM and Nawab NN 2010. Investigating phenotypic structures and allelochemical compounds of the fruits of Momordicha charantia L.

genotypes as a source of resistance against Bactrocera cucurbitae (Coquillett) (Diptera : Tephritidae). Crop Protection 29: 884-89

Gogi MD, Ashfaq M, Arif MJ, Sarfraz RM and Nawab NN 2010. Investigating phenotypic structures and allelochemical compounds of the fruits of Momordicha charantia L. genotypes as a source of resistance against Bactrocera cucurbitae (Coquillett) (Diptera : Tephritidae). Crop Protection 29: 884-890

Gomez KA and Gomez AA 1984. Statistical Procedures for Agricultural Research. John Wiley and Sons, New York. p 680 196

Gould JL and Gould CG 1988. The Honey Bee. New York: Scientific American Library.

Grewal SS and Sandhu GS 1983. Orientation and feeding response of the red pumpkin beetle to the detached leaves of cucurbits. Journal of research (PAU) 20(4): 560-562

Grewal, G. S. and A. S. Sidhu. 1978. Insect – pollinators of some cucurbits in Punjab. Indian J. Jgric. Sci., 48 (2):79- 83.

Gupta BL and Verma AK 1995. Host specific demographic studies of the melon fruit fly, Dacus cucurbitae Coquillett (Diptera : Tephritidae). Journal of Insect Science 89(1): 87-89

Gupta D 2004. Insect pests of cucurbit vegetables. In: Pest Management in Horticulture crops (LR Verma, AK Verma and DC Gautam, eds), Aristech Publishers Inc. p 437

Gupta D and Bhatia R 2000. Population fluctuations of fruit flies, Bactrocera spp. in submountainous mango and guava orchards. Journal of Applied Horticulture 2(1): 47-49

Gupta D and Verma AK 1992. Population fluctuations of the maggots of fruit flies, Bactrocera cucurbitae (Coquillett) and B. tau (Walker) infesting cucurbitaceous crops. Advances in Plant Sciences 5 (2): 518-523

Gupta JN and Verma AN 1978. Screening of different cucurbit crops for the attack of the melon fruit fly, Dacus cucurbitae Coquillett (Diptera : Tephritidae). Haryana Journal of Horticultural Sciences 7(1-2): 78-82

Gupta JN and Verma AN 1982. Effectiveness of fenitrothion bait sprays against melon fruit fly, Dacus cucurbitae. Coquillett in bitter gourd. Indian Journal of Agricultural Research 16(1): 41-46

Hameed SF and Adlakha RL 1973. Studies on the population and control of epilachna species on potato in Kullu Valley. Pesticides 7(4): 30-31

Hardy DE 1979. Review of economic fruit flies of the South Pacific region. Pacific Insects 20: 429-432

Hart WJ, Fujimoto MS, Kamakah D and Harris EJ 1967. Attraction of melon flies to lures on foliage or fibers at various heights. Journal of Economic Entomology 60(4): 1139-1142 197

Haseeb M, Qamar M and Sharma DK 2009. Seasonal incidence of brinjal hadda beetle, Henosepilachna vigintioctopunctata (F) (Coleoptera : Coccinellidae). In Aligarh, Utter Pradesh. Trends in Biosciences 2 (1): 31-32

Hassan E 1998. Insecticidal toxicity of neem seed kernel extract (NSKE) on Bactrocera trygoni (Frogg), (Diptera : Tephritidae) and repellency on persimmon fruit. Zeitshrift fur Plenzenkrankheiten and Pflanzenschutz 105(4): 411-416

Hasyim A, Muryati and de Kogel WJ 2008. Population fluctuation of adult males of the fruit fly, Bactrocera tau Walker (Diptera : Tephritidae) in passion fruit orchards in relation to abiotic factors and sanitation. Indonesian Journal of Agricultural Science 9(1): 29-33

Hasyim A, Muryati, Istianto M and de Kogel WJ 2007. Male fruit fly, Bactrocera tau (Diptera : Tephritidae) attractants from Elsholtzia pubescens Bith. Asian Journal of Plant Sciences 6(1): 181-183

Heslop-Harrison J and Heslop-Harrison Y 1985. Germination of stresstolerant Eucalyptus pollen. J Cell Sci 73: 135-157.

Hilje, L., H. S. Costa, and P. A. Stansly. 2001. Cultural practices for managing Bemisia tabaci and associated viral diseases. Crop Prot. 20: 801-812.

Hillingsworth R, Vagalo M, Tsatsia F 1997. Biology of melon fly, with special reference to the Solomon Islands. In: Management of fruit flies in the Pacific (AJ Allwood and RAI, eds), Proceedings of Australian Country. Industrial Agricultural Research 76(1): 104-144

Hirano K 1993. Populations dynamics of a phytophagous lady beetle, Epilachna vigintioctopunctata (Fabricius) (Coleoptera : Coccinellidae), living in spatio temporally heterogeneous habitats II. Seasonal changes in adult movement and spatial distribution of the population. Applied Entomology and Zoology 28 (2): 131-140

Hong LJ and Hui Y 2005. Population dynamics of Bactrocera dorsalis (Diptera : Tephritidae) in Yuanjiang dr hot Valley, Yunnan with an analysis of the related factors. Acta Entomologica Sinica 48 (5): 706-711

Hui YE and Jianghong LIU 2007. Population dynamics of oriental fruit fly, Bactrocera dorsalis (Diptera :Tephritidae) in Xishuangbanna, Yunnan Province, China. Frontiers of Agriculture in China 1 (1): 76-80 198

Hurd, P. D., E. G. Linsley, and A. E. Michelbacher. 1974. Ecology of the squash and gourd bee, Peponapis pruinosa, on cultivated cucurbits in California (Hymenoptera: Apoidea). Smithson. Contrib. Zool. 168: 1Ð17.

Hurd, P. D., E. G. Linsley, and T. W. Whitaker. 1970. Squash and gourd bees (Peponapis and Xenoglossa) and the origin of the cultivated Cucucbita. Evolution 25: 218Ð 234.

Hurtado HC, Aluja M and Liedo P 1995. Adult population fluctuation of Anastrepha species (Diptera : Tephritidae) in tropical orchard habitats of Chiapas, Mexico. Environmental Entomology 24 (4): 861-869

Hutchinson J and Dalziel JM 1954. Flora of West Tropical Africa, Crown Agents, Vol. 1 Part 1 (Revised by R. W. J. Keay). p 215.

Iftekar B and Khan NH 1980. Effect of temperature and relative humidity conditions on the occurrence of certain coccinellids on brinjal plant. Indian Journal of Environmental Research 1(11): 83-86

Inayatullah C, Khan L and Haq M-UL 1991. Relationship between fruit infestation and the density of melon fruit fly adults and puparia. Indian Journal of Entomology 53(2): 239-243

Inayatullah C, Khan L, Haq M-UL and Mohisin-Ata-Ul 1991. Weather based models to predict the population densities of melon fruit fly, Dacus cucurbitae Coq. Tropical Pest Management 37(3): 211-215

Ingoley P Metha PK, Chauhan YS. Singh N and Awasthi CP 2005. Evaluation of cucumber genotypes for resistance to melon fruit fly, Bactrocera cucurbitae Coq. under mid hill conditions of Himachal Pradesh. Journal of Entomological Research 29(1): 57-60

Inoue T, Nakamure K, Salmah S and Abas I 1993. Population dynamics of animals in unpredictably-changing tropical environments. Journal of Bioscience 18(4): 425-455

Iwaizumi R, Sawaki K, Kobayashi K, Maeda C, Toyokawa Z, Kawakmi T and Matsui M 1991. A comparative experiment of the attractiveness of the several kinds of the culure toxicants to the melon fly, Dacus cucurbitae (Coquillett). Research Bulletin of Plant Protection Service 27(1): 25-28

Jalaluddin SMD, NatarajanS K and Sadakathulla S 2001. Population fluctuation of the Guava fruit fly, Bactrocera correcta (Bezzi) in relation to host and abiotic factors. Journal of Experimental Zoology 4(2): 323-327

Jay SC 1986. Spatial management of honeybees on crops. Annu Rev Entomol 31: 49-65.

Jayaraman J 1981. Protein fractionation in gram flour. Laboratory Manual in Biochemistry. Willey Eastern Pvt. Ltd., New Delhi. p 79-80 199

Jeffrey C 1964. Key to the Cucurbitaceae of West Tropical Africa with a guide to localities and little known species. J W African Sc Assoc 9: 79-97.

Jeffrey C 1980. A review of the Cucurbitaceae. Bot J Linn Soc 81: 233-247.

Jesse, L. C., M. Gleason, A. Owens, J. Batzer, D. R. Lewis, and H. G. Taber. 2007. Use of bumble bees to extend the duration of row covers on muskmelon in Iowa. Poster (abstr.) D0401, Ann. Meeting Entomol. Soc. Am. (http:// esa.confex.com/esa/2007/ techprogram/paper_32418.htm).

Johri R and Johri PK 2003a. Seasonal biology, abundance and intensity of attack of red pumpkin beetle, Aulacophora foveicollis Lucas on cucurbit crops at Kanpur in Utter Pradesh. Journal of Applied Zoological Research 14(2): 144-147

Johri R and Johri PK 2003b. Food preference of red pumpkin beetle, Aulacophora foveicollis Lucas at Kanpur in Utter Pradesh. Journal of Applied Zoological Research 14(1): 80-81

Julier, H. E., and T. H. Roulston. 2009. Wild bee abundance and pollination service in cultivated pumpkins: farm management, nesting behavior, and landscape effects. J. Econ. Entomol. 102: 563-573.

Kanwar R 2006. Bioecology and management of hadda beetle, Henosepilachna vigintioctopunctata (Fabricias). M. Sc. Thesis,. Department of Entomology, CSK Himachal Pradesh Krishi Vishvavidyalaya, Palampur, India. p. 77

Kapoor VC and Agarwal ML 1983. Fruit flies and their natural enemies in India. In: Fruit flies of Economic Importance (Cavalloro, ed), Balkema Rotterdam. p 104-105

Kavadia VS, Hupta HCL and Pareek BL 1977. Comparative efficacy and residues of insecticides applied singly or in mixture with attractants against fruit fly affecting long melon (Cucumis utilissimus Duthie & Fuller). Indian Journal of Plant Protection 5(2): 183-187

Keck CB 1951. Effect of temperature on the development and activity of the melon fly. Journal of Economic Entomology 44(6): 100-102

Kerhoas C, Gay G, Duplan, JC and Dumas C 1986. Water content evolution in Cucurbita pepo during ageing: a NMR study. In:

Khan L, Inayatullah C and Haq M-UL 1992. Control of melon fly, Dacus cucurbitae (Diptera : Tephritidae) on melon in Pakistan. Tropical Pest Management 38(3): 261-264

Khan MZ and Hajela KP 1987. Studies on *Aulacophora foveicollis* Lucas (Coleoptera : Chrysomelidae) food preference and extent of damage. Indian Journal of Entomology 49(4): 457-459

Khan SM and Jehangir M 2000. Efficacy of different concentrations of sevin dust against red pumpkin beetle causes damage to musk melon. Journal of Biological Sciences 3(1): 183-185 200

Khan SM and Khattak SUR 2000. Chemical control of melon fruit fly, Bactrocera cucurbitae (Coq.) on musk melon (Cucumis sativus L.) by malathion and dipterex in D.I. Khan. Pakistan Journal of Biological Sciences 3(8): 1299-1300

Khan SM and Wasim M 2001. Assessment of different plant extracts for their repellency against red pumpkin beetle, (Aulacophora foveicollis Lucas), attacking musk melon. Journal of Biological Sciences 1(4): 198-200

Khandelwal RC and Nath P 1981. Evaluation of cultivars of water melon for resistance of fruit fly. Egyptian Journal of Genetics and Cytology 8(1): 46-56

Khattak SU, Afsar K, Hussain N, Khalil SK and Alamzeb 1990. Annual population and incidence of oriental fruit fly (Dacus dorsalis Hendel) in a fruit orchard at Peshawar, Pakistan. Bengladesh Journal of Zoology 18(1): 131-138

Kim, J., N. Williams, and C. Kremen. 2006. Effects of cultivation and proximity to natural habitat on ground-nesting native bees in California sunßower Þelds. J. Kans. Entomol. Soc. 79: 309-320.

Koul VK 1998. Insect pest management on cauliflower (Brassica oleracea L. var. botrytis L.). Ph. D thesis,. Department of Entomology, CSK Himachal Pradesh Krishi Vishvavidyalaya, Palampur, India. p 45

Kremen, C., N. M. Williams, and R. W. Thorp. 2002. Crop pollination from native bees at risk from agricultural intensiÞcation. Proc. Nat. Acad. Sci. 99: 16812-16816.

Krombein, K. V., P. D. Hurd, D. R. Smith, and B. D. Burks. 1979. Catalog of Hymenoptera in America North of Mexico, p. 1198. Smithsonian Institute Press, Washington, DC. Lexington.

Kumar A 2006. Studies on heterosis and inheritance of resistance to fruit fly in cucumber (Cucumis sativus L.). Ph. D Thesis,. Department of Vegetable Sciences, Dr YS Parmar Unversity of Horticulture and Forestry, Nauni, India. p 22

Kumar S and Kumar J 1998. Laboratory evaluation of some insecticides against strains of hadda beetle (Epilachna vigintioctopunctata Fab.) resistant to malathion and endosulfan. Pest Management and Economic Zoology 6(1-2): 133-137

Kumar S, Patel CB and Bhatt RL 1997. Studies on seasonal cyclicity of Bactrocera correctus Bezzi in mango and sapota orchards using methyl eugenol traps. Gujarat Agricultural University Research Journal 22(2): 68-74

Laskar N and Chatterjee H 2010. The effect of meterological factors on the population dynamics of melon fruit fly, Bactrocera cucurbitae (Coq.) (Diptera :Tephritidae) in the foot hills of Himalaya. Journal of Applied Science and Environmental Management 14 (3): 53-58 201

Latif A and Abdullah K 2005. Population dynamics and management of Bactrocera sp. (Diptera : Tephritidae) infecting mango. Pakistan Entomologist 27(2): 69-72

Lee H 1976. Information on the oriental fruit fly, Dacus dorsalis Hendel, damaging papaya. Journal of Agricultural Research of China 25(1): 156-162

Lee LWY, Hwang YB, Cheng CC and Chang JC 1992. Population fluctuations of the melon fly, D. cucurbitae in North Eastern Taiwan. Chinease Journal of Entomology 12(4): 285-292

Lee SL and Chen YL 1977. Attractancy of synthetic compounds related to methyl eugenol for oriental fruit fly and melon fly. Journal of Pesticide Science 2(2): 135-138

Levin, M.D. 1986. Using honey bees to pollinate crops USDA-ARS Etension Service Leaflet 549.

Liu YC 1983. Population studies on the oriental fruit fly, Dacus dorsalis Hendel in central Taiwan. In: Fruit Flies of Economic Importance (R Cavalloro, ed), AA Balkema, Rotterdam. p 62-65

Luo Y and Li Z (1999). Pollination Ecology of Chloranthus serratus (Thunb.) Roem. et Schult. and Ch. fortunei (A. Gray) Solms-Laub. (Chloranthaceae). Ann Bot 83: 489-499.

Ma S, Wang Y and Cui M (1997). A contribution to the reproductive biology of Chloranthus holostegius (Handel-Mazzeti) Pei & Shan. (Chloranthaceae) in Mile population (in Chinese). Acta Bot Yunnan 19: 415-422.

Mader, E., M. Vaughan, M. Shepherd, and S. H. Black. 2010. Alternative pollinators: native bees. The Xerces Society, Portland, OR. (http://www.attra.org/attra-pub/PDF/nativebee.pdf).

Madhura HS 2001. Management of fruit flies (Diptera :Tephritidae) using physical and chemical attractants. M. Sc. Thesis,. Department of Entomology, University of Agricultural Sciences, Bangalore, India. p 80

Makkar HPS, Blummel M, Borowy NK and Becker K 1993. Gravimetric determination of tannis and their correlation with chemicals and protein precitation method. Journal of the Science and Agriculture 61(2): 161-165

Mall NP, Pandey RS, Singh SV and Singh SK 1992. Seasonal incidence of insect pests and estimation of the losses caused by shoot and fruit borer on brinjal. Indian Journal of Entomology 54(3): 241-247

Mandal SK, Malam, Sattar A, Sah SB and Gupta SC 2006. Influence of weather parameters on the incidence of the melon fly, Bactrocera cucurbitae (Coquillett), in bitter gourd. Pest management and Economic Zoology 14(1&2): 45-48

Mann GS 1990. Bio-ecology and management of fruit flies. In summer institute on key insect pests of India. Their bio-ecology with special reference to integrated pest management. Punjab Agricultural University, Judhiana. p 253-257 202

Mann GS and Sohi AS 1983. Nomenclature status of genus Aulacophora Duponchel & Chevrolat and Aulacophora foveicollis (Lucas). Entomon 8 (1): 105-106

Mathewson, J. A. 1968. Nest construction and life history of the eastern cucurbit bee, Peponapis pruinosa (Hymenoptera: Apoidea). J. Kans. Entomol. Soc. 41: 255Ð261.

Mavi GS and Bajwa DS 1985. Comparative efficacy of some newer insecticides for the control of red pumpkin beetle, Aulacophora foveicollis Lucas on musk melon. Pesticides 19(2): 24-27

Mavi MS and Bajwa DS 1984. Comparative efficacy of some emulsifiable concentrates applied with "Fogair" for the control of red pumpkin beetle, Aulacophora foveicollis (Lucas) on musk melon. Journal of Entomological Research 8(2): 167-170

McGregor SE & Todd FE 1952. Cantaloupe production with honeybees. J Econ Entomol 45: 43-47.

McGregor, S. E. 1976. Insect pollination of cultivated crop plants. U.S. Department of Agriculture Agricultural Research Service, Washington, DC. Handbook No. 496.

Mehmood T, Khokhar KM and Shakeel M 2006. Comparative effect of different control methods on red pumpkin beetle, Aulacophora foveicollis on cucumber. Sarhad Journal of Agriculture 22 (3): 473-475

Mehta PK and Sandhu GS 1989. Studies on host preference and rate of feeding by red pumpkin beetle on different cucurbits. Vegetable Science 16(1): 66-74

Mehta PK and Sandhu GS 1990. Monitorin of red pumpkin beetle, (Aulacophora foveicollis) by cucurbitacins as kairomones through poison baits. Indian Journal of Agricultural Sciences 60(3): 225-226

Mehta PK and Sandhu GS 1992. Influence of cucurbitacins on the feeding activity of red pumpkin beetle, Aulacophora foveicollis (Lucas). Journal of Insect Science 5(2): 187-189

Mehta PK, Chandel RS and Kashyap NP 2000. Control of fruit fly, Bactrocera cucurbitae (Coq.) on cucumber in Himachal Predash. Pestology 24 (10): 53-55

Mehta PK, Sharma TN and Chandel RS 1998. Resistance in pea genotypes to leaf miner *Chromatomyia horticola* (Goureau). Indian Journal of Agricultural Sciences 68 (5): 271-273

Mensah B A. and Kudom A A. 2011. Foraging dynamics and pollination efficiency of Apis mellifera and Xylocopa olivacea on *Luffa aegyptiaca* Mill (Cucurbitaceae) in Southern Ghana. Journal of Pollination Ecology, 4(5): 34-38

Metcalf RL and Metcalf ER 1992. Fruit flies of the family tephritidae. In: Plant Kairomones in Insect Ecology and Control (RL Metcalf and ER Metcalf, eds), Chapman and Hall, Inc. London, United Kingdom. p 109-152 203

Metcalf RL, Mitchell WC and Metcalf ER 1983. Olfactory receptors in the melon fly, Dacus cucurbitae and the oriental fruit fly, Dacus dorsalis. Proceedings of National Acadamic Science USA 80: 3143-3147

Michelbacher, A. E., R. F. Smith, and P. D. Hurd. 1964. Bees are essential to pollination of squashes, gourds, and pumpkins. Calif. Agric. 18: 2Ð4.

Minter L. M., and Bessin R. T. 2014. Evaluation of Native Bees as Pollinators of Cucurbit Crops Under Floating Row Covers. Environ. Entomol. 43(5): 1354-1363.

Minter, L. 2012. Integrating sustainable pest management and pollinator conservation efforts in cucurbit production systems. Ph.D dissertation, University of Kentucky,

Mohasin M and De BK 1994. Control Henosepilachna vigintioctopunctata Fabr. On potato in West Bengal plains. Journal of the Indian Potato Association 21(1-2): 151-153

Morimoto, Y.; Gikunku, M. and P. Maundu. 2004. Pollinators to the bottlegourd (Lagenaria siceraria) observed in Kenya. International Journal of tropical insect science, 24(1):79-86.

Mukhopadhyay A and Mandal A 1994. Screening of brinjal (Solanum melongena) for resistance to major insect pests. Indian Journal of Agricultural Sciences 64(11): 798-803

Mulcahy DL, Bergamini MG & Ottaviano E, eds. 1986 Biotechnology and ecology of pollen. New York: Springer Verlag, 502-505.

Muthukumar M and Kalyanasundaram 2003. Influence of abiotic factors on the incidence of major insect pests in brinjal (Solanum melongena). South Indian Horticulture 51(1-6): 214-218

Nagia DK, Kumar S, Sharma P, Meena RP and Saini ML 1992. Mass multiplication of Henosepilachna vigintioctopunctata (Fabricus) (Coleoptera : Coccinellidae) on Physalis minima L. Plant Protection Bulletin, Faridabad 44(3): 24-25

Nagia DK, Kumar S, Sharma P, Meena RP, Sani ML and Goel SC 1992. Laboratory evaluation of insecticides for the control of Henosepilachna vigintioctopunctata (Fabricias) on brinjal (Solanum melongena L.) (Coleoptera : Coccinellidae). In: Proceedings of the National Symposium on Growth, Development and Control Technology of Insect Pests (SC Goel, ed.), SD Collage Muzuffarnagar. p 188-191

Naik VCB, Rao PA, Krishnayya PV and Rao VS 2008. Seasonal incidence and management of spotted leaf beetle, Henosepilachna vigintioctopunctata (F.) in brinjal. Journal of Applied Zoological Research 19(1): 1-4

Narayanan ES and Batra HN 1960. Fruit flies and their control, ICAR (Indian Council of Agricultural Research). New Delhi. p 68 Nath P 1966a. Varietal resistance of gourds to the fruit fly. Indian Journal of Horticulture 23(2): 69-77 204

Nath P 1966b. Observations on resistance of cucurbits to the fruit fly. Indian Journal of Horticulture 21(2): 173-175

Nath P and Thakur MR (1965). Evaluation of red pumpkin beetle resistance in gourds. Indian Journal of Horticulture 22 (2&4): 330-339

Nath P, Datta OP, Sundari V and Swamy KRM 1976. Inheritance of resistance to fruit fly in pumpkin. SABRAO Journal 8(2): 117-119

Nepi M and Pacini E 1993. Pollination, pollen viability and pistil receptivity in Cucurbita pepo L. Ann Bot 72: 527-536.

Njoroge, G. N.; Gemmill, B.; Bussmann, R.; Newton, L. E. and V.W. Ngumi. 2004. Pollination ecology of Citrullus lanatus at Yatta, Kenya. International Journal of Tropical insect science, 24(1):73-77.

Oke OA 2008. Effectiveness of two insecticides to control melon fruit fly (Bactrocera cucurbitae Coq.) in cucumber (Cucumis sativus L). crop at Anse Boileau Seychelles. European Journal of Scientific Research 22(1): 84-86

Okoli BE 1984. Wild and cultivated cucurbits in Nigeria. Econ Bot 38: 350-357

Ordway, E.; Buchmann S. L.; Kuehl R. O. and C. W. Shipman. 1987. Pollen dispersal in buffalo gourd, Cucurbita foetidissima (Cucurbitaceae) by bees of the genera Apis, Peponapis and Xenoglossa (Hymenoptera:Apidae, Anthophoridae). J. Kans. Entomol. Soc., 60(4):489- 503.

Orozco-Santos, M., O. Perez-Zamora, and O. Lopez-Arriaga. 1995. Floating row cover and transparent mulch to reduce insect populations, virus diseases, and increase yield in cantaloupe. Fla. Entomol. 78: 493-501.

Owens, A., L. Jesse, M. L. Gleason, and J. Batzer. 2008. Effectiveness of extended duration row covers for suppression of bacterial wilt in muskmelon. Phytopathology 98: S118.

Pal AB, Srinivasan K and Vani A 1983. Development of breeding lines of musk melon for resistance to fruit fly. Progressive Horticulture 15(1-2): 100-104

Pal AB, Srinivasan K, Bharatan G and Chandravadana MV. 1978. Location of source of resistance to the red pumpkin beetle, Rapidopalpa foveicollis Lucas against pumpkin germplasm. Journal of Entomological Research 2(2): 148-153

Papadoulos NT, Katsoyannos BI, Carey JR and Kouloussis NA 2001. Seasonal and annual occurance of Mediterranean fruit fly (Diptera : Tephritidae) in Northern Greece. Ecology and Population Biology p 94 Pareek BL and Kavadia VS 1986. Seasonal incidence of insect pests on cucurbits in Rajasthan. Annals of Arid Zone 35 (4): 300-311

Pareek BL and Kavadia VS 1987. Field evaluation of insecticides against hadda beetle, Epilachna vigintioctopunctata Fabr. infesting muskmelon. Indian Journal of Plant Protection 15(1): 105-107

Pareek BL and Kavadia VS 1988. Economic insecticidal control of two major pests of musk melon, Cucumis melo, in the pumpkin beetle, Rapidopalpa spp. and the fruit fly, Dacus cucurbitae in Rajasthan, India. Tropical Pest Management 34(1): 15-18 205

Pareek BL and Kavadia VS 1993a. Screening of musk melon varieties against red pumpkin beetle, Aulacophora foveicollis Lucas. Indian Journal of Entomology 55(3): 245-251

Pareek BL and Kavadia VS 1994a. Relative preference of fruit fly Dacus cucurbitae Coquillett on different cucurbits. Indian Journal of Entomology 56(1): 72-75

Pareek BL and Kavadia VS 1994b. Screening of musk melon varieties against fruit fly, Dacus cucurbitae Coquillett under field conditions. Indian Journal of Entomology 57(4): 417-420

Pavlakos J 1943. The biology and control of Aulacophora foveicollis Lucas in Greece. Zeitschrift für Angewandte Entomologie 30(1): 72-78

Pawar DB, Joi MB and Sonone HN 1984. Chemical control of red pumpkin beetle and fruit fly of musk melon by modern insecticides. South Indian Horticulture 32(5): 317-318

Pawar DB, Mote UN and Lawande KE 1991. Monitoring of fruit fly population in bitter gourd crop with the help of lure trap. Journal of Maharashtra Agricultural University 16(2): 281

Perring, T. M., R. N. Royalty, and C. A. Farrar. 1989. Floating row covers for the exclusion of virus vectors and the effect on disease incidence and yield of cantaloupe. J. Econ. Entomol. 82: 1709-1715.

Philippe JM (1991). La Pollinisation par les Abeilles : pose des colonies dans les cultures en floraison en vue d'accroître les rendements des productions végétales. Aix-en-Provence: Edisud.

Purseglove JW (1968). Tropical Crops: Dicotyledons. Vol. 1, London: Longman, 116-124.

Radford AE, Dickison WC, Massey JR and Bell CR (1974). Vascular Plant Systematics. New York and London: Harper and Row. 891pp.

Rafiq MA, Ghaffar A and Arshid M 2008. Population dynamics of whitefly (Bemisia tabaci) on cultivated crop hosts and their role in regulation its carry-over to cotton. International Journal of Agricultural Biology 10(4): 493-498

Raghuraman M and Veeravel R 1999. Influence of abiotic factors on the incidence of spotted leaf beetle, Henosepilachna vigintioctopunctata (F.) in brinjal. Pest Management in Horticultural Ecosystems 5(1): 17-20

Ragumoorthi KN, Rao PVS and Dhawan AK 1998. Neem products and plant extracts for managing Moringa fruit fly, Gitona distigma (Meigon). In: Ecological Agriculture and Sustainable Development Volume 2., Proceedings of International Conference on Ecological and Sustainable Agriculture, Chandigarh, India, 15-17 Nov. 1997. p 38 206

Rahaman MA and Prodhan MDH 2007. Effects of net barrier and synthetic pesticides on red pumpkin beetle and yield of cucumber. International Journal of Sustainable Crop Production 2(3): 30-34

Rahman A. 2016. Bioecology and management of carpenter bees in India. In: D.P.Abrol and R.K.Thakur (Edited) Recent advances in pollination biology. Kalyani Publishres New Delhi, India.

Rahman A.2016. Pollinators for sustainable agriculture in Assam. In: D P Abrol and R.K Thakur (edited) Recent advances in pollination biology. Kalyani publishers, New Delhi pp.

Rajagopal D and Trivedi P 1989. Status, bioecology and management of epliachna beetle, Epilachna vigintioctopunctata (Fab.) (Coleoptera : Coccinellidae) on potato in India. Tropical Pest Management 35(4): 410-413

Rajak DC 2000. Studies on populations of red pumpkin beetle on muskmelon (Cucumis melo L.). Agricultural Science Digest 20(1): 54-55

Rajak DC 2001. Host range and food preference of the red pumpkin beetle, Aulacophora foveicollis Lucas (Chrysomelidae : Coleoptera). Agricultural Science Digest 21(3): 179-181

Rajak DC and Singh HM 2002. Comparative efficacy of pesticides against red pumpkin beetle, Aulacophora foveicollis on musk melon. Annals of Plant Protection Sciences 10(1): 147-148

Rajapakse R and Ratnasekera D 2007. The management of the major insect pests Bactrocera cucurbitae (Diptera : Tephritidae) and Aulacophora spp. (Coleoptera : Scarabaeidae) in cucurbits in Sourhern Sri Lanka under three intensive systems: integrated, chemical and organic agriculture. Acta Horticulture 731: 303-307

Ramani S, 1997. Biosystematic studies on fruit flies with special reference to the fauna of Karnataka and Andaman and Nicobar. Ph. D thesis, Department of Entomology, University of Agricultural Sciences, Bangalore, India. p 214

Ramzan M, Singh D, Singh G, Mann GS and Bhalla JS. 1990. Comperative development and seasonal abundance of hadda beetle, Henosepilachna vigintioctopunctata (Fabr.) on some solanaceous host plants. Journal of Research (PAU) 27(2): 253-262

Ranganath HR, Suryanearayana MA and Vennakumari K 1997. Management of melon fly, Bactrocera cucurbitae Coquillett in cucurbits in South Andaman. Insect Environment 3(2): 32-33 207

Ranjitha AR and Viraktamath S 2005. Response of fruit flies to different types of traps in mango orchard. Pest Management in Horticultural Ecosystems 11(1): 15-25

Ranjitha AR and Viraktamath S 2006. Investigations on the population dynamics of fruit flies in mango orchard at Dharwad, Karnataka. Journal of Agricultural Sciences 19(1): 134-137

Ranjitha AR and Viraktamath S 2006. Montoring of fruit flies (Diptera : Tephritidae) in guava orchard at Dharward, Karnataka. Karnataka Journal of Agricultural Sciences 19(1): 45-49

Rao VR, Chitra KC and Rao PK 1989. Relative toxicity of syntheyic pyrethroids to Henosepilachna vigintioctopunctata (Fabricias). Indian Journal of Entomology 51(1): 51-54

Rath LK, Nayak VS and Dash D 2002. Non-preference mechanism of resistance in egg plant to epliachna beetle, Henosepilachna vigintioctopunctata. Indian Journal of Entomology 64 (1): 44-47

Rathod ST, Borad PK and Bhatt NA 2009. Bio-efficacy of neem based and synthetic insecticides against red pumpkin beetle, Aulacophora foveicollis (Lucas) on bottle gourd. Pest Management in Horticultural Ecosystems 15 (2): 150-154

Ravikumar CH and Viraktamath S 2006. Influence of weather parameters on fruit fly trap catches in Dharwad, Karnataka. Pest Management in Horticultural Ecosystems 12(2): 143-151

Ravindranath K and Pillai KS 1986. Control of fruit fly of bitter gourd using synthetic pyrethroids. Entomon 11(4): 269-272

Reddy PVR, Chitra KC and Rao PK 1990. Field evaluation of certain plant extracts for the control of brinjal spotted leaf beetle, Epilachna vigintioctopunctata Fab. Journal of Insect Science 3(2): 194-195

Richards AJ 1986. Plant Breeding Systems. London: George Allen and Unwin 529 pages.

Robinson WS, Nowogrodzki R and Morse RA 1989. The value of honey bees as pollinators of U.S. crops. Am Bee J 129: 411-423.

Ronald FLM and Jayma LMK. Distribution and host range of Bactrocera cucurbitae (Coquillet).http://extento.hawaii.edu/kbase/crop/type/bactro_c.htm [5th May, 2011] 208

Roy DC and Pande YD 1991. Seasonal incidence, host preference and feeding rate of red pumpkin beetle (Aulacophora foveicollis) in Tripura. Indian Journal of Agricultural Sciences 61(8): 603-607

Sambandam CN and Chelliah S 1972. Scheme for evaluation of cantaloupe and musk melon varieties for resistance to fruit fly Bactrocera (Dacus) cucurbitae (Coq.). Final Report, USDA, PL-480. Reaearch Project, FG-IN-339. p 67

Sandhu GS and Grewal SS 1985. Testing cucurbits through multiple-choice and no-choice tests under cages by infesting red pumpkin beetle adults. Indian Journal of Entomology 47(2): 169-172

Sapkota R, Dahal KC and Thapa RB 2010. Damage Assessment and management of cucurbit fruit flies in spring-summer squash. Journal of Entomology and Nemotology 2(1): 007-012

Sarada G, Maheswari TU and Purushotham K 2001. Seasonal incidence and population fluctuation of fruit flies in mango and guava. Indian Journal of Entomology 63(3): 272-276

Sardana RH, Tyagi A and Singh A 2005. Knowledge resource on fruit flies (Tephritidae : Diptera) in India. National Center for Integrated Pest Management, New Delhi. p 1

Sarfraz M, Dosdall LM and Keddie BA 2006. Dimondback moth-host plant intractions: implications for pest management. Crop Protection 25: 625-639

Sarfraz M, Dosdall LM and Keddie BA 2007. Resistance of some cultivated Brassicaceae to infestation by Plutella zylostella (Lepidopteran : Plutellidae). Journal of Economic Entomology 100 (1): 215-224

Sarver, M. J. 2007. Farm management for native bees: a guide for Delaware, p. 20. U.S. Dep. Agric. NRCS, Delaware Dep. of Argriculture. Dover, DE.

Sarwar G, Aslam M., Munawar M.S., Raja S and Mahmood R. 2008. Effect of honeybee (Apis mellifera L.) pollination on fruit setting and yield of cucumber (Cucumis sativus L.) Pak. Entomol. Vol. 30(2): 185

Sharma GC and Hall CV 1971. Influnce of cucurbitacins, sugars and fatty acids on cucurbit susceptibility to spotted cucumber beetle. Journal of Ambrican Society for Horticulture Science 96(5): 675-680

Sharma RK and Sinha SR 2009. Evaluation of some novel and eco-friendly insecticides against fruit fly of bitter gourd. Journal of Insect Science 22(2): 202-203 209

Sharma SS, Bhanot JP and Kalra VK 1999. Host preference, extent of damage and control of red pumpkin beetle, Raphidopalpa foveicollis (Lucas). Journal of Insect Science 12(2): 168-170

Shekarappa, Viraktamath SA, Patil BV and Yelshetty S 1998. Influence of weather parameters on the pheromone trap catches of fruit flies. In: Proceedings of First National Symposium on Integrated Pest Management in Horticulture Crops, Bangalore, India. p 27

Shinde CB and Purohit ML 1978. Population study of red pumpkin beetle, Aulacophora foveicollis Lucas – a serious insect pest of cucurbits. Science and Culture 44 (7): 335

Shivarkar DT and Dumber RB 1985. Bionomics and chemical control of melon fly. Journal of Maharashtra Agricultural University 10(3): 298-300

Shukla RP and Prasad VG 1985. Population fluctuations of the oriental fruit fly, Dacus dorsalis Hendel in relation to hosts and abiotic factors. Tropical Pest Management 31(4): 273-275

Shuler, R. E., T. H. Roulston, and G. E. Farris. 2005. Farming practices inßuence wild pollinator populations on squash and pumpkin. J. Econ. Entomol. 98: 790-795.

Siemens DH, Garner SH, Mitchell-Olds T and Callaway RM 2002. Cost of defense in the context of plant competition: Brassica rapa may grow and defend. Ecologia 83 (3): 505-517

Singh D, Ranzan M and Gill SPS 1977. Field incidence of cucurbit fruit fly (Dacus cucurbitae) on different cultivars of bitter gourd at Ludhina (Punjab). Indian Journal of Plant Protection 5(1): 93-94

Singh G and Mishra PN 1977. Note on the efficacy of some insecticidal dusts and sprays against red pumpkin beetle, Raphidopalpa foveicollis (Lucas). Indian Journal of Agricultural Sciences 47(5): 257-258

Singh RP and Srivastava BG 1983. Alcohol extract of neem (Azadirachta indica A. Jass) seed oil as oviposition deterrent for Dacus cucurbitae (Coq.). Indian Journal of Entomology 45(4): 497-498

Singh SB, Singh HM and Singh AK 2007. Seasonal occurrence of fruit flies in Eastern Utter Pradesh. Journal of Applied Zoological Research 18(2): 124-127

Singh SN, Chakrabarti AK and Peshwani KM 1977. Control of red pumpkin beetle, Raphidopalpa foveicollis Lucas. infesting Cucurbit seedlings in the field. Seed Research 5(1): 44-48

Singh SV and Kavadia VS 1989. Insecticidal control schedule for the pests attacking brinjal II, during fruiting stage. Indian Journal of Entomology 51(1): 69-75

Singh SV, Mishra A, Bisen RS, Malik YP and Mishra A 2000. Host preference of red pumpkin beetle, Aulacophora foveicollis and melon fruit fly, Bactrocera cucurbitae. Indian Journal of Entomology 62(3): 242-246

Singh YP and Singh PP 2002. Pest complex of egg plant (Solanum melongena) and their succession at medium high altitude hills. Indian Journal of Entomology 64(3): 335-342

Sinha AK and Krishna SS 1969. Further studies on the feeding behavior of Aulacophora foveicollis on cucurbitacin. Journal of Economic Entomology 62(2): 512-513

Sinha SN and Chakrabarti AK 1983. Effect of seed treatment with carbofuran on the incidence of red pumpkin beetle, Raphidopalpa foveicollis Lucas on cucurbits. Indian Journal of Entomology 45(2): 145-151

Spaulding, D., and T. Coolong. 2011. Acorn and butternut squash cultivar trial. In T. Coolong, J. Snyder, and C. Smigell (eds.), 2010 Fruit and Vegetable Crops Research Report, PR-608, p. 35. Agricultural Eperiment Station, University of Kentucky, Lexington, XY.

Stanghellini, M. S.; Ambrose, J. T. and J. R. Schultheis. 1997. The effects of honey bee and bumble bee pollination on fruit set and abortion of cucumber and water melon. Am. Bee J., 37(5):386-391.

Stanghellini, M. S.; Ambrose, J. T. and J. R. Schultheis. 1998. Seed production in watermelon: A comparison between two commercially available pollinators. Hort. Sci., 33:28-30.

Stanghellini, M. S.; Ambrose, J. T. and J.R. Schultheis. 2002 b. Diurnal activity, floral visitation and pollen deposition by honeybees and bumblebees on field grown cucumber and watermelon. Journal of Apicultural Research, 41(1-2):27-34. Stanghellini, M. S.; Schultheis, J. R. and J. T. Ambrose. 2002 a. Pollen mobilization in selected Cucurbitaceae and the putative effects of pollinator abundance on pollen depletion rates. J. Amer. Soc. Hort. Sci., 127(5):729-736

Su CY 1986. Seasonal fluctuations of Dacus cucurbitae in Southern Taiwan. Plant Protection Bulletin, Taiwan 28(2): 170-171

Sunandita and Gupta D 2007. A note on host range of fruit fly species infesting summer vegetable crops in mid hills of Himachal Pradesh. Journal of Insect Science 20(1): 106-107

Takeish H 1992. A study on the fruit flies (Diptera : Tephritidae) found in fresh fruits caused by passengers from the Thailand to Narita Airport, Japan. Research Bulletin of the Plant Protection Service, Japan 28: 75-78

Tan KH and Serit M 1994. Adult population dynamics of Bactrocera dorsalis (Diptera :Tephritidae) in relation to host phenology and weather in two villages of Penang Islands, Malaysia. Environmental Entomology 23 (2): 267-275

Tangmitcharoen S and Owens JN 1997. Floral biology, pollination, pistil receptivity, and pollen tube growth of teak (Tectona grandis). Ann Bot 79: 227-241.

Tepedino, V. J. 1981. The pollination efficiency of the squash bee (Peponapis pruinosa) and the honey bee (Apis mellifera) on summer squash (Curcubita pepo). J. Kans. Entomol. Soc. 54: 359-377.

Thakur JC, Khattra AS and Brar KS 1992. Comparatives resistance to fruit fly in bitter gourd. Haryana Journal of Horticultural Sciences 21(3-4): 285-288

Thakur JC, Khattra AS and Brar KS 1994. Stability analysis for economic traits and infeatation of melon fruit fly, (Dacus cucurbitae) in bitter gourd (Momordica charantia). Indian Journal of Agricultural Sciences 64(6): 378-381

Thomas C and Jacob S 1991. Bio-efficacy of carbofuran against epilachna beetle, Henosepilachna 28 punctata Fab. infesting bitter gourd (Momordica charantia L.) under field and laboratory conditions. Indian Journal of Entomology 53(3): 435-441

Thomas C and Jacob S 1994. Biological efficacy of carbofuran against the red pumpkin beetle, Raphidopalpa foveicollis Lucas infesting bitter gourd. Indian Journal of Entomology 56(2): 164-168

Thomas J, Faleiro R, Vidya CV, Satarkar VR, Stonehouse JM, Verghese A and Mumford JD 2005. Melon fly attraction and control by baits in Central Kerala. Pest Management in Horticultural Ecosystems 11(2): 110-112

Tsuchiya T and Gupta PK 1991. Chromosome Engineering in Plants: Genetics, Breeding, Evolution, part B. p 181-195 Fort Collins, Colorado USA.

Vaissiere, B. E. and R. Froissart. 1996. Pollination of cantaloupes under spunbonded row cover by honeybees (Hymenoptera: Apidae) in West Africa. J. Econ. Entomol, 89(5):1213-1222.

Vargas RI, Burns RE, Mau RFL, Stark JD, Cook P and Pinero JC 2009. Captures in methyl eugenol and cuelure detection traps with and without insecticides and with a farma tech solid lure and insecticide dispenser. Journal of Economic Entomology 102(2): 552-557

Vargas RI, Stark JD and Nishidia T 1990. Population dynamics, habitat, preference and seasonal distribution patterns of oriental fruit fly and melon fly (Diptera : Tephritidae) in an agriculture area. Environmental Entomology 19(6): 1820-1828

Vargas RI, Stark JD, Kido MH, Ketter HM and Whitehand LC 2000. Methyl eugenol and cuelure traps for suppression of male oriental fruit flies (Diptera : Tephritidae) in Hawaii: Effects of lure mixtures and weathering. Journal of Economic Entomology 93(1): 81-87 212

Vashistha RN and Chaudhury B 1971. Studies on resistance to red pumpkin beetle (Aulacophora foveicollis Lucas) in musk melon, bottle gourd and water melon. Progressive Horticulture 2(4): 47-58

Vaughan, M., M. Shepard, C. Kremen, and S. H. Black. 2007. Farming for bees: guidelines for providing native bee habitat on farms, p. 44. The Xerces Society, Portland, OR.

Verghese A, Nagarraju DK and Sreedevi K 2005. Comparision of three indigenous lures/ baits with three established attractants in case of fruit flies (Diptera : Tephritidae). Pest Management in Horticultural Ecosystems 11(1): 75-78

Wadhwani K and Khan NH 1983. Effects of sublethal doses of toxic baits on the reproductive potential of Dacus cucurbitae Coq. Indian Journal of Agricultural Sciences 53(11): 936-938

Weems HV, and Heppner JB 2001. Melon fly, Bactrocera cucurbitae Coquillett (Insecta : Diptera : Tephritidae) Florida Department of Agriculture and Consumer Services, Division of Plant Industry and TR Fasulo, University of Florida. University Florida Publication EENY. p 199

Whitaker TW and Davis GN (1962). Cucurbits: Botany, Cultivation and Utilisation, London: Leonard Hill,.

White IM and Elson-Harris MM 1992. Fruit flies of economic significance, their identification and bionomics. CAB International Wallingford, United Kingdom. P 125-135

Willis, D. S. and P. G. Kevan. 1995. Foraging Dynamics of Peponapis pruinosa (Hymenoptera: Anthophoridae) on pumpkin (Cucurbita pepo) in Southern Ontario. Canada Entomol., 127(2):167-175

Wong TTY, McInnis DO, Ramadan MM and Nishimoto JI 1991. Age related response of male melon flies, Dacus cucurbitae (Diptera : Tephritidae) to cuelure. Journal of Chemical Ecology 17(12): 2481

***Bitter gourd flowers***

**Figure 7.4: Insect pollinators cucumber and bitter gourd flowers**

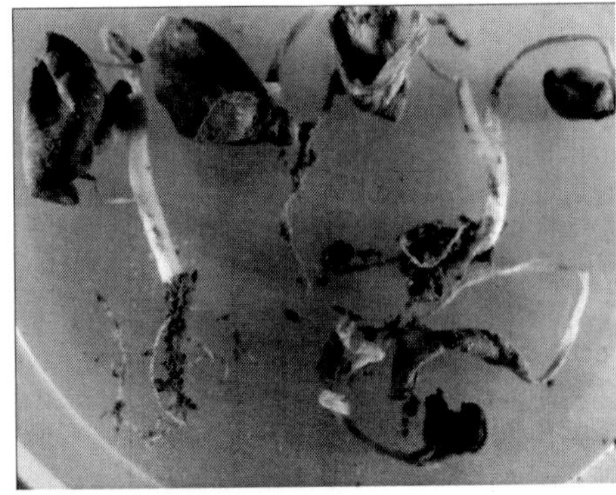

*Red pumpkin beetle danage Maggot fly damage*

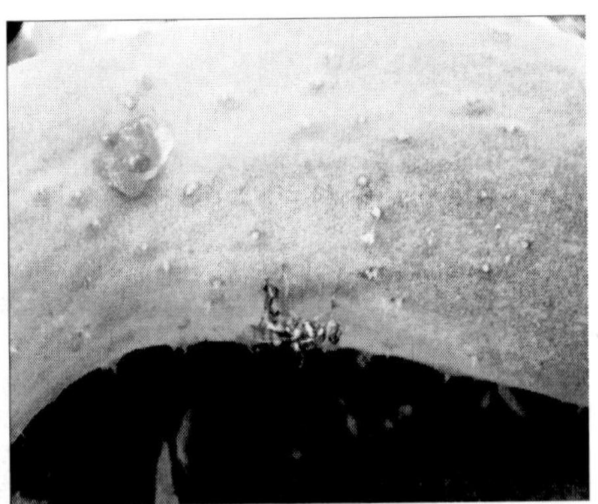

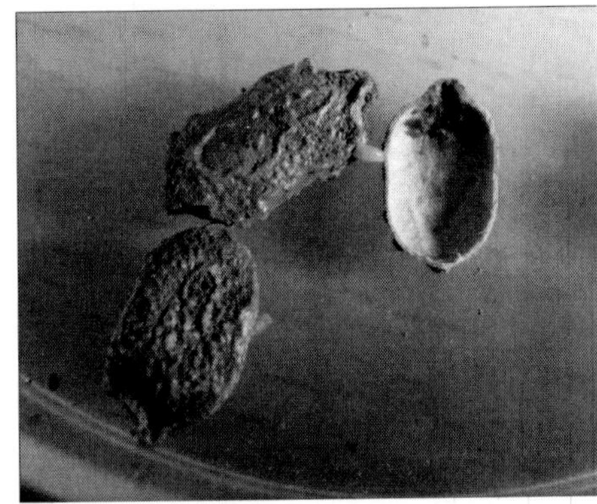

*Fruit fly damaging the cucumber fruit Maggots damaging Karela seeds*

**Figure 7.5: Insect pests of cucumber and bitter gourd flowers**

**CHAPTER**
**8**

# Malvaeceae

## Okra

The indiscriminate use of chemical insecticides for management of crop pests has resulted in reduction of the numbers of beneficial insects which provide pollination services. The pollinator decline is expected to result in the reduction of crop yields of which okra is no exception. This chapter attempts to integrate the effect of pollinators and pollination on the reproductive performance of okra (*Abelmoschus esculentus* (L) Moench) and safety of pollinators under pest management strategies.

## Okra, *Abelmoschus esculentus* (L.) Moench (Malvaceae)

Okra (*Abelmoschus esculentus* (L.) Moench) is an annual fruit vegetable crop grown in tropical and subtropical parts of the world (Pushpalatha, 2008). Okra is a native of West Africa (Joshi *et al.*, 1974; Kochhar, 1986). It is grown on a large scale in Africa, especially in the Sudan, Egypt, Ghana and Nigeria. It is also very important in other tropical areas including Asia, Central and South America and northern Europe (Joshi *et al.*, 1974; George, 1989, FAOSTAT, 2008). The Nile Basin seems to have been the route by which this plant spread through North Africa, the Eastern Mediterranean, Asia and to India. Okra reached the new world through Brazil and Dutch Guinea. African slaves brought okra to North America by way of New Orleans (Hamon *et al.*, 1990; Bish *et al.*, 1995). There are a number of varieties, both wild and cultivated. Some of these are *A. esculentus*, *A. caillei*, *A. moschatus*, *A. manihot*, *A. ficulneus* and *A. tetraphyllus*. The cultivated species in the genus *Abelmoschus* are: A. *manihot* L. and A. *moschatus* L. (Stevels, 1988; Siemonsma, 1991). The world okra production has been estimated at 4.8 million tons with India contributing 70% followed by Nigeria (15%), Pakistan (2%), Ghana (2%), Egypt (1.7%) and Iraq (1.7%) (Gulsen *et al.*, 2007). In West Africa, okra ranks second in vegetable production after tomato (Anon, 2002). On an average, 10 - 15 t /ha of yield can be obtained under good management conditions (NARP, 1993).

*Abelmoschus esculentus* has large bisexual flowers (Mc Gregor, 1976; Hamon and Koechlin, 1991). Its flower structure combines hermaphroditism and self-compatibility. Hence in okra, as the conditions for selfing become favourable, there is also an opportunity for cross-pollination (George, 1989; Al Ghzawi et al., 2003). Passive self pollination characterises *A. esculentus* flowers (Purewal and Rhandawa, 1957; Srivastava and Sachan, 1973; Chandra & Bhatnagar, 1975) despites the fact that the flowers are insect pollinated (George, 1989; Al Ghzawi et al., 2003). Few published data do exist on the relationships between insect foragers and *A. esculentus* flowers.

Okra is self-fertile, and when the anthers come in contact with the stigmas, self pollination may result; however, cross-pollination also occurs. Okra has alternate palmate broad leaves and the flowers have five large yellow petals with a large purple area covering the base (Martin, 1983; Hamon *et al.*, 1990). About 35-60 days after emergence, the plant begins to flower. The flower usually remains open for a day. Okra plants continue to flower and to fruit for an indefinite time, depending upon the variety, the season and soil moisture and fertility. Infact the regular harvesting stimulates continued fruiting, so much so that it may be necessary to harvest every day in climates where growth is especially vigorous.

No doubt okra is mostly self-fertilized; however, insects such as honeybees and bumble bees can cross-pollinate. Okra is self-compatible, and passive self-pollination can take place in its hermaphrodite flowers (Al-Ghzawi *et al.*, 2003). Its pollen grains are very large and echinate, 156 μm in diameter with spines over 20 μm in length (Vaissière and Vinson, 1994) so that pollination with both self and cross-pollen is possibly achieved by insects (Hamon and Koechlin, 1991; Al-Ghzawi *et al.*, 2003). Anthesis takes place at dawn, and the flower remains open all morning and closes by noon or early afternoon. Cross pollination can occur in okra up to a maximum of 42.2% (Mitidieri and Vencovsky, 1974). The extent of cross-pollination in a particular place depends upon several factors such as cultivar, competitive flora, insect population and season. Al Ghzawi *et al.* (2003) reported that, no significant differences were found between insect- and self-pollinated plants for the number of flowers.

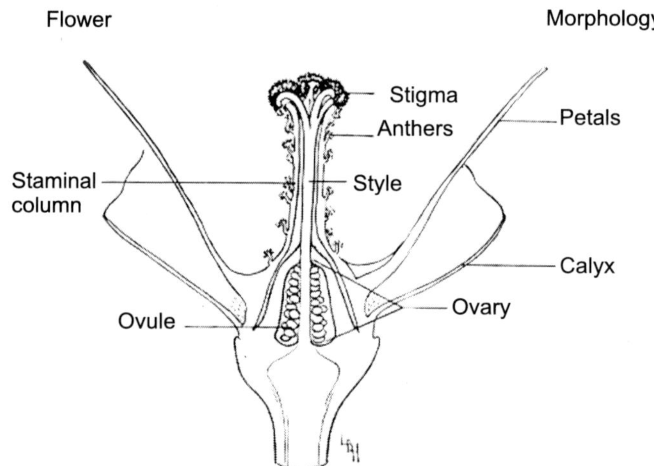


**Figure 8. Median section of okra flower (McGregor 1976)**

The single showy okra flower, as much as 2 inches across, resembles the cotton flower, with its wide corolla usually made up of five yellow to cream-colored petals (Figure 8.1). The erect sexual parts consist of a five to nine part style, each part with a capitate stigma, surrounded by the staminal tube bearing numerous filaments (Purewal and Randhawa 1947). The flower opens shortly after sunrise and remains open until about noon. The petals wilt in the afternoon and usually fall the following day. The anthers dehisce 15 to 20 minutes after the flower opens, and some of the pollen comes in contact with the stigma.

The single showy okra flowers resemble the cotton flowers are borne vertically on a peduncle 2.0 – 2.5 cm long. The flowers are about 2 inches in diameter, with five white to yellow petals with a red or purple spot at the base of each petal. The flowers were almost actinomorphic. The perianth consisted of 5 sepals and 5 distinct petals. The androecium consisted of very many monadelphous stamens which bears filaments(Purewal and Randhawa 1947). The gynoecium was a single pistil consisting of several carpels, and a superior ovary. The calyx, corolla and stamens are fused together at the base and fell off as one piece after anthesis.

## Pollination Ecology

Venkatramini (1952) observed that flower bud initiation, flowering, anthesis and stigma receptivity are influenced by genotype and climatic factors like temperature and humidity. From studies made on three okra varieties, flower buds are initiated at 22-26 days and the first flower opened 41-48 days after sowing. Once, initiated flowering continues for 40-60 days. Anthesis was observed between 6 a.m. and 10 a.m. Anthers dehisce before flower opening, and hence self pollination may occur at anthesis. Purewal and Randhawa (1947) reported that dehiscence of anthers is transvers and occurs within 5-10 minutes. 50-70%) of the stigma receptivity was also observed the day before flowering. Flowers opened only once in the morning and closed after pollination on the same day. The corolla withered in the following morning.

At anthesis, self-fertilization is far from complete and only a few ovules are then fertilized. Ovules continue to be fertilized during the morning and allogamy may be possible. The possibility of allogamy decreases rapidly and a pollen grain deposited on the style after midday has no impact on progeny. In okra, stigmas are exposed to allopollination at anthesis, and only anthers in the upper ring come into contact with stigmas. The moment of contact depends on the initial distance between these and the rate of elongation of the stamen sheath. In a given variety, the number of ovules available depends on the number of ovules per carpel and on the number of carpels per fruit. These are varietal characteristics. All the *Abelmoschus* species bear fruits with five carpels. Though pollinators are not must for pollination , the flowers are very attractive to bees and the plants are cross-pollinated. The cross pollination up to the extent of 4-19% (Choudhury et al.,1994; Shalaby, 1998) with maximum of 42.2% (Mitidieri and Vencovsky, 1998) has been reported. The extent of cross pollination in a particular place will depend upon the cultivar, competitive flora, insect population and season

## Nectar

Okra flowers produce nectar which attract various insects such as butterflies, houseflies and honeybees. Ant was observed to be the most prominent out of the various insects observed. Started visiting the flower immediately after opening is initiated. Pollen grain of okra are heavy and sticky and could not be readily transferred by wind, therefore insect are responsible for the transfer of pollen grains and consequently cross pollinated in okra plant.

## Insect visitor's of Okra

It is freely visited by honey bees and bumble bees, but the value of insect pollinator visitation is unknown (McGregor, 1976). In okra, there is a positive correlation between the number of seeds in a pod and its weight, and insect pollination increased the number of seeds present in a pod (Free, 1976). Nineteen percent increase in yield in Punjab,

India, was reported by Tanda (1984, 1985) due to intensive bee pollination. Mishra *et al.* (1987) found that the weight and length of capsules and seed number per capsule were significantly higher in open pollinated than in bagged flowers. Azo'o *et al.* (2011) reported that the mean number of matured seeds is higher in open-pollinated plants and smaller in self-pollinated plants. Cane and Tepedino (2001) advocate that if pollinators are in decline it should be possible to measure the results of their absence as reductions in fruit or seed production in natural or agricultural ecosystems.

In India for instance, honeybees *Apis cerana* and solitary bees *Halictus* spp. appeared as the main okra pollinators (Crane, 1991; Free, 1993) and preliminary observations indicated that okra flowers were fully visited by the two wild bee species between 7:00 and 10:00 am (Azo'o *et al.*, 2011).

In Cameroon, *Megachile spp.*, *Halictus spp.*, *Xylocopa spp.* and *Apis mellifera* were observed on preliminary trials as the main okra anthophilous insects (Njoya et al., 2005); but till-date there exist no precise data on their foraging activities. A positive impact of wild bees on the yields of eight okra genotypes was shown in Jordan. (Al Ghzawi et al., 2003).

Njoya et al (2005) reported that hand and insect pollination of okra flowers gave seed sets varying between 73–84% per pod which differed significantly ($p < 0.05$) from that of the bagged flowers which just resulted in 57% seed sets per flower. It was further interesting to find that a fecundated seed contains 91.5 µg nitrogen whereas an unfecundated seed has only 2.6 µg nitrogen; this means that a fecundated seeds contains 35 times much more nitrogen. As a consequence, cross-pollinated flowers rendered more fecundated seeds; with 311.1 µg more nitrogen per carpel than seeds from spontaneous-self pollinated flowers with more unfecundated seeds. An increase of 754.1 µg carbons and 192.2 µg of nitrogen per carpel were noted comparing seeds from forced self-pollinated flowers with those from spontaneous self-pollinated flowers indicating that cross pollination in okra is essential to achieve optimum yields both in seed quality and seed sets.

Although okra *Abelmoschus esculentus* (L.) is a self pollinated crop, but cross pollination has been reported in it (Purewal and Randhawa 1947., Frankel and Galun 1977). A number of insects visit and pollinate okra flowers (Mcgregor 1976, Frankel and Galun 1977). Choudhary et al (1973) observed *Apis cerana* and *Halictus sp* to be the chief pollinators of this crop. According to Tanda (1985) the most important insect pollinators of okra were the honeybees *A. mellifera* and *A. c. indica* which mainly foraged for nectar.\

Al-Ghzawi et al (2003) reported that in okra, higher numbers of seeds and seed weight per plant were obtained from the plants pollinated by insects than self-pollinated. Insect-pollinated plants showed a greater number of young pods and mature pods due to greater transformation of flowers into young pods. On the other hand, no significant differences were found between insect- and self-pollinated plants for the number of flowers. Insect-pollinated plants developed a greater percentage of flowers into young pods than selfed plants. Also, Insect pollination increased the number of seeds per mature pods in all genotypes and the development of fertilized ovules into seed.

Al-Ghzawi et al (2003) reported that in okra on an average seed bearing seeds per plant were 546.3 in insect pollinated plants and 215.3 in self-pollinated plants. The seed formation is maximum in insect pollinated plants and the least number of seeds in self-pollinated plants. It is quite evident that insects helped in pollination and formation of seeds in okra. The increase in yield of okra seed due to insect pollination over self-pollination was recorded to be 45%. All genotypes produced higher seed yield with insect pollinations than without insect pollinations

Insect pollination had a shorter flower life and flowering period (Barbier, 1978; Lerin, 1982; Williams, 1984). The plants pollinated by insects produced 7.1 g seed plant-1 as compared to 3.9 g obtained from non-pollinated plants. During observations, it was also noted that the retention and size of the seeds were greater in the insect pollinated plants than self-pollinated ones. The number of seeds per plant was greater in insect's pollinated plants than self-pollinated plants. Thus, the data indicate that okra is basically an insects-pollinated crop as has been reported by Lerin (1982).

Azo et al (2011) evaluated the impact of the floral activity of the insect visiting fauna on *Abelmoschus esculentus* (L.) Moench (Malvaceae) seed yields. The studies showed that two wild bees *Eucara macrognatha* and *Tetralonia fraterna* were predominant with 92.2% and 97.5% of the total insect visits. *Apis mellifera* was poorly represented by 3.3%. Insects foraged throughout the day light period. Their activity was highest in the morning hours (9:00 a.m.-10:00 a.m.). The installation of the nest of *E. macrognatha* and *T. fraterna* close to *A. esculentus* field is recommended to improve the seed production of okra.

Halder et al (2014) reported that the effective pollinators in case of okra (*Abelmoschus esculentus* L.) are *Hermetia illucens*. wild bees, *Eurema hecabe*, *Orthellia*sp etc. Interestingly, in non-urbanized area insect pollinators' diversity was higher.

Angbanyere and Baidoo (2014) reported honeybees were the main pollinators of okra flower, however their numbers were significantly different on the different treatments. Many of the other insects found on the plant were not involved in pollination of the flowers. There were no significant differences (P>0.05) in the numbers of flowers per plant and numbers of aborted flowers, but significant differences (P<0.05) were recorded in the numbers of pods per plant, yield per plant and the numbers of seed per pod.

Baskaran et al (1999) in Tamil Nadu, India found that okra (*Abelmoschus esculentus*) flowers were visited by 4 bee species namely the little bee (*Apis florea*), the Indian bee (*A. cerana indica*), the digger bee (*Anthophora occidentalis*) and the rock bee (*Apis dorsata*) with relative abundances of 39.84, 33.93, 13.49 and 12.74%, respectively. The dominance of the bees could be related to their colour preferences in flowers.

Malerbo-Souza et al (2001) reported that in Sao Paulo, Brazil flowers of okra cv. Chifre-de-veado were visited by insects between 7.00 and 18.00 h. Okra flowers began opening by 9.40-10.00 h, and closed by 14.45 and 15.20 h of the same day, when they started to shrivel. The nectar hoarding insects were Thysanoptera, Formicidae, Coleoptera and Hymenoptera (*Melipona* sp. and *Apis mellifera*). The crop showed 100% self-fertilization and did not require pollinators to produce fruits. However, flowers visited by insect fruits were heavier, longer and wider.

Njoya et al (2005) reported that *Megachile sp.* is the most efficient pollinator of okra as it had contacts with more than (56.1%) stigma upon landing followed by *Halictus spp.* which also acted as most important pollinator of okra as they frequently rolled on the anthers and consequently on the stigma of the same flower (86.3%) before taking off. *Xylocopa sp.* did not aid in pollination as it visited okra flower just to collect pollen and acted as pollen thief. *Apis mellifera* was mainly a nectar collector in okra flowers.

Abrol (1995) studied the foraging populations of various insects on okra (*Hibiscus esculentus*) [=*Abelmoschus esculentus*] in relation to some weather factors. *Apis dorsata*, *Apis cerana*, *Xylocopa fenestrata*, *X. pubescens*, *Lasioglossum* sp.,*Halictus* sp., *Chalicodoma flavipes* and *C. lanatum* visited okra flowers. Ants, thrips, coccinellids, *Syrphus* sp. and butterflies were also on okra blossoms in very low populations. Each bee species had a specific thermal requirement for initiation of flight activity. *Apis dorsata, Apis cerana, Bombus* sp. and *Xylocopa fenestrate* commenced flight activity at an air temperature of 16-19°C but this requirement was 21.5-24.0° for *C. flavipes, C. lanata, Halictus* sp. and *Lasioglossum* sp., and 24-26° for *P. smaragdula*. Maximum populations of pollinators were observed between 10.00 and 12.00 h. The activity of these pollinators was significantly and positively correlated with atmospheric temperature. *Xylocopa fenestrata, A. dorsata* and *Bombus* sp.and were most abundant throughout the day.

## Impact of pollination on qualitative and quantitative characters

Azo'o et al (2012) found that *E. macrognatha* and *T. fraterna* were effective pollinators of okra, *Abelmoschus esculentus* (L.) Moench (Malvaceae). Their visits increased the proportion of ovules that developed into seeds (pollination rate), which led to longer fruits. Overall, the pollination rate, the number of matured seeds per pod and the mean length of fruits were higher in the flowers that received a single visit of *E. macrognatha* than those that received a single visit of *T. fraterna* and those that were isolated for autonomous self-pollination. At the same time, the activity of *E. macrognatha* was higher than that of *T. fraterna* as concerns the reduction of aborted ovules. Indeed, a floral visit by each studied bee species brought a supplementary number of pollen grains on the stigma, which had already received a quantity of pollen grains from autonomous self-pollination. This additional pollen deposition had a significant impact on the pollination rate, the number of developed seeds per pod and the length of the fruit. Amongst the two wild bee species, the contribution of *E. macrognatha* is higher than that of *T. fraterna* on the pollination, the pod length and the seed number of okra. Therefore, *E. macrognatha* and *T. fraterna* could serve as backup or alternate pollinators of *A. esculentus* in Maroua. Then the conservation of the two wild-bee nests in areas surrounding okra plantations in bloom is recommended (Table 1).

**Table 8.1:** *Parameters related to the production of okra as a function of treatments (Source : Azo'o et al 2012).*

| **Parameter** | **Sngle bee visit of** | | **F** |
|---|---|---|---|
| | ***Eucara macrognatha*** | ***Tetralonia fraterna*** | |
| 100a | 100a | 100a | |
| 71.01 ± 7.44a | 81.43 ± 5.85b | 76.60 ± 7.40c | 16.66 |
| 79.83 ± 8.97a | 94.26 ± 6.99b | 84.13 ± 9.37a | 22.74 |
| 32.43 ± 7.01a | 21.13 ± 6.87b | 26.60 ± 7.01c | 15.68 |
| 15.08 ± 1.22a | 16.19 ± 1.19b | 15.21 ± 1.47b | 08.54 |

Means ± SD within a column followed by the same letter are not significantly different.

Mishra et al (1987) reported that flowers exposed to insects produced heavier and longer fruits than those isolated from insect visitors by nylon net or muslin cloth. The average seed set was significantly higher in open pollinated flowers as compared to muslin net or nylon net caged flowers. However, mode of pollination did not affect seed weight, per cent dry matter, protein and alcohol soluble sugar content.

**Table 8.2:** *Effect of modes of pollination on some qualitative and quantitative characters of okra (average of 9 cultivars)*

| Character | Mode of pollination | | | CD 0.05 |
|---|---|---|---|---|
| | Open pollination | Net caged | Cloth bagged | |
| Capsule weight (gm) | 18.847 (+25.03) | 14.355 (-4.77) | 15.074 | 2.313 |
| Capsule length (cm) | 14.13 (+ 17.95) | 12.06 (+0.67) | 11.98 | 1.21 |
| Number of seeds/ capsule [1] | 77.22 (+32.98) | 66.89 (+15.19) | 58.07 | 7.34 |
| Seed weight gm (50 seeds )[1] | 2.772 (+ 0.87) | 2.631 (- 4.26) | 2.748 | 0.102 |
| Dry matter (%) | 16.67 (24.07) | 16.19 (23.69) | 15.86 (23.45) | NS |
| Protein (%) | 17.46 (24.60) | 18.03 (25.12) | 16.70 (24.30) | NS |
| Alcohol soluble sugars(%)[3] | 9.90 (3.14) | 9.73 (3.11) | 9.54 ( 3.08) | NS |

**Source:** Mishra et al 1987.

Figure in parentheses are increase (+) or decrease (–) over selfing., Figures in parentheses are angular transformation. Figures in parentheses are $^{n}\sqrt{}$ transformation

**NS:** non-significant

## Pollination Recommendations and Practices

There are no clear recommendations on the number of honeybee colonies required for optimum fruit production, yet the available information conclusively indicates that cross-pollination by insects have a marked influence in improving the quality and yield of okra crop. Angbanyere and Baidoo (2014) reported that Insect pollination reduced the number of flowers aborted, increased number of fruits per plant, yield per plant and number of seeds per pod. They further found that insect pollination therefore improved the reproductive performance of okra. Some level of insect pollination is therefore necessary for better yield even in plants that could be self-pollinated.

## Insect pests of okra

Okra production has been estimated to be around 4 million tonnes in the tropical, subtropical and Mediterranean climates which contributes about 4% of the total vegetable consumption in the world (Figure 8.2). Average yield of okra is 6.5 to 7.5 t/ha which is far below the expected due to attack of several insect pests and diseases which cause extensive loss.

In India, Okra is ravaged by more than 30- 40 insect pests right from germination to harvest (Kumar *et al.*, 2002; Gulati, 2004, Jagtab *et al*., 2007). The okra crop is infested by a number of insect pests like *Amrasca devastans, Earias vittella, Bemisia tabaci, Helicoverpa armigera, Acrocercops bifasciata, Thrips tabaci, Aphis gossypii, Podagrica, Anomisflava, Sylepta derogata, Haritalodes derogata, Dysdercus koengii* and *Nezara viridula.* But *Amrasca devastans*, *E. vittela, H. armigera* and *B. tabaci*, are the notorious and major insect pests of okra (Dubey *et al*., 1999; Basu, 1995; Lohar, 2001). Insect pest infestation not only reduce the growth but also transmit pathogenic diseases (Sheedi, 1980; Dhaliwal *et al*., 1981).

Some major diseases affecting the production of okra include powdery mildew, okra mosaic virus, *Cercospora* leaf spot, fusarium wilt, verticillum wilt, etc. The aphids and jassids pierce and suck the cell sap from the tender leaves, flowers, twigs and tiny fruits. The powdery mildew and blight are also common diseases in okra crop and damaging in different stages when the plants are in 6-8 leaf stage. The infected plants results in reducing the quality of fruits and yield reduction. Blight causes yellowing and drying of leaves which starts from the margin and goes to the center of the leaves. Finally, it results in droping of leaves and hindering the photosynthesis as well as reducing yield. This challenging situation invites attention of entomologists to concentrate energies on integrated pest management (IPM) of okra and other crops (Huffaker and Carl, 1980; Dhaliwal *et al*., 1981; Mahadevan and Chelliah, 1986; Dubey *et al*., 1999; Kumawat *et al*., 2000).

*Earias* sp. alone has been reported to causes damage up to 71% (Pareek and Bhargava, 2003) whereas in general the overall pod damage is about 49% and 42% according to Kanwar and Ameta (2007) and Fletcher and Mishra (1990) respectively. Okra shoot and fruit borer results in 88-100% fruits damage in okra (Radake and Undirwade (1981). They lay eggs individually on leaves, floral buds and on tender fruits. After hatching the small brown caterpillars bore into the top shoot and feed inside before fruit formation. the resulting fruits are smaller and deformed(Mohan *et al.* 1983, Atwal, 1976).

Ahmad et. al. (2015) reported that Okra (*Abelmoschus esculentus*) is attacked by several insect pests such as aphid, leafhopper, spotted bollworm and dusky cotton bug and diseases like powdery mildew and okra blight. The spotted bollworms are the most dreadful pests causing serious turn down of the produce, in terms of quality as well as of quantity (Suman *et al.*1984). *Earias* spp. are distinguished from other pests of "okra" by their marked tendency for stem boring. The larvae enter the terminal bud of vegetable shoot move down the growing point. Severe attack, results in the wilting of top leaves and collapsing of the main stem. The larvae also bore onto the flower buds, flowers and fruits of the crop. *Earias* spp. is one of the most important pests of 'okra' which can cause tremendous losses (36-90%) in yield (Misra *et al*., 2002). Serious decline in production of "okra" due to the okra fruit and shoot borer has been reported by different workers; 8.4 to 73.2% variations in fruit infestation (Kumar and Urs, 1988), 32.06 to 40.84% loss in yield (Singh and Brar, 1994) and 20 to 51 reduction in yield (Krishnikumar and Srinivasan, 1987).

Use of chemical insecticides is in vogue for the control of this pest. Exclusive reliance on insecticides as a control strategy against this pest has resulted in several undesirable effects, suc as pesticide pollution, resurgence of secondary pests, insecticide resistance, killing of beneficial fauna and human health problems. There is a need to explore alternative approaches to reduce the sole dependence on insecticides. For this purpose, varietal control is the cornerstone in an ecological approach to pest control (IPM), due to its compatibility with other control tactics.

The use of crop varieties showing resistance to insects is being practiced since last six decades in modern agricultural production systems. Continuous rise in the food production costs in the developing countries has elevated the need to use the varieties having resistance against herbivore pests (Smith, 1986). Plant resistance to insects provides the indispensable foundation on which the structure of the IPM for different pests can be built. Insect resistant cultivars can be used as sole control method and also interact synergistically with biological, chemical and cultural control methods to reduce the spread of pest insects (Wiseman, 1990).

Plant characteristics have been recognized important resistance factors by both plant breeders and entomologists. Each plant species has a distinctive set of defense traits ranging from morphological to phytochemical factors that have behavioral and physiological effect for a potential herbivore consumer. Antibiotic and antixenotic effects manifested in insects may occur because of either presence or absence of these factors. Even low level of resistance in crop plants against pests plays important role in reducing the need for other control methods. The first plant organs contacted during the preliminary stages of host acceptance are surface hairs or trichomes. Insect species respond differentially to the presence of plant hairs. Pubescence as a resistance factor interferes with insect oviposition, attachment to the plant, feeding, colonization and ingestion. In some cases the trichomes possess associated glands that exude secondary plant metabolites. The effect of glandular trichomes may depend on the nature of the exudates, which may be of allelochemics such as alkaloids or terpenes (Elden and McCaslin, 1997; Gannon and Bach, 1996). Conversely, glabrous character of leaves may also result in reduction of population in certain cases (Lukefahr *et al.*, 1965). Some insects prefer leaves over bolls for oviposition in view of high density of hairs (Mehta, 1971) and pubescent accessions support more eggs than having glabrous leaves (Murugesan, 1982). In addition, increased leaf fiber, silica content (Bergvinson *et al.*, 1997a; Rojanaridpiched *et al.*, 1984), surface wax and high hemicellulose (Hedin *et al.*, 1993) have been identified as resistance mechanisms in different crops.

## Distribution of okra shoot and fruit borer

Butani and Jotwani (1984) reported that okra shoot and fruit borer, *Earias vittella* (Fabricius) is widely distributed and is recorded from Pakistan, India, Sri Lanka, Bangladesh, Burma (Myanmar), Indonesia, New Guinea and Fiji. This pest is common oriental species found from India and China to North Australia (Hill, 1983). Atwal (1976) reported that *Earias vittella* is widely distributed from in North Africa, Pakistan, India, and other countries and is a serious pest of okra and cotton.

## Biology of Okra shoot and fruit borer

Okra shoot and fruit borer is a holometabolous insect. So, it has four stages to complete its life cycle viz. egg, larva, pupa and adult.

## Egg

The eggs are about 0.5 mm in diameter, spherical and bluish green in color. The egg shell (chorion) has parallel longitudinal ribs forming a crown-like structure at the top. Eggs are laid singly or in small groups on young shoots, underside of the leaves, flower buds or young pods. Depending on the species, 82-378 eggs are laid in each 4-7 days and they hatch in 3-4 days in warm weather and 8-9 days under cold weather (Rahman and Ali, 1983).

## Larva

Newly hatched larva is 1-5 mm long; brownish-white has a dark head and prothoracic shield. The larvae undergo 4-5 molts. Larval duration varies from 9-20 days in warm weather and 50-60 days in winter (Rahman and Ali, 1983). Their main characteristics is that their body surface is irregularly spotted and spiny. Hence they are called as spiny bollworm or spotted bollworm. The fully developed (last instar) larvae are about 20 mm long, more or less spindle-shaped, greenish, dark grayish or brownish in color. The dorsal side or the back has a broad, whitish, longitudinal stripe with distinct dark spots. Two orange spots are found on the thoracic segments. Head and spiracles are black.

## Pupa

Pupa is shiny yellowish brown, about 12-14 mm long and found in a firm, yellowish-white to light grayish cocoon, shaped like an inverted boat. The pod borer pupates on top of the soil layer or on the plant, often on dried shoots and pods. Dark brown Pupa is enclosed in a dirty white to buff color cocoon. The pupal period lasts from a few days to more than months depending upon the climate. The average pupal period being 1-3 weeks (Rahman and Ali, 1983).

## Adult

The moth is yellowish-brown, about 12 mm long with a wingspan of about 20-25 mm. Color of fore wings is variable, depending on the season of the year (temperature), i.e. yellowish white with a pink shade or brownish-yellow and with a green (sometimes brown), more or less wedge-shaped longitudinal stripe. The dorsal side of the thorax has two green stripes. Hind wings are whitish. It has 11 generations in a year. The longest life cycle (49 days) was observed during January while the shortest life cycle (29 days) was found during July (Sharma *et al*., 1985).

## Host range of okra shoot and fruit borer

Gautam and Goswami (2004) stated that okra shoot and fruit borer (*Earias vittella)* feeds on many species of malvaceous plants. Satpute *et al*. (2002) studied on different hosts of *Earias vittella* and found that okra was most preferred host for the development of the pest, followed by cotton, artificial diet and mesta (*Hibiscus sp.*) based on average minimum pupal period, highest fecundity and maximum pupal and adult weight. Dongre and Rahaller (1992) were examined the relative food plant preference and induction of preference for feeding behavior in larvae of *Earias vittella* under dual choice conditions. Out of 5 food plants, *Abelmoschus esculentus* (okra) was the most and *Hibiscus rosa-sinensis* was the least preferred.

Butani and Jotwani (1984) found okra shoot and fruit borer as an oligophagous pest although okra and cotton are its main hosts. They also found it to feed on a large number of malvaceous plants, both wild as well as cultivated.

Khan and Verma (1946), Pearson (1958), Butani and Verma (1976), Atwal (1999), David (2001) reported that this pest has been infested to okra, cotton, hollyhock, safflower, Indian mallow, *Corchorus* sp., *Hibiscus* sp., *Malvas* sp., *Malvastrum* sp., *Sida* sp., *Theobrome* sp. and *Urena* sp.

Atwal (1976) mentioned that okra and cotton is the most favorite host of OSFB. Plant species including sonchal (*Malva parviflora*), gulkhaira (*Althaea officinalis*), holly hock (*Althaea rosea*) and some other Malvaceous plants are appear to be its alternate hosts. Rahman and Ali (1983) reported that when okra shoot and fruit borer were offered the choice of different parts of host plant they preferred okra fruit and shoot the best followed by cotton balls, ball. Flowers and buds of deshi cotton (*Gossypium arboretum*), buds and flower of kenaf and maize grains, flower of *Abutilon indicum*, flowers of *Hibiscus rosasinensis*, sarson (*Brassica campestris* var *sarson*), *Malvastrum tricuspidatum*, *Cassia fistula* and ears of pearl millet, pod of jute and soyabean.

## Nature of damage

Okra shoot and fruit borer, *Earias vittella* (Fab.) is one of the key insect pest of okra. This pest causes 36-90 % loss in the fruit yield of okra (Misra *et al.*, 2002). Shah *et al.* (2001) observed that the caterpillars of *E. vittella* bore into the tender shoots and developing floral buds causing drop of fruiting bodies and developing fruits making them unfit for human consumption. When the crop is only a few weeks old, the freshly hatched larvae bore into the tender shoots and tunnel downwards resulting withering of shoots and ultimately killing the growing points. As a result the apical dominance is lost and side shoots may arise and giving the plant a bushy appearance. With the formation of buds, flowers and fruits the caterpillars bore those and feed on the inner tissues. They move from bud to bud and fruit to fruit. The damaged buds and flowers wither and fall down without bearing any fruit. The affected fruits become deform in shape and remain stunted in growth (Butani and Jotwani, 1984; Acharya, 2010).

The larvae of okra shoot and fruit borer bore into the tender shoots, flower buds and fruits. As a result, the attacked shoot dries up while the flower buds and developing fruits dropped prematurely. Affected fruits remain on the plants become unfit for human consumption (Mohan *et al.,* 1983 and Atwal, 1976). The larvae of OSFB bore into the shoots and feed inside and also damage the seed (Karim, 1992).

The first symptom of attack by okra shoot and fruit borer was visible when the crop was three weeks old and the larvae bored into the shoots. Under severe attack, the top leaves wilted and the whole apex of the plant dropped down. In the reproductive stage of the crop, the larvae moved to the flower buds, small fruits and even mature pods and causing reduction of yield (Singh and Bichoo, 1989). Like other insects, the population of spotted bollworm is governed by their inherent capacity to increase, under the influence of various environmental factors.

The damage to the crop is done by two ways. First, the terminal portions of growing shoots are bored by the caterpillars, which move down by making tunnels inside. The shoots droop downward or dry up. Second, the larvae enter the fruits by making holes, rendering them unfit for human consumption (Misra *et al.*, 2002).

## Seasonal abundance of okra shoot and fruit borer

## Ecology

The insect was found to occur in high population during rainy season and its number drop in summer as the temperature increases. The development period of different stages prolonged during winter, the longevity, fecundity and coloration of the adult also fluctuate with environmental temperature and humidity (Schmutterer, 1961). Srinivasan and Gowder (1959) reported that 40-50% okra fruit were damaged due to attack of this pest in Madras. In another study Krisnaiah (1980) observed the attack of fruit borer to the extent of 35% in the harvestable fruit of okra. Rana (1983) observed the pick incidence of shoot and fruit borer of okra was observed in the last week of August with a range of 34 to 45% damage to fruits. The incidence of *Earias* spp. on okra was studied by Dhanwan and Sidhu (1984). He reported that the maximum damage occurred in fruits (67.7%) and buds (52.4%) in late October. The maximum in shoots (1.7%) and flowers (1.5%) occurred in mid-August. In spring, the maximum damage to fruits was 32.04% and increased larval population 1.4/plant was observed in late July. The population of *Earias spp* increased slowly upto mid September and rapidly thereafter. Dhamdhere *et al.* (1984) reported 25.9 to 40.9% damage to fruits in October.

Butani and Jotwani (1984) reported that there is no true hibernation but development and activity is considerably slowed down during winter. Khaliq and Yousuf (1986) also reported the increased incidence of *E. vittella* with the increasing temperature and humidity. In general, the population of OSFB fluctuates from month to month, season to season, even year to year. Dash *et al.* (1987) reported that the occurrence and seasonal abundance of noctuid *E. vittella* was maximum in shoots from July to October.

Dutt and Saha (1990) observed the lower activity of *E. vittella* during December-January and the higher activity was observed during the increasing temperature from February and a maximum peak in May-June. Khurana and Verma (1990) observed lower incidence (12.5%) of *E. vittella* during 1983 in a condition having mean maximum and minimum temperature of 34.30 C and 20.50 C, respectively with a mean RH of 73%, frequent rainfall between May and September. But they found comparatively higher incidence (20.5%) of the pest during 1987, in an environment condition with mean maximum and minimum temperatures of 36.3 C and 23.2 C, respectively having a mean RH of 64.8%.

Ali (1992) reported that the peak abundance and intensity of okra shoot and fruit borer/spotted bollworm in cotton field were in October-November and were more common during early to mid season on growing shoots, buds, pin bolls and developing bolls of cotton and during late season, particularly after January they tend to disappear.

Zala *et al.* (1999) found the activity of okra shoot and fruit borer, *Earias vittella* on shoot was started from the fifth week of July on four weeks old crop and continued till fourth week of September on twelve weeks old crop during 1996. In 1997 the infestation of *Earias vittella* on shoot was started on the first week of August on five weeks old okra crop till first week of October on thirteen weeks old crop. The maximum (26%) shoot damaged plant was observed during 1996 in second week of August on six weeks old okra crop, whereas it was 22% in the third week of August on seven weeks old okra crop during 1997.

Patel *et al.* (1999) reported the infestation of *E. vittella* on okra fruits appeared from the second week of August on six weeks old okra crop and continued till last harvest of fruit during 1996-1997. The intensity of fruit damage varied from 11.11% (second week of August) to 40.43% (fourth week of September) and 10.12% (third week of August) to 47.37% (first week of October) during 1996 and 1997, respectively. The larval activity started from fifth week of August in 1996 and 1997 and continued till the last harvest of the crop. The relevant observations were also reported by Mote (1977) and Kadam and Khaire (1995). Pareek *et al.* (2001) reported that the incidence of okra shoot and fruit borer started in first week of September and maximum fruit infestation recorded in the third week of October. Yadvendu (2001) recorded that the peak incidence of okra shoot and fruit borer and maximum fruit infestation in first and fourth week of September, respectively. Acharya (2002) and Dangi (2004) observed that the incidence of okra shoot and fruit borer commenced from the 4th week of August (6th week after sowing).

A field experiment was conducted by Sharma *et al.* (2010) to study the fluctuation of pest population of *Earias vittella* (Fab.) and their relation with prevailing weather condition at Horticulture Farm in Udaipur, India during Kharif 2005 and 2006. The results revealed that borer incidence commenced in the 29th standard week. The peak infestation of plants (91.6 %) was observed in 45th standard week. The maximum numbers of larvae (7.5 larvae/l0 plants) were recorded in the 42nd standard week.

## Management of okra shoot and fruit borer

Several biologically active compounds have been isolated from different parts of neem tree. Several vilasinin derivatives, salanins, salanols, salasnolactomes, vepaol, isovepaol, epoxyazadirachdone, gedunin, 7-deacetylgedunin have been isolated from neem kernels. Azadirachtin is the most potent growth regulator and antifeedant (Butterworth and Morgan, 1968; Warthen *et al.,* 1978). The triterpenoid azadirachtin (C35H44O16) was first isolated from the seeds of the tropical neem tree by Butterworth and Morgan (1968). Its definite structural formula, which resembles somewhat that of ecdysone, was finally explained in 1985 by kraus *et al.*, 1985 and by Bilton *et al.*, 1985. Azadirachtin is a limonoid alleliochemical (Butterworth and Morgan, 1968; Broughton *et al.,* 1986) present in the fruits and other tissues of the tropical neem tree (*Azadirachta indica*). The fruit is the most important aspect of neem that affects insects in various ways. The leaves, which may also be used for pest control, may reach a length of 30 cm.

A field experiment was conducted by Patil *et al.* (1991) in India for the control of okra fruit and shoot borer (*E. vittella*) with cypermethrin (15 gm/ha), fenvalerate (50gm/ha), acephat (375 gm/ha), quinaphos (250 gm/ha) and endosulfan. (250 gm/ha). All treatments reduced pod damage but cypermethrin treated plants were the least infested gave the best yield.

Misra (1989) studied the bio-efficacy of some insecticides against the pest complex of okra. The author reported that percent shoot infestation in insecticide treated plots varied from 1.74- 10.03% compared to 15.23% in untreated control plots. Adult pairs of males and females of *Earias vittella*, a pest of cotton and okra, were released in breeding chambers in different sets, containing the odours of the leaves of neem, *Azadirachta indica*; tulsi, *Ocimum basilicum*; eucalyptus, *Eucalyptus rosfrata*; lantana, *Lantana camara*; bulbs of garlic, *Allium sativum* and one control set with no odour. Adult longevity did not differ significantly among the treatments. All the treatments significantly reduced the egg output as compared to the control (172 eggs). Similarly, all the odours significantfy reduced egg hatching compared to the control (90.81%). The lowest number of eggs (128) and hatching (68.15%) were recorded with *Azadirachta* leaves odour (George, 1997).

Ambekar *et al.* (2000) evaluated the efficacy of neem-based formulations and synthetic insecticides against okra shoot and fruit borer. They found that all treatments significantly reduced fruit borer infestation over the untreated control. However, cypermethrin at 0.1% was the most effective and recorded the lowest infestation of 6.57%. Chinniah and Mohanasundaram (1999) studied the possible toxic effect or the safety of the neem derivatives to the predatory mites *Amblysieus* spp. The neem products *viz.,* neem cake extract (10%), neem seed kernel extract (5%) and neem oil (3%) proved safe by recording lower predator mortality.

## Mode of action of neem

### Settling Behavior

Crude neem extracts deters settling and reduces feeding in *M. persicae* (Griffiths *et al.* 1978 and 1989).

### Oviposition Behavior

The females of some lepidopterous insects are repelled by neem products on treated plant parts or other substrates and will not lay eggs on them under laboratory conditions.

### Feeding Behavior

Azadirachtin is a potent insect antifeedant. Antifeedancy is the result of effects on deterrent and other chemoreceptors. The antifeedant effects of azadirachtin have been reported for many species of insects. Reduction of feeding was also observed after topical application or injection of neem derivatives, including AZA and alcoholic neem seed kernel extract. This means that the reduction of food intake by insects is not only gustatory which means that sensory organs of the mouth parts but also non-gustatory regulate it. These two phagodeterrent/antifeedant effects were called primary and secondary (Schmutterer, 1985).

### Metamorphosis

Azadirachtin has different influence on the metamorphosis of the insects resulting in various morphogenetic defects as well as mortality, depending on the concentration applied.The IGR effect of neem derivatives such as methanolic neem leaf extract and azadirachtin in larvae and nymphs of insects was first observed in 1972 in Heteroptera (Leuschner, 1972) and Lepidoptera.

Molting, if it occurred, was incomplete and resulted in the death of the tested insects.Botanicals possess an array of properties including insecticidal activity and insect growth regulatory activity against many insect pests and mites (Rajasekaran and Kumaraswami, 1985; Prakash *et al.*, 1987; 1989 and 1990). Low mammalian toxicity, no reported development of resistance to their production so far, less hazardous to non-target organisms, no pest resurgence problem, no adverse effect on plant growth, negligible application risks, low cost and easy availability are the advantages of plant products over synthetic chemicals.

Ahmed (1984) enlisted 2121 plant species, possessing pest control property which include neem, sweetflag, cashew, custard apple, sugar apple, derris, lantana, tayanin, indian privet, agave, crow plant etc. 1005 species of plants having biological properties against insect pests including 384 species as antifeedants, 297 as repellents, 97 as attractants and 31 as growth inhibitors. About 413 different species/sub-species of insect pest have been listed by (Schmutterer, 1995) found to be susceptible to neem products. The listed species/sub-species belongs to different insect orders most of them were Lepidoptera (136) and Coleopteran (79).

## Management by botanicals

The use of neem based insecticides as a source of biologically active substances for pest control is increasing worldwide, and have recently gained popularity as components of integrated pest management (Banken and Stark, 1997). Maximum reduction in bollworm infestation (65.7%) was observed in garlic treated plot. Garlic extract and NSKE both at 10 per cent were found to be superior. Lowest bollworm incidence was observed with NSKE (10.3%), datura and neem oil emulsion (Anon., 1987).

Sardhana and Kumar (1989) studied the efficacy of neem oil, karanj oil (both at 0.5, 1.0 and 2.0%) and garlic oil (0.25, 0.5 and 1.0%) in comparison with monocrotophos (0.05%). Among the oils, neem oil and karanj oil offered effective control against okra fruit borers.

Weekly application of neem (*Azadirachta indica*) oil at 2% was effective for controlling *E. vittella* on okra (Sardana and Kumar, 1989). They observed that the plots having lower fruit damage and increased yields in treated plots monocrotophos at 0.05% and can therefore, be recommended for the use in an integrated control scheme for the rest. The most probable effect of neem in Lepidopterans is the disruption of the larval-pupal molt (i.e. pupation), which has been frequently reported (Schmutterer *et al.,* 1983; Koul and Isman, 1991). Repellent activity of neem against oviposition by Lepidopterous pests has also been reported for *Spodoptera litura* (Joshi and Sitaramaiah, 1979), *Cnaphalocrocis medinalis* (Saxena *et al*., 1981) and *E. vittella* (Sojitra and Patel, 1992). Extracts of neem and bakain caused maximum adverse effects on fecundity and hatching.

Numerous plant species have been reported to possess pest control properties but only a few seem to be ideally suited to practical utilization. Among these, neem (*Azadirachta indica* A. Juss) and bakain (*Melia azedarach* L) are the most promising plants from the entomological perspective (Schmutterer, 1990 and 1995). Neem oil produced non-toxic effects after spray and acted as antifeedant, growth inhibitor and oviposition deterrent against insect's pests of okra and cotton (Ahmed *et al.,* 1995). Patil (2000) conducted an experiment with 20 indigenous plant extracts to evaluate the antifeedant property against insect pest. *A. indica* exhibited maximum of 10-51 percent antifeedant followed by *A. calamus* (15.69%) and *A. squamosa* (17.31%) against third instar larvae of *Earias vittella.*

Morale *et al.* (2000) studied the effect of plant product against *E. vittella* of cotton under laboratory condition and revealed that neem oil 1%, karanj oil 1%, cotton seed oil 1%, neem seed extract (NSE aqueous) 5% and NSE (methanolic) 1% were significantly affected the larval period, larval mortality and fecundity of *E. vittella.* Lakshmanan (2001) reported effectiveness of neem extract alone or in combination with other plant extracts in managing lepidopteran pest's viz., *E. vittella, Chilo partellus* Swinhoe, *H. armigera* and *S. litura*.

Antifeedant effect of neem in combination with sweetflag and pongam extracts on okra shoot and fruit borer was studied by the Rao *et al.* (2002) which gave 43.12 to 80.00 percent mortality protection over control. Mudathir and Basedow (2004) found that different preparations of neem significantly reduced okra shoot and fruit borer infestation in okra.Singh *et al.* (2005) tested the efficacy of two botanicals and insecticides and reported that NSKE @ 1.5% was found superior after fenvalerate with respect to yield. NSKE (1.5%), NSKE (1%), karanj seed kernel extract (KSKE) (1.5%) and NSKE (1%) were superior by recording 58.27, 47.32, 44.25 and 41.5 q/ha yield, respectively as against 29.17 q/ha in untreated control.

## Management by chemicals

Borer is the major pest of okra, committing colossal losses to okra growers. Although various measures have been reported for controlling the pests, there is not a single such method that successfully be adopted to suppress the incidence and damage of the pests. This perhaps, is mainly due to the oligophagous nature of this pest that helps their year round population build up. However, a thorough search of review reveals that the approaches that had ever been made in controlling this pest comprise mainly the use of chemicals. Though various management approaches are practiced for this pest suggested by different workers, insecticides are found very effective which are cited below.

Badaya *et al.* (1999) studied the efficacy and economics of different insecticides for the control okra fruit and shoot borer, *Earias vittella* Fab. They found that among several insecticides against *Earias vittella* on okra Madhya Pradesh, India, fenvalerate at 0.02% was the most effective and profitable treatment, while dimethoate at 0.05% was relatively ineffective. A field experiment was conducted by Patil *et al.* (1991) in India for the control of okra fruit and shoot borer (*E. vittella*) with cypermethrin (15 gm/ha), fenvalerate (50 gm/ha), acephat (375 gm/ha), quinalphos (250gm/ha) and endosulfan (250 gm/ha). All treatments reduced pod damage but cypermethrin treated plants were the least infested gave the best yield.

Chaudhury and Dadheech (1989) reported that if the insecticidal protection was not given the okra shoot and fruit borer infested fruits were as much as 57.1% with a yield of 9.83 kg/plot. But the plots protected with alternate weekly sprays of 0.03% phosphamidon and 0.05% endosulphan provided yield of 15.65 kg/plot with 10% fruit infestation. Misra (1989) studied the bio-efficacy of some insecticides against the pest complex of okra. The athor reported that percent shoot infestation in insecticide treated plots varied from 1.74-10.03% compared to 15.23% in untreated control plots.

Sarkar and Nath (1989) conducted a field trial in Tripura, India, and indicated that decamethrin, malathion, endosulfan and carbaryl were effective to control okra shoot and fruit borer but fenvalerate (0.5 ml/L and 750 ml/ha) gave the greatest reduction in number of infected fruits. Haque (1993) conducted an experiment on the management of oka fruit and shoot borer and found that a foliar spray of Ripcord 10 EC @ 1.5ml/l of water showed significantly the lowest level shoot and fruit infestation (7.68%, 1.13% respectively).

## Seasonal incidence of sucking pests of okra

## Leafhoppers (*Amrasca biguttula biguttula*)

Hoppers lay pear shaped, elongated and yellowish white eggs in the veins on the under surface of leaves. Both the adults and nymphs suck the cell sap. Nymphs move diagonally when disturbed. There are 10 to 12 overlapping generations in a year. Plants loose their vitality and aff ected leaves turn yellow and curl upwards. When the infestation is high during summer, leaves turn brick red and show large necrotic spots, which is often confused with fungal infection.

## Management

TMTMApply neem cake @ 250 Kg/ha to soil immediately after germination and repeat after 30 days. In the initial stages of crop before fl owering spray systemic insecticide like Acephate 75 SP (1ml/l) or Imidacloprid (0.3 ml/l). Once the fruit harvest starts avoid systemic insecticides. Spray neem or pongamia soaps @ 0.5% or pulverized neem seed powder extract (NSPE) 4% at the lower surface of the leaves.

Senapathi and Khan (1978) from Bubaneshwar, India recorded weekly observations of the leafhopper on okra plant, from first week of July, 1975 to the last week of June,

1976. The results revealed that the insect occurred throughout the year on the crop and maintained high population level from November, 1975 to February, 1976 with its peak (8.33 per 100 sq. Cm area) in the month of December, due to low temperature and low relative humidity. Leafhopper population was low during early part of the March to third week of June which was due to prevailing warm weather

Mahamood *et al.* (1990) observed the appearance of pest, *A. biguttula biguttula* on okra in Pakistan and reported the activity of leafhopper until the end of crop season. Further, among various environmental factors, the only significant factor in both the years (1986-87) of the study was maximum and minimum temperature on the density of the pest. Neither relative humidity nor rainfall had significant influence on pest population. Sharma and Sharma (1997) conducted field studies in Haryana and recorded the number of nymphs of *A. biguttula biguttula* on the ventral surface of thirty leaves of ten okra plants at weekly intervals starting from the first week of July. The highest population was observed during the first week of August. Numbers were negatively correlated with the maximum temperature and positively correlated with the minimum temperature and average relative humidity.

## Aphids *(Aphis gossypii)*

This is a polyphagous pest, feeding in colonies and completely covers the shoot tips, buds and lower surface of leaves. Both nymphs and adults suck the sap. They also excrete honeydew on which sooty mould develops.

## Management

For management clip and destroy the infested shoots. During pre-flowering period spray systemic insecticides like Dimethoate 30 EC (2ml/l) or Acephate 75 SP or Acetamiprid.

## Petiole Maggot (*Melanagromyza hibisci* Spencer)

This is a dipteran fly and attacks the crop throughout the growth period. Generally drying of leaves is noticed. If petioles of such leaves are opened, yellow maggot and pupae can be noticed. Pest attacks immediately after germination and may cause mortality, or cracking of main stem.

## Management

Apply neem cake @ 250 kg/ha immediately after germination and repeat after 30 days followed by sprays of NSPE 4% or neem soap 1% at 10 days interval after flowering.

## Whitefly (*Bemisia tabaci*)

## Origin and Distribution of whitefly

Okra whitefly *(Bemisia tabaci* Genn.) is the most important insect pest of okra in Bangladesh and acts as the vector of Okra yellow vein clearing mosaic virus (OkYVCMV) on okra. *Bemisia tabaci* was first described in 1889 as a pest of tobacco in Greece and named as *Aleyrodes tabaci*, the tobacco whitefly (Gennadius, 1889). From 1926 to 1981, *B. tabaci* was reported as sporadic pest and was the most important vector of plant viruses in subtropical, tropical and temperate zones where winters are mild enough to permit year round survival (Cock, 1986). Worldwide distribution whitefly, *B. tabaci* is given below (Table 4)

**Table 8.3:** *Global distribution of whitefly as updated by CAB International Institute of Entomology, London (Cock, 1986)*

| Continent/Subcontinent | Countries |
|---|---|
| Europe | Cyprus, Denmark, Finland, Greece, Switzerland, Turkey, UK etc. |
| USSR(Former) | Azerbaijan SSR, Georgian SSR |
| Africa | Angola, Cape Verde Island, Egypt, Ethiopia, Ivory coast, Sierra Leone, Somalia, South Africa etc. |
| Asia | Afghanistan, Myanmar, China, India, Indonesia etc. |
| Pacific Islands and Australia | Australia and Hawaii etc. |
| North West Atlantic | Bermuda |
| America, Northern USA | California, Florida, Texas etc. |
| Canada | British Columbia, Quebec etc. |
| Central America and Caribbean | Barbados, Costa Rica, Puerto Rico etc. |
| South America | Argentina, Brazil, Venezuela, Colombia etc. |

There are more than 1000 whitefly species in the world. Twenty species occur in Hawaii; 12 of which have been accidentally introduced in the last 20 years. The sweet potato whitefly (*Bemisia tabaci*) is one of the most pestiferous of the group. This pest was first described as *Aleyrodes tabaci* from tobacco in Greece in 1889. In Hawaii, it was first found at Pearl City, Oahu in October 1982. When discovered it had already been in Hawaii for a considerable time. Field surveys conducted at the time turned up infestations in leeward Oahu (Pearl City), central Oahu (Kalihi), and windward Oahu (Kailua). As of October 1990, it occurs on the islands of Hawaii, Kauai, Maui, Molokai, and Oahu.

## Host range of whitefly

The sweet potato whitefly has an extremely wide host range. It attacks more than 500 species of plants (Greathead, 1986) from 63 plant families (Mound and Halsey, 1978). Mau and Tsuda (1991) recorded annona (cherimoya, atemoya, sugarapple), avocado, broccoli, cauliflower, Chinese cabbage, Chinese wax gourd, cucumber, *Dendrobium* (flowers), edible gourds, eggplant, fig, squash, sweet potato, togan, tomato, ung-choi, watermelon, yard long beans and zucchini. Although not yet reported in the state, other crop hosts include cabbage, chrysanthemum, beans, bitter melon, dishrag squash, pepper, pea, and radish as host plants of sweet potato white fly. Besides crop plants, weeds also serve as alternate hosts. some of the non crop hosts also serve as hosts castorbean, *Euphorbia,* ilima, *Ipomoea* spp., *Asystasia*, *Coccinia* sp., kikania-lai, *Momordica* sp., *Malva* sp., mountain apple, sowthistle, spurge, and *Xanthium* sp

## Biology of whitefly

Whiteflies have six life stages which include the egg, four nymphal stages, and the adult. The development period from egg to adult takes 15-70 days depending upon temperature and plant host.

## Egg

Female whiteflies deposit pear-shaped eggs into the mesophyll or inner tissue of the leaf from the lower surface. Eggs are attached to the leaf by a stalk-like process. Eggs are white when first laid, and become brown prior to hatching. They are generally laid on the underside surface of the younger, upper leaves of the plant. Females lay from 28-300 eggs depending on host and temperature. On cotton, a female lays 81 eggs, on average, at 80.060 F (26.70 C) and 72 eggs at 89.960 F (32.20 C) (Butler *et al.*, 1983). Egg densities can be as high as 1,200 eggs per square inch.

Under laboratory conditions on cotton leaves, eggs hatched in 5 days at 90.50F (32.5 0C) and 22.5 days at 62.10 F (16.70 C) (Butler *et al.*, 1983). Eggs did not hatch in temperatures above 96.80 F (360 C) in the same study. Similar incubations have been reported on sweet potato and potato (El-Helay *et .al.*, 1971; Azab *et. al.*, 1971) and eggplant (Avidov, 1956; Butler *et. al.*, 1983).

Low temperatures increase mortality. However, humidity is not a factor in egg mortality and egg incubation periods.

## Nymph

The first nymphal stage is called crawlers and the last stage is as the pupa. After hatching the crawlers move a short distance and settle to feed. Once settled, the subsequent three nymphal stages are scale-like and sedentary. Nymphs are creamy white to light green and oval in outline. The total nymphal period lasts about 2-4 weeks.

## Adult

Adult sweet potato whiteflies are small, approximately 1/25 inch in length, with a pale yellow body and two pairs of white wings and covered with a white waxy powder. At rest, wings are held in an inverted V position. Their compound eyes are red.

Adults usually emerge from their pupal cases in the morning hours and may copulate a few hours later. Oviposition occurs from 1 to 8 days after mating. Adult life span ranges from 6-55 days dependent on temperature. Females live only 10-15 days under southern continental U.S. summer conditions, but can live several months during the winter.

In this species, reproduction can occur with or without copulation. Unmated females can reproduce by parthenogenesis. Females lay 80 to more than 300 eggs in their lifetime. The plant host reportedly plays an important role in female fecundity. This species is similar in appearance with other whitefly species.

## Nature of damage

*B. tabaci* continues to be an economically important pest of greenhouse and field crops throughout equatorial areas of the world (De Barro, 1995). The insect breeds throughout the year and the female lays stalked yellow spindle shaped eggs singly on the lower surface of the leaf. Nymphs and adults suck the sap usually from the under surface of the leaves and excrete honeydew. Leaves appear sickly and get coated with sooty mold. The whitefly serves as the vector for the spread of yellow vein mosaic virus (YVMV) disease causing damage to okra crop. Berlinger (1986) reported that whitefly, *B. tabaci* damaging the plants in three means that were discussed below:

## Direct damage

The damage is caused by the piercing and sucking sap from the foliage of plants.. This feeding causes weakening and early wilting of the plant and reduces the plant growth rate and yield (Berlinger, 1986). It may also cause leaf chlorosis, leaf withering, premature dropping of leaves and plant death. Infestations of sweet potato whitefly nymphs are associated with the occurrence of irregular ripening of tomatoes and silver leaf of squash. .

## Indirect damage

Indirect damage results by the accumulation of honeydew produced by the whiteflies. This honeydew serves as a substrate for the growth of black sooty mold on leaves and fruit. The mold reduces photosynthesis and lessens the market value of the plant or yields it unmarketable (Berlinger, 1986).

Mandahar and Singh (1972) studied the effect of *Okra Yellow Vein Mosaic Virus* on host. They reported that infection of *Hibiscus esculentus* induced 62-82% reduction in total chlorophyll and 56.61% reduction in total photosynthesis, while the respiration of infected tissue was increased 8.33%. It was concluded that carbohydrates were transported from healthy to diseased leaves in which they accumulate and this may account in part for infected plants not to bear any fruit.

## Transmission of virus

The third type of damage is caused by the vectoring of plant viruses by Whiteflies (Cohen and Berlinger 1986). Whiteflies have been reported to cause over 40 diseases of vegetable and fiber crops worldwide. The sweet potato whitefly is considered the most common and important whitefly vector of plant viruses worldwide (Basu, 1995).

## Management of whitefly

High reproductive rate and multiple host sequences provide optimal conditions for whitefly population development which is the most difficult challenge for formulating control and suppression strategies. Egg mortality is usually minimal and even weather and predation does not have much effect. They have already developed resistance to chemical insecticides. Regular insecticide applications can result in resurgence of other pests. evidently, exploration of other options gain importance. Rosaiah (2001) found that neem oil 0.5% was significantly superior in reducing the whitefly population and shoot and fruit borer damage on okra followed by NSKE (5%). Several other investigators also found neem formulations effective in suppressing *B. tabaci* population (Jayaraj *et al.* 1986, Natarajan *et al.* 1986, Nandihalli *et al.* 1990). Srinivasamurthy and Sharma (1997) reported that some of the traditional pest management practices such as use of Neem oil (2%), castor seed oil (5%), *Madhuca latifolia* (J. F. Gmel.) oil (0.5%), tobacco leaf extract (0.5%), fish oil rosin soap (0.2%) and nicotine sulphate (0.2%) resulted in considerable reduction of whitefly population. Haider (1996) found that Ripcord was the very effective in managing the virus disseminating whitefly. Torres (2004) found that Thiamethoxam and Imidaclorid showed significant control of Whitefly in comparison with untreated plants up to 40 days after treatment in potted plants.

## Mites (*Tetranychus cinnobarinus, Oligonychus coffeae*)

The infestation of mites is mostly serious during warm and dry season. Nymphs and adults lacerate the leaves resulting in whitish grey patches and affected leaves become mottled, turn brown and fall. Puttaswamy and Channabasavanna (1980) found that increase in mite population on okra was correlated with low rainfall. Pest started building up during April and attained peak during May-June. Kumar and Sharma (1993) reported that population of *T. ludeni* on okra appeared in first week of April, with 27 mites/10 cm2 leaf area. Thereafter the population increased gradually and reached to a maximum of 126 mites/10 cm2 leaf area in the second week of June. Gulati (2004) reported that in summer crop mites appeared in the month of April, showed an increasing trend in May and then attained a peak in the month of June

## Management

Spray Dicofol 18.5 EC @ 2.5 ml or wettable sulphur 50 WP @ 3g/l thoroughly on the under surface of the leaves for good control. Neem soap/pongamia soap 1% may also be used as an alternative to synthetic acaricides.

## Solenopsis mealybug (*Phenacoccus solenopsis* Tinsley)

It is a polyphagous, invasive, recently reported from tomato, ladies finger, brinjal, chillies, pumpkin, etc. Both nymphs and adults cause damage by sucking the sap from the growing points. The infested plants loose its vitality and in heavy infestation gradually dried.

## Root-knot Nematodes (*Meloidogyne incognita, M. javanica)*

The root-knot nematodes typically cause root galls on the feeder roots, affecting the uptake of nutrition and water. The reduction in size of fruits is often accompanied by fresh flower drop.

## Management

Seed treatment with bio-pesticide *Pseudomonas fluorescens* @ 10g/kg seed. In case of severe incidence, apply Carbofuran 1.5 kg ai/ha.

**Table 8.4:** *Yield losses due to major insect pests and diseases in Okra (Source: Shivalingaswamy et al., 2002)*

| Insect pest | Damage % |
|---|---|
| Fruit borer *(H. armigera)* | 22 |
| Leafhopper (*Amrasca biguttula biguttula*) | 54-66 |
| Whitefly (*Bemisia tabaci*) | 54 |
| Shoot and fruit borer (*Earias vittella*) | 23-54 |
| Bud rot (*Choanephora cucurbitarium)* | 5-50 |

Fruit characteristics and varietal resistance

Physico-morphic characters of okra play an important role in selection of plants for for oviposition, infestation and establishment of the phytophagous pests. Mehta and Saxena (1970) reported that the oviposition was correlated with the moisture and intercrossing hairs on the plant surface of okra. Similarly, density of hairs on crown leaves (Agarwal and Katiyar, 1974) and leaf hairiness (Sharma and Agarwal, 1983) also had a significant effect on the selection of oviposition sites on the Noctuid, *Earias vittella* (Fab.).

Saini and Singh (1999) reported that was a significant and positive correlation of *Earias insulana* (Boisd.) with trichome density of different host plants. The resistance of okra to *Earias vittella* (Fab.) was correlated with increased fruit hair density (Kumbher *et al*., 1991). Conversely several workers reported that okra varieties showing greater fruit infestation had more hair density (Dahatonde, 1970; Patil, 1975) and soft skin (Teli and Dalaya, 1981). Chaudhuri *et al.* (2004) evaluated damage potential of *Earias vittella* (Fab.) larvae against okra fruits and reported that fruit damage was higher in open pollinated cultivars than hybrids. Hair density in fruits resulted in less pupation and longer larval period.

## Chemical Plant Factors

Chemical factors and food constituents of plants play a major role for the selection of any host by the pest. Essential oils, steroids and terpenoids in *Abelmoschus esculentus* fruit and *Gossypium hirsutum* leaf excited oviposition of *Earias fabia* on the plants (Mehta and Saxena, 1970). The developing seed of okra fruit functioned as mandatory requirement for both the sexes of *Earias vittella* (Fab.) during rearing in the laboratory to deposit highest number of total and viable eggs (Mani *et al.*,1986)..

## Varietal resistance

Varietal resistance is one of the most cost-effective and hazardless methods of pest confers environment an extra advantage. In eaerlier studies (Madave and Dumbre 1985) varietal resistance in okra has often been used against spotted bollworms. Madave and Dumbre (1985) studied the resistance of 14 varieties of "okra" to *E. vittella* and found that AE-76, Pusa sawani, Long green and White-velvet showed tolerance, while Indo-American Hybrid and Koparwadi local, were resistant to the pest.

## Natural enemies

Utilization of natural enemies, parasitoids and predators and botanicals needs to be encouraged as ecofriendly options for management of insect pests of okra (Bindu *et al.*, 2003; Singh and Brar, 2004; Paulraj and Ignacimuthu, 2005, Kapadia and Puri, 1991; Stelzel and Devetak, 1999; Biesinger and Haefner, 2005; Sardana *et al., 2005a;* Shivalingaswamy *et al.,* 2002; Telang *et al.,* 2004 ) Botanical insecticides, microbial control and biological control agents should be integrated for economic management of insect pests (Arora, *et al.*, 1996; Abro *et al.,* 2004; Memon *et al.*, 2004).

## Effect of biopesticides

Mudathir and Basedow (2004) found that different preparations of neem significantly reduced insect pest infestation in okra. Similarly, predators population was significantly higher in biosal treated okra compared with thiodan indicating that neem insecticide was comparatively less harmful to predators. Natural enemies play important role in the population regulation of phytophagous insects. Sardana *et al.* (2005b) evaluated impact of IPM programmes to reduce pest infestation on okra and tested different IPM modules. Module including bio-intensive, cultural and chemical treatments provided optimum control of pests. Tanwar *et al.* (2007) conducted trails to evaluate the impact of IPM on conservation of natural enemies and found that *T. chilonis* helped to reduce *H. armigera* population.

There were many species of spiders present in okra crop. Spiders are general predators present in agroecosystem feeding on phytophagous insects. Predatory spiders and coccinelids are the main defenders against insect pests in okra fields (Mishra and Mishra, 2002). Kumar *et al.* (2004) determined the predatory potential of spider against insect pest and recorded 13 species of spiders in okra crop. Kubar *et al.* (2006) observed many spider species bhbhin okra crop feeding on phytophagous insects.

It is concluded that proper application of bio-insecticide and natural enemies have good impact on plant yield parameter. The Biosal and tobacco extract were safe to predators. The natural enemies suppressed the insect pests and have good impact on agro-ecosystem. The crop protection expenses of bio-insecticides and natural enemies were less compared to chemical insecticide. The bio-insecticide plots yielded more compared to chemical insecticides plot.

## Impact of climatic factors on insect pests

Weather factors play an important role in the population and infestation of spotted bollworms on the shoots and fruits of 'okra'. *Earias* spp. start their activity on early stages of 'okra' crop and remain active till the harvest of the crop (Zala *et al.,* 1999). The shoot infestation starts in 2-3 week old crop (Kumar and Urs, 1988) and reaches to its peak (8.5%) before the fruiting stage (Shukla *et al.*, 1997). In case of cotton, the pest remains active on shoots from 6 to 11 weeks (Patel and Patel, 2000). The incidence of *Earias* spp. on 'okra' fruits starts with their very setting, i.e., 6 weeks after germination (Kumar and Urs, 1988). Fruit infestation increased progressively upto harvesting of crop and reaches to its peak (41.25%) before the crop harvest, in the first fortnight of June (Shukla *et. al.*, 1997). Fruit infestation reaches to its maximum within 3-4 weeks of the crop ememrgence (Mote, 1977). The crop suffered heavily in the 10th and 11th weeks during 1983-84 and 1984-85, respectively (Kumar and Urs, 1988). In cotton, the *E. vittella* started infestation on buds and bolls from 8-11th weeks (Patel and Patel, 2000).

## Interaction of hybrids and varieties against sucking pests of okra

Uttamaswamy *et al.* (1973) reported that okra varieties IV-7194 and AE-52 were tolerant to leafhopper damage despite of high leafhopper population low damage was noticed in resistant varieties. This may be due to non-preference for feeding, which may be further attributed to high hair density on lamina of these varieties. Bindra and Mahal (1979) screened ten okra varieties in the field for resistance to leafhopper. New selection, Sel-2-2 and Sel-6-2 were found to be promising as these either carried low leafhopper population or suffered low damage. Leafhopper injury index increased with the increase in leafhopper number. Ten nymphs of 3rd and 4th instars were considered as an optimum population level to screen varieties for leafhopper resistance, varietal difference were observed in leaf area and hair density on lamina side vein and mid rib. However, leafhopper population was found to decrease only with the increase in hair density on leaf lamina.

Singh *et al.* (1988) studied relationship of resistance in okra to *A. biguttula biguttula* with leaf hairiness and biochemicals during summer 1982-83 at Delhi. Susceptible and resistance genotypes had a percentage survivorship of 95.35 to 97.42 and 29.69 to 55.90 and development time of 6.15 to 6.94 and 9.99 to 11.84 days, respectively. The incidence of *A. biguttulla biguttula* was negatively correlated with hair density and its length on the lower surface of leaf.

## Use of organic compost extracts against insect pests of okra

Insect pests infestation is one of the major factors against cultivation of okra. Different phenologies of okra have been reportedly attacked by the insects. Generally, synthetic insecticides are the most effective means of controlling insect pests due to their quick action and long lasing effect. However, most of these synthetic insecticides have been responsible for development of insect resistance, environmental pollution, carcinogenic effect and destruction of beneficiary insects (Isman 2008). In view of the these problems, the need for cheap, cost effective and easily biodegradable products such as kaolin, botanical insecticides and organic composts have received a greater attention as promising alternative to synthetic insecticides(Akanbi et al 2007).

## References

A.O.A.C., 1990. In: Official Methods of Analysis, 15th ed., Sec. 959.08. AOAC.Inc., Suite 400, 220 Wilson Boulevard, Arlington, Virginia, 22201, USA.

Abdullah, N.M.M. and Singh, J. (2004). Effect of insecticides on longevity of the whitefly, Bemisia tabaci Gennadius on cotton. University Aden J. Natural Appl. Sci. 8(ii):261-268.

Abro, G.H., Memon, A.J., Syed, T.S. and Shaikh, A.A., 2004. Infestation of Earias spp. on cotton and okra grown as mono and mix crops. Pak. J. biol. Sci., 7: 937-942

Abrol, D. P. 1995 Studies on abundance, diversity and behaviour of insect visitors of okra (Hibiscus esculentus L. (Moench)) blossoms and their importance as pollinators. Korean Journal of Apiculture 10( 1):1-4

Acharya, M.C. 2002. Determination of economic threshold levels and integrated management of fruit borer, Earias vittella (Fab.) on okra, Abelmoschus esculentus (L.) Moench. M.Sc. Thesis, Maharana Pratap University of Agriculture and Technology, Udaipur, Rajasthan.

Acharya, M.C. 2010. Determination of Economic Threshold level and Integrated Management of Okra Fruit Borer Earias vittella (Fab.). J. Prot. Society. 2: 47-61.

Adebayo T A. 2003. Efficacy of mixture formulations of synthetic and botanical insecticides in the control of insect pests of okra and cowpea. Ph. D Thesis. Ladoke Akintola University of Technology, Ogbomoso.Nigeria.

Agarwal, R.A. and Gupta, G.P. 1986. Recent advances in cotton pest management. Plant Prot. Bull. 88(1-4): 375-378.

Agarwal, R.A. and K. N. Katiyar. 1974. Ovipositional preference and damage by spotted bollworm (Earias fabia) in cotton. Cotton Dev. 4: 28-30.

Ahmad S., Gupta S.C., Prakash N. and Mandal S.K., Weather factors and larval population of Earias vitella

Ahmad Tufail, Abraham Finan, Mohammedsaid Hassen, Welday Nebiat and Gebrehiwet Samuel 2015. Seasonal incidence of insects and diseases on okra (Abelmoschus esculentus) crop in zoba Anseba. International Research Journal of Biological Sciences. 4(1), 71-74.

Ahmad, F., Khan, F.R. And Khan, M.R., 1995, Comparative efficacy of some traditional and non- traditional insecticides against sucking insect pests of cotton. Sarhad J. Agriculture, 11 : 733-739.

Ahmad, S., S.C. Gupta, N. Prakash, and S.K. Mandal. 2000. Weather factors and larval population of Earias vittella Fabr. in summer sown 'okra' (Abelmoschus esculentus L.). J. Res., Bisra Agric. Uni. 12(2): 215-217.

Ahmed, H., Ali, S., Khan, M.A. and Habib, A. 1995. Efficacy of neem oil and synthetic growth regulators on control of okra insect pests as compared to chemical pesticides. J. Sustan. Agric Environ. 5(1-2): 232-245.

Ahmed, H., Ali, S., Khan, M.A. and Habib, A., 1995. Efficacy of neem oil and synthetic growth regulators on control of okra insect pests as compared to chemical pesticides. J. Sustan. Agric Environ., 5(1/2): 232-245.

Ahmed, K.U. 1995. Pal-Phul O Shak-Shabji (In Bengali) 5th ed. Mirpur, Dhaka, Bangladesh. p. 400.

Ahmed, S.S. 1984. Some promising plant species for use as pest control agent under traditional farming system. In: Proceedings on 2nd Neem Conference Reschhdzhouson FRG. pp. 565-580.

Akanbi W B, Adebayo T A, Togun O A, Adeyeye A S, and Olaniran O A. 2007. World Journal of Agricultural Sciences, 3(5): 642-652.

Akila, S. and Babu, P.C.S. 1994. Release of different dose of Trichogramma and its effect on internode borer, yield and quality of sugarcane. Sugarcane (London) 2: 22-23.

Alao F. O., Adebayo T. A., Olaniran O.A. and Akanbi, W.B. 2011. Preliminary Evaluation of the Insecticidal Potential of Organic Compost Extracts against Insect Pests of Okra (Abemoschus esculentus (L.) Moench). Asian Journal of Plant Science and Research, 2011, 1 (3):123-130

AL-Eryan, M.A.S., Zaitoon, A.A. And Rezk, H.A., 2001. The use of Coccinella 11-punctata (Coleoptera: Coccinellidae) against Aphis gossypii (Homoptera: Aphididae) on okra plant. Alexand. J. agric Res., 46: 107-114.

Al-Ghzawi A.M, Zaittoun ST, Makadmeh I, Al –Tawaha ARM 2003. The impact of wild bees on the pollination of eight okra genotypes under semi-arid Mediterranean conditions. Int J Agric Biol. 5:409-411.

Al-Ghzawi Abdul-Majeed A., Zaittoun Shahera T., Makadmeh Ibrahim and Al-Tawaha Olugbenga E. Ige and Eludire M. O. 2014. Floral Biology and Pollination Ecology of Okra (Abelmoschus esculentus L. Moench). American International Journal of Biology. 2( 2): 01-09

Ali M, Hossain M Z, Sarkern N C. 2000. Inheritance of Yellow Vein Euphytica .3: 205-209.

Ali, M.I. 1992. Seasonal occurrence of different cotton bollworms on cotton in Bangladesh. Bangladesh J. Entomol. 2(1 & 2).

Ambekar, J.S., Pawar, A.S. and Sakhare, M.V. 2000. Bio-efficacy of certain neem products against okra fruit borer. J. Maharashtra Agril. Uni. 25(1): 42-43.

Anand S. and Anil S., 2003. Effect of some weather parameters on the development of powdery mildew (Erysiphe polygoni) in black gram, Pl. Dis. Res., 18(1), 63-64

Angbanyere M. A and P. K. Baidoo 2014. The Effect of Pollinators and Pollination on Fruit Set and Fruit Yield of Okra (Abelmoschus esculentus (L.) Moench) in the Forest Region of Ghana. American Journal of Experimental Agriculture 4(9): 985-995.

Anitha K.R., 2007 Seasonal incidence and management of sucking pests of okra. M.Sc. dissertation, University of Agricultural Sciences, Dharwad. Karnatka, India.

Anjorin S.T., Jolaoso M.A. and Golu M.T., 2012. A survey of incidence and severity of pests and diseases of okra (Abelmoschus esculentus L. Moench) and egg plant (Solanum melongena L.) in Abuja, Nigeria, American J. Res. Comm., 1(11), 333-349

Anonymous. 1987. Annual Report for (1986-87). Central Institute for Cotton Research, Nagpur. P. 27.

Anonymous. 1993. Research and development of vegetable crops. Paper presented in the workshop on importance of okra as summer vegetable on March 9-10, IPSA, Gazipur.pp.1-07.

AOAC, 1975. In: Official Methods of Analysis, 12th ed., AOAC, Washington, DC.

Arora, R.K., Dhillon, M.K. and Singh, H., 1996. Management of pest complex in okra research summation. Annl. Agric. Biol. Res., 1(1/2): 37-45.

Aryo, O.J., M.E. Aken'ova and C.A. Fatokun. 1987. Plant character correlation and path analysis of pod yield in okra (Abelmoschus esculentus). Euphyta. 36(2): 677-686.

Atiri, G.I., Ivbijaro, M.F. and Oladele, A.D. 1991. Effects of natural synthetic chemicals on the incidence severity of okra mosaic virus in okra. Tropical Agril. 68 (ii):178-180.

Atwal AS 1976. Agricultural Pests of India and South- East Asia. Kalyani Publishers, New Delhi, Ludhiana: pp. 283-285. Aziz MA (2010). Sustainable management of spotted bollworms, Earias vittella (Fabricius) and Earias insulana (Boisduval), (Lepidoptera: Noctuidae) on Okra, Abelmoschus esculentus (L.) in Punjab, Pakistan. PhD (Ento.) Thesis, Uni. Agric. Faisalabad. 42P.

Atwal, A.S. 1976. Agricultural Pest of India and South-East Asia. Kalayani Publishers, New Delhi. pp. 283-285.

Atwal, A.S. 1999. Pests of Cotton. 4 th Edn., Kalyani Publishers, New Delhi. pp. 224-234. Avidov, Z. 1956. Bionomics of the tobacco whitefly (Bemisia tabaci Gennad.) in Israel. Katvim. 7: 25-41.

Atwal, A.S. and Singh, B. 1990. Pest population and assessment of crop losses, Publication. Indian Agriculture Research Institute, New Delhi.

Aust H.J. and Jurgen H.H., 1986. Microclimate in relation to epidemics of powdery mildew, Ann. Rev. Phtopathol., 29, 144-145

Azab, A.K., M.M. Megahed and D.H. El-Mirsawi. 1971. On the Biology of Bemisia tabaci (Genn.). Societe Entomol. D'Egypte Bull. 55: 305-315.

Azam, K.M., Razvil, Z. and Al-Raeesi, A.A. 1997. Management of whitefly Bemisia tabaci Gennadius and tomato leaf curl virus in tomato crops. Arab and Near East Plant Protect. Newsl. No. 25, 27.

Aziz MA, Hasan M, Ali A 2011. Impact of abiotic factors on incidence of fruit and shoot damage of spotted bollworms, Earias spp. on okra (Abelmoschus esculentus L.). Pak. J. Zool. 43: 863-868.

Azo'o Ela M, Ali M, Fohouo FNT, Messi J. 2012. The importance of a single floral visit of Eucara macrognatha and Tetralonia fraterna (Hymenoptera: Apidae) in the pollination and the yields of Abelmoschus esculentus in Maroua, Cameroon. African Journal of Agricultural Research. 7(18):2853-2857. Accessed 7 June 2011. Available: http://www.academicjournals.org/AJAR

Azo'o, M. E.; Fohouo, F. N. T.; Messi, J. 2011. Influence of the foraging activity of the entomofauna on okra (Abelmoschus esculentus) seed yield. International Journal of Agriculture and Biology 13 (5): 761-765

Badaya, A.K., Deshpande, R.R. and Ali. S.A. 1999. Efficacy and economics of different insecticides for the control of okra fruit and shoot borer, Earias vittella Fab. Bhartiya Krishi Anusandhan Patrika, 6(3-4): 181-185.

Balasubramanian, T. and S. Sadasivam. 1987. Changes in starch, oil, protein and amino acids in developing seeds of okra (Abelmoschus esculentus L.Moench). Plant foods for Human Nutrition. 37(1): 41-46.

Bale, J.S., G.J. Masters, I.D. Hodkinson, C. Awmack, T.M. Bezemer, V.K. Brown, J. Butterfield, A. Buse, J.C. Coulson, J. Farrar, J.E.G. Good, R. Harrington, S. Hartley, T.H. Jones, R.L. Lindroth, M.C. Press, I. Symrnioudis, A. D. Watt, and J.B. Whittaker., 2002. Herbivory in global climate change research: direct effects of rising temperature on insect herbivores. Global Change Biology. 8:1-16.

Banken, J.A.O. and Stark, J.D. 1997. Stage and age influence on the susceptibility of Coccinella septempunctata after direct exposure to neemix, a neem insecticide. J. Eco. Entomol. 90(5): 1102-1105.

Barbier, E., 1978. Pollination du colza par les abeilles. Rev. Fr. Apic., 365: 285–91

Baskaran, S.; Vigneswari, A.; Kasthuribai, M. 2001Composition and relative abundance of the flower visitors of bhendi,Abelmoschus esculentus Linn. Insect Environment 7 (1): 17

Basu, A. N. 1995. Bemisia tabaci Gennadius-Crop pest and principal whitefly vector of plant viruses. Oxford &IBH Pub. Co. Pvt. Ltd. New Delhi. pp. 183.

Basu, A.N., 1995. Bemesia tabaci (Genn.) crop pest and principal whitefly vector of plant viruses. West View Press, San Fransisco, pp. 183.

BBS (Bangladesh Bureau of Statistics) 2011. Ministry of Planning, Government of Peoples' Republic of Bangladesh.Yearbook of Agric. Stat. Bangladesh.

BBS (Bangladesh Bureau of Statistics). 2009. Ministry of Planning, Government of Peoples Republic of Bangladesh. Yearbook of Agric. Statistics of Bangladesh. Dhaka.

Begum, M.A. 2002. Prevalence and spread of Okra yellow vein clearing mosaic virus (OkYVCMV) in the field and its impact on growth and yield of okra. M.S. Thesis. Bangabandhu Sheikh Mujibur Rahman Agricultural University, Gazipur. pp 135.

Gopalakrishnan, T.R. 2007. Vegetable Crops, New India Publishing Agency, New Delhi.

Berger R.D., 1980. Measuring disease intensity, In: Teng, P.S., S.V. Krupa (eds.): Crop Loss Assessment, Proc. E.C. Stakman Commem. Symp. Misc. Publ. 7, Agric. Exp. Stn, 28-31

Bergvinson, D.J., J.T. Arnason and R.I. Hamilton. 1997a. Phytochemical changes during recurrent selection for resistance to the European corn borer. Crop Sci. 37:1567-1572.

Berlinger, M.J. 1986. Host Plant Resistance to Bemisia tabaci. Agric. Ecosystems Environ. 17: 69-82.

Berlinger, M.J., Taylor, R.A.J. Lebiush-Mordechi, S., Shalhevet, S. and Spharim, I. 2002. Efficacy of exclusion screens for preventing whitefly transmission of tomato yellow leaf curl virus of tomatoes in Israel. Bull. Entomol. Res. 92(5):367-373.

Bhatti J. A., M.A. Khan, M.A. Murtaza, M.Z. Majeed, and M.A.l. Javaid., 2007. Effect of abiotic factors on incidence and development of spotted bollworm (Earias spp.) on advanced genotypes of cotton under unsprayed conditions. J. Agric. Res. 45 (2): 145-150.

Biesinger, Z. and Haefner, J.W., 2005. Proximate cues for predator searching: a quantitative analysis of hunger and encounter rate in the ladybird beetle, Coccinella septempunctata. Anim. Behav., 69: 235-244.

Bilton, J.N., Broughton, H.B., Ley, S.V., Lidert, Z., Morgan, E.D., Rzepa, H.S. and Sheppard, R.N. 1985. Structural reappraisal of the limonoid insect antifeedant azadirachtin. J. Chem. Soc., Chem. Commun. pp. 968-971.

Bindu, P., Bharpoda, T.M., Patel, J.R. and Patel, J.J., 2003. Evaluation of various schedules based on botanical and synthetic insecticides in okra ecology. Indian J. Ent., 65: 344-346.

Blennerhassett, R.M. and B.M. El-Zeftawi. 1986. Productivity and growth of six okra cultivars. Australian J. Exp. Agric. 26(5): 625 – 629.

Borah, R.K. and Nath, P.D. 1995. Evaluation of an insecticide schedule on the incidence of whitefly, Bemisia tabaci Genn. and yellow vein mosaic in okra. Indian J. Virol. 11(ii):65-67.

Brickle, D.S., Tumiseed, S.G. and Sullivan, M.J. 2001. Ovicidal effects of insecticides on eggs of gram podborer (Helicoverpa armigera) and spotted bollworm (Earias vittella). Indian J. Agril. Sci. 63(12):853-855.

Broughton, H.B., Bilton, J.N., Ley, S.V., Lidert, Z., Morgan, E.D., Rzepa, H.S. and Sheppard, R.N. 1986. Structural reappraisal of the limonoid insect antifeedant azadirachtin. J. Chem. Soc., Chem. Commun. pp. 968-971.

Brown, J.K., Frohilch, D.R. and Rossel, R.C. 1995. The sweet potato or silverleaf whiteflies: Biotype of Bemisia tabaci or a species complex. Ann. Rev. Entomol. 40:511-34.

Butani DK, Jotwani MG 1984. Insects in vegetables. Periodical Expert Book Agency. Vivek-Vihar, Delhi (India). pp. 45-66.

Butani, Dhomo K. and Jotwani, M.G. 1984. Insects in vegetables. Periodical expert Book Agercy, India. p: 356.

Butani, K. and Verma, S. 1976. Insect pests of vegetables and their control. 10: pp. 31-37.

Butler,G.D.J.,T.J. Henneberry and T.E. Clayton. 1983. Bemisia tabaci (Homoptera: Aleyrodidae): Development, Oviposition, and Longevity in Relation to Temperature. Ann. Entomol. Soc. America. 76(2): 310-313.

Butterworth, J.H. and Morgan, E.D. 1968. Isolation of substance that suppresses feeding in locusts. J. Chem. Commun. pp. 23-24.

Cahill, M., Byrne, F.J., Gorman, K., Denholm, I., and Devonshire, A.L. 1995. Pyrethroid and Organophosphate resistance in tobacco whitefly, Bemisia tabaci (Homoptera:Aleyrodidae). Bull. Entomol. Res. 85: 181-187.

Cammel, M.E., and J.D. Knight., 1992. Effects of climatic change on the population dynamics of crop pests. Adv. Ecol. Res. 22: 117-162.

Cane JH, Tepedino VJ. Causes and extent of declines among native North American invertebrate pollinators: Detection, evidence and consequences. 2001. Conservation Ecolog. 2001; 5(1):1. Accessed 8 January 2011. Available: http://www.consecol.org/vol5/iss1/art1/(accessed on 17 November 2011)

CCSHAU. 1997. The Package of Practices for Kharif Crops. Hisar: India. CCS Haryana Agricultural University.

Chaudhuri, N., S. Pal, and S.K. Senapati. 2004. Fruit damage potential of Earias vittella (Fab.) as influenced by physical characters of okra fruit. Indian J. Entomol. 66(4): 372-374.

Chaudhury, D.S. 2001. Evaluation of synthetic and bio-pesticides against OSFB. Current Research Report. Mahatama Phula Agril. Univ. 2(1): 256-261.

Chaudhury, H. R. and Dadheech, L.N. 1989. Incidence of insects attacking okra and the avoidable losses caused by them. Ann. Arid Zone. 28 (3-4):305-307.

Chauhan, D.V.S. 1972. Vegetable production in India. 3rd Edition. Published by Ram Prasad Sons, Agra. pp. 28-30.

Chinniah and Mohanasundaram, M., 1999. Evaluation of neem derivatives for their toxic effect on safety to predatory mites, Amblyseius spp. (Acarina : Phytoseidae) in cotton ecosystem. Pestology, 123 : 45-48. aphid (*Aphis gossypii*). Indian J. Agric. Sci., 63 : 134- 135.

Chitra, K. C.; Rao, S. J.; Rao, P. K.1997. Efficacy of plant extracts for controlling cotton aphid (Aphis gossypii). Indian Journal of Agricultural Sciences 67 pp.134-135

Choi In Hu, Jang YongSeok, Kim GilHah and Kim JeongWha, 2004. Control effects of some insecticides on different stages of the stone leek leafininer , Liriomyza chinensis Kato (Diptera:Agromyzidae).Korean Appl. Entomol. 43(2): 169-173.

Cock, M.J.W.Ed. 1986. Bemisia tabaci, A Literature Survey on the Cotton Whitefly with an Annotated Bibliography. Food and Agriculture Organization of the U.N., C.A.B. Int. Inst. Biol. Cont. 121 pages.

Cohen, S. and M.J. Berlinger. 1986. Transmission and Cultural Control of Whitefly-borne Viruses. Agric. Ecosystems Environ. 17: 89-97.

Dangi, P.R. 2004. Incidence of Earias vittella (Fabricius) and its management in okra, Abelmoschus esculentus (L.) Moench. M.Sc. Thesis, Maharana Pratap University of Agriculture and Technology, Udaipur, Rajasthan.

Das SN, Subhraj R, Chatterjee ML, Ray S 2001 Bioefficacy, yield benefit and cost effectivety of some new molecules against okra fruit borer, Earias vittella (F) (Noctidae: Lepidoptera). Journal of Interacad 5: 346-351

Das, S.N., Subhraj, R., Chatterjee, M.L. and Ray, S., 2001. Bioefficacy, yield benefit and cost pest management options in okra effectivety of some new molecules against okra fruit borer, Earias vittella (F) (Noctidae: Lepidoptera). J. Interacad., 5: 346-351.

Dash, A.N., Mahapatra, H., Pradhan, A.C. and Patnaik, N.C. 1987. Effect of mixed and inter cropping on occurrence of some pest in Orissa. Environ. Ecol. 5(3): 526-530.

David, B.R. 2001. Elements of Economic Entomology. Population Book Report, Chennai. pp: 198-207.

De Barro, P.J. 1995. Bemisia tabaci Biotype B: a review of its biology, distribution and control. Division of Entomol .Technical paper no. 33. Common. Sci. Indus. Res. Org., Canberra, Australia.

Dhaliwal, J.S., Sidhu, M.S. and Singh, G., 1981. Effect of different dates of sowing on sorghum and the incidence of red leaf spot disease and major insect pests in Punjab. J. Res. Agric. Univ. Ludhiana, 18: 400-405.

Dhaliwal, J.S., Sidhu, M.S. and Singh, G., 1981. Effect of different dates of sowing on sorghum and the incidence of red leaf spot disease and major insect pests in Punjab. J. Res. Agric. Univ. Ludhiana, 18: 400-405.

Dhamdhere, S.V., Bhadur, I. and Mishra, U.S. 1984. Studies on occurrence and succession of pests of okra at Gwalior. Indian J. Plant Prot. 12: 9-12.

Dhandapani, N., 1986. Consumption digestion and utilisation of bhendi varieties by Earias vittella Fabricius. Madras Agric. J. 73 (12): 676-678.

Dhandapani, N., Kalyanasundaram, M., Swamiappan, M., Babu, P.C.S. and Jayaraj, S. 1992. Experiments on management of major pests of cotton with biocontrol agents in India. J. Appl. Entomol. 114: 52- 56.

Dhane, A.S., Ghodaki, B.S. and Babar, M.S. 2007. Field evaluation of Chelonus blackburni against shoot and fruit borer Earias vittella infesting okra. Indian J. Entomol. 69(1): 54-57.

Dhawan, A.K. and Sidhu, A.S. 1984. Incidence and relative abundance of different species of spotted bollworms on okra at Ludhiana, Punjab. J. Res. Punjab Agric.Univ. 21: 533-542.

Dilruba S, Hasanuzzaman M, Karim R, ahar K (2009). Yield response of okra to different sowing time and application of growth hormones. J. Hort. Sci. Ornamental Plants. 1:10-14.

Dinesh B.M., 2009 Studies on powdery mildew of sunflower caused by Erysiphe cichoracearum DC. M.Sc. thesis submitted in Dharwad Agri. Univ., 1-86

Dongre, T.K. and Rahalkar, G.W. 1992. Relative host plant preference and induction of preference for feeding behavior in spotted bollworm. Indian Ecol. 19(1): 61-64.

Dubey, V.K., Bhagat, K. and Yadu, Y.K., 1999. Insect pest succession studies on okra. J. appl. Zool. Res., 10: 144-145.

Dubey, V.K., Bhagat, K. and Yadu, Y.K., 1999. Insect pest succession studies on okra. J. appl. Zool. Res., 10: 144-145.

Dutt, A.K. and Saha, J.L. 1990. Insect pest management of winter planted cotton in coastal rice fallow of West Bengal. Tropical pest manage. 36(2): Pp. 89-92.

Elden, T.C. and M. McCaslin. 1997. Potato leafhopper (Homoptera: Cicadellidae) resistance in perennial glandular haired alfalfa clones. J. Econ. Entomol. 90 (3): 842-847.

El-Helaly, M.S., A.Y.El-Shazil, and F.H. El-Gayar. 1971. Biological studies on Bemisia tabaci Genn. (Homoptera: Aleyrodidae) attacking cotton in the coastal plain of Israel. Bull. Entomol. Res. 70: 213-219.

El-Nahry, F.I., M.I. El-Ghorab and R. Younes. 2005. Nutritive values of local varieties of fresh and sundried okra (Abelmoschus esculentus) pods and seeds. Plant Foods for Human Nutrition. 28 (3): 227-231.

Fabr. in summer sown okra (Abelmoschus esculentus L.), J. Res. Birsa Agri. Univ., 12(2), 215-217 (2000)

FAO 2007. Agricultural production data base. Food and Agricultural Organization. http//:apps.fao.org./faostat./ 340p.

FAOSTAT., 2008. Agricultural production data base. Food and Agricultural Organization (http://www.fao.org)

Farooq, A.K., Jalal-ud-din, A. Ghaffoor, and K.W. Khan. 2002. Evaluation of different cultivars of 'okra' (Abelmoscus esculentus L.) under the agro-climatic conditions of Dera Ismail Khan. Asian J. Plant Sci. 1(6): 663-664.

Farooq, A.K., Jalal-ud-din, A.Ghaffoor, and K.W. Khan. 2002. Evaluation of different cultivars of okra (Abelmoscus esculentus L.) under the agro-climatic conditions of Dera Ismail Khan. Asian J. Plant Sci. 1(6): 663-664.

Febvay, G., J. Bonnin, Y. Rhabe, R. Bournoville, S. Delrot and J.L. Bonnemain.1988. Resistance to different Lucerne cultivrs to the pea aphid Acyrthosiphon pisum: influence of phloem composition on aphid fecundity. Entomologia Experimentalis et Applicata. 48(2): 127-134.

Fletcher TB, Mishra CS 1990. Cotton boll worms in India. Report, Proc. Entom. Meeting. 2: 443-472.

Forehand, L.M., Orr, D.B. and Linker, H.M. 2006. Evaluation of a commercially available

Free JB. 1976. Insect pollination of tropical crops. Report of the Central Association of Beekeepers 12.

Gannon, A.J. and C.E. Bach. 1996. Effect of soyabean trichome density on Mexican bean beetle (Coleoptera: Coccinelidae) development and feeding preference 25(5): 1077-1082.

Gautam, R.K. and Goswami, N. 2004. Different food varies for amino acids in moth larvae. Uttar Pradesh Zool. 24 (2):183-186.

Gennadious, P. 1889. Disease of the tobacco plantation in the Triconia. The aleurodid of tobacco. Ellenike Ga. 5:1-3.

George, P.J.E. 1997 Effect of some plant odours in the breeding environment on fecundity and hatching of okra shoot and fruit borer, Earias vittella Fab. Under laboratory conditions. Crop Res. Hisar. 13 (3):157-161.

Gomez, K.A. and Gomez, A.A. 1976. Statistical Procedure for Agricultural Research (2nd Ed). A Willey Inter science Publication, New York. P. 680.

Gopal an, C., Rama Sastri, B.V. and Balasubramanian, S. 2007. Nutritive Value of Indian Foods, published by National Institute of Nutrition (NIN), ICMR.

Gowri, S., Ramachandrarao, G. and Nagalingam, B., 2002. Impact of neem formulations on Coccinellid predators of okra pest complex. Pestic. Res. J., 14: 242-243.

Gowri, S., Ramachandrarao, G. and Nagalingam, B., 2002. Impact of neem formulations on Coccinellid predators of okra pest complex. Pestic. Res. J., 14: 242-243.

Greathead, A.H. 1986. Host Plants. Chapter 3, pp. 17-25. In: Bemisia tabaci - a literature survey on the cotton whitefly with an annotated bibliography (Ed. M.J.W. Cock). CAB International Institute of Biological Control, Ascot, UK. P. 121.

Griffiths, D.C., Greenway, A.R. and Lioyd, S.L. 1978. The influence of repellent materials and aphid extracts on settling behavior and larviposition of Myzus piersicae (Sulzer) (Hemiptera: Aphididae). Bull. Ent. Res. 68: 613-619.

Griffiths, D.C., Pickett, J.A., Smart, L.E. and Woodcock C.M. 1989. Use of insect antifeedant against aphid vectors of plant virus disease. Pestic. Sci. 27: 269-276.

Gulati, R., 2004. Incidence of Tetranychus cinnabarinus infestation in different varieties of Abelmoschus esculentus. Ann. Pl. Protec. Sci.10:239-242.

Gulsen O, Karagul S, Abak K. 2007. Diversity and relationships among Turkish germplasm by SRAP and Phenotypic marker polymorphism. Biologia, Bratislava 62(1):41-45.

Gupta, G.P. and Sharma, K. 1997. Neem-based pest management strategy in cotton system. Pesticide Res. J. 9: 190–197.

Gupta, S. C., G. S. Prasad, and A. Sarfraz., 1998. Weather factors and incidence of Earias vittella Fabr, in 'okra', Abelmoschus esculentus (L.) Moench. J. Res. Bisra Agric. Uni. 10 (1): 12-15.

Haider, M.Z. 1996. Effectiveness of some IPM packages for the management of virus-disseminating whitefly on tomato. M.S. thesis in Entomology, IPSA, Salna, Gazipur, Bangladesh. 1-65 pp.

Halder, A.; Sarkar, S.; Dey, D.; Debnath, N. 2014 Pollinators and flower visitors of okra (Abelmoschus esculentus) in non-urbanized and urbanized areas.Environment and Ecology 32 (2) : 543-547

Haque, M.M. 1993. Tomato, Brinjal and Okra. In: Ahmed and Sahajahan, M. (ed). Homestead vegetable production, Bangladesh Agricultural Research Institute, Joydebpur, Gazipur. p. 61.

Harender, R. Bhardwaj, M. L. and Sharma, N.K. 1993. Performance of commercial okra( Hibiscus esculentus) varieties in relation to disease and insect pests. Indian J. Agric. Sci. 63(xi): 747-748.

Hedin, P., F.M. Davis and E.P. William. 1993. 2,4-dihydroxy-7-methoxy-(2H)-1-,4- benzin-3(4H)-one (N-O-ME-DIMBOA), a possible toxic factor in corn to the southwestern corn borer. J. Chem. Ecol. 19, 531-542.

Hill D.S, and Waller J M. 1988. Pests and Diseases of Tropical Crops, Logman, New York, pp: 414.

Hill, D.S. 1983. Agricultural Pests of the Tropics and their control. Cambridge University press. Cambridge, New York. pp. 64-367.

Hopper, K.R. 1986. Preference, acceptance and fitness components of Microplitis croceipes (Hymenoptera: Braconidae) attacking various instars of Heliothis virescence (Lepidoptera: Noctuidae). Environ. Entomol. 15: 274-280.

Huffaker, C.B. and Carl, B., 1980. New technology of pest control. John Wiley and sons, New York, USA, pp. 500.

Huffaker, C.B. and Carl, B., 1980. New technology of pest control. John Wiley and sons, New York, USA, pp. 500.

Husain, M.A., Trehan K.N. and Verma, P.N. 1939. Eonomics of field scale spraying against the whitefly of cotton (Bemisia gossypiperda, M. and L.). Indian J. Agric. Sci. 9: 109-126.

International Board for Plant Genetic Resources IBPGR. 1990. Report on International Workshop on Okra Genetic resources held at the National bureau for Plant Genetic Resources, New Delhi, India

Islam, M.N., Kumar, N.K. and Karim, M.A. 1999. Efficacy of different insecticides for the control of okra shoot and fruit borer , Earias vittella F. Annual Report.(1998-99). BARI, Joydevpur, Gazipur, Bangladesh: pp. 33-34.

Isman M B. 2008. Perspective. Botanical insecticides: for richer, for poorer Pest management science .64: 8-11.

Jackson, M.L. 1958. Soil chemistry analysis. Printice-Hall, Inc., Englewood Chiffs, New Jersy.

Jagtab, C.R., Shetgar, S.S. and Nalwandikar, P.K. 2007. Fluctuation in population of lepidopterous pest infesting okra in relation to weather parameters during Kharif. Indian J. Entomol. 69(3): 218-220.

Jayaraj, S., Rangarajan, G.A.V., Murugesan, S., Santharam, G. Vijayaraghavan, S. And Thangaraj, D., 1986, Outbreak of whitefly Bemisia tabaci (G.) on cotton in Tamil Nadu and its management. All Indian Group Discussion on Whitefly in Cotton, Guntur, April 29-30, (1986).

Jayaswal A.P. and Sundaramurthy V.T., 1992 Achievements in insect management in cotton. All India Coordinated Crop Improvement Project Achievements, 117-151

Jeyarani, S., Karuppuchamy, P. and Sathiah, N. 2008. Interaction between the egg larval parasitoid, Chelonus blackburni and Nucleopolyhedrovirus in Helicoverpa armigera (Hübner). pp.126-130. In: Ignachimuthu and Jayaraj, S. (Ed.). Recent trends in pest management. Elite publishing (Pvt.) Ltd. New Delhi.

Jhansi, K. 1984. Studies on Bracon hebetor (Say), a larval parasitoids of Maruca testulalis

Joshi, B. G.; Sitaramaiah, S., 1979: Neem kernel as an ovipositional repellent for Spodoptera litura (F.) moths. Phytoparasitica 7(3): 199-202.

Kabir, K.H., Rouf, F.M.A., Islam, M.N. and Malakar, P.K. 1994. Efficacy of different insecticides in controlling brinjal shoot and fruit borer, Leucinodes orbonalis Guen. Univ. 3. Zool., Rajshahi Univ. 13: 1-8.

Kadam, J.R. and Khaire, V.M. 1995. Raining and relative humidity: key factors to suppress Earias vittella Fab infestation on Okra crop. J. Ento. Res. 19(3): 201-205.

Kadam, J.R., and V.M. Khaire., 1995. Raining and relative humidity: key factors to suppress Earias vittella (Fabricius) infestation on 'okra' crop. J. Entomol. Res. 19(3): 201-205.

Kahlon TS, Chapman MH, Smith GE 2007. In vitro binding of bile acids by okra beets asparagus eggplant turnips green beans carrots and cauliflower. Food Chem. 103: 676-680.

Kanwar N, Ameta OP 2007. Assessment of loss caused by insect pests of okra, Abelmoschus esculentus (L.) Moench. Pestology. 31 (5): 45-47. Kochhar SL, (1986). Tropical Crops.The Macmillan Press

Kapadia, M.N. and Puri, N., 1991. Biology and comparative predation efficacy of three heteropteran species recorded as predators of Bemisia tabaci (Genn.) in Maharashtra. Indian J. Ent., 36: 555-556.

Kapadia, M.N. and Puri, N., 1991. Biology and comparative predation efficacy of three heteropteran species recorded as predators of Bemisia tabaci (Genn.) in Maharashtra. Indian J. Ent., 36: 555-556.

Karim, M.A. 26-29 January 1992. Vegetable production and marketing. Proceedings of a national review of planning workshop hold at BARI, Gazipur, Bangladesh.

Khaliq, A. and Yousuf, M. 1986. Effect of weather on the light-trap captures of some insect pests of cotton. J. Agric. Res. Lahore. 24 (4): 313-319.

Khan, L., 1987. Bionomics and control of the melon fruit fly, Dacus cucurbitae Coq. (Dipt. Tephritidae). Ph.D. Thesis, Dep. Agric. Entomol. Univ. Agric. Faisalabad (Pakistan).

Khan, M.A. and Mukhopadhyay, S. 1985. Effect of different pesticide combinations on the incidence of yollow vein mosaic virus of okra (Abelmoschus esculentus) and its whitefly vector Bemisia tabaci Genn. Indian J. Virol. 1(ii):147-151.

Khan, M.H. and Verma, P.M. 1946. Studies on Earias spp (the spotted bollworm of cotton) in the Punjub. The biology of the common parasites of Earias fabia stoll, E. insulana (Boisd) and E. cupreoviridis (WK). Indian J. Entomol. 7: 41-63.

Khoso, A.W., 1992. Growing vegetables in Sindh. Department of Agronomy, Sindh Agriculture University, Tandojam, pp.15-18.

Khoso, A.W., 1992. Growing vegetables in Sindh. Department of Agronomy, Sindh Agriculture University, Tandojam, pp.15-18.

Khurana, A.D. and Verma, A.N. 1990. Comparative damage caused by bollworms and yield of seed cotton during a dry and wet year in Haryana. J. Insect Sci. 3(2): 180-182.

King, E.G., Powell, J.E. and Coleman, R.J. 1985. A high incidence of parasitism of Heliothis spp (Lepidoptera: Noctuidae) larvae in cotton in south-eastern Arkansas. Entomophaga 30: 419-426.

Klein AM, Vaissiere BE, Cane JH, Steffan-Dewinter I, Cunnigham SA, Kremen C, et al. Importance of pollinators in changing landscapes for world crops. Proc R Soc. 2007. 274(2):303-313.

Koul, O. and Isman, M.B. 1991. Effects of azadirachtin on the dietary utilization and the development of the variegated cutworm Performa saucier. J. Insect. Physiol. 37: 591-598.

Kraus, W., Bokel, R., Klenk, A. and Pohnl, H. 1985. Constituents of neem and related species: A reversed structure of azadirachtin. Third Int. Conf. of Chemistry and Biotechnology of Biologically Active Natural products (Sofia, Bulgaria, 1985), Abstr. pp. 446-450.

Krishna, S.S. 1987. Nutritional modulation of reproduction in two phytophagous insect pests. Id. Aca. Sci. 96(3): 153-169.

Krishnaiah, K. 1980. Methodology for assessing crop losses due to pests of vegetables "Assessment of crop losses due to pest and diseases". USA Tech. Series no. 33: 259-267.

Krishnakumar, N.K., and K. Srivasan., 1987. Efficacy ad economics of pest control of 'okra' with conventional and synthetic pyrethroid insecticides. Indian J. Pl. Prot.15: 81-83.

Kubar, M.I., Khuhro, R.D., Rajput, L.B. and Khuhro, S.N., 2006. Predators associated with okra, Hiscus esculantus. Proceedings 3rd National Conference on Agriculture and Animal Sciences. Sindh Agriculture University, Tandojam, pp. 47-51.

Kubar, M.I., Khuhro, R.D., Rajput, L.B. and Khuhro, S.N., 2006. Predators associated with okra, Hiscus esculantus. Proceedings 3rd National Conference on Agriculture and Animal Sciences. Sindh Agriculture University, Tandojam, pp. 47-51.

Kumar K.K. and Urs K.C.D., 1988. Population fluctuation of Earias vitella (Fab.) on okra in relation to abiotic factor, Ind. J. Pl. Protec., 16(2), 137-142

Kumar, K.K., and K.C.D. Urs., 1988. Population fluctuation of Earias vittella (Fab.) on 'okra' in relation to abiotic factors. Indian. J. Pl. Protec. 16: 137-142.

Kumar, K.K., and K.C.D. Urs., 1988. Population fluctuation of Earias vittella (Fab.) on 'okra' in relation to abiotic factors. Ind. J. Plant. Protect. 16(2): 137-142.

Kumar, N. K. K. and Moorthy, P. N. K. 2000. Transmission of Yellow vein mosaic geminivirus to imidacloprid treated okra by the white fly, Bemisia tabaci Gennadious. Insect Environ. 6 (i): 46-47.

Kumar, P.S., Alagarsamy, N., Sevarkodiyone, S. and Baskaran, S., 2004. Predatory potential of spiders on some insect pests on bhendi. Bionotes, 6: 109-110.

Kumar, P.S., Alagarsamy, N., Sevarkodiyone, S. and Baskaran, S., 2004. Predatory potential of spiders on some insect pests on bhendi. Bionotes, 6: 109-110.

Kumar, S., S. Prasad, and R.N. Singh., 2002. Resurgence of two spotted mite due to acaricides and botanicals on 'okra'. Ann. Pl. Protec. Sci.10: 51-54.

Kumawat, R.L., Pareek, B.L. and Meena, B.L., 2000. Seasonal incidence of jassid and whitefly on okra and their correlation with abiotic factors. Ann. Biol., 16: 167-169.

Kumawat, R.L., Pareek, B.L. and Meena, B.L., 2000. Seasonal incidence of jassid and whitefly on okra and their correlation with abiotic factors. Ann. Biol., 16: 167-169.

Kumbher, T.T., A. Dumber and A.S. Kokate. 1991. Studies on the varietal resistance in okra, Abelmoschus esculentus L. (Moench) to shoot and fruit borer, Earias spp. Maharashtra. J. Hort. 5: 78-82.

Lakshmanan, K.K. 2001. Neem a natural pesticide. The Hindu, March 1. P.8.

Leuschner, K. 1972. Effects of an unknown substance on a shield bug. Naturwissenshaften. 59: 217.

Lohar, M.K., 2001. Biological control. In: Applied entomology Published by Dr. Kashif Raza under Kashif Publication, Hyderabad. pp.147-167.

lohar, M.K., 2001. Biological control. In: Applied entomology Published by Dr. Kashif Raza under Kashif Publication, Hyderabad. pp.147-167.

Lukefahr, M.J., D.F. Marin and J.R. Meyer. 1965. Plant resistance to five Lepidoptera attacking cotton. J. Econ. Entomol. 58: 516-518.

Madave, R.P., and R.B. Dumbre., 1985. Reaction of 'okra' varieties to shoot and fruit borer. J. Maharashtra Agril. Univ. 10: 276-277.

Mahadevan. N.R. and Schelliah, S., 1986. Influence of season and weather factors on the occurrence of the sorghum stem borer, Chilo partellus (Swin) in Tamil Nadu. Trop. Pest Manag., 32: 212-214.

Mahadevan., N.R. and Schelliah, S., 1986. Influence of season and weather factors on the occurrence of the sorghum stem borer, Chilo partellus (Swin) in Tamil Nadu. Trop. Pest Manag., 32: 212-214.

Mahapatro, G.K., and G.P. Gupta., 1998. Pesticide induced resurgence. Pestology, 22(12): 14-20.

Makhadmeh, I.M. and I.E. Khalil. 2004. Geometric characteristics and chemical composition of okra (Abelmoschus esculentus L.) grown under semi-arid conditions. International J. Food Properties. 7(1): 83-90.

Maleque, M.A. 1998. Judicious application of insecticides for the management of Brinjal shoot and fruit borer. M.S. Thesis, BSMRAU, Salna, Gazipur, Bangladesh. management of fruit borer, Earias vittella (Fab.) on okra, Abelmoschus esculentus (L.) Moench. M.Sc. Thesis, Maharana Pratap University of Agriculture and Technology, Udaipur, Rajasthan.

Malerbo-Souza, D. T.; Toledo, V. de A. A. de; Stuchi, A. C.; Toledo, J. de O. A. de. 2001. Pollination of okra flowers, Abelmoschus esculentus (L.) Moench. Acta Scientiarum 2001 23( 5): 1281-1285

Mandahar, C.L. and Singh, J.S. 1972. Effect of bhendi yellow vein mosaic on its host. Acta. Acad. Sci. Hungaricae 7: 187-191.

Mandal, S.K., S. A. Sattar, S.B. Sah, and S.C. Gupta., 2006. Prediction of 'okra' shoot and fruit borer (Earias vittella Fab.) incidence using weather variables at Pusa, Bihar. Int. J. Agric. Sci. 2(2): 467-469.

Mandal, S.K., S.B. Abdus Sattar, and S.C. Gupta., 2006. Prediction of 'okra' shoot and fruit borer (Earias vittella Fab.) incidence using weather variables at Pusa, Bihar. Intl. J. Agric. Sci. 2(2): 467-469.

Mani, H.C., P.H. Pathak and S.S. Krishna. 1986. Effects of larval food quality on egg number and viability in Earias fabia Stoll (Lepidoptera: Noctuidae). Zeitschrift fuer Angewandte Zoologie (Germany, F.R.). 73(1): 115-120.

Mani, M., Krishnamoorthy, A. and Gopalakrishnan, C. 2005. Biological control of lepidopterous pests of Horticultural crops in India. A. Review. Agric. Res. 26(1): 39-49.

Marcano, R. and Gonzalez, E. 1993. Evaluation of insecticides for the control of the whitefly Bemisia tabaci (Gennadius) in tomato. Bol.Entomol. Venezolana. 8 (ii): 123-132.

Martin FW, Roberts R. 1990 Milling and uses of okra seed meal at the household level. J Agric Univ. 16:21–27.

Mau, Ronald F.L. and Dick Tsuda. 1991. Sweetpotato whitefly Bemisia tabaci (Gennadius) (Homoptera: Aleyrodidae).

McGregor, S.E., 1976. Insect Pollination of Cultivated Crop Plants. Agric. Handbook No. 496. U.S. Deptt. Agri., Washington, D.C., VIII: 411

Mehta, R.C. 1971. Survival and egg production of the cotton Spotted Bollworm, Earias fabia Stoll. (Lepidoptera: Noctuida) in relation to the plant infestation. Appl. Entomol. Zool. 6: 206-209.

Mehta, R.C. and Saxena, K.N. 1970. Ovipositional responses of the cotton spotted bollworm, Earias fabia (Lepidoptera, Noctuidae) in relation to its establishment on various plants. Entomol. Exp. Appl. 13: 10-20.

Memon, A. J., G.H. Abro, and T.S. Syed., 2004. Varietal resistance of 'okra' against Earias spp. J. Entomol. 1(1): 1-5.

Miah, M.A.S. 1988. Effect of date of planting and insecticidal spray on control of yellow vein mosaic of Lady`s finger (Hibiscus esculenturs L.) abstracts of Theses (1966-1990). Department of plant pathology, Bangladesh Agricultural University, Mymensingh-2202, March, 1991.79 p. Maharashtra Agric. Univ. 2(2): 175.

Mishra RC, Kumar J, Gupta JK.1987. Effect of mode of pollination on fruit characteristics of okra. Abelmoschus esculentus (L.) Moench. Proc Indian Nat Sci Acad B 53. 1987;157-160.

Mitidieri, S and Vencovsky, K. (1998). Rev. Agric (Piracicaba Braz) 49(1):3

Mishra, N.C. and Mishra. S.N., 2002. Impact of biopesticides on insect pests and defenders of okra. Indian J. Pl. Protect., 30: 99-101.

Mishra, N.C. and Mishra. S.N., 2002. Impact of biopesticides on insect pests and defenders of okra. Indian J. Pl. Protect., 30: 99-101.

Misra, H.P., D.D. Dash, and D. Mahapatro., 2002. Efficacy of some insecticide against 'okra' fruit borer and leaf roller. Ann. Pl. Proec. Sci. 10: 51-54.

Misra, H.P., Dash, D.D. and Mahapatro, D. 2002. Efficacy of some insecticide against okra fruit borer and leaf roller. Ann. Pl. Proect. Sci. 10: 51-54.

Misra, P.N. 1989. Bio-efficacy of some newer insecticides against the pest complex of okra. Bangladesh Hort. 17(1):1-4.

Mohan, J., Krishnaiah, K. and Prasad, V.G. 1983. Chemical control of insect pest of okra. Indian J. Entomol. 45(2): 252-258.

Morale, R.S., Sarnaik, D.N., Satpute, U.S. and Sadawarte, A.K. 2000. Effect of plant products on growth and development of Helicoverpa armigera on cotton. Pestol. 24(6): 50-55.

Mota, W.F., F.L. Finger, D.J. H. Silva, P.C. Correa, L.P. Firme and L.L.M. Neves. 2005. Physical and chemical characteristics from fruits of four okra cultivars. Horticultura Brasileira, Brasília. 23(3): 722-725.

Mote, U. N. 1977. Seasonal incidence of bhindi fruit borer, Earias vittella F. Journal of Maharashtra Agricultural Universities 2(2): 175.

Mound L.A. and S.H. Halsey. 1978. Bemisia tabaci (Gennadius). pp. 118-124. In Whitefly of the World, A Systematic Catalog of the Aleyrodidae (Homoptera) with Host Plant and Natural Enemy Data. British Museum (Natural History ) and John Wiley & Sons, Chichester, New York, Brisbane, Toronto. 340 pages.

Mudathir, M. and Basedow, T. 2004. Field experiments on the effects of neem products on pests and yields of okra Abelmoschus esculentus in the Sudan. Mitteil. Deutsch. Gesellsch. allgem. angew. Ent. 14: 407-410.

Murugesan, S. 1982. The Spotted Bollworm, Earias vittella (Fab), Ecology and host resistance. Ph. D. Thesis Tamil Nadu Agriculture University, Coimbatore, India.

Naik, M.I., S. Lingappa, I.G. Hiremath, and B.K. Shivanna., 1997. Pheromone trap to monitor adult spotted bollworm, Earias vittella (F.) on cotton. Mysore J. Agric. Sci. 31(1): 33-35.

Nandihalli, B.S., Parameshwara Hugar And Patil, B.V. 1990. Evaluation of neem and neem products against cotton whitefly, B. tabaci (Genadius). Karnataka J. Agril. Sci., 3 : 58-61.

Naresh, V., A.K. Biswas, K. Roy, and M.W. Reza., 2003. Relative susceptibility of different varieties of 'okra' to the shoot and fruit borer, Earias vittella (Fabr.) and leaf-roller, Sylepta derogata (Fabr.). Pest Manag. Econ. Zool.11(2): 119-122.

Nath, P. and Saikia, A.K. 1993. Assessment of yield loss due to yellow vein mosaic of okra (Abelmoschus esculentus L. ) in Assam J. Agric. Sci. Soc., North East India. 6 (iv): 87-88.

National Agricultural Research Project (NARP). Horticultural crops. NARP, CSIR, Accra. 1993;3.

Nayar K.K., Ananthakrishnan T.N. and David B.V., 1976 General and Applied Entomology, Tata McGraw-Hill Publishing Company Ltd, New Delhi, 589

Ngo, N., Moore, S., Lawson, S., Dugger, P. and Richter, D. 1999. Ovicidal research with emamectin benzoate against tobacco budworm eggs. Beitwide Cotton Conferences, Orlando, Florida, USA, 3-7 January, 1999:1090-1098. Schedule of insecticidal application against insect pest complex of brinjal with special reference to brinjal shoot & fruit borer,- Leucinodes orbonalis Guen. Indian J. 50 (1): 16-19.

Njoya Moses Tita, Wittmann Dieter, Schindler Mathias 2005. Effect of Bee Pollination on Seed Set and Nutrition on Okra (Abelmoschus esculentus) in Cameroon. Deutscher Tropentag, October 11-13, 2005, Hohenheim "The Global Food & Product Chain-Dynamics, Innovations, Conflicts, Strategies"

Nonnecke, I.L., 1989. Vegetable Production A VI Book, Published by Van Nostrad. Reinhold. pp 608-612.

NRI (Natural Resource Institute) 1996. Insect pests of Nigeria Crops. Government Printing Press, Agric, Dept, Nigeria.pp: 240.

Olugbenga E. Ige and Eludire M. O. 2014. Floral Biology and Pollination Ecology of Okra (Abelmoschus esculentus L. Moench). American International Journal of Biology. 2( 2): 01 09

Oyenuga,V.A., 1968. Nigeria's Foods and Feeding Stuffs, their Chemistry and Nutrients value (third ed.), University Press, Ibadan.

Pareek, B.L., Kumawat, R.L. and Patni, S.K. 2001. Effect of abiotic factors on the incidence of okra insect pests in semi-arid conditions. Proceeding of the National Conference on plant protection- New Horizons in the Millennium, Feb. 23-25, Udaipur. pp. 1-8.

Parkash, O. 1988. Schedule of Insecticidal Application against insect pest complex of brinjal with special reference to brinjal shoot and fruit borer, Leucinodes orbonalis Guen. Indian J. Entomol. 50 (1):16-19.

Parvin, S. 2004. Performance of four okra varieties relation to Okra Yellow Vein Clearing Mosaic Virus (OkYVCMV) incidence and its impact on growth and yield. M. S. Thesis. Department of plant pathology BSMRAU, Salna. Gazipur.

Pasha, M.R., 1979. Pre-investment study Faisalabad District. Directorate of Industries and Mineral Development, Punjab. 38 pp.

Patel I.S. and Rote N.B., 1995, Seasonal incidence of sucking pest complex of cotton under rainfed condition of southern Gujarat, Gujarat Agric. Univ. Res. J., 21(1), 127-129

Patel, J.R. and N.C. Patel., 2000. Behaviour of Earias vittella in relation to weather factors on cotton H-6. Ind. J. Entomol. 62(4):363-366.

Patel, M.S., Yadav, D.N. And Rai, A.B., 1994, Neemark (Azadirachtin) as ovipositional deterrent against cotton pests. Pestol., 18 : 17-19.

Patel, N.C., Zala, S.P. and Patel, J.R. 1999. Impact of weather on magnitude of Earias vittella infesting okra. Indian J. Ento. 61(4): 351-355.

Patil, C.S., P.R. Moholkar, and D.S. Ajri., 1986. Studies on moth attraction and egg laying preference of shoot and fruit borer, (Earias vittella Stoll.), on different varieties of 'okra', (Abelmoschus esculentus L.).Current Res. Reporter, Mahatma Phule Agric. Uni. 2(2):289-291.

Patil, C.S., Pawer, S.A., Mote, U.N. and Khaire, V.M. 1991. Evaluation of cypermethrin (SIRcis and RIS-cis isomers) against okra fruit borer. Ann. AppI. Biol. 118 (suppl.):20-21.

Patil, R.S. 2000. Utilization of plant products in the management of okra shoot and fruit borer Earias vittella in okra crop. M.Sc. (Agri.) Thesis, University of Agricultural Sciences, Dharwad.

Paulraj, M.G. and Ignacimuthu, T., 2005. Predatory insect fauna in mixed cropping agro ecosystems in Northeastern Tamil Nadu. Insect Environ., 11: 79-82.

Pawar. A.D. and Prasad, J. 1988. Evaluation of some parasites in biocontrol of cotton bollworms in Haryana. Indian J. Plant Prot. 13(1): 21-24.

Pearson, E.O. 1958. The insect pests of cotton in tropical Africa. Int. J. Applied Entomol. 12: 71-74.

Pfadt, R.E., 1980. Fundamentals of applied entomology. Macmillan Company, New York, pp. 99-104, 24-126.

Prakash, A., Rao, J., Jevari, S.N. and Gupta, S.P. 1990. Rice agro ecosystem management by pesticides and its consequences. Nat con. Publ. in Growth Develop.Natural Res. Conservations. pp. 131-137.

Prakash, A., Rao, J., Pasalu, I.C. and Mathur, K.C. 1987. Rice Storage and Insect Management. B.R. Publishing Corporation, New Delhi. p.337.

Praveen, P.M. and Dhandapani, N., 2001. Eco-friendly management of major pests of okra (Abelmoschus esculentus (L.) Moench). J. Veget. Crop Prod., 7: 3-12.

Praveen, P.M. and Dhandapani, N., 2001. Eco-friendly management of major pests of okra (Abelmoschus esculentus (L.) Moench). J. Veget. Crop Prod., 7: 3-12.

Priya, B.S. and Misra, H.P. 2007. Biopesticides for the management of okra fruit borer, Earias vittella (Fabricius). Pest Mgt. Horti. Ecosystem 13(2): 176-179.

Pruthi, H.S. 1946. Report of the Imperial Entomologist. Abridged science Reports, Agricultural Research Institute, New Delhi. (1941-44). 99: 64-71.

Purewal, S. S. and Randhawa, G.S. (1947), Studies in Hibiscus esculentus (Lady's Finger). Chromosome and pollination studies. Indian J. Agri. Sci, 17, 129-136.

Pushpalatha BT 2008. Effect of seed priming on storability and field performance in okra (Abelmoschus esculentus (L.) Moench) Msc. Thesis University of Agricultural Sciences, Dharwad. Hortic. 2008;288:103–10.

Rahman , M.H. and H. Ali. 1983.Biology of spotted bollworm of cotton, Earias vittella Fab. Pakistan J. Biol. 13(1-2): 105-110.

Rahman MA, Uddin MM, Haque MA and Rahman MM 2015. Varietal Preference of Okra Shoot and Fruit Borer, Earias vittella (Fab.) under Field Condition in Bangladesh Academic Research Journal of Agricultural Science and Research 3(1): 8-12

Rai, S. 1985. Chemical control of Bhindi pests . Indian J. Entomol. 7 (2): 173-178.

Rajaram, V., Mathirajan, V.G. and Krishnasamy, S., 2006. IPM in Cotton under dry farming condition. Int. J. agric. Sci., 2: 557-558.

Rajaram, V., Mathirajan, V.G. and Krishnasamy, S., 2006. IPM in Cotton under dry farming condition. Int. J. agric. Sci., 2: 557-558.

Rajasekaran, B. and Kumarswami, Y. 1985. Antifeedent properties of certain plant products against Spodoptera litura (F.). In: Proceedings of National Seminar on Behavioral Physiology Appr. Mrmt. Crop pests, Tamil Nadu Agricultural University. pp. 25-28.

Ramiah, M., Vidhyasekaran, P. and Kandaswamy, T.K. 1972. Changes in photosynthetic pigments of bhendi infected by yellow vein mosaic disease. Madras Agril. J. 59:402-404.

Rana, B.S. 1983. Studies in the effectiveness of some insecticides against insect pests of bhendi, Abelmoschus esculentus (L.) Moench. M.Sc. Thesis, Udaipur University, Rajasthan.

Rao, N.S. and Raguraman, S., 2005. Influence of neem based insecticides on egg parasitoid, Trichogramma chilonis and green lace-wing predator, Chrysoperla carnea. J. Ecobiol., 17: 437-443.

Rao, N.S. and Raguraman, S., 2005. Influence of neem based insecticides on egg parasitoid, Trichogramma chilonis and green lace-wing predator, Chrysoperla carnea. J. Ecobiol., 17: 437-443.

Rao, S.N., Rajendran, R. and Raguraman, S. 2002. Antifeedant and growth inhibitory effects of neem in combination with sweet-flag and pungam extracts on Okra shoot and fruit borer, Earias vittella (Fab.). J. Entomol. Res. 26 (3): 233-238.

Rashid, M. M. 1999. Shabjibiggayan (In Bengali). Rashid Publishing House, 94 Old DOHS, Dhaka-1206. P. 49.

Reichmuth, C.H., Scholler, M. and Ulrichs, C.H. 1997. Stored Product Research in Cereals. Verlag Thomas Mann, Gelsenkirchen, 119 p.

Reiter, P. W., 1989. Vector Biology and Arborial Recrudescence. In:The Arboiruses: epidemiology and Ecology, Vol. 1. T P Monath, editor, CRC Press. Inc, Boca Raton, Fe : 245.

Richard, L.A. 1954. Diagnosis and improvement of saline and alkali soils. USDA Agri. Hand Book 60,Government point office, Washington DC.

Rojanaridpiched, C., V.E. Gracen, H.L. Everett, J.C. Coors, B.F. Pugh and P. Bouthyette. 1984. Multiple resistance factor in maize to European corn borer. Maydica 29: 305-315.

Rosaiah, R., 2001, Performance of different botanicals against the pests complex in bhendi. Pestol, 25 : 17-19.

Sabri, M.A., and F. Shahzad., 2002. Temperature threshold and thermal constant estimated for development of spotted bollworm (Earias vittella F.) of cotton. Pak. Entomol. 24(1): 45-47.

Sangwan, H.S. 1972. Some observations on Stenobracon deesae Cameron. Indian J. Entomol. 34: 172-173.

Sardana, H.R. and Krishnakumar, N.K. 1989. Effectiveness of plant oils against leafhopper, shoot and fruit borer in okra. Indian J. Entomol. 51 (2): 167- 171.

Sardana, H.R., Bambawale, O.M., Kadu, L.N. and Singh, D.K., 2005b. Development and validation of adaptable IPM in okra through farmers participatory approach. Annl. Pl. Protect. Sci., 13: 54-59.


Sardana, H.R., Bambawale, O.M., Kadu, L.N. and Singh, D.K., 2005b. Development and validation of adaptable IPM in okra through farmers participatory approach. Annl. Pl. Protect. Sci., 13: 54-59.


Sardana, H.R., Bambawale, O.M., Singh, D.K. and Kadu, L.N., 2005a. Monitoring of insecticide residues in IPM and non-IPM fields of okra and brinjal. Indian J. Pl. Protect., 33: 197-201.


Sardana, H.R., Bambawale, O.M., Singh, D.K. and Kadu, L.N., 2005a. Monitoring of insecticide residues in IPM and non-IPM fields of okra and brinjal. Indian J. Pl. Protect., 33: 197-201.


Sarkar, B.B. and Nath, B.C. 1989. Field evaluation of various insecticides against fruit borer and effect on yield of lady's finger seed. Pestol. 13(4): 19-21.

Sastry, K. S. M. and Singh, S. J. 1974. Yellow vein mosaic virus infection on growth and yield of lira crop. Indian Phytopath. 27: 294-297.

Satpathy. S and Rai, S. 1999. Efficacy of different pesticides and combination against jassid borer of okra. Vege. Sd. 26(1):78-81.

Satpute, N.S., Deshmukh, S.D., Mane, P.N., Rao, N.G.V. and Nimbalkar, S.A. 2002. Growth and development of spotted bollworm, Earias vittella (Fab.) reared on different hosts under laboratory conditions. PKV. Res. 26(1/2):47-50.

Savello, P.A., Mortin, F.W. and Hill, J.M. 1980. Nutritional composition of okra seed meal. Agril. Food Chem. 28: 1163-1166.

Saxena, R.C., Waldbauer, G.P., Liquido, N.J. and Puma, B.C. 1981. Effects of neem seed oil to the rice leaf folder, Cnaphalocrocis medinalis. In: Natural Pesticides from the Neem Tree (Azadirachta indica A. Juss) and Other Tropical Plants. Ed. by Schmutterer, H.; Ascher, K. R. S. Rottach-Egern, Germany: Eschborn: GTZ, 189–204.

Schmutterer, H. 1961. The most important pests of crops in the central rainlands of the Sudan and their control. Anz. Schadlinge. 34: 177-180.

Schmutterer, H. 1985. Properties and potential of natural pesticides from the neem tree, Azadirachta indica. Annu. Rev. Entomol. 35: 271-297.

Schmutterer, H. 1995. The neem tree, Azadirachta indica A. Juss and other eliaceous plants Sources of unique natural products for integrated pest management, medicines, industry and other purposes. VCH Publishers Inc., New York, USA. P. 253.

Schmutterer, H., Saxena, R.C. and Heyde, V.D.J. 1983. Morphogenetic effects of some partially-purified fractions and methanolic extracts of neem seeds on Mythimna separata (Walker) and Caphalocrocis medinalis (Guyenne). Z. Angew. Entomol. 95: 230-237.

Schoonoven, L.M. 1982. Biological aspects of antifeedants. Entomol. Exp. Appl. 31: 57-69.

Senapati B. and Mohanty G.B., 1980. A note on the population fluctuation of sucking pests on cotton, Madras Agric. J., 67(7), 624-630

Shah, B.R., Vyas, H.N. and Jhala, R.C. 2001. Population dynamics of major insect pests of okra Abelmoschus esculentus (L.) in relation to weather parameters. In: Abstracts National conference: Plant Protection –New Horizons in the Millennium (NCPP). Udaipur (Raj.) p4.

Shalaby, G.J., 1998. Natural cross-pollination in okra. Journal on Agriculture Science 3(1): 381-386.

Sharma, H.C. and R.A. Agarwal. 1983. Ovipositional behaviour of spotted bollworm Earias vittella (Fab.) on some cotton genotypes. Insect Sci. Applic. 4: 373-376.

Sharma, I.N., Lall, B.S., Sinha, R.P. and Singh, B.N. 1985. Biology of spotted bollworm Earias vittella. Bull. Entomol. 26(1): 38-41.

Sharma, R.P., Swaminathan, R. and Bhati, K.K. 2010. Seasonal incidence of fruit and shoot borer of Okra along with climatic factors in Udaipur region of India. Asia J. Agril. Res. 4 (4): 232-236.

Sheeba, M. and Narendran, T.C. 2007. On the identity of two confusing species of (Braconidae: Hymenoptera) parasitic on Opisina arenosella (Walker). J. Plant. Crops 35(3): 195-197.

Sheedi, M.A., 1980. Arthropods associated with okra Hibiscus esculentus L. crop. M.Sc. thesis, Sindh Agriculture University, Tandojam. pp. 58.

Sheedi, M.A., 1980. Arthropods associated with okra Hibiscus esculentus L. crop. M.Sc. thesis, Sindh Agriculture University, Tandojam. pp. 58.

Shivalingaswamy, T.M., Satpathy, S. Singh, B. and Kumar, A., 2002. Predator-prey interaction between jassid (Amrasca biguttula biguttula, Ishida) and a staphylinid in okra. Veget. Sci., 29: 167-169.

Shukla, A., Pathak, S.C. and Agrawal, R.K., 1996. Efficacy economics of some insecticides and plant products against the infestation of okra shoot and fruit borer Earias vittella (Fab.). Crop. Res. Hisar., 12: 367-373.

Shukla, A., S.C. Pathak and P.K. Agrawal., 1998. Field evaluation of 'okra' varieties for resistance to shoot and fruit borer, Earias vittella (Fab.). J. Insect Sci. 11(1): 60-61.

Shukla, A., S.C. Pathak, and R.K. Agrawal., 1997. Seasonal incidence of 'okra' shoot and fruit borer Earias vittella (Fab.) and effect of temperature on its infestation level. Adv. Pl. Sci.10:169-172.

Siemonsma J.S., 1982. Laculturedugombo (Abelmoschus spp.) legume fruit tropical avec reference special alacoted'ivoire thesis Univ. of Wageningen, the Netherlands,

Singh A.K., 2006. Evaluation of fungicides for the control of powdery mildew disease in coriander (Coriandrum sativum L.), J. Spices and Aromatic Crops, 15(2), 123-124

Singh, A.K. and Kumar, M., 2003. Efficacy and economics of neem based products against cotton jassid, Amrasca biguttulla Ishida in okra. Crop. Res. Hisar., 26: 271-274.

Singh, B.K., Singh, A.K. and Singh, H.M. 2005. Efficacy of certain synthetic insecticides and two botanicals against the okra fruit and shoot borer, Earias vittella (Fab.). Pest Manage. Economic Zool. 13(1): 99-103.

Singh, G. and K.S. Brar., 1994. Effects of dates of sowing on the incidene of Amrasca biguttula (Ishida) and Earias spp. on 'okra'. Indian J. Ecol. 21: 140-144.

Singh, J. and Brar, K.S., 2004. Mass production and biological control potential of coccinellids in India. In: Indian insect predators in biological control (ed. K. Sahayaraj) Daya Publishing House, Delhi, India, pp. 204-260.

Singh, M.P., Sharma, D. P. And Mishra, H.P., 1999, Efficacy of plant extracts against cotton whitefly, Bemisia tabaci (genn.). Annals of Plant Protection Sciences, 26: 46

Singh, R. 1987. Effect of okra fruit blocks, seeds and pericarp on post-embryoni development of Earias vittella (Fab.) in relation to some phytochemicals of selected okra genotypes. Proceed. Ind. Acad. Sci., An. Sci. 96 (4): 361-367.

Singh, Y. and Bichoo, S.L. 1989. Some biological and bionomical observation of Earias spp. Bull. Entomol. New-Delhi. 30 (1): 84-91.

Sinha, S.N., S.P. Sharma and A.K. Chakrabarti. 1978. Effect of spotted bollworm Earias spp. infestation on seed quality of 'okra', Abelmoschus esculentus. Seed Research (New Delhi) 6(2): 161-164.

Smith, C.M. 1986. Trends affecting research strategies in plant resistance to insects. Agric. Ecosystems & Environ. 18(1): 1-7.

Snedecor, G.W. and Cochran, W.G. 1967. Statistical methods. Iowa state Univ. Press. Ames, Iowa, 462p.

Sojitra, I.R. and Patel, J.R. 1992. Effects of plant extracts (including Azadirachta indica, Ricinus communis and Pongamia pinnata) on ovipositional behaviour of spotted bollworm (Earias vittella) infesting okra (Abelmoschus esculentus). Indian J. Agric. Sci. 62: 848–849.

Sparks, D. L. (ed.).1996. Methods of soil analysis, part 3: Chemical Methods, Soil Science Society of America. Madison, Wisconsin.

Srabani, Debnath, Nath, P. S. and Denath, S. 2002. Management of yellow vein mosaic disease of okra through insecticides, plant products and suitable varieties. Ann. Plant Protect. Sci. 10 (ii):340-342.

Srinivasan Murthy, K. And Sharma, D.P., 1997, Traditional pest management in cotton. Insect Environ., 3 : 73-74.

Srinivasan, P.M. and Gowder, R.B. 1959. A preliminary note on the control of the okra shoot and fruit borer, Earias fabia and E. insulana. Indian J. Agric. Sci. 30 (1): 55-57.

Steel, R.G.D., J.H. Torrie and D.A. Dickey., 1990. Principles and Procedures of Statistics. A biometric Approach, 3rd edition. WCB Mc Graw Hill Companies, inc., USA.

Stelzel, M. and Devatak, D., 1999. Neuroptera in agriculture ecosystem. Agric. Ecosyst. Environ., 74: 305-321.

Suman, C.L., S.D. Wahi and N. Jagan Mohan., 1984. Distribution pattern of 'okra' shoot and fruit borer (Earias vittella Fab.) under natural conditions. Indian L. Ent. 45: 362-364.

Swamiappan, M. and Balasubramanian, M. 1979. Studies on mass multiplication and potentiality of Chelonus blackburni Cam a braconid parasitoid of cotton boll worms. Entomo. 5: 73-75.

Tanda AS.1984. Insect pollination in okra (Abelmoschus esculentus). Proc. III Ent. Orient. Symp. Kariavattom, India. 25.

Tanda, A. S. 1985. Floral biology, pollen dispersal, and foraging behaviour of honeybees in okra (Abelmoschus esculentum). J. Apic. Res., 24(4):223-227.

Tanwar, R.K., Bamawale, O.M., Kumar, P.J., Dhandapani, A., Kanwar, V., Sharma, O.P. and Monga, D., 2007. Impact of IPM on natural enemies in irrigated cotton of North India. Entomon, 32: 25-32

Telang, S.M., Rathod, K.S. and Rathod, R.M., 2004. Parasitization by different parasites of Earias Vittella in okra. Indian J. Soils Crops, 14: 335-339.

Teli, V.S. and V.P. Dalaya, 1981. Studies on varietal resistance in okra (Abelmoschus esculentus (Linn.) Moench) to the shoot and fruit borer, Earias vittella Fabricius. South Ind. Hort. 29(1): 54-60.

Thakur, M.R., Arora, S.K., Bose, T.K. and Son, M.G. 1986. Vegetable crops in India. 6th edition. Calcutta, India. p. 678.

Thanavendan G. and Jeyarani S. 2009. Biointensive Management of Okra Fruit Borers using Braconid Parasitoids (Braconidae: Hymenoptera). Tropical Agricultural Research 21(1): 39– 50

Thomson, H.C. and Kelly, W.C. 1979. (3rd ed.) Vegetable crops. Mcgraw Hill Co. New York. P. 562.

Tindall, H.D. 1986. Vegetables in the Tropics. 1st edition, Macmillan Publishers, Hong Kong. pp. 325-327.

Torres, J.B. Ruberson, J.R. 2004. Toxicity of thiamethoxam and imidacloprid to Podisus nigrispinus (Dallas) (Heteroptera:Pentatomidae) nymphs associated to aphid and whitefly control in cotton. Neotropic. Entomol. 33(i):99-106.

Udayasekhara, R.P. 2005. Chemical composition and biological evaluation of Okra (Abelmoschus esculentus) seeds and their kernels. Plant Foods for Human Nutrition. 35 (4): 389-396.

Ukwungwu, M.N. and J.A. Odebiyi. 1985. Incidence of Chilo zacconius Blesznski on some rice varieties in relation to plant characters. Insect Sci. Appl. 6(6): 653-656.

Verma, S, Gupta, R.S., Vishwa nath, S.P., Kawal, D. and Sing, R. 1980. Control of pests of okra. Indian J. Entomol. 42: 822-823.

Vyas, S.H., and J.R. Patel., 1990. Relative susceptibility of some lady finger cultivars to Earias vittella Fabricius. Ind. J. Pl. Prtotect. 18 (1): 115-118.

Vyas, S.H., and J.R. Patel., 1991. Intensity and damage of Earias vittella Fab. on various cultivars of Bhendi. Guj. Agric. Uni. Res. J. 17(1): 140-141

Warthen, J.D., Uebel, J., Dutky, E.C., Luspy, S.R. and Finegold, W.R. 1978. Adult house fly feeding deterrent from neem seeds. USDA/SEA Agric. Res. Results, Northeast. 2: 1-11.

Winkeman, G.R., R. Amin, W.A. Rice and M.B. Tahir. 1986. Methods: Manual soils laboratory, Barani Agricultural Research and Development Project. Pakistan Agricultural Research Council, Pakistan. pp 30-33.

Wiseman, B.R. 1990. Plant resistance to the insects in the Southeastern United States: An overview. The Florida Entomologist: 351-358.

Wright, R.T., and T. Stuczynski. 1996. Atomic absorption spectro photometry. pp 69-90. In: D. L. Sparks (ed.). Methods of soil analysis, part 3: Chemical Methods, Soil Science Society of America. Madison, Wisconsin.

Yadvendu, T.C. 2001. Evaluation of newer insecticides against insect pests of okra (Abelmoschus esculentus). M.Sc. Thesis, Maharana Pratap University of Agriculture and Technology, Udaipur, Rajasthan

Yamaguchi, M. 1998. World Vegetables Principles, Production and Nutritive Values. Van Nostrand Reiduction. New York, USA. pp. 415.

Zaki, F.N., EL-Shaarawy, M.F. and Farag, N.A., 1999. Release of two predators and two parasitoids to control aphids and whiteflies. Anz. Schadlingsk., 72: 19-20

Zala, S. P., J. R. Patel, and N.C. Patel., 1999. Impact of weather on magnitude of Earias vittella infesting 'okra'. Ind. J. Entomol. 61(4):351-355.

*Infested fruit* *Infested shoot*

*Okra jassid* *typical cupping symptoms*

**Figure 8.2. Insect pests of Okra**

CHAPTER 9

# Leguminosae

## Leguminous Vegetables

Leguminous vegetables constitute about 5% of the cultivated crops used for human consumption. They have been cultivated for more than 6,000 years in different parts of the world. Majority of the population in India depends upon, leguminous vegetables as a source of protein diet. They are also a good source of vitamins such as folate and dietary minerals like calcium, iron, magnesium and zinc (Heiser 1990). Most leguminous vegetables are rich in phosphorus, calcium, iron, and a number of essential vitamins. Leguminous vegetables are used in several different ways as fresh pod, immature seeds, sprouted seeds and mature dry seeds. They are also canned and frozen for the market. Legumes contain antioxidants and several other health-promoting substances which can reduce the risk of some diseases such as cancer. Hence leguminous vegetables are a boon for human health as well as a commercial crop for domestic and export purposes.

There has been no increase in productivity levels of legume crops in the past 30 years. This is a serious concern in this time of exponential human population growth because legume crops provide essential dietary components that are not present in cereals.

The demand for food increasing day by day and by 2050, the production has to be doubled and this target will have to be achieved in the backdrop of decrease in cultivable land, increased water deficit, and increased uncertainty of climatic conditions. With the predicton of 2-4 rise in temperature in the next 100 years, conditions will become further challenging to grow crops successfully.

Leguminous crops fix atmospheric nitrogen in the soil, thus improving soil fertility. Legumes are unique compared to other vegetables in that they can obtain free atmospheric nitrogen through nitrogen-fixing bacteria, *Rhizobiwn* or *Pradyrhizobium,* in legume root nodules.

The nitrogen fixed in the root nodules is not only available to the plant but they also enrich the soil, in varying amounts, when the plants complete their life cycle

Large amounts of legume seed are required annually for planting but are always difficult to obtain when intervals between crop rotaions are shorter and improved high yielding varieties are not available. Therefore, good pollination of existing varieties is essential. Poor weather, inadequate irrigation, attack of insect pests and diseases have caused the lower yields.

Vegetable legumes are also highly susceptible to insect pests and diseases which cause up to 80% yield losses or more (Singh et al. 1990; Afun et al. 1991; Dreyer et al. 1994; Ulrichs and Mewis 2004). Besides, inadequate pollination is the most common cause of drastic reduction in yields. Therefore, for obtaining optimum yields, sustainable pest management strategies offer a safe alternative to the misuse and overuse of pesticides on food legumes in tropical Asia and Africa and provide safety to pollinators and other beneficial insects.

## Pollination of leguminous vegetables

The pollination requirements for legume crops vary depending upon their breeding system whether they are compatible and or self in compatible. . In case of self-pollinating varieties, bees have little role to play such as peas, soybeans, peanuts, snap beans and subterranean clover. But in case of self sterile varieties, some outside agent like bees is required for any seed production such as lespedeza, lima beans and vetches. Legume crops produce both nectar and pollen and can be very attractive to pollinators, especially Hymenoptera. Many species of legumes are good sources of honey and pollen for bees and large quantities of honey can be harvested.

A typical legume flower comprises a pattern consisting of sepals, and five petals, including one standard, two wings, and two keels, stamens and carpels. In addition, extra-floral nectaries may also play a role in pollinator attraction (Westerkamp and Weber 1999).

Legumes flowers attract a wide variety of insects which include honeybees, wild solitary bees, feral bees (Free 1993; Delaplane and Mayer 2000). Legumes play an important role in providing sustainable agriculture by providing ecological services such as foraging places and nesting sites.

The provision of floral resources for supporting beneficial insect pollinator populations could be a promising strategy to enhance environmental function of legumes. Therefore ther is a need to develop pollinator-friendly cultivars which requires rethinking of crop breeding strategies. The role of pollinators as agents of pollination and hybridization and breeding strategies needs to be understood (Richards 2001). Crops with floral attractiveness and rewards can be used which may further lead to improved crop yields as well as increased resistance to biotic and abiotic stresses. Therefore, there are good breeding and ecological reasons for improved targeting of crop pollinator relationships (CPR). Optimising CPR would be a good step for establishment of breeding strategies that increase yield and its stability by using pollinating bees that facilitate the development of ''pollinator-friendly'' cultivar(Morandin and Winston 2006). This programme not only helps conserving biodiversity but farmers continue to get high yield and yield stabilit (Westerkamp and Gottsberger 2000; Aizen et al. 2008).

## Cowpea, *Vigna* spp

Cowpea (V*igna unguiculata* L.Walp) belongs to the family Fabaceae is an important grain legume grown in tropics and subtropics of the Savanna regions. It has around 150-190 species. The cultivated cowpea has five groups namely; *unguiculata*, *sesquipedalis*, *textiles*, *melanophtalmus*, and *biflora* (Westphall, 1974; Marechal *et.al*., 1978, Ng and Marechal, 1985).

Cowpea has high protein content (23-29%). It has the ability to fix atmospheric nitrogen, which makes it to grow on and improve poor soil (Steele, 1972). It is the second most important crop grown among the pulses (Marechal, *et al* 1978). It is a self pollinated crop, however, insects visiting cowpea flowers play a role in the transfer of pollen from one flower to another.

The flowers are placed 5-60 cm at the distal end of long peduncles. They are arranged in racemose or intermediate inflorescence. Flowers are borne in alternate pairs, with usually only two flowers per inflorescence. They are conspicuous, borne on short pedicels. The corollas may be white, dirty yellow pink, pale blue or purple in colour. Flowers open in the early day and close in the mid of the day, wilt and collapse.

The inflorescence consists of two to eight whitish, yellowish, or violet papilionaceous flowers in pairs crowded together on the tip of a slender peducle or stem, 1 inch to 6 inches long. The 3/4- to 1-inch flower has a bent style, bearded on the inner curve immediately below the oblique stigma, and uniform anthers in two fused groups around the style. The flower has a single ovary with eight to 20 ovules. Extrafloral nectaries are located at the base of the corolla. Whether floral nectar is secreted is not clear, but most likely it is. The flower is attractive to bumble bees and various other insects that forage upon both the nectar and pollen.

## Pollination Requirements

Robbins (1931) stated that the cowpea is capable of self-fertilization, and this is probably the common occurrence, although the flowers are often visited by honey bees or bumblebees, attracted chicfly by the extrafloral nectar.

## Pollinators

Purseglove (1968) stated that a heavy insect is required to depress the wings of the flower and expose the stamens and stigma. The pollen is sticky and heavy, indicating that the plant is not wind-pollinated (Mackie 1946). Cross-pollination occurs in areas where bumble bees are numerous. Mackie and Smith (1935) stated that bumble bees are the primary pollinators. Warnock and Hagedorn (1954) revealed that insect visitation activity is beneficial in increasing the number of pod set, the number of seeds per pod, or both. Pollinators recorded included honeybees *Apis mellifera* Lin. *Apis cerana* Fab., *Apis dorsata* Fab.), Carpenter bee (*Xylocopa sp.*), Bumble bee (*Bombus spp.*), Oriental wasp (*Vespa orientalis* (Lin.), and various species of flies such as Tabanifly (*Tabanus spp*), Syrphid fly (*Eristalis sp.*), Housefly (*Musca domestica* Lin.), Fruit fly (*Bactrocera sp.*) etc.

## Pollination Recommendations and Practices

Cowpea is self pollinating crop and usually pollination occurs in cowpea flower before they open. The enclosure of the pistil and stamen within the keel enhances self pollination. Pollen grains are transferred from the anther to the stigma and thus pollination occurs. However, Cowpea flowers produce nectars which attract insects to them. Ige et al ( 2011) observed various insects such as Butterflies, Houseflies (*Musca domestica),* Honeybees (*Apis mellifera*) visiting the cowpea flower. They found that honeybee *Apis mellifera* was observed to be the most prominent out of the various insects observed, started visitation to the flower immediately after flower opening was initiated. The bee lands on the wing of the flower thereby depressing it and allowing the stamens and stigma to be pushed out of the corolla. The bee then adopts a supine feeding posture with the head positioned towards the nectar chamber and collects nectar. After nectar feeding, it gradually withdraws from the chamber to its landing position and then flies away on to the next flower or away from the vicinity.

However, during this process of feeding, the hairs on the bee brush the anthers and pollen adheres to them and these pollen grains are transferred to another flower. Pollen grains of cowpea are heavy and sticky and could not be readily transferred by wind, therefore insects are responsible for transfer of pollen grains and consequently cross pollination in cowpea plant. *Apis mellifera* has been previously reported as good pollen collectors and as such; effective pollinators (Nepi and Pacini, 1993, Luo and Li, 1999).

Fernand-Nestor Tchuenguem Fohouo et al (2009) reported that number and dry weight of seed/pod, the pod length and the percentage of normal seeds/pod from flowers left for open pollination were significantly higher than those of flowers protected from insects. The fruitification rates were 62 and 48 per cent, whil;e the percentages og healthy seeds were 97.61 and 76.17 percent respectively in unprotected and protected infloresces. The installation of *A.mellifera adansonii* could be recommened near cowpea (*Vigna unguiculata*) field to improve its seed and pod production.

## Pigeonpea *Cajanus cajan* (L.) Millsp.

**PIGEONPEA** *Cajanus cajan* (L.) Millsp., family Leguminosae

Pigeonpea (*Cajanus cajan* (L.) Millsp.) is a member of the family Leguminosae. It is an important pulse crop in the semi arid tropics. Globally pigeonpea is commercially grown in over a dozen countries on about 5.25 million hectares and India accounts for over 80% of the total area and production. It contains 23.00-28.60% protein which is rich in amino acids like cystine, tryptophan and methionine. It also contains 52.50 to 59.90% starch by weight (Jambunathan and Singh, 1982). Pigeonpea forms a large many branched shrub of about 1 to 3m in different agro-ecological zones of India (Williams, 1983). The yellow or orange papilionaceous flowers develop into green or brown, hairy, 4-7 seeded long beaked pods. Pigeonpea is a pubescent, many-branched shrub, 4 to 10 feet or more tall, with yellow or orange papilionaceous flowers that produce brown, hairy, four- to seven-seeded, long-beaked pods, 2 to 3 inches Iong by 1/2 inch thick. The 1 1/2 to 5-inch terminal and/or axillary raceme bears 5-inch terminal and/or axillary raceme bears several flowers which resemble pea flowers but are yellow, red or yellow purple in colour. The pollen is shed before the flower opens. Maximum number of flowers open between 1 a m and 3a.m and the flower remain open for 6 hours (Purseglove 1968).

Although flowers are often cross-pollinated, natural cross-pollination ranging from 0.1-48.0% has been reported by various workers (Howard *et al.*, 1919). Variable amounts of cross-pollination are associated with differences in locality, cultivar, and time of year and pollinator populations. Stigma is receptive upto 12.00 h and pollen germination only 24.00 h after dehiscence, thus creating a chance for foreign pollen to germinate. The foreign pollen tube grows much faster than self (Onim *et al.*, 1979a). Flowers provide subsistence for pollinators as these produce pollen and nectar. Insects collect these food substances and during the process, they transfer pollen from one flower to the other. Cross pollination provides chances of gene recombination and hybrid seed production; thus exploiting the heterosis and hybrid vigour in crop seed (Kevan and Baker, 1983).

Krauss (1932) and Wilsie and Takahashi (1934) considered the pigeon pea to be a self-pollinated species. However, a high degree of cross-pollination, from 5 to 40 percent, has been observed (Wilsie and Takahashi 1934, Purseglove 1968, Matta and Dave 1931, Krauss 1927, Shaw 1932, Abrams 1967). Bees visit the flowers in great numbers (Wilsie and Takahashi 1934), but whether they collect pollen, nectar, or both has not been reported.

Honeybees, megachilids and wasps are known for their active involvement in cross pollination and improving quantity and quality of pigeon pea yield (Figure 9.4). The problem with honeybees is that they visit pigeon pea more often for nectar than for pollen. When they are searching for nectar they learn to avoid the tripping apparatus, so that only about one percent of the visits results in tripping. When they are specifically looking for pollen they can be very efficient, however for high yields to occur it is often necessary to have more bees on the field than are required to make a honey crop. Problems that occur with wild bees are that they almost always go to pigeonpea primarily for pollen. Many species are very rapid in their movements and trip at about twice the rate of pollen collecting honeybees. But they are not available in large numbers on most fields to do a thorough pollination, and their reliability is very low as their numbers fluctuate from one season to another. It is desirable for farmers to cooperate with neighbors in their region especially in the acquisition of beekeeping services, which often require remuneration because of low honey yields. Pigeonpea flowers are often cross-pollinated; natural cross-pollination ranging from 0.1-48.0% has been reported by various workers (Howard *et al.*, 1919). Variable amounts of cross-pollination are associated with differences in locality, cultivar, and time of year and pollinator populations. Stigma is receptive upto 12.00 h and pollen germination only 24.00 h after dehiscence, thus creating a chance for foreign pollen to germinate. The foreign pollen tube grows much faster than self (Onim *et al.*, 1979a). Flowers provide subsistence for pollinators as these produce pollen and nectar. Insects collect these food substances and during the process, they transfer pollen from one flower to the other. Cross pollination provides chances of gene recombination and hybrid seed production; thus exploiting the heterosis and hybrid vigour in crop seed (Kevan and Baker, 1983). Honeybees, megachilids and wasps are known for their active involvement in cross pollination and improving quantity and quality of pigeonpea yield.

Studies conducted at Hisar revealed that 24 insect species belonging to Hymenoptera (17), Lepidoptera (2) and diptera (5) have been reported to visit pigeonpea blossoms. Among them, *Megachile* spp. was the most dominant constituting 42.66 per cent of total insect visitors (Salunke, 2003). Pollination studies conducted in caged and uncaged conditions in two cultivars viz. Paras and Manak revealed that significantly higher number of pods per plant were produced in open pollinated crop (58.4 and 53.9 in Paras and Manak, respectively) than without insect pollinated crop (42.1 and 38.7 in Paras and Manak, respectively). 1000 seed weight (g) was also significantly higher in open pollinated crop (65.39 and 65.06g in Paras and Manak, respectively) than without insect pollinated crop (63.46 and 62.94g in Paras and Manak, respectively). Vigour index was significantly higher in open pollinated crop than without insect pollinated crop. However, no differences in per cent germination and protein content of seed produced under open and without insect pollination were noticed (Table 9.1).

**Table 9.1:** *Effect of different modes of pollination on yield parameters in two cultivars of pigeonpea (Cajanus cajan (L.) Millsp.)*

| Yield parameters | Paras | | Manak | |
|---|---|---|---|---|
| | **Without insect Pollination (Caged)** | **Open pollination (Without cage)** | **Without insect Pollination (Caged)** | **Open pollination (Without cage)** |
| Number of pods /plant | 42.1 | 58.4 | 38.7 | 53.9 |
| Number of seeds/pod | 3.9 | 4.2 | 3.7 | 4.0 |
| 1000-seed weight (g) | 63.46 | 65.39 | 62.94 | 65.06 |
| Seed weight (g)/ plant | 11.4 | 15.3 | 10.1 | 13.1 |
| Seed yield (kg)/ plot (3×3m) | 1.4 | 1.9 1.3 1.7 | | |
| Seed yield (q)/ ha | 16.1 | 21.2 | 14.7 | 19.2 |
| CD(P = 0.05) | **Cultivars** | **Type of pollination Cultivars x Type of pollination** | | |
| | | | | |
| Number of pods/plant | N.S. | 5.60 | N.S. | |
| Number of seeds/pod | N.S. | N.S. | N.S. | |
| 1000-seed weight (g) | N.S. | 0.75 | N.S. | |

Similarly in an experiment conducted in Jorhat (Assam) in pigeonpea with *Apis cerana* bees with following five treatments such as. Open pollination (OP), Without bee pollination (WBP), Bee pollination 1(BP1) @ 3 hives/ ha, Bee pollination 2 (BP2) @ 4 hives/ ha, Bee pollination 3 (BP3) @ 5 hives/ ha. It was concluded that the effect of honeybee pollination on yield increase of Pigeon pea was significant. In all the plots pollinated by honeybees the yield was more than the without bee and open pollination. However, 5 colonies of *A. cerana* was sufficient to increase the yield of pigeon pea than other treatments (Table 9.2)

**Table 9.2:** *Effect of honeybee Apis cerana pollination of seed set and yields of pigeon pea*

| Treatments | Yield/plot(kg) | Yield/ha(q) | Yield increase over WBP (%) | Yield increase over OP (%) |
|---|---|---|---|---|
| WBP | 1.50 ± 0.36 | 6.0 ± 1.44 | – | – |
| OP | 2.0 ± 0.44 | 8.0 ± 1.62 | 33.3 | – |
| BP1 | 2.5 ± 0.37 | 10.0 ± 2.04 | 66.6 | 25.0 |
| BP2 | 3.0 ± 0.19 | 12.0 ± 0.74 | 100.0 | 50.0 |
| BP3 | 3.5 ± 0.29 | 15.0 ± 1.18 | 150.0 | 87.5 |
| S.Ed+ | 0.310 | 1.24 | – | – |
| **C.D.(P = 0.05)** | **0.675** | **2.7** | – | – |

Rathi and Sihag ( 1993) found that number of pods/plant,number of seeds/pod and weight of 100 seeds were similar for bee pollinated and open pollinated plots in *Cajanus cajan*. However, seed weight per plant, seed yield per plot and seed yield/ha. were significantly different. All these parameters were significantly higher in plots caged with *Chalicodoma lanata* as compared to plots caged without bees.

**Table 9.3:** *Yield gain under different conditions of pollination treatments in Cajanus cajan (Source: Rathi and Sihag 1993).*

| Yield parameters* | Pollination treatments | | | |
|---|---|---|---|---|
| | Plots caged with *C. lanata* | Open pollinated plots | Plots caged without bees | C.D (p = 0.05) |
| No of pods/plant | 82.43 ± 1.33 | 77.73 ± 0.80 | 67.13 ± 0.93 | 5.02 |
| No of seeds/pod[1]* | 3.16 ± 0.10 | 3.07 ± 0.03 | 2.57 ± 0.03 | 0.47 |
| 100-seed weight[1] (g) * | 6.40 ± 0.06 | 5.97 ± 0.06 | 5.47 ± 0.03 | 0.47 |
| Seed weight /plant[1] (g) * | 16.69 ± 0.57 | 14.20 ± 0.16 | 9.42 ± 0.23 | 2.46 |
| Seed yield/ plot [2](kg) | 1.78 ± 0.10 | 1.43 ± 0.10 | 0.91 ± 0.20 | 0.16 |
| Seed yield/ha [2] (kg × 100) * | 19.76 ± 0.70 | 15.90 ± 0.70 | 10.13 ± 1.90 | 1.91 |

1: Mean ± s.e of three replicates; 2 = Mean ± s.d of three replicates

$P \leq 0.05$ : significant (ANOVA)

**Pigeonpea ( *Cajanus cajan*)**

## Insect visitors

Pigeon pea is one of most important grain legume crops grown in the tropics and subtropics (Saxena et al., 2002). It is very rich in minerals, fats and vitamins as compared to ordinary peas (Madeley, 1995). The green pods and seeds are the most utilized forms in Africa, though dry seeds are increasingly gaining popularity (Damaris, 2007).

Pigeonpea flower is a protogynous hermaphrodite (Choudhary et al. , 2012), which opens early in the morning and closes before noon (Songok et al. , 2010) and produces nectar and pollen that attract insects (Reddy et al. , 2004; Martins, 2008; Pando et al. ,2011a).

The relationships between *C. cajan* and anthophilous insects have not been well studied. Martins (2007, 2013), Otieno et al. (2011) and Otieno (2013) reported that carpenter and leaf-cutter bees manipulate the flower by landing on the keel and grappling with it in the process of seeking nectar and pollen. The studies carried out by Pando et al. (2011 a) have shown that *Chalicodoma cincta Cincta* visits *C. cajan* flowers to harvest both nectar and pollen and increases its pollination possibilities.

The size of *X. olivacea* (Mensah and Kudom, 2011) has been reported to play a positive role in crop pollination. According to Pando et al. (2011 b ,2013), when collecting nectar and/or pollen, *X. olivacea* shakes flowers and this movement could facilitate the liberation of pollen by anthers for the optimal occupation of the stigma. This bee can be managed for pollination (Roubik, 1995).

Martins (2008) found that pollen loads on the *X. olivacea* are abundant, and the ability of this bee to disperse swiftly across large areas suggests that it can cross-pollinate the pigeon pea flowers regularly and efficiently. William ( 1977) reported that of the 48 insect species of insects visiting *Cajanus* flowers *Megachile* species accounted for 31% of the total number of insect visitors, the other visitors included *A. dorsata, A. florea.* and *Xylocopa spp.*

Grewal et al (1990) found that 11 species of bees belonging to 4 families were involved in pollination *Cajanus cajan* flowers in Punjab India. Of these, 5 species belonged to the Megachilidae, 3 to the Apidae, 2 to the Xylocopidae and 1 to the Anthophoridae. *Apis* constituted 50.2% of the total bees caught followed by Megachilidae( 30.7%).

Yogesh and Salunke (2009) recorded a total of 21 insect species belonging to Order Hymenoptera (14), Diptera (5) and Lepidoptera (2) visiting *Cajanus cajan* flowers at Hisar, India, out of which 12 were found collecting nectar and pollen while nine foraged only for nectar.

*Megachile* sp. was the most abundant visitors of the bloom and constituted 42.11 and 42.66% of the total visitors on the respective cultivar. Pando et al (2011) observed individuals from 18 species of insects visiting flowers of *C. cajan* in the two years. They found that Chalicodoma cincta cincta was the most frequent, followed by *Xylocopa calens*, *Apis mellifera adansonii*, *Megachile bituberculata*, *Camponotus flavomarginatus* and *Xylocopa albiceps*. Apart from bees, wasps were also recorded as likely predators.

Li et al 2012 recorded more than 46 species of insects visiting Pigeon pea [*Cajanus cajan* (L.) Millspaugh] flowers in the open field for pigeon pea production outside China. Of all the flower visiting insects *Megachile* spp. are were the major pollinators. The flower-visiting insects at the blooming stage, visited each primary branch at a frequency of 2.8 times per 10 min for the CGMS male sterile line and 5.2 times for the CGMS restorer line. This indicated the preference of flower-visiting insects to the flowers of restorer line. This difference between male sterile and restorer lines resulted in very similar dry seed yields of the male sterile line (383.7 g per plant) and the restorer line (357.0 g per plant). Evidently, enough pollens can been transported from the restorer line to the male sterile line by insect pollinators, even much less visiting frequency on the flowers of male sterile line compared to that of the restorer line.

Rajkhowa and Deka (2013) in Jorhat Assam, India found that the insect pollinators visiting the pigeon pea flower consisted of honey bees *Apis cerana*, *A. mellifera*, *A. dorsata*, carpenter bees *Xylocopa leucothorax*, *X. fenestrata* and other insects like *Polister habreu*, *Vespa magnifica, Papilio demoleus, Musca domestica* and *Monomorium indicum*. Out of all these insects, honeybee were more numerous and comprised more than 66 per cent of the total flower visitors. *A. cerana* was most numerouf 51.65% followed by *A. mellifera* 10.20% and *A. dorsata* 4.24% respectivrely. The other pollinators which were found as were also found as frequent visitors comprised of *X. fenestarata* (5.32%), *X. leucothorax* (4.28%), *M. indicum* (9.72%) and *P. demoleus* (8.52%). Of various meteorological factors temperature significantly influenced the foraging behaviour of honeybees.On the other hand, relative humidity showed non significant negative relationship.

Fernand-Nestor Tchuenguem et al (2014) studied the impact of the carpenter bee, *Xylocopa olivacea* Fabricius 1787, on the pod and seed yields of Cajanus cajan (L.) Millsp. 1900, in Nkolbisson and found that its foraging activity resulted in a significant increase in the Fr by 22.26%, as well as the number of seeds/pod by 34.98% and the percentage of normal seeds by 12.09% in the two study years. Placing *X. Olivacea* nests close to *C. cajan* fields is recommended to increase pod and seed production in Cameroon.

Tchuenguem Fohouo et al. (2014) reported that nineteen species of insects were found visiting the flowers of *Cajanus cajan. Chalicodoma cincta cincta* was the most prominent one, followed by *X. olivacea, Apis mellifera adansonii, Megachile bituberculata* and *Xylocopa albiceps*., respectively.

More et al (2015) studied the effect of different modes of pollination which comprised of open pollination *Apis cerana* on 4 frames, *Apis mellifera* on 6 frames and a colony of *A. florea* in seed production of pigeon pea and found that that 4-framed *Apis indica* was superior for number of pods per plant and weight of 1000 seeds, whereas 6-framed *Apis mellifera* was superior for number of seeds per plant, yield in kg per ha and per cent seed setting and a colony of *Apis florea* was superior for per cent crinkled seeds while per cent of germination was high under open pollination.

## Fruit/seed setting

Tchuenguem Fohouo et al. (2014) evaluated the impact of carpenter bee, *Xylocopa olivacea* Fabricius 1787, on the pod and seed yields of *Cajanus cajan* (L.) Millsp. 1900, in Nkolbisson. They found that *Xylocopa olivacea* mainly foraged for nectar and pollen and its foraging speed was 11.75 flowers/min. The fruiting rate (Fr), the number of seeds/pod and the percentage of normal seeds of open pollinated inflorescences were significantly higher than those protected from insects. Placing *X. olivacea* nests close to *C. cajan* fields is recommended to increase pod and seed production in Cameroon. The size of *X. olivacea* (Mensah and Kudom, 2011) has been reported to play a positive role in crop pollination. According to Pando et al. (2011b, 2013), when collecting nectar and/or pollen, *X. olivacea* shakes flowers and this movement could facilitate the liberation of pollen by anthers for the optimal occupation of the stigma. Roubik, (1995) found that this bee can be managed for pollination. Martins (2008) reported *X. olivacea* pollinated pigeon pea flowers regularly and efficiently, and this bee has the ability to disperse swiftly across large areas the suggests that it can cross-pollinate the pigeonpea flowers regularly and efficiently.

**Table 9.4:** *Fruiting rate, mean number of seeds yield per pod and percentage of normal seeds according to the treatments of C. cajan.*

| Treatments | Years | Flowers | pods | Fruiting/ rate | Seeds/pod | | Total seeds | Normal seeds | % normal seeds |
|---|---|---|---|---|---|---|---|---|---|
| | | | | | mean | sd | | | |
| 1 (unlimited visits) | 2008 | 3188 | 2801 | 87.86 | 5.93 | 0.72 | 15271 | 14503 | 94.97 |
| 2 (bagged flowers ) | 2008 | 3154 | 2544 | 80.66 | 4.96 | 0.65 | 11707 | 10191 | 87.05 |
| 3 (unlimited visits) | 2009 | 3313 | 3021 | 91.19 | 5.99 | 0.59 | 16818 | 15971 | 94.96 |
| 4 (bagged flowers ) | 2009 | 3118 | 2404 | 77.10 | 4.75 | 0.72 | 11211 | 9653 | 86.10 |
| 5 (C. c. cincta flowers) | 2008 | 67 | 66 | 98.51 | 6.44 | 0.50 | 425 | 413 | 96.94 |
| 6 (C. c. cincta flowers) | 2009 | 91 | 89 | 97.80 | 6.39 | 0.59 | 569 | 562 | 98.77 |

Source: Pando et al 2011

Pando et al (2011) studied the impact of the *Megachile* bee, *Chalicodoma cincta cincta* on pod and seed set of *Cajanus cajan* L. Mill sp, The treatments consisted of open pollination, bagged flowers and allowed visits by *C. c. cincta* only. *Chalicodoma cincta* cincta mainly foraged for nectar and pollen resource. The mean foraging speed was 10.33 flowers/min. Flowers visited by *C. c. Cincta* had higher fruiting rate as compared to those isolated from insect pollinators by bagging. Interestingly, seed formation was much higher in flowers visited by *C. c. Cincta* as compared to all other pollination treatments. The fruiting rate, the number of seeds/ pod and the percentage of normal seeds of obtained from open pollinated inflorescences were significantly higher than those protected from insects. *Chalicodoma cincta cincta* foraging resulted to a significant increment of the fruiting rate by 19.65 %, as well as the number of seeds/per pod by 24.33 % and the percentage of normal seeds by 11.52 %. Conservation of *C. c. cincta* nests close to C. cajan fields could be recommended to improve pod and seeds production in the region.

## Beans

**Scarlet runner bean:** The scarlet runner bean (*Phaseolus coccineus* L., family Leguminosae). The scarlet runner bean, is native to Mexico or Central America (Graham 1941) and of major importance in parts of Europe, is of minor importance in the United States, where it is sometimes called the Oregon lima bean. The showy blossoms are borne on long racemes. They are attractive to various species of bees for both nectar and pollen. Darwin (1874) showed that the blossom is ingeniously contrived to insure that insect visitation results in pollination. This is brought about by a pair of nectar holes at the base of the staminal tube, almost covered by a flap on one stamen. When the bee forces this flap aside, it rubs across the staminal column and transfers pollen from anthers to stigma.

## Pollination Requirements

Darwin (1889) covered scarlet runner beans to exclude pollinating insects and obtained only one-eighth and one-third as many beans as were obtained from open plants. In a more elaborate experiment, Free (1966) caged scarlet runner beans with and without honey bees to determine the need for insect pollination. Plants caged with bees set 1,479 g of pods per plant, compared with 180 g per plant where bees were excluded. Other factors varied correspondingly: Pods per plant, 96:11; grams of seed per plant, 356:50; number of seeds, 206: 30; and seeds per pod, 3.9 in bee cages compared to 2.7 in cages where bees were excluded. Only the mean weight per seed was unchanged. The pods in the cages without bees could not be harvested until later in the season than those in the bee cages. Tedoradze (1959) obtained similar benefits from bees. Mommers (1971) studied the pollination of runner beans under glass and reported that bees had no effect on production of beans. He concluded that varietal differences influenced the results obtained by Free (1966), and that this accounted for the differences reported.

## Pollinators

Henslow (1878) showed that bumble bees pollinated scarlet runner beans. Free and Racey (1968) showed that honey bees were as useful as bumble bees in pollination of scarlet runner beans grown in glass houses, although Free (1968) found that individual bumble bees worked faster and visited more flowers per plant than did honey bees. Bumble bees often pierce a hole in the base of the flower and steal the nectar without contributing to pollination. Subsequently, honey bees visit this hole and again pollination is bypassed (Jany 1950).

## Pollination Recommendations and Practices

Blackwall (1964) said that adequate pollination is a must for early yields and that lack of insect pollination is the most important single cause for low yield of beans. Blackwall also stated that where lack of pollination is a recurrent problem, it may be worthwhile to hire colonies of honey bees to supplement the local pollinator population. She recommended that small, rapidly expanding honeybee colonies need to b e used which has great need for pollen collection. Free (1966) stated honey bees are effective in pollinating scarlet runner beans, their use should increase yields where pollination is a limiting factor. Though he did not make recommendations for the use of honeybee colonies for the pollination of this crop but indicated that to obtain maximum yields and harvest the seeds high pollinator population should be maintained during early flowering.

## The lima bean *Phaseolus lunatus* L., family Leguminosae

### Flowering

The white or cream-colored, legume-type flower of the lima bean is borne at the end of a pedicel on a 2- to 4-inch-long raceme. Usually, only a small percentage of the flowers on a raceme set. In the flower itself, the keel is elongated into a spirally twisting beak. The style, which follows the spirals of the keel, is also twisted. It is bearded below the stigma. Knuth (1908) stated that the stigma, and a portion of the spirally twisted style with pollen adhering to the brush of hairs, projects from the tip of the keel petal when this is depressed and returns again when the pressure is removed. The anthers surround the style and shed their pollen upon it, but the stigma is never dusted before it is exserted. The cross-pollination can occur if visited by insect visitor before it is touched by its own pollen. Anthesis occurs between 7 and 8 a.m. The flower never closes, but the corolla is shed after a few days. Nectar is secreted at the base of the corolla and is the source of a fine quality honey. Bees also visit the plant for its pollen. Nectar secretion seems to be greatest when plants first come into bloom, and it remains intense for about a week - then tapers off. The lima bean (*Phaseolus lunatus*) flower is on the end of a pedicel on a 2–4 in (5–10 cm) raceme. The keel of the flower is elongated into a twisting beak. In the keel are the style, stigma, and the pollen-bearing anthers.

Anthers drop pollen on the portion of style below the stigma. When pressure is applied to the flower, as from a visiting bee, the stigma and a portion of the style project out of the keel; they retract when pressure is removed. Pollen from the pollen-dusted style rubs on to the end of the keel and contacts the stigma as it retracts. Cross-pollination is possible if the stigma is pollinated by a bee before it retracts. Not all flowers on a raceme set fruit. The problem may be high temperature, low humidity, or poor pollination (McGregor, 1976). The lima bean flower produces nectar and pollen, and bees will visit the plant if no richer forages are available.

### Lima Bean Pollination Requirements

Both self-pollination and cross-pollination can occur in lima bean. The number of pods, beans per pod, and total weight of beans increase in non-caged plants compared to plants caged to exclude bees; but other studies contradict this. Crossing promotes hybrid vigour in lima bean, but this sometimes happens without the help of bees. Production of hybrid seed increases in the presence of bees. Overall, cross-pollination is rarely necessary, but probably helpful (McGregor, 1976).

Amos (1943) reported about a 30 percent increase in yield due to the presence of bees. He stated that 30 colonies were within 1 mile of the test plots, but the acreage of beans in bloom was not given. There was increase was in numbers of pods, beans per pod, and total weight of beans but he gave no indication as to the pollinator density in the field. Allard (1954) caged plants with and without bees to obtain a measurement of bee activity in terms of hybrid beans produced. in plants caged with bees 2.38 percent hybrids seed was obtained wheras it was 1.06 percent from which bees were excluded. These observations indicate that the lima bean, like other beans, is capable of self- pollination but cross-pollination can and does occur. The amount of crossing and its value seem to depend upon the number and species of insect pollinators present or the degree of insect visitation to the crop. Boswell (1961) stated that the lima bean does not benefit from the intervention of insects; however, other works are not in complete agreement.

## Lima Bean Pollinators

Honey bees and bumble bees readily visit lima bean and promote cross-pollination. A few studies have indicated the role of bees in giving higher yields (McGregor, 1976). Thrips are most numerous in lima bean flowers and may pollinate the flowers; they may even cross-pollinate if crowding forces them to leave one flower and go to another (Free, 1993). There are few or no research-based recommendations for bee densities in lima bean. However, there is evidence that the presence of bees can only help improve yield. Supplying one honey bee hive per acre (2.5 ha) is a prudent measure for optimizing cross-pollination and yields. Additional benefit may be realized by conserving local bumble bee populations by leaving field margins undisturbed and by planting bee pasture plants in order to offset natural dearth periods.

Mackie and Smith (1935) and Allard (1954) gave thrips the credit for the cross-pollination of lima beans . Jones and Rosa (1928) reported that honey bees, bumble bees, and other insects visit the flowers and cause cross-pollination. Vansell and Reinhardt (1948) compared production from caged and open plots and also from areas of high and low general bee activity throughout the flowering period. They noted that 34.4 percent of the flowers set in the area where bee activity was high and continuous, but only 5 to 22 percent set where insecticides were applied. Because of the insecticide applications, most of the colonies were moved away during part of the flowering period, resulting in low pollinator activity. In fact, they recorded one bee per yard of row before the evacuation of the colonies began, but only one bee per 33 yards of row afterwards. Yet even with this low bee activity, they reported a greater yield in both number and weight of beans from open than from the caged plants. Wester and Jorgensen (1950) stated that bumble bees and honey bees were responsible for most of the lima bean crossing in the field. Anderson (1959) reported that lima bean plants caged with bees set more beans during early blooming period, and the beans were larger than in cages where bees were excluded. He reported that bees visited lima beans in abundance.

## Pollination Recommendations and Practices

Despite the rather promising results on the value of ample insect pollination on the increased production of lima beans by Amos (1943) and Vansell and Reinhardt (1948), Drayner (1956) on the other hand disagreed with the value of cross-pollination to increased production. The number of colonies need for pollination of lima beans is not recommended. Studies leading to such a recommendation are needed. In the interim, the meager data available would indicate that growers would materially benefit if they arranged for placement of numerous colonies in and adjacent to their lima bean fields at flowering time and took steps to protect these pollinators from harmful pesticides

## Bean (Common, Green, Snap)

### Flowering

The common bean (*Phaseolus vulgaris*) flower has an elongated, twisted keel containing the style and ten stamens. Inside the flower, the anthers drop pollen on to the style in the evening before it opens the next day. The next day if pressure is applied to the flower, by an insect, the style and stigma protrude from the keel and retract when the pressure is released. The style leaves pollen at the opening of the keel and the cross-pollination is possible if the stigma contacts a pollen-coated bee while it is extended. Otherwise, the stigma will be self-pollinated when it retracts and contacts its own pollen at the opening of the keel (McGregor, 1976).

### Green Snap Bean Pollination Requirements

Self-pollination is the norm with green snap beans, and it happens automatically before flower opening in the morning. However, it takes 8–9 h for pollen tubes to grow and fertilize seeds, during which time honey bees and bumble bees can visit the flower and cross-pollinate it. Tubes of foreign pollen probably grow faster than the plant's own pollen, so crossing is likely when the plant is cross-pollinated (Free, 1993). There is no yield or other economic advantage to the grower through cross-pollination. For the bean breeder, cross-pollination is actually a hazard to maintaining purity of varietal seed. To reduce unwanted hybridizing, one can separate varieties by at least 6–12 ft (1.8–3.7 m) and a tall, dense barrier.

### Green Snap Bean Pollinators

Bumble bees visit green snap beans more than do honey bees (McGregor, 1976). However, the overall benefit of bees to green snap bean production is probably small.

### Flowering

The small, greenish flower of beet (*Beta vulgaris*) occurs on a branched seed stem in clusters of two or three (Fig. 22.1). Flowers are perfect but selfing is rare because stigmas mature later than the anthers. Flowers open in the morning and anthers release pollen before noon. Stigmas begin opening in the afternoon and are not fully open until days 2–3 when the anthers of the same flower are no longer producing pollen. Once open, stigmas may stay receptive for more than two weeks. The pollen is wind-borne. Many insects visit beet flowers. Honey bees visit beet, but they readily move to more attractive food sources.

### Pollinators

**Pollinators** recorded are as under: European honeybee (*Apis mellifera* Lin.), Asiatic honeybee (*Apis cerana* Fab.), Rock bee (*Apis dorsata* Fab.), Cowpea borer (*Lampides boeticus* L.), Bumble bee (*Bombus spp.*), Tabanid fly (*Tabanus spp*), Syrphid fly (*Syrphus sp.*), Housefly (*Musca domestica* Lin.), Rice skipper (*Pelopidas mathias* (F.), Castor butterfly (*Ergolis merione* Cram.), Cabbage butterfly (*Pieris brassicae* Lin., *P. canidia* Lin.), Golden wasp (*Vespa magnifica* (Smith), Mustard sawfly (*Athalia lugens proxima* (Klug.)

**Broad bean,*Vicia faba* L.:** Rowlands (1960) recorded an average of 24 percent set in the open but only 7.3 percent set in the glass house He concluded that some plants are highly self-sterile than others. Free (1970) found that field bean flowers are not much attractive to insects because of low nectar content and sugar concentration. Free (1970) found that bee visitors to extra-floral nectaries were more numerous during midday compared to those visiting the flowers from 2 to 4 p.m., when pollen was available. He found that pollen was more attractive to bees than the floral nectar.

**Pollination Requirements:** Free (1966) found that insect pollination greatly benefitted the production of broad beans but had no marked effect on production of field beans. Free (1970), found that the pollination requirements of broad bean and field bean were similar although inadequate pollination limited seed production in broad as compared to field beans. Darwin (1889) showed that 17 broad bean plants exposed to insects produced 135 seeds whereas those covered with a net to exclude pollinators produced only 40 seeds,

## Pollinators

Bond and Hawkins (1967) and Free (1962) have studied the behavior of bees on field bean. Free (1962) concluded that the activity of the honey bees may vary. They may visit only the extrafloral nectaries, they may visit only the holes cut in the bases of the corollas by bumble bees, or they may visit the corolla for nectar and pollen. Only in the latter case are the bees of value to the beans as a pollinating agent. Bond and Hawkins (1967) placed two colonies per acre of field bean, and the bees collected sizeable amounts of pollen. However, these workers concluded that bumble bees were primarily responsible for the cross-pollination that occurred. Wafa and Ibrahim (1960) concluded that the carpenter bee, *Xylocopa aestuans* (L.), was the most important wild pollinating insect in the Gaza region. Free (1959) concluded that better visitation to the bean flowers resulted when the honey bee colonies were moved to the crop after it started to flower. Watts and Marshall (1961) showed about 26 percent increase in seed production due to the presence of bees. They also showed the value of bees in seing the seed early. In their plots caged with bees and also in their open plots, they found an average of nine, nine, and two pods, respectively on the bottom, mid- and top portion of the plant, but in the no-bee cage three, five, and four pods were in the same areas. Free (1970) concluded that only insects with long tongues could reach the nectar in the bean flower and that honey bees and short-tongued bumble bees that enter the flower probably obtain only pollen most of the time. However, some bumble bee species with short tongues, male carpenter bees (*Xylocopa aestuans*), and certain ants (*Cataglyphis bicolor* (F.)) bite holes in the base of the corolla, therby robbing nectar without affecting pollination. Honey bees do not make holes, but they will rob nectar from holes made by other insects.

**Pollination Recommendations and Practices:** Allen and Scriven (1957) concluded that one colony of honey bees per acre should give adequate coverage of field beans

**Mung bean/Green gram, *Phaseolus* spp:** The inflorescence is an axillary raceme, with 10 to 25 pale-yellow flowers, 1 1/2 to 2 cm long, and clustered at the top. Pollen is shed the afternoon before the flower opens the following morning. The flower fades the same afternoon. Only about half of the flowers (64 percent) open to permit possible cross-pollination.

**Pollination Requirements:** Purseglove stated that the flowers are fully self-fertile and entirely self pollinated, whereas, van Rheenen (1964) found that 2.8 to 3 percent crossing occurs when varieties are alternated.

***Phaseolus vulgaris* L.:** The bean flower is of the typical legume shape; usually whitish, but may be tinged to deep violet, purple, or red; and is l/2 to 3/4 inch long. The keel is prolonged in a spirally twisted beak. The style follows the spirals of the keel. There are the usual 10 stamens, the upper one free while the other nine are united into a tube enclosing the long ovary and part of the style. The blooms are loosely scattered along a 2- to 3-inch rachis (stem). Weinstein (1926) reported that many pollen grains germinate in the anther, but whether such pollen plays a part in fertilization is unknown. Knuth (1908) stated that although the anthers surround the style and the pollen is released before the flower opens, the pollen does not get onto the stigma before tripping occurs.

When pressure, for example, the weight of a large bee, is applied to the wing petals, the style, with pollen adhering to the brush of hairs on it, projects from the tip of the keel. When pressure is released, the style retracts into the keel, sometimes leaving some pollen exposed. When the stigma is first exposed, it is free of pollen, and may touch a pollen-coated bee during its brief period of exposure. If this happens, cross-pollination can occur. As the stigma retracts into the keel, the pollen released from anthers comes into contact and selfing occurs.

**Pollination Requirements:** The flowers of most bean cultivars are capable of self-pollination although cross-pollination can and does occur to varying degrees, depending upon the cultivar and the pollinator population (Hawthorn and Pollard 1954, Mackie and Smith 1935, Barrons 1939). Much of the research on bean pollination has been concerned only with the impact of the cross pollination on varietal seed contamination, not on quantity or quality of fruit set. Rutger and Beckham (1970) stated that *P. vualgaris* is primarily a selfing species, but studies of controlled insect pollination are needed. Free (1966) concluded that the pollination requirements of different species of beans could not be predicted so they must be studied individually.

Thorough studies on the possible benefits of maximum pollinator activity on the various types and cultivars would appear to be highly valuable.Knuth (1908) stated that the anthers surround the style and shed their pollen upon it, but the stigma is never dusted until the keel is depressed. Then the style, with its pollen-covered brush, springs out of the opening. Cross pollination can occur if the the stigma is touched by the bee coated with pollen from other plants, before being contacted by its own pollen.

However, Taylor (1919) stated that the flowers are entirely self-fertile as is abundantly proved by their productiveness when grown in greenhouses where bees are not present. Yet, he stated that they can be cross-fertilized by bees, although the percentage of crossing is not great. Mackie and Smith (1935) and Barrons (1939) indicated that cross- pollination may amount to more than 8 percent. The pollinator population associated with the crossing was never measured. Jones and Rosa (1928) stated that because the stigma and the anthers are enclosed within the keel, insects reach them only with difficulty; however, visitation occurs and crossing results. They considered the amount negligible. They also stated that selfing occurs freely when flowers are enclosed. Kristofferson (1921) reported 0 to 1.4 percent crossing, and Emerson (1904), 0 to 10 percent crossing. Free (1966) working with 'Processor' cv. of *P. vulgaris* concluded that "honey bees are unlikely to have much effect on the yield." His data, however, showed that the plants in the cages with bees produced 21 percent more seed weight, with 6 percent more pod weight than plants from which bees were excluded. Beans are visited only sparingly by honey bees, and beekeepers do not consider this crop of significance as a source of nectar or pollen.

In Kenya, Kasina et al. (2009) found that cross-pollinated common beans resulted to increased seed set and nutrition improvement of the dry harvested seeds. Since the marketing of French bean is based on individual pod weight as well as uniformity in pod shape, commercial yield might be affected by insect pollination. Adequate pollination is necessary , when all ovules are fertilized, the pods grow uniformly giving a better product as compared to poor pollination resulting in fewer developing ovules leading to curved pods that are rejected during the marketing (Bliss 1980). French beans are reportedly fully self-fertile (White and Izquierdo 1991), yet visits by large robust bees such as carpenter bees and bumble bees that can trip flowers could enhance pollination with auto as well as allo-pollen. Free (1966) found that the runner bean (*Phaseolus multiflorus*) plants, isolated from insects, produced fewer pods or seeds than those visited by honeybees.

## Foraging and pollination activities of *Xylocopa olivacea* (Hymenoptera, Apidae) on *Phaseolus vulgaris* flowers

*P. vulgaris* flowers have been reported to attract twenty-four species of insect. Among various pollinator species, X. *olivacea* was the most frequent visitor and foraged intensely for nectar collection. Kasina et al (2009) found that in Kenya, *Apis mellifera* was the most abundant species visting *P. vulgaris* flowers, followed by *Xylocopa olivacea* and *X. Inconstans.* Bumblebees in South Africa (Palmer, 1967) and *A. mellifera* in Western Kenya (Kasina et al., 2009) have been reported as the main floral visitor of this crop. The peak activity of *X. olivacea* on *P. vulgaris* flowers occurred between 10.00 and 13.00 h, which correlated with the highest availability period of nectar on *P. vulgaris* flowers. However, this activity decreased after 16.00 to17.00 h could be related to decreased temperature in the evening. Warm or sunny days encouraged good floral activity (Kasper et al., 2008). Similarly, rainfall has been documented as an environmental factor that can disrupt the floral insect activity (McGregor, 1976). The abundance of *X. olivacea* foragers on 1000 flowers and the positive and highly significant correlation between the number of *P. vulgaris* flowers at bloom, as well as, the number of *X. olivacea* visits indicates the attractiveness of *P. vulgaris* nectar with respect to this bee. In fact, weather during bloom was found to affect the abundance and foraging of pollinator insects (Bramel et al., 2004, Julianna and Rufus, 2010). Among the 24 insect species visiting *P. vulgaris* flowers, *X. olivacea* was the most abundant (28.7%), followed by *Coryna* sp. (17.92%), *Chalicodoma rufipes* (15.24%) and *Ceratina* sp. (7.65%).

During the collection of nectar on each flower, *X.olivacea* foragers regularly come into contact with the stigma. They were also able to carry pollen with their hairs, legs and mouth accessories from a flower of one plant to stigma of another flower of the same plant (geitonogamy), to the same flower (autogamy) or to that of another plant (xenogamy). The contribution of *X. olivacea* to *P. vulgaris* production through its pollination efficiency was significantly higher than that of all insects on the exposed flowers. The weight of *X. olivacea* played a positive role during nectar collection. *X. olivacea* shook flowers facilitating the liberation of pollen by anthers for the optimal occupation of the stigma. Our results confirmed those of Mensah and Kudom (2011) who revealed that the development of fruits from *L. aegyptiaca* flowers that have received a single visit of *X. olivacea* produced a mean weight of 428.7 g that was 1.5 times heavier than fruits from flowers visited by *A. mellifera* (286.76 g). This phenomenon was also reported by Vanderborght and Rasmont (1987) for *X. bariwal*, an efficient *P. coccineus* pollinator. Higher productivity of pods and seeds in unlimited visits when compared with bagged flowers showed that insect visits were effective in increasing cross-pollination It is concluded that *P. vulgaris* black seed outlets is a highly nectariferous bee plant that obtained benefits from the pollination by insects among which *X. olivacea* is of great importance. The comparison of pods and seeds set of unprotected flowers with that of flowers visited exclusively by *X. olivacea* underscores the value of this bee in increasing pods and seed yields as well as seed quality. The installation of X. *olivacea* nests at the proximity of *P. vulgaris* small black seed fields should be recommended for the increase of pods and seed yields of this valuable crop.

**Pollinators:** Mommers (1971 ) studied the pollination of beans grown under glass and concluded that bees do not influence development of *P. vulgaris*. Diwan and Salvi (1965) stated that beans in India are eagerly visited by *Apis dorsata, A. florea*, and *Trigona spp*., but the flowers are generally ignored by A. cerana. Those who have studied the problem agree that bumble bees and honey bees are the most frequent visitors to bean flowers in the United States. Darwin (1858) and Palmer (1967) indicated that bees are of benefit to beans, but the kind of bean they referred to is not clear. The data indicate that, if pollinating insects are of benefit, the bumble bee is the best pollinator. Honey bees are apparently next in importance. They visit the blossom for both pollen and nectar. The exact effect on the flower or value to the different cultivars of *P. vulgaris* has not been well established.

**Dependence of crops on bee pollination:** The bagged and un-bagged flowers were used to assess the effects of bee pollination on crop yield. The bagged flowers were assumed to have had spontaneous self-pollination (or non-bee pollination) while there was a strong possibility that the un-bagged flowers were visited by bees. The number of seeds per pod of the legumes (pulses) obtained from bagged and un-bagged flowers were found to be significantly different.

However, this difference was relatively lower for beans (25%) and green grams (27%) compared to the other legumes, while cowpeas showed the highest difference (43%). The results show that the mean number of seeds was lower than the maximum attainable in the un-bagged flowers, implying that bee pollination of the legumes in the study area was not sufficient.

## Impact on yield

Kingha et al (2012) evaluated the impact of the carpenter bee (*Xylocopa olivacea*) on pod and seed sets of *Phaseolus vulgaris* during 2009 and 2010 cropping seasons and found that the foraging activity resulted in a significant increment in fruiting rate by 48.43 and 78.18%, the number of seeds/pod by 19.38 and 18.58% and the normal seeds/pod by 15.67 and 38.25%, in 2009 and 2010, respectively. Hence, conservation of *X. olivacea* nests close to *P. vulgaris* crop fields should be recommended to improve pod and seed production in the region. Higher productivity of pods and seeds in unlimited visits when compared with bagged flowers showed that insect visits were effective in increasing cross-pollination. Webster et al. (1982), Wells et al. (1988) and Ibarra-Perez et al. (1997) revealed that *P. vulgaris* flowers set little pods in the absence of insect pollinators. But similar experiments conducted in England (Free, 1966) and in Brazil (Free, 1993) revealed that pollination by insects was not always needed. Darwin (1876) showed that self-pollination of *P. vulgaris* flowers produced as many pods and seeds as exposed plants. Thus, pollination requirements may differ between plant varieties.

**Table 9.5:** *Phaseolus vulgaris mean yields under different pollination treatments.*

| Treatments | Years | Flowers | Pod | Fruiting rate | Seeds/pod | | Total seeds | Normal seeds | % normal seeds |
|---|---|---|---|---|---|---|---|---|---|
| | | | | | mean | sd | | | |
| Ff (unlimited visits) | 2009 | 120 | 99 | 82.5 | 6.33 | 1.25 | 627 | 572 | 91.22 |
| Pf (bagged flowers) | 2009 | 120 | 32 | 26.67 | 5.16 | 1.76 | 165 | 129 | 78.18 |
| Ff (unlimited visits) | 2010 | 120 | 114 | 95 | 3.69 | 1.22 | 728 | 678 | 93.13 |
| Pf (bagged flowers) | 2010 | 120 | 24 | 20 | 5.17 | 1.74 | 124 | 70 | 56.45 |
| Fva (X. olivacea flowers) | 2009 | 29 | 15 | 51.72 | 6.4 | 1.18 | 96 | 89 | 92.71 |
| Fva (X. olivacea flowers) | 2010 | 144 | 132 | 91.67 | 6.4 | 1.34 | 851 | 778 | 91.42 |

Ff: free flower, Pf: protected flowers, Fva: flowers visited exclusively by *X. olivacea*.

Source: Kingha et al 2012.

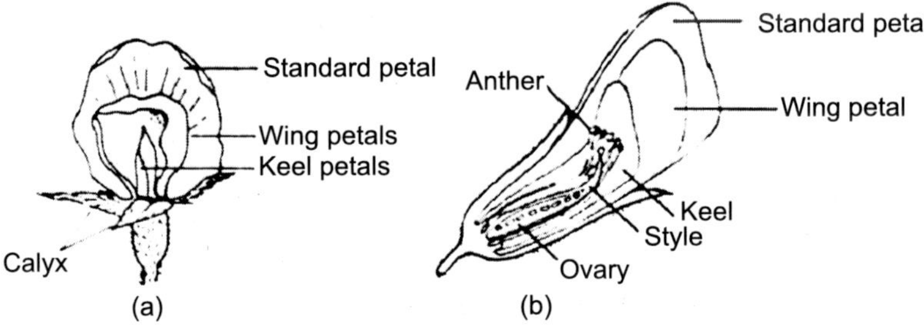


**Figure 9.1. Flower of Glycine max, soyabean a) front view (after Pohelman 1959), b) longitudinal section (after Purseglove 1968)**

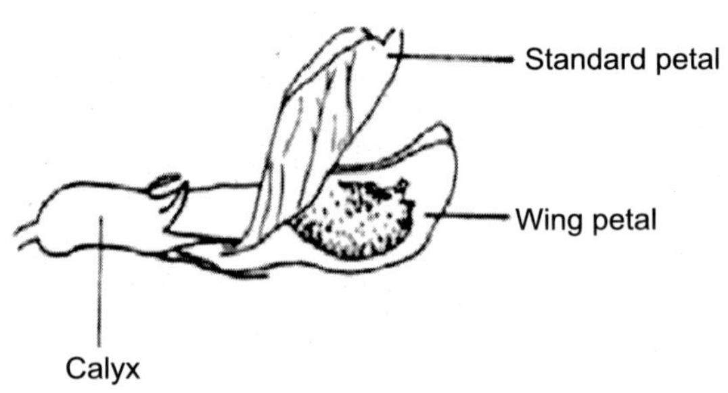

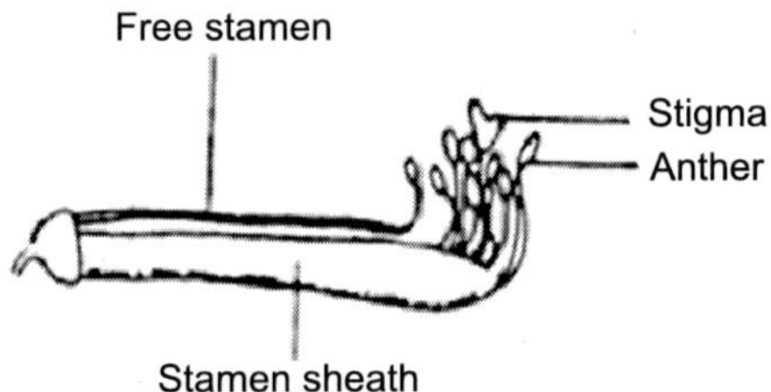


**Figure 9.2. Flower of Vicia faba, field bean; a, complete flower; b, petals and sepals**

## Insect visitors

Earlier studies have revealed that no seed set occurred from undisturbed bean flowers, whereas, flowers manipulated by wing petals had a great number of seeds (Darwin, 1858). Bumblebees have been reported as the most effective pollinator of beans collecting nectar and pollen in South Africa (Palmer, 1967) and North America (McGregor, 1976). Ibarra-Perez et al (1999) found that *Phaseolus vulgaris* flowers produced fewer seeds per pod in the absence of efficient pollinators in the United States of America. Kasina et al (2009) found that in Kenya *Apis mellifera* was the most abundant insect species visiting *P. vulgaris* flowers, followed by *Xylocopa olivacea* and *X. inconstans*.

**Table 9.6:** *Diversity of floral insects on P. vulgaris flowers in 2009 and 2010, number and percentage of visits of different insects Source: Kingha et al 2012.*

| Insects | | | 2009 | | 2010 | |
|---|---|---|---|---|---|---|
| **Order** | **Family** | **Genus,species, sub-species** | **n1** | **P1 (%)** | **n2** | **P2 (%** |
| | Apidae | *Xylocopa olivacea* (nectar) | 61 | 34.46 | 36 | 22.93 |
| | | *Xylocopa sp.* (nectar) | 5 | 2.82 | 3 | 1.91 |
| | | *Apis mellifera* adansonii (nectar) | 14 | 7.91 | 3 | 1.91 |
| | | *Amegila sp.* 1 (nectar) | 2 | 1.13 | 13 | 8.28 |
| | | *Amegila sp.* 2 (nectar) | – | – | 2 | 1.27 |
| | | *Braunsapis sp.* (nectar) | 11 | 6.21 | – | – |
| | | *Ceratina sp.* 1 (nectar+ pollen) | – | – | 24 | 15.29 |
| | | *Ceratina sp.* 2 (nectar) | – | – | 11 | 7.01 |
| Hymenoptera | | | | | | |
| | Halictidae | *Lasioglossum sp.* (nectar) | 1 | 0.56 | – | – |

| | Megachilidae | | | | | |
|---|---|---|---|---|---|---|
| | | *Chalicodoma rufipes* (nectar) | 28 | 15.82 | 23 | 14.65 |
| | | *Megachile sp.* 1 (nectar) | 1 | 0.56 | – | – |
| | | *Megachile sp.* 2 (nectar) | – | – | 1 | 0.64 |
| | | *Megachile sp.* 3 (nectar) | – | – | 1 | 0.64 |
| | | *Megachile sp.* 4 (nectar) | – | – | 5 | 3.18 |
| | | *Megachile sp.* 5 (nectar) | – | – | 2 | 1.27 |
| | Formicidae | *Camponotus flavomarginatus* (nectar) | 2 | 1.13 | – | – |
| | | *Camponotus sp.* (nectar) | 1 | 0.56 | – | – |
| | Total Hymenoptera | | 126 | 71.19 | 124 | 78.98 |
| Lepidoptera | Pieridae | *Eurema sp.* 1 (nectar) | 1 | 0.56 | 8 | 5.1 |
| | | *Eurema sp.* 2 (nectar) | 2 | 1.13 | – | – |
| | Lycaenidae | | | | | |
| | | (sp. 1) (nectar) | – | – | 2 | 1.27 |
| | | (sp. 2) (nectar) | – | – | 4 | 2.55 |
| | Hesperiidae | *Lambrix sp.* (nectar) | 2 | 1.13 | – | – |
| | Total Lepidoptera | | 5 | 2.82 | 14 | 8.92 |
| Diptera | Syrphidae | *Episyrphus sp.* (nectar) | 4 | 2.26 | - | - |
| Coleoptera | | | | | | |
| | Meloidae | *Coryna sp.* (eat flowers) | 42 | 23.73 | 19 | 12.1 |
| | Total | | | | | |
| | | 177 | 100 | 157 | 100 | |

n1: number of visits on 120 flowers in 11 days, n2: number of visits on 120 flowers in 09 days, p1 and p2: percentages of visits, p1= (n1 p2= (n2 X. olivacea visits for two years: $\chi 2 = 5.37$; t = 2.32; p < 10-2; sp: unidentified species.

Bliss (1980) recorded Carpenter bees as important pollinators of different cultivars of *P. vulgaris* in the Puerto Rico. Recently, Kingha et al (2012) also reported the Carpenter bees as efficient pollinators *P. vulgaris* grown for dry seeds in Cameroon. In the USA, bumble bees (*Bombus spp.*) has been found to be the most effective pollinators of *P. vulgaris* and their flower visits had a significant positive effect on seed production in two determinate varieties. In Kenya, honey bees are abundant visitors of French bean flowers (Kasina et al. 2009c), but exposure of bean plants to honey bee pollination did not result in higher pod yield of dry seeds in the UK nor in the USA (Free 1966; McGregor 1976; Delaplane & Mayer 2000).

Earlier studies have revealed *X. olivacea* foraging on *V. unguiculata* (Pando *et al.*, 2011), *Phaseolus vulgaris* (Kingha *et al.*, 2012) and *Phaseolus coccineus* (Fameni *et al.*, 2012). *X. olivacea* foraged *V. unguiculata* flowers for nectar and pollen. *X. olivacea* has been reported as the main floral visitors of this crop (Pando *et al.*, 2013) in Yaounde. This rank could be due to relative abundance of *X. olivacea* in different study areas, though the abundance and diversity of floral insects to a plant varies depending on the region (Roubik, 2000).

It is known that the anthophilous insects of a plant vary over time (Elfawal *et al.*, 1976; Tchuenguem, 2005). The peak of *X. olivacea* activity on flowers in the morning corresponds to the period of high availability of pollen and nectar on *V. unguiculata* flowers. The decrease in the activity of *X. olivacea* between 11.00 a.m. could be related to elevated temperature and the closure of flowers in the experimental field. Although, foragers preferred warm or sun rays for maximum floral activity (Kasper*et al.*, 2008), high temperature negatively affect insect activity on foraged flowers. Similarly, rainfall has been documented as an environmental factor that can disrupt the floral insect activity (McGregor, 1976).The abundance of *X. olivacea* foragers on 1000 flowers, as well as the positive and highly significant correlation between the number of *V. unguiculata* flowers at bloom and the number of *X. olivacea* visits indicates the attractiveness of *V. unguiculata* nectar and pollen with respect to this bee. In fact, weather conditions during bloom were revealed to affect the abundance and foraging of pollinator insects as *Apis mellifera Ligustica* and *Bombus impatiens* in Michigan (Julianna and Rufus, 2010).

During each of the two flowering periods of the host plant, *X. olivacea* intensely and regularly harvested nectar and pollen. This could be linked to the needs of individuals at the flowering period. The disruptions of visits by other insects reduced the duration of certain *X. olivacea.* This has obliged some bees to visit more flowers during a foraging trip in order to maximize their pollen loads. Similar observations were made for *Chalicodoma cincta cincta* (Hymenoptera: Megachilidae) foraging on *Cajanus cajan* (Fabaceae) flowers (Pando *et al.*, 2011b), *Xylocopa olivacea* workers foraging *Phaseolus vulgaris* of Small Black Seeds flowers (Kingha *et al.*, 2012), and *Xylocopa calens* foraging on flowers of *P. coccineus* (Pando *et al.*,2011a). *X. olivacea* foragers had a high affinity with respect to *V. unguiculata* when compared to the neighboring plant species, indicating their faithfulness to this fabaceae, a phenomenon known as "floral constancy" (Louveaux,1984; Backhaus, 1993; Basualdo *et al.*, 2000). Flower constancy or floral fidelity is an important aspect in the management of pollination and for this *X. olivacea* can provide benefits to pollination management of *V. unguiculata*

During the collection of nectar and pollen on each flower, *X. olivacea* individuals regularly come into contact with the stigma. They were also able to carry pollen with their hairs, legs and mouth accessories from a flower of one plant to stigma of another flower of the same plant (geitonogamy) and to the same flower (autogamy). The foragers might thus have influenced self- pollination and cross-pollination as early reported (Moffett *et al.*, 1975; Rao *et al.*, 1969). The significant contribution of *X. olivacea* in pod and seed yields of *V. unguiculata* lines with findings of Pando *et al.* (2013). This high productivity of pod and seeds in unlimited visits when compared with bagged flowers showed that insect visits were effective in increasing cross-pollination. Earlier, Pando *et al.*, (2013) at Yaounde and Kingha *et al.*, (2012) in Ngaoundere revealed that *V. unguiculata* and *P. vulgaris* flowers set little pods in the absence of insect pollinators. The contribution of *X. olivacea* to *V. unguiculata* production through its pollination efficiency was significantly higher than that of all insects visiting the exposed flowers. The weight of *X. olivacea* might have played a positive role during nectar and pollen collection. In fact, *X. olivacea* shook flowers when landing on plants, facilitating the liberation of pollen by the anthers for optimal occupation of the stigma. Similar experiments in England (Free, 1966) and in Brazil (Free, 1993) have shown that pollination by insects was not always needed. Thus, pollination requirements might differ between regions or countries.

## Soybean

*Glycine max* (L.) Merr., family Leguminosae

Soybean is one of the most important oilseed crop which are processed to obtain oil and a high protein meal. The soybean is a bushy, upright, annual, summer legume, 1 to 4 feet tall, with dense, alternate, and usually trifoliate oval leaves 2 to 4 inches across. There may be from 1 to 35 purple or white florets, three-eighths of an inch long, on each short raceme or flower cluster. A single plant may bear as many as 800 florets, but may set only 13 to 57 percent (van Shaik and Probst 1958). The floret has the characteristics and shape of many other legume flowers - a large standard petal, two small wing petals, and a keel petal that encloses the staminal column (fig. 174). The calyx is relatively large in proportion to the flower or even to the calyx of other legumes (Figure 9.1). Each floret is capable of producing a bean pod. Southern grown cultivars stop growing when flowering begins. Flowering usually continues for 4 to 6 weeks. There may be one-half million florets per acre. There are no extrafloral nectaries (Jaycox 1970).

Soybean flowers attract relatively few bees (Blickenstaff and Huggans 1962); however, bees do visit the flowers for both pollen and nectar. The anthers dehisce before the flower opens so that the stigma is in contact with and receptive to the pollen on the anthers. Pollination and fertilization is usually accomplished before the flower opens. Beekeepers in some areas frequently report honey crops from soybeans ( Hambleton 1936, Jaycox 1970, Pellett 1947). Beekeepers in other areas report that bees never visit the flowers, or if they do no honey is obtained.

## Pollination Requirements

The soybean is considered to be self-fertile and not benefited by insect pollination (Morse and Cartter 1937, Rubis 1970). Although there is no experimental evidence to support them, some soybean growers in Arkansas have indicated that bees increase production of beans, and they encourage the presence of apiaries near their fields. Tests with plants caged to exclude bees have shown no decrease in production over exposed plots (Woodhouse and Taylor 1913, Milum 1940, Piper and Morse 1910). Culter (1934) obtained an estimated 5 percent cross-pollination in open pollinated plots with an apiary nearby. Gordienko (1960) used honey bees in cages with two cultivars of soybeans. He fed the bees scented syrup to stimulate the floral visitation and obtained 29 percent hybrids on one cultivar and 44 percent on the other.

Brim and Young (1971) have reported that male sterility, although not cytoplasmic, has been found and that 99 percent of the seed set on the male-sterile plants was due to cross-pollination; however, the total seed set was extremely low.

Veatch (1930) showed that hybrid soybeans out-yielded the average parent variety up to 95.9 percent and the higher parent variety up to 71.5 percent. Weiss et al. (1947) showed a range of 14.2 to 71.3 percent heterosis over the best parent in the greenhouse, and 5.9 to 38.6 percent increase over the best parent in the field. Weiss (1949) showed an increase of 19.6 to 117 percent increase of the hybrid over the best parent. Wentz and Stewart (1924) showed increases of 60 to 397 percent. Weber et al. (1970) showed an average for all tests of the hybrids over the best parents of 13.4 percent.

Mention might be made that the increased yields for hybrids are all based upon a relatively low number of widely spaced plants. Because of the problems in making a sufficient number of hand pollinations to satisfactorily evaluate $F_1$'s in a normal-type planting, essentially no such studies have been conducted. The extremely large increases that have been reported are assumed to be partly a result of the extra space allotted to the plants.

Bradner (1969) stated that all of the breeding components have been described for the production of hybrid soybean seeds, except that a suitable pollen vector has not been found, although the honey bee "looks encouraging." Piper and Morse (1923) noted that soybean flowers were "much visited by bees." They also noted that three of their varieties, when in full flower at Jackson, Tenn., were very fragrant - the odor suggesting that of lilacs. Although they gave no indication that these varieties were more attractive to bees than other varieties, the information indicates that cultivars might be selected with aroma or attractiveness that when incorporated in a hybrid seed program might efficiently attract pollinating insects. Because of the potential value of hybrid soybeans, the breeders might watch for selections that show attractiveness to bees.

## Pollinators

Studies on pollinator complex of soybean is poorly understood. Jaycox (1970) in a survey on pollinating insects of soybean recorded only seven honey bees per 100 yards of row, but he gave no indication as to number of honey bee colonies required for pollination. Similarly, Blickenstaff and Huggans (1962) recorded very poor representation of pollinating insects in a collection 25,346 individual in a sweep net method. They found only one honey bee, three leafcutter bees, and nine halictids from this collection. Woodworth (1932) reported that thrips and honeybees were responsible for transfer of pollen. Caviness (1970) showed that thrips were not the effective pollinators but as honey bees were responsible for 7.7 percent crossing in soybeans. Culter (1934) and Gordienko (1960) showed that reported that honey bee populations can be built up on soybeans for pollination. Chiari et al (2005) evaluated the effect of the honeybee pollination in the production and quality of soybean seeds and found that honeybees were responsible for 95.5% of the pollination accomplished by insects. The pod number in covered treatment with honeybees was 61.38% higher than in the covered treatment without honeybees. Gill and O'Neal (2015) reported that honeybee *Apis mellifera* L. wild social and solitary bees and flower-visiting flies visit Soybean, *Glycine max* (L.) Merrill, flowers for nectar and pollen. the other species collected included *Agapostemon virescens* F. and *Lasioglossum* (Dialictus) species, *Melissodes bimaculata* Lepeletier (Hymenoptera: Apidae), and Toxomerus marginatus Say (Diptera: Syrphidae).

## Pollination Recommendations and Practices

There are no recommendations for the use of bees in pollination of soybeans. However, because of the interest in hybrid soybeans and the possibility of using pollinating insects in hybrid soybean production, a stocking rate of 1.5 colonies per hectare has been proposed based on foraging rates and bloom density.

## *Vicia* species

**Vetch** *Vicia* spp., family Leguminosae

Faba beans are a broad bean grown for human food, while "horse bean" and "field bean" with smaller, harder seeds are uszzed for animal feed. They have been reported originating from the Middle East around 8000 years ago before spreading to Western Europe. They form an important human food crop in China, the Middle East, the Mediterranean and in African countries. They are used for live stock nutrition in Australia and hold a potential for future market. Inadequate pollination is considered to be major factor resulting in low yields of the crop. The vetch inflorescence is usually a raceme. It bears few to many solitary flowers which are light blue to dark –purple in colour.

The flower, like that of most legumes, has the sexual column enclosed in the keel petal. If the column remains closed, only self-pollination can result (if the style and pollen are compatible) (Figure 9.2). If the column is freed, or tripped, and the stigma comes in contact with pollen from other flowers, cross- pollination can occur. Some species benefit by tripping even if cross- pollination does not occur. Repeated visitation by flower visiting pollinating bees has been reported to increase productiveness in some species. The flower needs to be tripped if bees are to collection pollen from it.

Nectar is usually secreted inside the corolla, sometimes on the outside of the base of the corolla. In many species, including at least *V. sativa, angustifolia*, and *pannonica*, nectar is also secreted in a nectary located on the leaf stipule (Knuth 1908, Herman 1960). Normally, when a bee visits the vetch flower for pollen or floral nectar, it settles on the wing petals, then forces its proboscis or "tongue" down the corolla. The pressure depresses the keel petal and forces the hairy style and pollen-laden anthers out. The pollen is thus shed onto the stigma, and some of it rubs off onto the bee, which then carries it to other flowers, resulting in cross-pollination.

However, Weaver (1956b) found that sometimes bees collect nectar from the side of rhe corolla without depressing the keel petal. Common and hairy vetches are an important source of excellent quality honey, and the production is usually sufficient for the storage of surplus quantities for the beekeeper. Hairy vetch is erratic in its nectar production and attractiveness to pollinating insects (Alex et al. 1950). However, common vetch produces a thick stipular nectar, which is more attractive to the bees than the floral nectar (Sculler 1956). Hungarian vetch has the reputation of supplying nectar in larger quantities for a longer period than any other cultivated plant (Schoth 1923).

## Pollination Requirements

Not too much is known about the pollination requirements of the different species and cultivars of *Vicia*. McKee and Pieters (1937) stated that seed production of some known species of vetches is greatly increased with bees. Several investigators have shown that production of hairy vetch (*V. villosa*) is greatly increased by bee visitation (Alex et al. 1950, Bieberdorf 1952, 1954, Thomas 1950, Weaver 1954, 1956a, b, 1957, Wojtowski and Mlyniec 1964).

Svetsugo and Kobayashi (1952) stated that fertilization of *V. villosa* is impossible when the plant is isolated from insects. Whether some of these workers considered the smooth- stemmed *V. dasycarpa* as a subspecies of *V. villosa* is not clear, but apparently the pollination requirements of the two are the same. Todd (1957) listed both hairy and purple vetch in the group of plants that produced more seed as a result of insect pollination. He stated that the brush arrangement of the vetch tripping requires repeated insect visits for pollination. He reported that most of Hungarian and narrowleaf vetches were largely self pollinated. Scullen (1956) stated that bees seldom visit Hungarian or the Willamette strain of common vetch, and that any pollination is probably done by bumble bees or other native bees.

McKee and Schoth (1925) stated that common vetch is self- pollinated, however, Schoth (USDA 1942) reported *V. villosa*, *V. benghalensis, V. pannonica*, and *V. sativa* as benefited by insect pollination. In the revised (USDA 1946) edition, only *V. villosa* and *V. pannonica* were listed as benefited by bees.

Knuth (1908) stated that *V. sativa* anthers dehisce in the bud so that automatic self-pollination is inevitable. However, he noted that the flowers were visited by pollinating insects. Further, he stated that the flower mechanism of *V. angustifolia* agrees with that of *V. sativa*, which would indicate that this species is capable of producing self-pollinated seeds. These flowers were visited by honey bees and other pollinating insects. In summary, therefore, the data indicate that hairy vetch (*V. villosa*) is greatly benefited by insect pollination. The pollination requirement of all of the species of vetch require clarification which offers an excellent opportunity for plant scientists to obtain data of considerable economic importance to our agriculture.

## Pollinators

Honey bees are the primary pollinators of hairy and smooth vetch. This has been established quite positively by von Schelhorn (1946), Weaver (1956a, b, 1957, 1965), and others. There is little information on the floral visitors of other species. Scullen (1956) stated that hairy vetch was worked freely by bumble bees and that they assisted in its pollination. He also stated that any pollination of Hungarian vetch was probably done by bumble bees or other native bees, but they seldom visit the flowers of the 'Willamette' cv. of common vetch. Where large- scale production of vetch seed is practiced in the United States, honey bees are used as a commercial practice. Alex et al. (1950) observed that fields with no colonies of bees within 2 miles produced 410 pounds seed per acre and fields with 0.1 to 1.0 colonies per acre within a mile produced 713 lb/acre; whereas fields with 1.5 to 3.0 colonies per acre within one-half mile produced 1,277 lb/acre. From these and other observations, they concluded that the bee saturation point had not been determined for hairy vetch.

Al-Ghzawi et al (2009) found 6 different species belonging to 6 genera and four families visiting the flowers of *Vicia villosa* spp. *dasycarpa*. The most frequent visitor species were *A. mellifera* and *Anthophora albigena* of family Apidae. Both made up more than 68.3% of the total number of the collected bees. The rest of flower visitors were shared in very low numbers except in the case of *Polistes dominulus* (Vispidae) which ranked as a third visitor in the abundance.

## Pollination Recommendations and Practices

There are few recommendations on the pollination of vetch. Alex et al. (1951) conservatively concluded that more than one colony of honey bees per acre increases hairy vetch seed production. Weaver (1956) recommended that one strong colony of honeybee is sufficient to provide adequate pollination for 3 to 5 acres of hairy vetch. The vetches are not much attractive to bees and for repeated visits to individual flowers strong population of honeybees is required.

Thus, the several colonies per acre previously mentioned seems to be a more realistic recommendation than the one colony per several acres. Of particular interest were the results of Drayner (1956), which showed that on continued inbreeding the ability to set selfed seed was progressively lost, but on hybridization self-fertility was restored. This is an area of benefit from pollination that has been largely overlooked and illustrates the value of bee pollination to such crops as vetches, which are considered to be self-fertile and even self-fertilizing. The results showed that the use of pollinating insects would be profitable and should be recommended for such self-fertile crops as the vetches.

## Pods and seed set

Al-Ghzawi et al (2009) studied the role of bee visitors on seed set and production of *Vicia villosa* spp. *dasycarpa* grown under semiarid conditions of Irbid, Jordan. They found that the percentage of pod set was significantly higher in open pollinated plants (14% ) as compared to those than the covered plants (2%)(Table 9.7).

**Table 9.7:** *Number of florets, pods and seeds per inflorescence, seed per pod, and percentage of formed pods and seeds for Vicia villosa spp. dasycarpa grown under uncovered and covered conditions during flowering stage of development.*

| Treatments | No./inflorescence | | | Seed/pod | Percentage of total florets | |
|---|---|---|---|---|---|---|
| | Florets | Pods | Seeds | | Pods | Seeds |
| Uncovered | 18a | 2.4a | 8.8a | 3.8a | 14.1a | 51.3a |
| Covered | 17a | 0.3b | 1.1b | 0.3b | 1.7b | 6.4b |

a, b: $P < 0.05$.

The resuls reveal that percentage of flower abscission was high and averaged more than 86%. Plant covering significantly reduced seed yield by reducing seed and pod number per plant and seed number per pod, but had no effect on individual seed weight (Table 9.8). In conclusion, preventing bees from visiting during flowering of *V. villosa* spp. *dasycarpa* decreased seed set, seed yield and yield components.

**Table 9.8:** Seed yield and number, pod number, seed number per pod and seed weight of Vicia villosa spp. dasycarpa plants grown under uncovered and covered conditions during flowering stage of development.

| Treatments | Seed yield (g.plant-1) | Seed no. | Pod no. | Seed no. | Seed weight |
|---|---|---|---|---|---|
| | | (no. plant-1) | | (no. pod-1) | (mg.seed-1) |
| Uncovered | 18.6a | 528a | 176a | 2.9a | 43a |
| Covered | 0.23b | 5b | 4b | 1.4b | 35a |

a,b: $P<0.05$.

It is clearly evident that plant covering significantly reduced seed yield by reducing all yield components except the individual seed weight. The covered plants had significantly lower seed and pod numbers per plant and seed number per pod. Individual seed weight was not significantly different between the covered and uncovered plants, which might be due to filling the remaining seeds on plants to large size after the covering treatment was released.

## Pests of leguminous vegetables and their management

Vegetable legumes suffer extensive damage from various pests and diseases. Several insect pests including bean flies, aphids, thrips, leafhopper, whitefly, leaf beetles, pod borers, pod bugs and bruchid beetles cause significant damage to food legumes in the field or in storage. Legume pod borer damage is a major pest causing up to 80% yield losses in Asia and Africa (Singh et al. 1990; Afun et al. 1991; Dreyer et al. 1994; Ulrichs and

Mewis 2004). Unfortunately, most of the farmers rely on chemical pesticides for control of various insect pests but without satisfactory control. Extensive use of pesticides has been made in various parts of the world. For example, Sodavy et al (2000) reported that Cambodian farmers sprayed up to 20 times per season with up to five different pesticides mixed together per spray on major vegetables including yard-long bean. Similarly, in Bangladesh, the bean is sprayed weekly or biweekly intervals or sometimes every day to control legume pod borer (Hoque et al. 2001). Sustainable pest management strategies offer a safe alternative to the misuse and overuse of pesticides on food legumes in tropical Asia and Africa.

## Insect and mite pests on vegetable legumes

### Bean flies

Bean flies are one of the most destructive pests of food legumes, especially during the seedling stage. They occur in Asia, Africa, Australia and Oceania. Larvae of these insects feed on legumes as internal feeders and leaf miners. The most important species are *Ophiomyia phaseoli* Tryon, *O. centrosematis* de Meijere, *O. spencerella* Greathead, *Melanagromyza sojae* Zehntner, *M. obtusa* Malloch (Diptera: Agromyzidae). *O. phaseoli* has the widest distribution and causes *the* maximum damage (Talekar 1990). another species *O. centrosematis is similar to O. phaseoli* but does not cause any serious damage. *O. spencerella* is more dominant than *O. phaseoli* on beans (Sariah and Makundi 2007). *M. sojae* is a common pest of soybean in Asia, Australia and Africa (Talekar 1990) and *M. obtusa* of pigeon pea throughout South and Southeast Asia (Shanower *et al*. 1998).

### Biology

**Eggs:** The eggs are laid in unifoliate or early foliate leaves by the adult females. The eggs are generally laid on the upper surface of the leaves near midrib close to the petiole (Talekar 1990). The eggs are milky white in colour, opaque or translucent and oval in shape. Afemale lays as much as 300 eggs during her life time. The egg period is 2 to 4 dais depending upon prevailing temperature. The egg laying behaviour varies in different species for example females of *O. phaseoli* lay eggs on the cotyledon of the newly germinated soybean plant(Talekar and Chen 1985) whereas *O. centrosematis* lays eggs just underneath the epidermis in the hypocotyl of young seedlings. in case *M. sojae* egg laying normally occurs at the basal part of the leaf lamina, near the petiole. *M. obtusa* has been reported to lays eggs singly on the green pods which are white and broad (Talekar 1990).

**Maggot:** The larvae of *O. phaseoli* has three instars which remain motionless beneath the leaf epidermis in a cavity made by its mother during egg laying. The larva starts feeding one or two hours after emergence. After reaching the midrib, yellowish white second instar larva constructs a tunnel in it and molts into the second instar. The third instar larva is light yellow-colored actively feeds in the petiole or stem. In younger plants, the larva feeds inside the stem in the junction of petiole and stem until its pupation in older plants (Talekar 1990). The larval period is about 8 to 9 days. The fully grown larva is sub-cylindrical, creamy white with dark black mouth parts. (Talekar 1990)

**Pupa:** The larva of *O. phaseoli* pupates on young plants beneath the epidermis of the stem near the soil surface. Under high larval density, pupation may also take place inside the stem pith. In mature plants, the pupation takes place at the junction of the leaf lamina and petiole. The pupa is yellow with a brownish tinge and is barrel shaped. The pupal period varies from one to three weeks depending on the temperature and altitude. Pupa is light yellow soon after pupation; it gradually turns to golden yellow and dark yellow toward adult emergence.

**Adult:** The adult *O. phaseoli* are metallic black in color, with light brown eyes. The adults could have on average life span of 3 weeks, however, they may live for 1.5 months. The copulating male usually lives for about 10 days, and the egg laying female lives for 1 to 3 weeks. The life span is only 7 to 12 days.

## Damage symptoms

The most serious damage by *O. phaseoli* adults starts at the unifoliate leaf stage. The unifoliate leaves show several feeding and oviposition punctures on the upper surface with corresponding light yellow spots, especially in the basal portion of the leaf. These feeding and oviposition wounds could predispose the plant for the entry of pathogens. The larvae start feeding in the leaf tissue, leaf lamina and then the petiole and stem. The larvae feed voraciously in the cortex, and continue downwards into the tap root. Because of the severe disturbances in the transport of water and nutrients, the plants, especially in the seedling stage, wilt and die. If the infestation starts in the later plant growth stages, the damage is limited to the leaf petioles, but not to the stem. The damage to the petiole leads to wilting of the leaves. Even stem damage at this stage may not result in plant mortality, but leads to stem swelling. *O. spencerella* also causes damage similar to *O. centrosematis* usually does not cause significant damage. The larva feeds inside the stem by tunneling that leads to weakening, and sometimes death of the plants (Talekar 1990).

Soon after emergence, the larva of *M. sojae* burrows through the mesophyll tissue into the closest vein moving downward in the leaf. Through the leaf vein and then the petiole, the larva continues feeding downwards into the stem and makes a tunnel into the pith, reaching the stem-root junction. It bores a little further into the thickened main root and moves upwards, thus widening the original tunnel (Talekar 1990).

The young larvae of *M. obtusa* attach themselves to the soft seeds inside the pods, and initially feed on the seed surface. They subsequently mine into the seeds, and the mines are often filled with fecal matter. Older larvae feed deep into the seeds, at times eating the embryo. Usually, a larva completes its development in one seed. The fully grown larvae come out of the seed, leaving behind a clear exit hole (Talekar 1990). Thus, the damaged pod is poorly filled.

## Management

1. Choose cultivars which are resistant or moderately, For instance, accessions or varieties having purple stems, thinner stems and smaller pith diameter are resistant to bean flies (AVRDC 1990; Talekar 1990). AVRDC soybean accessions (G11569, PI171444 and PI227687), mungbean accessions (V4281 and one of its progenies, VC4035-17), *Vigna glabrescens* accession (V1160) and *V. sublobata* were found to be resistant to *O. phaseoli, O. centrosematis* and *M. sojae* (AVRDC 1987; 1990; 1991; 1992). Consult the local extension agency for availability of resistant or tolerant varieties.
2. Earth up the plants three days after the appearance of the cotyledons above ground, so that most of the plants can overcome bean fly infestation. Production

of adventitious roots in the affected basal portion of the stem helps considerably in the resistance of the plant (Talekar 1990).

3. Although intercropping is not highly useful in reducing bean fly incidence, intercropping jute with mungbean simultaneously in alternate rows could reduce the bean fly infestation (AVRDC 1981 a&b). Growing pearl millet and mungbean as intercrops with pigeon pea reduce the damage of *M. obtusa* (Talekar 1990). Cultivation of onion as an intercrop with common bean (*Phaseolus vulgaris*) lowers bean fly infestation (Jemutai 2008).
4. Moth bean, chickpea, lentil, and cluster bean could be used as 'dead-end trap crops'—the bean fly adults lay eggs on these crops, but the eggs fail to hatch (Talekar and Lee 1988).
5. Plant late-maturing soybean varieties to reduce the yield losses due to *M. sojae* (AVRDC 1986). Early sowing of bean crops in Tanzania helps reduce bean fly infestations (Sariah and Makundi 2007).
6. Parasitoids such as *Opius phaseoli* Fischer against *O. phaseoli* and *Eucoilidea* sp. against *O. spencerella* and *O. centrosematis* are efficient parasitoids (Abate and Ampofo 1996). Protect the population of these parasitoids by avoiding the use of broad-spectrum pesticides.
7. Effective seed-dressing insecticides have been identified (Abate and Ampofo 1996; Jemutai 2008). Consult the local extension agency for appropriate local recommendations.

## Cowpea aphid

***Aphis craccivora* Koch** (Hemiptera: Aphididae)

This is a serious pest on cowpea, yard-long bean, hyacinth bean and peanut starting from the seedling stage to pod-producing stage. It acts as a direct pest and also transmits virus diseases. Although this aphid remains active throughout the year, it causes severe damage during the cool dry season.

## Biology

**Adult:** Unlike many insects, most aphids reproduce asexually in the tropics. They usually reproduce through parthenogenesis (development of embryo without mating with males) and are viviparous (give birth to nymphs directly rather than eggs). The early instar nymphs are grey in color, whereas the late instar nymphs and adult are black. Both wingless and winged forms occur. Winged forms are produced when population density is high, when host plants are of inferior quality, etc. The wingless forms are more common. Aphids mostly are found in groups. The nymphal period is about 5 to 8 days, and the total lifecycle is 11 to 14 days.

## Damage symptoms

*Aphis craccivora* prefers to feed on legume crops; it is commonly known as "cowpea aphid." Both the nymphs and adults possess piercing and sucking mouthparts. They occur in large numbers on the tender shoots, lower leaf surfaces, petioles, flowers and pods, and suck the plant sap. Slightly infested leaves exhibit yellowing. Severe aphid infestations cause stunting, crinkling and curling of leaves, delayed flowering, shriveling of pods, resulting in yield reduction. Young plants may be killed due to heavy infestation. *A. craccivora* also transmits. *Bean common mosaic virus* and *Cucumber mosaic virus* in a non-persistent manner (Damayanti *et al.* 2009). Large populations of the pest secrete substantial quantities of honeydew, which favors the growth of sooty mold on leaves and reduces the photosynthetic efficiency of the plants.

## Management

1. Avoid monoculture and follow crop rotation. The selected field should be located away from other legume crops.
2. Use entomopathogenic fungi (EPF) such as B*eauveria bassiana, Metarhizium anisopliae, Verticillium lecanii* and *Hirsutella thompsonii* at a concentration of 1×108 conidia ml-1 (Ekesi *et al.* 2000; Saranya *et al.* 2010).
3. Use neem oil, either alone or in combination with the EPF biopesticides (El-Hawary and Abd El-Salam 2008; Halder *et al.* 2013).
4. The ladybird beetles (*Menochilus sexmaculatus, Brumus suturalis, Harmonia dimidiate, Brumus suturalis* and *Coccinella septempunctata*) and green lacewings (*Chrysoperla carnea*) are efficient predators of aphids (Muniappan *et al.* 2012). These predators can be saved by avoiding the use of broad-spectrum pesticides.
5. *A. craccivora* can develop resistance to pesticides (Mokbel and Mohamed 2009). Use only those pesticides that have been recommended by local extension staff. Do not use the same compound or pesticide group continuously to avoid the development of pesticide resistance in insects referred to as puparium, although insects of this order (Hemiptera) do not have a perfect pupal stage (incomplete metamorphosis). Nymphal period is about 9 to 14 days during summer and 17 to 73 days in winter (David 2001). Adults emerge from puparia through a T-shaped slit, leaving behind empty pupal cases or exuviae.

**Adult:** The whitefly adult is a soft-bodied, moth-like fly. The wings are covered with powdery wax and the body is light yellow in color. The wings are held over the body like a tent. Adult males are slightly smaller in size than females. Adults live from 1 to 3 weeks.

## Damage symptoms

Both the adults and nymphs suck sap from the plants thereby reducing its vigour. In severe infestations, the leaves turn yellow and drop off. Large populations secrete substantial quantities of honeydew, which favors the growth of sooty mold on leaf surfaces and reduces the photosynthetic efficiency of the plants. If the infestation occurs during the pod formation stage, infected pods turn yellow and produce shriveled grains. In addition to direct damage, *B. tabaci* also acts as a vector for several viral diseases including *Mungbean yellow mosaic virus* (MYMV), *Cowpea mild mottle virus* (CPMMV), and *Bean golden yellow mosaic virus* (BGYMV) on crops such as mungbean, yard-long bean, common bean, cowpea and soybean (Mink and Keswani 1987; Bob *et al.* 2005; Morales *et al.* 2005; Brito *et al.* 2012).

## Management

1. Whitefly is a polyphagous insect; it has several host plants for feeding and survival ranging from cultivated crops to weeds. The field selected for vegetable legumes should be clean and not be located near any host plants and weeds.

2. Use of MYMV resistant or tolerant varieties is suggested to overcome the disease. For example, mungbean varieties ML 267, ML 613, NM 92 and VC 6372 (45-8-1) are found to be consistently resistant to MYMV across many locations (Shanmugasundaram *et al.* 2009). Mungbean genotypes such as ML 508, ML 537, ML 395, ML 505, and ML 543 are reported to be resistant against *B. tabaci* and MYMV (Chhabra and Kooner 1994a; Kooner 1998).
3. Grow seedlings for crops such as yard-long bean in insect-proof (50–64 mesh) net houses, net tunnels, greenhouses, or plastic houses.
4. If the seedlings are produced under open field conditions, use yellow sticky traps at the rate of 1-2 traps/50-100 m2 to trap the whiteflies. Hang the traps slightly above or at the canopy level for better trapping.
5. Maintain a high standard of weed control in seedling production areas and crop fields to reduce the availability of alternate host plants.
6. The fast growing plant crops like maize, sorghum, or pearl millet should be grown in the border of the field to act as barriers to reduce whitefly infestations. Reflective and yellow plastic or straw mulches may reduce landing of whiteflies.
7. Neem formulations (if available) can be applied as a soil drench or foliar application to control whitefly in vegetable legume seedlings. Seed treatment with neonicotinoid pesticide formulations (if available) also reduces whitefly populations and incidence of MYMV disease (Salam *et al.* 2009).
8. Natural enemies such as *Encarsia sophia* and *E. formosa* are efficient parasitoids of whiteflies (Bob *et al.* 2005). Protect the population of these parasitoids by avoiding the use of broad-spectrum pesticides.
9. Use only those systemic pesticides that have been recommended by the local extension service. Do not use the same compound or pesticide group continuously to avoid the development of pesticide resistance in insects.

## Leafhoppers

***Empoasca kerri* Puthi, *E. facialis* Jacobi,**

***E. fabae* Harri** (Homoptera: Cicadellidae)

*E. kerri* occurs mainly in Asia, whereas *E. facialis* occurs in Africa and *E. fabae* is widespread in Americas (Ranga Rao *et al.* 2013). The relatively dry and humid weather conditions generally favor the population build-up of leafhoppers.

## Biology

**Egg:** The adult females lay eggs along the midrib and lateral veins of the leaves, or into the petiole. The eggs are not visible to the naked eye. The egg period is about one week.

**Nymph:** The nymphs resemble the adults, but lack wings. Instead, they have slightly extended wing pads. They are pale green in color and tend to move sideways when disturbed. The nymphal period varies from 2 to 3 weeks, depending on the temperature.

**Adult:** The adults are wedge-shaped, pale green insects. They have fully developed wings, and are about 2.5 mm long.

## Damage symptoms

Both nymphs and adults suck the sap from the lower leaf surfaces through their piercing and sucking mouthparts, which leads to yellowing. When several insects suck the sap from the same leaf, yellow spots appear on the leaves, followed by crinkling, curling, bronzing (becoming reddish-brown) and drying. Leafhoppers also cause similar damage in cotton and potato.

## Management

1. Monitor the insects with yellow (570-580 nm) sticky traps placed at random in the field.
2. Okra, sesame, sorghum and pearl millet can be grown as inter-crops in vegetable legume (*e.g.*, mungbean) fields. This reduces the leafhopper damage significantly (Chakravorty and Yadav 2013).
3. Avoid the use of broad-spectrum pesticides to encourage the performance of natural enemies. Generalist predators such as ladybird beetles and green lacewings are efficient in preying on leafhopper nymphs and adults.
4. Use neem-based biopesticides at recommended doses. Neem seed kernel extract (NSKE) @ 5% can also be sprayed..
5. Use only those systemic pesticides recommended by the local extension service. Do not use the same compound or pesticide group continuously to avoid the development of pesticide resistance in insects.

## Common armyworm

***Spodoptera litura*** **Fabricius** (Lepidoptera: Noctuidae)

*S. litura* is a polyphagous and highly mobile insect and it is a pest of economic importance on many agricultural and horticultural crops. It is the predominant species on several vegetable legumes in tropical South and Southeast Asia. As they are nocturnal, the larvae feed actively during night hours. The larvae hide under crevices or plant debris in the field during day time.

## Biology

**Egg:** The eggs are laid in groups of 200–300, and covered with brown hairs from the body of the mother. The egg period is about 3 to 5 days.

**Larvae:** Upon hatching, the neonate larvae are translucent green with dark thorax and gregarious. The young larvae remain and feed in groups. However, they disperse when they grow older to feed individually. The grown-up larvae are green to pale greenish brown or black in color; they have stout, cylindrical bodies with prominent black spiracles. The body may have transverse and longitudinal grey and yellow bands. Mature larvae can grow to about 35–40 mm in length. When disturbed, the larvae curl into a ‘C’ shape with the head kept at the center. Larval period is about 15 to 30 days. The larva passes through six instars.

**Pupa:** Pupation takes place in the soil. Pupae are shiny reddish brown. The pupal period varies from 1 to 3 weeks.

**Adult:** The adult is a stout bodied moth with a wing span of about 40 mm. The adults are usually brown colored; the forewings have numerous criss-cross streaks in a cream or brown background. The hind wings are white with a brown patch along the border. The dense scales are present on the eighth abdominal segment of female moths.

## Damage symptoms

The neonate larvae feed on leaf surfaces and cause skeletonization, leaving behind the whitish membranous leaves only. Mature larvae feed on the whole leaves until only main veins are left. Sometimes, the larvae may also cut the seedlings or young plants at soil level, which may lead to complete destruction of the crop.

## Management

1. Castor (*Ricinus communis* L.) can be grown as a trap crop along the field border to attract egg-laying female adult moths. As eggs will be laid in masses, the egg masses and young larvae that still remain and feed in groups can be hand-picked and destroyed either on the trap crop or on the main crop.
2. Sex pheromones of *S. litura* are commercially available in many countries and can be used for monitoring as well as mass-trapping.
3. *Spodoptera litura* nucleopolyhedrovirus (SlNPV) is commercially available in some countries, and can be used to replace chemical pesticides. SlNPV is effective either alone or in combination with neem (Prasad and Wadhwani 2006). In addition, *Bacillus thuringiensis* formulations can be used to manage *S. litura* (Sharma *et al.* 2011).
4. The egg-parasitoids (*e.g., Trichogramma chilonis* Ishii) and larval parasitoids (*e.g., Campoletis chlorideae* Uchida) can be conserved and/or released (Bajpai *et al.* 2006) in vegetable legume fields at regular intervals to check the build-up of *S. litura*. If the natural enemies such as parasitoids and predators are present in the field, avoid using broad-spectrum chemical pesticides that are lethal to these natural enemies (Md. Abdullah *et al.* 2001). Instead, botanical pesticides such as neem that are safer for the parasitoids (Maria Packiam and Ignacimuthu 2012) can be used to augment the natural enemies.
5. Chemical pesticides can be applied as needed. Chemical pesticides may highly be effective against early larval stages, when the larvae remain in groups. Consult the local extension agency for appropriate recommendations.

## Beet army worm

***Spodoptera exigua* Hubner** (Lepidoptera: Noctuidae)

Like *S. litura, S. exigua* is also a polyphagous and nocturnal insect that mainly feeds on vegetable soybean, besides other vegetables such as tomato, eggplant, hot and sweet pepper, onion, etc.

## Biology

**Egg:** The eggs are laid in groups of 100–150 and covered with brown hairs from the body of the mother, similar to *S. litura*. The egg period is about 3 to 5 days.

**Larva:** The grown-up larvae are brownish green dorsally and pale yellow colored ventrally with a lateral white or yellow stripe. Larval period is about 2 to 3 weeks. The larva passes through five instars.

**Pupa:** Pupation takes place in earthen cocoons in soil. Pupae are light brown. The pupal period varies from 7 to 11 days.

**Adult:** Adults of *S. exigua* are medium-sized moths with a wing span of about 30 mm. The adults are usually brown colored; the forewings are mottled brown in color. The hind wings are grey with a brown line along the border. The adult period is about 10 days, and a single female on average lays about 500–600 eggs.

## Damage symptoms

The larvae cause defoliation. Sometimes, the larvae may also cut the seedlings or young plants at soil level. Occasionally the larvae may feed on the pods.

## Management

Similar to *S. litura.*

**Bean foliage beetles**

***Ootheca* spp.** (Coleoptera: Chrysomelidae)

*Ootheca mutabilis* (Schönherr) and *O. bennigseni* Weise are important foliage feeders on food legumes such as cowpea and common bean in eastern and southern African countries, especially Burundi, Kenya, Rwanda, Tanzania and Zambia (Abate and Ampofo 1996). They attack the foliage soon after germination, causing yield losses of 27- 100% (Singh and Allen 1980; Raheja 1981).

## Biology

**Egg:** The females lay eggs in clusters of about 60 eggs per cluster in soil close to the legume plants. Each female lays up to eight egg clusters. The eggs are elliptical and yellow in color. The egg period is about 2 to 3 weeks.

**Larva:** The larva of *Ootheca* spp. develops through three instars, followed by a pre-pupal stage. The larval period varies from 5 to 11 weeks depending on the temperature.

**Pupa:** The larva pupates in an earthen cell within the soil. The pupal stage lasts for about 16 days. Soil temperature plays an important role in its development.

**Adult:** The adults of *Ootheca* spp. undergo an obligatory diapause until the onset of the rainy season. The adult beetles are shiny, oval shaped and about 6 mm long. The adults vary in color ranging from orange to brown, and at times blackish.

## Damage symptoms

The adult beetles start feeding on leaves of newly planted legume crops, especially cowpea and beans. They feed on the foliage between the veins, and cause extensive defoliation. Damage to the growing shoots of young seedlings and heavy infestation completely destroy a crop. The adults occasionally feed on the floral parts. Additionally, larvae feeding on lateral roots cause wilting and premature senescence in bean plants (Abate and Ampofo 1996). Apart from direct feeding, the beetles also vector *Cowpea yellow mosaic virus* (Robertson 1963).

## Management

1. Choose tolerant or resistant cultivars available in the region. For instance, several bean lines are reported to be moderately resistant to *O. bennigseni* in Tanzania (Karel and Rweyemamu 1985). Consult the local extension agency for the availability of tolerant or resistant varieties.
2. Biopesticides such as neem are highly effective against *Ootheca* spp. Use biopesticides if they are available and recommended in the region.
3. Locally prepared botanical extracts such as aqueous extract from vernonia (*Vernonia lasiopus* var. *iodocalyx*) leaves are proven effective in reducing the infestation of *Ootheca* spp. (Paul *et al*. 2007).
4. Do not control the *Ootheca* infestation with broad-spectrum chemical pesticides. If necessary, spray a recommended pesticide after consulting with local extension staff.

## Thrips

***Megalurothrips distalis* Kany, *M. usitatus* (Bagnall), *M. sjostedti* (Tribom)** (Thysanoptera: Thripidae)

*M. distalis* and *M. usitatus* are widely distributed in South and Southeast Asia and Oceania; *M. sjostedti* predominantly occurs in Africa. They mainly feed on the flowers of legumes, and can cause 100% yield losses if left uncontrolled.

## Biology

**Egg:** The females lay eggs within the leaf tissues, especially in the terminal leaflets. The egg may not be visible to the naked eye. The egg period is about 2 to 3 days.

**Larva:** The larva of *M. distalis* is whitish translucent. *M. usitatus* larva is pale yellow in color, and turns to yellowish to orange-red, whereas the larva of *M. sjostedti* is whitish in the beginning and later turns to yellow to orange. The larval period varies from 1 to 2 weeks depending on the temperature. As pupation nears, the larva stops feeding and moves down to the soil surface either in the soil or under plant debris.

**Pupa:** The pupal stage lasts for five days to one week. Moisture plays an important role, because dryness in the soil could lead to the desiccation and death of the pupae.

**Adult:** The adults of *M. distalis* are brownish in color, whereas the adults of *M. usitatus* are dark brown . The adults of *M. sjostedti* are black in color. The adults of *Megalurothrips* are about 1.5-1.65 mm long. It is very difficult to confirm the exact identity of thrips with the naked eye; they must be identified under microscopes in laboratory conditions. Adults live from 1 to 3 weeks.

## Damage symptoms

Thrips remain hidden inside the flower buds and flowers. Both the larvae and adults feed on the tender leaves in the beginning. However, they prefer to feed mostly on flowers (inflorescence). Slightly infested leaves show silvery feeding scars. however, in case severe infestations on the flowers, the open flowers are discolored and distorted showing elongated brownish streaks; they dry out, and fall prematurely without forming pods. Infested pods are scarred and deformed.

## Management

1. Choose tolerant or resistant cultivars available in the region. For instance, mungbean lines SML 99, SML 100, SML 103, SML 112 and SML 117 are reported to be resistant to *M. distalis* (Chhabra and Kooner 1994b). Consult the local extension agency for the availability of tolerant or resistant varieties.
2. Grow vegetable legume (*e.g.*, yard-long bean and common bean) seedlings in insect-proof (50–64 mesh) net houses, net tunnels, greenhouses, or plastic houses to avoid early infestation, especially in the dry season.
3. Predators such as *Orius albidipennis* and *Paederus sabaeus* feed on *M. sjostedti* in West Africa (Tamo *et al.* 1993). Parasitoids such as *Ceranisus menes* Walker and *C. femoratus* Gahan (Hymenoptera: Eulophidae) are effective candidates against *M. sjostedti* and *M. usitatus* (Chang 1990; Tamo *et al.* 2012). However, they are unable to provide complete control. Use blue sticky traps to monitor thrips at regular intervals and determine when other pest management controls are required.
4. Entomopathogenic fungi could play a vital role in managing bean flower thrips, because of the moisture and humidity within the flower buds. For instance, 100% mortality of the adults of *M. sjostedti* was recorded at the highest concentration ($1\times108$ conidia $ml^{-1}$) of *Metarhizium anisopliae* at 8 days post-inoculation. At the same concentration, daily pollen consumption was significantly reduced at 2 days after treatment in infected adults (Ekesi and Maniania 2000).
5. Use mulch and reflective materials in vegetable legume fields to reduce the incidence of thrips.
6. Avoid using broad spectrum pesticides for the control of thrips as they result in resurgence of thrips, If necessary, spray a systemic pesticide.

## Pod bugs

***Clavigralla gibbosa* Spinola, *C. scutellaris* (Westwood), *C. tomentosicollis* (Stal.)** (Hemiptera: Coreidae)

Pod bugs, *C. gibbosa* and *C. scutellaris* are distributed in the tropics and sub-tropics, whereas *C. tomentosicollis* is predominantly present in Africa. They mainly feed on vegetable and grain legumes. These bugs comprise a serious group of pests causing damage to food legumes. They can cause yield loss of 20-100% in different parts of Asia and Africa.

## Biology

**Egg:** The females lay eggs mostly on the leaves and pods. On average, each *C. gibbosa* and *C. scutellaris* female lays about 60-200 eggs, whereas *C. tomentosicollis* lays about 120 eggs. The egg is brown in color. *C. scutellaris* eggs are smooth and shiny, and the eggs of *C. gibbosa* are roughly sculptured. Eggs are laid in clusters of 2-60. *C. gibbosa* generally lays clusters of 3-15 eggs, whereas *C. scutellaris* lays clusters of 18-20. The egg period varies from 2 to 5 days. However, it may be prolonged up to three weeks under low temperature conditions.

**Nymph:** The newly hatched nymphs of *Clavigralla* are reddish in color and later turn brown. Nymphs are gregarious. The nymphal period is about 1 to 3 weeks, depending on the temperature.

**Adult:** The *Clavigralla* bugs are brown-gray in color. In the field, *C. gibbosa* and *C. scuttellaris* are often mistaken for each other. However, *C. scutellaris* is more robust (about 12 mm long) than *C. gibbosa* and *C. tomentosicollis* (about 10 mm long). The adult longevity is more than three months for the *Clavigralla* bugs (Ranga Rao *et al.* 2013). Although the males and females are similar, the female is slightly larger with a rounded abdomen.

## Damage symptoms

The nymphs and adults have piercing and sucking mouthparts. They penetrate the pod walls and suck the sap from developing seeds inside. Occasionally they also feed on stems, leaves and flower buds. Feeding on the pods causes yellow blotches. Severe infestations lead to shriveled pods and seeds. The damage is serious during prolonged dry weather conditions.

## Management

1. Inter-cropping of legumes with millets such as sorghum could delay the infestation of pod bugs.
2. Parasitoids such as *Gryon fulviventris* (Crawford) (Hymenoptera: Scelionidae) against *C. tomentosicollis* in Africa, an unidentified *Gryon* sp. against *C. gibbosa,* and *Gryon clavigrallae* Mineo against *C. gibbosa* and *C. scutellaris* in Asia are found to be effective candidates for biological control (Bhagawat *et al.* 1994; Asante *et al.* 2000; Romeis *et al.* 2000). If the parasitoids are present in the field, avoid using broad-spectrum chemical pesticides that would be lethal to these natural enemies.
3. Entomopathogenic fungi could play a vital role in managing the *Clavigralla* bugs. For instance, high mortality of the adults of *C. tomentosicollis* was recorded when they were treated with *M. anisopliae* CPD 5 and *Beauveria bassiana* CPD 9 strains at 7 days post-inoculation (Ekesi 1999).
4. Do not spray broad-spectrum chemical pesticides. If necessary, spray a systemic pesticide after consulting with local extension staff.

## Bean bugs

***Riptortus pedestris* (F.), *R. clavatus* (Thunberg)** (Hemiptera: Alydidae)

Bean bugs are serious pests of food legumes and they feed mainly on soybean. They occur in South and Southeast Asia including Japan, Korea and Taiwan. They can cause significant reduction in pod and grain yields and quality.

## Biology

**Egg:** The females lay single eggs mostly on the bases of younger flowers and pods. On average, each female lays about 115 eggs. The egg period is 3 to 5 days.

**Nymph:** The nymphs are brownish black and hemispherical, resembling brown ants. The nymphal period is 2 to 3 weeks, and the nymphal stage is made up of five instars.

**Adult:** The *Riptortus* bugs brown in color, slender and longer (about 18 mm) than *Clavigralla* bugs.

## Damage symptoms

The damage symptoms are similar to *Clavigralla* bugs.

## Management

1. Delayed sowing from normal planting season of legumes helps avoid damage from bean bugs due to lower pest densities (Wada *et al.* 2006).
2. Parasitoids such as *Gryon japonicum* (Ashead), *Ooencyrtus nezarae* (Ishii), and *O. acastus* (Trjapitzin) and *G. nigricorne* (Dodd), are the major egg parasitoids of *Riptortus* bugs, and are effective candidates for biological control in Asia (Youn and Jung 2008). If the parasitoids are present in the field, avoid using broad-spectrum chemical pesticides that would be lethal to these natural enemies.
3. Inter-cropping of legumes (e.g. soybean) with trap crops (e.g. sesame and corn) can enhance the performance of parasitoids such as *G. japonicum* and *O. nezarae* and thus reduce the infestation of *Riptortus* bugs (Youn and Jung 2008).
4. Aggregation pheromones produced by the males of *R. pedestris* attract the adults and nymphs (Paik *et al.* 2009). Use synthetic aggregation pheromones of *R. pedestris*, if available.
5. Aggregation pheromones of *R. pedestris* also attract the egg parasitoid *O. nezarae*. Combined deployment of aggregation pheromone traps and non-viable eggs of *R. pedestris* for field multiplication of the parasitoid enhances the suppression of *R. pedestris* (Alim and Lim 2011).
6. Entomopathogenic fungi such as *B. bassiana* could be used to manage *Riptortus* bugs (AVRDC 1992).
7. Do not spray broad-spectrum chemical pesticides. If necessary, spray a systemic pesticide after consulting with local extension staff.

**Southern green stink bug *Nezara viridula* (L.)** (Hemiptera: Pentatomidae)

*N. viridula* is a highly polyphagous pest species and has a world wide distribution. It has been found feeding on 30 plant families (Todd 1989). Although it causes economic damage to many crop species, it mainly feeds on legumes.

## Biology

**Egg:** The barrel-shaped eggs are laid in clusters of 80-100 on the lower leaf surfaces and pods, in the upper regions of the crop. The eggs turn to deep yellow and finally reddish a few days before hatching. The egg period varies from 5 to 10 days.

**Nymph:** The newly hatched nymphs remain in groups near the egg-shells for a day or two and they do not feed. Although the second instar nymphs start feeding, they remain in groups until the third instar. The brightly colored nymphs develop through five instars and the larval period is about 3 to 7 weeks, depending on the temperature.

**Adult:** Although the nymphs are brightly colored with multiple color patterns, the adult is green. Adults live up to three weeks in hot summer weather, but they live longer over winter. In regions with a cold winter, the autumn generation will overwinter or diapause (Knight and Gurr 2007). Most diapausing adults turn to a reddish brown color. The adult is about 15 mm long.

## Damage symptoms

The damage symptoms are similar to other pod bugs, causing drying of shoots, shriveled pods and seeds. In addition, the bugs may carry the spores of fungal pathogens from plant to plant, and mechanically transmit plant pathogens while feeding (Muniappan *et al*. 2012).

## Management

1. Early maturing legume crops could be used as a trap crop for *N. viridula*. However, adults of *N. viridula* will quickly move from the trap crop to the main crop if they are at more attractive stages. Hence, chemical control of early instar *N. viridula* in trap crops is necessary to prevent movement into the main crop (McPherson and Newsom 1984).
2. Egg parasitoids such as *Ooencyrtus malayensis* (Ferriere), *Trissolcus basalis* Wollaston, *T. rudus* Le Xuan Hue, *T. mitsukurii* (Ashmead), *T. nakagawai* Watanabe, *Telenomus cyrus* (Nixon), *T. pacificus* (Gahan) and *T. comperei* Crawford are found in different countries in South and Southeast Asia (Muniappan *et al*. 2012). The Reduviid predator, *Sycanus collaris* Fab., also keeps this pest under control. Avoid using broad-spectrum chemical pesticides if these natural enemies are present in the field.
3. Entomopathogenic fungi can be used to manage *N. viridula*. However, they are mostly effective against the nymphal stages because *M. anisopliae* takes approximately 14 days from treatment to kill adult *N. viridula* (Sosa-Gomez and Moscardi 1998)—a length of infestation that most vegetable legume crops cannot tolerate (Knight and Gurr 2007).
4. Do not spray broad-spectrum chemical pesticides. If necessary, especially under exceedingly high populations of *N. viridula*, spray a systemic pesticide after consulting with local extension staff.

## Legume pod borer

***Maruca vitrata* (F.)** (Lepidoptera: Crambidae)

*M. vitrata* (syn. *M. testulalis*) is considered the most serious pest of food legumes in tropical Asia, sub-Saharan Africa, South America, North America, Australia and the Pacific (Sharma 1998). *M. vitrata* has been reported feeding on at least 45 different host plant species, including two non-leguminous hosts such *Sesamum* sp. and *Hibiscus* sp. in tropical Asia and sub-Saharan Africa. It is a major pest on at least nine cultivated legume species (*Vigna unguiculata* subsp. *sesquipedalis, V. radiata, V. mungo, Cajanus cajan, Lablab purpureus, Phaseolus angularis, P. vulgaris, Sesbania cannabina* and *S. grandiflora*) year-round. There are relatively few cultivated legumes that serve as the host plant for *M. vitrata* in sub-Saharan Africa, with cowpea (*V. unguiculata*) the predominant host. The majority of *M. vitrata* populations in Africa occur on perennial leguminous shrub or tree hosts, particularly during the main dry season (Tamo *et al*. 2002).

## Biology

**Egg:** The females lay eggs singly or in smaller groups of 2-16 on the floral buds, flowers and pods. Occasionally they are also laid on the leaves, leaf axils, terminal shoots and stem. Eggs are oval in shape and pale yellow in color. The egg period is about 2 to 4 days.

**Larva:** The larva is whitish to pale brown in color, with irregular but prominent brownish black spots on the dorsal side. The head capsule is light to dark brown. The larva has five instars and the larval period varies from 2 to 3 weeks depending on the temperature.

**Pupa:** Pupation occurs inside the silken cocoon, either in the soil or under plant debris. Sometimes pupation also takes place within the feeding web in the host plants. The pupal stage lasts for 5 to 10 days.

**Adult:** The adult moths of *M. vitrata* have light brown forewings with white patches, and white hind wings with an irregular brown border. Wings are 20-25 mm long. It often rests with the wings outspread. Adults live from 1 to 3 weeks.

## Damage symptoms

Infestation starts in the terminal shoots but later spreads to the reproductive structures (Jackai 1981). The larvae move from one flower to the other, and each may consume about four to six flowers before completing the larval stage. Infestation is highest on flowers, followed by floral buds, pods, and leaves (Karel 1985; Sharma 1998). The mature larvae, especially from the third instar, are capable of damaging pods, and occasionally the peduncle and stems (Taylor 1967). First instar larvae prefer to feed on flowers rather than pods or leaves. The larvae feed on floral buds, flowers, and pods by webbing. Often the damaged pods and inflorescence show frass.

## Management

1. Early sown crops may escape from *M. vitrata* damage, because the pod borer population tends to increase over the season.
2. Mixed cropping and intercropping can reduce the incidence of *M. vitrata*. For instance, intercropping of common bean with maize significantly minimizes pod borer damage on beans.
3. Grow Sunn hemp (*Crotalaria juncea*) as a trap crop against *M. vitrata*. The plant is highly attractive to the egg laying female moths, but highly unsuitable for the developing larvae of *M. vitrata*, and thus can be used as a 'dead-end' trap crop (Jackai and Singh 1983).
4. Sex pheromone lures of *M. vitrata* are commercially available and can be used for monitoring.
5. *Maruca vitrata* multiple nucleopolyhedrovirus (MaviMNPV) recently has been developed as a biopesticide in Taiwan and Benin (Tamo *et al.* 2012), and can be used to replace chemical pesticides. SlNPV is effective either alone or in combination with neem. In addition, B*acillus thuringiensis* formulations (AVRDC 1996 and 1997; Srinivasan 2008; Yule and Srinivasan 2013) and entomopathogenic fungi such as *B. bassiana* and *M. anisopliae* isolates and/or formulations (Ekesi *et al.* 2002; Sunitha *et al.* 2008; Sreekanth and Seshamahalakshmi 2012) can be used to manage *M. vitrata*.
6. Several parasitoid species have been reported for *M. vitrata* in tropical Asia and Africa (Ulrichs *et al.* 2001; Huang *et al.* 2003; Arodokoun et al. 2006; Muniappan *et al.* 2012). However, most are generalist parasitoids. Recently, AVRDC – The World Vegetable Center identified three relatively specific parasitoids *viz.*, an

egg-larval parasitoid, *Phanerotoma syleptae* Zettel, and two larval parasitoids, *Therophilus javanus* Bhat & Gupta and *T. marucae* van Achterberg & Long) of *M. vitrata* (Srinivasan *et al.* 2013), and developed pilot production methods (Yule and Srinivasan 2012). These parasitoids are commonly present in Southeast Asia, and they could be conserved and/or released in vegetable legume fields at regular intervals to check the build-up of *M. vitrata.*

7. Chemical pesticides can be applied as needed. Chemical pesticides may highly be effective against early larval stages, before the larvae entering into the reproductive parts of the plants. Avoid using the same pesticide group or compounds having the same mode of action, since *M. vitrata* is able to develop resistance quickly. Consult the local extension agency for appropriate recommendations.

## Lima bean pod borer

***Etiella zinckenella* Treitschke** (Lepidoptera: Pyralidae)

*E. zinckenella* widely occurs in Asia, sub-Saharan Africa, South America, North America, Australia, Europe and the Caribbean (Chhabra *et al.* 1993). Although it has been reported on several legume crops, it causes significant yield losses in lentil, peas and soybean.

## Biology

**Egg:** The females lay eggs singly or in smaller groups of 2-12 on immature pods and around the calyx. Eggs are shiny white. The egg period is about 4 to 5 days, although it could be as high as three weeks under low temperature.

**Larva:** The upper surface (dorsal side) of the larva is reddish-pink, whereas the sides and the lower surface of the body are pale green in color. The larval period varies from 3 to 6 weeks depending on the temperature. In parts of Russia and neighboring countries, it overwinters as larva. Toward pupation, the larva moves down to the soil surface for pupation either in the soil or under plant debris.

**Pupa:** Pupa is light green when newly formed, and later it turns to light brown. The pupal stage lasts for 1 to 3 weeks.

**Adult:** The adult is a grayish brown snout moth. The forewings have a distinct pale-white band along the costal margin. Hind wings are partially transparent with a dark marginal line (Chhabra *et al.* 1993). Wings are about 20 mm long. Adults live 2 to 3 weeks.

## Damage symptoms

Infestation of *E. zinckenella* usually starts late in the season. The larvae feed on floral buds, flowers, and pods by making rough and irregular incisions. The pods may have several entry holes. They feed on the seeds inside the pods. Often the damaged pods and inflorescence show light colored frass and loosely spun webs.

## Management

1. Choose tolerant or resistant cultivars available in the region. For instance, the soybean accession PI 227687 is consistently resistant to *E. zinckenella* (Talekar and Lin 1994). Consult the local extension agency for the availability of tolerant or resistant varieties.

2. Sex pheromone lures of *E. zinckenella* are commercially available, and can be used for monitoring.
3. Several parasitoid species have been reported for *E. zinckenella* in tropical Asia, Europe and North America. For instance, *Apanteles hanoii* Tobias & Long, *A. taragamae* Viereck, *Bracon* sp., *Trathala flavoorbitalis* (Cameron) and *Tropobracon luteus* Cameron have been recorded from *E. zinckenella* in Vietnam (Long and Hoa 2012). *Trichogrammatoidea bactrae* Nagaraja and *Trichogramma ostriniae* Pang and Chen were found to parasitize 80% eggs of *E. zinckenella* (AVRDC 1992). *Iconella etiellae* Viereck has been reported on *E. zinckenella* from western and central USA (Fernández-Triana *et al.* 2013). These parasitoids could be conserved and/or released in classical biological control programs to check the build-up of *E. zinckenella* in vegetable legume fields. Avoid using broad-spectrum chemical pesticides if these natural enemies are present in the field.
4. Biopesticides such as *B. thuringiensis* and neem are highly effective against *E. zinckenella* (Byrappa *et al.* 2012). Use biopesticides if they are available and recommended in the region.
5. Do not spray broad-spectrum chemical pesticides. If necessary, spray a systemic pesticide after consulting with local extension staff.

## Blue butterfly

***Lampides boeticus* (L.), *Euchrysops cnejus* (F.)** (Lepidoptera: Lycaenidae)

Blue butterflies occasionally cause serious yield losses in food legumes in Asia, sub-Saharan Africa, Europe, Australia and the Pacific. They mainly feed on legumes.

## Biology

**Egg:** The spherical shaped eggs are laid singly on floral buds, flowers and immature pods. Sometimes they are also laid on growing shoots and leaves. Eggs are pale blue or greenish white in color. The egg period is about 4 to 7 days.

**Larva:** The larva is relatively sedentary and it is yellowish green in color. The entire body of the larva has small and sparse setae. In general, the larva looks like a slug. The larval period varies from 2 to 4 weeks depending on the temperature.

**Pupa:** Pupation occurs on the leaves, twigs and pods, and to some extent in the soil or under plant debris. Pupa is reddish brown. The pupal stage lasts for a week.

**Adult:** The adult is a medium-sized butterfly. The wings are metallic blue to dusky blue in color. The tail of hind wings is black and tipped with white (Chhabra *et al.* 1993). Wings are about 3 cm long.

## Damage symptoms

The larva bores into the floral buds, flowers and green pods soon after emergence from eggs, and feeds on the inner contents. Damage on the pods is characterized by round holes.

## Management

1. Blue butterfly larvae occur infrequently in vegetable legumes. In addition, they are attacked by several parasitoids such as *Trichogramma chilotraeae* Nagaraja

& Nagarkatti, *T. bactrae* and *Cotesia specularis* Szepligeti. Hence specific control measures do not necessarily target blue butterflies.

2. If necessary, spray the most effective biopesticides based on *Paecilomyces lilacinus, Vetricillium lecani* and neem (Arivudainambi and Vijay Chandar 2009). These are superior to *B. thuringiensis* and other entomopathogenic fungi.

## Gram pod borer

***Helicoverpa armigera* Hübner** (Lepidoptera: Noctuidae)

The gram pod borer is a polyphagous and highly mobile insect and it is a pest of economic importance on many agricultural and horticultural crops. It is a major pest on a number of crops, including cotton, tobacco, corn, sorghum, sunflower, soybean, lucerne and pepper (Torres- Villa *et al.* 1996). It has been recorded on 180 cultivated and wild plant species in at least 45 families (Venette *et al.* 2003). Occasionally it causes significant yield losses in mungbean, vegetable soybean, chickpea and pigeon pea.

## Biology

**Egg:** The adults lay eggs singly and scattered, usually on or near the leaflets, floral buds or young pods. The eggs are spherical, about 0.5 mm in diameter, creamy white when laid, but later turn to brown and black just before hatching. The egg period is about 4 to 5 days depending on the temperature.

**Larva:** Upon hatching, the neonate larvae are creamy white with dark brown or black colored heads and prominent spines on the body. The grown-up larvae vary in color from pale green to brown or even black with lateral stripes on the body. Mature larvae can grow to about 4 cm in length. The larval period is about 15 to 25 days. The larva passes through six instars.

**Pupa:** Pupation takes place in soil, but sometimes also on plants. Pupae are dark brown. The pupal period varies from 6 to 33 days depending on the temperature. On average, it is about 10 days to two weeks.

**Adult:** Adults are stout-bodied moths with a wingspan of about 3.5–4 cm . Adult males are usually pale yellow in color with olive green or grey; adult females are reddish brown. The male forewings are pale yellow and olive green with light brown obscure transverse lines; female forewings are reddish brown with distinct brownish black transverse lines. The hind wings are white with a brown border.

## Damage symptoms

The neonate larvae feed on the surfaces of leaves or floral buds. However, the grown-up larvae prefer to feed on the contents of reproductive parts such as floral buds, flowers and young fruits. The larvae make the holes on these reproductive parts and feed inside by thrusting their head inside; hence the holes are circular and often surrounded by fecal pellets. Later, the larva feeds on most of the inner contents of the pod and hollows the pod out. Severely damaged pods rot and fall; partially damaged pods may become deformed.

## Management

1. Avoid growing legumes in the vicinity of other alternate host plants, because the *H. armigera* adults can easily migrate to the newly planted legume crop. It

may be too difficult to avoid this situation in countries where the land is highly fragmented. Erecting suitable physical barriers such as nylon nets or planting barrier crops that are non-host plants around the plots can reduce *H. armigera* damage on vegetable legumes. However, these measures cannot prevent the complete entry of the insects, as they are strong flyers.

2. Crop rotation should strictly be followed. If the growers plant vegetable legume (*e.g.*, mungbean) after mungbean or other alternate host plants such as tomato, chickpea, corn, cotton, etc, the damage will be higher from the emerging insects that already pupated in the soil during the previous crop cycle. This is more serious in locations where *H. armigera* diapause occurs during winter. Rotate the legume crop with a non-host cereal crop, cucurbit or cruciferous vegetable. In case of *H. armigera* sex pheromone traps can be used to monitor, mass-trap or disrupt the mating activities of male moths.
   (a) ***Monitoring*:** Sex pheromone traps baited with *H. armigera* pheromone lures can trap adult male moths to predict the population build-up in the field.
   (b) ***Mass-trapping*:** Sex pheromone traps baited with *H. armigera* pheromone lures can be used for trapping as many males as possible to reduce chances of females mating and producing viable eggs in the field. However, this has limited effect for polyphagous insects like *H. armigera*, since its population is always higher due to the availability of multiple host plants in crop production systems in the tropics.
   (c) ***Mating disruption*:** High concentration of full blends or, at times, only one component of the multi-component pheromone is placed in the field to permeate the area with the synthetic chemical. The high concentration of pheromone in the air overwhelms the male, making it impossible to locate a receptive female. Evidently, failure to mate results in failure to production of eggs which reduces the build-up of insect populations. Placing a high concentration of sex pheromone in a slow-release formulation on a 5- and 10-m grid in the field will result in a drastic reduction in male moths being attracted to virgin females, which adversely affects mating in *H. armigera* (AVRDC 1988).
4. Planting of African marigold (*Tagetes erecta* L.) as a trap crop reduces the incidence of *H. armigera*. *H. armigera* adults preferred marigold at flowering stage for oviposition. It is important to synchronize sowing/ transplanting of both crops so that their flowering coincides to attract *H. armigera* female adults.
5. The egg-parasitoids (*e.g., Trichogramma pretiosum* Riley) and larval parasitoids (*e.g., Campoletis chlorideae* Uchida) can be conserved and/or released in vegetable legume fields at regular intervals to check the build-up of *H. armigera*. If these parasitoids are present in the field, avoid using broad-spectrum chemical pesticides, which would be lethal to these natural enemies.
6. Commercially available biopesticides based on *B. thuringiensis*, *Helicoverpa armigera* nucleopolyhedrovirus (HaNPV) and neem can also be used against *H. armigera*. However, proper rotation should be followed while applying *B. thuringiensis* formulations to avoid development of resistance. For instance, *B. thuringiensis* subsp. *kurstaki*-based formulations could be rotated with *B. thuringiensis* subsp. *aizawai*-based formulations.

7. Chemical pesticides are widely used against this noxious insect in several parts of the world. Chemical pesticides are very effective against early larval stages, before the larvae enter into floral buds or pods. Pesticide spraying should be scheduled soon after noticing the eggs or early larval stages. As it would be laborious to monitor these stages, it could be coordinated with the sex-pheromone trap catches. However, proper pesticide rotation should be followed when using chemical pesticides. The effectiveness of these chemicals in the region and the registration status for the target legume crop should be checked before using.

**Spider mite *Tetranychus* spp.** (Acari: Tetranychidae)

Spider mites emerged as a serious pest of vegetable crops including vegetable legumes, eggplant, tomato, cucumber, and other field crops in South and Southeast Asia, Africa, Europe and Mediterranean countries. Low relative humidity favors the multiplication of mites wheras precipitation is the only important abiotic factor that restricts spider mite populations.

## Biology

Spider mites are very minute in size, and vary in color from green, greenish yellow, brown to orange red with two dark spots on the body. Eggs are round, white, or cream-colored; egg period is 2 to 4 days. Upon hatching, spider mites will pass through a larval stage and two nymphal stages (protonymph and deutonymph) before becoming adults. The lifecycle is completed in 1 to 2 weeks. There are several overlapping generations in a year. The adult lives up to 3 to 4 weeks.

## Damage symptoms

Spider mites usually extract the cell contents from the leaves using their long, needle-like mouthparts which results in resuction of chlorophyll content in the leaves, leading to the formation of white or yellow speckles on the leaves. In severe infestations, leaves will completely desiccate and drop off. The mites also produce webbing on the leaf and pod surfaces in severe conditions. Under high population densities, the mites move to the tip of the leaf or top of the plant and congregate using strands of silk to form a ball-like mass, which will be blown by winds to new leaves or plants, in a process known as "ballooning."

## Management

1. Several predators of spider mites occur in most countries. For instance, *Stethorus* spp., *Oligota* spp., *Anthrocnodax occidentalis* Felt, *Feltiella minuta* Felt, etc. are known to occur in Taiwan (Ho 2000). As application of broad-spectrum pesticides may kill predators and lead to outbreaks of spider mites, avoid spraying broad-spectrum pesticides.
2. Predatory mites such as *Phytoseiulus persimilis* Athias- Henriot and several species of *Amblyseius*, especially *A. womersleyi* Schicha and *A. fallacies* Garman can be used to control spider mites. They are more effective under protective structures and in high humidity conditions.
3. Green lacewings (*Mallada basalis* Walker and *Chrysoperla carnea* Stephens) also are effective generalist predators of spider mites. A third instar grub of *C. carnea* can consume 25–30 spider mite adults per day; however, it needs supplemental food for long-term survival (Hazarika *et al.* 2001).

4. Spray acaricides following local recommendations. Usually, the macrocyclic lactones (*e.g.* avermectins and milbemycins) are effective. However, continuous use may promote resistance in mites. Use proper pesticide rotations and follow window periods recommended by local extension staff.

## Bruchids

***Callosobruchus chinensis* L., *C. maculatus* Fab.** (Coleoptera: Bruchidae)

Bruchids have been reported from South and Southeast Asia, Africa, China, Taiwan and North America. They mainly attack grain legumes in storage, but also have been reported to attack corn, sorghum and cotton seeds (Chhabra *et al.* 1993).

## Biology

The eggs are glued on to the surface of the grain in storage, or on green pods if the infestation starts in the field. Although the eggs are laid singly, several eggs may be seen on a grain or pod. The eggs are elongate, oval and scale-like. When freshly laid, the eggs are translucent. However, they turn to pale yellow before hatching. The egg period is about one week, but may take more than two weeks at low temperature.

The young grub feeds inside the grains, and completes its development. The grub period is 2 to 5 weeks depending on the temperature. However, it hibernates during the winter, and hence it completes its grub development in 4 to 5 months (Chhabra *et al.* 1993). The mature grub is about 0.5 cm long and white in color with a brown head.

The pupa is oval in shape and white in color. Pupal stage lasts for 1 to 4 weeks depending on the temperature. The adult beetle is 3-4 mm in length. *C. chinensis* is reddish brown in color, whereas *C. maculatus* has grey wings with four black spots (Chhabra *et al.* 1993).

## Damage symptoms

The grub remains curved inside the grains and feeds, emptying out the seeds. When multiple generations occur repeatedly in the same seed lot, almost the entire grain is damaged, leaving a foul-flavored flour. Usually bruchid damage starts right from the field. Adults emerging from the initial infestation lead to a secondary infestation in storage, which is much more damaging (Talekar 1990). An initial infestation of 10% could lead to 100% grain loss in storage within 3 to 5 months under ordinary storage conditions (Chhabra *et al.* 1993).

## Management

1. Choose tolerant or resistant cultivars available in the region. For instance, TVu 2027 is a well-known bruchid resistant cowpea variety from IITA (Singh 1977). Two black gram accessions, VM2011 and VM2164, are highly resistant to *C. chinensis*, whereas two mungbean accessions (V2802, V2709) recently have been confirmed to possess complete resistance for *C. chinensis* and *C. maculatus* (Somta *et al.* 2007). Consult the local extension agency for the availability of tolerant or resistant varieties.
2. Storage of legume grains or seeds in air-tight containers is an effective way to eliminate bruchids, as they are unable to survive without air. Triple-bagging legume grains for storage can substantially reduce bruchid infestation (Kitch and

Ntoukam 1991a). This method recently has been promoted as Purdue Improved Cowpea Storage (PICS) technology (Baributsa *et al.* 2013). The PICS bag is a triple-layer plastic bag that serves as an air-tight (hermetic) way to store cowpea grain. Two high-density polyethylene (HDPE) inner bags fit inside an outer sack composed of woven polypropylene (PP). The inner HDPE liners have a wall thickness of 80 microns. The liners greatly hinder the movement of oxygen across the wall of the bag and the tough outer woven bag enables the bag to be easily handled.

3. Treating the legume grains with clays, sand, kaolin, and ash has been proven effective in controlling bruchid infestation in storage (Alice and Srikanth 2013). For instance, using attapulgite dust at the rate of 10-50 grams per kilogram of black gram seeds as a protectant reduced the *C. maculatus* infestation up to 135 days, without affecting seed viability (Pandey and Verma 1977). Treating cowpeas with activated kaolin at the rate of 10 grams per kilogram completely reduced the damage of *C. chinensis* for about 225 days (Swamiappan *et al.* 1976). When cowpea seeds are mixed with an equivalent volume of ash, they are protected from bruchids (Kitch and Ntoukam 1991b).
4. Vegetable oils (e.g. olive oil or mustard oil at the rate of 15 ml/kg of seed) can also be used to treat legume grains and seeds to protect from bruchid infestation. However, some vegetable oils may reduce seed viability.
5. Novel gadgets such as traps (e.g. pitfall trap and probe trap) can be used to monitor as well as mass-trap bruchids in storage. The gadget has an outer container and an inner perforated container with a rotating rod fixed with plastic brushes on all sides. The legume seeds with eggs are stored in the perforated container and the rod must be rotated one full circumference clockwise and anti-clockwise for 10 minutes three times a day. Due to the action of the rotating brush, the eggs are crushed and damage is prevented. The treatment does not affect seed germination (Mohan 2006).

## *Phaseolus vulgaris* pests- case study

Red kidney bean (*Phaseolus vulgaris*) is attacked by various insect pests such as *Epilachna vigintioctopunctata*, the flea beetle, *Longitarsus belgaumensis*, an aphid, *Smynthurodes betae*, and the bean fly, *Ophiomyia phaseoli* which cause considerable damage (Golob and Kilminster, 1982., Pajni and Jabbal, 1986., Rizvi and Singh, 1995., Flood et al 1995, Srivastava and Agarwal, 2004). The minor pests attacking the crop include , armyworm, black bug, grasshopper, thrips and leaf miners. Golob and Kilminster, 1982 reported *Zabrotes subfasciatus* as a serious pest of common beans, *P. vulgaris* in Brazil. Mexican bean beetle (*Epilachna varivestis*) has been recorded as the serious pest of beans in the eastern USA (Liebenberg, 2000).

Abrol et al (2006) reported eight species of insect pests (Figure 9.3) attacking common beans, *P. vulgaris which* included thrips, *Scirtothrips dorsalis*; aphids, *Aphis craccivora*; whitefly, *Bemisia tobaci*; hairy caterpillar, *Spilosoma obliqua*; stemfly, *Ophiomyia phaseoli*; pulse beetle *Callosobruchus chinensis*; bean gall weevil *Alcidodes signatus* and blister beetle *Cyaneolylta coerculea*. Of all these pests, bean gall weevil *Alcidodes signatus* and blister beetle, *Cyaneolylta coerculea* were the serious pests and recorded for the first from the crop and the country.

 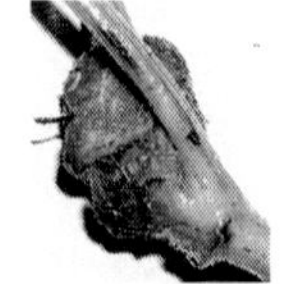 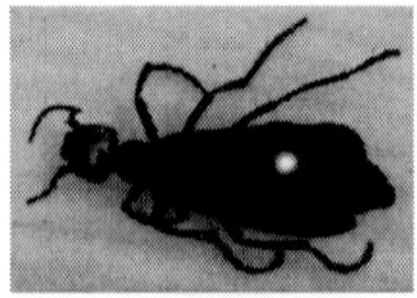

**Figure 9.3: A = *Alcidodes signatus* (adult); B = Showing gall formed by bean weevil *Al. signatus* on the internode of red kidney bean (*Phaseolus vulgaris*); C = Blister beetle, *Cyaneolylta coerculea* (Adult)**

**Integrated pest management approaches for insect and mite pests of vegetable legumes**

## Cultural control

1. Avoid vegetable legume monocultures and follow crop rotations.
2. Avoid growing legumes in the vicinity of other alternate host plants, because pests can easily migrate to the newly planted legume crop. Erecting suitable physical barriers such as nylon nets or planting barrier crops that are non-host plants around the plots can reduce pest damage on vegetable legumes. However, these measures cannot prevent the complete entry of insects as they are strong flyers.
3. Keep weeds under control in vegetable legume seedling production areas as well as in main fields to reduce the availability of alternate host plants for some of the major insect pests.
4. Grow trap crops, and focus pesticide spraying on the trap crop to manage bean flies, *S. litura, Riptortus* bugs, *N. viridula, M. vitrata* and *H. armigera*. Plant tall border crops like maize, sorghum, or pearl millet to reduce the infestation of whiteflies.
5. Early sowing or planting may reduce the damage of bean flies and *M. vitrata*.
6. Triple bagging, or treating legume grain or seed with clays, sand, kaolin, ash and vegetable oils may reduce the damage by bruchids in storage.

**Host plant resistance**

7. Choose resistant or tolerant cultivars for the major insect pests

**Mechanical control**

8. Grow seedlings for crops such as yard-long bean and common bean in insect-proof (50–64 mesh) net houses, net tunnels, greenhouses, or plastic houses.
9. Novel gadgets such as traps (e.g. pitfall trap and probe trap) can be used to mass-trap bruchid beetles, and to crush bruchid eggs in storage.

**Behavioral control**

10. To monitor insect populations, use yellow sticky traps to attract whitefly and leafhoppers and blue sticky traps to attract thrips.
11. Use reflective and yellow plastic or straw mulches to reduce the incidence of whiteflies and thrips on vegetable legume crops.

**Biological control**

12. Apply neem formulations and other systemic chemical pesticides, if recommended in the region, as soil drenches or through foliar application to control whitefly and aphids in vegetable legume seedling production.

13. Use neem, entomopathogenic fungi, *B. thuringiensis* based biopesticides, and botanical extracts, which will not interfere with the activities of natural enemies in vegetable legume production systems.
14. Install sex pheromone lures for insects such as *S. litura, S. exigua, M. vitrata, E. zinckenella* and *H. armigera*, and aggregation pheromone lures for *Riptortus* bugs in traps. Place the traps either at canopy level or at slightly above the canopy level for effective attraction. This will be highly effective when practiced in a community.

    **Chemical control**

15. Do not spray any broad-spectrum pesticides against early season sucking pests. This may disrupt the complex of natural enemies in the ecosystem and lead to a resurgence of sucking pests. If necessary, use systemic pesticides recommended by the local extension service. Do not use the same compound or pesticide group(s) continuously to reduce the development of pesticide resistance in insects. Applying chemical pesticides via seed treatment and soil drenching will avoid direct contact of pesticides with natural enemies.

## References

Abate T and Ampofo JKO. 1996. Insect pests of beans in Africa: their ecology and management. Annual Review of Entomology 41: 45-73.

Abrams, R.1967. Studies on natural cross-pollination in pigeonpeas (Cajanus cajan). Puerto Rico Univ. Jour. Agr. 51(1): 1-3.

Abrol D.P., Ramamurthy V. V. and Srivastava K. 2006. Bean Gall Weevil and Blister Beetle as New Pests on Red Kidney Bean (*Phaseolus vulgaris* L.) in India J. Asia-Pacific Entomol. 9(4): 1-4

Afun JVK, Jackai LEN and Hodgson CJ. 1991. Calendar and monitored insecticide application for the control of cowpea pests. Crop Protection 10: 363–370.

Alex, A. H., Thomas, F. L., and Warne, B. 1950. Importance of bees in vetch seed production. Tex. Agr. Expt. Sta. Prog. Rpt. 1306, 3 pp.

Al-Ghzawi A, Samarah N, Zaitoun S and Alqudah A. 2009. Impact of bee pollinators on seed set and yield of Vicia villosa spp. dasycarpa (Leguminosae) grown under semiarid conditions. Ital.J.Anim.Sci. 8, 65-74.

Al-Ghzawi, A., Zaitoun, S., Mazari, S., Schindler, M. and Wittmann, D., 2006. Diversity of bees (Hymenoptera, Apiformes) in extensive orchards in the high lands of Jordan. Arxius de Miscellània Zoologica 4:42-48.

Ali M, Umar Farooq and Shih YY. 2002. Vegetable Research and Development in the ASEAN Region: A Guideline for Setting Priorities. P.20-64. In: Kuo CG. (ed.). 2002. Perspectives of ASEAN Cooperation in Vegetable Research and Development: Proceedings of the Forum on the ASEAN-AVRDC Regional Network on Vegetable Research and Development (AARNET), Shanhua, Taiwan, 24-26 September 2001. Asian Vegetable Research and Development Center. AVRDC Publication No. 02-539, 151 p.

Alice J and Srikanth N. 2013. Non-destructive method to manage the most dreaded pest Callosobruchus maculatus (Fab.) in Black gram, Vigna mungo L. Journal of Stored Products and Postharvest Research 4(2): 30–34.

Alim MA and Lim UT. 2011. Refrigerated eggs of Riptortus pedestris (Hemiptera: Alydidae) added to aggregation pheromone traps increase field parasitism in soybean. Journal of Economic Entomology 104(6):1833-1839.

Allard, R. W. 1954. Natural hybridization in Lima beans in California. Amer. Soc. Hort. Sci. Proc. 64: 410-416.

Amos J. M. 1943. A measurement of the value of bees in the pollination of Lima beans. Amer. Bee Jour. 83: 240 - 241.

Anderson, E. J. 1959. Two part pollination program carried out on university farms. Penn. Beekeeper 34(3): 4 - 8.

Anonymous, 2005. Package of practices for Kharif crops. Directorate of extension Education, Sher- e –Kashmir University of Agricultural Sciences & Technoloy, Jammu, Jammu & Kashmir, India. 70p.

Arivudainambi S and Vijay Chandar A. 2009. Management of pulses blue butterfly, Lampides boeticus L. in green gram. Karnataka J. Agric. Sci. 22(3): 624-625.

Arnett, R.H., Jr., M.C. Thomas, P.E. Skelley and J.H. Frank. (editors). 2002. American Beetles, Volume II: Polyphaga: Scarabaeoidea through Curculionoidea. CRC Press. 861 pp.

Arodokoun DY, Tamò M, Cloutier C and Brodeur J. 2006. Larval parasitoids occurring on Maruca vitrata Fabricius (Lepidoptera: Pyralidae) in Benin, West Africa. Agriculture, Ecosystems and Environment 113: 320–325.

Arusha, Tanzania January 12-14, 2000., Bunda College of Agriculture, Lilongwe, Malawi

Asante SK, Jackai LEN and Tamo M. 2000. Efficiency of Gryon fulviventris (Hymenoptera: Scelionidae) as an egg parasitoid of Clavigralla tomentosicollis (Hemiptera: Coreidae) in northern Nigeria. 29(4): 815-821.

Asian Vegetable Research and Development Center. 1987. 1985 Progress Report. AVRDC Publication (86-205), Shanhua, Taiwan, R.O.C.

AVRDC, 1988. 1986 progress report. Asian Vegetable Research and Development Center, Shanhua, Taiwan.

AVRDC, 1990. 1989 Progress Report. AVRDC Publication (90-330), Asian Vegetable Research and Development Center, Shanhua, Tainan.

AVRDC, 1991. 1990 Progress Report. AVRDC Publication (91-338), Asian Vegetable Research and Development Center, Shanhua, Tainan.

AVRDC, 1992. 1991 Progress Report. AVRDC Publication (92-374), Asian Vegetable Research and Development Center, Shanhua, Tainan.

AVRDC, 2001. AVRDC Annual report 2001, Asian Vegetable Research and Development Center, Shanhua, Tainan, Taiwan. 57 p.

AVRDC. 1981a. AVRDC Progress Report 1979. Asian Vegetable Research and Development Center, Shanhua, Tainan, Taiwan, ROC, 107 p.

AVRDC. 1981b. AVRDC Progress Report 1980. Asian Vegetable Research and Development Center, Shanhua, Tainan, Taiwan, ROC, 110 p.

AVRDC. 1996. AVRDC 1995 Report. Asian Vegetable Research and Development Center, Shanhua, Tainan, Taiwan. 187 p.

AVRDC. 1997. AVRDC 1996 Report. Asian Vegetable Research and Development Center, Shanhua, Tainan, Taiwan, 172 p.

Azo'o Ela M and Messi J. 2012b. Yield responses of Cucumeropsis mannii (Cucurbitaceae) to the presence or the absence of the insect foraging activity at Nkolbisson in Cameroon. Journal of Animal & Plant Sciences 13, 1791-1799.

Azo'o Ela M, Madi A, Tchuenguem Fohouo FN and Messi J. 2012a.The importance of a single floral visit of Eucara macrognatha and Tetralonia fraterna (Hymenoptera: Apidae) in the pollination and the yields of Abelmoschus esculentus in Maroua, Cameroon. African Journal of Agricultural Research 7, 2853-2857.

Azo'o Ela M, Messi J, Tchuenguem Fohouo FN, Tamesse JL, Kekeunou S, Pando JB. 2010. Foraging behaviour of Apis mellifera adansonii and its impact on pollination, fruit and seed yields of Citrullus lanatus at Nkolbisson (Yaoundé, Cameroun). Cameroon Journal ExperimentalBiology 6, 41-48.

Azo'o Ela M, Tchuenguem Fohouo FN and Messi J. 2011. Influence of the foraging activity of the entomofauna on Okra (Abelmoschus esculentus) seed yield. International Journal of Agriculture & Biology 13, 761-765.

Backhaus W. 1993. Colour vision and colour choice behaviour of the honey bee. Apidologie 24, 309 331.

Bajpai NK, Ballal CR, Rao NS, Singh SP, Bhaskaran TV. 2006. Competitive interaction between two ichneumonid parasitoids of Spodoptera litura. BioControl 51(4): 419-438.

Baributsa D, Baoua I, Lowenberg-DeBoer J, Abdoulaye T and Murdock LL. 2013. Purdue Improved Cowpea Storage (PICS) Technology. Purdue Extension Publication E-262-W, p. 4. http://extension. entm.purdue.edu/publications/E-262.pdf

Barrett RP. 1990. Legume species as leaf vegetables. p. 391-396. In: Janick J, Simon JE. (eds.), Advances in new crops. Portland, OR: Timber Press

Barrons KC (1939). Natural crossing in beans at different degrees of isolation. Am. Soc. Hort. Sci. Proc., 36: 637-640.

Basualdo M, Bedascarrasbure E and DeJong D. 2000. Africanized honey bees (Hymenoptera: Apidae) have a greater fidelity to sunflowers than European bees. Journal of Economic Entomology 93, 304–307.

Beard, B. H., and Knowees, P. F. 1971. Frequency of cross-pollination of soybeans after irradiation. Crop Sci. 11: 489-492.

Beickenstaff, C. C., and Huggans, L. J. 1962. Soybean insects and related arthropods in Missouri. Mo. Agr. Expt. Sta. Bul. 803, 51 pp.

Benchasri S and Bairaman C. 2010. Evaluation of yield, yield components and consumers' satisfaction towards yard-long bean and cowpea in agricultural organic system. Bulg. J. Agric. Sci. 16: 705-712.

Bhagwat VR, Ariëns SJA and Shanower TG. 1994. Effect of host age and intercropping on parasitization of Clavigralla gibbosa eggs by Gryon sp. International Chickpea Newsletter 1: 38-39.

Bieberdorf, G. A 1954. Honey bees as vetch pollinating agents. Okla. Agr. Expt. Sta., Mimeo. Cir. M-250, 3 pp.

Bieberdorf, G. A. 1952. Honey bees and vetch pollination. Okla. Acad. Sci. Proc. 33: 36-37.

Blackwall, F. L. C. 1964. Runner beans: adequate pollination is a must for early yields. Grower 61: 1272-1273.

Bliss FA 1980 Common bean. In Fehr WR, Hadley HH (eds) Hybridization of crop plants. American Society of Agronomy & Crop Science Society of America, Madison, WI, pp 273-284.



Bliss FA 1980 Common bean. In Fehr WR, Hadley HH (eds) Hybridization of crop plants. American Society of Agronomy & Crop Science Society of America, Madison, WI, pp 273-284.



Bob MA, Odhiambo B, Kibata G and Ong'aro J. 2005. Whiteflies as pests and vectors of viruses in vegetables and legume mixed cropping systems in Eastern and Southern Africa – Kenya. pp. 129-140. In: Anderson PK, Morales FJ (eds.) Whitefly and Whitefly-borne viruses in the tropics: Building a knowledge base for global action. Centro Internacional de Agricultura Tropical (CIAT), Cali, Colombia. 351 p.

Bond DA and Fyfe JL 1962 Breeding field beans. Plant Breed. Inst. (England) report. 1960-1961, pp4-26

Booth, R.G, M.L Cox and R.B, Madge.1990. Guides to Insects of Importance to Man. 3. Coleoptera, C.A.B. International, Wallingford, 384pp.

Boswell, V. R. 1961. Flowering habits and production of seeds. U.S. Dept. Agr. Yearbook 1961: 57 - 64.

Bradner, N. R. 1969. Hybrid soybeans: fiction or fact? Soybean Digest 29(7): 16-17. 338

Bramel PJ, Kiran S, Reddy J, Ford-Lloyd B and Chancra S. 2004. Degree and distribution of pigeon pea landrace morphological diversity in traditional cropping systems in Andhra Pradesh. In: Bramel PJ,(ed.). Assessing the risk of losses of biodiversity in traditional cropping systems: A case study of pigeon pea in Andhra Pradesh. International Crop Research Institute for the semi-arid Tropics, Pantacheru Andhra Pradesh, India. 1-45 p.

Brim, C. A., and Young, M. F. 1971. Inheritance of male-sterile character in soybeans. Crop Sci. 11: 564-566.

Brito M, Fernández-Rodríguez T, Garrido MJ, Mejías A, Romano M, Marys E. 2012. First Report of Cowpea Mild Mottle Carlavirus on Yard-long Bean (Vigna unguiculata subsp. sesquipedalis) in Venezuela. Viruses 4: 3804-3811.

Byrappa AM, Kumar NG, Divya M. 2012. Impact of biopesticides application on pod borer complex in organically grown field bean ecosystem. J. Biopest. 5(2): 148-160.

Carvalho, R. P. L. DE and C. J. Resseto. 1968. Biology of Zabrotes subfasciatus (Boheman) (Coleoptera: Bruchidae). Revista Brasileira de Entomologia 13: 195-197.

Caviness, C. E. 1966. Estimates of natural cross-pollination in Jackson soybeans in Arkansas. Crop Sci. 6(2): 211.

Caviness, C. E. 1970. Cross-pollination in the soybean. In The Indispensable Pollinators, Ark. Agr. Ext. Serv. Misc. Pub. 127, pp.33-36.

Caviness, C. E. Walters, H. J., and Johnson, D. L. 1970. A partially male sterile strain of soybean. Crop Sci. 10(1): 107-108.

Chakravorty S and Yadav DK. 2013. Effect of intercrops on Jassid (Empoasca kerri Pruthi) infesting Green gram, *Vigna radiata* (L.) Wilczek. Annals of Plant Protection Sciences 21(2): 229-232.

Chang NT. 1990. Ceranisus menes (Walker) (Eulophidae: Hymenoptera), a new parasite of bean flower thrips, Megalurothrips usitatus (Bagnall) (Thripidae: Thysanoptera). Plant Protection Bulletin (Taipei) 32(3): 237-238.

Chhabra KS and Lal S, Kooner BS and Verma MM. 1993. Insect pests of pulses – Identification and control manual. Punjab Agricultural University and Directorate of Pulses Research, India. 88 pp.

Chhabra KS, Kooner BS. 1994a. Reaction of some promising mungbean genotypes towards whitefly, jassid and yellow mosaic virus. Pest Management Econ. Zool. 2: 11-14.

Chhabra KS, Kooner BS. 1994b. Reaction of certain summer mungbean genotypes towards thrips, Megalurothrips distalis (Karny). Indian J. Agric. Res. 28: 251-56.

Chiari,Wainer César , Vagner de Alencar Arnaut de Toledo, Maria Claudia Colla Ruvolo-Takasusuki, Arildo José Braz de Oliveira, Eduardo Shiguero Sakaguti, Valeria Maria Attencia, Fabiana Martins Costa and Marina Hitomi Mitsui. 2005. Pollination of Soybean (Glycine max L. Merril) by Honeybees (Apis mellifera L.). Brazilian Archives of Biology and Technology.48(1): 31-36

Coe. L. E. 1949. Pollination of hairy vetch in Oklahoma: the nearer the bees the better the harvest. Amer. Bee Jour. 89: 558-559.

Cordner, H. B. 1933. External and internal factors affecting blossom drop and set of pods in Lima beans. Amer. Soc. Hort. Sci. Proc. 39: 571 - 576.

Culter, G. H. 1934. A simple method for making soybean hybrids. Amer. Soc. Agron. Jour. 26: 252-254.

Damayanti TA, Alabi OJ, Naidu RA and Rauf A. 2009. Severe outbreak of a yellow mosaic disease on the yard-long bean in Bogor, West Java. HAYATI Journal of Biosciences 78-82.

Darwin C 1876. The effects of cross and self fertilization in the vegetable Kingdom. London: John Murray. p. 505.

Darwin, F. 1874. Bees visiting flowers. Nature 9: 189-190.

David BV. 2001. Elements of Economic Entomology (Revised and Enlarged Edition). Popular Book Depot, Chennai, India. 590 p.

Debouck D 1991. Systematics and morphology. In: Common beans: research for crop improvement, Van Schoonhoven, A and Voyset, O (eds), Cali, Colombie. pp 55-118.

Delaplane KS, Mayer DF 2000 Crop Pollination by Bees. CABI Publishing, Wallingford, UK and New York.

Denison RF, Kiers ET. 2004. Why are most rhizobia beneficial to their plant hosts, 1088 rather than parasitic. Microbes Infection 6, 1235-1239.

Drayner, J. M. 1956. Regulation of outbreeding in field beans. Nature 177: 489 490.

Dreyer H, Baumgartner J, Tamo M. 1994. Seed damaging field pests of cowpea (Vigna unguiculata L. Walp.) in Benin: occurrence and pest status. Int. J. Pest Manag. 40: 252–260.

Dryner JM 1956 Regulation of out breeding in field beans (Vicia faba). Nature 177(4506):489-490

Dryner JM (1959) Self and cross fertility in field beans (Vicia faba L.). J Agric Sci (England) 53:387-402

Ekesi S and Maniania NK. 2000. Susceptibility of Megalurothrips sjostedti developmental stages to Metarhizium anisopliae and the effects of infection on feeding, adult fecundity, egg fertility and longevity. Entomologia Experimentalis et Applicata 94(3): 229–236.

Ekesi S, Adamu RS, Maniania NK. 2002. Ovicidal activity of entomopathogenic hyphomycetes to the legume pod borer, Maruca vitrata and the pod sucking bug, Clavigralla tomentosicollis. Crop Protection 21(7): 589–595.

Ekesi S, Akpa AD, Onu I and Ogunlana M.O. 2000. Entomopathogenicity of Beauveria bassiana and Metarhizium anisopliae to the cowpea aphid, Aphis craccivora Koch (Homoptera: Aphididae). Archives of Phytopathology and Plant Protection 33(2): 171-180.

Ekesi S. 1999. Selection of virulent isolates of entomopathogenic hyphomycetes against Clavigralla tomentosicollis Stål. and evaluation in cage experiment using three cowpea varieties. Mycopathologia 148(3): 131–139.

El-Hawary FM and Abd El-Salam AME. 2008. Effect of neem and antitranspirant products against Aphis craccivora Koch and its biology. Egypt. Acad. J. Biolog. Sci. 1(2):189-196.

Enfawal MA, Bishr MA and Hassoub EK. 1976. Natural cross pollination in Egyptian Cotton (Gossypium barbadense L.) Journal of Agricultural Sciences 86, 205- 209.

Fameni ST, Tchuenguem Fohouo FN, Brueckner D. 2012. Pollination efficiency of Apis mellifera adansonii (Hymenoptera: Apidae) on Callistemon rigidus (Myrtaceae) flowers at Dang (Ngaoundéré, Cameroon). International Journal of Tropical Insect Science 32(1), 2-11.

FAO Food and Agriculture Organization. 2012. FAOSTAT. http:// faostat.fao.org [accessed 25 June 2014].

Fernández-Triana JL, Cardinal S, Whitfield JB, Hallwachs W, Smith MA and Janzen DH. 2013. A review of the New World species of the parasitoid wasp Iconella (Hymenoptera, Braconidae, Microgastrinae). ZooKeys 321: 65–87.

Fernand-Nestor Tchuenguem Fohouo, Albert Ngakou and beaudelaire Stephanie Kengini 2009. Pollination and yield response of cowpea (Vigna unguiculata L. Walp) to the foraging activity of Apis mellifera adansonii (Hymenoptera: Apidae ) at Ngaoundere(Cameroon). Afican journal of biotechnology 8(9): 1988-1996.

Fernand-Nestor Tchuenguem Fohouo, Joseph Blaise Pando and Joseph Lebel Tamesse 2014. Pollination efficiency of Xylocopa olivacea (Hymenoptera: Apidae) on Cajanus cajan (Fabaceae) flowers at Yaounde, Cameroon. International Journal of Tropical Insect Science 34(2): 138–148.

Flood, B., G, Hein and W. R. Beans. 1995. In "Vegetable Insect Management with Emphasis on the Midwest" (ed. R. Foster, B. Flood), Meister Publishing Co., Willoughby, Ohio, pp. 41-54.

Fluri P, Frick R 2005. Apiculture in Sweden: state and perspectives. Rev. Suisse. Agric., 37(2): 81-86.

Free J B 1966. The Pollination of the Beans Phaseolus Multiflorus and Phaseolus Vulgaris by Honeybees. Journal of Apicultural Research 5(2): 87-91

Free, J. B 1968. The behaviour of bees visiting runner beans (Phaseolus multiflorus). Jour. Appl. Ecol. 5: 631-638.

Free, J. B and Racey. P. A. 1968. The pollination of runner beans (*phaseolus multiflorus*) in a glasshouse. Jour. Apic. Res. 7: 67-69.

Free, J. B. 1966. The pollination of the beans *phaseolus multiflorus* and *phaseolus vulgaris* by honeybees. Jour. Apic. Res. 5: 87-91.

Garber, R. J., and Odband, T. E. 1926. Natural crossing in soybeans. Amer. Soc. Agron. Jour. 18: 967 - 970.

Garibaldi LA, Steffan-Dewenter I, Winfree R, Aizen MA, Bommarco R, Cunningham SA, Kremen C, Carvalheiro LG, Harder LD, Afik O, Bartomeus I, Benjamin F, Boreux V, Cariveau D, Chacoff NP, Dudenhöffer JH, Freitas BM, Ghazoul J, Greenleaf S, Hipólito J, Holzschuh A, Howlett B, Isaacs R, Javorek SK, Kennedy CM, Krewenka KM, Krishnan S, Mandelik Y, Mayfield MM, Motzke I, Munyuli T, Nault BA, Otieno M, Petersen J, Pisanty G, Potts SG, Rader R, Ricketts TH, Rundlöf M, Seymour CL, Schüepp C, Szentgyörgyi H, Taki H, Tscharntke T, Vergara CH, Viana BF, Wanger TC, Westphal C, Williams N, Klein AM (2013) Wild Pollinators Enhance Fruit Set of Crops Regardless of Honey Bee Abundance. Science 339: 1608-1611.

Gill KA and Neal O 2015 Survey of Soybean Insect Pollinators: Community Identification and Sampling Method Analysis. Environ Entomol. 44(3):488-98

Golob, P. and A. Kilminster. 1982. The biology and control of Zabrotes subfasciatus (Boheman) (Coleoptera: Bruchidae) infesting red kidney beans. Journal of Stored Products Research 18: 95-101.

Gordienko, V. 1960. Sexual hybrids of soya beans obtained by directed bee pollination. Pp.400-407. In Mel'nichenko, A. N., [Pollination of Agricultural Plants by Bees.] Izd-vo Minist. Sel'sko Khoz. SSSR. [In Russian.]

Graham PH, Ranalli P. 1997. Common bean (Phaseolus vulgaris L.). Field Crop Research 53, 131-146.

Greathead, D.J.1963. A review of the insect enemies of Acridoidea (Orthoptera). Transactions of the Royal Entomological Soc. London, 114(14): 437-517.

Grewal, G. S.; Gurdip Singh and Kandoria, J. L. 19990. Insect pollinators of pigeonpea (Cajanus cajan) around Ludhiana. Indian Journal of Agricultural Sciences. 60(3): 227-228

Gunn, C. R. 1971. Seeds of native and naturalized vetches of North America. U.S. Dept. Agr., Agr. Handb. 392,42 pp.

Halder J, Rai AB and Kodandaram MH. 2013. Compatibility of Neem Oil and Different Entomopathogens for the Management of Major Vegetable Sucking Pests. National Academy Science Letters 36(1): 19-25.

Hambleton J. I. 1936. Soybean for pollen and nectar. Gleanings Bee Cult. 64: 431.

Hazarika LK, Puzari KC and Wahab S. 2001. Biological control of tea pests. In: Upadhyay RK, Mukerji KG, Chamola BP. (eds.), Biocontrol potential and its exploitation in sustainable agriculture: Insect pests, Springer, USA. pp. 159-180.

Heiser CB. 1990. Seed to civilization: The story of food. Harvard University Press. 254 p.

Henslow, G. 1878. The fertilization of the scarlet runner bean by bumble- bees. Gar Chron., 10: 561.

Herman, F. J. 1960. Vetches in the united states native, naturalized and cultivate Dept. Agr., Agr. Handb. 168, 84 pp.

Heywood, V. H. and Ball, P. W. 1968. Leguminosae. In Tutin, T. G., Heyv Burges, N. A., and others, Flora Europaea 2: 80-136, Cambridge at the Ur

Ho CC. 2000. Spider-mite problems and control in Taiwan. Experime Acarology 24: 453-462.

Hoque M, Mollik SR, Nazimuddin M, Alam SN, Khorsheduzzaman M, Jasmine HS, Sultana NA, Rahman M, Ahmed B, Rajotte E and Luther GC. 2002. Survey of bean pod borer (Maruca testulalis), whitefly and aphids and their natural enemies on country bean (Lablab purpureus). Tenth Annual Report of the Integrated Pest Management Collaborative Research Support Program (IPM CRSP). http://www.oired.vt.edu/ipmcrsp/communications/annrepts/ annrep03/Bangladesh/8BangladeshY_10 ANNUAL _REPORT1. pdf#page=10. p. 123–124.

Huang CC, Peng WK and Talekar NS. 2003. Parasitoids and other natural enemies of Maruca vitrata feeding on Sesbania cannabina in Taiwan. Biol. Control. 48: 407–416.

Hungria M, Campo RJ and Chueire LMO. 2001. Symbiotic effectiveness of fast-growing rhizobia strains isolated from soybean nodules in Brazil. Biology and Fertility of Soil Soils 33, 387-394.

Ibarra-Perez FJ, Ehdaie B and Waines JG 1997. Estimation of outcrossing rate in common bean. Crop Sci., 37: 60-65.

Ibarra-Perez, FJ, Barnhart D, Ehdaie B, Knio KM, Waines JG 1999 Effects of insect tripping on seed yield of common beans. Crop Science 39: 428-433.

Ige, O. E., Olotuah, O.F and Akerele, V. 2011. Floral Biology and Pollination Ecology of Cowpea (*Vigna unguiculata* L. Walp). Modern Applied Science 5(4) : 74-82.

Jackai LEN and Singh SR. 1983. Suitability of selected leguminous plants for development of Maruca testulalis larvae. Entomologia Experimentalis et Applicata 34(2): 174–178.

Jackai LEN. 1981. Relationship between cowpea crop phenology and field infestation by the legume pod borer Maruca testulalis. Ann. Entomol. Soc. Am. 74: 402-408.

Jacob Remacle A. 1988. Comportement de butinage de l'abeille domestique et des abeilles sauvages dans les vergers de pommiers en Belgique. Apidologie 20, 271-285.

Jambunathan, R. and Singh, U. 1982. Grain quality of pigeonpea. Proc. Symp. On Increasing Pulse Production in India-Constraints and Opportunities. Oxford and IBH Publ., New Delhi. Pp.389-395.

Jany, E. 1950. [The piercing of scarlet runner flowers by bumble bees.] Ztschr. F. Angew. Ent. 32(2): 172-183. [in german.] Aa-45/54.

Jaycox, E. R. 1970. Ecological relationshtps between honey bees and soybeans. Amer. Bee Jour. 110(8): 306-307, (9): 343-345, (10): 383-385.

Jemutai J. 2008. Effect of seed dressing insecticide, intercropping and plant density on bean fly (Ophiomyia spp.) infestation in common bean (Phaseolus vulgaris L.). M.Sc. Thesis, Egerton University, Kenya. 52 pp.

Julianna KT and Rufus I. 2010. Weather during bloom affects pollination and yield of high bush blueberry. Journal of Economic Entomology 103, 557-562.

Karel AK and Rweyemamu CL. 1985. Resistance to foliar beetle, Ootheca bennigseni (Coleoptera: Chrysomelidae) in common beans. Environmental Entomology 14: 662-664.

Karel,AK, 1985. Yield losses from and control of bean pod borers, Maruca testulalis (Lepidoptera: Pyralidae) and Heliothis armigera (Lepidoptera:Noctuidae). Journal of Economic Entomology 78: 1323–1326.

asina JM, Hagen M, Kraemer M, Nderitu J, Martius C and Wittmann D 2009b Bee pollination enhances crop yield and fruit quality in kakamega, western Kenya. East African Agricultural and Forestry Journal 75(1):1-11.

Kasina M, Kraemer M, Martiusand C and Wittmann D 2009c Diversity and activity density of bees visiting crop flowers in Kakamega, western Kenya. Journal of Apicultural Research 48:134-139.

Kasina M, Kraemer M, Martiusand C, Wittmann D 2009a Farmers' knowledge of bees and their natural history in Kakamega district, Kenya. Journal of Apicultural Research 48:126-133.

Kasper ML, Reeson AF, Mackay DA and Austin AD. 2008. Environmental factors influencing daily foraging activity of Vespula germanica (Hymenoptera, Vespidae) in Mediterranean Australia. Insectes Sociaux 55, 288–296.

Keller LF and Waller DM 2002. Inbreeding effects in wild populations. Trends Ecol. Evol., 17: 230-241.

Kevan, P.G. and Baker, H.G. 1983. Insects as flower visitor and pollinators. Ann. Rev. Entomol. 28: 407-453.

Kiers ET, Rousseau RA, West SA and Denison RF. 2003. Host sanctions and the legume–Rhizobium mutualism. Nature 425, 78–81.

Killinger, G. B.1969. Pigeonpeas (Cajanus cajans (L.) druce)- a useful crop for Florida. Soil Crop Soc. Fla. Proc. 28: 162-167.

Kingha B. M. T, Tchuenguem Fohouo F.N, Ngakou A and Brückne D. 2012. Foraging and pollination activities of Xylocopa olivacea (Hymenoptera, Apidae) on Phaseolus vulgaris (Fabaceae) flowers at Dang (Ngaoundere-Cameroon). Journal of Agricultural Extension and Rural Development 4(6): 330-339.

Kitch LW and Ntoukam G. 1991a. Airtight storage of cowpea in triple plastic bags (triple-bagging). CRSP Technical Bulletin 3, Agronomic Research Institute of Cameroon, Maroua Research Center, CRSP Cowpea Storage Project. P. 11. http://www.entm.purdue.edu/entomology/research/cowpea/Extension%20bulletins/PDF%20publications/Tripple%20Bagging.pdf

Kitch LW and Ntoukam G. 1991b. Storage of cowpea in ash. CRSP Technical Bulletin 1. Bean–Cowpea CRSP, East Lansing, MI, USA. P. 11. http://www.entm.purdue.edu/entomology/research/cowpea/ Extension%20bulletins/PDF%20publications/Storage%20of%20 Cowpea%20in%20Ash.pdf

Klein AM, Vaissière BE, Cane JH, Steffan DI, Cunnigham SA, Kremen C and Tscharntke T (2007). Importance of pollinators in changing landscapes for world crops. Proc. Royal Soc., 274: 303-313.

Knight KMM and Gurr GM. 2007. Review of Nezara viridula (L.) management strategies and potential for IPM in field crops with emphasis on Australia. Crop Protection 26: 1–10.

Kooistra, E. 1968. Selection in runner beans (phaseolus concinnus) with special reference to the use of tuberous roots. Euphytica 17: 183-189.

Kooner BS.1998. Identification of sources of resistance to whitefly, jassids on Kharif mungbean. Abstr No. 1-A.33. In: Nat. Symp. on Management of Biotic and Abiotic Stresses in Pulse Crops. 26-28 June, 1998. IARI, New Delhi.

Krauss, F. G. 1927. Improvement of the pigeonpea, genetic analysis of cajanus indicus and the creation of new varieties through hybridization and selection. Jour. Hered. 18: 227-232.

Krauss, F. G. 1932. The pigeonpea (Cajanus indicus) its improvement, culture and utilization in hawaii. Hawaii Agr. Expt. Sta. Bul. 64, 46 pp.

Kuswanto, Budi Waluyo. 2011. The adaptation test on yard-long bean lines tolerant to aphids and high yield. Journal of Agricultural Science 33(2): 182-187.

Lambeth, V. N. 1950. Some factors influencing pod set and yield of the Lima bean. Mo. Agr. Expt. Sta. Res. Bul. 466, 60 pp.

LI, Zheng-Hong Liang N, Hong M A Saxena, K , Yang Tao , Liu Xiu-Xian , Zong Xu-Xiao 2012. Insect Pollinators in CGMS Hybrid Seed Production of Cajanus cajan. Acta Agronomica Sinica. 37(12): 2187–2193

Liebenberg, A.J. (Ed.) 2000. Dry bean production Department of Agriculture Directorate Agricultural Information Service, Private Bag X144, Pretoria 0001 South Africa

Long KD and Hoa DT. 2012. Notes on parasitoid assemblage reared from larvae of legume pod borers Maruca vitrata (Fabricius) and Etiella zinckenella Treitschke. Journal of Biology 34(1): 48-58.

Louveaux J. 1984. L'abeille domestique dans ses relations avec les plantes cultivées. In : ''Pollinisation et productions végétales'', Pesson P. & Louveaux J. (eds), INRA, Paris, 527-555.

Luo, Y. and Li Z. (1999). Pollination Ecology of Chloranches serratus (Thunb) Roemet Schult and Ch. Fortunei ( A. Gray) SolmsLaub. (Chloranthaceae). Ann. Bot. 83:489-499.

Mackie WW and Smith FL 1935. Evidence of field hybridization in beans. J. Am. Soc. Agron., 27: 903-909.

Magruder, R. 1948. Natural crossing of Lima beans in southern California in 1946. Amer. Soc. Hort. Sci. Proc. 51: 471-474.

Magruder, R. and Wester, R. E. 1940. Natural crossing in Lima beans in Maryland. Amer. Soc. Hort. Sci. Proc. 37: 731-736.

Magruder, R. and Wester, R. E. 1942. Prevention of field hybridization in the Lima bean. Amer. Soc. Hort. Sci. Proc. 60: 413-414.

Marechal, R., Mascherpa, J. M. and Stainier, F. 1978. Etude taxonomique d'un groupe d'especes des genres Phaseolus et Vigna (Papilonaceae) sur la base des donnees morphologiques et polliques, traitees pour l'analyse informatique. Boissiera 28: 1-273.

Margni M, Rossier D, Crettag P and Jolliet O. 2002. Life cycle impact assessment of pesticides on human health and ecosystem. Agriculture Ecosystem and Environment 93, 279–392.

Maria Packiam S and Ignacimuthu S. 2012. Effect of PONNEEM on *Spodoptera litura* (Fab.) and its compatibility with Trichogramma chilonis Ishii. Braz. Arch. Biol. Technol. 55(2): 291-298.

Martins D. J. 2008 Pigeon Pea (Cajanus cajan) in Mwanza, Western Tanzania. Initial Survey of Good Pollination Practices. Food and Agriculture Organization of the United Nations, Rome. Ultural Services Bulletin 118, 1–198.

Matta, D. N., and Dave, B. E. 1931. Studies in Cajanus indicus. India Dept. Agr., Mem. Bot. Ser. 19, 25 pp.

McGregor SE. 1976. Insect pollination of cultivated crop plants. Agricultural Research Service. United States Department of Agriculture, Agric. Handb, Washington, USA. 496:411.

Mckee, R. 1948. The legumes of many uses U.S. Dept. Agr. Yearbook 1918: 701-726.

Mckee, R. and Pieters, A. J. 1937. Miscellaneous forage and cover crop legumes. U.S. Dept. Agr. Yearbook 1937: 999-1031.

Mckee, R. and Schoth, H. A. 1925. Common vetch and its varieties. U.S. Dept. Agr. Bul. 1289, 20 pp.

Mckee, R. and Schoth, H. A. 1934. Vetch culture and uses. U.S. Dept. Agr. Farmers' Bul. 1740, 22 pp.

McPherson RM, Newsom LD. 1984. Trap crops for control of stink bugs in soybean. J. Georgia Entomol. Soc. 19: 470–480.

Md. Abdullah, Sarnthoy O, Isichaikul S, Tantakom S. 2001. Efficacy of Cypermethrin, neem extract and Bacillus thuringiensis for controlling insect pests of vegetable soybean. Kasetsart J. (Nat. Sci.) 35: 14-22.

Mensah Ben A and Kudom AA. 2011. Foraging dynamics and pollination efficiency of Apis mellifera and Xylocopa olivacea on Luffa aegyptiaca Mill. (Cucurbitaceae) in Southern Ghana. Journal of Pollination Ecology 4, 34-38.

Milum, V. G. 1940. Bees and soybeans. Amer. Bee Jour. 80: 22.

Mink GI and Keswani CL. 1987. First report of Cowpea mild mottle virus on bean and mung bean in Tanzania. Plant Dis. 71: 557.

Mlyniec, W. and Wojtowski, F. 1962. Use of bumble bees for research on flowering biology of hairy vetch. Ekol. Pol. Ser. B 8(1): 59-65. [In Polish, English summary.] AA-38/67.

Mlyniec, W.1962. The mechanism of pollination and generative reproduction in Vicia villosa Roth.] Genet. Polon. 3: 285- [In Polish.] Abstract in Plant Breeding Abs. 34: 669, p. 76, 1964.

Moffet JO, Stitth LS, Burkhart CC, Shipman CW. 1975. Honey bee visits to cotton flowers. Environmental Entomology 4, 203-206.

Mohan S. 2006. A device to remove insect eggs from stored pulse seeds. Patent Number 198434. http://www.allindianpatents.com/ patents/198434

Mommers, J. 1971. [The pollination of beans phaseolus under glass.] Maandschr. V. De bijent. 73(5): 79-80.

Morales F, Viana A, Palmieri M, Orozco M and Ruano RR. 2005. Whiteflies as vectors of viruses in legumes and vegetable mixed cropping systems in the tropical lowlands of Central America, Mexico and the Caribbean – Guatemala. pp. 188-196. In: Anderson PK, Morales FJ (Eds.), Whitefly and Whitefly-borne viruses in the tropics: Building a knowledge base for global action. Centro Internacional de Agricultura Tropical (CIAT), Cali, Colombia. 351 p.

More B.C., Latpate C.B. and Jadhav M.S 2015. Pollinators diversity with special reference to role of honeybees in seed production of pigeon pea, Cajanus cajan L.39, Issue : 2 119-122.

Morse, W. J., and Cartter, J. L. 1937. Improvement in soybeans. U.S. Dept. Agr. Yearbook 1937: 1154-1189.

Muniappan R, Shepard BM, Carner GR and Ooi PAC. 2012. Arthropod pests of horticultural crops in tropical Asia. Wallingford, Oxfordshire, CABI.

Murdock LL and Kitch LW. 1993. Proceedings and notes from the bean/cowpea CRSP. Cowpea Regional Economic Development Workshop for Subsaharan-Africa, Purdue University, West Lafayette, USA.113 p.

Nepi, M. and Pacini, E. 1993. Pollination, Pollen viability and pistil receptivity in Cucurbita pepo L. Ann. Bot. 72: 527-536.

Ng, N.Q. and Marechal, R. 1985. Cowpea taxonomy, origin and germplasm. In: Cowpea Research, Production and Utilization. Singh, S.R and K.O. Rachie (eds). John Wiley and Sons, Chichester, pp. 11-12.

Ngakou A, Megueni C, Makalao M, Nwaga D,Taine J and Ndjouenkeu R. 2008. Changes in the physic chemical properties of soil and harvested soybean seeds in response to soil solarization and bradyrhizobial inoculation. Archives of Agronomy and Soil Science 54(2), 189-202.

Ngakou A, Megueni C, Ousseini H, Massai A. 2009. Study on the isolation and characterization of rhizobia strains as biofertilizer tools or growth improvement of four grain legumes in Cameroon. International Journal of Biological and chemical Sciences 3(5), 1078-1089.

Ngakou A, Moctar M, Njintang YN and Tamo M. 2011. Some cowpea quality seed indicators as influenced by field application of selected biofertilizers and mycoinsecticide in three agroecological zones of Cameroon. Communication in Soil Science and Plant analysis 42(11), 1277-1289.

Ngakou A, Nwaga D, Ntonifor NN, Tamo M, Nebane CLN and Parh IA. 2007a. Contribution of Arbuscular Mycorrhizal Fungi (AMF), rhizobia and Metarhiziumanisopliae to cowpea production in Cameroon. International Journal of Agricultural Research 2(9), 754- 764.

Ngakou A. 2007. Potential of selection biofertilizers and a mycoinsecticide in managing Megalurothrips sjostidti and improving cowpea production in Cameroon. PhD Thesis, Departement of Biochemistry and Mirobiology, University of Buea, Cameroon. 197 p.

Nyirenda, G.K.C. 2000. Preliminary results of bean insect pests in Karonga Agricultural Development Division and some parts of Lilongwe and Shire Valley Agricultural Development Divisions in Malawi. Proceedings: Bean Seed Workshop

Onim, F.J.M., Eijnatten, C.L.M. and Pathak, R.S. 1979a. Some factors which encourage out-crossing in pigeonpea. Proc. Fourth Int. Symp. Polln., Maryland, pp. 219-225.

Paik CH, Lee GH, Oh YJ, Park CG, Hwang CY, Kim SS. 2009. Pheromone trap type and height for attracting of Riptortus clavatus (Thunberg) (Hemiptera: Alydidae) in soybean field. Korean J. Appl. Entomol. 48: 59-65.

Pajni, H. R. and A. Jabbal. 1986. Some observations of Zabrotes subfasciatus (Boh.) (Bruchidae: Coleoptera). Research Bulletin of the Panjab University Science 37: 11-16.

Palmer R (1967). Bees and bean pollination. S. Afr. Bee J., 39(4): 12-13.

Pandey GP, Varma BK. 1977. Attapulgite dust for the control of pulse beetle, Callosobruchus maculatus Fabricius on black gram (Phaseolus mungo). Bull. Grain Technol. 15: 188-193.

Pando J. B., Tchuenguem Fohouo F.-N. and Tamesse J 2013 Activite´ de butinage et de pollinisation de Xylocopa olivacea Fabricius 1787 (Hymenoptera: Apidae) sur les fleurs de Vigna unguiculata (L.) Walp. 1843 (Fabaceae) a` Yaounde´-Cameroun. Entomologie Faunistique 66, 47–59.

Pando J. B., Tchuenguem Fohouo F.-N. and Tamesse J. L. 2011a Pollination and yield responses of pigeon pea (Cajanus cajan L. Millsp.) to the foraging activity of

Chalicodoma cincta cincta (Hymenoptera: Megachilidae) in Yaounde′ (Cameroon). Journal of Animal and Plant Science 11, 1346–1357.

Pando JB, Tchuenguem Fohouo FN, Tamesse JL. 2011a. Foraging and pollination behaviour of Xylocopa calens Lepeletier (Hymenoptera: Apidae) on Phaseolus coccineus L. (Fabaceae) flowers at Yaoundé (Cameroon). Entomological Research 41, 185-193.

Pando JB, Tchuenguem Fohouo FN, Tamesse JL. 2011b. Pollination and yield responses of pigeon pea (Cajanus cajan L. Mill sp.) to the foraging activity of Chalicodoma cincta cincta (Hymenoptera: Megachilidae) in Yaoundé (Cameroon). Journal of Animal and Plant Science 11, 1346-1357.

Pando JB, Tchuenguem Fohouo FN, Tamesse JL. 2013. Activité de butinage et de pollinisation de Xylocopa olivacea Fabricius 1787 (Hymenoptera: Apidae) sur les fleurs de Vigna unguiculata (L.) Walp. 1843 (Fabaceae) à Yaoundé-Cameroun. Faunistic entomology 66, 47-59.

Pasquini M, Assogba-Komlan WF, Voster I. Shackleton CM, Abukutsa- Onyango MO. 2009. The production of African vegetables in urban and peri urban agriculture: a comparative analysis of case studies from Benin, Kenya and South Africa, Chapter 6. In: Shackleton CM, Pasquini MW, Drescher AW. African indigenous vegetables in urban agriculture. London: Earthscan.

Paul UV, Ampofo JKO, Hilbeck A, Edwards P. 2007. Evaluation of organic control methods of the bean beetle, Ootheca bennigseni, in east Africa. New Zealand Plant Protection 60:189-198.

Pedersen, M. W., Jones, L. G., and Rogers, T. H. 1961. Producing seed of legumes. U.S. Dept. Agr. Yearbook 1961: 171-181.

Pedune-Benin]. Production ecologiquement durable du niébé. 1999. Annual country report. Benin. International Institute of Tropical Agriculture, Cotonou, Benin.

Piper C.V. and Morse, W. J.1910. The soybean: history, varieties and field studies. U.S. Dept. Agr. Burl Plant Indus. Bul. 197, 84 pp.

Piper, C.V. and Morse, W.J. 1923. The soybean. 329 pp. McGraw-Hill Book Co., Inc., New York.

Prasad A, Wadhwani Y. 2006. Pathogenic virus and insect tissues: an effective way of pest control. Current Science 91(6): 803-807.

Pritsch, G. 1966. Investigations on the significance of honey bees (Apis mellifera L.) in assuring seed yield of winter vetches (Vicia villosa Roth).] Arch. f. Geflugelz. u. Kleintierk. 15(3): 205-211. [In German, English summary.

Raheja AK. 1981. Status of *Ootheca mutabilis* as a pest of cowpea in northern Nigeria. Samaru J. Agric. Res. 1: 111–118.

Rajkhowa Denisha and Deka M.K. 2013. Insect foragers and foraging behaviour of honey bee, apis cerana on pigeon pea. Indian Journal of Entomology, 75(3): 232-235 (2013)

Ranga Rao GV, Rameshwar Rao V, Ghaffar MA. 2013. Handbook on chickpea and pigeonpea insect pest identification and management. Information Bulletin No. 57, Patancheru, Andhra Pradesh 502 324, India: International Crops Research Institute for the Semi-Arid Tropics. 96 pp.

Rao CM, Naadre KR, Suryanarayana MC. 1996. Studies on the utility of honey bees on production of foundation seed of cotton cv NCMHH- 20. Indian Bee Journal 58, 13-15.

Rathi, A and Sihag, R.C., 1993 Foraging modes and foraging rates of different bee pollinators of pigeonpea (*Cajanus cajan* (L) Millsp). In: Pollination in Tropics, by G.K.Veeresh, R.Uma Shankar and K. Ganeshaiah(Eds.). Pub. IUSSI- Indian Chapter, Bangalore, pp: 93-95 (India)

Rathi, A and Sihag, R.C., 1993. Effect of the bee, Chalicodoma lanata F. on pollination and seed yield of pigeonpea (Cajanus cajan (L) Millsp). In: Pollination in Tropics by G.K.Veeresh, R.Uma Shankar and K Ganeshaiah (Eds.). Pub. IUSSI- Indian Chapter, Bangalore, pp: 175-177 (India)

Rathi, A. 1993. Field study of bee pollinator-pesticide interaction on pigeonpea (*Cajanus cajan* (L.) Millsp.). Ph.D. Thesis CCS Haryana Agricultural University, Hisar.

Rizvi, S.M.A, and A.K. Singh.1994.Succession of insect pests infesting Rajmash in eastern Uttar Pradesh. Indian J.Plant. Prot.22 (2):226-227.

Robbins, W. W. 1931. The botany of crop plants. ed. 3, 639 pp. P. Blakiston's son & co., inc., Philadelphia.

Robertson DG. 1963. Further studies on the host range of cowpea yellow mosaic virus. Trop. Agric. 40: 319–324.

Romeis J, Shanower TG, Madhuri K. 2000. Biology and field performance of Gryon clavigrallae (Hymenoptera: Scelionidae), an egg parasitoid of Clavigralla spp. (Hemiptera: Coreidae) in India. Bull Entomol Res. 90(3): 253-263.

Roubik D.W. (1995) Pollination of cultivated plants in the tropics. FAO Agric

Roubik DW. 2000. Pollination system stability in tropical America. Conservation Biology 14, 1235 1236.

Rubis, D. D. 1970. Breeding insect pollinated crops. In The lndispensable Pollinators. Ark. Agr. Ext. Serv. Misc. Pub. 127, pp.19-24.

Sabbahi R, De OD, Marceau J 2005. Influence of honey bee (Hymenoptera: Apidae) density on the production of canola (Crucifera: Brassicacae). J. Econ. Entomol., 98(2): 367-372.

Salam SA, Patil MS and Byadgi AS. 2009. IDM of Mungbean Yellow Mosaic Disease. Annals of Plant Protection Sciences 17(1): 157-160.

Sandsted, R. 1980. Energy inputs in snapbean production. p.127-128. In: Pimentel, D. (ed.), Handbook of energy utilization in agriculture. CRC Press, Inc., Boca Raton, FL.

Sansone, C. 2002. Blister Beetles, Texas Agricultural Extension fact sheet B-1401, Integrated Pest Management Guide for Texas Forage Crops. 41 p

Saranya S, Ushakumari R, Jacob S and Philip BM. 2010. Efficacy of different entomopathogenic fungi against cowpea aphid, Aphis craccivora (Koch). Journal of Biopesticides 3(1): 138-142.

Sariah JB and Makundi RH. 2007. Effect of sowing time on infestation of beans (Phaseolus vulgaris L.) by two species of the Bean Stem Maggot, Ophiomyia spencerella and Ophiomyia phaseoli (Diptera: Agromyzidae). Archives of Phytopathology and Plant Protection 40(1): 45-51.

Schelhorn, M. Von 1946. Biology of flowering and seed setting in Vicia villosa.] Zuchter 17/19: 22-24. [In German. ]

Schelhorn, M. VON. 1942. [Biological studies on the blossoms of hairy vetch.] Pflanzenbau 18(10): 311-320. [In German.]

Schoth, H. A. 1923. Hungarian vetch in Oregon. Oreg. Agr. Expt. Sta. Cir. 46, 4 pp.

Schoth, H. A. and Hyslop, G. R. 1925. Common vetch. Oreg. Agr. Expt. Sta. Bul. 213, 29 pp.

Selander, R.B. 1987.Behavioural observation in Cyaneolytta and a description of the triungulin larva of C. fryi (Col.: Melodae). J. Kansas Entomol. Soc. 60 (2): 288-304.

Sen, N. K., and Bhowal, J. G. A. 1962. A male-sterile mutant cowpea. Jour. Hered. 53(1): 44-46.

Shaik, P. H. Van, and Probst, A. H. 1958. Effects of some environmental factors on flower production and reproductive efficiency in soybeans. Agron. Jour. 50: 192-197.

Shanmugasundaram S, Keatinge JDH, Hughes, J. 2009. The Mungbean Transformation: Diversifying Crops, Defeating Malnutrition. IFPRI Discussion Paper 00922. http://www.ifpri.org/sites/default/files/ publications/ifpridp00922.pdf

Shanower TG, Lal SS, Bhagwat VR. 1998. Biology and management of Melanagromyza obtusa (Malloch) (Diptera:Agromyzidae). Crop Protection 17(3): 249-263.

Sharma HC. 1998. Bionomics, host plant resistance, and management of the legume pod borer, *Maruca vitrata* – a review. Crop Protection 17: 373–386.

Sharma OP, Bambawale OM, Gopali JB, Bhagat S, Yelshetty S, Singh SK, Anand R, Singh OP. 2011. Field Guide - Mungbean & Urdbean. Pp. 40. http://www.ncipm.org.in/A3P/UI/Home/publish/Field%20 Guide.pdf

Shaw, F. J. F. 1932. The breeding of cicer and cajanus. Imp. Burl Plant Genet., Herbage Plants Bul. 7: 14-16.

Sihag, R.C. and Rathi, A., 1994 Diversity , abundance, foraging behaviour and pollinating efficiency of different bees visiting pigeonpea (Cajanus cajan (L) Millsp.) blossoms. Indian Bee. J., 56, 187-201.

Sihag, R.C., 1993 Management of some sub-tropical megachilid bees for pollination of alfalfa (Medicago sativa L.) and pigeonpea (Cajanus cajan(L) Millsp). In :Pollination in Tropics, by G.K.Veeresh, R.Uma Shankar and K.Ganeshaiah (Eds.). Pub. IUSSI-Indian Chapter, Bangalore, pp: 229-232.

Sihag, R.C., 1994 Pollination of Asian crops. J. Entomol. Zool. 8: 158-60 (Japan)

Singh A, Baoule AL, Ahmed HG, Dikko AU, Aliyu U, Sokoto MB, Alhassan J, Musa M, Haliru B. 2011. Influence of phosphorus on the performance of cowpea (Vigna unguiculata (L) Walp.) varieties in the Sudan savanna of Nigeria. Agricultural Sciences 2(3), 313-317.

Singh BB, Chambliss OL and Sharma B. 1997. Recent advances in cowpea breeding. In: Singh BB, Mohan Raj DR, Dashiell KE, Jackaï LEN(Eds.), Advances in cowpea research. Copublication by International Institute of Tropical Agriculture (IITA) and Japan International Research Center for Agriculture and Science (JIRCAS), IITA, Ibadan, Nigeria, 30-49 p.

Singh BB, Hartmann P, Fatokun C, Tamo M, Tarawali S and Ortiz R. 2003. Recent progress on cowpea improvement. Chronica Horticulturae 43(2): 8–12.

Singh S. 1982. Ecology of the Agromyzidae (Diptera) associated with leguminous crops in India. Memoirs of the School of Entomol. No. 8, St. John's College, Agra, India. 126 pp.

Singh SR and Allen DJ. 1980. Pests, diseases and protection in cowpeas. In. Summerfield RJ, Bunting AH (eds.) Advances in legume science. Royal Botanic Gardens/Ministry of Agriculture, Fisheries & Food, Kew, London, pp. 419–443.

Singh SR, Jackai LEN, Dos Santos JHR and Adalla CB. 1990. Insect pests of cowpeas, pp. 43–90. In: Singh SR (ed.) Insect Pests of Tropical Legumes. John Wiley & Sons, Chichester, United Kingdom.

Singh SR. 1977. Cowpea cultivars resistant to insect pests in world germplasm collection. Tropical Grain Legume Bulletin 9: 1–7.

Sodavy P, Sitha M, Nugent R and Murphy H. 2000. Situation Analysis on Farmers' Awareness and Perceptions of the Effect of Pesticides on their Health. FAO COMMUNITY IPM PROGRAMME, Field document. p.16. http://www.communityipm.org/toxictrail/ Documents/ Cambodia-SituationAnalysis.pdf

Somta P, Ammaranan C, Ooi PAC and Srinives P. 2007. Inheritance of seed resistance to bruchids in cultivated mungbean (Vigna radiata, L. Wilczek). Euphytica 155:47–55.

Sosa-Gomez DR and Moscardi F. 1998. Laboratory and field studies on the infection of stink bugs, Nezara viridula, Piezodorus guildinii, and Euschistos heros (Hemiptera: Pentatomidae) with Metarhizium anisopliae and Beauveria bassiana in Brazil. J. Invertebr. Pathol. 71: 115–120.

Sreekanth M and Seshamahalakshmi M. 2012. Studies on relative toxicity of biopesticides to Helicoverpa armigera (Hubner) and Maruca vitrata (Geyer) on pigeonpea (Cajanus cajan L.). J Biopest. 5(2):191–195.

Srinivasan R, Yule S, Chang JC, Malini P, Lin MY, Hsu YC and Schafleitner R. 2013. Towards developing a sustainable management strategy for legume pod borer, Maruca vitrata on yard–long bean in Southeast Asia. In: Holmer R, Linwattana G, Nath P, Keatinge JDH, eds. 2013. Proceedings of the Regional Symposium on High Value Vegetables in Southeast Asia: Production, Supply and Demand (SEAVEG2012), 24-26 January 2012, Chiang Mai, Thailand. AVRDC – The World Vegetable Center, Publication No. 12-758. AVRDC – The World Vegetable Center, Taiwan. pp. 76-82.

Srinivasan R. 2008. Susceptibility of legume pod borer (LPB), Maruca vitrata to δ-endotoxins of Bacillus thuringiensis (Bt) in Taiwan. J Invertebr Pathol. 97: 79–81.

Srivastava, C and M. Agarwal. 2004. Relative toxicity of some insecticides tothe adults of rajma beetle Zabrotes subfasciatus (Boh.). Indian J. Entomol. 66 (1): 40-41.

Stoddard FL and Bond DA (1987). The pollination requirements of the faba bean. Bee World 68:144-152

Summers MC and Mondor EB. 2011. Rhizobium alter inducible defenses in broad bean, Vicia faba. Open Journal of Ecology 1(3), 52-62.

Sundramurthy, V.T. and K. Chitra. 1992. Integrated pest management in cotton. Indian J. Pl. Prot. 20: 1-17.

Sunitha V, Vijaya Lakshmi K, Ranga Rao GV. 2008. Laboratory evaluation of certain insecticides against pigeonpea pod borer, Maruca vitrata Geyer. J Food Legumes 21(2): 137–139.

Svetsugo, I., and Kobayashi, S. 1952. [Studies on Vicia villosa Roth, with special reference to the fertilization and seed production.] Crop Sci. Soc. Proc. 21(1/2): 150-151. [In Japanese, English summary.]

Swamiappan M, Jayaraj S, Chandy KC and Sundaramurthy VT. 1976. Effect of activated kaolinitic clay on some storage insects. Z. Angew. Entomol. 80: 385-389.

Talekar NS and Chen BS. 1985. The beanfly pest complex of tropical soybean. Pp.257-271. In: Soybean in tropical and sub-tropical cropping systems. Asian Vegetable Research and Development Center, Shanhua, Taiwan, Republic of China.

Talekar NS and Lee YH. 1988. Biology of Ophiomyia centrosematis (Diptera: Agromyzidae), a Pest of Soybean. Annals of the Entomological Society of America 6: 938-942.

Talekar NS and Lin CP. 1994. Characterization of resistance to Limabean pod borer (Lepidoptera: Pyralidae) in soybean. Journal of Economic Entomology 87(3): 821-825.

Talekar NS. 1990. Agronomyzid flies of food legumes in the tropics, Wiley Eastern Limited, Delhi, India. p.297.

Tamò M, Arodokoun DY, Zenz N, Tindo M, Agboton C and Adeoti R. 2002. The importance of alternative host plants for the biological control of two key cowpea insect pests, the pod borer Maruca vitrata (Fabricius) and the flower thrips Megalurothrips sjostedti (Trybom). In: Fatokun CA, Tarawali SA, Singh BB, Kormawa PM, Tamò M. (eds.) Proc. World Cowpea Conference III, Challenges and opportunities for enhancing sustainable cowpea production, September 4–7, 2000, International Institute of Tropical Agriculture, Ibadan, Nigeria, pp. 81–93.

Tamò M, Baumgärtner J, Delucchi V and Herren HR. 1993. Assessment of key factors responsible for the pest status of the bean flower thrips Megalurothrips sjostedti (Trybom) (Thysanoptera: Thripidae). Bulletin of Entomological Research 83: 251–258.

Tamò M, Srinivasan R, Dannon E, Agboton C, Datinon B, Dabiré C, Baoua I, Ba M, Haruna B and Pittendrigh B. 2012. Biological Control: a major component for the long-term cowpea pest management strategy. Pp. 249-259. In: Boukar O, Coulibaly O, Fatokun C, Lopez K, Tamò M. (eds.) Innovative research along the cowpea value chain. Proceedings of the Fifth World Cowpea Conference on Improving livelihoods in the cowpea value chain through advancement in science, 27 September – 1 October 2010, Saly, Senegal. 432 pp.

Taylor TA. 1967. The bionomics of Maruca testulalis Gey. (Lepidoptera: Pyralidae), a major pest of cowpeas in Nigeria. J. West Afr. Sci. Assoc. 12: 111–129.

Tchuenguem FFN, Djonwangwé D, Messi J and Brückner D 2009a. Foraging and pollination activities of Apis mellifera adansonii (Hymenoptera: Apidae) at Ngaoundéré (Cameroon). Cameroon J. Exp. Biol., 5(1): 1-9.

Tchuenguem FFN, Ngakou A and Kengni BS 2009b. Pollination and yield responses of cowpea (Vigna unguiculata L. Walp.) to the foraging activity of Apis mellifera adansonii (Hymenoptera: Apidae) at Ngaoundéré (Cameroon). Afr. J. Biotechnol., 9: 1988-1996.

Tchuenguem Fohouo Fernand-Nestor, Pando Joseph Blaise and Tamesse Joseph Lebel 2014. Pollination efficiency of Xylocopa olivacea (Hymenoptera: Apidae) on Cajanus cajan (Fabaceae) flowers at Yaounde´, Cameroon. International Journal of Tropical Insect Science 34(2): 138–148.

Tchuenguem Fohouo FN, Djonwangwe D, Messi J and Brueckner D. 2009a. Activité de butinage et de pollinisation de Apis mellifera adansonii Latreille (Hymenoptera: Apidae) sur les fleurs de Helianthus annus (Asteraceae) à Ngaoungéré (Cameroun). Cameroon Journal of Experimental Biology 5, 1-9.

Tchuenguem Fohouo FN, Fameni ST, Brueckner D. 2014. Foraging and pollination behaviour of Xylocopa olivacea (Hymenoptera: Apidae) on Phaseolus coccineus (Fabaceae) flowers at Ngaoundéré (Cameroon). International Journal of Tropical Insect Sciences 34(2), 127–137.

Tchuenguem Fohouo FN, Messi J, Brueckner D, Bouba B, Mbofung C and Hentchoya HJ. 2004. Foraging and pollination behaviour of the African honey bee (Apis mellifera adansonii) on Callistemon rigidus flowers at Ngaoundéré (Cameroon). Journal of the Cameroon Academy of Sciences 4, 133-140.

Tchuenguem Fohouo FN. 2005. Activité de butinage et de pollinisation d'Apis mellifera adansonii Latreille (Hymenoptera : Apidae, Apinae) sur les fleurs de trois plantes à Ngaoundéré (Cameroun): Callistemon rigidus (Myrtaceae), Sygygium guineense var. macrocarpum (Myrtaceae) et Voacanga africana (Apocynaceae). Thèse du Doctorat d'Etat, Université de Yaoundé I, Yaoundé, Cameroun. 103 p.

Tedoradze, S. G. 1959. [The role of bees in the shape-forming process of [scarlet runner] beans under georgian conditions.] Pchelovodstvo 36(9): 40-42. [in russian.]

Thiagarajan TR and Ahamad MH. 1993. Influence of a vesicular-arbuscular mycorrhizal fungus on the competitiveness ability of Bradyrhizobia spp. For nodulation of cowpea (Vigna unguiculata L. Walp.) in non-sterilized soil. Biology and Fertility of Soils 15, 294-296.

Thomas, F. L.1950. Vetch seed growers ask for honey bees. Amer. Bee Jour. 90: 315.

Thomas. F. L., and Warne, B. 1951. Bees for vetch pollination in texas. In lowa State Apiarist Rpt. 1950, pp. 31-33.

Todd JW. 1989. Ecology and behaviour of *Nezara viridula*. Annu. Rev. Entomol. 34, 273–292.

Torres-Villa LM, Rodrigues M and Lacasa A. 1996. An unusual behaviour in Helicoverpa armigera Hubner (Lepidoptera: Noctuidae): pupation inside tomato fruits. Journal of Insect Behaviour 9: 981-984.

Ulrichs C, Mewis I, Schnitzler WH and Burleigh JR. 2001. Effectivity of synthetic insecticides against Maruca vitrata F. and the parasitoid Bassus asper CHOU & SHARKEY in the Philippines. Mitteilungen der Deutschen Gesellschaft fur allgemeine und angewandte Entomologie 13(1-6): 279–282.

Ulrichs Ch and Mewis I. 2004. Evaluation of the efficacy of Trichogramma evanescens Westwood (Hym., Trichogrammatidae) inundative releases for the control of Maruca vitrata F. (Lep., Pyralidae). Journal of Applied Entomology 128(6): 426–431.

United States Department of Agriculture 1946. The dependence of agriculture on the beekeeping industry. A review. U.S. Dept. Agr. Burl Ent. and Plant Quar. E-584, revised, 39 pp.

Vansell, G. H., and Reinhardt, J. F. 1948. Do honey bees help pollinate baby lima beans? Gleanings Bee Cult. 76: 678-679.

Vaz GV, De Oliveira D and Ohashi OS. 1998. Pollinator contribution to the production of cowpea on the Amazon. Hortscience 33, 1157-1159.

Veatch, C. 1930. Vigor in soybeans as affected by hybridity. Jour. Amer. Soc. Agron. 22: 289-310.

Venette RC, Davis EE, Zaspel J, Heisler H, Larson M. 2003. Mini Risk Assessment: Old World bollworm, Helicoverpa armigera Hübner [Lepidoptera: Noctuidae]. University of Minnesota, St. Paul, MN 55108, USA. http://www.aphis.usda.gov/plant_health/plant_pest_ info/pest_detection/downloads/pra/harmigerapra.pdf

Wada T, Endo N and Takahashi M. 2006. Reducing seed damage by soybean bugs by growing small-seeded soybeans and delaying sowing time. Crop Protection 25(8): 726–731.

Warnock, S. J., and Hagedorn, D. J. 1954. Stigma receptivity in peas (pisum sativum b.). Agron. Jour. 46: 274.

Weaver, N. 1954. Pollination of hairy vetch by honeybees. Tex. Agr. Expt. Sta. Prog. Rpt. 1649, 5 pp.

Weaver, N. 1956a. The pollination of hairy vetch by honey bees. Jour. Econ. Ent. 49: 666-671.

Weaver, N. 1956b. The foraging behavior of honeybees on hairy vetch. i. foraging methods and learning to forage. Insectes Soc. 3: 538-549.

Weaver, N. 1957. The foraging behavior of honeybees on hairy vetch. ii. the foraging area and foraging speed. Insectes Soc. 4: 43-57.

Weaver, N. 1965. The foraging behavior of honeybees on hairy vetch. iii. differences in the vetch. Insectes Soc. 12: 321-325.

Weber, C. R., Emptg, L. T., and Thorne, J. C. 1970. Heterotic performance and combining ability of two- way f1 soybean hybrids. Crop Sci. 10: 159-160.

Webster BD, Ross RM and Evans T 1982. Nectar and nectary of Phaseolus vulgaris. L. J. Am. Soc. Hort. Sci., 107: 497-503.

Weinberger K, Pichop GN. 2009. Marketing of African indigenous vegetables along urban and peri-urban supply chains in sub-Saharan Africa. Chapter 7. In: Shackleton CM, Pasquini MW, Drescher AW. (eds.) African indigenous vegetables in urban agriculture. London: Earthscan.

Weiss M. G. 1949. Soybeans. Adv. Agron. 1: 77-157.

Weiss M. G. Weber, C. R., and Kalton, R. R. 1947. Early generation testing in soybeans. Jour. Amer. Soc. Agron. 39: 791-811.

Welch, J, E., and Grimball, E. L. 1951. Natural crossing in lima beans in s. carolina. Amer. Soc. Hort. Sci. Proc. 58: 254 - 256.

Wells WC, Isom WH, Waines JG 1988. Outcrossing rates of six common bean lines. Crop Sci., 28: 177-178

Wentz, J. B., and Stewart, R. T. 1924. Hybrid vigor in soybeans. Jour. Amer. Soc. Agron. 16: 534-540.

Wester, R. E and Jorgensen, H. 1951. Hybrid vigor in lima beans. Amer. Soc. Hort. Sci. Proc. 57: 305-309.

Wester, R. E., and Jorgensen, H. 1950. Emasculation unnecessary in hybridizing Lima beans. Amer. Soc. Hort. Sci. Proc. 55: 384 - 390.

Westphall, E. (1974). Pulses in Ethiopia: their taxonomy and agricultural significance. Field Crop Abstracts 24: 213-232.

Williams, I. H. 1977. Behavior of insects foraging on pigeonpea (Cajanus cajan L.). Millsp. in India. Trop. Agric.54:353-363.

Williams, I.H. 1983. The pollination of pigeonpea (Cajanus cajan (L.) Millsp.) in India. Second Int. Conf. Apic Torp., New Delhi. pp. 661-666.

Wilsie, C. P., and Takahashi M. 1934. Natural crossing in the pigeonpea. Jour. Agr. Res. 49: 923-927.

Wojtowski, F., and Mlyniec, W. 1964. [Behaviour of bees during pollination of hairy vetch.] Roczn. Wyz. Szk. Roln. Poznan. 19: 197-207. [In Polish, English and Russian summaries.] AA-245167

Woodhouse, E. J., and Taylor, C. S. 1913. The varieties of soybeans found in bengal, bikar and orissa and their commercial possibilities. India Dept. Agr. Mem. Bot. Ser. 5: 103-175.

Woodworth, C. M. 1922. The extent of natural cross-pollination in soybeans. Jour. Amer. Soc. Agron. 14: 278-283.

Woodworth, C. M. 1932. Genetics and breeding in the improvement of the soybean. Ill. Agr. Expt. Sta. Bul. 384: 297-404.

Yogesh Kumar and Salunke Satish 2009. Insect visitors of pigeon pea (Cajanus cajan (L.) Millsp.) blossom - foraging behaviour and abundance. Pest Management and Economic Zoology 17 No. 1 pp. 9-15

Youn HS and Jung C. 2008. Effect of trap cropping on the egg parasitism of Riptortus clavatus (Thunberg) in soybean fields. Journal of Asia- Pacific Entomology 11: 73–76.

Yule S and Srinivasan R. 2012. Phanerotoma philippinensis, a potential parasitoid for managing legume pod borer, Maruca vitrata. In: Abstracts of XXIV International Congress of Entomology (CD-ROM), 19-25 August 2012, Daegu, South Korea. O405TU05.

Yule S and Srinivasan R. 2013. Evaluation of biopesticides against legume pod borer, Maruca vitrata Fabricius (Lepidoptera: Pyralidae), in laboratory and field conditions in Thailand. J Asia–Pacific Entomol. 16(4):357–360.

***Pigeon pea field***

***Megachile lanata***

***Xylocopa fenestrata***

***Xylocopa latipes***

***Megachile hera***

***Apis mellifera***

**Figure 9.4. Insect pollinators of pigeon pea**

CHAPTER 10

# Alliaceae

## Allium cepa L., family Amaryllidaceae

Onions are grown in almost every country in the world. They are used in salads, as a raw or cooked vegetable, and as a condiment. In seed crop of onion (*Allium cepa* L.), cross pollination is common and self pollination can not occur because of protandry and sticky pollen. The onion (*Allium cepa* L.), a member of the Alliaceae family, is a biennial plant, requiring two growing seasons to complete the cycle from seed to seed. Bulbs are ready for harvesting after the first growing season, after which the plant will initiate the flowering process. Bulb onions are commercially grown as annual crops because the whole plant is harvested after the first growing season. The biennial cycle of onion seed production consists of two phases, the seed-bulb phase in the first year and the bulb-seed phase in the second year.

Onion seed is mainly produced in regions with reliable warm, dry summers with low atmospheric humidity to reduce the risk of diseases (Brewster, 2008). Warm, clear weather also promotes optimal insect pollination activity on blooming crops (Brewster, 2008).

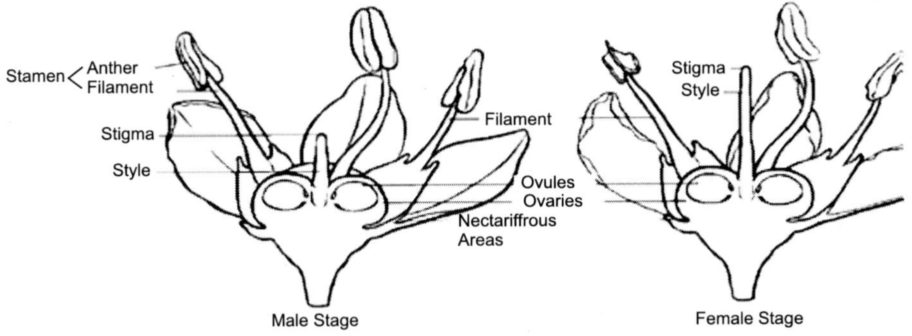


**Figure 10.1. Male and female flower of *Allium cepa***

Onion is a highly cross-pollinated crop and cross-pollination between cultivars is common (Free, 1993; Shanmugasundaram, 1998). Therefore, isolation distances between different onion cultivars is needed to prevent genetic contamination by foreign pollen. Recommendations on isolation distances ranges between 0.8 km and 5 km (Free, 1993). But Jones and Mann (1963) suggested that complete isolation would be virtually impossible because of the large foraging distances covered by some pollinator insects. They suggest that different localities should be dedicated for certain cultivars only and should be separated by at least 4.8 km.

The onion plant (*Allium cepa* L.) produces flower heads on single elongated inflorescence stalks (scapes) which are 1-2 meters long. The flower colour is white, green or striped, depending on the cultivar. Onion inflorescences or umbels carry between 200 and 600 small flowers (3-4 mm in length), depending on the cultivar and growing conditions. The flowers are made up of parts of three of each floral organ in five separate whorls, the outer perianth, inner perianth, outer stamens, inner stamens and the carpels (ovaries) (Figure 10.1). The superior carpels are fused with two ovules per carpel (Free, 1993b). The style elongates from 1 mm to 5 mm only after the stamens dehisced (Free, 1993b).

Only a few flowers open on the umbel at the onset of flowering which steadily increases until 50 or more florets open in one day at full bloom. Umbels are usually in bloom for a period of 30 days as the small flowers continue to open over a 2-3 week period. Individual Allium flowers are protandrous – the anthers dehisce before the stigma becomes receptive (Muller, 1883). When the anthers dehisce, all the pollen is shed within 24 to 36 hours, usually between 9 a.m. and 5 p.m. of the first day (McGregor, 1976). It has been found, under controlled conditions, that onion pollen remains viable for a period of 6 days (Mann and Woodbury, 1969). Temporal variation in pollen viability has also been recorded with pollen 2 to 3 times more viable when taken from flowers in the morning rather than in the evening (Nye et al., 1971). However, Mann and Woodbury (1969) found no difference in viability of pollen grains sampled in the morning vs. in the afternoon. Onion seeds are very small, black when ripe and born in silvery capsules (George, 1999). The size of the seeds vary between species and therefore also the number of seeds per gram of onion seed. Approximately 300 seeds per gram are generally recorded for onions (Brewster, 2008). The thousand-grain weight for onion seed is 3.6 g (George, 1999).

## Flowering

The onion (*Allium cepa*) inflorescence is an oval umbel 6–8 in (15.2–20.3 cm) across at the top and made up of 50–2000 florets. Each floret is about 1/8 in (3–4 mm) long and has six stamens divided into two whorls of three each, a style with a three-lobed stigma, and a three-celled ovary with two ovules in each cell. The anthers of the inner whorl open first and release pollen at irregular intervals, followed by those of the outer whorl. The period of pollen release lasts for 24–36 h and is completed before the stigma is receptive (Jones and Rosa 1928, Rodrigo et al. 1936).

Most of the pollen on the first day the flower opening is released between 9 a.m. and 5 p.m. There are nectaries at the bases of the inner whorl of stamens and the nectar accumulates between the ovary and the inner stamens (Knuth 1909, Roberts and Struckmeyer 1951). The style continues to grow while pollen is being shed. After all the pollen is released, the style reaches its maximum length and the stigma becomes receptive. Thus, the behaviour of individual florets discourages selfing, but because an umbel contains many florets at different stages of development it is possible for plant selfing to occur. The stigma remains receptive for up to 6 days, but the ability of pollen to germinate on its surface declines after the first day. Pollen is more viable in the morning than in the afternoon. Only a few florets on an umbel will open at the start of flowering. However, the number increases gradually until full bloom when there may be 50 or more open florets. A field of onion can be in various stages of bloom for over 30 days.

When flowering begins, only a few flowers open each day on an umber, but the number increases until at full bloom 50 or more florets may be open on a single day. They continue to open over a 2-week period, and 30 days or more may be involved in the flowering on all of the flower stalks. Moll (1954) showed that a flower may be pollinated as much as 6 days after it opens and found no significant difference in the percentage of set flowers pollinated 1 or 3 days after opening. Mann and Woodbury (1969) stated however, that pollen germination declined rapidly after the first day to zero percent by the sixth day. They concluded that the decline would be much more rapid under field conditions, making the age of the pollen an important factor in pollination. Nye et al. (1971) were in agreement in that they found that pollen taken from flowers opening in the morning was two or three times as viable as that taken from the flowers in the afternoon. The flowers are attractive to many species of bees and other hymenoptera, flies and other diptera, and numerous other orders of insects that feed upon the nectar, pollen, or both (Lederhouse et al. 1968, Bohart et al. 1970, Jones and Emsweller 1934, Shaw and Bourne 1936). The nectaries are shallow, and, unless the nectar is rapidly removed by insects, it can be easily seen glistening in the sunlight like a tiny jewel.

## Pollination Requirements

Pollination in the onion flower occurs when pollen is transferred from the dehiscing anthers of one floret to a receptive stigma of another floret. Effective transfer of pollen between florets on an umber or on an individual plant can transpire through the action of an outside agent, but self-pollination within the floret is impossible. Cross-pollination between plants is common and even obligatory in the fertilization of male-sterile onions used in hybrid seed production. The discovery of male sterility in onions (Jones and Emsweller 1936) made the production of hybrid onions possible under commercial conditions, and most of the onion seed produced now is hybrid seed. The procedure for utilization of male-sterility in the onion, which should be applicable to any crop plant in which male sterility is inherited in a similar way, was shown in detail by Jones and Clarke (1943).

In the production of hybrid seed, the grower plants a male-fertile row of a desired line to supply pollen to three to 10 rows of the male- sterile line (Franklin 1958), from which the hybrid seed will be obtained. Naturally, the greatest volume of hybrid seed possible is desired; therefore, the male-fertile or "bull" rows are kept at a minimum provided pollen is distributed sufficiently to set seed. Erickson and Gabelman (1956) showed that pollen dispersal from a point was logarithmic, with pollination at 7 feet from a source being only one-half that occurring at 1 foot. In order to secure maximum seed set, the grower encourages pollen dispersal to the maximum degree possible (Jones and Mann 1964).

## Onion Pollinators

Onion *(Alluim cepa* L.) flowers depends heavily on cross-pollinating insects quality and quantity of seed production (Singh and Dharmwal, 1970,Sihag,1986). The crop is protandrous with anthers shedding pollen before the stigma of the flower becomes receptive. Evidently, self pollination is not possible which necessitates the services of cross-pollinating insects for pollen transfer (Free, 1993). A large number of insect species have been reported visiting onion flowers (Jaboñski et al. 1982). Among all these insects, honeybees have, however, been reported as the most efficient and major pollinators of this crop because of their specific instinctive behavioural features for pollen and nectar collection and efficiency in pollen transfer (Free, 1993). Insect pollination is a complex phenomenon which depends upon several factors. Besides, physical features of flowers such as colour, shape, odour, potassium content in flowers of onion nectar (Waller, et al. 1972., Haggler, 1990), the diversity of environmental factors such as temperature, humidity, light, solar radiation, time of the day and nectar flow decisively shapes the behaviour of pollinating insects thereby influencing the cross-pollination and the production of the crop (Vischer and Seelay 1982, Corbet et al., 1993). Insect pollination has been reported to improve the quality and quantity of the onion seed production. Several investigators (Woyke, 1981., Kumar et al.1985., Rao and Suryanarayana, 1989) have reported that open pollinated plots gave significantly more and heavier seeds as compared to those deprived of insect visits.

Erickson and Gabelman (1956) concluded that in case of onion flowers, wind is not an important factor in pollination. Insects are primary vector for onion pollination.

When onion breeders want to get seed from a specific plant, they enclose the flowering umbels within a cage and introduce flies to transfer the pollen, or, if cross-pollination is desired, the umbels of the two lines are enclosed (Jones and Emsweller 1933). In large cage breeding work or pollination studies, honey bees are the primary agents used (Bohart et al. 1970, Moffett 1965, Shirck et al. 1945, Walsh 1965).

At least 276 species of insects are known to visit onion flowers; of these, honey bees, syrphid flies, halictid bees, and drone flies are the most important pollinators (Bohart *et al.*, 1970). When onion breeders want seed from a specific plant, they sometimes place a bag over the umbels and introduce flies to pollinate the florets. Mayer *et al.* (1993) tried to increase fly numbers by placing buckets of rotting meat around seed onion fields, but they had little success. Alfalfa leafcutting bees have not performed satisfactorily as onion pollinators (Mayer *et al.*, 1993). Only honey bees can be provided in the large and manageable quantities needed for commercial-scale seed production. Wind does not significantly aid pollination (Erickson and Gabelman, 1956).

Honey bees are usually adequate pollinators of open-pollinated onions because all of the plants produce nectar and pollen. However, in hybrid seed production using male-sterile plants (that do not shed pollen), only nectar-collecting bees will move freely from female to male plants, transferring pollen as they go. Pollen-collecting bees concentrate on the male rows and do not visit and cross-pollinate the male-sterile rows. Sometimes honeybees don't visit either open pollinated or hybrid flowers.

Onion seed yields, particularly hybrid seed yields, depend heavily on pollinator activity for good seed set. However, onion is not especially attractive to bees, and bees readily ignore it to concentrate on more attractive plants nearby (Gary *et al.*, 1972). Pollinator numbers are frequently low, even in fields well supplied with bee colonies, and pollination is often a limiting factor in onion seed production (Waller, 1983). The unpalatableness of onion nectar may be traced to its high levels of potassium or sugars (Mayer *et al.*, 1993). Nitrogen soil fertility does not seem to affect the quality of onion nectar. Some onion varieties are inherently more attractive to honey bees than others, and generally those most attractive produce the most seed (Hagler and Waller, 1991). This suggests that onion breeders may do well to select for stocks that are more attractive to bees.

Mayer *et al.* (1993) could not identify any cultural or bee management practices for increasing the attractiveness of onion flowers to honey bees. Thus, growers' only alternative is to use a heavy population of bees. However, simply placing hives in the field does not guarantee that bees will work the crop. Growers generally place 5–15 colonies of bees acre$^{-1}$ (12.4–37 per ha.) in or next to seed fields at the time of flowering. One recommendation calls for about two colonies acre$^{-1}$ (5 ha$^{-1}$) placed after flowering has begun, then adding another two acre$^{-1}$ at 3–4 day intervals to take advantage of the behaviour of inexperienced bees; however, when Mayer *et al.* (1993) tried this method they found no increase in the number of honey bees visiting onion.

In commercial production of seed, the provision of an adequate number of flies is impractical so the industry depends upon the honey bee as the primary pollinating agent. Bohart et al. (1970) reported more than 267 species of insect visitors on onion flowers, the most important of which were honey bees, small syrphid flies, halictid bees, and drone flies. Of these, only the honey bee can be manipulated and used in large-scale onion seed production. Kordakova (1956) and Sakharov (1956) gave major credit to the honey bee as a pollinator of onions in Russia.

Honey bees are effective pollinators of open-pollinated onions because both pollen and nectar are available on all umbels. In hybrid seed production where male-sterile plants are used, only the nectar collectors move freely from pollen-sterile to pollen-fertile plants, making the necessary transfer of pollen from male parent to female parent. Honey bees then become less than ideal pollinators of male-sterile onions. Pollen-collecting bees confine much of their activity to the pollen-producing rows without adequately visiting and cross-pollinating the male-sterile rows (Lederhouse et al. 1972). A strictly nectar collecting type of honey bee would be ideal because it would cross-visit and effectively pollinate the male-sterile flowers. In the absence of this perfect type of bee, the grower can only try to compensate by having more honey bees present in the field. Shasha's 30 conclusion, that too many bees may be detrimental, needs further study.

The lack of intense attractiveness of onions to bees, may cause the bees to neglect the crop, particularly if another highly attractive crop is in flower. The grower's only alternative is to make his crop as attractive as possible with best cultural practices and to use a heavy population of bees. Even then, the seed yielding potential of the crop may never be attained (Franklin 1970).

More research is needed on the factors that affect attractiveness of onions to honey bees (Sanduleac 1969, Singh and Dharamwal 1970). Franklin (1970) noted that mere placement of colonies of honey bees in the onion field does not guarantee that the bees will work the onion. Although Nye et al. (1971) reported an average of 100 bees per 100 feet of male fertile rows and a maximum of 40 per 100 feet on the male sterile rows, the number of honey bee visitors needed per onion plant, umber, or linear feet of row has not been determined.

Beekeepers occasionally obtain crops of onion honey with a characteristic onion flavour that disappears after a few weeks. Ewert (1942) reported that super phosphate and potassium fertilizers caused the nectar of onions to be richer in sugar, but the volume was not increased. The effect on the insect visitors was not reported. Waller (1970) and Waller et al. (1972) believed that a high level of potassium in the nectar might be an important clue to the reluctance of bees to visit onion flowers. Jula et al. (1965) calculated that onions produced 71 percent as much nectar per day as the highly attractive sainfoin. no longer releasing pollen, but style elongated and stigma receptive.

## Pollination Recommendations and Practices

As early as 1936, Shaw and Bourne (1936) indicated that growers of onion seed might find it useful to provide themselves with a supply of bees. They did not go into detail as to number of colonies, strength, or location. In a brief note without details, Hamilton (1946) stated that a grower produced much more onion seed than he had in the past after he rented eight colonies of bees. Sanduleac (1961 ) stated that bees increase production of onion and leek seed in Romania eight to 10 times, and he recommended about two colonies per acre.

The growers have learned through experience that the use of honey bees is essential and have generally adopted the practice of renting five to 15 colonies of bees per acre placed in or adjacent to their seed fields at flowering time. One suggestion has been to have about two colonies per acre delivered when flowering is well started. McGregor (1976) reported 2-15 colonies of honeybees per acre. Levin (1986) on the other hand recommended 3-4 colonies per acre. Mayer et al (1993) reported that 4-12 colonies per acre were required for optimum production of onion

Much information is needed on the factors that influence the activity of bees on onion flowers because, as Franklin (1970) pointed out, the mere placement of colonies in the field does not guarantee that the bees will work the onions. Continuous nectar foraging activity is the essential factor in hybrid onion fields especially during the peak period of flowering.

Onion (*Allium cepa* L.) seed yields, particularly hybrid seed, are heavily dependent on bee pollinator activity, especially honey bees (*Apis mellifera* L.), for pollination, seed set and adequate yields (Bohart et al. 1970; Caron et al. 1975; Franklin (1970), Erickson and Gabelman (1956), Lederhouse et al. (1968) and McGregor (1976). Onion seed yields in Washington range from 13.4 kg/ha to 2,241 kg/ha depending largely on the number of pollinating honey bees working in that field.

Mere placement of colonies of honey bees (12 to 30/ha) in or adjacent to onion seed fields does not guarantee they will work the field (Delaplane and Mayer 2000). In many cases high stocking rates have no effect on the number of foragers in the field. Lederhouse et al. (1968) and Waters (1972) suggested the high sugar concentration of onion nectar and Waller et al. (1972) suggested the high potassium ion concentration in the nectar makes onions unattractive to honey bees. Benedek and Gaal (1972) and Williams and Free (1974) noted honey bees foraging onion are primarily nectar collectors and onion is not favored as a source of pollen. Bohart et al. (1970) concluded that onion attractiveness was based on the nectar available and was, in general, independent of the presence or absence of pollen

Onion flowers produce nectar which accumulates in three cups between the lower ovary walls and the inner whorl of stamens (Brewster, 2008). Onion nectar is usually highly concentrated with sugar concentrations exceeding 40% (Free, 1993b; Hagler et al., 1990). Sugar concentrations were also found to differ considerably between hybrid parental lines (Silva and Dean, 2000). The sugar composition of onion nectar predominantly consist of relatively constant percentages of fructose and glucose between parental lines with almost undetectable levels of sucrose (Silva and Dean, 2000).

Onion flowers are often unattractive to honeybees (Gary et al., 1977, 1972), possibly due to the high potassium ion content (Waller et al., 1972; Hagler, 1990) or the high sugar concentration in the nectar (Lederhouse et al., 1972; Waters, 1972). Relatively large numbers of hives are required per hectare of onion seed crop and between 5 (open pollinated varieties) and 12 (hybrid varieties) hives are recommended to pollinate one hectare of onion crop (McGregor, 1976; Johannsmeier, 2001). This is in contrast to other vegetable seed crops dependent on managed honeybee hives, which generally only utilise an average of 2.4 hives per hectare in South Africa (carrot: 2-8 hives/ha; pumpkin: 0.2-4 hives/ha; pepper and tomato: 1-2 hives/ha; brassicas: 2-5 hives/ha) (Johannsmeier, 2001). However, the attractiveness of flowering hybrid onion crops to honeybees has not been found to be a problem in some parts of the world. Honeybee visitation has been found to be sufficient in Poland (Kordakov, 1956; Woyke, 1981). This is in contrast with observations made in North America (Mayer and Lunden, 2001). Despite the general avoidance that the honeybee has for onion flowers, it has been found to be an abundant and important forager on onion crops. (Witter & Blochtein, 2003; Howlett et al., 2005).

In addition, the nectaries of *Allium* flowers are shallow and easily accessible to many types of pollinating insects, giving rise to a generalized entomophilous pollination syndrome (Brewster, 2008). A diversity of insect visitors to onion seed crops have been identified by several authors in various parts of the world (Caron et al., 1975; Howlett et al., 2005; Lederhouse et al., 1968; Saeed et al., 2008; Sajjad et al., 2008; Williams and Free, 1974; Witter & Blochtein, 2003). Diptera and Hymenoptera species are often the most frequent visitors recorded on onion umbels (Free, 1993b; Williams & Free, 1974; Witter and Blochtein, 2003; Howlett et al., 2005; Saeed et al., 2008; Sajjad et al., 2008). Bees have been found to be the most abundant native pollinators within onion seed crops in some studies (Parker, 1982; Witter and Blochtein, 2003; Howlett et al., 2005), but flies were the more diverse group in other cases (Williams and Free, 1974; Saeed et al., 2008; Sajjad et al., 2008). In addition, several authors have investigated the efficiency of pollinator taxa other than honeybees in pollinating onion flowers (Moffett, 1965; Walsh, 1965; Williams and Free, 1974; Parker, 1982; Currah and Ockendon, 1983, 1984; Schittenhelm et al., 1997; Witter and Blochtein, 2003; Saeed et al., 2008) and the role pollinator diversity may play in providing pollination as an ecosystem service to onion seed crops. Parker (1982) found that Halictus farinosus Smith (Hymenoptera: Halictidae), a native bee in North America, was the most abundant and also the most efficient pollinator in an experimental onion crop with male-fertile and male-sterile cultivars and supplemented with one managed hive. Flowers visited by honeybees aborted more seed than flowers visited by Halictus bees, and seed abortion rates were higher in selfpollinated flowers (Parker, 1982).

Saeed et al. (2008) assessed the pollination potential of true flies and native bees in Pakistan on hybrid seed onion crops and found that bees showed greater pollination effectiveness than flies. However, they did not report the presence of any managed honeybee hives near the crop. In turn, Currah and Ockendon (1983, 1984) found no difference in pollination efficiency between honeybees and three species of blowflies in controlled experiments using large cages in terms of seed set and crossing level between two onion cultivars. In contrast, Walsh (1965) and Moffett (1965) found that houseflies were less efficient than honeybees in pollinating onion flowers in cages.

The bodies of honeybees have been found to carry more onion pollen than other insect visitors foraging on onion flowers (Kendall and Solomon, 1970; Free and Williams, 1972).

However, Parker and Hatley (1979) assessed the viability of pollen grains on the bodies of pollinator insects and found that pollen on the bodies of honeybees were the least viable in onion pollination. Parker (1982) concluded that native bees could be and were often more efficient pollinators of onions than honey bees.

Many authors found more honeybees on male-fertile rows than on male-sterile rows in onion seed crop fields (Williams & Free, 1974; McGregor, 1976; Woyke, 1981; Parker, 1982; Mayer and Lunden, 2001). Only nectar collecting bees move freely between the two lines while pollen collecting bees tend to concentrate their activity on the male-fertile lines (McGregor, 1976). However, Williams and Free (1974) found that onion flowers were not favoured by honeybees as a source of pollen but that most foragers were nectar-gatherers. This does not explain the large abundance of foraging bees on male-fertile rows. Rather, it is more likely that the greater attractiveness of male-fertile onion flowers could be ascribed to the greater sugar concentration of the nectar (Lederhouse et al., 1972) or their greater nectar production. The only solution to the unattractiveness of onion flowers to honeybees thus far has been to stock onion hybrid seed crops excessively with honeybees. Strategies to make onion hybrid seed crops more attractive to honeybees have been investigated by various authors, which include testing aromatic attractants like Citral, Geraniol and anise, preconditioning honeybees with onion flower scent compounds and testing honeybee response to simulated onion nectars containing various sugar and potassium concentrations (Woyke, 1981; Hagler, 1990; Silva et al., 2003).

Behavioural (competitive) interactions between honeybees and other pollinators in crop fields might be an important consideration when assessing the efficiency of wild and managed pollinators. Greenleaf and Kremen (2006) found in their investigation of the importance of behavioural interactions between honeybees and wild bees in field crops of sunflower hybrid seed that these interactions increased pollination up to 5-fold in some plots. These interactions caused honeybees to move more readily from male-fertile to male-sterile lines, thus increasing visitation and pollen transfer rates. They highlighted the importance of conservation of natural habitat at a landscape scale and the implementation of pollinator friendly crop management in order to increase the production of hybrid sunflower by increasing the frequency of these types of interactions.

Skrebtsov and Skrebtsova(1980) found that 21 species of insects visited onion flowers in the Volgo-Aktubinsk region of the USSR. They included both pollinators and natural enemies of insects. *Apis mellifera* L., was the important pollinator. Of the other insects, the most active were *Scolia flavifrons* F. (both a pollinator and a predator), syrphids (*Sphaerophoria menthastri* (L.), *Syrphus ribesii* (L.) and *Eristalinus aeneus* (Scop.) (*Eristalis aeneus*), which acted as pollinators in the adult stage and fed on aphids as larvae) and sometimes *Formica rufa* L., which acted mainly as a predator but may have served as a cross-pollinator at times. It was evident that onion seed-crops planted in the vicinity of other crops visited by insects attracted a variety of useful insect species. They recommended that care should be exercised in applying insecticides to them.

Kalmath and Sattigi (2004) found that among the six bee attractants,10% cacambe (a by product of sugarcane processing) attracted the most *Apis cerana* for 15 days after the first spray. Bee-Q was less effective, and sugar solution, sugarcane juice and molasses attracted lesser numbers of bees. Molasses was the least effective as it attracted no more bees to the blooming crop than water spray alone.

Caron et al (1975) in a study onduted at 3 areas of New York State, reported honeybees and drone flies (*Eristalis tenax*), were the most numerours pollinators on onion flowers. The most important pollinators were: *Apis mellifera*, *Eristalis tenax*, *Halictus ligatus*, *E. arbustorum*, *H. rubicundus*, *H. confusus*, *Dialictus imitatus*, *Bombus terricola*, *Hoplitis producta*, *D. nymphaearum*, *Syritta pipiens*, *B. impatiens*.

Sharma et al (2001) studied the pollination efficiency of *Apis dorsata* and *A. florea* as pollinators of onion and found that the number of loose pollen grains sticking to the bodies of individual foraging bees was 10 187 and 5625 for *A. dorsata* and *A. florea*, respectively. The results suggest that *A. dorsata* was a more efficient pollinator of onion compared with *A. florea*.

Bezabih and Gebretsadikan (2014) studied the effect of managed honeybee pollination on onion seed yield and quality through pollinator exclusion and found that open pollination treatments especially with honeybees increased onion seed quantity and quality.

Tanaka and Komochi (1982) used Adult *Musca domestica* in Japan to pollinate onions (*Allium cepa*) in an isolated netted room. In the pollination room, flies lived for 10-14 days. Since the onions flower for more than 1 month, pupae were introduced 2-3 times, at a rate of 10-20 pupae per plant.

Munawar et al (2011) studied the effect of honeybee and blowfly pollination on onion (*Allium cepa* L.) seed production at Islamabad. Pakistan and found that seed setting is highly dependent on insect pollination. The mean weight of seed produced from the plants caged with honeybees was higher (1.90±0.11 g) as compared to blow flies caged plants (1.55±0.89) and control (0.33±0.22).

Nascimento et al (1998) found that *Trigona spinipes* (Hymenoptera:Apidae, Meliponini), bees are highly effective pollinators of onion (*Allium cepa*) when placed in isolation cages. The use of this low cost but efficient pollinator for onion seed production in Brazil is recommended.

Schiavi et al (1984) found that that honeybees should be used as pollinators when several onion genotypes are grown together in large cages, but blowflies are to be preferred for selfing or single crossing. More seed (mean 6.7 g/plant) was produced by bee-pollinated plants than by fly-pollinated plants (mean 3.7 g/plant). The most efficient plot design for crossing 2 genotypes was to alternate the bulbs within a row.

Prasad et al (2000) found that the flowers of onion variety 'Patna red' caged with bees had a maximum 2.12 seeds per flower followed by open to all pollinators (1.56) and lowest in cloth. Jun (1998) compared the role of honey bee (*Apis mellifera*) and fly (*Lucilia caesar*), placed in a net house during flowering, on seed yield of onions and found that flies were superior pollinators than honeybees. Compared with open field pollination, seed yields were increased by flies and bees by 24 and 12% in 1994, and by 35 and 27% in 1995, respectively.

Chandel et al (2002) studied the effect of bee pollination on the quantity and quality of onion seed (cv. Nasik Red) under subtemperate conditions of Himachal Pradesh situated at an altitude of 1200 m above mean sea level. A number of insect visitors such as *Apis mellifera, A. cerana, A. dorsata, A. florea* and syrphids (*Episyrphus balteatus*, *Metasyrphus conferator* [*Eupeodes confrater*] and *M. corollae* [*Eupeodes corollae*]) visited onion flowers. *A. dorsata* proved to be the dominant visitor (7.4 bees $m^{-2}$2 $minutes^{-1}$) and most efficient pollinator covering on an average 7.5 flowers per umbel per visit during peak hours of their foraging activity (12.00-14.00 h) compared with *A. cerana* (5.4 flowers per umbel per visit). *A. florea* and *A. mellifera* visited onion flowers least and ranked fourth and fifth overall in visiting frequency, respectively. Syrphids covered a minimum of 1.6 flowers per umbel per visit, yet spent maximum time (8.8 seconds) on the flowers during 08.00-12.00 h. *A. dorsata* had the maximum foraging period (06.30-18.55 h) followed by *A. cerana* (06.45-18.30 h) and *A. mellifera* had the least foraging period (07.25-18.20 h) on onion seed crop. Induced bee pollination increased seed yield by 2.5 times and produced on an average 971 seeds per umbel compared to 406 in the control. The seeds from induced pollination field resulted in 90% germination compared to 69.5% germination from the control.

Rashad (1978) showed that honeybees constituted the highest percentage of insect visitors of onion (Allium cepa) flowers. The other important pollinators were *Megachile submucida, Chalicodoma flavipes, C. mucorea, Nomia unidentata, Ceratina tarsata* and *Andrena vetula*. In tests with artificial nest tubes at the natural nesting site, hollow stems of safflower and of bamboo reed were preferred by megachile bees.

Haslbachova et al (1981) found that honeybees represented 79% of all pollinators and were most active between 10.00 and 16.00 h on onion (*Allium cepa*) flowers. Nectar collection predominated, and in good and poor years pollen was collected by 18% and 50-55% of the bees, respectively.

Silva and Dean (2000) studied the Floral nectar characteristics of nine inbred lines of onion (*Allium cepa*) and their attractiveness to honey bees (*Apis mellifera*). Thyey found that Potassium and sugar concentrations of the nectar did not significantly correlate with the number of bee visits received by an umbel but average amount of nectar produced positively correlated with the number of bee visits.

They suggested that selection of flowers with high nectar production may lead to a higher rate of pollination of the onion seed crop. Witter and Blochtein (2003) studied the effects of pollination by bees and other insects on the seed production of onion cultivars Crioula Alto Vale and Bola Precoce in Brazil. The production of seeds with free insect visitation increased by more than 20% in relation to others without insects. Visitation of *A. mellifera* was most important for pollination of onion flowers as it transported more than 70% of onion pollen.

Woyke (1988) showed that at temperatures below 15°C, when the wind force was greater than 3-5 m/s, during rainfall, or when the plants were wet, the number of bee flights at full bloom fell by 2.3-16 times, compared with flight numbers in warm sunny weather. He suggested that the number of hives placed in seed onion fields should be increased by 2-4/ha above the number calculated from flower number and hive vigour.

Alves et al (1982) found that the most frequent visitors were *Trigona spinipes, Apis mellifera* and *Plebeia* spp. on onion flowers. There was an increase of 124.7% in the mean seed number on non-isolated inflorescences. Haslbachova and.Kubisova (1982) studied the number of bees visiting pollen-fertile and pollen-sterile lines of onions and found that the average number of bees foraging on 100 onion plants during a 12-h period was significantly higher in the open (16-30) than in the cages (3-8). In 13 of 16 cases regression analysis indicated a significant positive relationship between seed yield in the cages and number of bee visits.

Shasha'a et al (1973) studied the inter-relationships of pollinating insect activity and components of seed yield in four male-sterile and two male-fertile inbred lines of onions. They found that inbreds differed very significantly in their attractiveness to pollinating insects; the number of nectar-collecting honeybees on 100 umbels varied from 3.23 to 8.10, and the number of pollen-collecting bees varied from 0.13 to 4.14. Separating the correlation coefficients into components of direct and indirect effects revealed that pollination attractiveness was not the limiting factor in seed set for inbreds. Moreover, most of the variation in seed yield could be attributed to indirect effects on the components of yield, umbels per plant, flowers per umbel, percentage of fertilized flower, and seeds per fertilized flower.

Benedek and Gaal (1972) found that the fertilization rate and seed yield of uncaged onion inflorescences were greater than those of caged inflorescences; the germination rate of seeds from uncaged heads was also significantly higher. The weight of 1000 seeds and the number of seeds per capsule did not differ significantly in caged and uncaged plants.. On the onion plants, honeybees foraged almost entirely for nectar, though they became heavily dusted with pollen during the process. Because of this, and because of their large foraging area in the onion field, honeybees would seem likely to be effective cross-pollinators of onion.

Kumar et al 1989 found that seed set and seed weight were higher on the plot caged with bees (4-frame nucleus of *Apis cerana*), and estimated seed yield was 275 kg/ha compared with 98 kg/ha on the open plot. On a plot caged to exclude insects seed yield was 73 kg/ha, and for bagged umbels 3 kg/ha. The introduction of honeybee colonies to onion fields at the start of flowering is recommended.

Nye et al 1971 found that bee activity and seed yields decreased as the distance from the pollen rows increased. The viability of onion pollen from flowers in the morning was 2 to 3 times greater than in the afternoon.

Carlson 1974 found that of the five varieties grown in experimental plots, Red Wetherfield, followed by Yellow Sweet Spanish, attracted the highest number of honey bees (*Apis mellifera*) during the flowering period. The honey bees remained longer on the umbels of both Red Wetherfield and Australian Brown than on those of the other varieties.

Singh and Dharamwal (1970) noted the increase of 21.80 per cent in yield and better seed germination with insect pollination of onion crop. Further, they reported that the open pollinated onion crop showed 72.00 to 79.00 per cent increase in seed set compared to the caged or bagged plants. There was 12 to 300 times more increase in seed set due to insect pollination. Kumar et al. (1989) recorded 90.47 % seed set and 77.8 % germination in bee pollinated onion crop.

**Table 10.1:** *Effect of bee visitation on qualitative characters on onion*

| Treatment | Per cent crinkled seed | Per cent seed setting | 1000 seed wt. (g) | Germination (%) | Emergence (%) |
|---|---|---|---|---|---|
| Open pollination | 14.8 | 85.3 | 5.5 | 78.4 | 72.0 |
| Apis cerana indica 4F | 5.8 | 94.3 | 6.6 | 89.0 | 85.0 2 |
| Apis mellifera 2F | 6.1 | 93.9 | 6.2 | 88.0 | 84.0 |
| A. florea 2F | 4.8 | 95.2 | 7.1 | 90.0 | 86.0 |
| Bee-Q | 11.1 | 89.0 | 5.2 | 79.0 | 73.0 |
| SSS | 11.7 | 88.3 | 5.1 | 79.0 | 73.0 |
| BQ+SSS | 10.3 | 89.7 | 6.0 | 79.0 | 73.0 |
| Pollination without insect | 25.4 | 74.6 | 4.9 | 80.0 | 74.0 |
| SE + | 0.4 | 2.6 | 0.2 | 2.9 | 3.4 |
| CD at 5% | 1.3 | 8.3 | 0.7 | 8.8 | 10.2 |

**Source:** Mupade et al 2009

Mupade, et al (2009) found that different bee attractants (Table 10.1) were significantly superior to plots without bees and open pollinated in respect of crinkled seed. Among the treatments, 2 framed *A. florea*, 4 framed *A. cerana indica* and 2 framed *A. mellifera* recorded lower percentage of crinkled seed (4.77, 5.84 and 6.13 per cent), which were at par with one another. The treatments Bee-Q + SSS, Bee-Q spray and SSS pollination were at par with each other and superior to open pollination and pollination without insects. Open pollination though inferior to the former treatments, but superior to pollination without insects in decreasing crinkled seed. Mohan Rao and Suryanarayana (1989) who reported that the chaffy seeds were less in bee pollinated treatments (78.3) as compared to treatments of pollination without insects (190.2).

Kumar et al. (1989), who noted that per cent of seed set in bee pollination (90.5) was maximum as compared to open pollination (71.8), net caged (60.7) and cloth bagged (14.0) condition. Prasad et al. (2000) observed the seed set as 70.7, 58.3, 58.9 and 21.5 per cent in flower caged with bees, open to all pollinators, net caged and cloth bagged, respectively.

Lorenzone et al. (1993), who observed that the 1000 seed weight was significantly higher, wherever the bees were the frequent visitors. Earlier reports of Singh and Dharamwal (1970), Benedek and Gall (1972), Kumar et al. (1989) and Chandel et al. (2004), reported increase in seed germination in bee pollinated crop.

Onion umbels deprived of insect visits did not set any seed. Wind and gravity had no effect on pollination. Several species of insects have been identified identified visiting blooming onion umbels, the number of species depending on the locality (Currah, 1990). Honeybees are the most important pollinators of onion and hives ranging from 12 to 37/ ha have been recommended, starting from the time when 50% of umbels are showing open florets. Bees forage for nectar, not pollen, from onion flowers, and they are active at air temperatures above 16°C. They move more frequently from umbel to umbel when the sky is clear and cloud free, therefore increasing the chances of cross-pollination in sunny weather.

Sajjad et al (2008) reported that the community of pollinators visiting onion (*Allium cepa* L.) flowers was composed of two bees and eight true flies. Among bees, *Apis dorsata* proved to be an abundant pollinator while among true flies *Episyrphus balteatus* (Syrphidae) had the maximum abundance.

Estimates of increased seed set due to pollinators have been made in different parts of the world. Singh (1997) obtained an increase of 3.5 to 98.8 times in the onion seed yield due to assured pollination by bees. Similarly, in Poland, Woyke 1981 found that plots caged without bees, with bees and not caged produced 2, 210 and 669 seeds per inflorescence, respectively. Similar results were obtained by Kumar et al (1989) who obtained an estimated yield of 73, 275, 97 kg/ha, caged with bees, than from plots caged without bees and open plots, respectively. Rao and Sunyanarayana (1989) reported that non-availability of pollinators resulted in only 17% fruit setting and free availability of pollinators increased fruiting up to 73%.

**Diversity of insect visitors on onion umbels:** Rasekh Adel *et al.*, 2013 found a variety of insects visiting onion flowers. Among the community of pollinators, hymenopterans (*Andrena* sp. (Andrenidae), *Halictus patellatus* (Morawitz 1873), *Halictus quadricinctus* (Fabricius1777), (Halictidae), *Xylocopa valga* (Gerstaecker1872), *Amegilla* sp. (Apidae), and dipterans (*Episyrphus balteatus* deGeer, *Eupeodes corollae*, *Sphaerophoria scripta*, *Eristalinus aeneus* (Syrphidae), *Musca domestica* L. (Muscidae) were the most frequent visitors. Part of the community of pollinators of onion has already been published (Rasekh Adel *et al.*, 2012).

Rasekh Adel *et al*., (2013) found that the average increased seed production in the open pollinated treatment was 74%. In the absence of insect pollinators, the onion seed yield per hectare was 130 kg as compared to 500 kg/ ha contributed by insects. The results of this study was in agreement with other works (e.g., Jablonski *et al.,* 1982; Tolon and Duman, 2003; Ahmad *et al*., 2003; Kumar *et al.*, 1989; Rao and Lazar, 1983), and confirmed that the presence of insects during flowering period has substantial impact on onion seed setting. In this study, the number of seeds and weights of 1000 seeds in treatment without pollinating insect were significantly less than those of the free pollinated treatment. Similarly, Ahmad *et al*., (2003) found that reduced or limited visiting of pollinators caused 50-61% reduction in seed production of onion. Also, Rao and Lazar (1983) observed only 9.8% seed setting in onion in cages. Moreover, Kumar *et al*., (1989) recorded three times more yield in open pollinated treatment than the closed one.

Ahmad *et al*., (2003) claim that proper visiting of pollinators could increase seed production up to 50% by minimizing the side grown attractive crops. Also, completing the spraying schedule before onset of flowering or application of insecticides in the evening could minimize the hazardous effect of insecticides on pollinators. In addition, spraying onion plants with some insects attractant may result in a higher rate of insect visiting flowers and consequently increase the percentage of flower set and lower number of aborted flowers.

Ahmad *et al*., (2003) who found no significant difference in attraction of the four varieties of onion to pollinators. However, varieties possessing more concentration of nectar in flowers are more attractive to insect pollinators which in turn could result in a better seed setting. Several studies have reported honeybees as the most frequent visitors of onion bloom (Bohart *et al*., 1970; Priti, 1998). However, in this study the abundance of honey bees was not high when compared with leaf cutting bees (Megachilidae) and yellow jacket wasps (Vespidae). This may be explained by the fact that onions are not favorites of honey bees because onion nectar is high in potassium, and the nectar tends to increase in viscosity as temperature increases, so bees prefer to forage on alternate nectar sources if available (Voss *et al*., 1999).

## Pests of Onion

Onion is an important vegetable crop and is attacked by several pests which are as given below:

**I. Pests of Onion (*Allium cepa* L., Alliaceae)**

Onion ranks second only to tomato among cultivated vegetables. Onion originated in the region of Afghanistan, Iran, and Pakistan (George, 2011). Main onion producing countries in Asia are China, India, Japan, Pakistan, and South Korea. Several types of onions are grown all over the world and some are used both green and dry. Onions have become popular because of their medicinal value. Onion thrips that transmit virus diseases in addition to direct damage by army worms, onion fly, and leaf miners are the most serious insect pests of onions in tropical Asia. Onion maggot under North Indian conditions and earwig under South Indian conditions gain importance occasionally.

**Major pests of onion include:** Onion thrips *Thrips tabaci* (Thysanoptera Thripidae), **Onion** maggot *Delia antiqua* (Anthomyiidae: Diptera) and Earwig *Anisolabis stali* (Forficulidae: Dermaptera).

**Potential major pests are** Tobacco caterpillar *Spodoptera litura* Noctuidae Lepidoptera**,** Cutworm *Agrotis ipsilon* Noctuidae Lepidoptera

**1. Onion thrips: *Thrips tabaci* (Thysanoptera: Thripidae)**

**Distribution and status:** Found world-wide and is found throughout India as a major pest of onion and garlic (*Allium fistulosum*).

**Host range:** Onion, garlic, cotton (*Gossypium spp.*), cabbage, cauliflower, potato, tobacco, tomato, cucumber (*Cucumis sativus* L.), brinjal, tea, pear, pine apple, chillies, tomato, radish, grapes etc.

**Damage symptoms**

Adults as well as by nymphs lacerate the leaf tissue and feed on the plant juice. The insects are just visible to the unaided eye and are seen moving briskly on the flowers and leaves of onion and garlic plants. They usually congregate at the base of a leaf or in the flower. Leaves of attacked plants turn silvery white, curl, wrinkle and gradually dry from tip downwards. The plants do not form bulbs nor do the flowers set seed. Leaf tip discoloration and drying is the main symptom.

**Bionomics**

The adults are slender, yellowish brown and measure about 1 mm in length. Males are wingless, the females have long narrow strap like wings. Nymphs resemble the adults but are wingless and smaller in size. This pest is active throughout the year and breeds on onion and garlic from November to May, migrates to cotton and other summer host plants and breeds till September. In October, it is found on cabbage and cauliflower. The adult female 'lives for 2-4 weeks and lays 50-60 kidney-shaped eggs singly in slits made in leaf tissue with its sharp ovipositors. Egg period 4-9 days. Nymphs pass through four stages and are full-fed in 4-6 days. They descend therafter to the ground and pupate at a depth of about 25 mm. The pre-pupal and pupal stages last 1-2 and 2-4 days, respectively. Several generations are completed in a year.

**Management**

**Monitoring:** Sticky traps are commonly used for detecting thrips population. Various colours were found attracting different species of thrips in various geographical areas; widely used ones are yellow and bright blue traps.

**Cultural**

Good crop management practices can be readily manipulated to the disadvantage of thrips. Field sanitation techniques such as removing alternate weed host on bunds and destruction of culls of onion and garlic are helpful to some extent in bringing down thrips population. Avoid successive planting of onion and garlic or other preferred host. Therefore, planting in upwind direction could be helpful in escaping infestation from old planting to some extent in the initial stages.

**Planting Date**

Transplanting dates may be adjusted to make the onion varieties tolerant to thrips attack to avoid chemical applications

**Mulching**

Thrips being colour-sensitive, coloured mulches may be employed for their control.

Reflective plastic silver colour and aluminium painted black mulches were effective in repelling the thrips in seed crop.

**Irrigation**

Sprinkler irrigation reduces thrips population considerably compared to drip and surface irrigation. Insects would be washed off or drowned in the water accumulated in the leaf bases. Due to continuous retention of moisture in the soil, the pupae get rotten. In case of garlic, sprinklers are not that effective as in case of onion mainly due to the closer inner leaf alignment that protects the thrips from splashes of water.

**Barriers**

Two rows of maize or inner row of wheat and outer row of maize surrounding the onion plots.

**Biological Control**

These include lady beetles, minute pirate bugs, ground beetles, big-eyed bugs, lacewings, hover flies, predatory mites and spiders. So far limited success has been achieved in biological control of *T. tabaci* in many parts of the world. A predator, minute pirate bug, *Orius tantillus* and anthocorid bug, found effectively feeding on thrips on garlic. *Metarhizium ansopliae, Paecilomyces fumosoroseus, Verticillium lecani and Frankliniella* also killed *T. tabaci sp.* Mortality of thrips was highest with *B. bassiana* at 260 C and 75% RH. The eulophid parasitoids *Ceranisus vinctus* from the Philippines and *Goetheana* (*Dasyscapus*) *parvipennis* from Java have been reported from *Thripstabaci. Goetheana parvipennis* from Ghana was introduced and established in the West Indies for control of *T. tabaci* and *Selenothrips rubrocinctus*. The phytoseiid mites, *Amblyseius mckenziei, Amblyseius cucumeris*, and *Typhlodromips sessor* and the fungal pathogens *Neozygites parvispora* and *Verticillium lecanii* are used in glasshouses for control of *T. tabaci* in Europe. Conserve predators like *Scymnus nubilis,Orius albidipenis, Chrysopa* sp,and predatory thrips vis., *Aelothrips collartris*

**Plant Resistance**

Commercial varieties, N-2-4-1 and Pusa Ratnar were found resistant to T. *tabaci* in Punjab but the former was susceptible to thrips in Maharashtra. In Bihar, Pusa Red and N-53 had lowest thrips population while Arka Niketan had highest infestations. Grow resistant varieties *viz.,* White Persian, Grano, Sweet Spanish and Crystal Wax. Use neem coated urea to reduce the infestation of the pest. Install sky blue colour sticky traps@ 25/ha.

**Botanicals**

Neem was found effective when mixed with insecticides.

**Chemical**

Many insecticides like dimethoate (0.06%) and Methyl Demeton (0.05%) are recommended for thrips control. Cypermethrin at 60 g ai/ ha offers good control of this pest. Carbosulfan (0.05%) and Lambda cyhalothrin also effectively suppress thrips population.

The secret of thrips control in onion lies in the placement of the insecticide. The product should be applied in such a way that it targets the base of leaves where majority of thrips are hidden. Unlike other crops, onion and garlic require higher pressure and spray volume for good coverage of foliage. Addition of a spreader or sticker (0.05-1.0%) is useful for retention and spread of spray fluid on erect leaves of onion. The same insecticides should not be applied repeatedly to minimize insecticide resistance.

In seed crop, control of thrips is often difficult. The umbels will provide excellent hiding site and it is difficult to reach them with insecticide.

From the beginning, seed crop should 1be monitored regularly for thrips and should be controlled well before flowers open.

Avoid sprays after flower opening. If necessary, safer insecticide should be used late in the evening when no bees are foraging.

2. **Onion Maggot: *Delia antiqua* (Anthomyiidae: Diptera)**

**Distribution and status:** Widely distributed in France, Germany, Canada, USA, Japan, erstwhile USSR and England. This pest also attacks onions in northern India.

**Host range:** Onion and garlic

**Damage symptoms**

The maggots bore into the bulbs, causing the plants to become flabby and yellowish. It causes withering in the field and rotting in storage. Damage leads to the invasion of *Bacillus carolovorus* which causes soft rot of onion

**Bionomics**

The flies are slender, greyish, large-winged. The maggots are small, white and about 8 mm. in length. The female lays elongate, white eggs near the base of the plant, in cracks in the soil. The eggs hatch in 2-7 days. The maggots crawl up, enter the leaf sheath and reach the bulb. They feed there and become full-grown in 2-3 weeks. The maggots then crawl out of the bulb and pupate in the soil. After 2-3 weeks, the adults emerge and start the new generation. In the third generation, the pest often attacks the onions shortly before the harvest. It initiates the process of rotting of the onions in storage.

**Management**

**Cultural**

Avoid planting in soils that are high in undecomposed organic matter.

Avoid close spacing while planting.

Control of D. *antique* is exceptional in areas where crop rotation was followed in the previous year.

**Host Plant Resistance:** No resistant varieties are available against maggot.

Apply 10 kg of carbaryl 4G or phorate lOG to the soil followed by light irrigation. Spray methy–demeton 25 EC or dimethaote 30 EC 1.0 L in 500 – 750 L of water per ha. Application of Pirimiphos Methyl or Deltamethrin at 2ml/lit. Fipronil 80WG (@25 g ai/kg) and Cyromazine (75 WP @ 50 g ai/kg), an insect growth regulator as seed treatment were recommended.

**Natural enemies:** A nematode, *Heterotylenchus aberrans*, parasitized adult flies in France and Denmark. A parasitoid, *Aphaereta pallipes*, has been reported from Canada. Ground beetle is a good predator and establishing grassy refuse strips in onion crop will enhance the beetle population and reduce the maggot population.

**Common names:** Beet army worm, onion army worm

**Scientific name:** *Spodoptera exigua* (Hübner)

**Order:** Lepidoptera

**Family:** Noctuidae

**Distribution:** Europe, Asia, United States

**Hosts:** Onion, garlic, leek, cruciferous crops

**Life cycle:** Clusters of eggs are laid on leaves and covered with white feltlike scales. Eggs hatch in about 3 days. Larvae feed clustered together after hatching, but soon disperse. There are usually six larval instars which complete development in 9–14 days. Pupation occurs in the soil and the pupal period lasts for about 10 days

**Damage:** Damage is caused by larvae feeding on the leaves. In onions and shallots, the larvae enter the leaves as early instars and feed from inside the leaves, causing characteristic "windowpaning," leaving only the outer cuticle of the leaf.

**Pest Management**

Pheromone traps can be placed at 125 cm height and dispenser should be regularly replaced for effective trapping. NPV and Bt can also be employed for control of the pest. Insecticides like Cypermethrin, Quinolphos, Acephate, etc. can be used if infestation is more.

**Natural enemies:** A nucleopolyhedrosis virus has been isolated from *S. exigua* in Indonesia and commercial formulations are available for field use. This has helped to reduce dependence on chemical insecticides. The parasitoids *Diadegma semiclausum*, *Eriborus argenteopilosus*, *Charops brachypterum*, and *Euplectrus* cf. *chapadae* were recorded in Indonesia (Shepard and Barrion, 1998). A fungus, *Nomuraea rileyi*, and a microsporidian (protozoan) were also reported from Indonesia. *Chelonus insularis*, *Cotesia marginiventris*, *Meteorus autographae*, and the tachinid *Lespsia archippivora* have been recorded in the U.S. Predators found attacking the eggs and larvae in the U.S. were minute pirate bugs, *Orius* sp.; big-eyed bugs, *Geocoris* spp.; damsel bugs, *Nabis* spp.; and a predatory shield bug, *Podisus maculiventris.*

**Other remarks:** There are several larval color phases, ranging from green to nearly black. Sometimes the dark-colored larvae are confused with other army worm species. This pest has become resistant to many chemical pesticides.

**Common name:** Black onion aphid

**Scientific name:** *Neotoxoptera formosana* (Takahashi)

**Order:** Hemiptera

**Family:** Aphididae

**Distribution:** United States, Mexico, Brazil, Chile, France, Italy, Netherlands, Spain, Japan, China, Taiwan, Korea, Indonesia, Australia, Papua New Guinea, New Zealand

**Hosts:** Onion, chives, leek, garlic, shallot

**Life cycle:** This pest completes its nymphal stages in 7 days on shallots.

**Damage:** This aphid forms dense colonies and can reduce the value of the crop due to feeding and because of the physical presence of the aphids, especially on leaf onions.

**Natural enemies:** No information is available on the natural enemies of this pest but it is likely that generalist predators, such as ladybeetles and lacewing larvae, feed on this aphid.

**Other remarks:** *N. formosana* is difficult to separate from a closely related black aphid, *N. oliveri* (Essig)

3. **Earwig: *Anisolabis stali* (Forficulidae: Demaptera)**

   **Distribution and status:** Throughout India

   **Host range:** Onion, garlic, cabbage, cotton, sorghum and groundnu

   **Damage symptoms:** Nymphs bore into the bulb and make cavities which lead to withering of plants.

**Bionomics:** Adult is brown to black with forceps like caudal cerci and white jointed legs. Adult female lays 21-139 eggs, egg period 14 days. Nymphs white in the early stage turn brown in the later stage, nymphal period 50 -54 days.

**Control**

Application of malathion 5% @ 25 kg/ha prior to sowing in areas where the attack of earwig is more endemic. Repeat the above application on 40 days of sowing and incorporate in the soil during earthing up.

4. **Red Spider Mite (*Tetranychus cinnabarinus*)**

**Symptoms:** Tetranychid mites are found through out the world on major vegetable crops. This is emerging as a regular mite pest in onion and garlic.

**Biology**

The spider mite completes its life cycle from egg to adult in about a week. All stages of this mite are present throughput the year. Eggs are spherical, shiny, straw colored, and hatch in 3 days. They are only about 1/254 inch in diameter. The eggs are laid singly on the underside of the leaf surface.

They are attached to silken webs spun by the adults. Larvae are slightly larger than the egg, pinkish in colour and and have three pairs of legs. This stage lasts a short time, perhaps a day. There are two nymphal stages, the protonymph and deutonymph. The nymphal stage is slightly larger than the larval stage, reddish or greenish, and having 4 pairs of legs. This nymphal stage lasts about 4 days. Adult females are about 1/50 inch long, reddish, and more or less elliptical in shape. The males are slightly smaller and wedge shaped. They have a black spot on either side of their relatively colorless bodies. The life span of adult female is 200 days and lays 200 eggs.

## Damage symptoms

Adults and nymphs feed primarily on the undersides of the leaves. The upper surface of the leaves becomes stippled with little dots which reflect the feeding punctures. The mites tend to feed in "pockets" often near the midrib and veins. Silk webbing produced by these mites is usually visible. The leaves become bleached, discoloured and may fall off

## Management

Spraying miticides like Dicofol and Ethion.

Even a thorough water spray also washes off the mites from the plant.

## Eriophyid Mite (*Aceria tulipae Keifer*)

Garlic plants infested with mite show symptoms like stunting, twisting, curling and yellow mottling and reduce the plant stand as well as yield. Mites also attack the stored garlic bulbs. Most of the mites are located at the basal portion of cloves and make them dry and desiccated. Mite is capable of transmitting viruses such as wheat yellow streak mosaic virus in wheat, garlic mosaic virus and onion mosaic caused by Allium 1 virus.

## Management

Flood irrigation or heavy rains may reduce field populations. Avoid planting successive onion or garlic crops.

Wettable Sulfur 0.3% + Dimethoate (0.03%) as pre-sowing and post-sowing treatment is recommended.

Seed dressing and plant stand treatments with 0.5% Sulfur and 0.1% Dimethoate are also effective in controlling mites.

Hot water treatment of seed garlic at 55 and 600 C for 10 minutes was also effective but affects clove sprouting capacity.

Fumigation with Methyl Bromide at 32 g/m3 for 2 h at 800 F or 3 g/m3 Aluminium Phosphide for 72 h in airtight plastic box has been recommended for the control of mite in stored garlic.

## Bulb Mite (*Rhizoglyphus robini Claparede*)

This pest infests the bulbs by penetrating the outer layer of tissue. The plants wilt and rot as opportunistic fungi and bacteria gain entry inside the bulbs. Bulb mites can reduce plant stands, stunt plant growth, and promote rot of bulbs in storage. On seeded onions, they can cut off the radical before the plants become established.

## Management

Avoid planting successive onion or garlic crops.

Flood irrigation or heavy rains may reduce mite levels in the soil.

In garlic, sow clean seed cloves.

Hot water treatment of seed garlic before planting may reduce mite infestation

Soil drenching with Dicofol at 2 ml/1 or soil application of powdered sulfur effectively reduce the mite population considerably. Fumigation with Methyl Bromide @16 g/m3 for 3 h or 10 g/m3 for 6 h is recommended for control of mobile phases of the mite in stored garlic.

## Biology

The developmental stages in the life cycle are egg, larva, protonymph, heteromorphic deutonymph, tritonymph and adult. Females lay up to 700 eggs. Larval period is 3 to 7 days.

**Adult:** The longevity of female is 31 days and male 62 days.the mature bulb mites have four pair of legs and bary in size from 0.5 to 0.9 mm long. Their bodies are shiny, transparent, smooth and have reddish brown appendages. It takes 12.2 days from egg to become adult at 25°C.

## Damage symptoms

Bulbs infested with bulb mites may rot and fail to produce new growth, or new growth may be off color, stunted, and distorted. The bulb mites in general are not serious pests of bulbs but by injuring causes huge number of mites to get established in bulb and rotten. Their injury may facilitate the attack of other pests, such as the bulb scale mite or bulb flies. The bulb mites can attack healthy new roots and corms, especially in greenhouses.

Other major pests are:

**Common name:** Western flower thrips

## References

Abrol D. P 2009. Planned honeybee pollination for improvement in horticultural crop production. Final Technical Report Horticulture Mission Mode Project-1(HTMM-1, 2.28). ICAR New Delhi pp40

Abrol D. P. 1988 Effect of climatic factors on pollination activity of alfalfa-pollinating subtropical bees Megachile nana Bingh and Megachile flavipes Spinola (Hymenoptera, Megachilidae). Acta Oecologica 9: 371-377.

Abrol DP 1997. Impact of insect pollination on carrot seed production. Insect Environment 3 : 61.

Abrol, D. P. 2012. Pollination Biology: Biodiversity Conservation and Agricultural Production. Springer Dordrecht HeidelbergLondon New York.

Abrol, D.P 1991 Path analysis of environmental factors influencing daily flight activity of Apis dorsata F. Acta Oecologia, Ecologia Generalis 12 : 819-824.

Abrol, D.P 2006. Factors influencing flight activity of Apis florea F, an important pollinator of Daucus carota L Journal of Apicultural Research and Bee World 45(2): 2–6

Abrol, D.P. 1992 Bioenergetics in bee flower interrelationship- An analysis of foraging behaviour. Korean J.Apic. 7(1): 39-66.

Abrol, D.P. 1997 Impact of insect pollination on carrot seed production, Insect Environ., 3, 61.

Abrol, D.P. 1998 Environmental factors influencing flight activity of honeybees Apis cerana F. and A. mellifera L. (Hymenoptera: Apidae). Indian Bee J. 60(2): 71-75.

Abrol, D.P. 2006. Diversity of pollinating insects visiting litchi flowers (Litchi chinensis Sonn.) and path analysis of environmental factors influencing foraging behaviour of four honeybee species. J. Apic. Res., 45(4): 180–187.

Abrol, D.P. and Kapil, R.P. 1986. Factors affecting pollination activity of Megachile lanata Lepel: Proc. Indian Acad. Sci. (Anim. Sci.) 95: 757-769.

Abrol, D.P. and Kapil, R.P. 1991. Foraging strategies of honeybess and solitary bees as determined by nectar-sugar components.Proc. Indian Nat. Sci. Acad. B, 57(2): 127-132.

Abu-Hammour K. 2008 Pollination of MedicinalPlants (Nigella sativa and Coriandrum sativum) and Cucurbita pepo in Jordan. Doctor-Ingenieur thesis. University of Bonn. Germany. 90 pp.

Agati, G. 1952. Indagini ed osservazioni sulla biologia fiorale della cipolla. Riv. Ortoflorofruttic. ital. 36: 67-77.

Ahmad Munir and Aslam Muhammad 2002. Pollinators visiting carrot (daucus carota l.) seed crop. Journal of Research (Science), Bahauddin Zakariya University, Multan, Pakistan. 13(1): 31-35

Ahmad, M and Aslam, M. 2002. Pollinators visiting carrot (Daucus carota L.) seed crop. Journal of Research (Science), 13: 31–35.

Ahmad, S., Muhammad, T., Karim, A., Jabbar,A., and Ahmad, Gh. 2003. Effect of pollinators and insecticides on seed setting of onion (Allium cepa L.). Journal of Biological Science, 3 (3): 305-308.

Aizen, M.A., Feinsinger, P., 1994. Habitat fragmentation, native insect pollinators, and feral honey bees in Argentine "Chaco Serrano". Ecological Applications 4, 378–392.

Aizen, M.A., Garibaldi, L.A., Cunningham, S.A. and Klein, A.-M., 2008. Long-term global trends in crop yield and production reveal no current pollination shortage but increasing pollinator dependency. Current Biology 18, 1572–1575.

Aizen, MA, Garibaldi, LA, Cunningham, SA and Klein, AM. 2009. How much does agriculture depend on pollinators? Lessons from long-term trends in crop production. Annals of Botany, 103: 1579–1588.

Alam, M.Z. and Quadir, M.A. 1986. Role of honey bees in fruit and seed setting of bottle gourd, Lagenaria siceraria (Mol.) Staudl.Punjab Vegetable Grower,21: 32-34.

Alam, M.Z., Quadir, M.A. and Ali, M.S. 1987. Pollination behaviour of honeybee (Apisindica) and the influence on seed production of cauliflower. Bangladesh Hortic.,15(1): 25-30.

Alam, M.Z.; Quadir, M.A.; Ali M.S. 1987 Pollination behaviour of honeybee Apis indica F. and its influence on seed production of cauliflower, Bangladesh Horticulture 15 (1): 25-30.

Ali, M., Dowker, B.D., Currah, L. and Mumford, P.M. 1984. Floral biology and pollen viability of parental lines of onion hybrids.Ann. Appl. Bio., 104: 167–174.

Allen-Wardell, G., Bernhardt, P., Bitner, R., Burquez, A., Buchmann, S.L., Cox, P.A., Dalton, V., Feinsinger, P., Ingram, M., Inouye, D.W., Eugene, C., Kennedy, K., Kevan, P.G., Koopowitz, H., Medellin, R., Nabhan, G.P., Pavlik, B., Tepedino, V.J., Torchio, P.F., 1998. The potential consequences of pollinator declines on the conservation of biodiversity and stability of food crop yields. Conservation Biology 12, 8–17.

Allsopp, M.H., 2009. Bee disease threatens South African apiculture, Plant Protection News, No. 79, Agricultural Research Council: Plant Protection Research Institute.

Al-Sahaf, F. H. 2002. Effect of planting method and rose water spray on seed production in onion (Allium cepa L.). Emirate Journal of Agricultural Science, 14: 14-23. Aminpour, R. and Mortzavi Bak, A. 2004.

Altieri, M.A., 1995. Agroecology: the science of sustainable agriculture. Westview Press, Boulder, CO.

Altieri, M.A., 1999. The ecological role of biodiversity in agroecosystems. Agriculcture Ecosystems and Environment 74, 19–31.

Alves, S. B.; Vendramim, J. D.; Silveira Neto, S. and Nakano, O. 1982. Insect pollination of onion, Allium cepa L. Solo 74 (1/2): 18-22

Amaral, E., Mitidieri, J. and Vencousky, R. 1963. Studies on the activities of Apis mellifera L. while visiting the flowers of Cucumis sativus L. Olericultura,3: 181-193.

Anderson, D.L., Sedgeley, M., Short, J.R.T. and Allwood, A.J. 1982. Insect pollination of mango in northern Australia.Australian J. agric. Res.,33: 541-548.

Aouar-Sadli M., Louadi K., Doumandji S. 2008 Pollination of the broad bean (Vicia faba L. var. major) (Fabaceae) by wild bees and honey bees (Hymenoptera: Apoidea) and its impact on the seed production in the Tizi-Ouzou area (Algeria). African Journal of Agricultural Research 3(4): 266-272.

Association of Official Seed Analysts. Rules for testing seeds. Assoc. Offic. Seed Anal. Proc. Vol. 49, no. 2. 1959.

Backman, W.W. and Waller, G.D. 1971. Honeybee responses to sugar solutions of different composition.J. apic. Res.,16: 165-169.

Baker, H.G., Cruden, R.W. and Baker, I. 1971. Minor parasitism in pollination biology and its community function: the case of Ceiba acuminate.Bio. Sci.,21:1127-1129

Banaszak J. 1980 Studies on methods of censusing the numbers of bees (Hymenoptera, Apoidea). Polish Ecological Studies 6: 355-366.

Baswana K. S. 1984 Role of insect-pollinators on seed production in coriander and fennel. South Indian Horticulture 32: 117-118.

Bell C. R. 1971 Breeding systems and floral biology of the Umbelliferae or evidence for specialization in unspecialized flowers. In: Heywood V. H. (Ed.) The biology and chemistry of the Umbelliferae. Academic Press. London. pp. 93-107.

Belluscci, S. and Marques, M.D. 2001. Circadian activity rhythm of the foragers of a eusocial bee (Scaptotrigona affdepilis, Hymenoptera, Apidae, Meliponinae) outside the nest.Biol. Rhythm Res.,32 (2): 117–124.

Benachour K., Louadi K. and Terzo M. 2007 Role des abeilles sauvages et domestiques (Hymenoptera: Apoidea) dans la pollinisation de la fève (Vicia faba L. var. major) (Fabaceae) en région de Constantine (Algérie). Annales de la Société Entomologique de France 43(2): 213-219.

Bendifallah L Louadi K and Doumandji S 2013. Bee fauna potential visitors of coriander flowers Coriandrum sativum L. (Apiaceae) in the Mitidja area (Algeria). J. Apic. Sci. 57( 2): : 59-69

Bendifallah L., Louadi K. and Doumandji S. and Michez D. 2011 Role des abeilles solitaires (Hymenoptera: Apoidea) dans la pollinisation de la fève Vicia faba L.var. Major) (Fabaceae) en région de Mitidja (Algérie). In: Proceedings of the 9 th International Conference on Pests in Agriculture. Montpellier, France. 25-27 October 2011. pp. 75-80.

Benedek, P. 1976. Effect of environmental factors on the foraging rates of honey bees in red clover fields. Zeitschrift fur Angewandte Entomologie,81(1): 14-20.

Benedek, P. and Gaal, E. 1972. The effect of insect pollination on seed onion with observations on the behaviour of honey bees on crop.J. Apic. Res., 11:175-180.

Bernhardt, P., 2000. Convergent evolution and adaptive radiation of beetle-pollinated angiosperms. Plant Systematics and Evolution 222, 293–320.

Beyleveld, A.J., 1967. Ostrich farming in South Africa. Agrekon 6, 14–16.

Bezabih, G.; Gebretsadikan, K. 2014 Managed honeybees (Apis mellifera L.) increase onion (Allium cepa) seed yield and quality. Livestock Research for Rural Development 26 (1) 2014.

Biesmeijer, J.C., Roberts, S.P.M., Reemer, M., Ohlemuller, R., Edwards, M., Peeters, T., Schaffers, A.P., Potts, S.G., Kleukers, R., Thomas, C.D., Settele, J. and Kunin, W.E., 2006. Parallel declines in pollinators and insect-pollinated plants in Britain and the Netherlands. Science 313, 351–354.

Blanche, K.R. and Cunningham, S.A., 2005. Rainforest provides pollinating beetles for Atemoya crops. Journal of Economic Entomology 98, 1193–1201.

Blanche, K.R., Ludwig, J.A. and Cunningham, S.A., 2006. Proximity to rainforest enhances pollination and fruit set in orchards. Journal of Applied Ecology. 43, 1182–1187.

Bogoyavlenseii, S. H., and Akimenko, A. L. 1966. [Coriander as a nectiferous and entomophillous culture.] In Achievements of Science and Advanced Experiment in Beekeeping, pp. 119-125.

Bohart G. E and Nye W. P. 1960 Insect pollinators of carrots in Utah. Agricultural Experimental Station Bulletin. N°.49. Utah State University. Logan. UT. 16 pp.

Bohart, G. E., Nye, W. P., and Hawthorn, L. R. 1970. Onion p0llination as affected by different levels of pollinator activity. Utah Agr. Expt. Sta. Bul. 482, 60 pp.

Bohart, G.E. and Nye, W.P. 1968 "Insect pollinators of carrot in Utah", Utah Agri. Expt. Sta. Bull., 419, 16.

Bohart, G.E., Nye, W.P. and Hawthorn, L.R., 1970. Onion pollination as affected by different levels of pollinator activity. Utah Agricultural Experiment Station Bulletin, No. 482, 66.

Borneck, R. and Merle, B. 1989. Essaie d'une evaluation de l' incidence economique de l' abeillie pollinisatrice dans l' agriculture europenne. Apicata 24: 33-38.

Borthwick, H. A. 1931. Carrot seed germination. Amer. Hort. Sci. Soc. Proc. 28: 310-314.

Bose, T.K. and Som, M.G. 1990 "Vegetable crops in India", Kalyani Publishers, Naya Prakash, India, p. 409.

Boyd, J. and Banzhaf, S., 2007. What are ecosystem services? The need for standardized environmental accounting units. Ecological Economics 63, 616–626.

Braak, J. P., and Kho, Y. O. 1958. Some observations on the floral biology of the carrot (Daucus carota L.). Euphytica 7(2): 131-139.

Brantjes, N.B.M. and Leemans, J.A.A.M. 1976. Silane oties (Carryophyllaceae) pollinated by nocturenal Lepidoptera and mosquitoes. Acta Botanica Neerlandica, 25: 281-285.

Brewer, J.W. 1974. Pollination requirements for watermelon seed production. J. apic. Res., 13(3): 207-212.

Brewster, J.L., 2008. Onions and other vegetable Alliums, 2nd Edition. CAB International, Oxon.

Brosi, B.J., Daily, G.C., Shih, T.M., Oviedo, F., Duran, G., 2008. The effects of forest fragmentation on bee communities in tropical countryside. Journal of Applied Ecology 45, 773–783.

Buchmann, S. L. and Nabhan, G. P. 1996. The forgotten pollinators. Island Press, Washington D C.

Burill, R.M and Dietz, A. 1982. The responses of honeybees to variations in solar radiation and temperature. Apidologie 12 : 318-328.

Buscher, B.E., 2008. Conservation, neoliberalism, and social science: a critical reflection on the SCB 2007 annual meeting in South Africa. Conservation Biology 22, 229–231.

Carlson, E. C. 1974 Onion varieties, honeybee visitations, and seed yield. California Agriculture 28 (9): 16-18

Caron, D. M.; Lederhouse, R. C.; Morse, R. A. 1975. Insect pollinators of onion in New York State. Hortscience 10 (3) 273-274

CEPF, 2003. Ecosystem profile: The Succulent Karoo hotspot, Namibia and South Africa, Africa. Critical Ecosystem Partnership Fund.

Cervancia, C.R. and Bergonia, E.A. 1991. Insect pollination of cucumber (Cucumis sativus L.) in the Philippines.The Sixth International Symposium on Pollination, Tilburg, Netherlands, 27-31 August 1990, 278-282.

Chacoff, N.P. and Aizen, M.A., 2006. Edge effects on flower-visiting insects in grapefruit plantations bordering premontane subtropical forest. Journal of Applied Ecology. 43, 18–27.

Chandel R S, Thakur R K, Bhardwaj N R and Pathania N 2004. Onion seed crop pollination a missing dimension in mountain horticulture. Acta Horticulturae 631 : 79-86.

Chandel, R. S., Thakur, R. K, Bhardwaj, N. R., and Pathania, N. 2004. Onion seed croppollination: a missing dimension in mountain horticulture. Acta Horticulturae, 631: 79-86.

Chandel, R. S.; Thakur, R. K.; Bhardwaj, N. R.; Pathania, N 2002. Onion seed crop pollination: a missing dimension in mountain horticulture. Issues and advances in transplant production and stand establishment research, proceedings of the XXVI International Horticultural Congress, Toronto, Canada, 11-17 August, 2002.

Chaudhary O. P., Singh J. 2007 Diversity, temporal abundance, foraging behaviour of floral visitors and effect of different modes of pollination on coriander (Coriandrum sativum L.). Journal of Spieces and Aromatic Crops 16(1): 8-14.

Chaudhary, N. 2004. Factors influencing inter and intra specific preferences of honey bee for three concurrently flowering vegetable/ spice crops. Ph.D Thesis, CCSHAU, Hisar. 96+xxviii pp.

Chaudhary, N. and Sihag, R.C. 2003. Diversity, foraging behaviour and foraging efficiency of different pollinators of visiting onion (Allium cepa L.) blossoms.Korean J. Apic., 18(2): 103-108.

Choi, S.Y. 1987. Foraging activity of insects on weigh onion flowers (Allium sistulosum). Korean J. Apic.,2(2): 67-74.

Chopra, R.N. 1933 Indigenous drugs of India, The Art Press, Calcutta, p. 3, 61.

Cirudarescu, G. 1971. Pollinators of lucerne and factors influencing their effectiveness in South Eastern parts of Birse depression. Analele Universitati Bucuresti. Biologica Animale, 20: 77–81.

Clark, W.C. and Dickson, N.M., 2003. Sustainability science: the emerging research program. Proceedings of the National Academy of Sciences 100, 8059–8061.

Collison, C.H. 1973. Nectar secretion and how it affects the activity of honey bees in the pollination of hybrid pickling cucumbers, Cucumis sativus L. M.Sc. Thesis, Michigan State University, East Lansing, USA.

Collison, C.H. 1976. The interrelationship of honeybee activity, foraging behaviour, climatic conditions and flowering in the pollination of pickling cucumber, Cucumis sativus L. Ph.D. Thesis, Michigan State University, USA, 268pp.

Collison, C.H. and Martin, E.C. 1979. Behaviour of honey bees foraging on male and female flowers of Cucumis sativus.J. apic. Res.,18: 184-190.

Colville, J.F., Picker, M.D. and Cowling, R.M., 2002. Species turnover of monkey beetles (Scarabaeidae: Hopliini) along environmental and disturbance gradients in the Namaqualand region of the succulent Karoo, South Africa. Biodiversity and Conservation 11, 243–264.

Comin, D. 1946. Onion production. 186 pp. Orange-Judd Publishing Co., Inc., New York

Corbet, S.A. 1978 A bee's view of nectar. Bee World 57: 25-32.

Crailsheim, K. 1991. Interadult feeding of jelly in honey bees (Apis mellifera L.) colonies.J. Comp. Physiol.161B: 55-60.

Corbet, S.A., Fussell M., Ake, R., Fraser, A., Gunsen, C., Salvage, A and Smithy, K. 1993. Temperature and the pollinating activity of social bees. Ecological Entomology 18: 17-30.

Costanza, R., D'Arge, R., de Groot, R.S., Farber, S., Grasso, M., Hannon, B., Limburg, K., Naeem, S., O'Neill, R. V, Paruelo, J., Raskin, R.G., Sutton, P. and van den Belt, M., 1997. The value of the world's ecosystem services and natural capital. Nature 387, 253–260.

Costanza, R., D'Arge, R., de Groot, R.S., Farber, S., Grasso, M., Hannon, B., Limburg, K., Naeem, S., O'Neill, R. V, Paruelo, J., Raskin, R.G., Sutton, P. and van den Belt, M., 1998. The value of the world's ecosystem services and natural capital. Ecological Economics 25, 3–15.

Cowling, R.M. and Richardson, D.M., Pierce, S.M. (Eds.), 1997. Vegetation of southern Africa. Cambridge University Press, Cambridge.

Cowling, R.M., 1986. A description of the Karoo Biome Project. South African National Scientific Programmes Report No. 122. Pretoria.

Cowling, R.M., Egoh, B., Knight, A.T., Farrell, P.J.O., Reyers, B., Rouget, M., Roux, D.J., Welz, A., Wilhelm-Rechman, A. and O'Farrell, P.J., 2008. An operational model for mainstreaming ecosystem services for implementation. Proceedings of the National Academy of Sciences 105, 9483–9488.

Cowling, R.M., Esler, K.J. and Rundel, P.W., 1999. Namaqualand, South Africa – an overview of a unique winter-rainfall desert ecosystem. Plant Ecology 142, 3–21.

Cowling, R.M., Roux, P.W. and 1987. The karoo biome: a preliminary synthesis Part 2 - vegetation and history. South African National Scientific Programmes Report, No. 142, Pretoria.

Cowling, R.M., Roux, P.W., Pieterse, A.I.H., 1986. The karoo biome: a preliminary synthesis. Part 1 - physical environment. South African National Scientific Programmes Report, No. 142, Pretoria.

Crane, E. 1990. Bees and Beekeeping.Bath Press Ltd., Avon., 68-69pp.

Cremo, M.A. and Goswani, M., 1995. Divine nature: a spiritual perspective on the environmental crisis. Bhaktivedanta Book Trust International.

CSIRO 1991. The insects of Australia: A textbook for students and research workers. 2nd edn. Carlton, Victoria, Melbourne University Press.

Cunningham, SA, FitzGibbon, F and Heard, TA. 2002. The future of pollinators for Australian agriculture. Australian Journal of Agricultural Research, 53: 893–900.

Cupido, C.F., 2005. Assessment of veld utilisation practices and veld condition in the Little Karoo. MSc thesis, Stellenbosch University.

Currah, L. and Ockendon, D.J., 1983. Onion pollination by blowflies and honeybees in large cages. Annals of Applied Biology 103, 497–506.

Currah, L. and Ockendon, D.J., 1984. Pollination activity by blowflies and honeybees on onions in breeders' cages. Annals of Applied Biology 105, 167–176.

Daily, G.C. and Matson, P.A., 2008. Ecosystem services: From theory to implementation. Proceedings of the National Academy of Sciences 105, 9455–9456.

Daily, G.C., 1997. Nature's services: societal dependence on natural ecosystems. Island Press, Washington, D.C.

Daily, G.C., 2000. Management objectives for the protection of ecosystem services. Environmental Science & Policy 3, 333–339.

De Groot, R.S., Wilson, M.A. and Boumans, R.M.J., 2002. A typology for the classification, description and valuation of ecosystem functions, goods and services. Ecological Economics 41, 393– 408.

De Marco, P. and Coelho, F.M., 2004. Services performed by the ecosystem: forest remnants influence agricultural cultures' pollination and production. Biodiversity and Conservation 13, 1245–1255.

De Ruijter, A., 2002. Pollinator diversity and sustainable agriculture in the Netherlands, in: Kevan, P., Imperatriz-Fonseca, V.L. (Eds.), Pollinating bees: the conservation link between agriculture and nature. Ministry of Environment, Brasilia, pp. 67–70.

Dean, W.R.J. and Milton, S.J., 1995. Plant and invertebrate assemblages on old fields in the arid southern Karoo, South Africa. African Journal of Ecology 33, 1–13.

Dean, W.R.J. and Milton, S.J., 1999. The Karoo: ecological patterns and processes. Cambridge University Press, Cambridge.

Delbrassine S. and Rasmont P. 1988 Contribution à l'étude de la pollinisation du Colza, Brassica napus L. var. Oleifera (Moench) Delile, en Belgique. Bulletin

Delplane, K.S. and Mayer, D.F. 2000. Crop pollination by bees. CABI Publ., Wallingford, Oxon, UK.

Deodikar G B and Suryanarayana M C 1977 Pollination in the service of increasing farm production in India. Adv. Pollen Spore Res. 2: 67–87.

Deodikar G. B., Suryanarayana M. C. 1977. Pollination in the service of increasing farm production. In: Nair P. K. K. (Ed.) Advance in Pollen-Spore Research.

Deodikar, G.B. and Suryanaryana, M.C. 1972. Crop yields and bee pollination. Indian Bee J. 34:53-64.

Devi M, Sharma H. K and Rana K 2015. Measuring diversity and density of insect pollinators on coriander, Coriandrum sativum L by different sampling methods International Journal of Farm Sciences 5(3) : 179-189, 2015

Devi S, Gulati R, Kanika T and Poonia A. 2015 The pollination biology of onion (Allium cepa L.)- A Review Agri. Review, 36 (1): 1-13

Devillers, J., Dore, J.C., Tisseur, M., Luzeau, S. and Maurin, G. 2004. Modelling the flight activity of Apis mellifera at the hive entrance.Computers and Electronics in Agriculture, 42(2): 87–109.

Dewey. D.R.; Lu, K.U. 1959 A correlation and path analysis of crest wheat grass seed production. Agron. J. 51 : 515 –518.

Dewey. D.R.; Lu, K.U. 1959. A correlation and path analysis of crest wheat grass seed production. Agron. J. 51 : 515 –518.

Diederichsen A. 1996 Coriander (Coriandrum sativum L.). Promoting the conservation and use of underutilized and neglected crops. 3. Inst. Plant Genetics and Crop Plant Research. Gatersleben/ International Plant Genetic Ressources Institute. Rome. 83 pp.

Donaldson, J.S., Nänni, I., Zachariades, C. and Kemper, J., 2002. Effects of habitat fragmentation on pollinator diversity and plant reproductive success in renosterveld shrublands of South Africa. Conservation Biology 16, 1267–1276.

Driver, A., Desmet, P.G., Rouget, M., Cowling, R.M. and Maze, K., 2003. Succulent Karoo ecosystem plan. Biodiversity component technical report, Cape Conservation Unit Report No CCU 1/03. Kirstenbosch, Cape Town.

Duncan, A. A. 1965. Vegetable seed production in onion. Oreg. Agr. Ext. Serv. FS 88, Leaflet.

Duncan, D. B. 1955, Multiple range and multiple F tests. Biometrics 11(1): 1-42, 1955.

Eardley, C.D., 1996. The genus Scrapter Lepeletier & Serville (Hymenoptera: Colletidae). African Entomology 4, 37–92.

Ehrlich, P.R. and Raven, P.H., 1964. Butterflies and plants: a study in coevolution. Evolution 18, 586–608.

Eickwort G. C. and Ginsberg H. S. (1980) Foraging and mating behaviour in Apoidea. Annual Review of Entomology 25: 421-446.

El-Berry A R, Gawaad A A A, Moustafa M A F, El-Gayar F H, Berrry A R El, AbdelGawaad A A and Gayar F H El 1974 Pollinators other than honey bees visiting certain medicinal plants in Egypt. Zeitschrif fur Angewandte Entomologie. 76: 113–119.

El-Berry A. A., Moustafa M. A., Abdel-Gawaad A. A., El-Bialey S. 1974 Pollinators other than honey bees visiting certain vegetable plants in Egypt. Journal of Applied Entomology 77(1-4): 106-110.

Ellis P. R., Wheatley G. A. and Hardman J. A. 1978. Preliminary studies of carrot susceptibility to carrot fly attack. Annals of Applied Biology 88(1):159-170.

Emlen, J.M. 1966. The role of time and energy in food preference.American Naturalist, 100: 611-617.

Enzie J. V. 1943. Experiments in the production of carrot seed. N. Mex. Agr. Expt. Sta. Bul. 308, 11 pp.

Erickson, EH and Peterson, CE. 1979. Asynchrony of floral events and other differences in pollinator foraging stimuli between fertile and male-sterile carrot inbreds. Journal of the American Society for Horticultural Science, 104: 639–643.

Erickson, EH, Peterson, CE and Werner, P. 1979. Honey bee foraging and resultant seed set among male-fertile and cytoplasmically male-sterile carrot inbreds and hybrid seed parents. Journal of the American Society for Horticultural Science, 104: 635–638.

Erickson, H. T., and Gabelman, W. H. 1956. The effect of distance and direction on cross-pollination in onions. Amer. Soc. Hort. Sci. Proc. 68: 351-357.

Esty, D., Levy, M.A., Srebotnjak, T. and de Sherbinin, A., 2005. 2005 Environmental sustainability index: benchmarking national environmental stewardship. Yale Center for Environmental Law and Policy, New Haven.

Ewert, E. 1942. Nectar production of cooking onions. Deut Bienenzucht 49: 234- 235. [In German.]

Ewies, M.A. and El-Sahhar, K. 1977. Observation on the behaviour of honeybee on onion and their effect on seed yield.J. Apic Res., 16: 194-196.

Fahn, A. 1949. Studies in the ecology of nectar secretation.Palest. J. Bot. Ser,4: 207-224. FAO.2002. Production Year Book, 56.

Finch S. 1993. Integrated pest management of the cabbage root fly and the carrot fly. Crop Protection 12(6):423-430.

Fisher, B. and Turner, R.K., 2008. Ecosystem services: Classification for valuation. Biological Conservation 141, 1167–1169.

Fisher, B., Turner, R.K. and Morling, P., 2009. Defining and classifying ecosystem services for decision making. Ecological Economics 68, 643–653.

Flemion, R., and J. Olsen 1950. Lygus bugs in relation to seed production and occurrence of embryoless seeds in various Umbelliferae species. Boyce Thompson lust. Contrib. 16: 39-46. 1950.

Flemion, R., and K T. Hendrickson 1949. Further studies on the occurrence of embryoless seeds and immature embryos in the Umbelliferae. Boyce Thompson Inst. Contrib. 15: 291-297. 1949.

Franklin, D. F. 1948. Some problems in carrot seed production. Seed world 63(8): 8-9,44.

Franklin, D. F. 1953. Growing carrot seed in idaho. Idaho agr. Expt. Sta. Bul. 294, 35 pp.

Franklin, D. F. 1958. Effect on hybrid onion seed production of using different ratios of male-sterile and pollen rows. Amer. Soc. Hort. Sci. Proc. 71: 435-439.

Franklin, D. F. 1970. Problems in the production of vegetable seed. In the Indispensable pollinators, ark. Agr. Ext. Serv. Misc. Pub. 127, pp. 112-141.

Free J. B. 1993 Insect pollination of crops. Ed. Academic Press. Harcourt Brace Jovanich Publ. London. 684 pp.

Free, J. B. 1960. The behaviour of honey bees visiting the flowers of fruit trees. J. Anim. Ecol.,29: 385-395.

Free, J.B. and Williams, I.H., 1972. The transport of pollen on the body hairs of honeybees (Apis mellifera L.) and bumblebees (Bombus spp. L.). Journal of Applied Ecology 9, 609–615.

Free, J.B., Williams, I.H., Longden, P.C. and Johnson, M.G. 1975. Insect pollination of sugar beet (Beta vulgaris L.) seed crops.Ann. Appl. Biol.,81: 127-134.

Free, JB. 1970. Insect pollination of crops, London: Academic Press.

Frisch, K.Von. 1950. Bees, Their Vision Chemical Sense and Language.Cornell Univ. Press, Ithaca, New York.

Gallai, N., Salles, J., Settele, J. and Vaissière, B.E., 2009. Economic valuation of the vulnerability of world agriculture confronted with pollinator decline. Ecological Economics 68, 810–821.

Gary, N. E., Witherell, P. C., and Marston, J. 1972. Foraging range and distribution of honey bees used for Carrot and onion pollination. Environmental ent. 1(1): 71 - 78.

Gary, N.E. 1975. Activities and behaviour of honey bee. In: The Hive and Honeybee. Dadant and sons, Illinois, Hamilton, 237-240 pp.

Gary, N.E., Witherell, P.C. and Marston, J.M., 1972. Foraging range and distribution of honey bees used for carrot and onion pollination. Environmental Entomology 1, 71–78.

Gary, N.E., Witherell, P.C., Lorenzen, K. and Marston, J.M., 1977. The interfield distribution of honey bees foraging on carrots, onions and safflower. Environmental Entomology 6, 637–640.

Gegear, R.J. and Thomson, J.D. 2004. Does the flower constancy of bumble bees reflect foraging economics. Ethology.,110: 1-2.

George, R.A.T., 1999. Vegetable seed production, 2nd Edition. CAB Interantional Publishing, New York.

George, RAT. 1999. Vegetable seed production, 2nd edn, Eastbourne,, UK: Antony Rowe.

Gess, F.W. and Gess, S.K., 1993. Effects of increasing land utilization on species representation anddiversity of aculeate wasps and bees in the semi-arid areas of southern Africa., in: La Salle, J., Gauld, I.D. (Eds.), Hymenoptera and Biodiversity. CAB International, Wallingford, England, pp. 83–113.

Gess, S.K. and Gess, F.W., 2004a. Distributions of flower associations of pollen wasps (Vespidae: Masarinae) in southern Africa. Journal of Arid Environments 57, 17–44.

Gess, S.K., 1992. Biogeography of the masarine wasps (Hymenoptera: Vespidae: Masarinae), with particular emphasis on the southern African taxa and on correlations between masarine and forage plant distributions. Journal of Biogeography 19, 491–503.

Gess, S.K., 1996. The pollen wasps: ecology and natural history of the Masarinae. Harvard University Press, Cambridge.

Gess, S.K., 2001. The Karoo, its insect pollinators and the perils which they face. International Pollinator Initiative (IPI) Case Study. (www.ecoport.org)

Gess, S.K., Gess, F.W., 2004b. A comparative overview of flower visiting by non-Apis bees in the semi-arid to arid areas of southern Africa. Journal of the Kansas Entomological Society 77, 602–618.

Gilbert, F.S. 1980. Flower visiting by hoverflies (Syrphidae). Ann. Res. Ecol. Syst., 6: 139-170.

Gill, R.J., Ramos-Rodriguez, O. and Raine, N.E., 2012. Combined pesticide exposure severely affects individual- and colony-level traits in bees. Nature 491, 105–8.

Gill, S.S., Kaushik, H.D. and Sharma, S.K. 2001. Pollination efficiency of three Apis spp. foraging on phalsa (Grewia subinaequalis D.C.). Ann. Ent., 19(1): 19-21.

Gjorgji, M. and Rukie, A. 1997. The effects of pollination and the use of bees on number and quality of onion seed 1. Balkan Symposium on vegetables and potaoes, ISHS.Acta Hortic.,462: 12.

Gladis T 1997. Bees versus flies?—rearing methods and effectiveness of pollinators in crop germplasm regeneration. KW Richards VII International Symposium on Pollination. Lethbridge, Alberta, Canada 23–28June, 1996. Pp. 235 – 238.

Glukhov M. M. 1955 Honey plants. In: Mc Gregor S. E. (Ed.) Insect pollination of cultivated crop plants. Ed. Agriculture Handbook. Serv. Rech. Agri. U.S. Gov. Printing Off. Washington. 411pp.

Glukhov, M.M. 1955. Honey Plants. Izd-vo Seikhoz Zit-ry (In Russian): 512 PP. Izd. 6,

Goldman, R.L., Tallis, H., Kareiva, P., Daily, G.C., Goldman, L. and Dailyt, G.C., 2008. Field evidence that ecosystem service projects support biodiversity and diversify options. Proceedings of the National Academy of Sciences 105, 9445–9448.

Goyal, NP, Singh, M and Kandoria, JL. 1989. Role of insect pollination in seed production of carrot, Daucus carota Linn. Indian Bee Journal, 51: 89–93.

Grace J. and Nelson M. 1981 Insects and their pollen loads at a hybrid Heracleum site. New Phytologist 87: 413-423.

Greenleaf, S.S. and Kremen, C., 2006a. Wild bees enhance honey bees' pollination of hybrid sunflower. Proceeding of the National Academy of Sciences 103, 13890–13895.

Greenleaf, S.S. and Kremen, C., 2006b. Wild bee species increase tomato production and respond differently to surrounding land use in North California. Biological Conservation 133, 81–87.

Guy Hyperlink "http://link.springer.com/search?facet-creator = % 22 Guy + Boivin % 22" Boivin 1999. Integrated Management for Carrot Weevil integrated Pest Management Reviews 4(1):21-37.

Haglar, J.R., Cohen, A.C. and Loper, G.M. 1990. Production and decomposition of onion nectar and honey bee foraging activity in Arizon.Env.Ent.,19(2): 327-331.

Hagler, J.R 1990. Honey bee (Apis mellifera L.) response to simulated onion nectars containing variable sugar and potassium concentrations. Apidologie 21:115-121.

Hagler, J.R., Cohen, A.C. and Loper, G.M., 1990. Production and composition of onion nectar and honey bee (Hymenoptera: Apidae) foraging activity in Arizona. Environmental Entomology 19, 327–331.

Hamilton R. P. 1946. Onions need bees. Gleanings Bee Cult. 74: 23.

Hammert A., Harper D. A. T. and Ryan P. D. 2001 P.A.S.T : Palaeontological Statistics Software Package for Education and Data Analysis. Palaeontological Electronica 4 (1): 1-9.

Harris, L.F. and Johnson, S.D., 2004. The consequences of habitat fragmentation for plant–pollinator mutualisms. International Journal of Tropical Insect Science 24, 29–43.

Harrison, J.F. and Fewell, J.H. 2002. Environmental and genetic influences on flight metabolic rate in the honeybee, Apis mellifera.Comparative Biochem.Physio.,133(2): 323–333.

Haslbachova, H and Kubisova, S. 1982. Pollinating activity of honey bees (Apis mellifera) on onions (Allium cepa). I. Flight of bees and its influence on the amount of seed produced. Sbornik UVTIZ, Zahradnictvi 9 (2): 119-126

Haslbachova, H.; Kubisova, S. and Roznovsky, M. 1981. Studies of honeybee (Apis mellifera) activity on onion (Allium cepa). Acta Universitatis Agriculturae, Brno, A, Facultas Agronomica 29 (3/4): 313-321

Hawthorn, L. R. 1951. Studies of soil moisture and spacing for seed crops of carrots and Onions U.S. Dept. Agr. Cir. 892, 26 pp.

Hawthorn, L. R., Bohart, G. E, toole, E. H., and others. 1960. Carrot seed production as affected by insect pollination. Utah agr. Expt. Sta. Bul. 422, 18 pp.

Hawthorn, L. R., Bohart, G. E., and Toole, E. H. 1956. Carrot seed yield and germination as affected by Different levels of insect pollination. Amer. Soc. Hort. Sci. Proc. 67: 384- 389.

Hawthorn, R.L., Bohart, G.E., Toole, E.H., Nye, W.P. and Levin, M.D. 1960. Carrot seed production as affected by insect pollination, Bull. Utah Agric. Exp. Sta., 422, 18.

Heal, G., 2000. Valuing ecosystem services. Ecosystems 3, 24–30.

Heard, T.A. and Exley, E., 1994. Diversity, abundance and distribution of insect visitors to macadamia flowers. Environmental Entomology 23, 91–100.

Heeger E. F. 1989 Handbuch des Arznei- und Gewürzpflanzenbaues. Unveränderte Auflage. Frankfurt. 1956 pp.

Heinrich, B. 1975. Energetics of pollination. Ann. Rev. Ecol. Syst. 6: 139-170.

Heinrich, B. and Raven, P.H. 1972. Energetics and pollination ecology.Science, 17: 597-602.

Helmholtz Association of German Research Centres. 2008. Economic Value of insect pollination. Science Daily.

Hesse, A.J., 1938. A revision of the Bombyliidae (Diptera) of southern Africa. Part 1. Annals of the South African Museum 34, 1–1053.

Hilton-Tailor, C. and Le Roux, A., 1989. Conservation status of the Fynbos and Karoo biomes, in: Huntley, B.J. (Ed.), Biotic Diversity in southern Africa. Oxford University Press, Cape Town, pp. 202–223.

Hmamouchi M. 1997. Plantes alimentaires, aromatiques, condimentaites, medicinales et toxiques au

Hoffman, M.T., Cousins, B., Meyer, T., Petersen, A. and Hendricks, H., 1999. Historical and contemporary land use and the desertification of the Karoo, in: Dean, W.R.J., Milton, S.J. (Eds.), The Karoo: Ecological Patterns and Processes. Cambridge University Press, Cambridge, pp. 257–273.

Howlett, B.G., Donovan, B.J., McCallum, J.A., Newstrom, L.E. and Teulon, D.A.J., 2005. Between and within field variability of New Zealand indigenous flower visitors to onions. New Zealand Plant Protection 58, 213–218.

Howlett, BG, Walker, MK, Newstrom-Lloyd, LE, Donovan, BJ and Teulon, DA. 2009. Window traps and direct observations record similar arthropod flower visitor assemblages in two mass flowering crops. Journal of Applied Entomology, 133: 553–564

Hussein M. H. and Abdel-Aal S. A. 1982 Wild and honeybees as pollinators of 10 plant species in Assiut area. Egypt. Zeitschrift fur Angewandte Entomologie 93: 342-346.

Hwang, H.J., Suk, J.K. and Ha, I.J. 1998. Effect of Pollinating insect on seed yield in seed production of onion (*Allium cepa* L.).J. Hortic. Sci., 40(2): 27-30.

Ilyas M. 1980 Spices in India: III. Economic Botany 34: 236-259.

Inouye, D.W. 1980. The effect of proboscis length and corolla tube lengths on patterns and rates of flower by bumble bees.Oecologia, 45: 197-201.

Ivanova K. V., Stoletova E. A. 1990 Istorija kul'tury i vnutrividovaja klassifikacija koriandra posevnogo (Coriandrum sativum L.) Sbornik Naucnych Trudov Prikladnoj Botaniki, Genetiki i Selekcii 133: 26-40.

Jablonski, B., Skowronek, J., Woyke, H. W., and Doruchowski, R. W. 1982. Biologia kwitnienia, nektarowanie, zapylanie i owocowanie meskosterylnych linii cebuli (Allim cepa L.). Pszczelnicze-Zeszyty- Naukowe, 26: 57-104.

Jadhav, L.D. 1981. Role of insects in the pollination of onion (Allium cepa L) in the Maharastra state, India.Indian bee J., 43: 61-63.

Jain, K.L. and Kapil, R.P. 1980. Foraging rhythm of megachilid bees in relation to the flowering of *Medicago sativa* L. and *Parkinsonia aculeata* L. Indian Bee. J.,42: 35-38.

Jamieson, C.A. and Austin, G.H. 1956. Preference of honey bee for sugar solution. Proc.10th int. Congr.Entomol. Montreal, 4: 1059-1065.

Johannsmeier, M.F., 2001. Beekeeping in South Africa, 3rd Edition. Agricultural Research Council-Plant Protection Research Institute, Handbook No. 14, Pretoria.

Johannsmeier, M.F., 2005. Beeplants of the south-Western Cape. Agicultural Research Council- Plant Protection Research Institute, Handbook No. 17., Pretoria.

Johnson, S.D., 2004. An overview of plant–pollinator relationships in southern Africa. International Journal of Tropical Insect Science 24, 45–54.

Jones O. T. and Coaker T. H. 1977. Oriented responses of carrot fly larvae, Psila rosae, to plant odours, carbon dioxide and carrot root volatiles. Physiological Entomology 2(3):189-197.

Jones, D. F., and Clarke, A. E. 1943. inheritance of male-sterility in onion and the production of hybrid seed Amer. Soc. Hort. Sci. Proc. 43: 189-193.

Jones, H. A and Emsweller, S. L. 1934. The use of flies as onion pollinators. Amer. Soc. Hort. Sci. Proc. 31 (sup.): 160-164.

Jones, H. A and Emsweller, S. L. 1936. A male-sterile onion. Amer. Soc. Hort. Sci. Proc. 34: 582-585.

Jones, H. A and Mann, L. K. 1964. Onions and their allies. 286 pp. Leonard Hill Book CO., London.

Jones, H. A Poole, C. F., and Emsweller, S. L. 1936. Bolting habit in the onion. Abstract in Amer. Soc. Hort. Sci. Proc. 33: 490.

Jones, H. A., and Emsweller, S. L. 1933. Methods of breeding onions. Hilgardia 7: 625-642.

Jones, H.A. 1933 Vegetable breeding at the university o California. Proc. Am. Soc. Hortic. Sci., 29:572-581.

Jones, H.A. and Clarke, A.E., 1943. Inheritance of male sterility in the onion and the production of hybrid seed. Proceedings of the American Society for Horticultural Science 43, 189–194.

Jones, H.A. and Emsweller,S.L. 1933. Methods of breeding onions.Hilgardia, 7: 625-642.

Jones, H.A. and Mann, L.M., 1963. Onions and their allies, botany, cultivation and utilization. Leongard Hill, London.

Jones, H.A. and Rosa, J.T., 1928. Truck crop plants, 1st Edition. McGraw-Hill, New York.

Jula, F., Pirvu, E., and Illyes, C. 1965. [The melliferous value of sainfoin (onobrychis viciifolia) and onion (Allium cepa L.), under the soil and climatic conditions around CLUJ.] Lucr. Stiint. lnst. Agron. Cluj 21: 99-106. [In Romanian, English and Russian summaries.]

Jula, F., Privu, E. and Ilyes, G. 1965. The melliferous value of sainfoin (Onobrychis vichfolia) and onion (Allium cepa L.) under the soil and climatic condition around.Clij. Lucr.Stint.Inst.Agron.Cluj., 21: 99-106.

Jun Hwang Hae; Kyu Suh Jun; Jong Ha In; WooRyu Young 1998. Effect of pollinating insects on seed yield in seed production of onion (Allium cepa L.). RDA Journal of Horticulture Science : 27-30

Kalmath, B. S and Sattigi, H. N 2004 Efficacy of chemicals in attracting honey bees and other insect pollinators to onion bloom. Pest Management and Economic Zoology 12(1): 101-104

Kapil, R.P. and Brar, H.S. 1971. Foraging behaviour of Apis florea F. on Brassica campestris var toria.In: Proceedings of XXIIth International Congress of Apiculture, Moscow, 335–339.

Kapil, R.P. and Jain, K.L. 1980. Biology and utilization of insect pollinators for crop production.Final Tech. Report, Department of Zoology, Haryana Agricultural University, Hisar, India, 81p.

Kapil, R.P. and Kumar, S. 1974. Foraging activity of *Apis dorsata* F. on *Brassica juncea* Czern and Coss.J. Bombay Natural History Soc., 71(2): 327–331.

Kearns, CA, Inouye, DW and Waser, NM. 1998. Endangered mutualisms: the conservation of plant-pollinator interactions. Annual Review of Ecology and Systematics, 29: 83–112.

Kendall, D.A. and Solomon, M.E., 1970. Insect pollination of fruit trees, Republication of the Agriculture Horticultural Research Station, University of Bristol.

Kevan, PG and Baker, HG. 1983. Insects as flower visitors and pollinators. Annual Review of Entomology, 28: 407–453.

Kirtikar, K.R. and Basu, B.D. 1935. Indian medicinal plants, Lolit Mohan Basu, Allahabad, p. 61.

Klein, A.-M. and Steffan-Dewenter, I. and Tscharntke, T., 2003b. Fruit set of highland coffee increases with the diversity of pollinating bees. Proceedings of the Royal Society B: Biological Sciences 270, 955–961.

Klein, A.M., Steffan-Dewenter, I. and Tscharntke, T., 2003a. Pollination of Coffea canephora in relation to local and regional agroforestry management. Journal of Applied Ecology 40,837– 845.

Klein, A-M, Vaissiere, BE, Cane, JH, Steffan-Dewenter, I, Cunningham, SA, Kremen, C and Tscharntke, T. 2007. Importance of pollinators in changing landscapes for world crops. Proceedings of the Royal Society B: Biological Sciences, 274: 303–313

Kluser, S. and Peduzzi, P., 2007. Global pollinator decline: a literature review. UNEP/ GRID, Geneva.

Kochetov, A.S. 2004. Use of Carpenter bees for cucumber pollination in film greenhouses. Kartofel Ovoshchi, 6: 28.

Kordakov, Z.M., 1956. Honeybees and pollination of seed plants of the common onion, in: Gubin, A.F. (Ed.), Pollination of Agricultural Crops. Gos. Izd-vo Selkhoz Litry, pp. 163–171.

Kordakova, Z.M. 1956. Honey bees and pollination of seed plants of the common onion. In: Pollination of agricultural plants (eds. Krishchunas, I.V. and Gubin, A.F.). Moskva, Gos. Izd-vo Selkhoz Lit-r: 163-171.

Kosior, A., Celary, W., Olejniczak, P., Fijal, J., Król, W., Solarz, W. and Plonka, P., 2007. The decline of the bumble bees and cuckoo bees (Hymenoptera: Apidae: Bombini) of Western and Central Europe. Oryx 41, 79–88.

Koul A. K., Hamal I. A., Gupta S. K. 1989 Pollination mechanism in Coriandrum sativum Linn. Proceedings of the Indian Academy of Sciences Plant Science 99: 509-515.

Koul A. K., Koul P., Hamal I. A. 1986 Insects in relation to pollination of some umbellifers. Bulletin of the Botanical Survey of India 28: 39-42.

Koul, P, Koul, AK and Hamal, IA. 1989. Reproductive biology of wild and cultivated carrot (Daucus carota L.). New Phytologist, 112: 437–443. [CrossRef]

Koul, P, Sharma, N and Koul, AK. 1993. Pollination biology of Apiaceae. Current Science, 65: 219–222.

Kremen, C., Williams, N.M. and Thorp, R.W., 2002. Crop pollination from native bees at risk from agricultural intensification. Proceedings of the National Academy of Sciences 99, 16812–16816.

Kremen, C., Williams, N.M., Bugg, R.L., Fray, J.P. and Thorp, R.W., 2004. The area requirements of an ecosystem service: crop pollination by native bee communities in California. Ecology Letters 7, 1109–1119.

Kuhlmann, M., 2009. Patterns of diversity, endemism and distribution of bees (Insecta: Hymenoptera: Anthophila) in southern Africa. South African Journal of Botany 75, 726–738.

Kumar and Jaiswal 2012. Effect of honeybee (Apis mellifera L.) pollination on yield and quality in coriander Indian Journal of Entomology 74(3): 281-284.

Kumar J, Mishra R C and Gupta J K 1989. Effect of honey bee pollination on onion (Allium cepa L.) seed production. Indian Bee Journal 51 : 3-5.

Kumar, J., Mishra, R.C and Gupta, J K. 1985. The effect of mode of pollination on Allium species with observations on insects as pollinators. J. Apic. Res. 24(1):62-66.

Kumar, J., Mishra, R.C. and Gupta, J.K. 1989. Effect of honeybee pollination on onion (Allium cepa L) sedd production. Indian Bee J., 51:1, 3-5.

Kumar, J., Rao, K.V.K., Gupta, P.R and Dogra, G.S. 1989. Temporal distribution of bees (Hymenoptera: Apoidea) on spring flowering crops in mid hills of Himachal Pradesh, India. Indian Bee J. 51(2): 55-58.

Kumar, K. 2002. Studies on insect pollinators of sunflower (*Helianthus annuus* L.). M.Sc. Thesis, Haryana Agricultural University, Hisar, Haryana, India.

Kumar, R. 2004. Role of *Apis mellifera* L. in pollination of cucumber (*Cucumis sativus* L.) under poly-house conditions. M.Sc. Thesis, Haryana Agricultural University, Hisar, Haryana, India.

Kumar, S. 2002. Studies on the activities of insect pollinators of cucurbit crops. M.Sc. Thesis, Haryana Agricultural University, Hisar, Haryana, India.

Kuti, J.O. and Rovelo, C. 1992. Effect of pollination techniques on fruit set and seed yield of egusi melon (Citrullus colocynthis). Report Cucurbit Genetics Cooperative,15: 80-81.

Kwon, Y.J. and Saeed, S. 2003. Effect of temperature on the foraging activity of Bombus terrestris L. (Hymenoptera: Apidae) on greenhouse hot pepper (Capsicum annuum L.) Appl. Entomol. Zool.,38(3): 275-280.

Lamborn, E and Ollerton, J. 2000. Experimental assessment of the functional morphology of inflorescences of Daucus carota (Apiaceae): testing the 'fly catcher effect'. Functional Ecology, 14: 445–454

Lamm, R.D., 2006. The culture of growth and the culture of limits. Conservation Biology 20, 269–271.

Lawrence JF and Britton EB 1991. Coleoptera. In : The insects of Australia: A textbook for students and research workers. Vol II. 2nd edn. CSIRO Division of Entomology (ed), Melbourne, Melbourne University Press. Pp. 1137.

Le Baron, F.C. 1962. Onion seed sample costs and production. Calif. Agr. Ext. Sen., Cost Data Sheet 22, Leaflet.

Le Maitre, D.C., Colvin, C. and Maherry, A., 2009a. Water resources in the Klein Karoo: the challenge of sustainable development in a water-scarce area. South African Journal of Science 105, 39–48.

Le Maitre, D.C., O'Farrell, P.J., Milton, S.J., Atkinson, D., De Lange, W.J., Egoh, B., Reyers, B., Colvin, C., Maherry, A. and Blignaut, J.N., 2009b. Assessment and evaluation of ecosystem services in the Succulent Karoo biome. Council for Scientific and Industrial Research, Pretoria.

Leadley, P., Pereira, H.M., Alkemade, R., Fernandez-Manjarres, J.F., Proenca, V., Scharlemann, J.P.W. and Walpole, M.J., 2010. Biodiversity scenarios: projections of 21st century - change in biodiversity and associated ecosystem services. Secretariat of the Convention on Biological Diversity, Montreal.

Lederhouse, R.C., Caron, D.M and Morse, R.A. 1968.Onion pollination in New York. New York's Fd Life Sci. 1: 8-9.

Lederhouse, R.C., Caron, D.M. and Morse, R.A., 1972. Distribution and behavior of honey bees on onion. Environmental Entomology 1, 127–129.

Lerer, H., Bailey, W.G., Mills, P.F. and Panki, W.M. 1982. Pollination activity of Megachile rotundata (Hymenoptera; Apidae).Env.Ent.,11: 997-1000.

Lerer, H.; Bailey, W.G.; Mills, P.; Pankiw, M 1982 Pollination activity of Megachile rotundata F. (Hymenoptera: Apoidea). Environmental Entomology 11: 997-1000.

LI, C.C. 1956 The concept of path coefficients and its impact on population genetics. Biometrics 12: 997-1000.

Lindsey A. H. 1984 Reproductive biology of Apiaceae. I. Floral visitors to Thaspium and Zizia and their importance in pollination. American Journal of Botany 71: 375-387.

Lorenzone M C A, Martinho M R, Freire J A H and Reis F P 1993. Effect of Pollination by honey bees on onion hybrid seed production. Presquisa-Agropecuana-Brasieira 28 : 229-235.

Losey, J. and Vaughan, M. 2006. The economic value of ecological services provided by Insects. Bioscience, 56 (4): 311-323.

Luk'janov I. A., Reznikov A. R. 1976 Koriandr. In: Smoljanova A. M., Ksendza A. T. (Eds.) Efirnomaslicnye kul'tury. Kolos. Moskva. Russia. pp. 9-57.

MacArthur, R. H. and Pianka, E. R. 1966. On the optimaluse of a patchy environment. American Naturalist, 100: 603-609.

Macgillivray, J. H. 1948. Effect of irrigation on the yield of onion seed. Amer. Soc. Hort. Sci. Proc. 51: 423-427.

Magruder, R., Wester, R. E., and others. 1941. Descriptions of types of principal american varieties of onions. U.S. Dept. Agr. Misc. Pub. 435, 87 pp.

Mann, L. K 1952. Anatomy of the garlic bulb and factors affecting bulb development. Hilgardia 21(8): 195-235.

Mann, L. K and Little, T. M. 1957. Growing garlic in California. Calif. Univ., Dept. Veg. Crops, Veg. Crops Ser. 89, 10 pp.

Mann, L. K and Woodbury, G. W. 1969. The effect of flower age, time of day and variety on pollen germination of onion, Allium cepa L. Amer. Soc. Hort. Sci. Proc. 94: 102-104.

Mann, L.P. and Woodbury, G.W., 1969. The effect of flower age, time of day and variety on pollen germination of onion, Allium cepa L. Journal of the American Society of Horticultural Science 94, 102–104.

Manning, J.C. and Goldblatt, P., 1996. The Prosoeca peringueyi (Diptera: Nemestrinidae) pollination guild in southern Africa: long-tongued flies and their tubular flowers. Annals of the Missouri Botanical Garden 83, 67–86.

Maroufi K., Farahani H. A. and Darvishi H. H. 2010 Importance of Coriander (Coriandrum sativum L.) Between the Medicinal and Aromatic Plants. Advances in Environmental Biology 4: 433-436.

Martin, E.C. 1979. The use of bees for crop pollination. In: The Hive and Honey bee. (ed.) Dadant and Sons, 579-614 pp.

Mathias M. E. 1994 Magic, myth and medicine. Economic Botany 48: 3-7.

Mayer D. F. and Lunden, J. D. 2001. Honey bee management and wild bees for pollination of hybrid onion seed. Acta Horticulturae, 561: 275-278.

Mayer D.F. and Lunden J.D. 2001. Honey Bee Management and Wild Bees for Pollination of Hybrid Onion Seed Proc. 8th Pollination Symp. Eds. P. Benedek & K.W. Richards Acta Hort. 561. 275-278.

Mayer, C., 2004. Pollination services under different grazing intensities. International Journal of Tropical Insect Science 24, 95–103.

Mayer, C., Soka, G. and Picker, M.D., 2006. The importance of monkey beetle (Scarabaeidae: Hopliini) pollination for Aizoaceae and Asteraceae in grazed and ungrazed areas at Paulshoek, Succulent Karoo, South Africa. Journal of Insect Conservation. 10, 323–333.

Mayer, D.F. and Lunden, J.D., 2001. Honey bee management and wild bees for pollination of hybrid onion seed. Acta Horticulturae. 561, 275–278.

McGregor, S.E. 1976. Insect pollination of cultivated crop plants.Agriculture Handbook, United States Department of Agriculture, No. 496.

McGregor, S.E., 1976. Common vegetables for seed and fruit, in: McGregor, S.E. (Ed.), Insect Pollination of Cultivated Crop Plants. United States Department of Agriculture, Washington.

Meer, Q. P., Van der, and Bennekom and J. L. Van. 1968. Research in pollen distribution in onion seed fields. Euphytica 17: 216- 219.

Meer, Q. P., Van der, and Bennekom and J. L. Van. 1969. Effect of temperature on the occurrence of male sterility in onion (Allium cepa L.). Euphytica 18: 389394.

Meine, C., Soulé, M., Noss, R.F. and Soule, M., 2006. “A mission-driven discipline”: the growth of conservation biology. Conservation Biology 20, 631–651.

Meléndez-Ramirez, V., Magaña-Rueda, S., Parra-Tabla, V., Ayala, R. and Navarro, J. 2002. Diversity of native bee visitors of cucurbit crops (Cucurbitaceae) in Yucatán, México.J Insect Conser.,6: 135-147.

Meloyan, A.A. 1975. Nectar sugar production by winter vetch (Vicia villosa) cultivars of different origins.Zapiski Lenin Gradskogo Selskokhozyais tvennago instituta, 271: 61-63.

Mevetty, P.B.E., Pinnisch, R. and Scarth, R. 1989. The significance of floral characteristics in seed production of four summer rape cultivar.A-lines with pollination cytoplasm. Can. J. Pl. Sci., 69(3): 915-918.

Millenium Ecosystem Assessment, 2005a. Ecosystems and human well-being: biodiversity synthesis. Island Press, Washington, DC.

Millenium Ecosystem Assessment, 2005b. Ecosystems and human well-being: current state and trends. Island Press, Washington, DC.

Millenium Ecosystem Assessment, 2005c. Biodiversity regulation of ecosystem services, in: Ecosystems and human well-being: current state and trends. Island Press, Washington, DC., pp. 297–329.

Milton, S.J., Dean, W.R.J., Plessis, M.A. and Siegfried, W.R., 1994. A conceptual model of arid rangeland degradation. Bioscience 44, 70–76.

Misra, R.C., Kumar, T., Veeresh, G.K., Shaankar, R.U. and Ganeshaiah, K.N. 1993 “Status of research in pollination biology in Himachel Pradesh”, Proc. Intern. Symp. Poll. in Tropics, August 8-13, India, 279 - 295.

Mitchell, R.J. 1992 Testing evolutionary and ecological hypotheses using path analysis and structural equation modeling. Functional Ecology 6:123-129.

Mitchell, R.J.; Paton, D.C 1990 Effects of nectar volume and concentration on sugar intake rates of Australian honeyeaters (Meliphagidae). Oecologia 83:238246.

Moffett, J.O., 1965. Pollinating experimental onion varieties. American Bee Journal 105, 378.

Mohan Rao G and Suryanarayana M C 1989. Effect of honey bee pollination on seed yield in onion (Allium cepa L.). Indian Bee Journal 51 : 9-11.

Mohan Rao G and Suryanarayana M C 1989. Effect of honey bee pollination on seed yield in onion (Allium cepa L.). Indian Bee Journal 51 : 9-11.

Mohr, N.A. and Jay, S.C. 1990. Nectar Production of selected cultivars of *Brassica compestris* L. and *Brassica napus* L.J.agril. Res., 29(2): 95-100.

Mommers, J. 1977. The concentration and composition of nectar in relation to honeybee visits to fruits trees Apidologie,8: 357-361.

Morandin, L.A., Winston, M.L., 2005. Wild bee abundance and seed production in conventional, organic, and genetically modified Canola. Ecological Applications 15, 871–881.

Morris, H. 1968. Pollination of cucumbers.Glean. Bee Cult.,1: 14-16.

Morse R.A. and Calderone, NW. 2000. The value of honey bees as pollinators of U. S. crops. Bee Culture, 128:1–15.

Mortaza Rasekh Adel, Hussein Sadeghi Namaghi and Mojtaba Hosseini 2013. Effects of insect pollinators on onion seed production quality and quantity J. Crop Prot. 2013, 2 (4): 395-402

Moskva, Gos. Izd-vo Selkhoz Lit-ry, pp. 163-171. [In Russian.] le Baron, F. C. 1962. Onion seed, sample costs and production. Calif. Agr. Ext. Sen., Cost Data Sheet 22, Leaflet.

Mucina, L. and Rutherford, M.C. (Eds.), 2006. The vegetation of South Africa, Lesotho and Swaziland, Strelitzia 19. South African National Biodiversity Institute, Pretoria.

Muller, H. 1983. The fertilization of flowers. Macmillian, London.

Munawar, M. S.; Shazia Raja; Shahid Niaz; Ghulam Sarwar 2011. Comparative performance of honeybees (Apis mellifera L.) and blow flies (Phormia terronovae) in onion (Allium cepa L.) seed setting. Journal of Agricultural Research (Lahore) 2011 Vol. 49 No. 1 pp. 49-56

Munawar, M.S. and Muzaffar, N.(1999). Role of honeybee (Apis mellifera L.) pollination on onion (Allium cepa) seed production. In: 36th limondia Congress September, 12-18, Vancouver-Canada, 276pp.

Mupade, R.V. Kulkarni, S.N and Kamte, G.S.2009. Effect of bee pollination on qualitative characters of onion. Indian J Plant Prot. 37(1&2): 186-187

Myers, N., Mittermeier, R.A., Mittermeier, C.G., de Fonseca, G.A.B. and Kent, J., 2000. Biodiversity hotspots for conservation priorities. Nature 403, 853–858.

Nascimento, W. M.; Pessoa, H. B. S. V.; Araújo, M. de T.1998. Utilization of Trigona spinipes as a pollinator in onion (Allium cepa L.) breeding programmes in Brazil. Journal of Applied Seed Production 1998 Vol. 16 pp. 47-49

Nemirovich-Danchenko, E.N. 1964. Concerning the nectar yield and floral biology of cucumbers.Izv. Tomsk. Otd.Vses. Bot. Obshch.,5:127-132.

Nogueira, C.R.H. and Calmona, R.C. 1993. Insect pollination of cucumber (Cucumis sativus var. Aodai Melhorada). Naturalia Sao Paulo., 18: 77-82.

Nuenz, J.A. 1977. Nectar flow by melliferous flora and gathering flow by Apis mellifera ligustica. Journal of Insect Physiology, 23: 265–276.

Nunez, J. 1997 Circadian variation of flight activity in colonies of Apis mellifera lingustica. J.Insect Physiol. 23 : 387-392.

Nye, W. P.; Waller, G. D. and Waters, N. D. 1971. Factors affecting pollination of onions in Idaho during 1969. Journal of the American Society for Horticultural Science 96(3): 330-332

Nye, W.P. and Anderson, J.L. 1974. Insect pollinators frequenting strawberry blossoms and the effect of honey bee on yield and fruit quality. J. Am. Soc. Hortic. Sci., 99: 40–44.

O'Farrell, P.J., De Lange, W.J., Le Maitre, D.C., Reyers, B., Blignaut, J.N., Milton, S.J., Atkinson, D., Egoh, B., Maherry, A., Colvin, C. and Cowling, R.M., 2011. The possibilities and pitfalls presented by a pragmatic approach to ecosystem service valuation in an arid biodiversity hotspot. Journal of Arid Environments 75, 612–623.

O'Toole, C., 1993. Diversity of native bees and agroecosystems, in: LaSalle, J., Gauld, I.D. (Eds.), Hymenoptera and Biodiversity. CAB International, Wallingford, UK.

Oh, H.W. and Woo, K.S. 1990. A study of foraging and pollen collecting activity of honey bee (Apis mellifera L.) in the spring.Korear J. Apic, 5(1): 1-22.

Oldroyd, B.P., 2007. What's killing American honey bees? PLoS Biology. 5, e168.

Osgood, C.E. 1974 Relocation of nesting populations of Megachile rotundata F. an important pollinator of alfalfa. J. Apic. Res. 13 : 67-74.

Paci, P. 1956. Research on the floral biology of the carrot. Riv. Ortoflorofruttic. Ital. 40: 414-423. [ln italian, english summary.]

Pankratova, E. P 1958. [data on the biology of blossoming and pollination Of carrots.] Dokl. Tskha 36: 118 - 123. [in Russian.] Aa-727/62.

Pankratova, E. P. 1957. The effect of bee pollination on the harvest of carrot Seed. Dokl. Tskha 30 (part 2): 332-336. [in Russian.] aa-396/61.

Pankratova, E.P. 1964 "Data on the biology of blossoming and pollination of carrots", Dokl. TSKhA, 36, 118-123.

Parker, F.D. and 1982. Efficiency of bees in pollinating onion flowers. Journal of the Kansas Entomological Society 55, 171–176.

Parker, F.D. and Hatley, C.L. 1978. Onion pollination viability of onion pollen and pollen diversity on insects body hairs. Proc.IV Int. Symp. Poll. M D Agri. Exp. Stn. Spec. Publ. Misc., 1: 201-206.

Parker, F.D., Hatley, C.L., 1979. Onion pollination: viability of onion pollen and pollen diversity on insect body hairs., in: Proceedings of the Fourth International Symposium on Pollination. Maryland agricultural experiment station special miscellaneous publication 1, pp. 201–206.

Pathak, C.S., 2000. Hybrid seed production in onion. Journal of New Seeds 1, 89–108.

Pedersen, M. W., F. K Todd, and F. V. Lieberman 1950. A portable field cage. U. S. Dept. Agr. Bur. Ent. and PI. Quar. ET-289.

Percival M S 1950 Pollen presentation and pollen collection. New Phytol. 46: 142–173.

Percival M S 1955 The presentation of pollen in certain angiosperms and its collection by Apis mellifera. New Phytol. 54: 353– 368.

Pereboom, J.J.M. and Biesmeijer, J.C. 2003. Thermal constraints for stingless bee foragers: the importance of body size and coloration. Oecologia, 137 (1): 42–50.

Picker, M.D., Midgley, J.J., 1996. Pollination by monkey beetles (Coleoptera: Scarabaeidae: Hopliini): flower and colour preferences. African Entomology 4, 7–14.

Pierre J., Le Guen J., Esnault R., Debbagh S. 1997 Méthode d'étude de la fréquentation de diverses féveroles par les insectes pollinisateurs. In: Proceedings of Mediterranean Food Legumes. Rennes, France. 20-22 February 1988. pp. 199-206.

Pierre J., Suzo M. J., Moreno M. T., Esnault R. et Le Guen J. 1999. Diversité et efficacité de l'entomofaune pollinisatrice Hymenoptera : Apidae) de la féverole (Vicia faba L.) sur deux sites en France et en Espagne. Annales de la Société Entomologique de France 35 (n.s. suppl.): 312-318.

Pimm, S.L., Russell, G.J., Gittleman, J.L., Brooks, T.M., 1995. The future of biodiversity. Science 269, 347–350.

Poole C. F. 1937. Improving the root vegetables. U.s. Dept. Agr. Yearbook 1937: 300-325.

Prakash V. 1990 Leafy Spices. CRC Press Inc., Boca Raton, pp. 31-32.

Prasad Rabindra; Chand Hari and Singh, R. 2000. Effect of honeybee, Apis mellifera pollination on the yield and germination of onion, Allium cepa seeds. Shashpa 2000 Vol. 7 No. 1 pp. 53-55

Prasad, R., Chand, H. and Singh, R. 2000. Effect of boney bee, Apis mellifera pollination on the yield and germination of onion, *Allium cepa* seeds.Shashpa, 7: 1, 53-55.

Priti C and Sihag, R.C. 1997. Diversity, visitation frequency, foraging behaviour and pollinating efficiency of insect pollinators visiting cauliflower (Brassica aleracea) blossoms.Indian Bee J., 59(4): 230-237.

Priti, C. 1998. Abundance and Pollination efficiency of insect visitors of onion bloom. Indian Bee Journal, 60 (2): 75-78.

Pyke, G.H., Pultiam, H.R. and Charnov, E.L. 1977. Optimal foraging: Selective review of theory and tests. Qtly. Rev. Biol., 52: 137-154.

Rabindra Prasad, Hari Chand and Singh R 2000. Effect of honey bee Apis mellifera pollination on the yield and germination of onion (Allium cepa) seeds. Shashapa 7 : 53-55.

Rader, R, Howlett, BG, Cunningham, SA, Westcott, DA, Newstrom-Lloyd, LE, Walker, MK, Teulon, DAJ and Edwards, W. 2009. Alternative pollinator taxa are equally efficient but not as effective as the honeybee in a mass flowering crop. Journal of Applied Ecology, 46: 1080–1087.

Rao, G. M. and Lazar, M. 1983. Studies on bee behavior and pollination in onion (Allium cepa L.). Proceedings of the 2nd International Conference on Apiculture in Tropical Climates, New Delhi, 3: 580-589.

Rao, G. M. and Suryanarayana, M. C. 1989. Effect of honeybee pollination on seed yield in onion (Allium cepa L.). Indian Bee Journal, 50: 9-11.

Rao, G.M. and Suryanarayana, M.C. 1988. Studies on pollination of watermelon Citrullus lanatus (Thunb) Manst.Indian Bee J., 50: 5-8.

Rasekh Adel, M., Sadeghi Namaghi, H., and Hosseini, M. 2012. Biodiversity of Apoidea (Insecta: Hymenoptera) associated with onion and alfalfa fields in Mashhad and Chenaran Regions. Iranian Journal of Plant Protection Science, 43 (1): 191-199 (in Persian with abstract in English).

Rashad, S. E. 1978. Utilization of non-Apis bees as crop pollinators (EG-ARS-66); First Annual Report 48 pp.

Rees, W.E., 1992. Ecological footprints and appropriated carrying capacity: what urban economics leaves out. Environment and Urbanization 4, 121–130.

Ricciardelli D, Albore G and Ambrosio M 1979 Preliminary observations on pollination of Coriandrum sativum by honey bees and other insects. Apicoltore Moderno. 70: 151–157.

Ricciardelli D'Albore C. G. and D'Ambrosio M. 1979 Aminoacids composition variations of pollens gathered by honeybees during one year activity. In: Proceedings of the 27th International Apicultural Congress of Apimondia. Athens, Greece, 14-20 September 1979. pp.14-20.

Ricciardelli D'Albore C. G. and Intoppa F. 2000 Flori e api. La flora visitata dalle Api e dagli altri Apoidei in Europa. Calderini Edagricole. Bologna. 110 pp.

Ricciardelli D'Albore G. C. 1986 Les insectes pollinisateurs de quelques ombellifères d'intérêt agricole et condimentaire (Angelica archangelica L., Carum carvi L., Petroselinum crispum A.W. Hill., Apium graveolens L., Pimpinella anisum L., Daucus carota L., Foeniculum vulgare Miller v. Azoricum thell.). Apidologie. 17: 107-124.

Richard W., Rusta, R.W., Vaissièreb, B.E. and Westrichc, P. 2003. Pollinator biodiversity and floral resource use in Ecballium elaterium (Cucurbitaceae), a Mediterranean endemic.Apidologie, 34: 29-42.

Richards, K.W., 1993. Non-Apis bees as crop pollinators. Revue Suisse Zoology 100, 807–822.

Ricketts, T.H., 2004. Tropical forest fragments enhance pollinator activity in nearby coffee crops. Conservation Biology 18, 1262–1271.

Ricketts, T.H., Daily, G.C., Ehrlich, P.R. and Michener, C.D., 2004. Economic value of tropical forest to coffee production. Proceedings of the National Academy of Sciences 101, 12579–12582.

Rao, G.M. and Lazer, M. 1980. Studies on beee behavoiur and pollination on Allium species with observation on insect as pollinators. J. Apic. Res., 24: 62-66.

Robinson, Richard W. 1954 Seed germination problems in the Umbelliferae. Bot. Rev. 20(9): 531-550. 1954.

Rodel, G. 1990 "Pollination of carrot: recent work", Bull. Semances, 111, 47-51.

Rodel, G., Grosse, T.J.P., and Bonnel, A. 1991 "Pollination of male sterile carrot inbred lines (Daucus carota L.) by honey bees (Apis mellifera L.)", Proc. Eucarpia on carrot breeding, June 18-20, France, 29-135.

Rodet G, Torre Grossa JP and Bonnet A 1991. Foraging behavior of Apis mellifera L. on male-sterile and male-fertile inbred lines of carrot (Daucus carota L.) in gridded enclosures. C van Heemert and A de Ruijter VI International Symposium on Pollination, Tilburg, Netherlands. Pp. 371 – 375.

Rodger, J.G., Balkwill, K. and Gemmill-Herren, B., 2004. African pollination studies: where are the gaps? International Journal of Tropical Insect Science 24, 5–28.

Roubik, D. W. 1995. Pollination of cultivated plants in the tropics. FAO, Rome.

Ruszkowaski, A. and Bilinski, M. 1984. Proba wykorzystania trzmiek (Bombus Later) I miesiarki lucernowki (Megachile rotundata F.) dozaphlania materialu hodowlanego ogorkow. Pszczelnicze Zeszyty Naukowe, 28: 175-184.

Saeed, A., Muhammad, T., Karim, A., Jabbar, A. and Ahmad, G. 2008. Effect of pollinators and insecticides on seed setting of onion (Allium cepa L.) Ann. Appl. Boi.,130(3): 497-506.

Saeed, S., Saggad, A., Kwon, O. and Kwon, Y. J. 2008. Fidelity of Hymenoptera and Diptera pollinators in onion (Allium cepa L.) pollination. Entomological Research, 38: 276-280.

Saeed, S., Sajjad, A., Kwon, O. and Kwon, Y.J., 2008. Fidelity of Hymenoptera and Diptera pollinators in onion (Allium cepa L.) pollination. Entomological Research 38, 276–280.

Sajjad, A., Saeed, S. and Masood, A., 2008. Pollinator community of onion (Allium cepa L.) and its role in crop reproductive success. Pakistan Journal of Zoology 40, 451–456.

Sakharov, M.K. 1956. Bees and onion seed production. In: Pollination of Agriculture Plants (eds. Krishchunas, I. V. and Gubin, A. F.). Moskva, Goa Izdvo Selkhoz Lit-ry, 172-173 pp.

Sanduleac, E. 1961. The pollination of vegetable seed plants.Apiculture,14: 25-26.

Sarviva, J.A.P.G. 1985. Bioclimatic factors affecting flights of honey bee. Apis mellifera L. (Hymenoptera: Apidae) Ph.D. Thesis, Colorado State University Fort Collins. USA, xi+148pp.

Scheiner, S.M.; Mitchell, R.J.; Callahan, H. 2000 Using path analysis to measure natural selection. Journal of Evolutionary Biology 13:423-433

Schiavi, M.; Masera, R.; Nervo, G.; Mazzola, P.; Soressi, G. P.1984. Different efficiency of blow-flies and honey bees in controlled pollination of onion (Allium cepa L.). 3rd Eucarpia Allium Symposium, Wageningen, 4-6 September 1984. 1984 pp. 115

Schittenhelm, S, Gladis, T and Rao, VR. 1997. Efficiency of various insects in germplasm regeneration of carrot, onion and turnip rape accessions. Plant Breeding, 116: 369–375.

Schlessman M. A. 1982 Expression of andromonoecy and pollination of Tuberous lomatiums (Umbelliferae). Systematic Botany 7: 134-139.

Schmid-Hempel, P. 1984. The importance of handling time for flight directionality in bees. Ecol. Sociobiol. 15: 303- 309.

Schoener, T.W. 1971. Theory of feeding strategies.Ann. Rev. Ecol. Syst. 2: 369-404.

Schuphan, W. and Weiller, H. 1967 "Pollination of the carrot crop for better seed production", Qual. Plant. Mater. Veg., 15, 81-101

Seeley, T.D., Camazine, S. and Sneyd, J. 1991. Collective decision making in honey bees: How colonies choose among nectar sources. Behav. Ecol. Sociobial,28: 277-290.

Sergeeva D. S., Sil'zenko V. M. 1984 Resistance of coriander to low temperatures. Fiziologiya i Biokhimiya Kul'turnykh Rastenii 16: 52-55.

Seyedebrahimi, S., Ebadi, R., Mobli, M. and Hatami, B. 2004. Honey bee (Apis mellifera L.) foraging on some onion cultivars and its relation with their nectar quantity and quality. J. Sci Tech. Agri. Natural Resources, 8: 1, 235-246.

Shanmugasundaram, S., 1998. Onion cultivation and seed production. Shanhua, Tainan, AVRDC.

Sharma, P.L. and Sharma, B.R. 1968 "Insect pollination in carrot crop", Indian J. Hort., 25, 216.

Sharma, S. K.; Singh, J. R.; Kaushik, H. D. 2001. Efficiency of two wild honey bees (Apis spp.) as pollinators of onion (*Allium cepa* L.). Haryana Journal of Horticultural Sciences 30 No. 1/2 pp. 134-135

Sharma, S.K. and Singh, J.R. 1999. Pollination efficiency of Apis dorsata F. and A. florea F. on carrot (Daucus carota L.).Indian Bee J., 61(1): 1-4.

Sharma, S.K. and Singh, J.R. 2000. Efficiency of two Apis spp. pollinating sunflower (Helianthus annus L.).Ann. Ent.,18(1): 15-17.

Sharma, S.K., Singh, J.R. and Kaushik, H.D. 2001. Efficiency of two wild honey bees (Apis spp.) as pollinators of onion (Allium cepa L.). Haryana J. hortic. Sci., 30(1&2): 134-135.

Sharma, S.K., Singh, J.R. and Ombir 2000. Comparative efficiency of three Apis spp. on Brassica campestris var. 'sarson' B.S.H.1. Crop Res., 19(1): 125-127.

Shasha'a, N. S. 1972. Limitation studies of seed set in the onion (Allium cepa L.) (Liliaceae). Ph. D. dissertation, Utah State Univ., Logan. 1972. [Unpublished.]

Shasha'a, N. S.; Nye, W. P. and Campbell, W. F. 1973. Path-coefficient analysis of correlation between honey bee activity and seed yield in Allium cepa L. Journal of the American Society for Horticultural Science 98 (4) 341-347

Shelar D G and Suryanarayana M C 1981 Preliminary studies on pollination of coriander (Coriandrum sativum L.). Indian Bee J. 43: 110–111.

Shepherd, M., Buckmann, S.L., Vaughan, M. and Hoffmann-Black, S., 2003. Pollinator conservation handbook. The Xerces Society, Oregon.

Shuel, R.W. 1955. Nectar secretion.Am. Bee J., 95: 229-233.

Sihag R C 1986 Insect pollination increases seed production in cruciferous and umbelliferous crops. J. Apic. Res. 25: 121–126. Synge A D 1947 Pollen collection by honey bees (Apis mellifera). J. Anim. Ecol. 16: 122–138b

Sihag, R. C. 1986. Insect pollination increases seed production in cruciferous and umbelliferous crops. J. Apic. Res. 25: 121-126.

Sihag, R.C. 1984. Influence of environmental factors on the pollination activity of bees. Env. Ecol., 2:149-152.

Sihag, R.C. 1985. Floral biology, mellitophily and pollination ecology of onion crops. In: Recent advances in pollen research (ed. Varghese, T.M.). Allied Publishers, New Delhi.241-268 pp.

Sihag, R.C. 1991 Some economic estimates of use of pesticides and bee pollination for crop yield, Korean J. Apic., 6, 89-94.

Sihag, R.C. and Khatkar, S. 1999. Foraging pattern of three honeybee species on eight cultivars of oilseed crops. I. Diurnal foraging. Int. J.Trop. Agric., 17(1-4).

Silva, E.M. and Dean, B.B. and 2000. Effect of nectar composition and nectar concentration on honey bee (Hymenoptera: Apidae) visitations to hybrid onion flowers. Journal of Economic Entomology 93, 1216–1221.

Silva, E.M. and Dean, B.B. and Hiller, L.K., 2003. Honey Bee (Hymenoptera: Apidae) foraging in response to preconditioning with onion flower scent compounds. Journal of Economic Entomology 96, 1510–1513.

Sinacori A., Sinacori M., Oliveri E. and Carrubba A. 2009 Api e pronubi selvatici nella produzione di Coriandrum sativum L. Apoidea 6(3): 155-161.

Singh G. and Gangwar B. 1991 Management practices for higher productivity of coriander in India - a review. Agricultural Review Karnal 12: 15-21.

Singh H, Swaminathan R. and Hussain T.2010. Influence of certain plant products on the insect pollinators of coriander. Journal of Biopesticides 3(1): 208 -211.

Singh J P and Dharamwal S S 1970. The role of honey bees in seed setting of onion at Pant Nagar, Dist. Nainital (U P), India. Indian Bee Journal 32 : 23-27.

Singh, B. and Hameed, S.F. 1995. Physical factors and foraging activity of insects visiting rye bloosoms. In: Pollination Biology: Environment Factors and pollination (ed. Sihag, R.C.), Rajendra Scientific Publishers, Hisar. 111-116pp.

Singh, M. 1983 "Role of insect pollination in seed production of carrot, Daucus carota Linn.", Thesis Abst. Punjab Agri. Univ., Ludhiana, India, 9, 327-328.

Singh, Y. 1997. Role of honeybees in farm production, agricultural growth and rural reconstruction in India. Indian Bee Journal, 59 (1): 24-30.

Sinha, S.H. and Chakrabacti, A.K. 1992 "Insect pollination in carrot seed crop", Seed Res., 20, 37-40.

Skiredj A., Elattir H. and El Fadl A. 2002 Fiches techniques des cultures aromatiques et condimentaires. L'absinthe, l'ail, le céleri, le cerfeuil, la ciboulette, la coriandre, les épinards, l'estragon, le fenouil de florence, la menthe, la niora, le persil, le sésame, la verveine. Bulletin mensuel d'information et de liaison du PNTTA. Transfert de technologie en agriculture. 24 pp.

Skrebtsov, M. F. and Skrebtsova, N. D. 1980. Pollinators of onion seed crops. Zashchita Rastenii No. 10 pp. 22

Skrebtsova, N.D. 1957. Amount of pollen on the body of bees. Pchetorodstvo, 34(4): 39-42.

Slate, W. L.1927. Report of the director. Conn. Agr. Expt. Sta. Bul. 291: 91-111.

Snedecor, G. W. 1956 Statistical methods. 5th ed. Ames, Iowa. Iowa State College Press. 534 p.

Sokal, R.R and Rholf, F.J 1981. Biometry (2nd Edition) WH Freeman & Co; San Francisco.

Sonnet M., Jacob-Remacle A. 1987 Pollinisation de la légumineuse fourragère Hedysarum coronarium L. en Tunisie. Bulletin de la Recherche Agronomique de Gembloux 22: 19-32.

Southwick E. E., and Southwick, L. 1992. Estimating the economic value of honey bees (Hymenoptera: Apidae) as agricultural pollinators in the United States. Journal of Economic Entomology, 85: 621-633.

Spurr CJ, Mendham NJ and Brown PH 2001. The effect of Rutherglen bug (Nysius vinitor) on yield and quality of hybrid carrot seed. TRI Ltd (ed), Proceedings of the 10th Australian Agronomy Conference 2001. Hobart, The Regional Institute Ltd.

Stanghellini, M.S., Ambrose, J.T. and Schultheis, J.R. 1997. The effects of honeybee and bumble bee pollination on fruit set and abortion of cucumber and watermelon. Am. Bee J.,137: 386-391.

Stephen, W.A. 1970. Honey bees for cucumber pollination. Am. Bee J. 110: 132-133.

Steyn, T., 2009. Bysiekte in tot 85% van korwe. Die Burger, 14 July, p.6.

Struck, M., 1994a. Flowers and their insect visitors in the arid winter rainfall region of southern Africa: observations on permanent plots. Insect visitation behaviour. Journal of Arid Environments 28, 51–74.

Struck, M., 1994b. Flowers and their insect visitors in the arid winter rainfall region of southern Africa: observations on permanent plots. Composition of the anthophilous insect fauna. Journal of Arid Environments 28, 45–50.

Szabo, T.I. and Smith, M.V. 1972. The influence of light intensity and temperature on the activity of alfalfa leaf cutter bee.Megachile rotundata under field conditions.J. Apic. Res., 11: 157–165.

Tainton, N. (Ed.), 1999. Veld management in South Africa. University of Natal Press, Pietermaritzburg.

Tanaka, M. and Komochi, S.1982. Use, reproduction and storage of houseflies as pollinators for onion breeding. Research Bulletin, Hokkaido National Agricultural Experiment Station 1982 No. 135 pp. 73-81

Tepedino V. J., Bowlin W. R. and Griswold T. L. 2011 Diversity and pollination value of insects visiting the flowers of a rare buckwheat (Eriogonum pelinophilum: Polygonaceae) in disturbed and " natural" areas. Journal of Pollination Ecology 4(8): 57-67.

Thapa R. B. 2006 Honeybees and other insect pollinators of cultivated plants. Journal of the Institute of Agriculture and Animal Science 27: 1-23.

Thompson, D. J. 1962 natural cross-pollination in carrots. Amer. Soc. Hort. Sci. Proc. 81: 332 - 334.

Thompson, M., Vlok, J., Cowling, R.M., Cundill, S.L. and Mundau, N., 2005. A land transformation map for the Little Karoo. GeoterraImage, Pretoria.

Tilman, D., Fargione, J., Wolff, B., D'Antonio, C., Dobson, A., Howarth, R., Schindler, D., Schlesinger, W.H., Simberloff, D. and Swackhamer, D., 2001. Forecasting agriculturally driven global environmental change. Science 292, 281–284.

Tolon, B. and Duman, Y. 2003. The effect of pollination by honeybee (Apis mellifera) on onion (Allim cepa) seed production and quality. XXXVIII Apimondia Ineternational Apicultural Congres, 554.

Tornier I and Drescher W 1991. Development of a method for mass production of the pollinator and aphid predator Episyrphus balteatus Deg. (Diptera, Syrphidae). C van Heemert and A de Ruijter VI International Symposium on Pollination, Tilburg, Netherlands 27–31 August, 1990. Pp. 464 – 468.

Treherne, R.C. 1923. The relation of insects to vegetable seed production. Rep. Queb. Soc. Prot. Pl. 15: 47-49.

Trofimec, N.K. 1940. The biology and fertilization in Allium. Vest. Sots Rasteniev. Rasten. 5:76-86.

Uvah I. I. I. and Coaker T. H. 1984 Effect of mixed cropping on some insect pests of carrots and onions. Entomologia Experimentalis et Applicata 36(2): 159-167

Van Der Meer, Q. P. and van Bennekom, J. L. 1972. Influence of the environment on the percentage of self-fertilization in onion and some consequences for breeding. Euphytica, 21: 45-453.

Van der Meer, Q.P. and Van Bennekom, J.L. 1968. Research in pollen distribution in onion seed fields.Euphytica, 7: 216- 219.

Van der Walt, W.J., 1999. Changing face of South African seed trade. Journal of New Seeds 1, 81–97.

Van der Walt, W.J., 2002. A review of the South African seed industry. Journal of New Seeds 4, 57–70.

Vansell, G.H. 1934. Relation between the nectar sugar concentration in fruit blossoms and the visits of honey bee. J. econ.Ent.,27: 943-945.

Vavilov, N.I., 1951. The origin, variation, immunity and breeding of cultivated plants. Chronica Botanica 13, 1–6.

Vischer, P.K and Seeley, T.D. 1982. Foraging strategies of honeybee colonies in a temperate deciduous forest. Ecology 63: 1790-1801.

Vitousek, P.M., Mooney, H.A., Lubchenco, J. and Melillo, J.M., 1997. Human domination of Earth's ecosystems. Science 277, 494–499.

Vlok, J. and Schutte-Vlok, A.L., 2010. Plants of the Klein Karoo. Umdaus Press, Pretoria.

Voss R. E., Murray M., Bradford K., Mayberry K. S. and Miller, I. 1999. Onion Seed Production in California. University of California Division of Agriculture and Natural Resources Publication, 8008: 1-10.

Wackernagel, M., 1994. Ecological footprint and appropriated carrying capacity: a tool for planning toward sustainability. PhD thesis, The University of British Columbia.

Wäckers, FL, Romeis, J and van Rijn, P. 2007. Nectar and pollen feeding by insect herbivores and implications for multitrophic interactions. Annual Review of Entomology, 52: 301–323.

Waddington, K.D. 1982. Information used in foraging. In: The Biology of Social Insects. (eds. Breed, M.D., Michner, C.D and Evans, H.E.). Boulder Westview Press.24-27 pp.

Wallace, K.J., 2007. Classification of ecosystem services: Problems and solutions. Biological Conservation 139, 235–246.

Waller, C.D. 1970. Attracting honey bees to alfafa with citral, geraniol and anise.J. Apic. Res.,9, 9-12.

Waller, G. D., Carpenter, E. W and Ziehl, O. A. 1974. Potassium in onion nectar and its probable effects on attractiveness of onion flowers to honey bees. J. Am. Soc. Hortic. Sci 97:535–539.

Waller, G.D., Carpenter, E.W. and Ziehl, O.A., 1972. Potassium in onion nectar and its probable effect on attractiveness of onion flowers to honey bees. Journal of the American Society of Horticultural Science 97, 535 539.

Walsh, R.S. 1965. Pollination of onion plants by honey bee. New Zeal. Beekeeper, 27(2):18- 20.

Wangensteen H., Samuelsen A. B., Malterud K. E. 2004 Antioxydant activity in extracts from coriander. Food Chemistry 88: 293-297.

Waser N. M., Olerton J. 2006 Plant-pollinator Interactions. From specialization to generalization. University Chicago Press. Chicago. 441pp.

Watanabe, M.E., 1994. Pollination worries rise as honey bees decline. Science 265, 1170.

Waters, N.D. 1972. Honey bee activity in blooming onion fields in Idaho. Amer Bee J., 112:218-220.

Waters, N.D., 1972. Honeybee activity in blooming onion fields in Idaho. American Bee Journal 112, 218–219.

Welch, J. E., and Grimball, E. L., Jr. 1947. Male sterility in the carrot. Science 106: 594.

Whitaker, T. W., Sherf, A. F., Lange, W. H., and others. 1970. Carrot production in the united states. U.s. Dept. Agr., agr. Handb. 375, 37 pp.

Whitaker, T.W., Sherf, A.F. and Lange, W.H. 1970 "Carrot production in United States", U.S. Dept. Agri. Handbook, 375, 37

Whitehead, V.B., 1984. Distribution, biology and flower relationships of fideliid bees of southern Africa (Hymenoptera, Apoidea, Fideliidae). South African Journal of Zoology 19, 87–90.

Wilkaniec, Z., Giejdasz, K. and Proszynski, G. 2004. Effect of pollination of onion seeds under isolation by the red mason bee (Osmia rufa L.) (Apoidea, Megachilidae) on the setting and quality of obtained seeds. Journal of Apicultural Science, 48 (2): 35-41.

Wilkaniec, Z., Warakomska, Z. and Giejdasz, K. 2002. Rooeliny ywicielskie pszczoy samotniczej Osmia rufa L. (Apoidea, Megachilidae) okreoelone na podstawie analizy py³ku z komór lêgowych. Zesz.Nauk. PTPN, 93: 199-205.

Williams, I.H. and Free, J.B. 1974. The pollination of onion (Allium cepa L.) to produce hybrid seed. J. Appl. Ecol., 11:409-412.

Williams, I.H. and Free, J.B., 1974. The pollination of onion (Allium cepa L.) to produce hybrid seed. Journal of Applied Ecology 11, 409–417.

Williams, N.M., Minckley, R.L. and Silveira, F.A., 2001. Variation in native bee faunas and its implications for detecting community changes. Conservation Ecology 5, 7.

Wills, A.B. and North, C., 1978. Problems of hybrid seed production. Acta Horticulturae. 83, 31–36.

Winfree, R. 2008. Pollinator-dependent crops: an increasingly risky business. Current Biology, 18: R968–R969

Witter, S. and Blochtein, B. 2003. Effect of pollination by bees and other insects on the production of onion seeds. Pesquisa Agropecuaria Brasileria, 38 (12): 1399-1407.

Witter, S. and Blochtein, B. 2003. Effect of pollination by bees and other insects on the production of onion seeds. Pesquisa Agropecuária Brasileira 38: 1399-1407

Wolf, S., Lensky, Y. and Paldi, N. 1999. Genetic variability in flower attractivness of honey bee (Apis mellifera L.) within the genus Citrullus. Hort. Sci.,34(5): 860-863.

Woyke, H. 1988. The influence of weather on insect foraging intensity in seed onions.

Woyke, H. W. 1981. Some aspects of the role of the honeybee in onion seed production in Poland. Acta Horticulture, 111: 91-98

Wykes, G.R. 1952. The preferences of honey bee for solutions of various sugar which occur in nectar. J. exp. Biol., 29: 511-519.

Wykes, G.R. 1953. The sugar contents of nectars. Biochem. J., 53: 294-296.

Yucel, B. and Duman, I. 2005. Effect of foraging activity of honey bee (Apis mellifera L.) on onion (Allium cepa) seed production and quality.Pakistan J. Biol. Sci., 8(1): 123-126.

Zych M. 2007 On flower visitors and true pollinators: The case of protandrous Heracleum sphondylium L. (Apiaceae). Plant Systematics and Evolution 263: 159-179.

CHAPTER

11

# Planned Pollination in Oilseed Crops

## INTRODUCTION

Edible vegetable oils are a high-value agricultural commodity and demand for high-quality seed oils continues to grow as the world population increases (Wittkop *et al.*, 2009). With insect pollinators enhancing yields in almost 70% crops worldwide declines poses a genuine threat to global food security (Klein *et al.,* 2007). Insect pollination enhances yield in oilseed rape by approximately 25% and selecting varieties with higher nectar secretion is likely to have positive implications for both pollinator populations and agricultural production. Furthermore, oilseed rape is economically important for commercial beekeepers as it is the main source of nectar in spring (Bommarco *et al.*, 2012). Within intensively managed agricultural landscapes, natural or semi-natural components provide important nesting and foraging sites for wild pollinators and proximity to such habitats increases pollinator species richness, crop visitation rates and thus pollination success (Garibaldi *et al.,* 2011). Maintaining pollinator diversity can ensure resilience of pollination as an ecosystem service due to species' showing differential response to environmental change (i.e. response diversity). Although, pollination improves the yield of most crop species and contributes to one-third of global crop production, it is alarming in the scenario of agricultural intensification due to loss of floral resources and diminishing pollination services (Klatt *et al.*, 2014; Vanbergen *et al.,* 2013; Wratten *et al.*, 2012).

When the natural process of pollination is mediated through the use of bees in a planned manner such a intervention can significantly influence the quality and quantity of yield of crops. This practice has become an essential component of modern cultivation system to enhance the productivity of crops to maintain the quality and quantity of food commodity.

Oilseed crops have a special significance in India due to their contribution in Indian economy. India occupies a prominent position, both in regard to acreage and production of oilseeds. Cultivation of oilseeds constitutes second largest agricultural commodity after cereals in India, sharing 14 per cent of the country's gross cropped area. These crops are cultivated on about 16.5 million hectares, with total production of 10 million tonnes (Abrol and Shankar 2012).

The important oilseed crops grown in India are rapeseed and mustard, sesame, linseed, safflower, sunflower, Niger and linseed (Guidry, 1964). Most of the oilseed crops are either fully dependent upon cross pollinating agents for seed production or benefit greatly by insect pollination. Inadequate pollination results not only in reduced yields but also in delayed yield and a high percentage of culls or inferior fruits. Though self pollination can set fruits but if allowed insect pollination, there can be number of advantages as far as quality or quanitity of seed/fruit production is concerned. Insect pollination increases either: (a) the proportion of fruits set; or (b) the quality of fruits set, because fruit size depends on the number of seeds set, or the size of seeds. Seed size is generally greater after cross-pollination by insects (Charlesworth and Charlesworth 1987; Richards, 1997). Number of oilseed crops such as oil-seed rape (Free and Ferguson 1983; Williams *et al.*, 1987), flax and linseed, sunflowers (Free and Williams 1976) belong to this group of pollination systems.

The level of productivity of oilseeds in India is far behind the average productivity in the world. One of the factors responsible for low productivity of oilseed crop is due to the failure of proper seed setting due to lack of pollination (Rao *et al.*, 1980; Free, 1993; Abrol, 2007, 2008, 2009). Insect pollinators not only enhance the yield of the crop but also contribute uniform and early setting. Therefore, cross pollination of entomophilous crops by honeybees is considered as one of the most effective and cheapest methods for triggering the crop yields. Applied pollination, pollinator management and managed pollination or interventions are the common efforts recently being practiced for optimization/maximization of production in cross pollinated crops and to bring the pollinators to the target crops (Fig. 11.1–11.2).

Among the various pollinating agents, insect pollinators play a predominant role in increasing the yield of oilseed crops (Mishra *et al.*, 1988). Important oil crops that need pollination interventions for enhancing producity are Rapeseed-mustard, Sunflower, Safflower, Sesame, Niger, Taramira and Linseed. The pollination requirements, type of pollinators, their behaviour and efficiency in the pollination and enhanced production of different crops is briefly described below:

## Rapeseed, Mustard and Canola *(Brassica spp.)*

Rapeseed mustard group is mostly cross-pollinated, although some varieties are self-fruitful. Selfing has been reported to reduce seed yield, seed size, and yield in subsequent generation (Delaplane and Mayer 2000). Self-incompatible plants require pollen transfer from plant to plant (Wallace *et al.*, 2002). Pollination interventions in the form of planned pollination (Fig. 11.3–11.4) has been found to greatly benefit quality and quantity of seed production in crucifers. For instance, in male sterile oilseed rape (*B. napus*), yields of the species were increased by honey bee pollination (Westcott and Nelson 2001). Similarly, in sarson (*B. campestris*), pollination by insects, increased the seed yield, caused formation of well-shaped, larger grain, and more viable seed (Abrol, 2009). Impact of insect pollination on production of different oil crop is presented in table 11.1.

Many species of *Brassica,* such as rape (*Brassica napus*), sarson (*Brassica campestris* var. sarson), toria (*Brassica campestris* var. toria), Indian mustard or rai (*Brassica juncea*), white mustard (*Brassica alba*), and black mustard (*Brassica nigra*) are cultivated widely as oilseed crops throughout the world. Most of these crops bloom during February-March for over a month. The bright yellow fragrant flowers are produced in long terminal racemes which are highly attractive to honeybees and other natural insect pollinators. Mohammed (1935) in West Punjab at Lyallpur recorded 117 visiting species of insects belonging to seven orders; *Andrena ilerda*, *Apis florea*, *Halictus* spp. comprised 82 % of insects and were the most important in this order. He also observed that stigma of the crops remained receptive for two days after opening and in 5 days their fertility was lost whereas, pollen remained viable for 7 days. It was observed in Australia that among *Apis mellifera* foragers 72 % collected nectar only 25 % both nectar and pollen and 3 % collected pollen only (Langridge and Goodman 1975). The pollen was collected between 10.00 to 14.00 hours.

Mohammed (1935) reported that enclosing individual plants in muslin bags, seed set was 12.3 % in toria, 20.3 % in brown sarson, and 91.0 % in yellow seeded sarson. He further reported that in case of hand pollinated toria flowers, there was 100 % pod formation. In a similar study, Sihag (1986) studied seed set in toria plots caged and not caged and left open for natural cross pollination; there developed 25 and 1060 pods per plant, 6.3 and 18.2 seeds per pod, and 4 and 385 seeds per plant. Similar observations on sarson gave 59 and 95 % pods containing 3.5 and 12.7 seeds per pod (Mishra *et al.*, 1988). Observations made in field plots of both sarson and toria showed that when crop was caged to exclude insects yield was 68 g; when caged with *Apis cerena* gave 219 g and in open field near honeybee colonies where other insect pollinators were also present yield was 244 g (Latif *et al.*, 1960). In general, there is quite a high degree of self incompatibility in these crops and good yields are obtained by cross pollination with the aid of natural fauna in most places where good weather prevails and probably yield can be further increased with the help of honey bee colonies to the extent of 10-30 % (Rahman, 1940; Singh, 1954) covering for inclement weather. Singh and Singh (1992) observed that bee pollinated (BP) plants in field cages as compared with self pollinated (SP) plants in individual bags produced 3 times heavier pods, 4 times more seed per pod, 50 times more seeds and 84 times more seeds yield per plant.

Singh and Singh (1992) studied the impact of bee pollination on seed yield, carbohydrate composition and lipid composition of mustard seed *Brassica campestris* L. and found that bee pollinated plants produced 3 times heavier pods, 4 times more seeds per pod, 50 times more seeds per plant, 11 times more pods per plant and 84 times more seed yield per plant than self pollinated plants. Carbohydrate content was inversely proportional to lipid content. In the seeds of self pollinated plants the total carbohydrate content was about twice that of seeds of bee pollinated plants. Triglycerides constituted the majority of neutral lipids. In the seeds of BP plants triglycerides constituted about 74% of the total non-polar lipids, which was about 20 times more than in SP plants. Sterol was the least abundant of all the lipids and phosphatidylcholine was absent from all seeds. High lipid content was directly related to high seed yield, and concentration of total lipid increased according to type of pollination in the order SP< OP < HP < BP.

Deaodikar and Suryanarayana (1972) found enhanced yield in oilseed crops when pollinated by bees (Table 11.2). Langridge and Goodman (1975) also found significant differences in production of rapeseed in open and caged plots. Plots receiving unrestrained visits had significantly higher plant density, seed yield, seed/plant, weight of 1000 seeds, weight of seeds per plant, germination and percentage of oil content (Table 11.3) as compared to those isolated from insect visits.

From the seed yield data (Table 4) it is revealed that % pod setting was significantly higher in plants which had access to insect pollination (95.21 %) than those which were net caged ( 70.27 %) or muslin bagged (58.81 % ). Similarly, number of seeds per pod was also significantly higher in open pollinated flowers 912.62 seeds /pod) in comparison to nylon caged (5.55 seeds/pod) and muslin bagged (3.05 seeds/pod). Pods obtained from open pollinated flowers had significantly more healthy seeds (84.78 %) than those from net caged (41.45 %) or muslin bagged (48.37 %) flowers; the latter two, however, did not differ from each other. The average weight of 100 healthy dry seeds from muslin bagged flowers was significantly higher (542.20 mg) than the weight of seeds from open pollinated (478.50 mg), and net caged flowers (459.40 mg). Per cent oil content of healthy seeds from muslin bagged (41.64 %) flowers was significantly higher than that from open pollinated (39.02 %) flowers. However, oil content of seeds from muslin bagged flowers was equal to that from net caged flowers. When all the yield parameters were taken into consideration for calculating oil content (mg/pod), it was found that increase in total oil yield from open pollinated and net caged plants, was 9.76 and 1.55 times more, respectively, than from muslin bagged plants.

Singh *et al.* (2000) found that there was significant improvement in qualitative and quantitative parameters of rape when pollined by *Apis cerana himalaya* (Table 11.5).

Atmowidi *et al.* (2007) analyzed the diversity of pollinator insects and its effect to seed set of mustard (*Brassica rapa*) planted in agricultural ecosystem in West Java. At least 19 species of insects pollinated the mustard, and three species, i.e. *Apis cerana*, *Ceratina* sp., and *Apis dorsata* showed a high abundance. The higher abundance and species richness of pollinators occurred at 08.30-10.30 am and the diversity was related to the number of flowering plants. Insect pollinations increased the number of pods, seeds per pod, seed weights per plant, and seed germination.

Chhuneja *et al.* (2007) studied the role of *Apis mellifera* in the seed production of *Brassica campestris* var. toria and found that the mean number of siliqua per plant, number of seeds per 20 siliquae, thousand seed weight, seed yield per plant and oil content was significantly higher upto 50 and 75m distances from bee colonies. It was further reported that one colony of *Apis mellifera* with 10 bee strength was sufficient for effective pollination of one hectare of crop area. Araneda *et al.* (2010) evaluated the yield of *B. napus* cv. Artus pollinated by *A. mellifera* in Chile and found that the parameter least affected by bee intervention was the grains per silique variable. In contrast, siliqua per plant and 1000 grain weight parameters presented significant differences, contributing to a yield greater than 5 t ha$^{-1}$; which represented a figure 50.34% higher than in the treatment without bees. He further concluded that the inclusion of bees in crops is fully justified as a production tool.

Sharma and Abrol (2014) found that toria crops benefit greatly from cross-pollination. They found that the number of siliqua per plant in toria under open pollination were twice (279.33 siliqua/plant) than those formed under caged condition (128.99 siliqua/plant). Seed weight was lower under caged conditions than under open pollination, and more and better quality seeds were obtained from open pollination. The combined effect was that open pollination yielded 1.80 times more seed than plants having no access to insect pollinators (table 11.6). Sabir *et al.* (2000) found that honeybees maximized seed yield, 1000 grain seed yield and germination percentage in *Brassica campestris*. Earlier, Sabbahi *et al.* (2005) found a significant improvement in the seed yield when honeybees were present. Munawar *et al.* (2009) reported significant increases in a range of plant parameters when caged with bees as compared to plants without bees. Abrol (2009) studied that pollination requirement of different oilseed crops and expected increase in yield due to bee pollination which has been summarized in Table 11.7.

For rapeseed, although the wind is the principal vector in terms of the distances over which pollen is transported (Hoyle *et al.,* 2007), bees are the principal pollinators, being the most abundant insects in the cultivars and varieties of rapeseed used for seed production (Westcott and Nelson 2001; Pordel *et al.*, 2007). Thus the introduction of hives of *A. mellifera* in controlled pollination helps to increase the production of crops such as rapeseed (*Brassica napus*) (Sabbahi *et al.*, 2006), since this plant is characterized by the production of abundant pollen and good quality nectar at relatively high concentrations of sugar, in flowers with a color and structure which are attractive to insects, particularly bees (Smith, 2002; Sabbahi *et al.*, 2006).

The studies clearly establish that most of the oilseed crops are cross pollinated and adequate pollination is vital for any significant seed production. An increase in seed yield as a result of insect pollination has been reported in mustard (Mohammed, 1935; Latif *et al.*, 1960). Latif *et al.* (1960) found that *Apis cerana* colonies near sarson and toria fields increased the seed yield by 60%. Similar increase in field has been reported in oil seed rape (Langride and Goodman 1975; Kisselhegn, 1977) and *Brassica campestris* var Jambuck (Kubisova *et al.*, 1980). Bisht *et al.* (1980) found that flowers of rapeseed visited by *Apis* species had higher pod set, more number of seeds per pod and weight of seed was also higher than those deprived of pollinator visit. Mishra *et al.* (1988) found that in *Brassica campestris* per cent pod setting, number of pods per plant and proportion of healthy seeds was significantly higher in open pollinated flowers than in net caged and muslin bagged ones. Similarly, average weight of seeds and oil content was higher in open pollinated flowers. *Apis cerana* was the most common pollinating species (Fig. 11.5). The other pollinators observed included *A. mellifera*, *A. dorsata*, bumblebee, solitary bee, syrphid flies (Fig. 6-10) etc. In a similar study, Prasad *et al.* (1989) found that pollination *of Brsassica juncea* by *A. cerana* resulted in more siliqua setting, increased length of siliqua, seed weight increased yield and had a pronounced effect on oil contents and germination (Table 11.8).

Mishra and Kaushik (1992) reported that yield and oil content was higher in honeybee pollinated crops in comparison to self pollinated. The results showed that higher percentage of pod setting was obtained in open pollinated crops of *Brassica nigra*, *B.campestris*, *B. carinata, B. napus* and *Eruca sativa* having 80.00, 84.42, 71.90 and 70.30 per cent, respectively. Average weight of 1000 healthy seeds varied between 3.0 to 5.1 g under open pollination, whereas it was 1.8 to 3.7 g under self pollination. Langridge and Goodman (2003) in a trial on oilseed rape (*Brassica napus*) cv. Midas in northern Victoria, found that no increase in yield of seed, oil content, or percentage germination was obtained from plots where bees and larger insects had access compared with plots enclosed in cages of 2.5 mm mesh to exclude these insects. The weight of 1000 seeds was slightly greater in the enclosed plots. Bees and a Syrphid, *Melangyna viridiceps*, were the predominant insects visiting the flowers. The bees stored some surplus honey and pollen and built up colony population (Fig. 11.11). Self pollination and possibly wind pollination appear to be the norm for this cultivar.

Evidently, pollination is one of the most important natural factors enhancing crop production. Among the various pollinating agents, insect pollinators play a predominant role in increasing the yield of oil seed crop. Almost one third of total cropped area under oil seeds has been reported to be entomophilous (Mishra *et al.*, 1988). The yield of rape seed and can be doubled through pollination by insects pollinators not only enhance the yield of the crop but also contributes uniform and early pod setting. Being a cross pollinated crop rapeseed attracts a large number of insect pollinators.

## Sunflower (*Helianthus annus* L.; Family Compositae)

In case of Sunflower (*Helianthus annuus* L.) there is strong evidence that insufficient pollination can significantly minimize the yields (Free, 1999). Sunflower benefits from insects that visit its flowers for pollen or nectar. Honeybee (*Apis mellifera* L.) pollination increases the seed yield by 30% and oil content by more than 6% in hybrid varieties (Furgala *et al.*, 1979; Jyoti and Brewer 1999). Honeybees are known as the main pollinators of sunflower in most parts of the world (Fig. 11.12–11.13). For example, in

Viamão (Brazil), 96% of the sunflower insect visitors were *A. mellifera* L. (Hoffmann, 1994). Other bees, especially the non- *Apis* bees are also known to frequent sunflower but are largely perceived as unreliable and ineffective pollinators mainly due to their low activity (Radford *et al.*, 1979). However, De Grandi-Hoffman and Watkins (2000) noted a possible indirect role of non-*Apis* bees, shown by their ability to enhance pollination by *Apis mellifera* (Fig. 11.14– 11.16).

Recently, Greenleaf and Kremen (2006) showed clearly that non-*Apis* bees are beneficial in sunflower pollination in their behavioral interactions with honeybees. Honeybees are forced to move to many flowers when several non-*Apis* bees forage same time and in same floral heads. This improves efficiency of honeybees to pollinate sunflower. Other than bees, different insect species visit flowers of sunflower.

In an attempt to determine the effect of honeybee visits on sunflower seed production in Manitoba, Furgala (1954b) placed 2.5 colonies per hectare along the east border of sunflower fields. The seed production per hectare in these two fields was as much as recorded in four fields located 4.8 km away from the honeybee colonies. Furthermore, he noted that seed yield in the farmers' field decreased progressively as the distance increased from the East to the west.

Insect pollination of sunflower has also been studied in Pakistan (Manzoor-ul-Haq and Faiz 1980). Wakhle *et al.* (1978) on the basis of chemical analysis of sunflower seeds obtained from different pollination treatment showed that there was a significant increase in oil (6.5 %) as also oil and protein contents together (7%) in seeds resulting from bee pollination when compared with self pollinated seeds (Table 11.9).

Evidently bee pollination is beneficial not only quantitative but also qualitative improvement in sunflower seeds. In a similar study Singh *et al.* (2000) reported that seeds in BP and CP treatments have higher oil percentages and higher oil and protein contents compared to those in SP or even OP treatments. Seeds under OP, BP and CP treatments showed nearly 6, 7 and 8.5 per cent increase in oil and protein contents over SP treatment, respectively (Table 10). Oz et al. (2009) determined the efficiency of pollination with honeybee (*Apis mellifera*) on sunflower hybrid seed production under (1) in cages with honeybees, (2) hand pollination (in cages) and (3) in cages without honeybees. They found that seed set ratios were 98-99% for pollinations in cages with honeybees or by hand as artificial whereas, this ratios was reduced to a level of 4-5% by pollination in cages without honeybees. Pollination in cages with honeybees and by hand increased seed yield per head by about 206 and 226%, respectively, compared with pollination in cages without honeybees. They further added that the use of honeybees for sunflower hybrid seed production improved seed set ratio, 100 seed weight, number of filled seed per head and seed yield per head. Similarly, Rajasri *et al.* (2012) in a two years study on sunflower hybrid NDSH1 found that the crop covered with insect proof net without honey bees recorded significantly lower seed setting % in both the years of study with a mean of 6.33% compared to other methods (Table 11.11).

All other pollination methods where honey bees were introduced were found to be significantly superior to open pollination (16.52 %). Highest mean seed setting percent of 74.13 % was recorded with bee pollination combined with hand pollination (Fig. 11.17) followed by bee pollination with eight frames and four frames (50.89 and 33.26% respectively). Honeybees + hand pollination gave significantly higher seed yield (14.4 q/ha) than open pollination (6.4 q/ha) and with crop caged with net (2.3 q/ha) there by the yield was increased about 526 % with honey bee + hand pollination and 178 % with open pollination compared to control plots. There were about 330 and 191 per cent increased yield of sunflower due to introduction of honey bees of 8 frames and 4 frames colonies respectively. The supplemental honeybee pollination + hand pollination significantly increased the percentage of seed setting and seed yield compared with open. Pollination and crop in cages without honeybees. The Hybrid sunflower seeds obtained from honey bee pollination coupled with hand pollination showed significant superiority with higher percentage of germination accounting for 99 per cent followed by open pollination (96%) and Bee pollination (95%) compared to control plot without bees (93%).

## Safflower (*Carthamus tinctorius* L.; family Asteraceae)

Seed production in safflower is directly related with success of pollination because the plants show self-pollination in absence of pollinators (Knowles, 1969). Classen (1950) reported zero to 100 per cent cross-pollination. In most of the plants, cross-pollination ranged from 5-40%. Pollinators contribute to various degrees of pollination of the flower (Kadam and Patankar 1942; Levin and Butler 1966; Butler *et al.*, 1966; Levin *et al.*, 1967). Safflower is usually considered to be a self pollinated crop. Insects, particularly bees, are the major agents of pollination (Boch, 1961; Eckert, 1962; Rubis *et al.,* 1966). Temperature and humidity affect seed setting of bagged flowers (Patil and Chavan 1958). Pandey and Kumari (2007) found that there was 85.99% seed setting in the open pollinated heads followed by 38.15% (in muslin cloth) and 35.54% (in butter paper) bagged condition (Table 11.12).

## Sesame (*Sesamum indicum* L; family Pedaliaceae)

Sesame (*Sesamum indicum* L.) is one of the oldest and important oilseed crops cultivated on about 7.8 million hectares of the total world crop area (FAOSTAT, 2012). Kamel *et al*. (2013) recorded 29 insect pollinators associated with sesame in which *Apis mellifera* and *Ceratina tarsata* were the most dominant species during different years. In case of sesame, the plants exposed to insects gave 25 % higher yield as compared with those covered with cages to exclude insects. Both opened and bee pollination treatments were effective in increasing the seed yield of sesame from 22 to 33% compared to 'pollination without insects' (Panda *et al*., 1988) and quantitative and qualitative parameters had also been improved in opened pollination (Mahmoud, 2012) (Fig. 11.18– 11.21). *Apis cerena, A. dorsata, A. florea* were the most common insects on the crops and they began foraging at 0600, 0700, 0800 h and visited 6.3, 5.0 and 1.9 flowers per minute in open condition and flowers pollinated by them gave 51, 53, 4 seeds per pod, respectively. Seed weight was also heavier as compared with the controls (Rao *et al*., 1981). While foraging 72 % of *A. cerena* collected nectar 23 % pollen only and 4 % both nectar or pollen. At one time or the other they touched anthers to bring about cross pollination and thus increased the yield (Phadke *et al.,* 1967). *Apis mellifera* is also known to work on flowers very actively. Abrol (1991) reported *A. dorsata* as an important pollinator of sesame and found that field activities of bees were made intense at higher temperature, light intensity and radiation. Extensive studies were made in Punjab on the genetics and pollination requirement of this crop (Sikka and Gupta 1949).

## Linseed/ Flax (*Linum usitatissimum* L., Family Linaceae)

Flax is considered to be normally self-pollinated although some crossing (1 to 6 per cent) occurs (Dillman, 1938; Dillman and Goar 1937; Masuo, 1958), mostly among the large flowered types. Rubis (1970) worked with a male sterile line having disk-form flowers, and stated that he obtained practically no cross pollination of the male-sterile lines with lines having tubular flowers; however, good seed set was obtained, indicating heavy cross-pollination, with other lines that had large disk form flowers.

Several tests have shown that bee pollination improves seed yields in fiber flax. Bezdenezhoykh (1956) in Russia reported that honey bees in cages increased seed production of fiber flax 22.5 percent over plants in cages without bees. Gubin (1945) also studied the effect of bee pollination of fiber flax in Russia and reported that bees increased seed production 22.5 to 38.5 percent. Luttso (1957), also in Russia, reported that bee pollination increased seed production by 29 percent, the number of seeds per capsule by 18 percent, and the weight per seed by 11 percent in comparison to fields without bee pollination. Likewise, Smirnov (1956) showed a 19 percent increase in the number of seeds per capsule, a 22 percent increase in the total weight of seeds, and a 2.2 percent increase in the weight per seed. He also reported that bee-visited plants set up the crop and ceased blooming earlier than plants from which bees were excluded. Kozin (1954) reported a sizeable increase in seeds per boll and seed weight when 40 colonies were placed near a fiber flax field, but he did not indicate the size of the field.

The studies on pollination of crops clearly establish that cross pollination not only improves quantitatively but improves the quality of oil crops as well. Even self compatible varieties give enhanced yields and improve qualitatively when cross pollinated (Table 13) (Fig. 11.22– 11.25). Therefore, pollination in oil crops assumes utmost significance for boosting production of oil crops in a world deficient of oil/oilseeds.

Clearly most of the oilseed crops are cross pollinated and exclusively depend upon bees and other pollinators for pollination services.

## Pollination management

### Problems associated with bee pollination

Pollinator-plant interaction is a very complex phenomenon and is influenced by many overlapping effects. The protection of pollinators, including honeybees is as essential as the protection of crops from the insect-pest damage. The use of pesticides for pest control on the one hand and the role of honeybees (*Apis* spp.) for crop pollination on the other have become essential components of modern agriculture. Unfortunately, these two practices are not always compatible, as honeybees are susceptible to many of commonly used pesticides (Johansen, 1977., Russell *et al.,* 1998., Cunningham *et al.,* 2002; Sundararaju, 2003), used for the control of insect pests (Poehling, 1989; Stark *et al.,*1995). The major constraint confronting pollinator-plant interaction is the indiscriminate and excessive use of pesticides for controlling insect-pests (Bisht *et al.*, 1980, Rana and Goyal 1991; Zhong *et al.*, 2004). The loss of honeybees directly effect beekeeping through loss of honey production and indirectly the crop production due to inadequate pollination. Reduction of population of these beneficial insects due to insecticides, therefore, incurs significant environmental, ecological and economic costs (Bai and Reddy 1977; Pimentel *et al.,* 1980; Crane and Walker 1983; Praksah and Kumaraswami 1984; Khan and Dethe 2004).

Furthermore, modern agricultural practices have resulted in the reduction of wild insect pollinators and disturbed insect flower- relationship by way of disappearance of wastelands and uncultivated strips of land, destruction of certain food sources by weed control and overall changes in the environment. Wild bees are also damaged by pesticides.

Poisoning may result from contaminated food as well as from florets, leaves, soil, or other material used by the bees in nesting. The toxicity of a specific insecticide to honey bees and wild bees is not always the same, and even among wild bees some materials are more toxic to one species than to another. Poisoning may result from contaminated food as well as from florets, leaves, soil, or other material used by the bees in nesting. Pesticide application practices that may reduce bee poisoning should be adopted as given below:

1. Apply pesticides only when needed.
2. Choose the pesticide with the lowest hazard rating for bees, particularly the lowest residual toxic effect, from the list of pesticides available for a particular pest control program.
3. Liquid or granule applications are less hazardous than dusts. Microencapsulated forms of pesticides have a significantly longer residual life than other application forms. The minute capsules can be carried back to the colony in the same manner that pollen is carried, and can kill brood and young adult bees.
4. Ground application is less hazardous than aerial application, particularly when applied in close distances to apiaries. Fine sprays are less toxic than coarse sprays
5. Where practicable, apply pesticides when bees are not active on the crop. For pesticides considered a low hazard when they have dried, early morning may be suitable. For pesticides with a residual toxic effect of a few hours, apply in the late afternoon or early evening.
6. The time of day when a pesticidc is applied should be chosen to minimise the risk of spray drift occurring either over apiaries or over plants being foraged by bees.n Treatments can and should be applied only when bees are not foraging for nectar or pollen. If bee-attractive plants need to be treated while in bloom, they should be treated at night or in the early morning or late evening when the bees are not flying.
7. Where practical, beekeepers should be given prior notice, preferably a minimum of 48 hours, of a pesticide application to allow apiaries to be moved from the area or their entrance closed.

## Number of Colonies Required for Pollination

Several investigators have attempted to determine the number of colonies of honeybees required for increased yields in rape and their recommendations vary from place to place and crop (Table 11.14). For instance, Hammer (1963 & 1966) recommended three colonies/ha; Radchenko (1964), two; Downey and Bolton (1961), one; White (1970), two; and Vesely (1962) three to four colonies per hectare. Although hoverflies appear to play some role in pollination of rape we consider the honeybee the more efficient pollinator. The ideal pollinator population and proper distribution of colonies for most efficient pollination needs to be determined for various oil crops.

## Pollination Recommendations

Honey bees are the most effective agents involved in cross pollination of rapeseed mustard and other oil crops. Indiscriminate use of pesticides/fungicides often kills a large number of pollinators. In certain cases, a single crop over a vast area is cultivated. This also reduces number of wild honey bee colonies in those areas. Therefore, the importance of managed and migratory beekeeping in the field is being realized as an important input to increase the production of oilseed crops. For effective pollination and increased yield of oilseed crops, the efficiency of a bee colony as a pollinator would depend upon certain factors such as colony strength, number and time of placement of colonies, distribution of colonies in the field/orchards, time and placement of colonies and weather conditions.

**Table 11.1:** *Impact of animal pollination on production of oil crops*

| Crop species | Commodity | World production (Mt) | Animal pollination | References |
|---|---|---|---|---|
| *Brassica napus* | Rapeseed, Oilseed rape | 46 770 903 | increase | Free, 1993; Adegas and Noqueira Couto 1992; Abel and Wilson 1999; Bürger, 2004; Manning and Boland 2000; Abel *et al.*, 2003; Morandin and Winston 2005 |
| *Helianthus annuus* | Sunflower | 26 460 824 | increase | Bichee and Sharma 1988; Crane, 1991; Free, 1993; DeGrandi- Hoffman and Martin 1993; Moreti *et al.*, 1996; Heard, 1999; De Grandi Hoffman and Watkins 2000; Dag *et al.*, 2002; Greenleaf and Kremen 2006b |

**Table 11.2:** *Yield in self pollinated and bee pollinated crops*

| Crop Botanical Name | Reported range of seed increase from bee pollinated over self pollinated | |
|---|---|---|
| | Percent more | Times more |
| Brassica napus L. (rape) | 12.8-139.3 | 1.128-2.39 |
| Brassica campestris L. var. toria (Toria) | 66.0-220.9 | 1.66-3.20 |
| Brassica campestris L. var. dichotoma (sarson) | 222 | 3.22 |
| Brassica juncea Czern &Coss. (Rai,Indian mustard) | 18.4 | 1.184 |
| Brassica alba Boiss. (white mustard) | 128.1-151.8 | 2.28-2.51 |

(Source: Deodikar and Suryanaryana 1972)

**Table 11.3:** *Rapeseed crop with and without pollinators*

| Attribute | Open plots | Closed plots | Significance of difference |
|---|---|---|---|
| Plant density | 459 | 406 | – |
| Seed yield | 725 | 453 | $P < 0.01$ |
| Seed/plant | 1.61 g | 1.10 g | $P < 0.01$ |
| Wt of 1000 seeds | 1.78 g | 1.88 g | $P < 0.01$ |
| No. seeds/plant | 927 | 588 | $P < 0.01$ |
| % germination | 95.3 | 97.1 | Ns |

(Source: Langridge and Goodman 1975)

Note: All data on left of the table are averages per plot, except oil content, which was an average of five composite samples.

**Table 11.4:** *Yield and oil potential of Brassica campestris var sarson under three conditions of pollination.*

| Parameter | Open pollination (X+S.E) | Net caged (X+S.E) | Muslin bagged (X+S.E) | Critical difference | Transformation |
|---|---|---|---|---|---|
| % pod setting | 95.21 + 2.46 | 70.27 + 4.38 | 58.81 + 6.57 | (10.09) | Angular |
| Seeds/pod | 12.72 + 0.82 | 5.55 + 1.18 | 3.95 + 0.85 | (0.73) | √n+1 |
| % healthy seeds | 84.78 + 2.76 | 41.45 + 8.25 | 48.37 + 12.28 | 18.41 | Angular |
| Seed weight in mg/100seeds | 478.50 + 21.83 | 459.40 + 23.41 | 542.20 + 11.00 | (1.31) | √n |
| % oil content | 39.02 + 0.29 | 40.76 + 0.90 | 41.64 + 0.27 | 1.81 | – |
| Oil yield (mg) pod | 19.03 | 3.02+ | 1.95 | | |
| Increase in oil yield over muslin bagged | 9.76 | 1.55 | 1.0 | | |

(Source: Mishra *et al.*, 1988)

**Table 11.5:** *Quantitaive and qualitative effects of Apis cerena himalaya pollination on rapeseed.*

| Yield parameters | Control (PWI) | Open Pollinated (BP) | Bee Pollinated (BP) | CD at 5 % | % increase over PWI | % increase over OP |
|---|---|---|---|---|---|---|
| Siliqua set (%) | 20.67 (26.39) | 68.42 (56.01) | 70.23 (57.24) | (5.00) | 239.77 | 2.65 |
| Siliqua length (cm) | 3.94 (2.11) | 4.16 (2.12) | 4.52 (2.24) | (0.07) | 14.72 | 8.65 |
| No. of seeds per Siliqua | 9.12 (3.14) | 11.65 (3.52) | 13.73 (3.80) | (0.24) | 50.55 | 17.85 |
| Wt. of 1000 seeds (gm) | 2.01 (1.66) | 2.31 (1.75) | 2.36 (1.76) | NS | 17.41 | 2.16 |
| Seed yield (q/ha) | 1.96 | 12.37 | 13.45 | 2.90 | 586.22 | 8.73 |
| Seed germination (%) | 55.00\| (47.93) | 81.00 (65.98) | 83.00 (66.02) | (5.21) | 5.90 | 2.50 |
| Oil content (%) | 33.25 (35.18) | 38.75 (38.48) | 40.33 (39.41) | (0.81) | 21.30 | 4.10 |
| Figures in parentheses are transformed values | | | | | | |

(Source: Singh *et al.*, 2000)

**Table 11.6:** *Seed yield and yield parameters in toria as influenced by varied mode of pollination*

| Yield attribute | Bee pollination | Open pollination | Closed pollination |
|---|---|---|---|
| Siliqua per plant | 258.33 + 0.25 | 276.33 + 0.16 | 122.67 + 0.32 |
| Seed per siliqua | 22.10 + 0.11 | 29.50 + 0.22 | 15.67 + 0.17 |
| Pod length (cm) | 6.23 + 0.006 | 7.37 + 0.18 | 5.26 + 0.01 |
| 1000 seed weight (g) | 5.65 + 0.01 | 5.97 + 0.04 | 2.15 + 0.01 |
| Seed yield (q/ha) | 12.77 + 0.02 | 13.73 + 0.01 | 7.50 + 0.02 |

(Source: Sharma and Abrol 2014)

**Table 11.7:** *Expected increase in crop yield due to bee pollination*

| Crop | Pollination requirements | Expected percent increase in yield due to cross-pollination |
|---|---|---|
| Mustard | Often cross pollinated | 13.00-222.00 |
| Safflower | Often cross pollinated | 4.00-114.00 |
| Sunflower | Often cross pollinated | 21.00-3,400 |
| Sesame | Often cross pollinated | 24.00-40.00 |
| Niger | Often cross pollinated | 17.00-45.00 |
| Linseed | Mostly cross pollinated | 2.0-49.00 |

(Source: Abrol, 2009)

**Table 11.8:** *Effect of different pollination treatments on oil and protein contents and germination of seeds in rai cv Pusa bold in Pusa, Bihar*

| Treatment | Oil content±SE (%)* | Protein content ± SE (%)** | Germination ± SE (%)** |
|---|---|---|---|
| Caged with bee | 31.80 (34.03 ± 0.11) | 17.80 (4.29 ± 0.00) | 93.30 (9.64 ± 0.02) |
| Uncaged | 32.10 (34.51 ± 0.11) | 18.44 (4.29 ± 0.01) | 98.00 (9.64 ± 0.02) |
| Caged | 26.30 (30.85 ± 0.13) | 6.39 (4.05 ± 0.01) | 73.00 (8.54 ± 0.03) |
| CV (%) | (0.9) | (0.6) | (0.1) |
| SE(m) | (3.09) | (0.01) | (0.0) |
| LSD (5%) | (0.2) | (0.02) | (0.01) |

(Source: Prasad *et al.*, 1989)

Average of 7 replications, * Figures in parenthesis are values transformed into angles, ** Figures in parenthesis are values transformed in square root

**Table 11.9:** *Average value of moisture, oil and protein contents in sunflower seeds variety EC 68414*

| Type of treatment | Number of samples | Moisture | Oil* | Proteins* | Oil & Proteins * |
|---|---|---|---|---|---|
| SP | 6 | 8.43 (0.142) | 32.44 (2.187) | 17.83 (2.48) | 49.72 (3.94) |
| OP | 7 | 10.197 (1.66) | 33.978 (4.0.34) | 25.55 (3.13) | 55.47 (4.33) |
| BP | 9 | 9.75 (2.048) | 38.81 (3.41) | 17.90 (2.952) | 56.71 (4.35) |
| CP | 12 | 8.533 (0.0681) | 38.94 (3.045) | 20.33 (4.051) | 58.29 (4.62) |

(Source: Wakhle *et al.*, 1978)

*The figures are percentages of oil and protein contents on moisture free basis

Value in brackets indicate standard deviation

**Table 11.10:** *Quantitaive and qualitative effects of Apis cerana himalaya pollination on sunflower*

| Yield parameters | Control (PWI) | Open Pollinated (BP) | Bee Pollinated (BP) | CD at 5 % | % increase over PWI | % increase over OP |
|---|---|---|---|---|---|---|
| Seed set (%) | 2.93 (9.84) | 60.50 (51.13) | 71.58 (57.81) | (2.20) | 2343.00 | 18.31 |
| Wt. of 1000 seeds (gm) | 50.32 | 64.29 | 54.88 | 3.26 | 9.06 | 17.15 |
| No. of seeds per gm | 20.87 (4.63) | 16.58 (4.13) | 17.04 (4.19) | 2.77 (0.37) | | |
| Seed yield (q/ha) | 2.06 | 16.60 | 18.55 | 1.47 | 800.49 | 11.75 |
| Seed germination 9%) | 69.88 (56.78) | 86.50 (68.80) | 91.13 (74.55) | (11.94) | 30.41 | 5.35 |
| Oil content (%) | 35.96 (36.85) | 46.79 (43.16) | 43.70 (41.38) | (1.32) | 21.52 | 7.07 |
| Figures in parentheses are transformed values. | | | | | | |

(Source: Singh *et al.*, 2000)

**Table 11.11:** *Effect of honeybees on seed setting % and seed yield of hybrid sunflower, NDSH1*

| Treatments | Seed set % | | | Seed yield (q $ha^{-1}$) | | | Percentage increase in yield over control |
|---|---|---|---|---|---|---|---|
| | 2006-07 | 2007-08 | Mean | 2006-07 | 2007-08 | Mean | |
| Open pollination | 23.88 | 10.23 | 16.52 | 4.03 | 8.70 | 6.4 | 178 |
| Covered with net | 11.88 | 1.16 | 6.33 | 2.06 | 2.50 | 2.3 | – |
| Caged with 4 frame colony | 19.87 | 47.1 | 33.26 | 3.34 | 10.08 | 6.7 | 191 |
| Caged with 8 frame colony | 51.6 | 50.18 | 50.89 | 8.64 | 11.15 | 9.9 | 330 |
| Bee + hand pollination | 87.8 | 60.46 | 74.13 | 15.3 | 13.54 | 14.4 | 526 |
| F-test | Sig | Sig | | Sig | Sig | | |
| CD | 4.90 | 3.27 | | 0.71 | 2.29 | | |
| CV | 6.20 | 4.6 | | 5.59 | 17.64 | | |

(Source : Rajasri *et al.* (2012)

**Table 11.12:** *Self incompatibility test in different experimental conditions*

| Types of Pollination | Total number of flowers | Number of filled seeds | Number of unfilled seeds | Percentage of seeds setting |
|---|---|---|---|---|
| Naturally pollinated heads | 54.75+7.70 | 47.08+13.78 | 7.67+6.07 | 85.99 |
| Butter paper bagged head | 54.76+5.50 | 19.46+8.88 | 35.3+3.38 | 35.54 |
| Muslin cloth Bagged heads | 51.53+5.35 | 19.66+4.67 | 31.87+0.68 | 38.15 |

(Source: Pandey and Kumari 2007)

**Table 11.13:** *Seed yield and yield parameters in different oil seed crops as influenced by bee pollination*

| Crops (variety) | Yield attributes | Pollination | |
|---|---|---|---|
| | | With bees | Without bees |
| Mustard (var. M-27) | Pod set (%) | 71.90 | 48.60 |
| | Seeds /pod | 10.80 | 5.90 |
| | 1000 seeds weight (gms) | 5.00 | 2.00 |
| | Seed yield (q/ha) | 13.90 | 1.20 |
| | Oil content (%) | 36.40 | 32.60 |
| Niger (var. M-15) | Seed set (%) | 45.80 | 28.40 |
| | Seed yield (q/ha) | 3.79 | 3.11 |
| | Total florets | 45.75 | 33.42 |
| | Oil content (%) | 35.20 | 32.00 |
| | Average seed set % | 23.80 | 14.18 |
| | seed weight (g)/head | 0.80 | 0.07 |
| Sesame (var. Kalika) | Average yield (q/ha) | 8.47 | 5.64 |
| | Oil content (%) | 36.80 | 36.40 |
| Sunflower (var. modern) | Seed yield(g)/head | 23.80 | 11.60 |
| | No. of seeds /head | 540.80 | 418.25 |
| | Filled seed/head (%) | 62.40 | 58.65 |
| | Seed yield (q/ha) | 27.54 | 9.78 |

**Fig. 11.1:** ***Apis mellifera*** **apiary nearby the mustard field**

**Fig. 11.2: Pollination intervention in mustard**

**Fig. 11.3: Planned pollination in mustard**

**Fig. 11.4: Conservation of Non-*Apis* bees for oilseed pollination**

**Fig. 11.5:** ***Apis cerana*** **foraging on mustard flowers**

**Fig. 11.6:** ***Apis mellifera*** **foraging on mustard**

**Fig. 11.7:** ***Apis dorsata on mustard***

**Fig. 11.8:** ***Bombus haemorrhoidalis collecting nectar from crucifer***

**Fig. 11.9: Solitary bee foraging on mustard**

**Fig. 11.10: Syrphid fly collecting nectar from mustard**

**Fig. 11.11:** ***A. mellifera*** **collecting nectar and pollen in bee hive**

**Fig. 11.12: Pollination intervention in sunflower**

**Fig. 11.13:** ***A. mellifera*** **pollinating sunflower**

**Fig. 11.14: Megachile bee foraging on Sunflower**

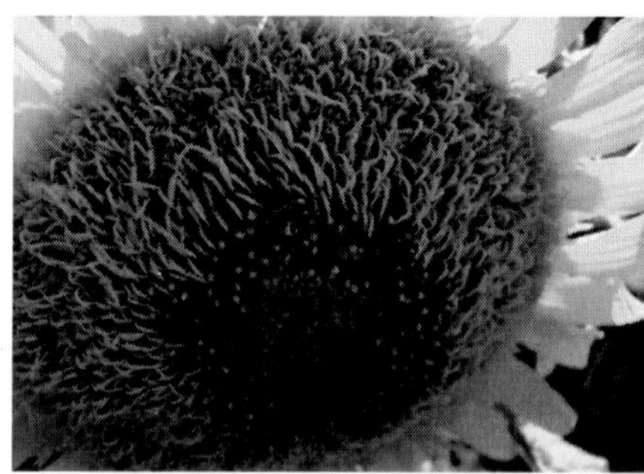

**Fig. 11.15:** ***Pithetes smargdula*** **working on sunflower**

**Fig. 11.16: View of nesting materials for Non-*Apis* bees**

**Fig. 11.17: Hand pollination in sunflower**

**Fig. 11.18:A view of sesame cultivation on Hills**

**Fig. 11.19: *Apis dorsata* foraging on sesame flower**

**Fig. 11.20:Nomia bee pollinating sesame flower**

**Fig. 11.21: *Bombus haemorrhoidalis* on sesame flowers**

**Fig. 11.22: Field view of Linseed cultivation**

**Fig. 11.23: Flowering in Linseed crop**

**Fig. 11.24:*Apis mellifera* pollinating linseed flower**

**Fig. 11.25:*Apis cerana* foraging on linseed**

## Conclusions and Future Strategies

The present chapter clearly indicates that pollination interventions in terms of supplementing oilseed field with honeybee colonies can boost oilseed production. A more comprehensive strategy for management of crop pollination is needed to ensure a reliable source of pollinators, both managed as well as native pollinators. The honeybee pollination not only responsible to enhance the yield in self incompatible varieties but even the self fertile varieties give many fold increased yield due to intervention of managed pollinators. This requires understanding of the biology and ecology of pollinating insects, as well as providing appropriate nesting habitat, and ensuring the availability of alternative sources of "forage" to sustain populations when the target crops are not in bloom. A number of other bee and insect pollinators, such as orchard bees, bumblebees, and squash bees, which are not affected by either the mites or the Africanized bees, are considered as likely candidates for management and use in commercial agriculture. An additional role can be played by native or wild pollinators, provided that attention is given to curtailing of population losses caused by both inadvertent insecticide poisoning and habitat destruction (Kevan *et al.*, 1990). Pollinator management, and managed pollination are the common efforts recently being practiced for maximization of production in cross pollinated crops and to bring the pollinator to the target crop. There is strong evidence that pollinators are declining as a result of local and global environmental degradation (Kluser and Peduzzi 2007). Recent declines in both wild and domesticated pollinators, and parallel declines in the plants which rely upon them possess a serious threat to global food security and sustainable agriculture (Potts *et al.,* 2010). Agriculture has become more pollinator dependent because of a disproportionate increase in the area cultivated with pollinator dependent crops. If the trend toward favoring cultivation of pollinator-dependent crops continues, the need for the service provided by declining pollinators will greatly increase in the near future.

Kevan *et al.* (1990) highlighted the pivotal role played by managed and native insect pollinators in sustainability of agricultural practices. However, wild insect pollinators and the pollination services they provide have declined in agricultural landscapes in some regions (Biesmeijer *et al.*, 2006; Potts *et al.*, 2010). Several factors associated with increased farming intensity to support growing human populations can limit the suitability of farm environments for insect pollinators, including reduction of natural areas, habitat fragmentation, and scarcity of flowering food and nesting resources (Carvell *et al.*, 2007). Monoculture plantings of crops lack floral diversity and can limit the provision of resources for pollinators throughout the season. Compared to more diverse landscapes, the lack of resources in agricultural landscapes can reduce insect pollinator diversity (O'Toole, 1993) and potentially decrease wild bee contributions to crop pollination (Potts *et al.*, 2010). Enhancement of structurally resource-poor environments through the establishment of habitats containing flowering plants and grasses can support beneficial insects in agricultural landscapes (Long *et al.*, 1998; Kells *et al.*, 2001; Sheffield *et al.,* 2008). These agricultural restoration programs are expected to provide greatest support for bees in simple landscapes with the greatest floral contrast to the background landscape (Scheper *et al.* 2013). Wildflower plantings can provide pollen and nectar resources when the crop is not in bloom and, depending on the bee species biology, may also provide nesting habitat (Carreck and Williams 2002; Kremen *et al.*, 2004; Heard *et al.*, 2007). Hoverflies are efficient pollinators of other crops, such as mango (Dag and Gazit 2001) and oilseed rape (Jauker and Wolters 2008), and the larvae of aphidophagous species are also biological control agents of many soft-bodied arthropods (Bugg *et al.*, 2008; Smith *et al.*, 2008), thus providing an additional ecosystem service to crops.

## References

Abel, C.A. and Wilson, R.L., 1999. The use of diverse plant species for increasing Osmia cornifrons (Hymenoptera: Megachilidae) in field cages. J. Kansas Entomol. Soc. 71, 23–28.

Abel, C.A., Wilson, R.L. and Luhmann, R.L., 2003. Pollinating efficacy of Osmia cornifrons and Osmia lignaria subsp. lignaria (Hymenoptera: Megachilidae) on three Brassicaceae species grown under field cages. J. Kans. Entomol. Soc. 38, 545–552.

Abrol, D.P. and Shankar, U., 2012. Pollination in oil crops-recent advances and future strategies. In: Gupta S.K. (Ed.) Technological Innovations in Major World Oil Crops, Volume 2, Springer Verlag. Pp. 221-267.

Abrol, D.P., 1991. Conservation of pollinators for promotion of agricultural production in India. Journal of Animal Morphology and Physiology 38(1-2), 123–139.

Abrol, D.P., 2007. Foraging behaviour of Apis mellifera and Apis cerana as determined by the energetics of nectar production in different cultivars of Brassica campestris var. toria. Journal of Apicultural Science 51(2), 5-10.

Abrol, D.P., 2008. Plant insect interaction in crucifers In: Gupta, S.K. (Ed.), Biology and Breeding of Crucifers. CRC Press, Taylor and Francis, USA pp. 129-149.

Abrol, D.P., 2009. Bees and Beekeeping in India. 2nd Edition Kalyani Publishers, New Delhi. Pp.720.

Abrol, D.P., 2009. Honeybee Diseases and their Management. 2nd edition. Kalyani Publishers, Ludhiana Pp 750.

Adegas, J.E.B. and Nogueira Couto, R.H., 1992. Entomophilous pollination in rape (Brassica napus L. var. oleifera) in Brazil. Apidologie 23, 203-209.

Araneda, X.D., Ulloa, R.B., Carrillo, J.A., Contreras, J.L. and Bastidas. M.T., 2010. Evaluation of yield component traits of honeybee pollinated (Apis mellifera L.) rapeseed canola (Brassica napus L.). Chilean Journal of Agricultural Research 70(2), 309-314.

Atmowidi, T., Buchori, D., Manuwoto, S., Suryobroto, B. and Hidayat, P., 2007. Diversity of Pollinator Insects in Relation to Seed Set of Mustard (Brassica rapa L.: Cruciferae). HAYATI Journal of Biosciences 14(4), 155-161.

Bai, A.R.K. and Reddy, C.C., 1977. Laboratory toxicity of some insecticides to Apis cerana indica. Journal of Apicultural Research 16 (3), 161-162.

Bezdenezhoykh, S.M., 1956. Pollination of fiber flax and training of bees. In: Krishchunas, I.V., Gubin, A.F. (Eds.) Pollination of agricultural plants. Gos Izd-vo. Selkhoz Litry, Moskva, pp 21–24.

Bichee, S.L., Sharma, M., 1988. Effect of different modes of pollination in sunflower Helianthus annuus L. (Compositae) sown on different dates by Trigona iridipennis Smith. Apiacta 23, 65-68.

Biesmeijer, J.C., Roberts, S.P.M., Reemer, M., Ohlemuller, R., Edwards, M., Peeters, T., Schaffers, A.P., Potts, S.G., Kleukers, R., Thomas, C.D., Settele, J. and Kunin, W.E., 2006. Parallel Declines in Pollinators and Insect-Pollinated Plants in Britain and the Netherlands. Science 313, 351.

Bisht, D.D., Naim, M. and Mehrotra, K.N., 1980. Studies on the role of honeybees in rapeseed production. In: Proc. 2nd Int. Conf. Apiculture in Tropical Climate. Indian Agricultural Research Institute, New Delhi, p. 491–496.

Boch, R., 1961. Honey bee activity on safflower (Carthamus tinctorius L.), Canad. Jour. Plant. Sci.41, 559-562.

Bommarco, R., Marini, L. and Vaissiere, B.E., 2012. Insect pollination enhances seed yield, quality, and market value in oilseed rape. Oecologia 169, 1025–1035.

Bugg, R.L., Colfer, R., Chaney, W.E., Smith, H.A. and Cannon, J., 2008. Flower flies (Syrphidae) and other biological control agents for aphids in vegetable crops. Agriculture and Natural Resources Publication 8285, 1–25.

Bürger, C., 2004. Effects of landscape structure on bee diversity and pollination at different spatial scales. Ph.D. thesis. Institute of Agroecology, University of Göttingen, Germany.

Butler, G.D.Jr., Werner, E.G. and Leven, M.D., 1966. Native bees associated with safflower in Southercenteral Arizona, Kans Ent. Soc. Jour. 39(3), 434-436.

Carreck, N.L. and Williams, I.H., 2002. Food for insect pollinators on farmland: insect visits to flowers of annual seed mixtures. Journal of Insect Conservation 6, 13–23.

Carvell, C., Meek, W.R., Pywell, R.F., Goulson, D., Nowakowski, M., 2007. Comparing the efficacy of agri-environment schemes to enhance bumble bee abundance and diversity on arable field margins. Journal of Applied Ecology 44, 29–40.

Charlesworth, D., Charlesworth, B., 1987. Inbreeding depression and its evolutionary consequences. Annual Review of Ecology and Systematics 18, 237-268.

Chhuneja, P. K., Singh, J., Gatoria, G.S. and Blossom. 2007. Assessing the role of honey bee (Apis mellifera L.) in the seed production of Brassica campestris var. toria in the Punjab. Indian Journal of Crop Science 2(2),34-40

Classen, C.E., 1950. Natural and controlled crossing in safflower (Carthamus tinctorius L.), Argon. Jour. 42, 381-384.

Crane, E. and Walker, P., 1983. Pollination of tropical and subtropical crops by bees. In: Crane, E., Walker, P. (Eds.) The impact of pest management on bees and pollination. Cardiff : IBRA. pp. 5-21.

Crane, E., 1991. Apis species of tropical Asia as pollinators and some rearing methods for them. Acta Hort. 288, 29-48.

Cuningham, S.A., Fitzgibben, F. and Heard, T.A., 2002. The future of pollinators for Australian Agriculture. Agriculture 53(8), 893-900.

Dag, A., Gazit, S., 2001. Mango pollinators in Israel. Journal of Applied Horticulture 2, 39-43.

Dag, A., Lior, E. and Afik, O., 2002. Pollination of confection sunflowers (Helianthus annuus L.) by honey bees (Apis mellifera). Am. Bee J. 142, 443- 445.

De Grandi Hoffman, G. and Martin, J.H., 1993. The size and distribution of the honey bee (Apis mellifera L.) cross-pollinating population on male sterile sunflowers (Helianthus annuus L.). J. Apic. Res. 32, 135-142.

De Grandi-Hoffman, G., Watkins, J.C., 2000. The foraging activity of honey bees Apis mellifera and non-Apis bees on hybrid sunflowers (Helianthus annuus) and its influence on cross-pollination and seed set. J. Apic. Res. 39, 37-45.

Delaplane, K.S. and Mayer, D.F., 2000. Crop pollination by bees. CABI Publishing, New York, NY, USA.

Deodikar, G.B. and Suryanarayana, M.C., 1972. Crop yields and bee pollination. Indian Bee J. 34, 53-64.

Dillman, A.C., 1938. Natural crossing in flax. J. Am. Soc. Agron. 30, 279-286.

Dillman, A.C., Goar, L.G., 1937. Flaxseed production in the far western states. U.S. Dept. Agr. Farmers' Bul. 1792, 22 pp.

Downey, R.K., Bolton, J.L., 1961. Production of [Polish and Argentine] rape in Western Canada. Canada Dept. Agr. Res. Br. Pub. 1021, 19 pp.

Eckert, J.E., 1962. The relation of honey bees to Safflower, Amer. Bee Jour. 102, 349-350.

FAOSTAT., 2012. Food and Agriculture Organization of the United Nations. Roma, Italy.

Free, J.B. and Ferguson, A.W., 1983. Foraging behaviour of honeybees on oilseed rape. Bee World 64, 2-24.

Free, J.B., 1993. Insect pollination of crops (2nd ed.). Academic Press, Harcourt Brace Jovanovich Publ., London. 684 p.

Free, J.B., 1999. Pollination in the tropics. Beekeeping and Development 51, 6-7.

Free, J.B., and Williams, I.H., 1976. Pollination as a factor limiting the yield of field bean (Vicia faba L.). J. Agric. Sci., Camb. 87, 395-399.

Furgala, B., 1954b. Honey bees increase seed yields of cultivated sunflowers. Glean. Bee Cult. 82, 532-534.

Furgala, B., Noetzel, D.M. and Robinson, R.G., 1979. Observations on the pollination of hybrid sunflowers. Proc. IVth Int. Symp. on Pollination, Md Agric Exp Sta Spec Misc Publ 1, 45-48.

Garibaldi, L.A., Carvalheiro, L.G., Leonhardt, S.D., Aizen, M.A., Blaauw, B.R., Isaacs, R., Kuhlmann, M., Kleijn, D., Klein, A.M., Kremen, C., Morandin, L., Scheper, J. and Winfree, R., 2014. From research to action: enhancing crop yield through wild pollinators. Front Ecol Environ 12(8), 439–447.

Greenleaf, S.A. and Kremen, C., 2006. Wild bees enhance honey bees' pollination of hybrid sunflower. Proc. Nat. Acad. Sci. USA. 103, 13890-13895.

Gubin, A.F., 1945. Cross pollination of fibre flax. Bee World 26, 30-31.

Guidry, N.P., 1964. Graphic summary of world agriculture. U. S. Dept. Agr. Misc. Pub. 705, pp. 64.

Hammer, O., 1963. Summer Rape as a Competitor Affecting the Pollination of Clovers. Dansk froavl no. 14, 7 pp.

Hammer, O., 1966. Some problems of competition between summer rape and clover, in relation to pollination. In: 2nd Int. Symp. Pollination, London, 1964. Bee World, 47(Suppl. 1), 99–106.

Heard, M.S., Carvell, C., Carreck, N.L., Rothery, P., Osborne, J.L., Bourke, A.F.G., 2007. Landscape context not patch size determines bumble-bee density on flower mixtures sown for agrienvironment schemes. Biology Letters, 3, 638-641.

Heard, T.A., 1999. The role of stingless bees in crop pollination. Ann. Rev. Entomol. 44, 183-206.

Hoffmann, M., 1994. Observações sobre a polinização entomófila de Helianthus annuus L. em Viamão, Rio Grande do Sul. An Soc Entomol Bras, Londrina 23(3), 391-397. [In Portuguese].

Hoyle, M., Hayter, K. and Cresswell, J.E., 2007. Effect of pollinator abundance on self-fertilization and gene flow: application to GM Canola. Ecol. Appl. 17, 2123- 2135.

Jauker, F. and Wolters, V., 2008. Hover flies are efficient pollinators of oilseed rape. Oecologia 156, 819-823.

Johansen, C.A., 1977. Pesticides and pollinators. Ann. Rev. Ent. 22, 177-192.

Jyoti, J. and Brewer, G.J., 1999. Effect of honeybee (Hymenoptera:Apidae) pollination on sunflower hybrids. Proc. 21st Sunflower Research Workshop. Nat Sunflower Assoc. Jan 14-15, pp. 103-107.

Kadam, B.S., Patankar, V.K., 1942. Natural cross-pollination in safflower, Indian Jour. Genet. 2, 69–70.

Kamel, S.M., Blal, A.H., Mahfouz, H.M. and Said, M., 2013. The most common insect pollinator species on sesame crop (Sesamum indicum L.) in Ismailia Governorate, Egypt. Arthropods, 2(2), 66-74.

Kells, A., Holland, J., Goulson, D., 2001. The value of uncropped field margins for foraging bumblebees. Journal of Insect Conservation 5, 283–291.

Kevan, P.G. and Eisikowitch, D., 1990. The effects of insect pollination on canola (Brassica napus L. cv. O.A.C. Triton) seed germination. Euphytica 45, 39-41.

Khan, R.B. and Dethe, M.D., 2004. Median Lethal Time of New Pesticides to foragers of honeybee. Pestology 28(1), 28-29.

Kisselhagen, S., 1977. Seed yield is increased if there are bees on the rape. Tidsskrift for Biavl 111, 66–67.

Klatt, B.K., Holzschuh, A., Westphal, C., Clough, Y., Smit, I., Pawelzik, E. and Tscharntke, T., 2014. Bee pollination improves crop quality, shelf life and commercial value. Proc. R. Soc. B 281, 20132440.

Klein, A.M., Vaissière, E.B., Cane, J.H., Steffan-Dewenter, I., Cunningham, S.A., Kremen, C. and Tscharntke, T., 2007. Importance of pollinators in changing landscapes for world crops. Proceedings of the Royal Society B 274, 303-313.

Kluser, S. and Peduzzi, P., 2007. Global Pollinator Decline: A Literature Review (Geneva: UNEP/GRID).

Knowles, P.F., 1969. Centers of plant diversity and conservation of crop germplasm, Safflower, Econ. Bot. 23, 324-329.

Kozin, R.B., 1954. Effect of bees on the seed crop of flax. Pchelowdstvo, Mosk. 6, 41-43.

Kremen, C., Williams, N.M., Bugg, R.L., Fay, J.P. and Thorp, R.W., 2004. The area requirements of an ecosystem service: crop pollination by native bee communities in California. Ecology Letters, 7, 1109–1119.

Kubisova, S., Nedbalova, V. and Plesnik, R., 1980. Cinnost vcely medonosne (Apis mellifera L.) na repce (Brassica napus L. var. napus) Pol'nohospoddrstvo 26, 744-754.

Kumari, A. and Pandey, A.K., 2005. Pollination mechanism and behaviour of pollinators in safflower (Carthamus tinctorius L.). J. Ind. Bot. Soc. 82,56–61.

Langridge, D.F. and Goodman, R.D., 1975. A study on pollination of oilseed rape (Brassica campestris). Austral. J. Exptl. Agric. Anim. Husband.15, 285–288.

Langridge, D.F. and Goodman, R.D., 2003. Honeybee pollination of oilseed rape, cultivar Midas. Australian Journal of Experimental Agriculture and Animal Husbandry 22(115), 124–126.

Latif, A. and Qayyum, A., Abbas, M., 1960. The role of Apis indica in the pollination of toria and sarson (Brassica campestris var. Toria and dichotoma). Bee World 41,283–286.

Levin, M.D. and Butler, G.D., 1966. Bees associated with Safflower in South Central Arizona, Jour. Eco.Ent. 59, 654-657.

Levin, M.D. and Butler, G.D., Rubis, D.D., 1967. Pollination of safflower by insects other than honeybees. Jour. Econ. Ent. 60, 1481-1482.

Long, R.F., Corbett, A., Lamb, C., Reberg-Horton, C., Chandler, J. and Stimmann, M., 1998. Beneficial insects move from flowering plants to nearby crops. California Agriculture 52, 23–26.

Luttso, V.P., 1957. The pollination of flax by honeybees. Dokl. Tskha 30, 327-331.

Mahmoud, M.F., 2012. Insects associated with sesame (Sesamun indicum L.) and the impact of insect pollinators on crop production. Pestic. Phytomed. (Belgrade), 27(2), 117-129.

Manning, R. and Boland, J., 2000. A preliminary investigation into honey bee (Apis mellifera) pollination of canola (Brassica napus cv. Karoo) in Western Australia. Aust. J. Exp. Agr. 40, 439-442.

Manzoor-ul-Haq, and Fiaz, M., 1980. Insect pollination of sunflower (Helianthus annuus) in Pakistan. J. Apic. Res. 19, 83-88.

Masuo, Y., 1958. On natural crossing in flax. Crop Sci. Soc. Japan Proc. 27: 321-323.

Mishra, R.C. and Kaushik, H.D., 1992. Effect of cross-pollination on yield and oil content of Brassica campestris L. var. sarson with pollination efficiency of honeybees, Apis spp. Ann. Entomol. 10,33–37.

Mishra, R.C., Kumar, J., and Gupta, J.K. 1988. The effect of mode of pollination on yield and oil potential of Brassica campestris L. var. sarson with observation on insect pollinators. J. Apic. Res. 27, 186–189.

Mohammed, A., 1935. Pollination studies in tara (Brassica napus L. var. dichotsma Prain) and sarson (Brassica campestris L. var. sarson, Prain). Indian J. Agric. Sci. 5, 125-154.

Morandin, L.A., Winston, M.L., 2005. Wild bee abundance and seed production in conventional organic and genetically modified canola. Ecol. Appl. 15, 871-881.

Moreti, A.C. de C.C., da Silva, R.M.B., da Silva, E.C.A., Alves, M.L.T.M.F., Otsuk, I.P., 1996. Aumento na produção de sementes de girassol (Helianthus annuus) Pela ação de insetos polinizadores. Sci. Agric. 53, 280-284.

Munawar, M. S., Raja, S., Siddique, M., Niaz, S., Amjad, M., 2009. The pollination by honeybee (Apis mellifera L.) increases yield of canola (Brassica napus L.) Pakistan Entomologist 31(2), 103-106.

O'Toole, C., 1993. Diversity of native bees and agroecosystems. Hymenoptera and Biodiversity, In:LaSalle, J., Gauld, I.D) (eds.), pp. 169-196. CAB Int., Wallingford, UK.

Oz, M., Karasu, A., Cakmak, I., Goksoy, A.T., Turan, Z.M., 2009. Effects of honeybee (Apis mellifera) pollination on seed set in hybrid sunflower (Helianthus annuus L.). African Journal of Biotechnology 8(6), 1037-1043.

Panda, P., Sontakke, B.K. and Sarangi, P.K., 1988. Preliminary studies on the effect of bee (Apis cerana indica F.) pollination on yield of sesamum and niger. India Bee J. 50, 63-64.

Pandey, A.K., Kumari, A., 2007. Pollination ecology of safflower (Carthamus tinctorius Linn.). In: Seventh International Safflower Conference, Waaga Wagga Australia, pp 1–10.

Parkash, R. and Kumaraswami, T., 1984. Toxicity of some insecticide spraying on hymenopertrans pollinators in new Brunswick. Canadian Entomologist 112, 259-269.

Patil, J.A. and Chavan, V.M., 1948. Selfing methods in Safflower, Ind. oilseeds Jour. 2, 10-12.

Phadke, K.G., Pant, N.C. and Sinha, R.B.P., 1967. The foraging behaviour of Apis indica F (Hymenoptera: Apidae) visiting flowers of sesame (Sesamum orientate L.). Indian J Ent 29, 157–163.

Pimentel, D., Andow, D., Dyson-Hudson, R., Gallahan, D., Jacobson, S., Irish, M., Kroop, S., Moss, A., Schreiner, I., Shepard, M., Thompson, T. and Vinzant, B., 1980. Environmental and social costs of pesticides: a preliminary assessment. Oikos, 34(2),127-140.

Pordel, M.R., Hatami, B., Mobli, M. and Ebadi, R., 2007. Identification of insect pollinators of three different cultivars of winter canola and their effect on seed yield in Isfahan. J. Sci. Technol. Agric. Natur. Resour. 10(4B),413-426.

Potts, S.G., Biesmeijer, J.C., Kremen, C., Neumann, P., Schweiger, O. and Kunin, W.E., 2010. Global pollinator declines: trends, impacts and drivers. Trends in Ecology and Evolution 25, 345-353.

Prakash, R. and Kumaraswami, J., 1984. Toxicity of some insecticides to the Indian bee, Apis cerana Feb.. Indian Bee Journal 46, 15–17.

Prasad, D., Hameed, S. F., Singh, R., Yazdani, S.S. and Singh, B., 1989. Effect of bee pollination on the quantity and quality of rai crop (Brassica juncea Coss). Indian Bee Journal 51(2), 45-47.

Radford, B., Nelson, R. and Rhodes, J., 1979. Agents of pollination in sunflower crops in the central Darlingdowns, Queensland. Aust J Exp Agric Anim Husb 19, 565-69.

Rahman, K.A., 1940. Insect pollinators of toria (Brassica napus L. var dichotoma) and Sarson (Brassica campestris L. var Sarson) at Lyallpur. Indian J. Agric. Sci. 10, 422-447.

Rajasri, M., Kanakadurga, K., Durga Rani, V. and Choudhary, A., 2012. Honey bees–potential pollinators in hybrid seed production of sunflower. International Journal of Applied biology and pharmaceutical technology 3(2), 216-221.

Rana, B.S., Goyal, N.P., 1991. Field toxicity of methyl demeton and dimethoate to the foragers of honeybee (Apis cerana indica F.) visitors to Brassica chinensis L. Indian Bee Journal. 53(1-4), 73-77.

Rao, G.M., Lazar, G.M. and Suryanarayana, M.C., 1981. Foraging behavior of honeybees in sesame (Sesamum indicum L.). Indian Bee J. 43, 97-100.

Richards, A.J., 1997. Plant breeding systems, 2nd Edn. London: Chapman and Hall.

Rubis, D. D., Levin, M.D. and Mc Gregor, S.E., 1966. Effects of honey bee activity and cages on attributes of thin hull and normal Safflower lines, Crop Sci. 6, 11-14.

Rubis, D.D., 1970. Breeding insect pollinated crops. Arkansas Agricultural Extension Service 127, 19-24.

Russell, D., Meyer, D. and Bukowski, J., 1998. Potential impact of micro encapsulated pesticides on New Jersey Apiaries. American Bee Journal 38(3), 207-210.

Sabbahi, R., De Oliveira, D. and Marceau, J., 2005. Influence of honey bee (Hymenoptera: Apidae) density on the production of canola (Crucifera: Brassicacae). Journal of Economic Entomology 98(2), 367-372.

Sabbahi, R., De Oliveira, D. and Marceau, J., Houle, E., 2006. Floraison du canola (Crucifera: Brassicaceae) et le rendement des colonies d'abeilles domestiques (Hymenoptera: Apidae). Avalaible at http://www.agrireseau.qc.ca/apiculture/documents/Rendem_apicole-Rachid_revise.pdf

Sabir, A.M., Suhail, A. and Hussain, A., Saeed, A., 2000. Insecticidal mortality, forgaing behaviour and pollination role of honeybee (Apis mellifera L.) on brassica (Brassica campestris L.). Pakistan Journal of Zoology 32(4), 369-372.

Scheper, J., Holzschuh, A. and Kuussaari, M., Potts, S.G., Rundlof, M., Smith, H.G., Kleijn, D., 2013. Environmental factors driving the effectiveness of European agri-environmental measures in mitigating pollinator loss – a meta-analysis. Ecology Letters 16, 912–920.

Sharma, D., Abrol, D.P., 2014. Effect of insecticides on foraging behaviour and pollination role of Apis mellifera L. (Hymenoptera: Apidae) on toria (Brassica campestris var. toria) crop. Egyptian Journal of Biology 16, 79-86.

Sheffield, C.S., Westby, S.M. and Smith, R.F., Kevan, P.G., 2008. Potential of bigleaf lupine for building and sustaining Osmia lignaria populations for pollination of apple. The Canadian Entomologist, 140, 589–599.

Sihag, R.C., 1986, Insect pollination increases seed production in cruciferous and umbelliferous crops. J. Agric. Res. 25, 121-126.

Sikka, S.M. and Gupta, N.P., 1949. Pollination studies in Sesamum orientale L. Indian J. Genetics 9, 33-191.

Singh, M.P. and Singh, K.I., Devi, C.S. 2000. Role of Apis cerena Himalaya pollination on yield and quality of rape seed and sunflower crops. In proceedings of 4th Asian Apicultural Association, International Conference, Kathmandu, Nepal March 23-28, 1998.

Singh, R.P. and Singh, P.N., 1992. Impact of bee pollination on seed yield, carbohydrate composition and lipid composition of mustard seed. Journal of Apicultural Research 31, 128-133.

Singh, R.P. and Singh, P.N.,1992. Impact of bee pollination on seed yield, carbohydrate composition and lipid composition of mustard seed. J.Apic Res. 31, 128-133.

Singh, S., 1954. Insect pollination and breeding of fruit varieties. Indian J Hort Sci 11, 123-125.

Smirnov, V.M., 1956. Cross-pollination of fl ax by bees. Pchelovodstvo Mosk 9,53–55.

Smith, H.A., Chaney, W.E., Bensen, T.A., 2008. Role of syrphid larvae and other predators in suppressing aphid infestations in organic lettuce on California's Central Coast. Journal of Economic Entomology 101, 1526–1532.

Smith, W., 2002. Honey bees on canola. New South Wales Agriculture, Department of Primary Industries, Orange, New South Wales, Australia.

Stark, D.J.D., Jepson, P.C. and Mayer, D.F., 1995. Limitation to use of topical toxicity data for production of pesticide side effects in the field. Journal of Economic Entomology 88 (5), 1081-88

Sundararaju, D., 2003. Occurrence of bee fauna and extent of pollination in insecticide sprayed ecosystem of cashew. J. Palynol 39, 121–125.

Vanbergen, A.J. and Baude, M., Biesmeijer, J.C., Britton, N.F., Brown, M.J.F., Bryden, J., Budge, G.E., Bull, J.C., Carvell, C., Challinor, A.J., Connolly, C.N., Evans, D.J., Feil, E.J., Garratt, M.P., Greco, M.K., Heard, M.S., Jansen, V.A.A., Keeling, M.J., Kunin, W.E., Marris, G.C., Memmott, J., Murray, J.T., Nicolson, S.W., Osborne, J.L., Paxton, R.J., Pirk, C.W.W., Polce, C., Potts, S.G., Priest, N.K., Raine, N.E., Roberts, S., Ryabov, E.V., Shafir, S., Shirley, M.D.F., Simpson, S.J., Stevenson, P.C., Stone, G.N., Termansen, M., Wright, G.A., 2013. Threats to an ecosystem service : Pressures on pollinators. Frontiers in Ecology and the Environment 11 (5), 251-259.

Vesely, V., 1962. The economic effectiveness of bee pollination on winter rape (Brassica napus l., var. Oleifera metz.). Min. Zemedel. Lesn. A vodniho hospodar. Ust. Vedtech. Inform. Zemedel. Ekon 8, 659–673.

Wakhle, D.M., Shakuntala Nair, K. and Phadke, R.P., 1978. Effect of bee-pollination on the oil and protein content in the seeds of sunflower (Helianthus annuus L.). Indian Bee J. 40, 1-2.

Wallace, H.M., Maynard, G.V. and Trueman, S.J., 2002. Insect flower visitors, foraging behaviour and their effectiveness as pollinators of Persoonia virgata R. Br. (Proteaceae). Austral J Entomol 41, 55- 59.

Westcott, L. and Nelson, D., 2001. Canola pollination: an update. Bee World 82, 115-129.

White, B., 1970. Pollination of commercial rape seed crops. Australasian Beekeeper 72, 99–100.

Williams, I.H., Martin, A.P. and White, R.P., 1987. The effect of insect pollination on plant development and seed production in winter oil-seed rape (Brassica napus L.). J. Agric. Sci., Camb. 109, 135-139.

Wittkop, B., Snowdon, R.J. and Friedt, W., 2009. Status and perspectives of breeding for enhanced yield and quality of oilseed crops for Europe. Euphytica 170,131–140.

Wratten, S.D., Gillespie, M., Decourtye, A., Mader, E. and Desneux, N., 2012. Pollinator habitat enhancement: benefits to other ecosystem services. Agriculture, Ecosystems and Environment 159, 112-122.

Zhong, H.E., Mark, L., Sheve, P. and Cate, B., 2004. Minimizing the import if the mosquito adulterine noted or honeybees, Apis mellifera (Hymenoptera: Apidae): Aerial Ultra volume application using a high-pressure nozzle system. Journal of Economic Entomology 97 (1), 1-7.

CHAPTER 12

# Integrated Pest Management in Oilseed Crops

## INTRODUCTION

Oilseed crops have been the backbone of several agricultural economies from antiquity and play a prominent role in agricultural industries and trade throughout the world. A major amount of edible oil comes from the crop resources. The quantum increase in production through various methods of improvements and technological interventions has also offered a congenial and suitable environment for the ravages caused by insect pests. Among several factors, insect pests are a significant constraint limiting production of oilseeds. Oil from the seeds of plants belonging to the genus *Brassica,* family Cruciferae, have been utilized by man for thousands of years (Prakash and Hinata, 1980), but it is only during the last thirty years that oilseed *Brassica* crops have become internationally important (Lamb, 1989). The two most important commercial oilseed crops are *Brassica campestris* var. *oleifera,* which has a worldwide distribution, and *B. napus* var. *oleifera*, which tends to be restricted to Europe and North Africa (Weiss, 1983). The other oilseeds crops are Groundnut, sunflower, soybean, safflower and sesame yielding high grade edible oil. The non-edible oil producing species include castor and linseed.

### Scenario of oilseed crops in the world

There are about 40 different oilseeds crops in the world whose oil can be consumed (Lennerts, 1983), but only a few are significant. Oilseed crops are of three distinct categories: the annual or biennial group which includes soybeans, sunflower, groundnuts and rapeseed; the perennial tree crops include coconuts and oil palms. Cotton and corn germ belong to third group in which embryo is a by-product. Other oilseeds are of minor importance in the world trade but they do play a significant role in local markets or serve as raw materials for special products (Sharma et al., 2012). *Brassica* species are the second largest oilseed crop after soybean (*Glycine max* (L.) Merr.) in the world oilseed production, surpassing peanut (*Arachis hypogaea* L.), sunflower (*Helianthus annuus* L.) and cottonseed ( *Gossypium hirsutum* L.) during the last two decades (FAO, 2010; Raymer, 2002, Sharma et al., 2014 ).

Of the 37 species in the *Brassica* genus, the four most widely cultivated species for oilseed and vegetable production are *Brassica rapa* L., *Brassica juncea* (L.) Czernj and Cosson, *Brassica napus* L. and *Brassica carinata* A. Braun (Raymer, 2002; Rakow, 2004; Sovero, 1993). Rapeseed is the most favored vegetable oil. The leading producers are Canada, China, India, Germany and Ukraine along with Poland, UK and Australia which contribute about 77% of the total rapeseed mustard production of the world (FAO, 2010). Germany has the highest productivity of rapeseed (37.60 q/ha) followed by United Kingdom (32.98 q/ha) and Czech Republic (29.38 q/ha).

Oil palm is an alternative and potential long term source of edible oil which is expected to contribute significantly towards meeting the growing edible oil demand. It can yield, on an average, four tonnes of vegetable oil per hectare as compared to less than half a tonne from other annual oilseed crops. Palms are grown predominantly in the tropical areas of the world as perennial trees and are an important source of vegetable oil. Production of palm kernel was around 11.74 mmt; however, with respect to oil production, palm oil dominates the world vegetable oil economy with production touching 42.41 mmt during 2008–2009. It contributes about 36% of the total world oil production. The major palm oil producing countries are Indonesia, Malaysia, Thailand, Nigeria and Colombia which together contributed about 93% of total world palm oil production in the year 2008-2009. Cotton is cultivated in 70 countries worldwide with total coverage of about 34 mha and production of around 41.26 mmt of cotton seed and 4.72 mmt of cotton seed oil in 2008–2009 (FAO, 2010). Over a quarter of the world cotton area is in India, followed by USA (16%), China (14%) and Pakistan (8%). The remaining production comes from Turkey, Australia, Greece, Brazil and Egypt. The cotton is a dual purpose crop, producing both seed and fibre as valuable primary agriculture products. Groundnut (*Arachis hypogaea* L.) is a leading oilseed crop in India and an important oilseed crop of tropical and subtropical regions of the world. The seeds are rich source of edible oil (43-45%) and protein (25-28%). Groundnut (*A. hypogaea* L.) is the most important oilseed and a highly explored agricultural commodity. Groundnut is grown in about 84 counties on an area of 24.52 mha with the production of 34.15 mmt and average productivity of 13.92 q/ha. Among the groundnut producing nations, India sows the highest area (around 28% of total world area) under this crop and is the second largest producer, next only to China. In India, groundnut occupies about 6.85 mha area with a production of about 7.34 mmt. China, India, Nigeria, USA and Myanmar contributed about 85% of the total world production of groundnut in 2008. China (31.02 q/ ha), USA (38.29 q/ha), Nicaragua (36.10 q/ha), Turkey (34.33 q/ ha), Egypt (33.99 q/ ha) and Syrian Arab Republic (31.87 q/ha) are the countries having very high average yields and thus contribute significantly to the world pool of groundnut.

Among the oilseed crops, sunflower (*H. annus* L.) occupies fourth position in area and production in the world after soybean, rape seed mustard and groundnut. World harvest of sunflower was 33.03 mmt from an area of 18.98 mha during 2008. Russian Federation, Ukraine, Argentina, India and China are the top five countries in the world with respect to production (21.49 mmt) and area under crop (15.93 mha). Europe alone accounts for about 50% of world area and total production. Asia and Africa share about 21 and 6% of the total sunflower area in the world, respectively. Average yield of the crop is significantly very high in countries like Crotia (31.03 q/ha), Austria (29.73 q/ha) and Switzerland (29.00 q/ha). USA, Romania, Bulgaria and Hungary are the other countries which have high potential for this crop. Olive, *Olea europaea* L., is a sub-tropical, evergreen tree and one of the world's oldest cultivated crops originated in the Mediterranean region. The olive tree is the sixth most important oil crop in the world. World olive oil production is around 2.9 mmt, the Mediterranean countries being the major contributors (International Olive Council 2009, http://www.internationaloliveoil.org/downloads/ production1_ang. PDF). Spain has largest area (2.60 mha) and highest production (6.22 mmt) followed in production by Italy, Greece, Turkey, Tunisia and Greece. These five countries accounted for 80.84% of the total world production (18.32 mmt) in 2008. Egypt has the highest productivity (63.60 q/ha) followed by Peru (55.46 q/ha) and Australia (54.40 q/ha). USA, Chile, Mexico, Argentina and Slovenia are some of the other countries having higher productivity levels. Sesame (*Sesamum indicum* Lin.) is known as the 'queen of oilseeds' and one of oldest oilseed crops grown in India. India ranks first in area under cultivation representing 30% of the world production. Castor (*Ricinus communis* L) is one of the industrially important non-edible oilseed crops of the world. India ranks first among the major castor producing countries (Brazil and China) in the world with 68% of area and 76% of world castor production.

## Scenario of oilseed crops in India

India is among the largest oil economies in the south East Asian region and occupies a distinct position in terms of diversity in annual oilseed crops. The prevailing agro-ecological conditions have been favorable for growing several important annual oilseeds, including edible (namely, groundnut, rapeseed-mustard, soybean, sunflower, safflower and sesame) and non-edible oilseeds (namely, castor and linseed). In addition, a wide range of other minor oilseeds and oil bearing tree species add to the diversity as well as oilseed production in the country. On the world map, India occupies a prominent position in acreage and production of oilseed crops. It constitutes the second largest agricultural commodity next to cereals in India, sharing 14 per cent of the country's gross cropped area and accounting for nearly 5 per cent of the gross national product and 10% of the value of all agricultural products. Today, oilseed crops are cultivated on about 16.5 million hectares, with total production of 10 million tones which constitutes approximately one-tenth of the total cultivated area in India. The major oilseed crops growing states are Madhya Pradesh (24.1 %), Gujarat (11.6%), Andhra Pradesh (11.5%), Rajasthan (11.4%), Maharashtra (10.9%), Karnataka (9.5%), Uttar Pradesh (6%) and Tamil Nadu (4.5%). Approximately 15 million peoples are engaged in the production and processing of oilseeds (Hegde, 2000). On the basis of importance the oilseed crops grown country are groundnut, rapeseed and mustard, sesame, linseed, safflower, castor, sunflower and niger. More than half of the world's diet of fats and oils comes from oilseeds-coconuts, cotton, oil palm, olives, peanuts, rape, soybeans, and sunflower (Guidry, 1964).

India contributes a large share to the global castor production (76.9 per cent) and also a substantial one in the production of sesame (31.2 per cent) and groundnut (25.1 per cent). The country is the largest producer of castor and sesame and second largest producer of groundnut and rapeseed-mustard only next to China.

Castor productivity in India is more than the world average mainly due to higher productivity in Gujarat, India, where castor is cultivated under irrigated conditions otherwise it is grown in semi-arid and arid regions. Castor is grown in all most all states of the country, Gujarat, Andhra Pradesh, Tamilnadu, Rajasthan and Karnataka account for about 51, 29, 6, 5 and 3 per cent of the total area under the crop, respectively. Gujarat state of India alone contributes to 83% of the total country's production.

## Constraints in oilseed crop production

Oilseed crops are affected by a large magnitude of insect pests of which, about a dozen are of serious concern. These include leaf miner, thrips, and tobacco caterpillar; head borer in sunflower; aphids and *H. armigera* in safflower; leaf webber and capsule borer in sesame and, aphids, saw fly, cabbage butterfly, and painted bug in rapeseed-mustard, semilooper and Bihar hairy caterpillar in castor and bud fly and *H. armigera* in linseed. These crops are often attacked by polyphagous pests like red hairy caterpillar, Bihar hairy caterpillar, root grubs, termites, cutworms and serpentine leafminer. The yield losses due to these pests could range from 50-90% across oilseed growing areas. Soybean crop is severely attacked by about a dozen of insect pests causing damage ranging from 20 to 100 per cent. However, with implementation of IPM schedule, 30-35 per cent additional yield can be obtained (Sharma et al., 2014). In order to develop high yielding genotypes having resistance to insect-pests, the detailed knowledge of the insects, the plant and their interaction is must. In general, the available oilseed crop cultivars are susceptible to major insect pests however, some have shown escape to some insect pests. Although the use of insecticides is quite effective but their excessive and injudicious use have resulted into imbalance in agro-ecosystem and health hazards. Therefore, the answer lies in the development of resistant/tolerant varieties, exploitation of biological control agents and use of bio-rational approaches against the major insect pests of oilseed crops. One of the viable options is the adoption of IPM involving several approaches like seed treatment with *Trichoderma* sp, release of *Trichogramma,* use of pheromone traps and use of need based application of pesticides etc. New technologies have made significant impact in the field of IPM.

Climate change, food insecurity, and energy demand are major concerns for modern agriculture and their impact is increasing rapidly. A rapidly growing human population has resulted in a demand for increased food production. As the land resources became limited, efforts were made to increase productivity by fighting the losses inflicted by insects, weeds, and plant pathogens through the interventions of pesticides. At the same time, pesticides proved to be more dangerous due to their indiscriminate and excessive use, contaminating food and the environment resulting in pest resistance, pest resurgence, and pest outbreaks. Consequently, the need arose for eco-friendly integrated pest management strategies to produce food safe from the negative impact of pesticide residues (Abrol and Shankar 2014). The IPM principle does not preclude chemical pesticide use, but rather uses it as one of the tools in the management package, to be used prudently and integrated with other tools. Thus, the concept of IPM of insects contains three basic elements: (i) maintaining insect populations below levels that cause economic damage; (ii) the use of multiple tactics to manage insect populations; and (iii) the conservation of environment quality. Excessive and indiscriminate use of pesticides endangers the health of farm workers and consumers of agricultural products worldwide (Goodell, 1984). Many of the steps of IPM depend upon accurate and readily available information for proper decision-making. Approaches to be used within IPM strategies include (i) the greater use of monitoring to determine pest abundance more accurately, (ii) the use of trap crops to attract and to concentrate pests thereby reducing pest density and damage to the main crop, and (iii) the promotion of conservation biological control with particular emphasis on minimum tillage and avoidance of insecticides during flowering (Cook and Denholm 2008). The important insect pests and their management strategies of different oilseed crops are briefly discussed below-

## Important insect pests of Oilseed crops

### Insect pests of Brassica

Many species of the insect pests that attack the oilseed *Brassica* crops also attack other cruciferous vegetables such as cabbage and radish. The common pests of crucifers as described by Bonnemaison (1965) before oilseed rape became an important crop in Europe and North America. Crucifers crops are susceptible enough to attack by insect pests in the early seedling stages, that is from emergence until they have 4-5 leaves stage. The number of insect pests associated with the rapeseed and mustard crops has increased from 24 (Rai, 1976) to about 40 (Bakhetia and Sekhon 1989; Sachan and Gangwar 1980) owing to changed cropping patterns and introduction of new crops like *Brassica napus* and B. *carinata*. The mustard aphid, *Lipaphis erysimi* (Kaltenbach) is the key pest responsible for bulk of yield losses. The most serious pests of the oilseed *Brassica* crops are members of the order coleoptera, particularly in Europe and North America. On the Indian subcontinent and in New Zealand, aphids cause severe damage (Bakhetia and Sekhon 1989; Lammerink and Banfield, 1980; Singhvi et al., 1973), although in India insect pests from other orders are also important (Bakhetia and Labana 1978). In Asia, insecticides are applied to control aphids and the diamondback moth, *Plutella xylostella,* in oilseed *Brassica* crops (Lee, 1968), but relatively little has been published on their pest status. Similarly, little has been published on the pests of oilseed *Brassica* crops in South America, Australia, and Africa. It is important to correctly identify the pest causing damage so appropriate control measures may then be taken well in time.

The flea beetle, *Phyllotreta cruciferae* is not usually a significant pest in Europe (Augustin et al., 1986), although it is the dominant flea beetle pest of oilseed *Brassica* crops in Canada (Burgess, 1977) and in India (Bakhetia and Labana 1978). Root maggots, *Delia* spp., are significant pests of rape only in parts of Alberta, Canada (Griffiths, 1986) and in Finland (Tiittanen and Varis 1960), although *Delia radicum* is a widespread pest of other crucifers throughout most of Europe and North America (Bonnemaison, 1965). The externalities triggered by the chemical control of insect pests have necessitated the search for alternative control methods in judicious way. Among these, role of resistant varieties comprise the foremost strategies of pest management because of their well documented advantages. In India, in the recent past, extensive work has been done on various aspects of HPR to the key pest (*Lipaphis erysimi*) but the work on the other major insect-pests of rapeseed and mustard has been negligible.

## Mustard aphid, *Lipaphis erysimi* Kalt. and *Brevicoryne brassicae*

This aphid is active throughout the year however, main activity period coincides with the growing period of cruciferous crops from September until March. During lean season the aphid shifts to wild or cultivated off-season crucifers in damp places in orchards and kitchen gardens (Sidhu and Singh, 1964). Different off-season hosts have been recorded for this pest like Asvagandha (*Withania somnifera*) from May to July, cauliflower (*B. oleracea*) from August to March, cabbage (*B. oleracea* var. *capitata*) and Indian mustard and from March to April, radish (*Raphanus sativus)* (Chandra and Kushwaha 1987). Mustard aphids (*Lipaphis erysimi* Kalt and *Brevicoryne brassicae* L.) is a key pest of rapeseed-mustard (Fig. 12.1–12.3), causing a 35% to 75% reduction in yield (Rohilla et al., 1987; Rana, 2005) and a 6% reduction in oil content (Singh et al., 1987). A minimum temperature in the range of 4-13.8°C and a mean maximum temperature 14.0-30.0°C and 48-80% relative humidity, favor quick aphid multiplication (Sekhon and Bakhetia 1994; Bishnoi *et al.*, 1992). Rainfall has been reported to have profound effect on aphid as mild rainfall helps in aphid multiplication when the crop is in flowering stage (Singh *et al.*, 1990) while the rainfall totally wipes off the aphid at pod stage of this crop (Mahal *et al.*, 1988).

## Flea beetle, *Phyllotreta cruciferae* Gorge

This pest sometimes becomes very serious causing economic damage to the crop and it has been reported to cause serious damage to the rapeseed and mustard crops in Uttar

Pradesh and West Bengal while moderate incidence has been reported from everywhere from rapeseed and mustard growing areas of this country (Rai, 1976) and other parts of the world (Burgess, 1977; Tansey et al., 2009). Adult beetles are present in the fields all the year round except from November to January, when these hibernate in the soil or plant debris. Further, Makhmoor and Shinde (1986) had observed a positive correlation between flea beetle incidence and nitrogen application.

## Painted bug, Bagrada hilaris, B. cruciferarum

This pest causes damage to rapeseed and mustard crops at seedling and maturity/harvesting stage of the crop (fig 4). Both adults and nymphs suck the sap from the leaves, young plants or even for the seed and in earlier case it devitalize the plant completely. The seedling stage of the crop is the most sensitive which may even be killed, if heavy incidence of this pest occurs at this time (Vora et al, 1985). On full grown plants, the nymphs and adult congregates on the pods and suck the sap.

## Mustard sawfly, *Athalia lungens* Klug

This pest preferably feeds on mustard plant in choice situation otherwise is a serious pest of all the cruciferous crops in seedling stage (Vora et al, 1985). The grub is the main damaging stage of this pest which feed on leaves (Fig 12.5 and 12.6) and causes shot holes and sometime riddle whole leaf in severe damage situations. In case of less damage, the plant recovers quickly and the shot holes are of no economic importance. In some months this pest aestivates as grub/pupa and again become active when the temperature starts declining. It completes 3 overlapping generation from September to January.

## Pea leafminer, *Chromatomyia horticola*

This pest is a polyphagous pest and may cause damage up to 15.2% on yield basis (Singh, 1991; Singh et al., 1984). It also affects plants of compositae and leguminoseae in addition to family Cruciferae. The maggots mine the leaves with typical gallery formation. The larvae bore the mesophyll and spongy parenchymatous tissues of the leaves. They usually don't consume the entire leaf surface. The mines run both on upper and lower leaf surface. During summer and autumn months, it aestivates and the fly again emerges in December with a peak activity period being February to April, with 4-5 generations in a year. The population of this pest generally can't sustain the high temperature conditions beyond 30°C.

## Management Strategies

*Cultural control:* Lesser nitrogen applications normally restrict fecundity, faster development and multiplication of aphids.

*Natural enemies:* There are many natural enemies which play role in the management of rapeseed and mustard insect pests at one or the other stage of the crop growth. Mustard aphid is one of the most important pests of this crop. It is parasitized by *Diaretiella rapae* up to 93 per cent (Chandra and Khushwaha 1987) (Fig. 12.7). The parasitization is quite high in January and February. Pea leafminer has been reported to be parasitized by *Chrymotomyia formosa, Diglyphus isaea, Opius exigua* and Sohegion sp. up to 86 per cent by Kumar, 1985. Many coccinellids like C. *septempunctata* (Fig. 12.8– 12.9), *M. sexmaculata, Brumoides suturalis, Hippodamia variegata, Scymnus nubilus* and *Verania viacta* have been found to be active in the aphid infested fields (Atwal, 1971; Kalra, 1988, Rai et al., 2010). Syrphids, *Ischiodon scutellaris*(Fig. 12.10) is the only syrphid feeding on mustard aphid (Cherian, 1934; Kumar et al. 2011) (Fig. 12.9). In Punjab, *Diaeretiella rapae* parasitized only 8.2% aphids in February which increased to 100% in April (Atwal et al., 1969). A bacteria, *Serratia marcescens* have been reported to cause disease to *Athalia proxima* larvae by Bogawat et al., 1966. On the other hand Entomophthora sp. and *Cephalosporium* sp. have been reported to cause disease in Aphid, *Isipaphis erysimi* (Singh, 1982). The mustard aphid is known to be preyed upon or parasitized by a large number a predators and parasites like coccinellids, syrphids, *Chrysoperla carnea, Aphidius* sp, etc. There are only a few reports on the biology and efficacy of syrphids and coccinellids in the laboratory and the population dynamics of the aphid in the field. In India, the rate of parasitism of mustard aphid, *Lipaphis erysimi* (Kaltenbach) on rape-seed mustard crop varied from 20% (Thakur et al., 1989) to 51% (Dogra et al., 2003) in Himachal Pradesh to 60-97% in Maharashtra (Katole et al., 1998)

*Mechanical Control:* Mechanical removal of twigs and leaves harboring hairy caterpillar is very effective and is advised on community basis. In case of the mustard aphids too, heavily infested twigs removed without disturbing the whole plant can check the proliferation of the mustard aphid.

## Insect pests of Groundnut

Insect pests form the major constraints in its production of groundnut. Red hairy caterpillars: *Amsacta albistriga* Walker, *A. moorei*, leaf miner: *Aproaerema modicella* Deventer, bihar hairy caterpillar: *Spilosoma obliqua* Walker, tobacco caterpillar: *Spodoptera litura* (Fabricius) Aphids: *Aphis craccivora* Koch, and leaf hopper: *Empoasca kerri* Pruthi, are important insect pests of groundnut crop in India. In the recent years incidence of thrips on groundnut crop is increasing and known to cause yield loss to the tune of 14 to 40 per cent (Ghewande, 1987) and also responsible for spreading a viral disease called bud necrosis in groundnut. The leaf miner is considered as the most important insect pest of groundnut in South India and particularly under rainfed situations (Ayyar, 1963; Nair, 1975; Reddy, 1988). The pest initially appears as a leaf miner causing short blister like mines. Older larvae fold the leaflets and feed within. As a result, the leaflets turn brown, shrivel and dry up. Severely infested crop gives a burnt up appearance and yield losses can reach up to 76 per cent (Anonymous, 1986).

## Management Strategies

*Cultural control:* Mixed sowing of cereals with groundnut is a common practice and is known to reduce the incidence of insect pests like thrips, jassids and leafminer on groundnut. Mulches ant summer ploughing can also protect crops in many different ways. Summer ploughing is known to reduce the grubs' population significantly (>70%).

Resistant varieties: It has been reported that the bunch varieties had higher population of, and showed more injury owing to, leaf-webber, aphid and grey weevil, whereas the spreading and semi-spreading types harboured lower populations (Brar and Sandhu, 1976). This difference may be due to more of foliage and more of succulence of the bunch types which provide conditions favourable for providing with better shelter and preferred food. High level of resistance to jassids has been noticed in EC 5240. Successful trasfer of jassid resistance from EC 5040 and EC 5043 into suitable lines having good agronomic traits and high yield has been achieved.

*Biological control*: More than 30 species of hymenopterous parasitoids attack the eggs, larvae and pupae of *A. modicella* on groundnut and soyabean in India. Gadad and Hegde (2014) evaluated the efficacy of different biopesticides and biorationals against groundnut thrips and leaf miner at Dharwad, India. Results revealed that spinosad 45 EC @ 0.20 ml/l, *Lecanicillium lecanii* @ 6 g/l and *L. lecanii* @ 4 g/l found effective in reducing thrips population, whereas spinosad 45 EC @ 0.20 ml/l, buprofezin (25 EC @ 1.0 ml/l), *Beauveria bassiana* @ 6 g/l and *Metarrhizium anisopliae* @ 6 g/l found effective against leaf miner. Considering the cost of cultivation and the gross profit in different treatments, the benefit cost ratio (BCR) and net profit was calculated. Among the different treatments, spinosad and buprofezin recorded higher B:C ratio net profit and found superior over all other treatments in the trial. Among the biopesticides *L. lecanii* @ 6 g/l recorded higher B:C ratio and net profit as compare to other biopesticides treatments.

*Trichogramma evanescens minutum* has often been recorded parasitizing the eggs of *Amsacta albistriga* and *Diacrisia obliqua,* (Krishnamurthi and Usman 1954). *A. albistriga* eggs are frequently parasitised by *T. eyanescens minulum* with an effective recovery of 59.4 per cent (Sundaramurthy *et al.,* 1976; Gunathilagraj and Babu 1987). However, only 10.3 per cent parasitization have been recorded in *A. modicella* eggs in this crop (Muthaiah and Kareem 2000). Leafminer, *Aproerema modicella* have been parasitized by an Eulophid, *Stenomesius japonicum,* and a chalcid, *Brachymeria wittei* for the first time by Muthaiah, 1991. *Goniozus indicus* and *B. wittei* have been found parasitizing 16.7 and 20 per cent larvae of *A. modicella* in Tamil Nadu by Muthaiah and Kareen, 2000. Twenty six per cent *Helicoverpa armigera* pupae were found parasitized up to by *Carcelia illota* and *Chelonus* sp. in Tamil Nadu (Muthaiah and Kareem 2000).

Coccinellids, *Menochiles sexmaculatus and* C. *septempunctata* have been found responsible for 10-97 per cent decrease in population of *Aphis craccivors* in groundnut grown in Gujrat (Patel et al., 1976). Patel and Vyas 1984, while studying feeding potential of *M. sexmaculata* reported it to be consuming 474 aphids during 6 days of larval development. *Ischiodon scutellaris, Xanthogramma scutellaris* and *Chrysoperla carnea* were also recorded predating upon *A. craccivora* in Gujrat. Thrips on the other hand were being preyed upon by C. *septempunctata, Orius* sp. and *M. sexmaculata*. In another very important observation, Sahariyaraj, 1999 reported 92.73 and 94.61 per cent reduction in the population of *H. armigera* and S. *litura,* respectively by *Rhinicoris marginatus.*

*Bacillus thuringiensis* have been found to be effectively managing the population of *A. modicella* up to 36.4 per cent by Brar and Sandhu 1982; Chandla et al., 1988. *H. cousanguinea,* a very important pest of groundnut in the sub-tropics was effectively managed by *Bacillus popillae* var. holotrichiae @ 5 billion spores per square meter (Vyas *et al.,* 1991). Fungi, *Metarrhizium anisopliae* was found infesting *H. armigera* by Kalra and Kulshreshtha, 1961. *A modicella* was found diseased by *Beauveria bassiana* up to 30.80 per cent. Many caterpillars infesting groundnut have been reported to be infected by NPV. SNPV was found to be causing 100 per cent infection in S. *litura* larvae on castor followed by groundnut with 98.17 per cent infection under laboratory conditions (Gururaj *et al.,* 1999). A protozoon, *Nosema* sp. was parasitizing S. *litura* to the extent of 13.8 per cent in Andhra Pradesh (Sridhar and Prasad, 1996).

*Botanicals:* Neem seed extract is found as effective against *A. modicella* and *A. craccivora* as phosphamidon.

## Insect Pests of sesame

Sesame crop is attacked by several insect pests which cause considerable damge (Satyagopal et al 2014). The main insect pests include sesame shoot webber and pod borer, *Antigastra catalaunalis;* sesame gall fly, *Asphondylia sesame;* sphingid moth, *Acherontia* styx, and hairy caterpillar, *Spilosoma obliqua.*

**Sesame shoot webber and pod borer, *Antigastis catalaunalis:*** This pest is distributed throughout the sesame growing area of India. It causes terminal destortion of plants through rolling and webbing of a few top leaves. Early stage infestation (10-15 days old plants) results in plant death without producing any branches or shoots. Sometime larva enters into the shoot and the pods (Fig. 12.11). One larva can destroy 2-3 young plants and 4-6 flower buds. It leads to 10-70 per cent plant infestation, 1-43.5% pod infestation and 8.86-71.53 per cent yield lose. - It completes 14 generation in a year. This pest is active from March-December.

## Sesame gallfly, *Asphondylia sesami*

Female lays its eggs in flower buds. Incubation period is 2-4 days. Maggots feed on floral parts, especially gynaeceum. The plant transforms into malformed gall like buds which fail to turn into seed capsules and infested buds wither and fall into ground. The pest is active from flower initiation stage of the plant. Intensity of the pest is 1- 21 per cent.

## Sphingid moth, *Acherontia styx*

This is massive caterpillar (7.5-10 cm) feeds voraciously on foliage. Robust moths are known as honey robbers as they rob honey from the bee hives. The pest is active from mid March - early December but maximum infestation is in September. Full grown larva is green with oblique yellow streak along the sides. Larva can be recognized by presence of curved horn on dorsal side. Three colors of larvae are there which are green, yellow and brown.

## Hairy caterpillars, *Diacrisia obliqua, Amsacta moorei*

This pest is highly polyphagous and defoliate a number of kharif crops (Fig. 12.12). The larvae have long hairs throughout their body. Young caterpillars are gregarious. The insect feed on all parts except stem. There is only one generation of *A. moorei* and 3-8 in *D. obliqua* in a year.

## Pod sucking bug, *Nezara virudula, Aphanus sordidus* and Thrips

Bugs suck sap from pods of standing crop (fig 13) and even from the crop kept on thrashing floors while thrips infestation results in malformation of stamens, injury to the ovary and development of a dark pigment on fruit wall instead of usual green color.

## Management strategies

Cultural control: Sowing the crop during June and July will escape from leaf webber damage and delayed sowing resulted in a significantly higher level of damage to leaves, flowers and pods and poor yield (Patra, 2001). Intercropping with pigeon pea (Nath et al., 2003) and black gram, green gram, cluster bean, sorghum and pearl millet (Ahirwar et al., 2009; Ahuja et al., 2009) was found to significantly reduce the leaf webber damage.

Mechanical control: Collection and destruction of infested parts of plants further minimize the damage of caterpillar. If possible manual collection and destruction of larvae will reduce the population build up.

Resistance to sesame leaf-roller: It has been reported that the branching pattern and maturity of the plants are positively correlated with their resistance to leaf-roller. Profuse branching and early maturing varieties provide conditions which are more conducive to pest development. The late maturing varieties NP 29, C 1036 and Chanda 3 have been reported to the less susceptible to the pest.

Biological control: Conservation of existing natural enemies (spiders, coccinellid beetles, predatory stink bugs, preying mantids, black ant) and parasitoids (Braconids and Ichneumonids) through ETL based (2 webbed leaves/sq. m or 10% damage) application of botanical insecticides and safer chemicals. Augmented release of parasitoids *viz., Trathala flavoorbitalis* (Behera, 2011) and *Apanteles* sp and the predators like *Chrysoperla carnea* also would reduce the population build up. Sesame leaf webber and pod borer, *Antigastra catalaunalis* is parasitized by five different parasitoids. Cheema and Singh (1987) have recorded parasitization up to 20% in Punjab, India whereas, Kalra (1989) recorded parasitism up to 42 per cent. In Gujrat, *Diadegma* sp. has been reported parasitizing leaf webber and pod borer up to 50% (Patel and Bhalani, 1990). In earlier studies, Khan and Hussain, 1964 found Lucern caterpillar, *Laphygina exigua* parasitized by Mermethid nematode, *Mermis indica* up to 92.8% and Rao et al., 1980 had recorded parasitization up to 93% of *A. styx* by *T. australiacum.*

Bio-pesticides: Spray of neem oil 1% or Neem Seed Kernel Extract 5% at the early stage of infestation. Application of bio-inoculants (*Azospirillum*) induces the insect resistance among the treated plants and recorded the minimum leaf damage (Anandh et al., 2010) by increasing levels of phosphorus and potassium level in the plants (Selvanarayanan, 2013). Chemical control Spray of carbaryl 0.1% or dusting carbary l 5% reduces the damage. Two sprays of quinalphos 0.05% at 30 and 45 days after sowing will control the pest effectively.

## Insect pests of Sunflower

Besides, other factors, attack of insect pests is a major constraint in sunflower seed production which cause considerable economic loss (Atwal and Dhaliwal 2009). A large number of insect pests have been reported causing extensive damage to sunflower. In India, more than fifty insect species have been found feeding on sunflower of which seedling pests, cutworm (*Agrotis* spp) and grasshopper (*Atractomorpha crenulata*), soil insects (termites and whitegrubs), sucking pests such as leaf hopper (*Amrasca biguttula biguttula*), whitefly (*Bemesia tabaci*), thrips (*Scirtothrips dorsalis*), defoliators like Bihar hairy caterpillar (*Spilosoma obliqua*), tobacco caterpillar (*Spodoptera litura*), green semilooper (*Plusia orichalcea*), cabbage semilooper (*Trichoplusia sppi*) and capitulum borer (*Helicoverpa armigera*) (Fig. 12.14) are of economic importance (Abrol et al., 2007). Besides, non insect pests such as rabbits, parakeets, doves, house sparrows, crows, rats etc. have been reported to cause severe damage on sunflower (Goel and Kumar1990; Sandhu et al. 1973). A survey of insect pests attacking sunflower was conducted in different locations of Jammu region. For this purpose, 20 plants were selected randomly in the crop and number of insect pests/plant of each species recorded. The mean of these observations constituted reading for each pest. The studies revealed the occurrence of several insect pests which have been identified as *Altica cyanea* Weber (Coleoptera: Chrysomelidae); *Agonoscelis nubila* Fabricius (Hetero-ptera:Pentatomidae); *Eysarcoris guttger* Thunberg; *E.ventralis*

Westwood (Heteroptera:Pentatomidae); *Reptortus linearis* Fabricius (Heteroptera : Alydidae); *Monolepta signata* Olivier (Coleoptera: Chrysomelidae); *Dolycoris indicus* Stal (Heteroptera: Pentatomidae). All these insect pests were found to cause considerable damage up to 30% from seedling pests, 46% from sucking pests and 50% from inflorescence pests. The number of insect pests attacking sunflower/plant (n = 20) on an average was 3.0+1.2 (*A. cyanea*); 3.6+0.8 (*Agonoscelis nubila*); 2.6+1.0 (*Eysarcoris guttger*); 4.2+0.8 (*Reptortus linearis*); 6.0+1.2 (*Monolepta signata*); 3.2+1.0 (*Eysarcoris ventralis*); 1.8+0.8 (*Dolycoris indicus*). Perusal of the literature reveals that *Agonoscelis nubila, Eysarcoris guttger*; *Reptortus linearis* and *Monolepta signata* have not been recorded attacking sunflower from India and elsewhere. Though several other species of the genus *Agonoscelis, Eysarcoris, Riptortus* and *Monolepta* have been recorded attacking several other crops in different parts of the world (Goel and Kumar 1990). Evidently, their occurrence with sunflower constitutes new record from India and elsewhere. Though *A. cyanea* has earlier been recorded from different parts of the world attacking several other crops (Goel and Kumar 1990; Sandhu et al. 1973; Sethi et al. 1978) but its occurrence with sunflower in the present study constitutes first record of its association with the crop from India and elsewhere. In earlier studies, *Dolycoris indicus* has been recorded attacking sunflower in India (Atwal and Dhaliwal 2004). However, its occurrence with sunflower in present study constitutes first record from Jammu and Kashmir State.

## Management Strategies

**Cultural practices:** Mixed cropping of sunflower with red gram reduced the population of *H. armigera* on sunflower to some extent. In BSH-I or Modern variety of sunflower at 4:2 or 2:4 ratio, the number of borers were comparatively lower than that in solo crop of hybrid (Singh et al., 2003). For control of cutworms, growing of sunflower on ridges reduced the damage to seedlings to a very significant extent. In case of sucking pests, increased spacing encouraged the jassid build-up, while increased dosage of N fertilization favored heavy multiplication of jassids.

**Resistance sources:** More than 100-germplasm lines and several cultivars have been evaluated in the field against *H. armigera.* The lines were categorized on the basis of egg or larval counts and damage to the capitulum.

**Biological control:** Various biological control agents attack these pests. *Spilosoma obliqua* larvae havebeen reported to be parasitized by *Charops* sp. upto 38 per cent and up to 15.12% by *Apanteles obliquaei* and *A. ruides* (Shetgar et al., 1990). *Helicoverpa armigera,* a dreaded pest of sunflower is parasitized by two species of tachinid viz. *Carcelia illata* and *Exorista xanthaspis* upto 24.54% in Gujrat (Patel and Talati, 1987). Timely spray of Ha*NPV* @ 450 LE/ha twice can protect sunflower crop from head borer. Two species of *B. thuringiensis* when evaluated against fourth instar larvae of S. *obliqua* under laboratory conditions gave 100 per cent mortality within 24 hrs. (Chahal et al., 2001). On the basis of an extensive study in Northern districts of Karnataka, various treatments NPV, *C. carnea* and endosulfan alone or in various combinations with NPV, the *chrysopids* hold a great promise for effective management of this pest on sunflower.

**Botanicals:** Many plant products like neem, sweet flag etc. have been found effective against *S. litura* and *H. armigera.*

## Safflower

This crop is mainly attacked by Safflower aphid, *Uroleucon carthemi* and pod borer, *Helicoverpa armigera.* There are a few natural enemies which have been reported feeding on these insect pests.

## Management Strategies

**Cultural practices:** Early sowing can be a suitable component in integrated aphid management since it escapes the peak pest incidence. In Karnataka and Maharasthra, September sown crop experienced less damage.

**Resistance varieties:** Woody, spiny and early maturing types were mostly tolerant while late non- spiny types were more susceptible.

**Mechanical:** Collection and destruction of larvae of *Helicoverpa* and severely affected plants.

**Biological control:** Lady bird beetles *(Coccinellids), Chrysoperla* sp. and syrphids are the main predators of safflower aphid. Pawar et al., 1985 have recorded *Chrysoperla sp.* feeding on safflower aphids. Another parasitoid, *Endaphis aphidimyra* has been reported parasitizing aphids up to 32% (Warhade and Shivpuje 1987). A cecidomyiid, *Pseudendaphis* sp. has been reported to cause 60 per cent parasitization in Karnataka during February to May (Bhumannavar and Thontadarya 1979a, b).

**Chemical control:** Among several seed dressers evaluated, carbosulfan 25 DS @ 20 g/kg of seed to check the initial build up and the later population was effectively managed by peripheral (1.8 m all around the field) with NSKE (5%) or dimethoate (0.05%). Several plant products either aqueous petroleum ether or absolute alcohol extract were been found effective against aphids under laboratory conditions. Kernel extract from neem seed 5.7% and custard apple seed were effective in managing the aphid. Three years of experimentation with spraying of NSKE (5%) at 45 days after sowing followed by dimethoate (0.05%) spray at 15 days later in Karnataka has confirmed its supremacy over 2 sprays of dimethoate (Singh et al., 2003).

## Insect pests of Olive

The main bottle neck in the olive fruit production is the ravages caused by several insects, diseases, nematodes and weed pests throughout the world. Among the insect pest, olive fruit fly (*Bactrocera oleae*), olive psyllid (*Euphyllura olivine* Costa) (Fig. 12.15), olive moth (*Prays oleae*), olive black tinged bug (Fig. 12.16) and black scale (*Saissetia oleae* Bern) have been reported to be of major importance (Daku *et al.*, 2000) and green stink bug (*Nezara virudula* L.) in India (Kaul *et al.*, 2007a). While studying the pest complex in Himachal Pradesh, Thakur *et al.* (1997) observed seventeen insects infesting cultivated olive and among sap suckers, the olive psylla to be a regular pest. Farahbakhch and Moini (1975) presented a list of all the known pests and fungus diseases reported from olive growing countries in Iran, which include 95 species of insects and 6 species of mites.

Olive fruit fly is one of the key pests causing substantial damage in olive growing regions all over the Mediterranean area. The major potential economic losses caused by this pest include pre-harvest fruit drop and reduction of olive fruit and oil quality due to increased acidity caused by fungi entering the fruit through the *B. olea* maggots exit hole. The olive psylla also known as jumping plant lice, *E. olivina* is considered as one of the most important flowering predator for olive trees causing 60-100 per cent loss (Zouiten and Hadrami, 2001). This species is widely distributed in many parts of the world where olive is grown (Mathur, 1975). The first report of Olive psylla, *Euphyllura pakistanica* from India was made by Thakur et al. (1989) on cultivated olive. In this connection, incidence and management of olive psylla, *E. pakistanica* was also discussed by kaul et al., 2007b. Black scale damages olive trees in two ways: by sucking the sap and by secreting honeydew which serves as a substrate for the development of sooty mould.

## Management Strategies

Tilling the area under the tree and in entire olive orchard after harvest helps in killing the pupae of olive fruit fly. Fruit fly population can also be monitored with pheromone traps. Furthermore, pruning the crowded and bushy branches can help increase the mortality of black scale and olive psylla. Use of resistant varieties and if necessary spray bio-pesticides at evening hours to conserve the natural enemy's fauna to work upon the olive pests. In addition to above, the migrating and mobile pests can be suppressed by the collective action of various groups including agrochemical industry, pesticide dealers, government institutions, NGOs, and groups of farmers (Knight & Norton 1989).

## Insect pests of Castor

It is an industrially important oilseed crop which is damaged by several insect pests out of which castor semilooper, *Achaea janata;* tobacco caterpillar, *Spodoptera litura;* castor shoot and capsule borer, *Conogathus punctiferalis;* leafhopper, *Empoasca flavescense;* whitefly, *Trialeurodes ricini* and red hairy caterpillar, *Amsacta* spp. are of greater economic importance. In recent years serpentine leafminer, *Liriomyza trifolii* (Burgess) is also becoming serious on the crop. Besides, several hairy caterpillars like *Spilosoma obliqua* (Walker), spiny caterpillar, *Ergollis merione* (Fig 17), *Euproctis* spp, *Pericalia ricini* Fabricius, *Amsacta albistriga* Walker, *A. moorei* Butl. and castor slug caterpillar, *Parasa lepida*, castor gall fly, *Asphondylia ricini*, and red spider mite, *Tetranychus telarius* L. are also assumed regional importance and are sporadic pests. In recent years, castor inflorescence thrips (*Scirtothrips dorsalis* Hood.) also attained a pest status by infesting the crop at flowering stage causing considerable loss to the crop especially in Gujarat. The polyphagous pest *H. armigera* also cause considerable damage to castor crop by boring castor capsules apart from feeding on foliage (Basappa, 2003).

## Management Strategies

In the castor ecosystem, insect pests are attacked by good number of natural enemies among them, egg parasitoid, *Trichogramma chilonis* Ishii parasitized Castor semilooper up to 100 per cent and Another parasite, *Telenomus* spp. parasitise this pest from 10.6 to 98.4 per cent (Kulshreshta *et al.,* 1967) which has been recorded up to 80% in endamic areas of Andhra Pradesh by Thobbi et al., 1976. A larval parasitoid, *Microplitis ophiusae* has been reported parasitizing 70-80 per cent population of *A. janata* in Coimbatore (Ayyar, 1927). An ecto-parasite, *Eurytoma* sp. parasitized pupae of castor gall fly, *A. licini* (Rao, 1952). Insect predators, insectivorous birds and some of the microbial agents such as *Bacillus thuringiensis* cause mortality up to 52.3 per cent to castor semilooper (Kulshreshtha et al., 1965) and *Nosema rileyi* causing disease up to 100 per cent in castor semilooper in Maharashtra (Phadke and Rao, 1978).

**Cultural practices:** Sowing of cucumber crop in the borders or in the main crop of the castor preferably earlier than the main crop attracts the migrating larvae which can be destroyed mechanically. Similarly, wild castor and *Calotropis* twigs can be used to attract the migrating larvae and the congregated larvae can be easily killed mechanically or otherwise by using insecticides.

**Resistance sources:** RC 1066, RC 488 and Aruna showed multiple resistances, thus paving the way for evolving new varieties with higher level of resistance in desirable agronomic background.

**Mechanical control:** Collection and destruction of eggs and gregarious larvae of the hairy caterpillar can be very effective.

**Biological control:** Among the 27 species parasitizing *A. janata*, *Trichogramma chilonis, T. achaeae* and *Telenomus* spp, are egg parasitoids and are favored over others for inundative releases because of their easy multiplication, low cost and multi-parasitic nature. It has been found that the releases of 50, 000 parasitized eggs/ha/week during the egg laying period could keep the pest population under check.

**Chemical control:** Spraying of castor with pesticide should be done for controlling semilooper before 5 1arvae/plant for efficient plant protection.

## Insect Pests of Linseed

Linseed crop is attacked by about a dozen insect-pests among which linseed bud fly, *Dasyneura lini* (fig 18) and *Helicoverpa armigera* (fig 19) are the key pest (Abrol and Shankar, 2011).

## Linseed bud fly, *Dasyneura lini*

This insect appears as a serious pest of linseed in some parts of India including Andhra Pradesh, Madhya Pradesh, Bihar, Uttar Pradesh, Delhi and Punjab. The adult is small diptern fly, delicate, mosquito like orange colored insect. The damage is caused by maggots, which feed on the flower buds and prevent their proper opening. Consequently, the seed dose not set properly. Due to their feeding, galls are produced and there is no pod formation. The incidence of this pest goes up to 20 per cent.

## Management Strategies

Linseed bud fly is parasitized up to 50.7 per cent in their larval stage during February and one larva may parasitize 3-4 larvae in a bud (Ahmad and Mani, 1939). Coccincllids, C. *septempunctata* and *M. sexmaculata* are predators of midge and a predation up to 10-15 per cent have been reported from Kanpur by Malik, 2000; Malik and Singh 1997. Similarly, planting of resistant varieties are the promising option to mitigate the linseed bud fly attack. Installation of Pheromone traps to early detection and mass destruction of adults of *Helicoverpa armigera* and erection of bird perches for black drongo (fig 20) to predate on *Helicoverpa* larvae. Biological control agents such as *Campoletes chloridae* is promising in suppressing the *Helicoverpa* larval population on linseed crop (Abrol and Shankar 2011).

**(1)Mustard Aphid, *Lipaphis erysimi* on flowers**

**(2) *Lipaphis erysimi* infestation on mustard siliqua**

**(3) *Brevicorny brassicae* infestation on mustard pods/crucifers**

**(4) Painted bug, Bagrada on mustard**

**(5) Saw fly larva feeding on mustard foliage**

**(6) Adult saw fly foraging on mustard flowers**

**(7) Parasitoids probing for aphid parasitization**

**(8) *Coccinella septumpunctata* feeding on mustard aphids**

***(9) Coccinella grub feeding on mustard aphids***

**(10) Adult syrphid fly foraging on mustard flower for nectar**

**(11) Sesame pod borer**

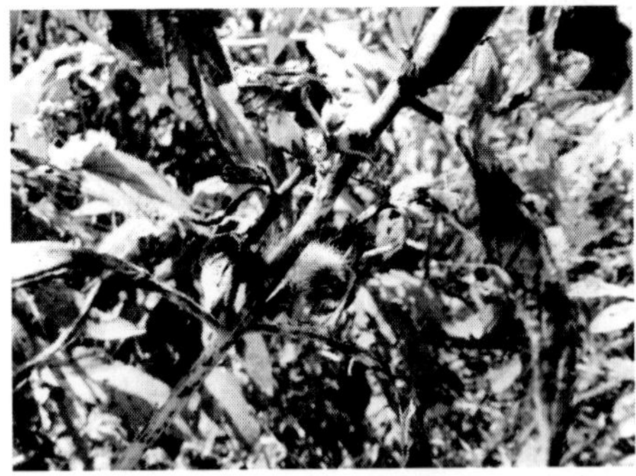

**(12) Bihar hairy caterpillar**

**(13) Green Stink bug nymphs infesting sesame pods**

**(14) *Helicoverpa armigera* larva feeding on sunflower head/ capitulum**

**(15) Olive psylla infestation on Olive**

**(16) Olive tinged bug**

**(17) Bihar hairy caterpillar and Castor Spiny caterpillar on castor**

**(18) Linseed pod damaged by linseed budfly**

**(19) *Helicoverpa armigera* damaging the linseed buds**

**(20) Black drongo looking for larval predation in linseed**



## Conclusions and future strategies

The multitude of insect pests occurring in different locations in different oilseed crops cause heavy losses in the production of oilseeds in already oilseed deficit world. The pests occurring at different stages of crop growth are being tackled by chemical control methods, which drastically effect the population of beneficial insects such as natural enemies, predators, parasitoids and pollinators. It is therefore recommended to adopt the most widely acceptable tools of integrated pest management for sustainable production of oilseeds, by regular monitoring of the pest situation alongwith the weather condition of a particular area so that timely protection from the pest and the prophylactic measures if any may be initiated, followed by conservation and augmentation of beneficial species of parasites (Powell, 1986; Blande et al., 2008), predators and microorganisms, by selective use of chemical pesticides, and training of field functionaries/farmers for their awareness about biological control agents coupled with IPM demonstrations in farmers field.

## References

Abrol, D.P. and Shankar, U., 2011. All India Co-ordinated Project on Linseed. Annual reports-2010-11. SKUAST-Jammu.

Abrol, D.P. and Shankar, U., 2014. Pesticides, Food Safety and Integrated Pest Management. In: Pimental, D., Peshin, R., (eds.) Integrated Pest management- Pesticide problems, Springer. pp. 167-199.

Abrol, D.P., Kotwal, D.R. and Hafeez, A., 2007. New insect pests attacking sunflower (*Helianthus Annuus* L.). Journal of Research SKUAST–J 6(1):117-119.

Ahirwar, R.M., Banerjee, and S., Gupta, M.P., 2009. Seasonal incidence of insect pests of sesame in relation to abiotic factors. *Annals of Plant Protection Sciences* 17(2): 351-356.

Ahmad, T. and Mani, M.S., 1939. Two new chalcidoid parasites of linseed midge, *Dasyneura lini. Indian J. Agril. Sciences* 9: 531-539.

Ahuja, D.B., Rajpurohit, T.S., Singh, M., Solanki, Z.S., Kalyan, R.K., Ahuja, U.R., Dhandapani, A. and Sundria, M.M., 2009. *Indian Journal of Agricultural Sciences* 79(10): 808-812.

Anandh, G.V., Selvanarayanan, V. and Tholkappian, P., 2010. Influence of arbuscular mycorrhizal fungi and bio-inoculants on host plant resistance *Antigastra catalaunalis* Duponchel in sesame *Sesamum indicum* Linn. *Journal of Biopesticides* 3(1): 152-154.

Anonymous., 1986. Proceedings of VII Annual Rabi-Summer Groundnut Workshop, Mahatma Phule Agricultural University Rahuri, Ahmednagar, pp 4-5.

Atwal, A.S., Chaudhary, J.P. and Ramzan, M., 1971. Mortality factor on the natural population of cabbage aphid, *Lipaphis erysimi* (Kalt.) (Aphididae: Homoptera) in relation to parasites, predators and weather conditions. *Ind. J. Agric. Sci.* 41: 507-510.

Atwal, A.S., Chowdhary, J.P. and Ramzan, M., 1969.Some preliminary studies in India on the bionomics and rate of the parasitism of Diaeretiella rapae (Braconidae: Hym.) a parasitoid of aphid. Journal of Research, PAU Ludhiana 6: 177-182.

Atwal, A.S. and Dhaliwal, G.S., 2009. Agricultural pests of South Asia and their management. Kalyani Publishers, New Delhi.

Augustin, A., Tulisalo, U. and Korpela, S., 1986. Flea beetles (Coleoptera, Chrysomelidae, Halticinae) on rapeseed and sugarbeet in Finland. J. Agric. Sci. Finl. 58:69-82.

Ayyar, T.V.R., 1927. The parasitic hymenoptera of economic importance noted from South India. Bull. Entomol. Res. 18: 73.

Ayyar, T.V.R., 1963.Handbook of Economic Entomology for South India, Government Press, Madras, p. 516.

Bakhetia, D.R.C. and Labana, K.S., 1978. Insect resistance in *Brassica* crops. Crop Improvement 5:95-103.

Bakhetia, D.R.C. and Sekhon, B.S., 1989. Insect-pests and their management in rapeseed mustard. J. Oilseeds Research **6** (2) : 269-299.

Basappa, H., 2003. Integrated Pest Management in Castor. Directorate of Oil seeds Research, Rajendranagar, Hyderabad-500 030, India. VIII, 52 p.

Basappa, H., 2004. Integrated pest management in sunflower: an Indian scenario. In: Proc. 16th International Sunflower Conference, Fargo, ND USA.pp.853-859.29 Aug-02 September, 2004.

Behera, P.K., 2011. Trathala flavo-orbitalis - natural enemy of sesame shoot webber and capsule borer in Coastal Odisha. Insect Environment 17(3): 133-134.

Bhumannavar, B.S. and Thontadarya, T.S., 1979a, Estimation of safflower crop losses due to the aphid *D. caarthami* Theobold (Hemiptera:Aphididae) under rainfed conditions. Pestology 3 (6): 18-19.

Bhumannavar, B.S. and Thontadarya, T.S., 1979 b. Effectiveness of different insecticides against safflower aphid, *D. compoitae* Theobald (Hemiptera Aphididae) on Safflower in Karnataka. Pesticides 13 (9): 28-30.

Bishnoi, O.P., Singh, H. and Singh, R., 1992. Incidence and multiplication of mustard aphid (Lipaphis erysimi) in relation to meteorological variables. Indian J. Agric. Sci. 62(10): 710-712.

Blande, J.D., Pickett, J.A. and Poppy, G.M., 2008. Host foraging for differentially adapted brassica-feeding aphids by the braconid parasitoid Diaeretiella rapae. Plant Signal Behav. 3(8): 580-582.

Bonnemaison, L., 1965. Insect pests of crucifers and their control. Ann. Rev. Entomol. 10:233-56.

Brar, K.S. and Sandhu, G.S., 1982. Field biology of white grub, *Holotrichia consanguinea* Blanchard (Scarabaeidae: Coleoptera) in Punjab. J. Soil. Biol. Eco. 2: 32-35.

Brar, K.S. and Sandhu, G.S.,1976. Major pests of oilseed crops in Northern India and their Control. Oilseeds J. 6:16.

Burgess, L., 1977. Flea beetles (Coleoptera: Chrysomelidae) attacking rape crops in the Canadian prairie provinces. Can. Entomologist 109:21-32.

Chahal, V.P.S., Singh,G. and Chahal, P.P.K., 2001. In: Biopesticides : Emerging Trends BET 2001, Chandigarh, India, pp.136.

Chandla, V.K., Misra, S.S., Bhala, O.P. and Thakar, J.R., 1988. White grub *Brahmina coriacea* (Hope) infesting potato in Shimla hills. Seeds and Forms 14: 12-13.

Chandra, S. and Kushwaha, K.S., 1987. Impact of environmental resistance on aphid complex of cruciferous crops under the agroclimatic conditions of Udaipur II. Biotic components. Ind. J. Entomol. 49: 86-113.

Cheema, J.S. and Singh, G., 1987. Biology of sesame leaf webber and capsule borer, *Antigastra catalaunalis*(Duponchel) (Pyralidae: Lepidoptera) in Punjab. J.Res. Punjab Agril. Univ. 24(1): 65-74.

Cherian, M.C., 1934. Notes on some south Indian Syrphids. J. Bombay nat. Hist. Soc. 37: 697-699.

Cook, S.M. and Denholm, I., 2008. Ecological approaches to the control of pollen beetles in oilseed rape. IOBC/wprs Bulletin 38: 110-113.

Daku, L., Norton, G.W., Pfeiffer, D.G., Luther, G.C., Pitts, C.W., Taylor, D.B., Tedeschini, J., and Uka, R., 2000. "Farmers' knowledge and attitudes towards pesticide use and olive pest management practices in Vlora, Albania: A Baseline Survey." IPM CRSP Working Paper 001-1, Virginia Polytechnic Institute and State University.

Dogra, I., Devi, N. and Raj, D., 2003. Parasitization of mustard aphid, *Lipaphis erysimi* (Kalt.) by *Diaeretiella rapae* in mid hill zone of Himachal Pradesh (India). J. Entomol. Res. 27(2): 145-149.

FAO., 2010. FAO agricultural outlook 2010–19.

Farahbakhch, G. and Moini, M., 1975. Olive pests in Iran-Plant Pests and Diseases Research Institute, National Agricultural Research Organisation, Ministry of Agriculture and Natural Resources, Tehran, Iran, 73 pp.

Gadad, H. and Hegde, M., 2014. Evaluation of Biopesticides and Biorationals against Thrips and Leafminer in *Rabi*/Summer Groundnut. Research and Reviews: Journal of Agriculture and Allied Sciences 3(3):25-31.

Ghewande, M.P., 1987. Virus diseases of groundnut and their management. In symposium on diseases of oilseed and pulse crops and their control. Vyas, S.C.C., Sgroff, V.N., (Eds.) Jawaharlal Nehru Krishi Vishwa Vidyalaya Indore (M.P) pp. 62-67.

Goel, S.C. and Kumar, A., 1990. Insect pests and predators associated to sunflower in winter of northern India. Indian Journal of Entomology 52:39-45.

Goodell, G., 1984. Challenges to international pest management research and extension in the third world: do we really want IPM to work? Bulletin of the Entomological Society of America 30: 18–26.

Guidry, N.P., 1964. Graphic summary of world agriculture. US Dept Agric Misc Publ 705:64.

Gunathilagaraj, K., Babu, P.C.S., 1987. FAO Plant Protection Bulletin 35 (2): 63- 64.

Gururaj, G.; Kulkarni and Hugar, P.S. 1999. Effect of host plant on POBs yield and mortality of Spodoptera litura larvae infected with NPV. Karnataka J. Agril. Sci.12(1-4): 60-63.

Hegde, D.M., 2000. Oilseeds situation. Directorate of Oilseeds Research, ICAR, Hyderabad.

Kalra, A.N. and Kulshreshtha, J.P. 1961. Studies on the biology and control of Lachnosterna consanguinea Bl. a pest of sugarcane in Bihar Bull. Ent. Res. 52: 1577- 1587.

Kalra, V.K., 1988. Population dynamics of various predators associated with mustard aphid, L. erysimi Kalt. J. Bio. Cont. 2: 77-79.

Kalra, V.K., 1989. Natural parasitism of sesame leaf webber and pod borer, *Antigastra catalaunalis* Duponchel. Indian Journal of Plant Protection 17(1): 9-11.

Katole, S.R., Nimbalkar, S.A., Yadgirwar, P.V., Kadu, N.R. and Kolhe A.V., 1998. Note on parasitization of *Diaeretiella rapea* M'Intosh on mustard aphids, *Lipaphis erysimi* Kalt. PKV Res. J. 22: 214-216.

Kaul, V., Tiku, A.K., Shankar, U. and Monobrullah, Md., 2007a. Green stink bug (Hemiptera: Pentatomidae) recorded as a new pest of olive in India. Journal of Asia-Pacific Entomology 10(1): 81-83.

Kaul, V., Malik, G.H., Shankar, U. and Monobrullah, Md., 2007b. Incidence and management of olive psylla, *Euphyllura pakistanica*. Indian Journal of Entomology 69(4): 331-340.

Khan, M.Q. and Hussain, M., 1964. Natural control of Lapbygma exigua by a Mermis sp. (Nematoda). Indian J. Entomol. 26:124-125.

Knight, A.L. and Norton, G.W., 1989. "Economics of Agricultural Pesticide Resistance in Arthropods". Annual Review of Entomology 34: 293-313.

Krishnamurti, B. and Usman, S. 1954. Some insect parasites of economic importance noted in Mysore State. Indian Journal of Entomology 16: 327-344.

Kulashreshta, J.P., Sanghi, P.K., Thobbi, V.V. and Kulkarni, L.G., 1967. Mass multiplication of Telenomus sp. (Hymenoptera: Scelionidae) an egg parasite imported in India for biological control of castor semilooper, *Achea janata* L. and some observations on its biology. Indian J. Entomol. 29: 290-294.

Kulshreshtha, J.P., Sanghi, P.K. and Ravindranath, V., 1965. Microbial control of castor semilooper *Achea janata*. Indian J. Entomol. 27: 353-354.

Kumar, A., 1985. Estimation of damage caused to *Chromatomyia horticola* population on *Brassica campestris* by hymenopteran parasites. Ann. Entomol. 10:49-53.

Kumar, G., Prasad, D.K. and Pavan, S., 2011. Hover-flies (Diptera : Syrphidae) of the Coromandel Coast in Andhra Carnatic, Peninsular India. Bionotes 13(2): 78-86.

Lamb, R.J. and 1989. Entomology of oilseed *Brassica* crops. Ann. Rev. Entomol. 34:211-229.

Lammerink, J., Banfield, R.A., 1980. Effect of aphid control by disulfoton on seed yield components and seed quality of oilseed rape. NZ J. Exp. Agric. 8:45-48.

Lee, H.S., 1968. Evaluation of some granulated insecticides against the aphids and diamond-back moth on rapeseed by soil treatment. Plant Prot. Bull. 10:69-70 (In Chinese).

Lennerts, L., 1983. Oelschrote, oelkuchen, pfl anzliche Oele und Fette, Herkunft, Gewinning, Verwendung, Bonn 1983. Alfred Strothe, Hannover.

Mahal, M.S., Singh, R. and Singh, B., 1988. Dispersion of *Lipaphis erysimi* (Kalt) and development of a sequential sampling plan for its supervised control on raya, *Brassica juncea* Coss. J. Insect Sci. 1: 27-38.

Makhmoor, H.D. and Shinde, C.D. 1986. Preliminary study on the effect of different doses of N. P. and K. and their combination in mustard in on the incidence of *Athalia proxima* Klug., *Crocidolomia binotalis* Zell. and *Phytomyza atriconis*. Res. Devel. Rep. 3:66-71.

Malik, Y.P., 2000. Assessment of yield loss due to host pest interaction in linseed, *Linum usitatissimum*. Indian J. Agril. Sci. 70:53-54.

Malik, Y.P. and Singh, S.V., 1997. Evaluation of linseed germplasm against bud fly, *Dasyneura lini* Barnes and some important disease. *J. Insect Sci. 10:70-71.*

Mathur, R.N., 1975. Psyllidae of the Indian subcontinent. Indian Council of Agricultural Research, New Delhi, 429 pp.

Muthiah, C., 1991. Some new parasites of groundnut leaf miner (*Aproaerema modicella* Dev.). Indian Journal of Plant Protection 19(1):105.

Muthiah, C. and Kareem, A.A., 2000. Survey of groundnut leafminer and its natural enemies in Tamil Nadu, India. International Arachis Newsletter 20: 62-63.

Nair, M.R.G.K., 1975. Insect and Mites of Crops in India, ICAR New Delhi. pp 404.

Nath, P., Bhushan, S. and Singh, A.K., 2003. Effect of intercropping on the population of insect pests of sesamum. Journal of Applied Zoological Researches 14(1): 76-79.

Patel, A.A. and Bhalani, P.A., 1990. A new record of *Diadegma* sp. as a larval parasite of the sesamum leafroller in Gujarat. Indian Journal of Entomology 51(4):474.

Patel, A.G. and Vyas, H.N., 1984. Studies on predatory capacity of ladybird beetle, Menochilus sexmaculatus (Fab.) against *Aphis craccivora* Koch under laboratory condition. Pesticides 18(11): 8-9.

Patel, A.J. and Talati, G.M., 1987. Biology of *H. armigera* (Hb.) as a pest of sunflower. Gujarat Agril. Univ. Res. J. 12(2): 54.

Patel, R.C., Yadav, D.N. and Patel, J.R., 1976. Natural control of groundnut aphid, *Aphis craccivora* Koch in Central Gujrat. Current Sci. 45(1): 34-35.

Patra, A.K., 2001. Response of sesame (*Sesamum indicum*) varieties to dates of sowing during rainy season. Indian Journal of Agricultural Sciences 71(8): 523-524.

Pawar, C.S. and Bhatnagar, U.S. and Jadhav, D.R., 1985. Some predatory insects and their parasites. Sci. Cult. 51 (1): 101-102.

Phadke, C.H. and Rao, V.G., 1978. Studies on entomogenous fungus, *Nomuraea rilyei* (Farlow). Curr. Sci. 47: 511-512.

Powell, W., 1986. Enhancing parasitoid activity in crops. In: Insect Parasitoids. 13th Symp. Royal Ent. Soc. London, 18-19 Sept. 1985 (Waage, J., Greathead, D., eds.), Acad. Press, London, pp. 19-340.

Prakash, S. and Hinata, K., 1980. Taxonomy, cytogenetics and origin of crop brassicas, a review. Opera Bot. 55:1-57.

Rai, B.K., 1976. Pests of oilseed crops in India and their control, Indian Council of Agricultural Research, New Delhi, pp. 121.

Rai, S., Singh, N.N. and Shankar, U., 2010. Functional response of grub and adult of *Coccinella septempunctata* (L.) and *Coccinella transversalis* (Fab.) on mustard aphid, *Lipaphis erysimi* (Kalt). Archives of Plant protection and Phytopathology 43(18): 1829-1835.

Rakow, G., 2004. Species origin and economic importance of Brassica. In: Pua, E.C., Douglas, C.J., (eds) Biotechnology in agriculture and forestry, vol 54, Brassica. Springer, Berlin, pp 3–11.

Rana, J.S., 2005. Performance of *Lipaphis erysimi* (Homoptera: Aphididae) on different *Brassica* species in a tropical environment. J. Pest Sci. 78: 155-160.

Rao, A.S., 1952. A gall midge pest of castor in India. Current Sci., 21: 347-348.

Raymer, P.L., 2002. Canola: an emerging oilseed crop. In: Janick, J., Whipkey, A., (eds.) Trends in new crops and new uses. ASHS, Alexandria, VA, pp 122–126.

Reddy, L.J., 1988. The Groundnut, ICAR Publication. pp. 572.

Rohilla, H.R., Singh, H., Kalra, V.K. and Kharub, S.S. 1987. Losses caused by mustard aphid, *lipaphis erysimi* (Kalt.) in different brassica genotypes. Proc.7$^{th}$ Int. Rapeseed Cong. Poland. 5:1077-1083.

Sachan, J.N. and Gangwar, S.K., 1980. Vertical distribution of important insect-pest of cole crops in Meghalaya as influenced by the environment factors. Indian J. Entomol. 42:414–421.

Sahayaraj, K., 1999. Field evaluation of *Rhynocoris marginatus* (Fab.) against two groundnut defoliators. International Arachis News Letter 19: 41-42.

Sandhu, G.S., Brar, K.S. and Bakhetia, D.R.C., 1973. Soil systemic insecticides for the control of leafminer on mustard. Pesticides 11(7):19-21.

Satyagopal, K., S.N. Sushil, P. Jeyakumar, G. Shankar, O.P. Sharma, D. Boina, S.K. Sain, D. Chattopadhyay, Ram Asre, K.S. Kapoor, Sanjay Arya, Subhash Kumar, C.S. Patni, C. Chattopadhyay, A. Pandey, R. Pachori, A.Y. Thakare, K. Basavanagoud, A.S. Halepyati, M.B. Patil and A.G. Sreenivas. 2014. AESA based IPM package for Mustard/Rapeseed. pp 49.

Sekhon, B.S. and Bakhetia, D.R.C., 1994. Relationship between population of mustard aphid, Lipaphis erysimi (Kalt.) and yield losses in Brassica napus. Journal of Insect Science 7: 67-69.

Selvanarayanan, V., 2013. Bio-inoculants to boost resistance in crop plants against insects: field success and struggles in India. IOBC/WPRS Bulletin, 89: 373-378.

Sethi, M.A., Singh, K.M. and Prasad, H., 1978. Incidence of insect pests on different varieties of sunflower. Indian J. Ent. 40:101-103.

Sharma, A.N., Gupta, G.K., Verma, R.K., Sharma, O.P., Bhagat, S., Amaresan, N. and Saini, M.R., Chattopadhyay, C., Sushil, S.N., Asre, R., Kapoor, K.S., Satyagopal, K., Jeyakumar, P., 2014. Integrated Pest Management for Soyabean. Directorate of Plant Protection, Quarantine & Storage, CGO Complex, NH IV, Faridabad- 121001, pp 41.

Sharma, P., Saini, M.K., Deep, S., Kumar, V., 2012. Biological Control of Groundnut Root Rot in Farmer's Field. Journal of Agricultural Science 4(8):48-59.

Shetgar, S.S., Bilapate, G.G., Patel, V.V., Londhe, G.M., 1990. A note on the natural enemies of Bihar hairy caterpillar, Spilosoma oblique (Walker). Indian J. Entomol. 52(1): 158.

Sidhu, H. S., Singh, S., 1964. Biology of the Mustard Aphid, *Lipaphis erysimi* in the Punjab. Indian Oilseeds J. 8(4): 348-359.

Singh, H., 1982. Studies on insect-pest complex in *Brassica campestris* L. var. brown 'sarson'. Thesis, Doctor of Philosophy, Entomology, submitted to Haryana Agricultural University, Hisar, 192 pp.

Singh, H., 1991. A success story of five-fold increase of oilseeds production in Haryana. Haryana Agricultural University Hisar, India.

Singh, H., Rohilla, H.R., Kalra, V.K. and Yadava, T.P., 1984. Response of Brassica cultivars sown on different dates to attack of mustard aphid, *Lipaphis erysimi* (Kalt.). J. Oilseeds Res. 1: 49-56.

Singh, H., Singh, Z. and Yadav, J.P., 1987. Post harvest losses in Rapeseed caused by aphid pest. Porc. 7th Intl. cong. Poland, 5: 1138-1142.

Singh, H., Singh, Z. and Yadava, T.P., 1990. Influence of abiotic factors on alate production in mustard aphid Lipaphis erysimi (Kaltenbach). J. Aphidol. 4: 71-74.

Singh, S., Singh, N., Bhat, N. S., Singh, S.B., 2003. Development and validation of IPM in oilseed based production systems. In: Proceedings and recommendations of the NATP interactive Workshop on Integrated Pest Management, February 26-28, 2003. Recent Advances in Integrated Pest Management (Singh, A., Trivedi, T.P, Sardana, H.R., Sharma,O.P., Sabir, N., eds.) pp. 91-98.

Singhvi, S.M., Verma, N.D. and Yadava, T.P., 1973. Estimation of losses in rapeseed (*Brassica campestris L. var. toria)* and mustard *(Brassica juncea* Coss.) due to mustard aphid (*Lipaphis erysimi* Kalt.). Horyana Agric. Univ. J. Res. 3:5-7.

Sovero, M., 1993. Rapeseed, a new oilseed crop for the United States. In: Janick, J., Simon, J.E., (eds.) New crops. Wiley, New York, pp 302–307.

Sridhar, V. and Prasad, V.D., 1996. Life table studies on natural population of *Spodoptera litura* on groundnut. Ann. Pl. Protec. Sci. 4: 142-147.

Sundaramurthy, V.T., Jayraj, S. and Swamiappan, M., 1976. *Telenomus malonus* Nixon, A potential egg parasite of groundnut red hairy caterpillar, *Amsacta albigastra* Walker. Curr Sci. 45(21):775.

Tansey, J., Dosdall, L.M., Keddie, B.A., 2009. *Phyllotreta crucifera* and *Phyllotreta striolata* responses to insecticidal seed treatments with different models of action. J. Appl. Entomol. 133:201-209.

Thakur, J.R., Atwal, R.K. and Gupta, P.R., 1997. Insect pests of olive in Himachal Pradesh. Pest Management and Economic Zoology 5(1): 75-79.

Thakur, J.R., Sharma, P.C. and Gupta, P.R., 1989. Occurrence of olive psylla, *Euphyllura pakistanica* on olive in India: A new record. Tropical Pest Management 35(3): 331.

Thakur, J.R., Sharma, P.C. and Gupta, P.R., 1989. Occurrence of olive psylla, *Euphyllura pakistanica* on olive in India: A new record. Tropical Pest Management 35(3): 331.

Thobbi, V.V., Singh, B.U., Amin-ul-Hussain, Sarves, S.O.H. and Rao, N.G.P., 1976. Biological Control of Castor Semilooper. Indian Farming 26(6): 15-17.

Tiittanen, K. and Varis, A.L., 1960. The treatment of seeds of swede, turnip, and turnip rape in the control of flea beetles (*Phyllotreta* spp.) and cabbage root flies (*Hylemyia brassicae* Bouch and H. *floralis* Fall.). Valt. Maatalouskoetoiminnan Julk. 181:1-11.

Vora, V.J., Bharodia, R.K., and Kapadia, M.N. 1985. Pests of oilseed crops and their control on rapeseed and mustard. Pesticides 19:38-40.

Vyas, R.V., Yadav, D.N. and Patel, R.J., 1991, Studies on mass production of *Bacillus popillae* var. *holotrichiae*, a promising pathogen of white grub in Gujarat, Gujarat Agricultural University Research Journal 17: 30-37.

Weiss, E. A., 1983. Oilseed processing and products, In: Oilseed crops, 528-596, ISBN 0-632- 05259-7, London: Longman.

Zouiten, N. and Hadrami. I. E., 2001. Olive tree psyllid: State of knowledge and perspective of control. Agricultures 4(10): 225-232.

CHAPTER
13

# Quarantine Control of Pests

## INTRODUCTION

The word quarantine is derived from an Italian word which means 40. The word owes its origin from the appearance of Black Death disease in Europe which required at least 40 days showing its symptoms. The ships carrying the passengers suspected to be infected with the disease were not allowed to disembark and confined for at least 40 days so that symptoms if any could appear during this period to avoid further spread of the disease (MacKenzie, 2001). The same principle later became applicable to both plants and animals of unknown or suspected health status. However, in some countries the term quarantine has acquired wider dimensions in relation to prevention and spread of harmful organisms from one country to another. Different terms are used to refer quarantine measures in different countries as most of the European countries use the terms 'plant quarantine' and 'plant health' to address the same issue. However, in North America, the term plant protection is used to cover the legislative and regulatory measures to prevent the entry and spread of unwanted harmful organisms by the human activities.

Quarantine is based on the principle that prevention is better than cure with the objective of reducing their risk factor against the introduction of exotic pests and diseases and if already present checks its further spread and multiplication. Quarantine can, therefore, be defined as a legal enforcement of measures aimed to prevent pests and pathogens from spreading, or prevent them from multiplying further in case they have already found entry and have established in a new restricted area. Though quarantine measures may not guarantee an everlasting protection against the entry of exotic pests but will certainly check or delay introduction of these unwanted organisms and their subsequent establishment in hitherto clean areas. The legislative measures such as international and national quarantine and modified campaigns need to be enacted to prevent the entry of new pests and diseases and their further spread. As a consequence various countries have promulgated bee laws/ legislations to avoid the risk of introduction of bee diseases and other related problems.

Quarantine has different meaning in different countries as to what areas are to be covered. They include eradication campaigns, surveys, risk assessment and all closely related topics. Plant health also includes certification and marketing of commodities, control of migratory pests and prevention of the misuse of plant pests. All these activities have been included under a new technical term referred to as 'phytosanitation'. The plant health also embraces issues like use of pesticides, their registration and regulation. Plant health deals with a great variety of organisms inflicting damage to plants which include fungi, bacteria, viruses, insects, mites, nematodes and members of many other categories of organisms all collectively referred to as 'plant pest'.

Plant health deals with endless problems and circumstances which can involve almost any branch of science. It involves plant pathology, entomology, nematology including other invertebrate pests. it also involves science of pesticides and their chemistry, commerce and law contributing both to its interest and to its intellectual challenge. The logic is to achieve good plant health through administrative and regulatory systems.

Our lives are dependent upon plants, their sustainable and safe cultivation, plant products and availability of natural resources. However, all the resources such as food, environment and renewable natural resources are threatened by the invasion of alien species. The problem is being further complicated by expanding global trade, tourism and changes in climate which help spread of alien species. Lack of proper coordination and resources in different countries is another important factor having serious implications on the economy and ecology.

Due to easy and fast mode of transport the invasive alien species have been reported to pose serious threat to agriculture, food security and biological diversity. The International Plant Protection Convention (IPPC) aims to provide effective action plan for preventing the spread and introduction of pests of plants and plant products and their effective control. The IPPC is of great relevance to all those countries that are parties to the policies and programmes of Convention on Biological Diversity CBD to prevent the introduction of, control or eradicate those alien species which threaten ecosystems, habitats or species.

Most countries in the world lack resources to identify the risks associated with the spread of invasive species. Lack of knowledge about ecological threats, the pressure of global trade and travel and fragmented national authority are main causes for risk analysis and control measures. However, countries that are contracting parties to IPPC enjoy benefits of well developed network as both IPPC and CBD have close cooperation to address the significant problem of invasive alien species.

## Early history of plant quarantine

The destruction of barberry (*Berberis vulgaris*) plants is an interesting example of early quarantine history. It was found that wheat and other cereals experienced severe attack of blast near the areas having berberry plantations in northern Europe. The notion was so strong that berberry spread the blast or black rust resulting in a legislation in 1660 in Rouen, France for the destruction of barberry plants. The berberry plants were the cause of the problem was confirmed by Anton De Bary, the renowned mycologist in 1865 who found that these plants acted as an alternate host to the parasitic fungus *Puccinia graminis*, which was the cause of the problem. The barberry plant was necessary for the fungus to complete its life cycle. This was a great finding which resulted in enactment. between 1726-1772 in New England, Connecticut, Massachusetts and Rhode Island for its complete eradication to save wheat and other cereals. This was followed by several other countries such as Germany in Schaumburg-Lippe (1805) and by Bremen in 1815 (Fulling, 1942, 1943). This was further followed by many other European states such as Denmark (1869, 1904), Prussia (1880), Austria (1882), Norway (1916), Bavaria (1920) and Hungary (1920).

The plant Quarantine Act was passed by congress in USA in 1912 which was followed by North Dakota in 1917. Immediately after this 13 states followed the act in 1918 recommending complete eradication of barberry from all the wheat growing areas of USA,

## The Risk from Alien Pests

The movement of plants has been a part of movement of human being carried with them, consciously or unconsciously as a source of food, and forage for livestock they carried to sustain them (Diamond, 1998). This facilitated the spread of cultivated plants within the ancient empires of the middle east, the, Greeks and Romans. These transportations also carried with them the pests, and diseases. The trade routes served as pathways for the transfer of plants and their associated pests. The establishment of silk routes between Asia and Europe facilitated the transport of many plants from east to west such as apples, oranges and from west to east such as barley and wheat. Similarly, trade routes between China, South-East Asia and the Middle East and East Africa also facilitated the transport of crops such as rice, sugarcane, soybean and bananas along with their Pests. Although trade included plant products and occasionally seeds but by the early $19^{th}$ century seeds became the most traded commodity. Besides seeds of food plants, other materials transported included seeds of ornamental plants to Europe and America from all over the world. . By the middle of the 19th century the long-distance transport of growing plants became a regular phenomenon which increased the potential risk of pests spreading with their host plants. The invention of the Wardian case in 1830 by Nathaniel Bagshaw Ward was a turning point in the eay transport of living plants to long distances in long sea journeys which could take months together thereby helping the pests to successfully survive with their host plants (Hobhouse, 1992).

Furthermore with the introduction of steam ships which dramatically decreased the length of sea voyages and now modern system of air freight has further aggravated the fast transport and spread of plants from one continent to another overnight. The use of modern refrigeration system where the plants can be kept in good condition associated with pests during transport.

In the last 50 years tourism is another factor which has taken new dimension for the spread of plants along with associated pests to new areas with the movement of people to new and unexplored areas. Many a times the organisms which are not damaging in their native areas become highly destructive in new areas where they establish themselves. Evidently, enforcement of plant health and quarantine. Is the most needed and urgent basic task rto prevent the spread of pests and diseases.

Recent advances in trade among different nations with ultra modern facilities of cold chain such as refrigerated cold storage facility for fresh fruits and vegetables as increased the risk manifold. Though several countries have quarantine restriction yet they have not been fully successful to avoid accidental introductions.

The spread of alien invasive species has serious consequences for damage to crops and the environment especially aggravated by the threat of climate change. The exotic pests are not only threat to agriculture, horticulture or forestry but inflict major damage to environment as well as biodiversity.

The different countries have adopted their own regulations to prevent the entry and spread of exotic pests. The government of India issued a plant quarantine regulation order (Regulation of Import into India) Order, 2003 (PQ Order, 2003) under the destructive insects and pests act, 1914. This regulation is implemented by National Plant Protection Organization under the directorate of Directorate of Plant Protection Quarantine and Storage (DPPQS). More than 700 pests are being regulated under the plant quarantine based on pest risk analysis. The entry of exotic pests is prevented through checking of plant materials at various airports/sea ports. Inspite of serious efforts made to prevent their entry, some of the exotic pests have made entry into the country causing considerable economic losses.

## Pest risk analysis

Pest risk analysis has been one of the most important steps in plant protection and as a means of regulatory decisions in almost all countries of the world. With the advancement of scientific information it has undergone a sea change to support better analysis and remedial measures thereof. It helps to decide which pests require surveillance and what are the products to be imported from other countries for effective management.

Despite the importance of pest risk analysis and critical needs for it, there has been little opportunity for formal education in pest risk analysis. Most pest risk analysts are entomologists, plant risk analysis from the perspective of regulatory science, rather than from a purely academic perspective.

Advanced or specialized methods developed in other fields (e.g. engineering, ecological modelling, toxicology, economics, etc.) may be modified and adapted to do pest risk analysis. The pest risk analysis community has come a long way in creating better models, refining methods and incorporating diverse areas of expertise, but we are still improving. Some organizations are beginning to incorporate more complex mathematical models into their analyses, while others are attempting to streamline and simplify their work. We believe that those who undertake pest risk analysis should understand the basic concepts of risk, and how to analyse risk – and then provide the foundation and building blocks for the conduct of pest risk analysis. Pest risk analysis is a multi-disciplinary field – entomology, plant pathology, botany, weed science, ecology, economics, geography, mathematics, statistics and even social sciences can play important roles in conducting pest risk analysis. Just as no one type of method, process or scheme will fit every type of analysis, there is no one type of expert best suited for every type of analysis.

## Spread of pests in trade

Carl Linnaeus was the first scientist to express concerns about the spread of plant pests with their hosts (Usinger, 1964). He was aware that a beetle was causing damage to pea seeds in north America introduced from English colonies of England. In 1752 he opined that the seed beetles could be introduced to England with the imported pea seed and that from England they might spread and cause damage to peas in continental Europe. He also advocated many scientific methods of controlling many kinds of insects such as fumigation with smokes, oily barriers on the trunks of fruit trees against the winter moth (*Operophtera brumata*), and biological control with a carabid beetle against caterpillars and with *Coccinella* ladybirds, ichneumons and braconid wasps against, aphids. By the middle of the 19th century the causes of disease in plants became much more clear. Better land and sea transport were enabling people to travel much more widely and rapidly which has not been possible previously also facilitated the exchange of goods, including plants and plant products.

## History of Plant Quarantine in India

The quarantine measures are of great relevance and most important for country like India whose economy is dependent largely on Agriculture. The importance of quarantine measures was realized in India in the early 20$^{th}$ century, when in 1905 Indian government ordered the compulsory fumigation of cotton bales to prevent the introduction of dreaded Mexican cotton boll weevil(*Antonymous grandis*). Comprehensive Plant Quarantine Act, known as Destructive Insects and Pests Act, (DIP Act) become operative On February 3, 1914. DIP Act has been revised and amended several times over the years. It was however; periodically reviewed to meet the growing requirements of liberalized trade under the WTO.

In 1946, the Directorate of Plant Protection, Quarantine and Storage were set up, under the ministry of Food and Agriculture, Government of India. In 1946, Plant quarantine activity started in the Botany Division at Indian Agricultural Research Institute (IARI) New Delhi with the initiation of plant introduction scheme.

In October 1949, the Directorate started its quarantine activities at Bombay seaport. The first plant Quarantine and Fumigation station in India was formally inaugurated On December 25, 1951.

In August, 1976 the National Bureau of Plant Genetic Resource (NBPGR) was created. In 1978, the Division of Plant Quarantine was created with Entomology, Plant Pathology and Nematology as its sections. In October, 1988, the Plants, Fruits and seeds (Regulation of Import into India) order, 1989 popularly known as PFS order came into force.

## The plant quarantine order in India

Government of India in the year 2003 approved a new Plant Quarantine regulation policy (Regulation of Import into India, Order, 2003) to improve the export certification process and standards to ensure phytosanitary certification in such a way to regulate pests and vectors, including alien species for importing countries. In India, the plant quarantine operations are carried out by the Directorate of Plant Protection, Quarantine and Storage which works under the control of the Ministry of Agriculture.

The development of the new Plant Quarantine order reflects following objectives:

It aims to prevent the introduction and spread of exotic pests that are destructive to the country by regulating the import of plants and plant products through adequate policy and statutory measures, and to support India's agricultural exports through credible export certification to facilitate safe global trade in agriculture by assisting producers, exporters and importers by providing proper technical guidance and phytosanitary certifications.

Before promogulation of new order on 18 November 2003, the plant quarantine statutes and regulations were governed by the earlier order of 1914 (Destructive Insects and Pests Act, 1914) which was mainly concerned with prevention of introduction and spread of destructive pests affecting various crops, seeds and fruits(Regulation of Import into India) Order, 1989.

Some other Rules were promulgated for regulating import of live insects (1941), fungi (1943) and cotton (1972). In 1988, new seed policy was formulated to help the farmers to get the best available seeds and planting, material. The Order of 1989 (Regulation of Import into India, 1989, prohibiting and regulating the import into India of plants, plant materials has now been replaced by the new Order of 2003

## New quarantine order in India

The government of India notified Plant Quarantine (Regulation of Import into India) (Second Amendment) Order, 2014 which included several amendments over earlier Destructive Insects and Pests Act, 1914 (2 of 1914) and Plant Quarantine (Regulation of

Import into India) vide order, 2003. The new order is promogulated with the following objectives to Liberalize trade in agriculture as a consequence of the WTO Agreements. Though it offers economic opportunities, yet on the other hand posed fresh challenges and amplified pest risks. India with diversity of agroclimatic zones and surplus food production has the potential to expand its trade which requires strict phytosanitary regulations and their stringent compliance.

To overcome these problems, there was an urgent need to fill the gaps in order of 1989, by regulating the import of germplasm of genetically modified organisms, transgenic plants and agents like fungi and live insects.

Evidently safe conduct of global trade in agriculture required fulfilment of certain legal obligations in accordance with international agreements. The need was felt to incorporate certain provisions to safe guard the interests of the farmers and national biodiversity from threats of destructive pests, invasive and alien pests.

Agricultural imports are now classified as prohibited plant species, restricted species where import is permitted only by authorized institutions. The restricted species are permitted only with additional and strict declarations of freedom from quarantine pests and subject to specified treatment certifications and the; and (d) plant material imported for consumption or industrial processing permitted with normal phytosanitary certification.

A permit is required for import of seeds, propagating material and mushroom spawn cultures. More than 144 agricultural commodities are specified in the new order specifically listing as many as 590 quarantine pests and 61 weed species. The entry points where the restrictions are enforced has increased to 130 as against 59 previously. The structure of certification fees and inspection charges are rationalized according to new order. The new order is in conformity International Plant Protection Convention in several ways.

## Action plan for pest risk analysis

Pest risk analysis plays a key role for plant quarantine in India. In this context an action plan was prepared for the establishment of a national pest risk analysis unit. The functions of which include organizing of pest risk analysis trainings, establishing groups, holding workshops attended by national and international experts for prioritize commodities and crops for pest risk analysis.

Some 36 commodities (Strawberry, banana, kiwi, musk melon, corn, pearl millet, sorghum Lettuce,garlic, Broccoli, potato, Chinese cabbage, Mustard, sunflower, safflower, linseed, castor, rape seed watermelon, pears, mandarin, cashew nut, apple, grape, citrus fruits, lentil, red beans, chickpea, jute, black gram, green gram, cotton, wheat, rice, barley, maize, baby) were selected for which a pest database is under development. Detailed pest risk analysis has started with the objective of completing 13 commodities each year.

## Control of invasive alien species in India

The national integrated pest management (IPM) programme is the responsible for preventing and controlling the threat posed by invasive alien species within the country. Several agencies like State governments, non-governmental, private sector organizations, research institutions are increasingly involved in the surveillance and detection of pests and diseases. International cooperation has helped in dealing with migratory locust, a pest of great concern. India maintains active coordination with neighbouring countries and FAO and for surveillance, early detection and control measures for locust.

A very amazing and unexplained cyclic problem of rodent surge exists in hilly areas of northeast region which coincides with the periodic gregarious flowering of bamboo tress. Research and preventive control measures are underway to devise strategies for their management.

Research on Impact of climate change and threat posed by invasive species is of particular concern in India. For example, the attack of white woolly aphid on sugarcane in parts of peninsular India caused serious losses in more than 200 000 ha of sugarcane during 2002, whereas the attack of this pest was not reported in earlier studies.

Several new pests are emerging which requires proper forewarning and management.

## Development of phytosanitary agreements

During 1865 to 1875, the French wine industry was threatened by the spread of American vine louse (*Viteus vitifolii*) in French vine orchards formerly (*Phylloxera vastatrix*), which later spread to other parts of Europe. The loss equivalent to £50 million sterling was reported in France alone (Large, 1940). Several countries like Austria, France, Germany, Italy, Portugal, Spain and Switzerland signed the first international agreement September, 1878 during 1878 to prevent the spread of this pest (Anon., 1914). This agreement had the provision that 1) material being traded should be free from pest, and power to inspect materials and take remedial measures for items not complying with the requiste terms and conditions.

Another example is of the Colorado beetle (*Leptinotarsa decemlineata* ) which has been responsible for the destruction of the potato (*Solanum tuberosum*), as a plant pest for over the past 150 years.

It was first was first described from the Colorado area in 1824 by Thomas Say hence the name Colorado beetle. It started spreading to other areas and by the middle of the 19th century it had become a recognized pest of potatoes in the USA and elsewhere on continent of Europe, especially near major ports but also, notably, near Torgau in Germany, south of Berlin.

## Testing and certification of field crop seeds/fruits

Testing of seeds for purity was started first in Denmark and Germany in 1869. This gave rise to the development of certification of seed potatoes. Regional seed testing congresses were held at Hamburg in 1906, and at Munster and Wageningen in 1910. After the First World War a larger congress, comprising 16 countries, was held in Copenhagen in 1921, followed in 1924 by a congress of 28 countries at Cambridge, England. (Thompson, 1974; Ainsworth, 1981). Similarly, work on certification of virus disease of fruit trees in Europe and America were initiated at the British research stations of East Malling and Long Ashton starting with vegetatively propagated apple and plum rootstocks. By 1955 a considerable number of virus-tested clones of tree fruits and their rootstocks were available.

Implementation of plant quarantine regulations is done by Ministry of Agriculture through The Directorate of Plant Protection, Quarantine and Storage (DPPQS) as its nodal agency which have recently been revised and known as the Plant Quarantine. DPPQS deals with the commercial import of consignments through its network of 35 Plant Quarantine Stations spread across the country such as seaports, airports, land frontiers which deals with grains, plants and plant products for consumption.

Besides the twenty eight plant quarantine stations, there are seven stations working units under the major stations. National Bureau of Plant Genetic Resources (NBPGR) undertakes the quarantine processing of all germplasm including transgenic planting material under exchange for research purposes.

For intra-country or internal quarantine of plants/ planting material, domestic quarantine regulations have been issued from time to time for inter-state movement of agricultural commodities under the Destructive Insects and Pests (DIP) Act of 1914.

Presently, there is a provision to restrict the inter-state movement of nine pests such as fluted scale, San José scale, coffee berry borer, codling moth, *Banana bunchy top virus*, *Banana mosaic virus*, potato cyst nematode, potato wart and apple scab. Keeping in view the recommendations and the present status of quarantine, the areas which need special attention in order to upgrade our quarantine system are broadly categorized as follows:

## International Quarantine

Strengthen pest risk analysis (PRA) for import of agricultural commodities, be it plant or animal and other risk factors associated with GMOs, biocontrol agents and other microorganisms. For the purpose, upgrade Plant Quarantine Stations in terms of trained manpower and infrastructure. Continue the review of the national quarantine legislation in the light of Agreement and the status of our compliance with SPS/ WTO. Initiate pre-import inspection and strengthen the post-entry quarantine (PEQ) growing and inspection of imported material. Establish a National Biosecurity Centre to deal with biosecurity in a comprehensive manner to achieve food safety, protect animal and plant life and health and also protect the environment.

## Domestic Quarantine

Review the status of existing domestic quarantine for establishment of interstate quarantine check-posts to monitor movement of pests of significance. Review the national regulatory framework and develop a mechanism for distribution or sale of pest-free seeds/ plants/ planting material within the country, be it seed distribution for multiplications testing under All India Coordinated Research Projects, inland supply of germplasm by NBPGR or seed distribution by the National/ State Seed Corporations/ private organizations. Strengthen state certification mechanism to ensure the supply of pest-free nursery material. Develop an early warning system and rapid response to avoid further spread and establishment of the recently introduced pests.

## Research and HRD

Reprioritize research in areas that need immediate emphasis such as survey and surveillance of diseases/pests, development of digitized keys and development of diagnostic protocols using serological and molecular techniques for detection of exotic pests/ transgenes. Develop a strong programme of awareness generation on importance of biosecurity for various stakeholders.

## Issues concerning Plant Quarantine in India

There are several issues related to exchange of plant/ planting material. The methodology adopted is specific for materials to be exchanged in bulk or small samples needed for research purposes. The national legislation needs to be in harmony with the norms laid down by the international plant protection convention. India in its latest order has complied with the standard norms by bringing out the P Q Order.

The various schedules *viz.*, V (Restricted species where import is permitted only by authorized institutions), VI (Restricted species whose import is permitted only with additional declarations or specified treatment certifications) and and VII (Plant material whose import for consumption / industrial processing is permitted with a normal Phytosanitary Certificate) of the PQ Order.

However there are certain difficulties in proper implementation of PQ order as officials at the ports of entry do not have the authority to search and seize the personal baggage of the incoming travellers, and the custom officials who search are not professionally trained to understand the importance of pests. They need proper orientation programmes to make the quarantine programmes most successful.

The provisions under the DIP Act, 1914 and its amendments for domestic quarantine to restrict the inter-state movement of nine invasive pests need to implemented to check their further spread into neighbouring states within the country. Establishment of inter-state barriers or check-posts could be a possible solution on the movement of seeds and planting materials of their host species. Besides, some of pests which have made entry into India during different years also need to be considered for inclusion under domestic quarantine, they include sunflower downy mildew (*Plasmopara halstedii*) in 1984, American serpentine leaf miner (*Liriomyza trifolii*) in 1991, spiraling white fly (*Aleurodicus dispersus*) in 1993 and biotype 'B' of white fly *Bemisia tabaci* in 1999.

The significant interceptions made, which are yet not reported from India, include insects like *Anthonomus grandis* in *Gossypium* spp. from USA, *Acanthoscelides obtectus* in *Cajanus cajan and Phaseolus vulgaris* from Brazil, Colombia, Italy and Nigeria, *Ephestia elutella* in *Macadamia* nuts and *Vigna* spp. from USA, nematodes like *Heterodera schachtii* from Denmark, Germany and Italy, *Quadrastichodella eucalyptii* in *Eucalyptus* from Australia, fungi like *Claviceps purpurea* in seeds of wheat, barley, etc. *Ditylenchus dipsaci, D. destructor, Rhadinaphelenchus cocophilus*, etc. in soil clods and plant debris, and, *Peronospora manshurica* on soybean from several countries, *Fusarium nivale* on wheat barley and *Aegilops sp.* from Germany, Italy, Hungary, Sweden and UK, Uromyces betae on sugarbeet from Belgium, Germany, Italy, UK and USA, bacteria like *Xanthomonas campestris pv.* campestris on Brassica spp. from Canada, France, Pakistan, Sweden, Taiwan, UK and USA and viruses like *Barley stripe mosaic virus* on wheat from USA, *Cowpea mottle virus* on cowpea from Philippines, *Raspberry ring-spot* virus on soybean from AVRDC (Taiwan), Sri Lanka, Thailand, USA etc. Such interceptions signify the success of quarantine as otherwise these pests could have entered the country and played havoc with the plant biodiversity and agriculture had they not been detected and stopped during quarantine processing.

## Impact of Pest Incursions into India

In recent years inspite of hard efforts made to prevent the entry of exotic pests, several pests such as papaya mealy bug, coconut eriophid mite, eucalyptus gall wasp, erythrina gall wasp, coffee berry borer, eucalyptus gall wasp, erythrina gall wasp, spiralling whitefly, serpentine leaf miner, sunflower downy mildew, anthurium blight, crown gall, parthenium, phalaris etc., have made entry into India and are still causing economic damage. Some of the serious plant pathogens which are emerging as serious problems and spreading fast include *Groundnut bud necrosis virus* (GBNV), bunchy top of banana, soybean bud blight, mungbean/ urd bean leaf curl, tomato bud blight and potato stem necrosis, and *Tobacco streak virus* (TSV) in sunflower and other crops. Late blight disease of potato caused by *Phytophthora infestans* has been reported causing an estimated loss of over US$ 236 million in India. Besides, Foliar blight in wheat, bract mosaic in banana, sheath blight in maize and paddy are other pathogens of serious concern causing serious losses.

## Movement of Domestic Quarantine pests

India ranks second in world in the production of fresh vegetables. Department of Agriculture, Ministry of Agriculture, Government of India has notified nine introduced pests under domestic quarantine regulation by restricting their movement to other parts of India. These pests include Banana bunchy top, Banana mosaic virus, Fluted scale and Sanjose scale which have spread to most part of India due to inadequacy in implementation of domestic quarantine regulations whereas Potato wart, Golden cyst nematode of potato, Apple scab, Codling moth and Coffee berry borer are restricted to some extent due to altitude/climatic requirements of the pests. Recently, the Golden Cyst Nematode of Potato (*Globodera rostochiensis and G. pallida*) which was hitherto confined only to Nilgiri Hills of Tamil Nadu has now been reported from Shimla, Himachal Pradesh where seed potatoes are grown (Ganguly *et.al.* 2010). There is a risk of this pest spreading to other potato growing areas of India like Uttar Pradesh and West Bengal.

## Quarantine Pests prevented from entering India

The impact of invasive pests and losses caused by them are not yet seriously evaluated in India. Their entry could be prevented if proper awareness and vigil is in place. Due to wide diversity of climates and crops available in India, the risks of invasive pest incursions are much more. Inspite of difficulties, Plant Quarantine Organization of India has intercepted and prevented the entry of exotic pests which include poty virus in Oil Palm (Reddy, *et al.*, 1996), *Xanthomonas campestris* pv. *dieffenbachiae* on Anthurium (Sathyanaryana, *et al.,* 1998), Crown gall (*Agrobacterium tumefaciens*) pathogen in Rose (Sathyanaryana, *et al.,* 2002), palm seed weevil (*Caryobruchus gleditsiae*) on Sabal palm seeds (Kumarasamy, *et al.*, 2002), Cymbidium mosaic virus on Dendrobium (Sathyanarayana, *et al*., 2003)Garlic bulb canker (*Embellisia allii*) on garlic bulbs (Latha, *et al.,* 2006), many nematodes on various ornamental and fruit plants (Latha, *et al.,* 1997;1999), *Hypothenemus obscures* (Tropical nut borer) on Butia seeds (Latha *et al*. 2011) and *Fusarium oxysporum f.sp. callistephi* on Marigold, Plumeria rust (*Coleosporium domingense*), downymildew (*Peronospora farinosa*) on beet *Peronospora manschurica* on soybean, *Xanthomonas vesicatoria* on Chilli and tomato seeds, Chrysanthemum white rust (*Puccinia horiana*) on Chrysanthemum cut flowers, *Odontoglossum ringspot virus* on Orchids, were intercepted on agricultural commodities imported for commercial purpose. Khetarpal and Gupta (2007) reported that a number of exotic pests such as *Uromyces betae* on Sugarbeet, *Fusarium nivale* on Wheat, *Cowpea mottle virus* on Cowpea, *Tomato black ring virus* on French bean, *Heterodera sachactii* on Sugarbeet, *Anthonomous grandis* on Cotton and *Quadrastichodella eucalypti* on Eucalyptus and many such quarantine pests were intercepted on Germplasm materials.

## Exotic pests of concern on Vegetable crops

India is the second position in world in production of fresh vegetables stands at in the world followed by China. Many seeds of vegetables are imported from various countries in huge quantities to meet demand for seeds of high yielding varieties / hybrids. India is still free from many exotic pests of concern to horticultural crops. Vegetable seeds like Coriander, Chilli, Tomato, Onion, Okra, Egg plant, Beet, Carrot, Cucumber, Ridge gourd, Bottle gourd, Bitter gourd, Cabbage, Broccoli, Cauliflower, Brussels sprout, etc., are imported regularly. The seed-borne pathogens accompanying such commodities are a real threat to biosecurity of the nation. Major plant disease threats through imported vegetable seeds include Cocoa and all species of Sterculiaceae and Bombaceae From Africa, Sri Lanka, West Indies, Coffee beans from Africa, South America, Sri Lanka, Rubber from South America, West Indies.

Some of the examples where plant pathogens have been reported causing serious damage include : Fusarium wilt(Fungi *Fusarium oxysporum f.sp. cucumerinum*) of Cucumber is host specific and infects only Cucumber. It is externally and internally seed borne and can survive on stored seeds for more than a year. Similarly, Downy mildew of Beet (*Peronospora farinosa*) is able to disseminate as seed contaminant through imported seeds. The mycelium and oospores can be isolated from the contaminated seeds. The infection can cause weight loss in beet and loss in sugar content in sugar beet ranging from 1.5 to 2.3 per cent. The major hosts are Beet and Sugarbeet, Sugarcane from Australia, Fiji, Papua New Guinea and Sunflower from Argentina, Peru

1. **Prohibited for general public:** Coconut plants and seeds, coffee plants and seeds, cotton seeds and unginned cotton, forest tree seed (*Castanea, Pinus, Ulmus*), groundnut seeds and cuttings, potato, sugarcane, tobacco seeds and wheat seeds.
2. **Plants/seeds which require post entry quarantine:** Cocoa, citrus, coconut, groundnut, potato, sugarcane, sunflower, tobacco and wheat.
3. **Additional declarations required for freedom from pest include:** All species of *Allium* (onion, garlic, leek, chive, shallot, etc.) from Smut (*Urocystis cepulae*), Cocoa and all species of the family Sterculiaceae and Bombaceae from Pod rot (*Monilia rorei*), Mealy pod (*Trachysphaeria* and *fructigena*), Witches' broom (*Crinipellia perniciosus*) Swollen shoot virus, Sugarcane (all species of *Saccharum*) from Leaf scald (*Xanthomonas albineans*), Gummosis (*Xanthomonas vasculorum*), Sereh, downy mildew, chlorotic streak and Fiji, Potato (all species of *Solanum*) Wart (*Synchytrium endobioticum*) and freedom of parent crop from virus diseases

## Spread of invasive species

Range expansion over large distances is often exhibited by invasive pest species (Porter *et al.*, 1991; Bale *et al.*, 2002). Anthropogenic causes are largely responsible for spread of invasive species into newer areas but climate plays a major role in the initial survival and establishment which is essential for subsequent spread in the invaded areas. India has witnessed several incursions from invasive species in the last 25 years and there are several quarantine pests waiting in the wings to gain entry into India either due to anthropogenic or climate induced reasons and therefore vigilance is warranted. In the new areas, the invasive species generally lack native specialist parasitoids unless they have moved along with pest like in the case of leaf miners (Cornell and Hawkins, 1993; Godfray *et al.*, 1995, Muniappan, 2008). Natural enemies that are generalists may not get adapted to search the niches occupied by the invading species and are unlikely to prevent the rapid initial establishment and subsequent population build-up. Diamond back moth can travel large distances to new cruciferous crop plantings and escape from its natural enemies

(Talekar and Shelon, 1993). Warming induced northward shift in the potential distribution of European corn borer, *Ostrinia nubilalis* is predicted. Altitudinal range expansion and increased overwintering survival of corn earworms *Heliothis zea* (Boddie) and *Helicoverpa armigera* (Hubner) may cause heavy yield loss and put forth major challenge for pest management in maize (Diffenbaugh *et al.* 2008). In north India, *H. armigera* is predicted to become a major pest on cotton, pulses and vegetables due to warming (Sharma *et al.*, 2005; 2010). Effect of changing climate on geographical distribution, range expansion and shifts in pest status of insect species in different regions is likely to alter the structure, diversity and functioning of ecosystems (Timoney, 2003).

Natural enemies will often be able to track climate induced changes in range expansion of pests as the expansion is most often gradual and incremental unlike the long range expansion in the case of invasive species. Some natural enemies are also highly mobile. The braconid parasitoid, *Cotesia flavipes* Cameron, could travel beyond 2000 km from its release site when introduced into Africa from its native Asia for the control of the stem borer, *Chilo partellus* (Swinhoe) in maize and Sorghum (Assefa *et al*., 2008). With few exceptions, parasitoids generally show reduced movement relative to their prey (Anton *et al*, 2007).

Greater understanding of the thermal requirements of insect pests, parasitoids and predators is required to predict the likely impacts of global warming on individual species responses and their trophic interactions. Natural enemies, both parasitoids and predators, are widely perceived to be more sensitive to climate change than their herbivore hosts. Herbivores quickly bounce back to previous densities prior to a wipe-out compared to natural enemies as a result of higher intrinsic rates (Thomson et al., 2010). However, each species responds differently to a steady or abrupt rise in temperatures and this understanding is crucial to predict its response to a changing climate.

## Quarantine control of honeybee diseases

There have always been efforts to acquire races of bees of bees better qualitative and quantitative characters for good pollination and more honey production. This has resulted in importation of stocks from one country to another. but such importations without proper quarantine measures has been a cause of concern in many countries where unwanted races/strains have threatened the indigenous pollinators and accidently brought parasites, pests and diseases to new areas for example introduction of *Apis mellifera* eliminated *Apis cerana japonica* in Japan (Gatoria et al 2000, Abrol 2001)

## Import of pollinators

Dochkova (1984) reported that iImport of *Megachile rotundata* F. in Bulgaria from France and Canada was found to be very be effective for pollination of Lucerne (Dochkova,1984). Populations of 1200-2000 female bees per hectare resulted in seed yield increasers of 75-550 kg/ha. Aballay et al (1986) reported the successful establishment of Bombus ruderatus (F.) imported from Newzealand into Chile for pollination of red clover (*Trifolium pretense* L.). Successful introduction of the of bumble bees for pollination of red clover in Newzealand, led Donovan (1990) to introduce two megachile bee species namely *Megachile rotundata* and *Nomia melanderi* Cockerell as pollinators of Lucerne. Similarly, Richards and Krunic (1990) importrd *M. rotundata* bees from Canada for pollination of alfalfa (*Medicago sativa* L.) in Yugoslavia which also proved worthwhile. during 1996,a total of 2500 prepupae were imported M. rotundata were imported from Newzealand to Australia. The immature stages of pollinators were screened for various diseases and reared under quarantine for adult emergence and for later use as pollinators. Keven and Laverty (1990) reported that parasitic mite like *Varroa jacobsoni* can survive on imported pollinators that may come in contact with infested honeybees visiting these flowers in their native country. It is therefore essential that import of pollinators is made through strict quarantine procedures after ensuring them from such parasites.

## Conclusion

The responsibility of prevention of pest incursions into the country lies with all stakeholders. This is possible by proper inspection, sampling and testing followed by appropriate action at the event of interception of the pest from further spreading thereby risking the food security. If the proper action is initiated at the right time exorbitant expenditures incurred to control exotic pests can be avoided. As more than 50 per cent of population is directly or indirectly dependent upon agriculture as the main source of livelihood, the promotion of safe trade, use of certified seeds and timely interventions to eradicate the pests would be the key principles to save the biodiversity in India.

## References

Aballay, F.E.; Ruiz,M.; Maraboli, A and Arrretz,V.P. 1986. Analysis of establishment in chile of *Bombus ruderatus* (F.) introduced foe red clover (*Trifolium pretense* L.) pollination. *Investigation Agricola* 9(1):31-36.

Abrol, D.P. 2001. Import export policy of honeybees and other polinators.*Indian Bee Journal* 63:55-58.

Ainsworth, G.C. 1981 *Introduction to the History of Plant Pathology*. Cambridge University Press, Cambridge.

Anon. 1914 *Actes de la Conférence Internationale de Phytopathologie*. Institut International d'Agriculture, Rome, pp. 237–241.

Anonymous, 1948. Illustrated Guide to Pests and Diseases Covered by External Quarantine. Ministry of Agriculture, USSR, Moscow, pp.84-85.

Anton, C., Zeisset, I., Musche, M., Durka, W., Boomsma, J.J. and Settele, J. 2007. Population structure of a large blue butterfly and its specialist parasitoid in a fragmented landscape. *Molecular Ecology* 16, 3828–3838.

Assefa, Y., Mitchell, A., Conlong, D.E. and Muirhead, K.A. 2008 Establishment of *Cotesia flavipes* (Hymenoptera: Braconidae) in sugarcane fields of Ethiopia and origin of founding population. *Journal of Economic Entomology* 101, 686–691.

Bale, J.S., Masters, G.J., Hodkinson, I.D., Awmack, C., Bezemer, T.M., Brown, V.K., Butterfield, J., Buse, A., Coulson, J.C., Farrar, J., Good, J.E., Harrington, R., Hartley, S.E., Jones, T.H., Lindroth, R.L., Press, M.C., Symrioudis, I., Watt, A.D. and Whittaker, J.B. (2002) Herbivory in global climate change research: direct effects of rising temperature on insect herbivores. *Global Change Biology* 8, 1–16.

Bazdyreva, V.I. 1932. Cottony-cushion scale (Icerya purchasi Mask. and measures for its control. Zashchita Rastenii, **3**: 35-64.

Bohm, H. 1958. Zum Vokommen der Mitterlmeer fruchtfliege Ceratitis capitata Wied., Pflanzenschutzberichte, **21** (9/10): 129-158.

Braun, H. 1959. Synchytrium endobioticum. Kartoffelbae,**10**: 234-237.

Butler, E. J. 1913. Diseases of rice. Agric. Res. Inst. Pusa Bull., **34**.

Butler, E. J. 1919. The rice worm (*Tylenchus angustus*) and its control. Mem. Dept. Agric. India, Bot. Ser., **10**(1): 1-37.

Caltagirone, L.E. 1956. Observationes sobre Arrenoclavus Koehleri Blanchard (Hym., Calc., Encyrtidae). Agric. Tech. II. of Gor. Santiago, Chle.

Chitwood, B. and Buhrer, 1946. The life history of the golden nematode of potato (Heterodera rostochiensis Wall) under conditions in Long Island, New York, Phytopathology, **36**: 3.

Clausen, C.P. 1931. Insects injurious to Agriculture in Japan. USDA cir 168. 116 pp.

Clausen, C.P., King, J.L. and Teranichi, C. 1927. Parasites of Popillia japonica in Japan and Korea and their introduction into the United States. Bull. 1429, Washington.

Cornell, H.V., and Hawkins, B.A. 1993 Accumulation of native parasitoid species on introduced herbivores: a comparison of hosts-as-natives and hosts-as-invaders. *American Naturalist* 141, 847-865.

Curl, L.F. 1952. Insects: pink bollworm, The Yearbook of Agriculture, USDA, Washington, 505-511.

Diffenbaugh, N.S., Krupke, C.H., White, M.A., and Alexander, C.E. 2008 Global warming presents new challenges for maize pest management. *Environmental Research Letters* 3, 044007.

Dochkova, B. 1984. Studies on the economic importance of *Megachile rotundata* . *Rastewevudni Nauki*. 21(4):116-124.

Donovan, B.J. 1990. Selection and importation of new pollinators to New Zealand. *New Zealand Entomologist*. 13:26-32.

Gatoria,G.S .; Chhuneja,P.K and Garch,S.2000. Quarantine control of honeybee diseases. *Indian Bee Journal*.62(1-2) :1-10.

Glass, F.X. 1960. Apple maggot fly emergence in Western New York. New York Agr. Exp. Sta. Bull. **759**: 29.

Godfray, H.C.J., Agassiz, D.J.L., Nash, D.R. and Lawton, J.H. 1995 The recruitment of parasitoid species to two invading herbivores. *The Journal of Animal Ecology* 64, 393–402.

Jacob, J. P., Devaraj, R. and Natarajan, R. (2007) Outbreak of the invasive gall inducing wasp Leptocybe invasa on eucalypts in India. *Newsletter. Asia–Pacific Forest Invasive Species Network* 8, 4–5.

Jhala, R.C., Bharpoda, T.M. and Patel, M.G. 2008 *Phenacoccus solenopsis* Tinsley (Hemiptera: Pseudococcidae), the mealy bug species recorded first time on cotton and its alternate host plants in Gujarat, India. *Uttar Pradesh Journal of Zoology* 28, 403-406.

Joshi, S., Poorani, J. and Chandish R. Ballal 2013. Invasive species of mealybugs in India. http://www.nbaii.res.in/news/pestalert1.pdf (accessed on 20.08.2013)

Kahn, R.P. 1983. Safe guarding the international exchange of plant germplasm. **In**: 10th International Congress of Plant Protection vol 2. Proceedings of a conference held at Brighton, England, 20-25 November 1983.: 866-872.

Kazas, I.A., Gopkavenko, A.S. and Poichenko 1960. Phylloxera and measures of its control. Simferopol.

Keven, P.G and Laverty, T.M. 1990. A brief survey and caution about importing alternative pollinators to Canada. *Canadian Beekeeping*.15 (8):176-177.

Khetarpal, R.K. and Gupta, K. 2007. Plant Biosecurity in India – Status and Strategy. As. *Biotechnological Development Reviews,* **20**: 83-107.

Kothekar, V.S. 1970. Quarantine and Dangerous Pests. **In**: A Hand Book of Pests,Disease and Weeds of Quarantine Significance. Amerind Publishing Co. Pvt. Ltd., New Delhi, pp. 13-103.

Kumar, P.K.V., Prakasan, C.B. and Vijayalakshmi, C.K., 1990 Coffee berry borer Hypothenemus hampei (Coleoptera : scolytidae): First record from India. *Journal of Coffee Research* 20, 161-164.

Kumarasamy, M., Latha, S., Sathyanarayana, N. and Reddy, O. R. 2002. Interception of palm seed weevil (*Caryobruchus gelditisiae (L.))* on seeds of *Sabaluresana* imported from Germany. *Indian Journal of Plant Protection,* **30** (1): 67 – 68.

Large, E.C. (1940) *The Advance of the Fungi*. Jonathan Cape, London.

Latha S, Kumarasamy M, Verma Shiv Sagar, Murali R, Gyaneshwer Banchhor and Rahul Mishra 2011. Interception of Tropical Nut Borer Hypothenemus obscurus – A New Host Record on Ornamental Palm Seed Butia rigida. *Indian Journal of Plant Protection* 39: 75-76.

Latha S. and Sathyanarayana N.2012. An overview of the status and the potential impact of the exotic pathogens on Indian horticulture. *Pest Management in Horticultural Ecosystems*, 18 (1):88-93.

Latha, S., Babu, D. V. N., Sathyanarayana, N., Reddy, O. R. and Renu Sharma. 1997. Nematodes intercepted from imported fruit plants. *Indian Journal of Plant Protection,* **25** (1): 60 – 61.

Latha, S., Babu, D. V. N., Sathyanarayana, N., Reddy, O. R. and Sharma. 1999. Plant parasitic nematodes intercepted from imported ornamental plants. *Indian Phytopathology,* **52** (3): 283 – 284.

Latha, S., Sathyanarayana, N. and Reddy O.R. 2006. "Interception of *Embellisia allii* and a virus in Garlic bulbs imported from China". *Indian Phytopathology.* **59** (3): 387.

Lindgren, D.L. and Vincent, L.E. 1959. Biology and control of Trogoderma granarium Everts. J. Eco. Ento. 52(2): 312-319.

MacKenzie, D. 2001 All fall down. *New Scientist*, 172, 34-37

Mani, M., Sunil Joshi, Kalyanasundaram, M., Shivaraju, C., Krishnamoorthy, K, Asokan, R. and Rebijith, K.B. 2013 A new invasive Jack Beardsley mealybug, *Pseudococcus jackbeardsleyi* (Hemiptera: Pseudococcidae) on papaya in India. *Florida Entomologist* 96(1), 242-245.

Mathys, G. and Baker, E.A. 1980. An appraisal of the effectiveness of quarantines. Annual Review of Phytopathology. **18**: 85-89

Muniappan, R., Meyerdirk, D. E., Sengebau, F. M., Berringer, D. D., and. Reddy, G. V. P. 2006 Classical biological control of the papaya mealybug, Paracoccus marginatus (Hemiptera: Pseudococcidae) in the republic of Palau. *Florida Entomologist* 89, 212–217.

Muniappan, R., Shepard, B.M., Watson, G.W., Carner, G.R., Sartiami, D., Rauf, A., and Hammig, M.D. (2008) First report of the papaya mealybug, *Paracoccus marginatus*

(Hemiptera: Pseudococcidae), in Indonesia and India. *Journal of Agricultural and Urban Entomology* 25, 37-40.

Palaniswami, M.S., Pillai, K.S., Nair, R.R. and Mohandas, C. 1995 A new cassava pest in India. *Cassava Newsletter* 19,6-7.

Pickett, A.D. 1937. Studies on the genus Rhagoletis (Trypetidae) with special references to R. pomonella Walsh. Can. J. of Sci., **15**(3): 55-75.

Porter, J.H., Parry, M.L., and Carter, T.R. 1991 The potential effects of climate change on agricultural insect pests. *Agricultural and Forest Meteorology* 57, 221-240.

Reddy, O. R., Sathyanarayana, N., Vani, S., Babu, D. V. N. and Latha, S. 1996. Virus disease of Oil palm seedlings and it's management. In: *National Seminar on Oil palm production and processing.* 13 – 14 February, 1996, Mysore, Karnataka, India. 28 – 31.

Richards, K.W and Krunic, M.D. 1990. Introduction of alfalfa leaf cutter bee to pollinate alfalfa in Yugoslavia. *Entomologist.* 109(3):130-135.

Sathiamma, B., Radhakrishnan Nair, C.P., and Koshy, P.K. 1998 Outbreak of a nut inflesting eriophyid mite *Eripphyes guerreronis* (K.) in coconut plantations in India. *Indian Coconut Journal* 29(8), 1–3.

Sathyanarayana, N., Latha, S. and Reddy, O. R. 2002. Interception of *Agrobacterium tumefaciens* – A crown gall bacterium on imported rose plants. In: *Proceedings of National Seminar on Resources Management in Plant Protection During Twenty First Century,* organized by Plant Protection Association of India, Hyderabad, India. 177 - 179.

Sathyanarayana, N., Latha, S. and Reddy, O. R. 2003. Interception of Cymbidium mosaic virus on imported Dendrobium plants. In: *Indian Journal of Plant Protection,* **31** (2): 118 – 119.

Sathyanarayana, N., Reddy, O. R., Latha, S. and Rajak, R. L. 1998. Interception of *Xanthomonas campestris pv. dieffenbachiae* on Anthurium plants from the Netherlands. *Plant Disease,* **82**(2): 262.

Sharma, H.C. (ed.). 2005 *Heliothis/Helicoverpa Management: Emerging Trends and Strategies for Future Research.* New Delhi, India: Oxford & IBH, and Science Publishers, USA. 469 pp.

Sharma, H.C., Srivastava, C.P., Durairaj, C., and Gowda, C.L.L. 2010 *Pest management in grain legumes and climate change.* In: Climate Change and Management of Cool Season Grain Legume Crops (Yadav, S.S., McNeil, D.L., Redden, R. and Patil, S.A., eds.). Dordrecht, The Netherlands: Springer Science + Business Media. pp.115-140.

Singh, S.P. and Jalali, S.K 1990 Spread of *Heteropsylla cubana* Crawford, subabul psyllid to Karnataka. *Myforest* 26 (2), 153-156.

Smith, L.B. and Hadley, C.H. 1952. Insects: Japanese beetle. The Yearbook of Agriculture, Washington, 567-573.

Startsev, N.V. and Arapova, L.I. 1966. Koloradskii Zhuk v Belorussii (The Colorado beetle in Belorussia). Zashchita Rastenii, **9**.

Sudarshan Ganguly, Singh, M. And Ganguly, A. K. 2010. Record of Potato Cyst Nematode, *Globodera rostochiensi* and *G. Pallid* in Shimla, Himachal Prades, India. *Indian Journal of Nematology,* **40** (1): 96 – 102.

Talekar, N.S. and Shelton, A.M. 1993 Biology, ecology and management of the diamondback moth. *Annual Review of Entomology* 38, 275–301.

Thompson, J.R. 1974 The International Seed Testing Association 1924–74. *Seed Science and Technology* 2, 267–283

Thomson, L.J., Macfadyen, S. and Hoffmann, A.A. 2010 Predicting the effects of climate change on natural enemies of agricultural pests. *Biological Control* 52, 296–306.

Timoney, K.P. 2003 Changing disturbance regime of the boreal forest of the Canadian Prairie provinces. *Forest Chronicle* 79, 502-516.

Ulrich, J. 1957. Physiologic specialization of Synchytrium endobioticum. FAO Plant Protection Bull. **12**: 181-187.

Usinger, R.L. 1964 The role of Linnaeus in the advancement of entomology. *Annual Review of Entomology* 9, 1–16.

Vasseur, R. and Schvester, D. 1955. Essais toxicologicues de laboratorei contre la mouchemediterranne des Fruits (Ceratitis capitata Wied) (Dipt., Trypetidae). Ann. Epiphyt, **2**:6

Viraktamath, C.A., Tewari, G.C., Srinivasan, K. and Gupta, M. 1993 American serpentine leaf miner is a new threat to crops. Indian Farming 43, 10-12.